Birkhäuser

Operator Theory: Advances and Applications

Volume 236

Founded in 1979 by Israel Gohberg

Manuel Cepedello Boiso
Håkan Hedenmalm
Marinus A. Kaashoek
Alfonso Montes Rodríguez
Sergei Treil
Editors

Concrete Operators, Spectral Theory, Operators in Harmonic Analysis and Approximation

22nd International Workshop in Operator Theory and its Applications, Sevilla, July 2011

Birkhäuser

Editors
Manuel Cepedello Boiso
Departamento de Análisis Matemático
Universidad de Sevilla
Sevilla
Spain

Håkan Hedenmalm
Department of Mathematics
The Royal Institute of Technology
Stockholm
Sweden

Marinus A. Kaashoek
Dept. of Mathematics & Computer Science
Free University Amsterdam
Amsterdam
The Netherlands

Alfonso Montes Rodríguez
Departamento de Análisis Matemático
Universidad de Sevilla
Sevilla
Spain

Sergei Treil
Department of Mathematics
Brown University
Providence, RI
USA

ISSN 0255-0156 ISSN 2296-4878 (electronic)
ISBN 978-3-0348-0647-3 ISBN 978-3-0348-0648-0 (eBook)
DOI 10.1007/978-3-0348-0648-0
Springer Basel Heidelberg New York Dordrecht London

Mathematics Subject Classification (2010): 47-06; 34B45, 47A10, 47B38, 93D15

Printed on acid-free paper

Springer Basel is part of Springer Science+Business Media (www.birkhauser-science.com)

Contents

Operator Theory:
Advances and Applications, Vol. 236, ix–xvi

Israel Gohberg Memorial Session

This text is composed by the Editors from the speeches given at the Memorial Session preceding the Conference Dinner.

OPENING

Rien Kaashoek (VU University, Amsterdam)

Ladies and gentlemen, dear friends and colleagues.

Israel Gohberg passed away on October 12, 2009. I welcome you at this special session dedicated to his memory. I am speaking on behalf of the IWOTA Steering Committee. IWOTA stands for International Workshop on Operator Theory and its Applications. As many of you know, professor Israel Gohberg was the first and only president of the Steering Committee. Together with William Helton he initiated in 1974 the IWOTA idea and as president he determined from the very beginning until the end of his life the general principles and the main directions of the IWOTA meetings.

I am very grateful to Alfonso Rodrigues, the chief organizer of the present IWOTA, for dedicating the conference to the memory of Israel Gohberg and organizing this special session. On the web site of the conference the following words were used: *Being one of the main promoters of the IWOTA series Gohberg's kind courageous humor will not be forgotten.*

This evening, just before the conference banquet, we have this special session to commemorate Gohberg's great mathematical legacy, his outstanding work and his wonderful personality. We do that in the presence of his wife Bella Gohberg, who also will contribute to this session, his sister Feija, and his two daughters Zvia and Yanina. I am very happy that they are here. I hope that they will experience the great influence their husband, brother and father, had on us, his former colleagues, students, co-workers, friends, and on the field we are working in.

The session begins with a mathematical oriented contribution, a short review of Israel Gohberg's research work by Leiba Rodman, the first Ph.D. student of Gohberg after his emigration in 1974 from Kishinev, Moldavia to Israel. Next the program consists of personal reminiscences on Israel Gohberg by Nikolai Nikolski, a colleague from the very beginning, Yuli Eidelman, who was his co-worker in the field of numerical analysis, and Henri Landau, a life long friend of Gohberg and the Gohberg family. The session will be concluded by Mrs. Bella Gohberg who will speak on behalf of the family. Her speech will be in Russian and will be translated by her oldest daughter Zvia Faro.

I wish all of you an inspiring session.

The review of Gohberg's research work presented by Leiba Rodman appears as the first paper in these proceedings.

REMEMBERING ISRAEL GOHBERG – PERSONAL PIECES OF PAST

Nikolai Nikolski (*University Bordeaux* 1/*Steklov Institute of Mathematics, St. Petersburg*)

I think I first met Israel Gohberg at the 1966 Moscow ICM. This was a time marked by the famous Gohberg-and-Kreĭn treatises on operator theory: the first volume "Introduction to the theory of non-selfadjoint operators" had already triumphantly appeared, the second one was announced on editors prospects, and many people (including the authors) spoke of a third (never published). It was a time of great enthusiasm towards operator theory and its value in mathematics. In particular, in the Soviet Union, the (until then classified) results of M. Keldysh in perturbation theory were published, results which played a role in reducing the "fluttering effect" of the first supersonic aircrafts. The Gohberg–Kreĭn books revolutionised the subject and allowed it to enter an age of maturity.

However, my distant acquaintance with Gohberg's mathematics started earlier, during my undergraduate university years. I was delighted to read two book size survey articles (1957 and 1958) by Israel Gohberg and Mark Kreĭn in "Uspehi Matematicheskih Nauk" ("Soviet Surveys – Uspehi"); they formed part of my entering graduate study examination in 1963. Their completeness, novelty and rigor have had a life-lasting influence on my own understanding of how to do and how to expose mathematics.

During the 1966 ICM Congress, I had many occasions to meet Israel – with an admiration and respect for one of the brilliant young leaders of mathematics in the second half of the twentieth century. In particular, I asked him the favor to be a referee (an "official opponent", following the Russian wording) for my Ph.D. defence, which I had prepared about this time at Leningrad State University. Having his agreement and returning to my University, I approached the President of the Scientific Council who was responsible for the defence. For a non-Russian reader, I would like to mention that the Russian (then Soviet) system of scientific degrees was (and maybe is) much more sophisticated and effective than the existing one in the West. First, there existed two degrees "Candidate of Science" (a qualification for a first job as a mathematician – university assistant or docent (associate professor)) and "Doctor of Science" (making accessible full professor positions). Defences were possible in sessions of special Scientific Councils assigned by the president of the university. In my time, for my Ph.D. (and later for full doctor) defences at the Leningrad University the Council numbered more than 40 members, representing all mathematical disciplines; the quorum was at 2/3 of the Council list. A dissertation passed many stages before the defence, in particular, a so-called "external report" – an "official evaluation" of the thesis by another mathematical institute (a university mathematical department or an institute of the Academy of Sciences). If a Council decided to accept a thesis for a defence, it would assign two (for full doctor degree – three) "official opponents" – approved experts in the field, who would give 15–20 minute speeches during the defence explaining to the Council why it should/should not award the candidate the desired degree. Similar speeches would be given by the thesis advisor (in a Ph.D. case) and by the thesis author himself. The session ended with a secret vote by the Council members, and the desired degree was attributed if no fewer than 2/3 votes were positive (with fractions rounded for the candidate's advantage). So, it was a spectacular, valuable and testing procedure, sometimes, with an unexpected outcome (notwithstanding the fact that a thesis cannot be submitted before a candidate has at least two publications in refereed journals). On the whole, this was one of the rare systems of the Russian-Soviet academic life which were undoubtedly effective.

Returning to my case, I approached the President of the Council (an outstanding applied mathematician and a fighter pilot during World War 2) and asked him to include Israel Gohberg on the list of referees. He showed his surprise: "How come you are such a big boy and yet do not know the '2/5 rule'?" I had never heard before about that and asked for an explanation. He said (it was just a private conversation in a corridor) that there are five active participants in a Ph.D. defence (the author, a thesis scientific advisor, two opponents, and an external referee) and communist party advisors required that the team contained no more than two Jews. He added that he had no intention to try to cut through the barrier in my case. I already had in the team Victor Havin (my beloved advisor) and Boris Mityagin of Moscow as a referee (arranged previously and, in fact, having proposed the entire subject of my research – invariant subspaces of weighted shift operators). It was a dead end and I was forced to accept this anti-semitic "2/5 rule". (My second referee became yet another prominent mathematician.) Perhaps elsewhere there were exceptions from the "rule" but, being alone against a hard system, as the Soviet system was, and having absolutely no information about other similar cases, one had no choice. I sent my apologies to Israel, and he understood the circumstances.

Later on, I visited several times Kishinev for Gohberg's lovely seminar (with A. Markus, I. Feldman, N. Krupnik and others). I subsequently met Israel after the "iron curtain" started to be raised and I was permitted to go abroad. The first such meeting happened during the memorable Lancaster NATO ASI in operator theory in 1984. During this meeting the following small but indicative exchange happened showing the influence and moral authority that Gohberg had on the community. Having in my talk frequently repeated the expression "Hankel and Toeplitz operators" and being tired to pronounce it tens of times, I decided to shorten it to "Ha-plitz operators". Many people appreciated my neologism but Israel approached me after the talk to say that this curious invention may be incorrect because it distorts proper names. After reflection I accepted and laid aside forever this jokey terminology. (However apparently some other people still use this shortening.)

From the beginning of the 1990's meetings with Gohberg became regular. The mathematical conversations were always deep and influential. Now, I understand well how much I have gathered from so generous and rich a personality as Israel Gohberg was.

Reminiscences of Israel Gohberg

Yuli Eidelman (Tel-Aviv University)

There were two periods in my acquaintance with Israel Gohberg. The first one began in my early childhood. My father and he were great friends. They had a lot in common, in their outlook of life and in their fanatical passion for mathematics. Uncle Iz'a was mentioned often in the family discussions. From about the age eleven, I remember his association with my father and the deep impression Israel made on me, as a large and kind man with much humor, who called me "druzhische (my friend)".

In 1966 my family went on vacation in Odessa. Israel was also there and spent quite a lot of time with us; some of which was at Mark Kreĭn's summer house in Arkadia; I knew Kreĭn's grandson. It was only many years later that I realized what a historical place I have visited, what mathematical masterpieces had been created there. In 1967 Israel attended the very first Voronezh Winter Mathematical School and visited our home during his trip. These schools became a notable event in soviet mathematical life. The friendship between Israel and my father, which began in '50s lasted until my father passed away in 2005.

I did not cross paths with Israel until 27 years later, as a very new immigrant I came to look for work at Tel-Aviv University. Israel greeted me very warmly but in the same time explained the situation to me very clearly. He recommended to change my direction of research, and suggested that I work in numerical methods for structured matrices. Such a topic seemed to surprise some of his colleagues as Israel was renowned as an outstanding expert in pure mathematics. In particular my father said that if some time ago somebody would have told him that Gohberg is involved in such a field of interest he would not have believed it. However, I was not the first to get involved in this direction of his work. On the moment of my appearance in Tel-Aviv my two predecessors in this field were Israel Koltracht and Vadim Olshevsky; they continued their careers in the USA. It was not easy for me to make the switch to numerical issues, but in time I got used to it and also my previous programming experience turned out to be useful in this.

My joint work with Israel was devoted to an interesting and important class of structured matrices. This class contains the band matrices, the semiseparable matrices, inverses of band and of semiseparable matrices, and other interesting examples. Various algorithms for such matrices are the subject of my interest till now. Not only my involvement in this direction has grown but also that of other researchers, which points to how topical it has become. In working so closely together, I have witnessed Israel's great intuition, how he follows the "scent" to use an expression by Iz'a Koltracht. One could absolutely rely on Israel's innate ability to understand what is worth paying attention to, what is worth making an effort to achieve, and what is not worth pursuing. We had a series of papers on our results: by ourselves and later jointly with our colleagues Vadim Olshevsky, Tom Bella, Israel Koltracht and Pavel Zhlobich from Storrs (Connecticut, USA), Dario Bini, Luca Gemignani and Paola Boito from Pisa, and Eugene Tyrtyshnikov from Moscow. Together with our colleague Iulian Haimovici we started to work on a large monograph devoted to a systematic presentation of the results in the field of our interest. Now we must finish this work without Israel.

Once Israel said about our long-standing acquaintance and my present age that now every young guy is 50 years old (in Russian "seichas kazhdomu soplyaku 50 let"). He was a very wise man, I got from him valuable advice not only on mathematics but also on life.

It was not an easy task to work together with Israel, my colleagues surely can confirm this. But the stronger requirements he applied to himself, his work was a very important part of his life.

Our collaboration continued 15 years, till the very end. A few days before Israel passed away there was a day when he felt better, and in our talk by phone we discussed as usually our issues. He said that it will be difficult to continue our discussions in the university but we will go on at his home. But what happened has happened. And I'd like to say that I and many of us were very happy to meet such a person in our life. We must remember him.

Recollections

Henry J. Landau (Bell Labs, Murray Hill, NJ)

Now at the end of this great conference with its breadth – eight parallel sessions! – depth and variety, I think that Israel Gohberg is the only person who could understand and appreciate every talk, and I picture how he would have enjoyed it. My great luck was to have known him for many years as collaborator and friend so I wanted to say a few words about him, although a much better account is the wonderful book, *Israel Gohberg and Friends*, by Harm Bart, Thomas Hempfling, and Rien Kaashoek. Thinking about him as I so often do, I said to my wife how much pleasure he took in mathematics. "In everything," she said, which is the essence of it all. We see him in the sea, the image of delight, or with his friends, or in his office with the entire wall behind his desk lined with books and journals he created, or with his marvelous family. It was Cora Sadosky who expressed it best when she spoke of his inclusiveness, in mathematics, with people, the warmth with which he brought us all together, pulling us in from every continent like a magnet. But I would like here to give just a glimpse of the hardships, which without respite he had to overcome, that were hidden behind his unfailing optimism.

Nowadays, living as we do, we believe in fairness, in recognizing high merit, but in totalitarian regimes it is entirely different. The worst people come to the top and rule arbitrarily, by fear. And so it was exactly on Israel's 12th birthday, two months after the Soviets took over Tarutino, the small town where he lived with his parents and younger sister, that three of the secret police came to their house and took away his father, for no reason of course. He was never seen again. Decades later the family received notification that he had been "free of guilt" and died in a gulag, but for the present the mere fact that he had been taken would have ostracized them, and so had to be hidden. A year later the Nazis invaded, and their mother set out to get the children away, in the total chaos of war. Try to imagine having to leave everything familiar behind, taking only what you can carry, not knowing at any point where you are and whether to move left or right, but knowing that everything hinges on making the correct choice. I will tell only one incident, which stops the heart. Israel's mother was a midwife so presenting herself as a nurse to accompany the wounded, managed somehow to keep the children with her, and started on their way east on such trains as she could manage to board, sometimes riding inside, sometimes in open wagons, each destination uncertain. Their train halted in a small town where it was supposed to remain all day, so she ventured out to try to barter something for a little food, but when she returned the train was gone!! She was told that the town had another station where perhaps the train had been shunted so, desperate, she gave away her only warm coat to be taken to it – providentially, the train was there and they were reunited. Ultimately they found themselves in Kyrgyzstan, in the north near China on a collective farm, all working in the fields.

School was always their focus however, and somehow even there Israel's talent was noticed; he got a scholarship and ultimately was sent alone to Kishinev, five days' travel away from his family. There, working independently, he found his way to research problems and had results published (in Doklady!) while still a student. Pervasive antisemitism blocked the way to any job, and even with his qualifications it needed intervention by a well-placed official to find him a post in a two-year institute for training elementary school teachers, where he had to teach in Moldavian, a dialect he didn't know. Minimal though this position was, it required character to resist the prevailing baseness, and he never forgot those who helped him in these early years[1]. In that period, Mark Grigorievich Kreĭn, a great mathematician of our time, was lecturing at a military institute in nearby Odessa, so Israel, always enterprising and daring, managed to slip by the guards to find him. They soon became collaborators and close friends (for 24 years! with one room in Kreĭn's small apartment designated as "Iz'ia's room" for his visits) working as equals on a host of fundamental questions in operator theory. In time, a group of outstanding students and an international reputation crystalized itself around him. He became professor and member of the Moldavian Academy, but with official antisemitism rapidly

[1]Among those who are now perhaps less known but who deserve to be remembered are M.S. Shumbarsky, his teacher in high school in Kyrgyzstan; G.Ya. Sukhomlinov, G.A. Bykov, and especially D.L. Picus who taught him later, and T.A. Itskovich in Kishinev; V.A. Andrunakevich and I.F. Volkov, his former professors, who recommended him to Mrs. A.N. Krachun, then Minister of Education for Moldavia; and V.G. Ceban, his colleague in Beltsy and Kishinev.

rising again he foresaw that, despite these advantages, the children's lives there would become unbearably distorted. Acting purely on his intuition about the West, since everything they heard about it was only hostile propaganda, he persuaded his entire familiy to apply for emigration. A full storm of enmity and vilification from all officialdom then broke over them. Both his wife and sister lost their jobs as doctors, his children were denounced at school, friends feared to be seen with them, but they remained steadfast and ultimately, with the help of the international mathematical community, were able to leave. Once again he was in a new world without a language – immediately invited to Stony Brook, he memorized his lectures phonetically – but soon began the deep currents of work that we see today.

There was an ironic but touching postscript to this period. Exactly thirty years after expelling him from his positions, the university in Kishinev and the Moldavian Academy invited him back for an honorary doctorate. It was problematic but the whole family went. The official ceremony did not mention the past, although Israel accepted in Moldavian and Hebrew, but the meaningful event took place later in an informal meeting with the faculty. Israel said frankly that people had not acted well but that he understood the pressures, and then there came an outpouring of emotion from the others, who clearly had so treasured their memories of his time with them. They spoke of his openness, his help to them, and the joy of mathematics that he brought them – just as all of us remember him now.

He lived squarely in the real world: he could repair shoes, select a good watermelon, grow grapes, make wine. There are two expressions I associate with him. When faced with a question, particularly one involving people, his verdict would often be "it's not simple", but if it came to something in his power to do, it was invariably "no problems!" Then there is his phenomenal work among us: over 500 papers and books, many in collaboration as befitted his inviting nature, each marked by something unmistakably his, an important idea or a step in a new direction. He traveled extensively, and wherever he went operator theory and friendship blossomed together – there are wonderful tributes to him from all over the world in the book that I mentioned. But, tragically, his health began to decline. His heart weakened and his kidneys began to fail; in time, he was spending five hours a day, five days a week, in dialysis. Yet never did he so much as mention it. He only joked: two friends meet on the street. "How are things?" says one. "In one word?" asks the other. "Yes." "Good." "And in two words?" "Not good." His mind was always on celebration of life. Accordingly, for their 50th wedding anniversary he and Bella decided that as their original wedding had been in a drab office in Moscow they should have a real one in Israel. And so they did, with over 200 guests! There is a photograph showing their portraits on the two occasions, fifty years apart, the light in their faces exactly the same.

Is it possible to live with such commitment and integrity, such courage, such harmony among one another, and between life and work? Israel and his family give us a proof of existence. And also, I believe, of uniqueness.

Closing remarks

Mrs. Bella Gohberg (Ra'anana, Israel)

Dear Conference Committee, friends and IWOTA participants,
I have attended many conferences together with my husband Israel Gohberg (Z″L), but IWOTA 2011 is different, because he is not here with us. This is the first time that I am present at the conference in his honor without him.

I have mixed feelings tonight. On one hand, I am extremely happy to be here with Israel's sister Feya and my daughters. I feel a lot of warmth, attention and am glad to meet our friends and Israel's colleagues.

On the other hand it is very sad that my husband (Z″L) is not here to enjoy it with us. I vividly remember IWOTA 2008 in Williamsburg. It was a happy occasion – the celebration of his 80th birthday. We are grateful that all our family could share the celebration with him.

Israel was an amazing person, who loved life and was an eternal optimist. He loved mathematics; it was his hobby, his profession, his inspiration and muse. He followed his passion traveling around the world and wherever he went he published new papers, formed new collaborations and met new friends.

Israel (Z″L) was not a healthy person, but this never slowed him down or stopped him. Many of his colleagues were not aware of his health condition, because he never complained and continued working till his last day. He inspired others and defeated his disease. Doctors classified Israel as a medical miracle, an example of the power of mind over body. Israel's (Z″L) courage and sense of humor helped him in the most critical and even hopeless of situations.

Israel's wisdom, kindness and generous heart attracted collaborators, colleagues and friends. Although in his professional world Professor Gohberg resided in the abstract multi/infinite-dimensional spaces, he was a humble down to earth person, devoted family man, reliable colleague and loyal friend.

Losing Israel was very painful and extremely difficult for our family. We are grateful to his colleagues and friends for the kind words, letters and support during this hard time. Thank you for the wonderful In Memoriams, journals and publications in his honor.

They say that the person dies when he is forgotten, but those who are remembered keep living. Thank you for keeping his memory alive.

I would like to express my gratitude to the conference committee for inviting me to participate at this event. Special thanks to Professor Alfonso Montes Rodriguez for his generosity, warm hospitality and for organizing my travel arrangements and accommodations.

I wish you all a long and healthy life, professional success and all the best.

Operator Theory:
Advances and Applications, Vol. 236, xvii–xxii

Research Work of Israel Gohberg

Leiba Rodman

Abstract. This presentation was given at IWOTA 2011, Seville, Spain, during the special evening session in memory of Israel Gohberg. It is based on a memorial paper, joint with A.E. Frazho and M.A. Kaashoek [12].

Mathematics Subject Classification (2010). 15-02, 47-02.

Keywords. State space method, Gohberg–Semencul formula.

1. Introduction

I am honored to talk here about research work of Israel Gohberg. Many of us knew, and were greatly influenced by, Israel Gohberg (1928–2009), a brilliant and charismatic mathematician. His mathematical legacy is fundamental and extensive: more than 450 mathematical articles, 40 Ph.D. students, 27 co-authored books. These are very impressive numbers. But the numbers are not the whole story, and perhaps even not the main story. There are people who would crank up a paper every two weeks or so without regard to quality or interest. Gohberg was definitely not one of them. On the contrary, every research paper by Israel Gohberg displays creative ideas and novel insights.

Gohberg's contributions are mainly in analysis, operator theory, linear algebra, numerical analysis, and control theory. His work is leading in the following research areas:

- singular integral equations and their discrete analogues;
- Toeplitz operators and equations;
- the theory of nonselfadjoint operators;
- spectral theory of matrix and operator functions;
- factorizations of matrix and operator functions;
- inversion problems for structured matrices.

Gohberg's numerous honors and awards include:

- Foreign Member of the Netherlands Academy of Arts and Science (1985);
- Landau Prize (1976);

- Rothschild Prize in Mathematics (1986);
- Humboldt Research Prize (1992);
- Hans Schneider Prize in Linear Algebra (1994);
- M.G. Kreĭn prize of the Ukrainian National Academy of Sciences (2007);
- SIAM Fellowship (2009);
- honorary doctorates: Technical University Darmstadt (1997), Technical University Vienna (2001), University of Timisoara (2002), University of Kishinev (2002), University of Balti (2002), Technion (2008).

Obviously, it is impossible to cover much of Gohberg's research work during this short 20 minutes presentation. Bona fide applications were always part of his research outlook. I have decided therefore to mention here only two areas of seminal, lasting, and influential contributions of Israel Gohberg that are related to applications in control theory. My apologies to non-mathematicians in the audience: I cannot talk meaningfully about Gohberg's research work without resorting to mathematical lingo.

2. The state space method

The *state space model* of a linear time invariant differential equation is given by

$$\dot{x} = Ax + Bu, \qquad y = Cx + Du, \tag{2.1}$$

where A, B, C, D are matrices of suitable sizes, and where u, x, and y are the *input*, *state*, and *output* of the model. This is the most basic model of control systems. The function $G(s) = D + C(sI - A)^{-1}B$ is known as the *transfer function* of the state space model. The realization theory for this model, in other words construction of A, B, C, D by given rational matrix function $G(s)$, has been developed by Kalman, Gilbert, and others in the 1960's.

A large part of research of Israel Gohberg and his circle of close collaborators and friends in the last thirty years has been devoted to expanding this state space method into various areas, including:

- interpolation with applications to H^∞ (worst case) control [4, 5];
- integral equations [6, 7, 9];
- canonical systems, direct and inverse spectral problems [1, 2, 3, 15, 16].

A key application of the state space method is the *factorization principle* formulated in [6, 10], see also the books [8, 9]. Minimal cascade decompositions of systems, or alternatively minimal factorizations of the respective transfer functions, can be explicitly described in one-to-one fashion in terms of certain invariant subspaces of their corresponding state space operators. In the case of systems written in the form (2.1) with invertible D, these are pairs of complementary subspaces, one invariant for A, the other invariant for $A - BD^{-1}C$. This *geometric principle of factorization* will be illustrated with one basic example of canonical factorization.

Let $G(s)$ be a rational $m \times m$ matrix function with no poles on the imaginary axis and at infinity. Consider factorization problem:

$$G(s) = G_-(s)G_+(s), \tag{2.2}$$

where $G_+(s)$ and $G_-(s)$ are biproper rational $m \times m$ matrix functions such that:

- $G_+(s)$ and $G_+(s)^{-1}$ have all their poles in the open right half-plane $\mathbb{C}_{\text{right}}$;
- $G_-(s)$ and $G_-(s)^{-1}$ have all their poles in the open left half-plane $\mathbb{C}_{\text{left}}$.

Such factorizations are called *(right) canonical factorizations.*

If G admits a canonical factorization, then $G(s)$ is nonsingular for each s on the imaginary axis and at infinity.

Since G has no poles on the imaginary axis and at infinity, it admits a realization

$$G(s) = D + C(sI_n - A)^{-1}B$$

such that the matrices A, B and C are partitioned:

$$A = \begin{bmatrix} A_1 & 0 \\ 0 & A_2 \end{bmatrix}, \quad B = \begin{bmatrix} B_1 \\ B_2 \end{bmatrix}, \quad C = \begin{bmatrix} C_1 & C_2 \end{bmatrix},$$

so that A_1, A_2 are matrices of sizes $n_1 \times n_1$, $n_2 \times n_2$ ($n_1 + n_2 = n$), and $\sigma(A_1) \subset \mathbb{C}_{\text{left}}$, $\sigma(A_2) \subset \mathbb{C}_{\text{right}}$, i.e., A_1 and $-A_2$ are stable.

Assume that $G(s)$ is nonsingular for every s on the imaginary axis and at infinity (necessary condition for existence of canonical factorization). Then D is nonsingular, and the matrix $A^\times := A - BD^{-1}C$ has no eigenvalues on the imaginary axis. Consider the subspace

$$M = \left\{ \begin{bmatrix} x \\ 0 \end{bmatrix} \mid x \in \mathbb{C}^{n_1} \right\}$$

It is the spectral subspace consisting of all eigenvectors and generalized eigenvectors of A corresponding to the eigenvalues in $\mathbb{C}_{\text{left}}$. Let $M^\times$ be the $A^\times$-invariant subspace consisting of all eigenvectors and generalized eigenvectors of $A^\times$ corresponding to the eigenvalues in $\mathbb{C}_{\text{right}}$. Then the factorization principle states that G *admits a right canonical factorization if and only if* M *and* $M^\times$ *are complementary subspaces in* $\mathbb{C}^n$.

Moreover, as it turns out, the factorization, when it exists, can be further described in terms of formulas involving solutions to certain Riccati equations. Indeed, G admits a right canonical factorization if and only if the Riccati equation

$$RB_2D^{-1}C_1R - R(A_2 - B_2D^{-1}C_2) + (A_1 - B_1D^{-1}C_1)R - B_1D^{-1}C_2 = 0$$

has a solution R satisfying the spectral constraints

$$\sigma\big(A_1 - (B_1 - RB_2)D^{-1}C_1\big) \subset \mathbb{C}_{\text{left}},$$
$$\sigma\big(A_2 - B_2D^{-1}(C_1R + C_2)\big) \subset \mathbb{C}_{\text{right}}$$

(such solution is necessarily unique). If the above conditions are satisfied, a right canonical factorization $G(s) = G_-(s)G_+(s)$ of G is given by

$$G_-(s) = D_1 + C_1(sI_{n_1} - A_1)^{-1}(B_1 - RB_2)D_2^{-1},$$
$$G_+(s) = D_2 + D_1^{-1}(C_1R + C_2)(sI_{n_2} - A_2)^{-1}B_2.$$

and

$$G_-^{-1}(s) = D_1^{-1} - D_1^{-1}C_1(sI_{n_1} - A_1^{\times})^{-1}(B_1 - RB_2)D^{-1},$$

$$G_+^{-1}(s) = D_2^{-1} - D^{-1}(C_1R + C_2)(sI_{n_2} - A_2^{\times})^{-1}B_2D_2^{-1},$$

where

$$A_1^{\times} = A_1 - (B_1 - RB_2)D^{-1}C_1, \qquad A_2^{\times} = A_2 - B_2D^{-1}(C_1R + C_2),$$

and where $D = D_1D_2$ is a factorization of D with invertible D_1 and D_2.

The factorization principle bears the distinctive characteristics of Gohberg's results: it is complete (if and only if statement), explicit (formulas are given), well motivated, rooted in applications, connects diverse ideas and approaches (algebra of factorizations vs geometry of invariant subspaces), and inspired many further developments. As an example of such further developments, let me mention non-canonical factorizations (more general than (2.2)), which also can be described in terms of invariant subspaces. In this case however the factorization principle is more involved because the subspaces in question are not anymore direct complements to each other.

3. Gohberg–Semencul formula

Toeplitz matrices model many types of processes in which certain shift invariance, either in time or in space, is present. This includes time series analysis and signal processing. The Gohberg–Semencul formula for inverting Toeplitz matrices was originally published in 1972 [17] (for matrices with scalar entries), and extended in [13, 14] for matrices with non-commutative entries, in particular block Toeplitz matrices.

The original version of the formula for complex matrices is now presented. Let T_n be an $n \times n$ complex Toeplitz matrix. Consider the equations

$$e_0 = T_n \begin{bmatrix} \alpha_0 & \alpha_1 & \cdots & \alpha_{n-2} & \alpha_{n-1} \end{bmatrix}^{tr}, \tag{3.1}$$

$$e_{n-1} = T_n \begin{bmatrix} \beta_0 & \beta_1 & \cdots & \beta_{n-2} & \beta_{n-1} \end{bmatrix}^{tr}, \tag{3.2}$$

where e_0 is the vector in $\mathbb{C}^n$ with one in the first position and all the other entries zero, and e_{n-1} is the vector in $\mathbb{C}^n$ with one in the last position and all the other entries zero. If there exists a solution to $T_n\alpha = e_0$ and $T_n\beta = e_{n-1}$, where $\alpha_0 \neq 0$, then T_n is invertible, $\alpha_0 = \beta_{n-1}$, and

$$T_n^{-1} = \frac{1}{\alpha_0}\left(L_1U_2 - L_3U_4\right),$$

where L_1 and L_3 are the lower triangular Toeplitz matrices, U_2 and U_4 are the upper triangular Toeplitz matrices given by the following formulas:

$$L_1 = \begin{bmatrix} \alpha_0 & 0 & \cdots & 0 \\ \alpha_1 & \alpha_0 & \cdots & 0 \\ \vdots & \vdots & \ddots & \vdots \\ \alpha_{n-1} & \alpha_{n-2} & \cdots & \alpha_0 \end{bmatrix}, \qquad L_3 = \begin{bmatrix} 0 & 0 & \cdots & 0 \\ \beta_0 & 0 & \cdots & 0 \\ \vdots & \vdots & \ddots & \vdots \\ \beta_{n-2} & \beta_{n-3} & \cdots & 0 \end{bmatrix},$$

$$U_2 = \begin{bmatrix} \beta_{n-1} & \beta_{n-2} & \cdots & \beta_0 \\ 0 & \beta_{n-1} & \cdots & \beta_1 \\ \vdots & \vdots & \ddots & \vdots \\ 0 & 0 & \cdots & \beta_{n-1} \end{bmatrix}, \qquad U_4 = \begin{bmatrix} 0 & \alpha_{n-1} & \cdots & \alpha_1 \\ 0 & 0 & \cdots & \alpha_2 \\ \vdots & \vdots & \ddots & \vdots \\ 0 & 0 & \cdots & 0 \end{bmatrix}.$$

The Gohberg–Semencul formula led to many developments, including:

- inversion of structured matrices;
- displacement structure;
- continuous time analogues;
- non-commutative analogues;
- applications in signal processing, filtering;
- inversion of multilevel and multivariable Toeplitz matrices;
- numerical algorithms for fast inversion.

The importance and influence of the Gohberg–Semencul formula in mathematics and engineering cannot be overestimated. As an indication, note that the Google scholar shows 496 hits for the formula and 549 hits for Gohberg–Semencul (as of July 2011), mostly in engineering literature.

Reviews of the mathematical work of Gohberg, authored by Kaashoek and by Kaashoek and Lerer, are given in Part 2 of [11].

References

[1] D. Alpay and I. Gohberg, *Discrete analogs of canonical systems with pseudo-exponential potential. Definitions and formulas for the spectral matrix functions*, pp. 1–47, OT 161, Birkhäuser, Basel, 2006.

[2] D. Alpay and I. Gohberg, *State space method for inverse spectral problems*, pp. 1–16, Progr. Systems Control Theory, 22, Birkhäuser Boston, Boston, MA, 1997.

[3] D. Alpay and I. Gohberg, *Discrete analogs of canonical systems with pseudo-exponential potential*, pp. 31–65, OT 165, Birkhäuser, Basel, 2006.

[4] J.A. Ball, I. Gohberg, and L. Rodman, *Interpolation of rational matrix functions*, OT 45, Birkhäuser Verlag, 1990.

[5] J.A. Ball, I. Gohberg, and L. Rodman, *The state space method in the study of interpolation by rational matrix functions*, Mathematical system theory, 503–508, Springer, Berlin, 1991.

[6] H. Bart, I. Gohberg, and M.A. Kaashoek, *Minimal factorization of matrix and operator functions*, OT 8, Birkhäuser Verlag, Basel and Boston, 1979.

[7] ______, *Fredholm theory of Wiener–Hopf equations in terms of realization of their symbols*, Integral Equations and Operator Theory **8** (1985), 590–613.

[8] H. Bart, I. Gohberg, M.A. Kaashoek, and A.C.M. Ran, *Factorization of matrix and operator functions: the state space method*, OT 178, Birkhäuser Verlag, Basel, 2008.

[9] H. Bart, I. Gohberg, M.A. Kaashoek, and A.C.M. Ran, *A state space approach to canonical factorization with applications*, OT 200, Birkhäuser Verlag, Basel and Boston, 2010.

[10] H. Bart, I. Gohberg, M.A. Kaashoek, and P. Van Dooren, *Factorizations of transfer functions* SIAM J. Control Optim. **18** (1980), 675–696.

[11] H. Bart, T. Hempfling, and M.A. Kaashoek (eds.), *Israel Gohberg and friends. On the occasion of his 80th birthday*, Birkäuser Verlag, Basel, 2008.

[12] A.E. Frazho, M.A. Kaashoek, and L. Rodman, *Israel Gohberg in memoriam*, IEEE Control Systems Magazine **30** (2010), 135–139.

[13] I.C. Gohberg and G. Heinig, *Inversion of finite Toeplitz matrices consisting of elements of a noncommutative algebra*, (Russian), Rev. Roumaine Math. Pures Appl. **19** (1974), 623–663. Translation: pp. 7–46, OT 206, Birkhäuser Verlag, Basel, 2010.

[14] I.C. Gohberg and G. Heinig, *The inversion of finite Toeplitz matrices*, (Russian), Mat. Issled. **8** (1973), 151–156, 183. Translation: pp. 1–6, OT 206, Birkhäuser Verlag, Basel, 2010.

[15] I. Gohberg, M.A. Kaashoek, and A.L. Sakhnovich, *Canonical systems with rational spectral densities: explicit formulas and applications*, Math. Nachr. **194** (1998), 93–125.

[16] I. Gohberg, M.A. Kaashoek, and A.L. Sakhnovich, *Pseudo-canonical systems with rational Weyl functions: explicit formulas and applications*, J. Differential Equations **146** (1998), 375–398.

[17] I.C. Gohberg and A.A. Semencul, *The inversion of finite Toeplitz matrices and their continual analogues*, (Russian), Mat. Issled. **7** (1972), 201–223, 290.

Leiba Rodman
Department of Mathematics
College of William and Mary
Williamsburg, VA 23187, USA
e-mail: `lxrodm@math.wm.edu`

Operator Theory:
Advances and Applications, Vol. 236, 1–10

Some New Refined Hardy Type Inequalities with Breaking Points $p = 2$ or $p = 3$

S. Abramovich and L.-E. Persson

Abstract. For usual Hardy type inequalities the natural "breaking point" (the parameter value where the inequality reverses) is $p = 1$. Recently, J. Oguntuase and L.-E. Persson proved a refined Hardy type inequality with breaking point at $p = 2$. In this paper we show that this refinement is not unique and can be replaced by another refined Hardy type inequality with breaking point at $p = 2$. Moreover, a new refined Hardy type inequality with breaking point at $p = 3$ is obtained. One key idea is to prove some new Jensen type inequalities related to convex or superquadratic funcions, which are also of independent interest.

Mathematics Subject Classification (2010). 26D15.

Keywords. Inequalities, refined Hardy type inequalities, convex functions, superquadratic functions, Jensen type inequalities.

1. Introduction

First we consider the following well-known Hardy-type inequality: If the function f is non-negative and measurable on $(0, \infty)$, then

$$\int_0^b \left(\frac{1}{x}\int_0^x f(y)\,dy\right)^p x^\alpha dx \le \left(\frac{p}{p-1-\alpha}\right)^p \int_0^b f^p(x)\,x^\alpha \left(1-\left(\frac{x}{b}\right)^{\frac{p-\alpha-1}{p}}\right) dx, \tag{1.1}$$

where $p \geq 1$ and $\alpha < p - 1$. The constant on the right-hand side is sharp. A simple proof of this inequality was recently presented in [12], where also some historical remarks can be found. In particular, for $\alpha = 0$ and for $b = \infty$ (1.1) is the classical Hardy inequality, stated by G.H. Hardy in 1920 (see [2]) and where it was finally proved in 1925 (see [3]). Moreover, for $b = \infty$ (1.1) coincides with the first weighted version also proved by Hardy himself in 1928 (see [4]). Further development of Hardy-type inequalities can be found in the books [6], [7] and [8]. Even if the inequality (1.1) is sharp, it is possible to refine it by inserting a second positive term in the left-hand side of (1.1). Such a result was first proved by C.O. Imoru in 1977 (see [5], and also its generalization in [10]). In all cases the "breaking point", i.e., the point where the inequality reverses is $p = 1$. In 2008 J.A. Oguntuase and L.-E. Persson proved the following refined Hardy inequality with "breaking point" $p = 2$ (see [9] and cf. also [11]): Let $p \geq 1$, $\alpha < p - 1$ and $0 < b \leq \infty$. If $p \geq 2$, and the function f is non-negative and locally integrable on $(0, b)$ and $\int_0^b x^\alpha f^p(x)\,dx < \infty$, then

$$\int_0^b \left(\frac{1}{x}\int_0^x f(y)\,dy\right)^p x^\alpha dx + \frac{p-1-\alpha}{p}\int_0^b \int_t^b \left|\frac{p}{p-\alpha-1}\left(\frac{t}{x}\right)^{1-\frac{p-\alpha-1}{p}} f(t) - \frac{1}{x}\int_0^x f(\tau)\,d\tau\right|^p x^{\alpha-\frac{p-\alpha-1}{p}}\,dx \cdot t^{\frac{p-\alpha-1}{p}-1}dt$$

$$\leq \left(\frac{p}{p-\alpha-1}\right)^p \int_0^b f^p(x)\,x^\alpha \left(1-\left(\frac{x}{b}\right)^{\frac{p-\alpha-1}{p}}\right)dx. \tag{1.2}$$

If $1 < p \leq 2$, then (1.2) holds in the reversed direction. In particular, for $p = 2$ we have equality in (1.2).

In this paper we derive a new refined Hardy type inequality different than (1.2) but again, with a natural breaking point at $p = 2$ (see Theorem 3.2). We also present and prove another new refined Hardy type inequality with breaking point at $p = 3$ (see Theorem 3.4).

One key idea is to prove some new Jensen type inequalities connected to functions of the type $F(x) = x\varphi(x)$, where φ is convex/concave or supequadratic/subquadratic. These results are of independent interest. Next we present and prove some new inequalities for such functions (see Propositions 3.1 and 3.3), which, in their turn, are crucial for the proofs of the new Hardy type inequalities.

The paper is organized as follows: In order not to disturb our discussions later on, the new Jensen type inequalities and other preliminaries are collected in Section 2. The main results concerning refined Hardy type inequalities are stated and discussed in Section 3 and the proofs are given in Section 4.

2. Preliminaries

Our first important Lemma reads:

Lemma 2.1. *Let $K(x) = x\varphi(x)$, where $\varphi(x)$ is convex on $[0, b)$. Then*

$$K(y) - K(x) \geq \varphi(x)(y-x) + C_\varphi(x)\,y\,(y-x), \tag{2.1}$$

holds for $x \in [0, b)$, $y \in [0, b)$, C_φ is the constant in the definition of convexity and the inequality

$$\int_\Omega K(f(s))\,d\mu(s) - K\left(\int_\Omega f(s)\,d\mu(s)\right) \tag{2.2}$$

$$\geq \int_\Omega C_\varphi(x)\,f(s)\,(f(s)-x)\,d\mu(s) = \int_\Omega C_\varphi(x)\,(f(s)-x)^2\,d\mu(s)$$

holds, where f is any non-negative μ-integrable function on the probability measure space (Ω, μ) and $x = \int_\Omega f d\mu$. The inequalities (2.1) and (2.2) hold in the reverse direction when φ is concave.

Example. The inequalities (2.1) and (2.2) are satisfied in particular by $K(x) = x^p$, $p \geq 2$. For $1 < p \leq 2$ the reverse inequalities hold. They reduce to equalities for $p = 2$.

Proof of Lemma 2.1. Multiplying by y the inequality satisfied by any convex function

$$\varphi(y) - \varphi(x) \geq C_\varphi(x)(y - x), \tag{2.3}$$

by simple manipulations we get that $K(x) = x\varphi(x)$ satisfies (2.1) when φ is convex.

By putting $y = f(s)$ and $x = \int_\Omega f d\mu$ in (2.1) and integrating with respect to the probability measure μ we arrive at (2.2). Moreover, if φ is concave, then (2.3) holds in the reverse direction so the same proof as above shows that in fact (2.1) and (2.2) both hold in the reverse direction when $K(x) = x\varphi(x)$, where $\varphi(x)$ is concave. □

Remark 2.2. Inequality (2.2) may be regarded as a new type of Jensen inequality, for the functions $K(x) = x\varphi(x)$, where φ is convex/concave.

Next we define the crucial concept of superquadratic and subquadratic functions (see [1]).

Definition 2.3. Let $\varphi : [0, b) \to \mathbb{R}$. The function φ is superquadratic if for all $x \in [0, b)$ there exists $C_\varphi \in \mathbb{R}$ such that

$$\varphi(y) - \varphi(x) \geq C_\varphi(x)(y - x) + \varphi(|y - x|) \tag{2.4}$$

for all $y \in [0, b)$.

The function φ is subquadratic if $-\varphi$ is superquadratic and the reverse inequality of (2.4) holds.

Remark 2.4. Inequality (2.4) holds for all $\varphi(x) = x^p$, $x \geq 0$, $p \geq 2$. It holds in the reverse direction if $0 < p < 2$ and it reduces to equality for $\varphi(x) = x^2$.

The following result is useful (see [1, Lemma 2.1]):

Lemma 2.5. *Let φ be a superquadratic function with $C_\varphi(x)$ as in (2.4).*

(i) *Then $\varphi(0) \leq 0$.*
(ii) *If $\varphi(0) = \varphi'(0) = 0$, then $C_\varphi(x) = \varphi'(x)$ whenever φ is differentiable at $x > 0$.*
(iii) *If $\varphi \geq 0$, then φ is convex and $\varphi(0) = \varphi'(0) = 0$.*

We are now ready to formulate the similar result as in Lemma 2.1 but when φ is superquadratic instead of convex.

Lemma 2.6. *Let $K(x) = x\varphi(x)$, where $\varphi(x)$ is superquadratic on $[0, b)$. Then*

$$K(y) - K(x) \geq \varphi(x)(y-x) + C_\varphi(x)\, y(y-x) + y\varphi(|y-x|) \tag{2.5}$$

holds for $x \in [0,b)$, $y \in [0,b)$, $C_\varphi(x)$ is defined by (2.4) and

$$\int_\Omega K(f(s))\,d\mu(s) - K\left(\int_\Omega f(s)\,d\mu(s)\right) \geq \int_\Omega \left[C_\varphi(x) f(s)(f(s)-x) + f(s)\varphi(|f(s)-x|)\right] d\mu(s) \tag{2.6}$$

holds, where f is any non-negative μ-integrable function on the probability measure space (Ω, μ) and $x = \int_\Omega f d\mu$. The inequalities (2.5) and (2.6) hold in the reverse direction when φ is subquadratic.

Example. The inequalities (2.5) and (2.6) are satisfied in particular by $K(x) = x^p$, $p \geq 3$. For $1 < p \leq 3$ the reverse inequalities hold. They reduce to equalities for $p = 3$.

Proof of Lemma 2.6. Multiplying (2.4) by y, by simple manipulations we get that $K(x) = x\varphi(x)$ satisfies (2.5) when φ is superquadratic.

Next we consider (2.5) with $x = \int_\Omega f d\mu$ and $y = f(s)$, and integrate with respect to the probability measure μ and obtain (2.6). Furthermore, if φ is subquadratic, then (2.4) holds in the reverse direction and the same proof as above shows that (2.5) and (2.6) hold in the reverse direction in this case. □

Remark 2.7. Inequality (2.6) may be regarded as a new Jensen type inequality for the functions $K(x) = x\varphi(x)$ where φ is superquadratic/subquadratic.

3. The main results

In this section we state the main results concerning Hardy type inequalities related to $K(x) = x\varphi(x)$, where $\varphi(x)$ is convex/concave and then to $K(x) = x\varphi(x)$, where $\varphi(x)$ is superquadratic/subquadratic. The proofs are given in the next section.

The following result is crucial for the proof of Theorem 3.2 and of independent interest.

Proposition 3.1. *Let $0 < b \leq \infty$, $u : (0, \infty) \to \mathbb{R}$ be a non-negative weight function such that $\frac{u(x)}{x^2}$ is locally integrable on $(0, \infty)$ and let the weight function v be defined by*

$$v(t) = t\int_t^b \frac{u(x)}{x^2}\,dx, \quad t \in (0, b). \tag{3.1}$$

If the function φ *is integrable and convex on* $[0,b)$ *and* $K(x) = x\varphi(x)$, *then the inequality*

$$\begin{aligned} &\int_0^b K(f(x))\frac{v(x)}{x}dx - \int_0^b K\left(\frac{1}{x}\int_0^x (f(t)dt)\right)\frac{u(x)}{x}dx \\ &\geq \int_0^b\int_t^b f(t)\left(f(t)-\frac{1}{x}\int_0^x f(\tau)\,d\tau\right)C_\varphi\left(\frac{1}{x}\int_0^x f(\tau)\,d\tau\right)\frac{u(x)}{x^2}dxdt \\ &= \int_0^b\int_t^b \left(f(t)-\frac{1}{x}\int_0^x f(\tau)\,d\tau\right)^2 C_\varphi\left(\frac{1}{x}\int_0^x f(\tau)\,d\tau\right)\frac{u(x)}{x^2}dxdt \end{aligned} \tag{3.2}$$

holds for all non-negative locally integrable functions f *and with* C_φ *defined by* (2.3). *If* φ *is concave, then* (3.2) *holds in the reverse direction.*

Example. From Proposition 3.1 for $\varphi(x) = x^{p-1}$, $p \geq 2$ (therefore $C_\varphi(x) = \varphi'(x) = (p-1)x^{p-2}$), choosing $u(x) = 1$, we find that

$$\begin{aligned} &\int_0^b \left(1-\frac{x}{b}\right) f^p(x)\frac{dx}{x} - \int_0^b \left(\frac{1}{x}\int_0^x f(t)\,dt\right)^p \frac{dx}{x} \\ &\geq (p-1)\int_0^b\int_t^b \left(f(t)-\frac{1}{x}\int_0^x f(\tau)\,d\tau\right)^2\left(\frac{1}{x}\int_0^x f(\tau)\,d\tau\right)^{p-2}\frac{dx}{x^2}dt. \end{aligned} \tag{3.3}$$

The inequality (3.3) holds in the reverse direction for $1 < p < 2$.

By using (3.3) we are now ready to derive our new refined Hardy type inequality with breaking point $p = 2$.

Theorem 3.2. *Let* $p \geq 2$, $k > 1$, $0 < b \leq \infty$, *and let the function* f *be non-negative and locally integrable on* $(0,b)$. *Then*

$$\begin{aligned} &\left(\frac{p}{k-1}\right)^p \int_0^b \left(1-\left(\frac{x}{b}\right)^{\frac{k-1}{p}}\right) x^{p-k} f^p(x)\,dx - \int_0^b x^{-k}\left(\int_0^x f(t)\,dt\right)^p dx \\ &\geq \frac{(p-1)}{p}(k-1)\int_0^b\int_t^b \left(f(t)\frac{p}{k-1}\left(\frac{t}{x}\right)^{1-\frac{k-1}{p}} - \frac{1}{x}\int_0^x f(\sigma)\,d\sigma\right)^2 \\ &\cdot\left(\frac{1}{x}\int_0^x f(\sigma)\,d\sigma\right)^{p-2} x^{\left(1-\frac{k-1}{p}\right)(p+1)} t^{\frac{k-1}{p}-1}\frac{dx}{x^2}dt. \end{aligned} \tag{3.4}$$

Moreover, the double integral of the right-hand side of (3.4) *is non-negative. If* $1 < p \leq 2$, *then the inequality* (3.4) *holds in reverse direction. Equality holds when* $p = 2$.

Next we formulate a result which is similar to Proposition 3.1 but where $\varphi(x)$ is superquadratic instead of convex.

Proposition 3.3. *Let* $0 < b \leq \infty$, $u : (0,\infty) \to \mathbb{R}$ *be a non-negative weight function such that* $\frac{u(x)}{x^2}$ *is locally integrable on* $(0,\infty)$ *and let the weight function* v *be* *defined by*

$$v(t) = t\int_t^b \frac{u(x)}{x^2}dx, \quad t \in (0,b). \tag{3.5}$$

If the function φ is integrable and superquadratic on $[0,b)$ and $K(x) = x\varphi(x)$, then the inequality

$$\begin{aligned}
&\int_0^b K(f(x))\frac{v(x)}{x}dx - \int_0^b K\left(\frac{1}{x}\int_0^x (f(t)dt)\right)\frac{u(x)}{x}dx \qquad (3.6)\\
&\geq \int_0^b\int_t^b f(t)\left(f(t) - \frac{1}{x}\int_0^x f(\tau)\,d\tau\right)C_\varphi\left(\frac{1}{x}\int_0^x f(\tau)\,d\tau\right)\frac{u(x)}{x^2}dxdt\\
&\quad + \int_0^b\int_t^b f(t)\varphi\left(\left|f(t) - \frac{1}{x}\int_0^x f(\tau)\,d\tau\right|\right)\frac{u(x)}{x^2}dxdt\\
&= \int_0^b\int_t^b \left(f(t) - \frac{1}{x}\int_0^x f(\tau)\,d\tau\right)^2 C_\varphi\left(\frac{1}{x}\int_0^x f(\tau)\,d\tau\right)\frac{u(x)}{x^2}dxdt\\
&\quad + \int_0^b\int_t^b f(t)\varphi\left(\left|f(t) - \frac{1}{x}\int_0^x f(\tau)\,d\tau\right|\right)\frac{u(x)}{x^2}dxdt,
\end{aligned}$$

holds for all non-negative locally integrable functions f and with C_φ defined by (2.4). *If the function φ is subquadratic, then* (3.6) *holds in the reverse direction.*

By using (3.6) we are now ready to state our next new refined Hardy type inequality with breaking point $p = 3$.

Theorem 3.4. *Let $p \geq 3$, $k > 1$, $0 < b \leq \infty$, and let the function f be non-negative and locally integrable on $(0,b)$. Then*

$$\begin{aligned}
&\left(\frac{p}{k-1}\right)^p \int_0^b \left(1 - \left(\frac{x}{b}\right)^{\frac{k-1}{p}}\right) x^{p-k} f^p(x)\,dx - \int_0^b x^{-k}\left(\int_0^x f(t)\,dt\right)^p dx \qquad (3.7)\\
&\geq \frac{p-1}{p}(k-1)\int_0^b\int_t^b \left(f(t)\frac{p}{k-1}\left(\frac{t}{x}\right)^{1-\frac{k-1}{p}} - \frac{1}{x}\int_0^x f(\sigma)\,d\sigma\right)^2\\
&\quad \cdot \left(\frac{1}{x}\int_0^x f(\sigma)\,d\sigma\right)^{p-2} x^{\left(1-\frac{k-1}{p}\right)(p+1)} t^{\frac{k-1}{p}-1}\frac{dx}{x^2}dt\\
&\quad + \int_0^b\int_t^b \left(f(t)\,t^{1-\frac{k-1}{p}}\right)\left(\left|f(t)\frac{p}{k-1}\left(\frac{t}{x}\right)^{1-\frac{k-1}{p}} - \frac{1}{x}\int_0^x f(\sigma)\,d\sigma\right|\right)^{p-1}\\
&\quad \cdot x^{\left(1-\frac{k-1}{p}\right)p} t^{\frac{k-1}{p}-1}\frac{dx}{x^2}dt.
\end{aligned}$$

Moreover, each double integral of the right-hand side of (3.7) *is non-negative. If $1 < p \leq 3$, then the inequality* (3.7) *holds in the reverse direction. Equality holds when $p = 3$.*

4. Proofs

Proof of Proposition 3.1. Let us choose in (2.2) the probability measure $d\mu(t) = \frac{1}{x}dt$, $0 \le t \le x$. Then

$$\begin{aligned} &\frac{1}{x}\int_0^x K(f(t))\,dt - K\left(\frac{1}{x}\int_0^x f(t)\,dt\right) \qquad (4.1)\\ &\ge C_\varphi\left(\frac{1}{x}\int_0^x f(t)\,dt\right)\frac{1}{x}\int_0^x f(t)\left(f(t)-\frac{1}{x}\int_0^x f(\tau)\,d\tau\right)dt\\ &= C_\varphi\left(\frac{1}{x}\int_0^x f(t)\,dt\right)\frac{1}{x}\int_0^x \left(f(t)-\frac{1}{x}\int_0^x f(\tau)\,d\tau\right)^2 dt. \end{aligned}$$

Multiplying (4.1) by $\frac{u(x)}{x}$ and integrating over $0 \le x \le b$, we get that

$$\begin{aligned} &\int_0^b\int_0^x K(f(t))\,dt\frac{u(x)}{x^2}dx - \int_0^b K\left(\frac{1}{x}\int_0^x f(t)\,dt\right)\frac{u(x)}{x}dx \qquad (4.2)\\ &\ge \int_0^b\int_0^x \left(f(t)-\frac{1}{x}\int_0^x f(\tau)\,d\tau\right)^2 C_\varphi\left(\frac{1}{x}\int_0^x f(\tau)\,d\tau\right)\frac{u(x)}{x^2}dtdx. \end{aligned}$$

Now, by using (3.1) and Fubini's theorem, we find that

$$\begin{aligned} \int_0^b\int_0^x K(f(t))\frac{u(x)}{x^2}dtdx &= \int_0^b \frac{K(f(t))}{t}\left(t\int_t^b \frac{u(x)}{x^2}dx\right)dt \qquad (4.3)\\ &= \int_0^b K(f(t))\frac{v(t)}{t}dt = \int_0^b K(f(x))\frac{v(x)}{x}dx. \end{aligned}$$

(4.2) and (4.3) lead to (3.2). In the case that $K(x) = x\varphi(x)$ and φ is concave then (4.1), (4.2) and (4.3) hold in the reverse direction, which leads to the reverse of (3.2). The proof is complete. □

Proof of Theorem 3.2. We denote the right-hand side of (3.3) by R and replace the parameter b by $b^{\frac{k-1}{p}}$ and $f(x)$ by $f\left(x^{\frac{p}{k-1}}\right)x^{\frac{p}{k-1}-1}$. Then

$$\begin{aligned} R = &\int_0^{b^{\frac{k-1}{p}}}\int_t^{b^{\frac{k-1}{p}}}\left(f\left(t^{\frac{p}{k-1}}\right)t^{\frac{p}{k-1}-1} - \frac{1}{x}\int_0^x f\left(\tau^{\frac{p}{k-1}}\right)\tau^{\frac{p}{k-1}-1}d\tau\right)^2 \qquad (4.4)\\ &\cdot (p-1)\left(\frac{1}{x}\int_0^x f\left(\tau^{\frac{p}{k-1}}\right)\tau^{\frac{p}{k-1}-1}d\tau\right)^{p-2}\frac{dx}{x^2}dt. \end{aligned}$$

We now make the substitutions

$$y = x^{\frac{p}{k-1}} \quad \text{and} \quad s = t^{\frac{p}{k-1}} \Leftrightarrow x = y^{\frac{k-1}{p}} \quad t = s^{\frac{k-1}{p}}$$

from which it follows that

$$t = b^{\frac{k-1}{p}} \Rightarrow s = b, \quad x = b^{\frac{k-1}{p}} \Rightarrow y = b, \quad dt = \tfrac{k-1}{p}s^{\frac{k-1}{p}-1}ds, \quad \tfrac{k-1}{p}ds = t^{\frac{p}{k-1}-1}dt,$$

$$dx = y^{\frac{k-1}{p}-1}\tfrac{k-1}{p}dy, \quad dy = \tfrac{p}{k-1}x^{\frac{p}{k-1}-1}dx, \quad \text{and} \quad t^{\frac{p}{k-1}-1} = s^{1-\frac{k-1}{p}}.$$

By using these substitutions we get from (4.4) that

$$R=(p-1)\left(\frac{k-1}{p}\right)^{p+2}\int_0^b\int_s^b\left(\frac{p}{k-1}f(s)\left(\frac{s}{y}\right)^{\left(1-\frac{k-1}{p}\right)}-\frac{1}{y}\int_0^y f(\sigma)\,d\sigma\right)^2$$
$$\cdot\left(\frac{1}{y}\int_0^y f(\sigma)\,d\sigma\right)^{p-2} y^{\left(1-\frac{k-1}{p}\right)(p+1)} s^{\frac{k-1}{p}-1}\frac{dy}{y^2}ds\,. \tag{4.5}$$

Now we make the same changes on the left-hand side of (3.3), denoted by L, that is, we replace b by $b^{\frac{k-1}{p}}$ and $f(x)$ by $f\left(x^{\frac{p}{k-1}}\right)x^{\frac{p}{k-1}-1}$ and by the substitution $y=x^{\frac{p}{k-1}}$ we get that

$$L=\int_0^b\frac{k-1}{p}\left(1-\left(\frac{y}{b}\right)^{\frac{k-1}{p}}\right)y^{p-k}\left(f(y)\right)^p dy \tag{4.6}$$
$$-\left(\frac{k-1}{p}\right)^{p+1}\int_0^b y^{-k}\left(\int_0^y f(s)\,ds\right)^p dy.$$

Therefore from (4.4)–(4.6), after dividing L and R by $((k-1)/p)^{p+1}$, we get (3.4).

The reverse of the crucial inequality (3.4) holds for $1<p\le 2$ because in this case the function $\varphi(x)=x^{p-1}$, $x>0$, is concave. Hence the proof follows in the same way also in this case.

This completes the proof of the theorem. □

Proof of Proposition 3.3. Let us choose in (2.6) the probability measure $d\mu(t)=\frac{1}{x}dt$, $0\le t\le x$. Then

$$\frac{1}{x}\int_0^x K(f(t))\,dt-K\left(\frac{1}{x}\int_0^x f(t)\,dt\right) \tag{4.7}$$
$$\ge C_\varphi\left(\frac{1}{x}\int_0^x f(t)\,dt\right)\frac{1}{x}\int_0^x f(t)\left(f(t)-\frac{1}{x}\int_0^x f(\tau)\,d\tau\right)dt$$
$$+\frac{1}{x}\int_0^x f(t)\,\varphi\left(\left|f(t)-\frac{1}{x}\int_0^x f(\tau)\,d\tau\right|\right)dt$$
$$=C_\varphi\left(\frac{1}{x}\int_0^x f(t)\,dt\right)\frac{1}{x}\int_0^x\left(f(t)-\frac{1}{x}\int_0^x f(\tau)\,d\tau\right)^2 dt$$
$$+\frac{1}{x}\int_0^x f(t)\,\varphi\left(\left|f(t)-\frac{1}{x}\int_0^x f(\tau)\,d\tau\right|\right)dt.$$

Multiplying (4.7) by $\frac{u(x)}{x}$ and integrating over $0\le x\le b$, we find that

$$\int_0^b\int_0^x K(f(t))\,dt\frac{u(x)}{x^2}dx-\int_0^b K\left(\frac{1}{x}\int_0^x f(t)\,dt\right)\frac{u(x)}{x}dx \tag{4.8}$$
$$\ge\int_0^b\int_0^x f(t)\left(f(t)-\frac{1}{x}\int_0^x f(\tau)\,d\tau\right)C_\varphi\left(\frac{1}{x}\int_0^x f(\tau)\,d\tau\right)\frac{u(x)}{x^2}dtdx$$

$$+\int_0^b\int_0^x f(t)\,\varphi\left(\left|f(t)-\frac{1}{x}\int_0^x f(\tau)\,d\tau\right|\right)\frac{u(x)}{x^2}dtdx$$
$$=\int_0^b\int_0^x\left(f(t)-\frac{1}{x}\int_0^x f(\tau)\,d\tau\right)^2 C_\varphi\left(\frac{1}{x}\int_0^x f(\tau)\,d\tau\right)\frac{u(x)}{x^2}dtdx$$
$$+\int_0^b\int_0^x f(t)\,\varphi\left(\left|f(t)-\frac{1}{x}\int_0^x f(\tau)\,d\tau\right|\right)\frac{u(x)}{x^2}dtdx.$$

Now, by using (3.5) and Fubini's theorem, we obtain that

$$\int_0^b\int_0^x K(f(t))\frac{u(x)}{x^2}dtdx=\int_0^b\frac{K(f(t))}{t}\left(t\int_t^b\frac{u(x)}{x}dx\right)dt \tag{4.9}$$
$$=\int_0^b K(f(t))\frac{v(t)}{t}dt=\int_0^b K(f(x))\frac{v(x)}{x}dt.$$

By combining (4.8) and (4.9) we obtain (3.6).

If φ is subquadratic, then the crucial inequality (2.6) holds in the reverse direction. Hence, the proof follows in the same way also in this case. □

Proof of Theorem 3.4. First we note that by using Proposition 3.3 with $\varphi(x)=x^{p-1}$, $p\geq 3$ (therefore $C_\varphi(x)=\varphi'(x)=(p-1)\,x^{p-2}$), choosing $u(x)=1$, we obtain that

$$\int_0^b\left(1-\frac{x}{b}\right)f^p(x)\frac{dx}{x}-\int_0^b\left(\frac{1}{x}\int_0^x f(t)\,dt\right)^p\frac{dx}{x} \tag{4.10}$$
$$\geq\int_0^b\int_t^b f(t)\left(f(t)-\frac{1}{x}\int_0^x f(\tau)\,d\tau\right)(p-1)\left(\frac{1}{x}\int_0^x f(\tau)\,d\tau\right)^{p-2}\frac{dx}{x^2}dt$$
$$+\int_0^b\int_t^b f(t)\left(\left|f(t)-\frac{1}{x}\int_0^x f(\tau)\,d\tau\right|\right)^{p-1}\frac{dx}{x^2}dt$$
$$=\int_0^b\int_t^b\left(f(t)-\frac{1}{x}\int_0^x f(\tau)\,d\tau\right)^2(p-1)\left(\frac{1}{x}\int_0^x f(\tau)\,d\tau\right)^{p-2}\frac{dx}{x^2}dt$$
$$+\int_0^b\int_t^b f(t)\left(\left|f(t)-\frac{1}{x}\int_0^x f(\tau)\,d\tau\right|\right)^{p-1}\frac{dx}{x^2}dt.$$

By now making the same steps and variable substitutions as in the proof of Theorem 3.2 but now with (4.10) as the crucial inequality instead of (3.3) we obtain the proof of Theorem 3.4. We leave out the details. □

References

[1] S. Abramovich, G. Jameson and G. Sinnamon, Refining Jensen's inequality, Bull. Math. Soc. Sci. Math. Roumanie (N.S.) 47 (95), (2004), 3–14.

[2] G.H. Hardy, Note on a theorem of Hilbert, Math. Z. 6 (1920), 314–317.

[3] G.H. Hardy, Notes on some points in the integral calculus, LX. An inequality between integrals, Messenger Math. 54 (1925), 150–156.

[4] G.H. Hardy, Notes on some points in the integral calculus, LXIV. Further inequalities between integrals, Messenger of Math. 57 (1928), 12–16.

[5] C.O. Imoru, On some integral inequalities related to Hardy's, Canad. Math. Bull 20(3) (1977), 307–312.

[6] V. Kokitashivili, A Meshki and L.-E. Persson, Weighted norm inequalities for integral transform with product kernels, Mathematics Research Development Series, Nova Science Publishers Inc., New York, 2010.

[7] A. Kufner, L. Maligranda and L.-E. Persson, The Hardy inequality. About its history and related results, Vydavatelsky Servis, Pilzen, 2007.

[8] A. Kufner and L.-E. Persson, Weighted inequalities of Hardy type, World Scientific Publishing Co., Inc., River Edge, NJ, 2003.

[9] J.A. Oguntuase and L.-E. Persson, Refinement of Hardy's inequalities via superquadratic and subquadratic functions, J. Math. Anal. Appl. 339 (2008) 1305–1312.

[10] J.A. Oguntuase and L.-E. Persson, Refinement of Hardy's inequality for "all" p. Banach and function spaces II, Yokohama Publ., Yokohama, 2008, 129–144.

[11] J.A. Oguntuase and L.-E. Persson, Hardy type inequalities via convexity – the journey so far, Aust. J. Math. Anal. Appl. 7 (2010), no 2, Art. 18, 19pp.

[12] L.-E. Persson and N. Samko, What should have happened if Hardy had discovered this?, J. Inequal. Appl., 2012:29.

S. Abramovich
Department of Mathematics
University of Haifa
Haifa, 34987 Israel
e-mail: `abramos@math.haifa.ac.il`

L.-E. Persson
Department of Mathematics
Lulea University of Technology
SE-971871, Lulea, Sweden

and

Narvik University College
P.O.Box 385
N-8505, Narvik, Norway
e-mail: `larserik@ltu.se`

Operator Theory:
Advances and Applications, Vol. 236, 11–41

Non-negative Self-adjoint Extensions in Rigged Hilbert Space

Yury Arlinskiĭ and Sergey Belyi

Abstract. We study non-negative self-adjoint extensions of a non densely defined non-negative symmetric operator $\dot{A}$ with the exit in the rigged Hilbert space constructed by means of the adjoint operator $\dot{A}^*$ (bi-extensions). Criteria of existence and descriptions of such extensions and associated closed forms are obtained. Moreover, we introduce the concept of an extremal non-negative bi-extension and provide its complete description. After that we state and prove the existence and uniqueness results for extremal non-negative bi-extensions in terms of the Kreĭn–von Neumann and Friedrichs extensions of a given non-negative symmetric operator. Further, the connections between positive boundary triplets and non-negative self-adjoint bi-extensions are presented.

Mathematics Subject Classification (2010). Primary 47A10, 47B44; Secondary 46E20, 46F05.

Keywords. Non-negative symmetric operator, self-adjoint bi-extension, non-negative self-adjoint bi-extension, extremal bi-extension.

1. Introduction

In order to describe the main ideas and results of the current paper, we first recall the notion of the rigged Hilbert spaces. A triplet $\mathcal{H}_+ \subset \mathcal{H} \subset \mathcal{H}_-$ is a rigged Hilbert space constructed upon a symmetric operator $\dot{A}$ in a Hilbert space $\mathcal{H}$ if $\mathcal{H}_+ = \mathrm{Dom}(\dot{A}^*)$ with an inner product defined by

$$(f,g)_+ = (f,g) + (\dot{A}^* f, \dot{A}^* g), \quad f,g \in \mathrm{Dom}(A^*). \tag{1.1}$$

and $\mathcal{H}_-$ is the space of all anti-linear functional on $\mathcal{H}_+$ that are continuous w.r.t. $\|\cdot\|_+$. An extension theory of symmetric operators in rigged Hilbert spaces was thoroughly covered in [7]. One of the objects of this theory is a self-adjoint bi-extension $\mathbb{A}$ of a symmetric operator $\dot{A}$ whose definition is given below in Preliminaries section. Throughout this entire article, by a non-negative operator in a rigged Hilbert space we understand an operator $\mathbb{T}$ such that $(\mathbb{T}f, f) \geq 0$ for all $f \in \text{Dom}(\mathbb{T})$. In this paper we put our main focus on non-negative bi-extensions of a non-negative symmetric operator. The theory of extensions of non-negative symmetric operators originates in the works of von Neumann, Friedrichs, and Kreĭn (see survey [12]). That is why most of the main results of the paper are given in terms of the Kreĭn–von Neumann and Friedrichs extensions of a given non-negative symmetric operator that are described in details in Section 3. The existence conditions for non-negative bi-extensions are presented in Section 4 and rely on the concepts if disjointness and transversality of self-adjoint extensions that were introduced in Preliminaries. Here we also give a descriptions of the non-negative self-adjoint bi-extensions and associated closed quadratic forms. Section 5 is solely dedicated to extremal self-adjoint bi-extensions and contains existence and uniqueness results. The connections between non-negative self-adjoint bi-extensions and boundary triplets is established in Section 6.

The results of the current paper complement and enhance the classical results of the theory of extensions of non-negative symmetric operators as well as some new developments of this theory in rigged Hilbert spaces discussed in [7], [8]. Applications of these results may be used in solving realization problems for Stieltjes and inverse Stieltjes functions in infinite-dimensional Hilbert spaces similarly to finite-dimensional cases treated in [13] and [14].

2. Preliminaries

For a pair of Hilbert spaces $\mathcal{H}_1$, $\mathcal{H}_2$ we denote by $[\mathcal{H}_1, \mathcal{H}_2]$ the set of all bounded linear operators from $\mathcal{H}_1$ to $\mathcal{H}_2$. Let $\dot{A}$ be a closed, densely defined, symmetric operator in a Hilbert space $\mathcal{H}$ with inner product $(f, g), f, g \in \mathcal{H}$.

Consider the rigged Hilbert space (see [15], [31]) $\mathcal{H}_+ \subset \mathcal{H} \subset \mathcal{H}_-$, where $\mathcal{H}_+ = \text{Dom}(\dot{A}^*)$ and $(f, g)_+$ is defined by (1.1). Note that by the second representation theorem [20] we have

$$\text{Dom}(I + \dot{A}\dot{A}^*)^{1/2} = \mathcal{H}_+, \ \text{Ran}(I + \dot{A}\dot{A}^*)^{1/2} = \mathcal{H},$$

and

$$(f, g)_+ = ((I + \dot{A}\dot{A}^*)^{1/2} f, (I + \dot{A}\dot{A}^*)^{1/2} g), \quad f, g \in \mathcal{H}.$$

The Hilbert space $\mathcal{H}_+$ admits the following (+)-orthogonal decomposition

$$\mathcal{H}_+ = \text{Dom}(\dot{A}) \oplus \mathfrak{N}_{-i} \oplus \mathfrak{N}_i,$$

where $\mathfrak{N}_\lambda := \ker(\dot{A}^* - \lambda I)$, $\text{Im}\,\lambda \neq 0$ is the defect subspace of $\dot{A}$. Denote

$$\mathfrak{M} = \mathfrak{N}_{-i} \oplus \mathfrak{N}_i$$

and let

$$P^+_{\text{Dom}(\dot{A})}, \ P^+_{\mathfrak{N}_{-i}}, \ P^+_{\mathfrak{N}_i}, \ P^+_{\mathfrak{M}}$$

be (+)-orthogonal projections in $\mathcal{H}_+$ onto $\text{Dom}(\dot{A})$, $\mathfrak{N}_{-i}$, $\mathfrak{N}_i$, and $\mathfrak{M}$, respectively.

Recall that $\mathcal{H}_-$ can be identified with the space of all anti-linear functional on $\mathcal{H}_+$ and continuous w.r.t. $\|\cdot\|_+$. Let $\mathcal{R}$ be the *Riesz–Berezansky operator* (see [7]) which maps $\mathcal{H}_-$ onto $\mathcal{H}_+$ such that $(f,g) = (f,\mathcal{R}g)_+$ and $\|\mathcal{R}g\|_+ = \|g\|_-$ for all $f \in \mathcal{H}_+$, $g \in \mathcal{H}_-$. Clearly

$$\mathcal{R}\restriction\mathcal{H} = (I + \dot{A}\dot{A}^*)^{-1}.$$

Definition 2.1. Let $\mathbb{A}$ be a linear operator with $\operatorname{Dom}(\mathbb{A})$ dense in $\mathcal{H}_+$ and with values in $\mathcal{H}_-$. Then the adjoint operator $\mathbb{A}^*$ is defined as follows:

$$\begin{aligned}&\operatorname{Dom}(\mathbb{A}^*) = \{u \in \mathcal{H}_+ : \exists\, \psi \in \mathcal{H}_- \mid (u,\mathbb{A}f) = (\psi, f) \text{ for all } f \in \operatorname{Dom}(\mathbb{A})\},\\ &\quad\mathbb{A}^*u = \psi.\end{aligned}$$

It is easy to see $\mathcal{R}\mathbb{A}^* : \mathcal{H}_+ \supseteq \operatorname{Dom}(\mathbb{A}^*) \to \mathcal{H}_+$ is the $(+)$-adjoint operator to $\mathcal{R}\mathbb{A}$ acting in $\mathcal{H}_+$.

Definition 2.2. An operator $\mathbb{A} : \mathcal{H}_+ \supset \operatorname{Dom}(\mathbb{A}) \to \mathcal{H}_-$ is called a *generalized self-adjoint* if $\operatorname{Dom}(\mathbb{A})$ is dense in $\mathcal{H}_+$ and $\mathbb{A}^* = \mathbb{A}$.

Definition 2.3. A generalized self-adjoint operator $\mathcal{H}_+ \supset \operatorname{Dom}(\mathbb{A}) \to \mathcal{H}_-$ is called *self-adjoint bi-extension* of a symmetric operator $\dot{A}$ if $\mathbb{A} \supset \dot{A}$.

The formula (see [9], [7])

$$\mathbb{A} = \dot{A}^* + \mathcal{R}^{-1}\left(\mathcal{S} - \frac{i}{2}P^+_{\mathfrak{N}_i} + \frac{i}{2}P^+_{\mathfrak{N}_{-i}}\right)P^+_{\mathfrak{M}} = \dot{A}^* + \mathcal{R}^{-1}\left(\mathcal{S} - \frac{1}{2}\dot{A}^*\right)P^+_{\mathfrak{M}} \tag{2.1}$$

establishes a one-to-one correspondence between the set of all self-adjoint bi-extensions of $\dot{A}$ and the set of all $(+)$-self-adjoint operators $\mathcal{S}$ in $\mathfrak{M}$.

Let $\mathbb{A}$ be a self-adjoint bi-extension of $\dot{A}$ and let the operator $\widehat{A}$ in $\mathcal{H}$ be defined as follows:

$$\operatorname{Dom}(\widehat{A}) = \{f \in \mathcal{H}_+ : \widehat{A}f \in \mathcal{H}\}, \quad \widehat{A} = \mathbb{A}\restriction\operatorname{Dom}(\widehat{A}).$$

The operator $\widehat{A}$ is called a *quasi-kernel* of a self-adjoint bi-extension $\mathbb{A}$ (see [31]). We say that a self-adjoint bi-extension $\mathbb{A}$ of $\dot{A}$ is *twice-self-adjoint* or *t-self-adjoint* (see [7]) if its quasi-kernel $\widehat{A}$ is a self-adjoint operator in $\mathcal{H}$.

For the existence, description, and analog of von Neumann's formulas for bounded self-adjoint bi-extensions and $(*)$-extensions see [7] and references therein. In what follows we suppose that $\dot{A}$ has equal deficiency indices. Recall that two self-adjoint extensions A_1 and A_0 of $\dot{A}$ are called *disjoint* if

$$\operatorname{Dom}(A_1) \cap \operatorname{Dom}(A_0) = \operatorname{Dom}(\dot{A}) \tag{2.2}$$

and *transversal* if

$$\operatorname{Dom}(A_1) + \operatorname{Dom}(A_0) = \operatorname{Dom}(\dot{A}^*).$$

Note that it immediately follows from von Neumann formulas that two transversal self-adjoint extensions are automatically disjoint.

The following statements for two self-adjoint extensions A_1 and A_0 of $\dot{A}$ are evident:

$$A_1, A_0 \text{ are disjoint} \iff \overline{\operatorname{Ran}}\left((A_1-\lambda I)^{-1}-(A_0-\lambda I)^{-1}\right)=\mathfrak{N}_\lambda,$$
$$A_1, A_0 \text{ are transversal} \iff \operatorname{Ran}\left((A_1-\lambda I)^{-1}-(A_0-\lambda I)^{-1}\right)=\mathfrak{N}_\lambda$$

for at least one $\lambda \in \rho(A_1)\cap\rho(A_0)$.

Thus, if the deficiency numbers of $\dot{A}$ are finite (and equal), then two self-adjoint extensions of $\dot{A}$ are transversal if and only they are disjoint.

Let $\dot{A}$ be a closed densely defined symmetric operator and let A_1 be its self-adjoint extension. It has been shown in [2], [9] that any self-adjoint bi-extension $\mathbb{A}$ of $\dot{A}$ such that $\mathbb{A}\supset A_1$ is *generated* by a disjoint to A_1 self-adjoint extension A_0 of $\dot{A}$ via the formulas

$$\begin{aligned}\operatorname{Dom}(\mathbb{A})&=\operatorname{Dom}(A_1)+\operatorname{Dom}(A_0),\\ \mathbb{A}f&=\dot{A}^*f-\mathcal{R}^{-1}\dot{A}^*\mathcal{P}_G f,\quad f\in\operatorname{Dom}(\mathbb{A}),\end{aligned}$$

where $\mathcal{P}_G$ is a skew projection operator in $\operatorname{Dom}(\mathbb{A})$ onto G parallel to $\operatorname{Dom}(A_1)$ and G is defined from the $(+)$-orthogonal decomposition

$$\operatorname{Dom}(A_0)=\operatorname{Dom}(\dot{A})\oplus G. \tag{2.3}$$

The operator $\mathcal{S}$ corresponding to $\mathbb{A}$ in (2.1) is of the form

$$\begin{aligned}\mathcal{S}f&=\frac{1}{2}\dot{A}^*f,\qquad f\in\operatorname{Dom}(A_1)\ominus\operatorname{Dom}(\dot{A}),\\ \mathcal{S}g&=-\frac{1}{2}\dot{A}^*g,\quad g\in\operatorname{Dom}(A_0)\ominus\operatorname{Dom}(\dot{A}).\end{aligned}\tag{2.4}$$

In particular,

$$\mathbb{A}g=(\dot{A}^*-\mathcal{R}^{-1}\dot{A}^*P^+_{\mathfrak{M}})g,\quad g\in\operatorname{Dom}(A_0).$$

The following formula immediately follows from (2.3)

$$(\mathbb{A}f,f)=(A_1f_1,f_1)+(A_0f_0,f_0)+2\operatorname{Re}(A_1f_1,f_0), \tag{2.5}$$

where $f=f_1+f_0$, $f_l\in\operatorname{Dom}(A_l)$, $(l=0,1)$.

Let $\mathbb{A}$ be a self-adjoint bi-extension of $\dot{A}$. We define a *dual* extension $\mathbb{A}'$ on $\operatorname{Dom}(\mathbb{A})$ by the formula

$$(\mathbb{A}'f,g)=(\dot{A}^*f,g)+(f,\dot{A}^*g)-(\mathbb{A}f,g),\quad f,g\in\operatorname{Dom}(\mathbb{A}). \tag{2.6}$$

We note that $\dot{A}^*\in[\mathcal{H}_+,\mathcal{H}]\subset[\mathcal{H}_+,\mathcal{H}_-]$ and the *generalized adjoint* of $\dot{A}^*$ takes the form [7]

$$\left(\dot{A}^*\right)^*=\dot{A}^*-\mathcal{R}^{-1}\dot{A}^*P^+_{\mathfrak{M}}. \tag{2.7}$$

It follows from (2.1) that if

$$\mathbb{A}=\dot{A}^*+\mathcal{R}^{-1}\left(\mathcal{S}-\frac{i}{2}P^+_{\mathfrak{N}_i}+\frac{i}{2}P^+_{\mathfrak{N}_{-i}}\right)P^+_{\mathfrak{M}}$$

is a self-adjoint bi-extension of $\dot{A}$, then $\mathbb{A}'$ is of the form

$$\mathbb{A}' = \dot{A}^* + \mathcal{R}^{-1}\left(-\mathcal{S} - \frac{i}{2}P^+_{\mathfrak{N}_i} + \frac{i}{2}P^+_{\mathfrak{N}_{-i}}\right)P^+_{\mathfrak{M}}.$$

So, if $\mathbb{A}$ is a self-adjoint bi-extension of $\dot{A}$, then $\mathbb{A}'$ is a self-adjoin bi-extension of $\dot{A}$ as well. It was also shown in [2] that if $\mathbb{A}$ is a t-self-adjoint of $\dot{A}$, then $\mathbb{A}'$ is also a t-self-adjoint bi-extension of $\dot{A}$. Moreover, if $\hat{A}$ is a quasi-kernel of $\mathbb{A}$ and $\mathbb{A}$ is generated by a disjoint to $\hat{A}$ self-adjoint extension A, then the quasi-kernel of $\mathbb{A}'$ coincides with A and $\mathbb{A}'$ is generated by $\hat{A}$. Clearly, $(\mathbb{A}')' = \mathbb{A}$.

Notice that from (2.6) and the inequality

$$2|(\dot{A}^*f, f)| \le 2||f||\,||\dot{A}^*f|| \le ||f||^2 + ||\dot{A}^*f||^2 = ||f||^2_+,$$

we get

$$-||f||^2_+ \le (\mathbb{A}f, f) + (\mathbb{A}'f, f) \le ||f||^2_+.$$

3. The Friedrichs and Kreĭn–von Neumann extensions

Let $\tau[\cdot,\cdot]$ be a sesquilinear form in a Hilbert space $\mathcal{H}$ defined on a linear manifold $\mathrm{Dom}(\tau)$. The form τ is called symmetric if $\tau[u,v] = \overline{\tau[v,u]}$ for all $u, v \in \mathrm{Dom}(\tau)$ and non-negative if $\tau[u] := \tau[u,u] \ge 0$ for all $u \in \mathrm{Dom}(\tau)$.

A sequence $\{u_n\}$ is called *τ-converging* to the vector $u \in \mathcal{H}$ [20] if

$$\lim_{n\to\infty} u_n = u \quad \text{and} \quad \lim_{n,m\to\infty} \tau[u_n - u_m] = 0.$$

The form τ is called *closed* if for every sequence $\{u_n\}$ τ- converging to a vector u it follows that $u \in \mathrm{Dom}(\tau)$ and $\lim\limits_{n\to\infty} \tau[u - u_n] = 0$. The form τ is *closable* [20], i.e., there exists a minimal closed extension (the closure) of τ. We recall that a symmetric operator $\dot{B}$ is called *non-negative* if

$$(\dot{B}f, f) \ge 0, \quad \forall f \in \mathrm{Dom}(\dot{B}).$$

If τ is a closed, densely defined non-negative form, then according to First Representation Theorem [23], [20] there exists a unique self-adjoint non-negative operator T in $\mathfrak{H}$, associated with τ, i.e.,

$$(Tu, v) = \tau[u,v] \quad \text{for all} \quad u \in \mathrm{Dom}(T) \quad \text{and} \quad \text{for all} \quad v \in \mathrm{Dom}(\tau).$$

According to the Second Representation Theorem [23], [20] the identities hold:

$$\mathrm{Dom}(\tau) = \mathrm{Dom}(T^{1/2}), \quad \tau[u,v] = (T^{1/2}u, T^{1/2}v).$$

Let $\dot{B}$ be a non-negative symmetric operator in a Hilbert space $\mathcal{H}$. It is known [20] that the non-negative sesquilinear form $\tau_{\dot{B}}[f,g] = (\dot{B}f, g)$, $\mathrm{Dom}(\tau) = \mathrm{Dom}(\dot{B})$, is closable. Following the M. Kreĭn notations we denote by $\dot{B}[\cdot,\cdot]$ the closure of $\tau_{\dot{B}}$ and by $\mathcal{D}[\dot{B}]$ its domain. By definition $\dot{B}[u] = \dot{B}[u,u]$ for all $u \in \mathcal{D}[\dot{B}]$. Because $\dot{B}[u,v]$ is closed, it possesses the property: if

$$\lim_{n\to\infty} u_n = u \quad \text{and} \quad \lim_{n,m\to\infty} \dot{B}[u_n - u_m] = 0,$$

then $\lim_{n\to\infty} \dot{B}[u - u_n] = 0$. For a densely defined $\dot{B}$, the *Friedrichs extension* B_F of $\dot{B}$ is defined as a non-negative self-adjoint operator associated with the form $\dot{B}[\cdot,\cdot]$ by the First Representation Theorem. If $\dot{B}$ is densely defined then, clearly,

$$\mathrm{Dom}(B_F) = \mathcal{D}[\dot{B}] \cap \mathrm{Dom}(\dot{B}^*), \quad B_F = \dot{B}^* \upharpoonright \mathrm{Dom}(B_F).$$

The Friedrichs extension B_F is a unique non-negative self-adjoint extension having the domain in $\mathcal{D}[\dot{B}]$. Notice that by the Second Representation Theorem [20] one has

$$\mathcal{D}[\dot{B}] = \mathcal{D}[B_F] = \mathrm{Dom}(B_F^{1/2}), \quad \dot{B}[u,v] = (B_F^{1/2}u, B_F^{1/2}v), \quad u,v \in \mathcal{D}[\dot{B}].$$

If $\dot{B}$ is non-densely defined, then its Friedrichs extension B_F is a non-negative linear relation of the form (see [28])

$$B_F = \left\{ \left\langle x, (\dot{B}_0)_F x \right\rangle,\ x \in \mathrm{Dom}((\dot{B}_0)_F) \right\} \oplus \langle 0, \mathfrak{B} \rangle,$$

where $(B_0)_F$ is the Friedrichs extension of the operator $\dot{B}_0 := P_{\overline{\mathrm{Dom}}(\dot{B})}\dot{B}$ in the subspace $\overline{\mathrm{Dom}}(\dot{B})$ and $\mathfrak{B} = \mathcal{H} \ominus \mathrm{Dom}(\dot{B})$.

The Kreĭn–von Neumann extension is defined as follows [1], [16]:

$$\dot{B}_K = ((\dot{B}^{-1})_F)^{-1},$$

where $\dot{B}^{-1}$ is the linear relation inverse to the graph of $\dot{B}$.

Theorem 3.1 ([1]). *The following relations describing $\mathcal{D}[B_K]$ and $B_K[u]$ hold:*

$$\begin{aligned} \mathcal{D}[B_K] &= \left\{ u \in \mathcal{H} : \sup_{f \in \mathrm{Dom}(\dot{B})} \frac{|(\dot{B}f,u)|^2}{(\dot{B}f,f)} < \infty \right\}, \\ B_K[u] &= \sup_{f \in \mathrm{Dom}(\dot{B})} \frac{|(\dot{B}f,u)|^2}{(\dot{B}f,f)}, \quad u \in \mathcal{D}[B_K]. \end{aligned} \tag{3.1}$$

We note the equalities for an arbitrary non-negative self-adjoint operator B in a Hilbert space $\mathcal{H}$:

$$\mathrm{Ran}(B^{1/2}) = \left\{ g \in \mathcal{H} : \sup_{f \in \mathrm{Dom}(B)} \frac{|(f,g)|^2}{(Bf,f)} < \infty \right\},$$

$$\|B^{[-1/2]}g\|^2 = \sup_{f \in \mathrm{Dom}(B)} \frac{|(f,g)|^2}{(Bf,f)}, \quad g \in \mathrm{Ran}(B^{1/2}),$$

where $B^{[-1]}$ is the Moore–Penrose inverse. The Kreĭn–von Neumann extension of a non-densely defined non-negative operator $\dot{B}$ is an operator (not just a linear relation) if and only if the domain $\mathcal{D}[B_K]$ is dense in $\mathfrak{H}$. According to [1] a non-negative operator $\dot{B}$ is called *positively closable* if from $\lim_{n\to\infty} \dot{B}\varphi_n = g$ and $\lim_{n\to\infty} (\dot{B}\varphi_n, \varphi_n) = 0$ follows $g = 0$ ($\{\varphi_n\} \subset \mathrm{Dom}(\dot{B})$). Notice that a densely defined $\dot{B}$ is positively closable. A theorem of Ando and Nishio [1] states that $\dot{B}$

admits non-negative self-adjoint extensions, which are operators, if and only if $\dot{B}$ is positively closable.

A non-negative self-adjoint extension $\widetilde{B}$ of $\dot{B}$ is called *extremal* [3], [5], [6] if the relation

$$\inf\left\{\left(\widetilde{B}(u-\varphi), u-\varphi\right) : \varphi \in \mathrm{Dom}(\dot{B})\right\} = 0$$

holds for every $u \in \mathrm{Dom}(\widetilde{B})$. A characterization of the Kreĭn–von Neumann extension B_K is obtained in [5] and [6]: *the Kreĭn–von Neumann extension B_K is the unique extremal non-negative self-adjoint extension of $\dot{B}$ having maximal domain of its closed associated sesquilinear form.*

Theorem 3.2. *Let $\widetilde{B}$ be a non-negative self-adjoint extension of $\dot{B}$. Then*

$$B_K \le \widetilde{B} \le B_F \tag{3.2}$$

in the sense of quadratic forms. More precisely

$$\begin{aligned} &\mathcal{D}[\dot{B}] \subseteq \mathcal{D}[\widetilde{B}] \subseteq \mathcal{D}[B_K], \\ &\widetilde{B}[u] \ge B_K[u] && \text{for all} \quad u \in \mathcal{D}[\widetilde{B}], \\ &\widetilde{B}[v] = \dot{B}[v] && \text{for all} \quad v \in \mathcal{D}[\dot{B}]. \end{aligned}$$

Besides,

$$\mathcal{D}[\widetilde{B}] = \mathcal{D}[\dot{B}] \dot{+} (\mathcal{D}[\widetilde{B}] \cap \mathcal{N}_z), \tag{3.3}$$

where $\mathcal{N}_z$ is the defect subspace of $\dot{B}$, $z \in \mathbb{C} \setminus [0, +\infty)$.

For a densely defined non-negative $\dot{B}$ inequalities (3.2) in the equivalent form

$$(B_F + I)^{-1} \le (\widetilde{B} + I)^{-1} \le (B_K + I)^{-1}$$

and equality (3.3) for $z < 0$ were established by M. Kreĭn [23]. For a sectorial operator $\dot{B}$ with vertex at zero and for sectorial linear relations all statements of Theorem 3.2 can be found in [5] and [6].

The next theorem gives a descriptions of all closed forms associated with non-negative self-adjoint extensions of $\dot{B}$.

Theorem 3.3 ([5]). *If $\widetilde{B}$ is a non-negative self-adjoint extension of a non-negative symmetric operator $\dot{B}$, then the form*

$$(\widetilde{B}u, v) - B_K[u, v], \quad u, v \in \mathrm{Dom}(\widetilde{B})$$

is non-negative and closable in the Hilbert space $\mathcal{D}[B_K]$. Moreover, the formulas

$$\begin{aligned} &\mathcal{D}[\widetilde{B}] = \mathcal{D}[\tau], \\ &\widetilde{B}[u, v] = B_K[u, v] + \tau[u, v], \ u, v \in \mathcal{D}[\widetilde{B}] \end{aligned}$$

give a one-to-one correspondence between all closed forms $\widetilde{S}[\cdot,\cdot]$ associated with non-negative self-adjoint extensions $\widetilde{B}$ of $\dot{B}$ and all non-negative forms $\tau[\cdot,\cdot]$ closed in the Hilbert space $\mathcal{D}[B_K]$ and such that $\tau[\varphi] = 0$ for all $\varphi \in \mathcal{D}[\dot{B}]$.

In addition, the closed forms associated with extremal extensions are closed restrictions of the form $B_K[\cdot,\cdot]$ on the linear manifolds $\mathcal{M}$ such that

$$\mathcal{D}[\dot{B}] \subseteq \mathcal{M} \subseteq \mathcal{D}[B_K].$$

The next theorem can be found in [29], [30], [6], [19].

Theorem 3.4. *Let $\dot{B}$ be a bounded non-densely defined non-negative symmetric operator in a Hilbert space $\mathcal{H}$, $\mathrm{Dom}(\dot{B}) = \mathcal{H}_0$. Let $\dot{B}^* \in [\mathcal{H}, \mathcal{H}_0]$ be the adjoint of $\dot{B}$. Put $\dot{B}_0 = P_{\mathcal{H}_0}\dot{B}$, $\mathcal{N} = \mathcal{H} \ominus \mathcal{H}_0$, where $P_{\mathcal{H}_0}$ is an orthogonal projection in $\mathcal{H}$ onto $\mathcal{H}_0$. Then the following statements are equivalent*

(i) *$\dot{B}$ admits bounded non-negative self-adjoint extensions in $\mathcal{H}$;*

(ii) $\sup\limits_{f\in\mathcal{H}_0} \dfrac{||\dot{B}f||^2}{(\dot{B}f, f)} < \infty$;

(iii) $\dot{B}^*\mathcal{N} \subseteq \mathrm{Ran}(\dot{B}_0^{1/2})$.

Let $\dot{B}$ be a non-negative closed symmetric operator. Consider the symmetric contractions

$$\dot{S} = (I - \dot{B})(I + \dot{B})^{-1},$$

defined on $\mathrm{Dom}(\dot{S}) = (I + \dot{B})\mathrm{Dom}(\dot{B})$. Notice that the orthogonal complement $\mathfrak{N} = \mathcal{H} \ominus \mathrm{Dom}(\dot{S})$ coincides with the defect subspace $\mathfrak{N}_{-1}$ of the operator $\dot{B}$. There is a one-to-one correspondence given by the Cayley transform

$$B = (I - S)(I + S)^{-1}, \quad S = (I - B)(I + B)^{-1},$$

between all non-negative self-adjoint extensions B (linear relations in general) of the operator $\dot{B}$ and all self-adjoint contractive (*sc*) extensions S of $\dot{S}$. As was established by M. Kreĭn in [23], [24] the set of all *sc*-extensions of $\dot{A}$ forms an operator interval $[S_\mu, S_M]$. Following M. Kreĭn's notations we call the endpoints S_μ and S_M by the *rigid* and the *soft* extensions, respectively. They possess the properties

$$\begin{aligned} &\inf_{\varphi\in\mathrm{Dom}(\dot{S})} ((I + S_\mu)(f - \varphi), (f - \varphi) = 0, \\ &\inf_{\varphi\in\mathrm{Dom}(\dot{S})} ((I - S_M)(f - \varphi), (f - \varphi) = 0, \end{aligned} \tag{3.4}$$

for all $f \in \mathcal{H}$. The operator interval $[S_\mu, S_M]$ can be parameterized as follows

$$S = \frac{1}{2}(S_M + S_\mu) + \frac{1}{2}(S_M - S_\mu)^{1/2}X(S_M - S_\mu)^{1/2}, \tag{3.5}$$

where X is a self-adjoint contraction in the subspace $\overline{\mathrm{Ran}(S_M - S_\mu)} (\subseteq \mathfrak{N})$.

Notice that for each $S \in [S_\mu, S_M]$ the equalities (3.4) imply

$$\begin{aligned} &\inf_{\varphi\in\mathrm{Dom}(\dot{S})} ((I + S)(f - \varphi), (f - \varphi) = ((S - S_\mu)f, f), \\ &\inf_{\varphi\in\mathrm{Dom}(\dot{S})} ((I - S)(f - \varphi), (f - \varphi) = ((S_M - S)f, f), \; f \in \mathcal{H}. \end{aligned} \tag{3.6}$$

Using the relation (see [23])

$$\inf_{\varphi\in\operatorname{Dom}(\dot S)}((I+S)(f-\varphi),(f-\varphi)=||P_\Omega(I+S)^{1/2}f||^2,$$

where

$$\Omega=\{g\in\mathcal{H}:(I+S)^{1/2}g\in\mathfrak{N}\},$$

from (3.6) we get the equalities

$$\begin{aligned}(I+S)^{1/2}\Omega&=\operatorname{Ran}((S-S_\mu)^{1/2}),\\ ||(I+S)^{[-1/2]}f||&=||(S-S_\mu)^{[-1/2]}f||^2,\quad f\in\operatorname{Ran}((S-S_\mu)^{1/2}).\end{aligned}\tag{3.7}$$

Let L be a bounded non-negative self-adjoint operator in the Hilbert space $\mathcal{H}$ and let $\mathcal{M}$ be a subspace in $\mathcal{H}$. The Kreĭn shorted operator $L_\mathcal{M}$ [23], [1] is given by the following definition

$$L_\mathcal{M}=\max\{X\le L\mid \operatorname{Ran}(X)\subseteq\mathcal{M}\}.$$

It is shown in [23], that

$$L_\mathcal{M}=L^{1/2}QL^{1/2},\tag{3.8}$$

where Q is an orthoprojection operator onto the subspace $\operatorname{Ran}(Q)=(L^{1/2})^{-1}(\mathcal{M})$. Moreover, [23]

$$(L_\mathcal{M}f,f)=\inf_{\varphi\in\mathcal{H}\ominus\mathcal{M}}(L(f-\varphi),f-\varphi),\ f\in\mathcal{H}.\tag{3.9}$$

Thus, from (3.6) we have

$$(I+S)_\mathfrak{N}=S-S_\mu,\ (I-S)_\mathfrak{N}=S_M-S.$$

The next result describes the sesquilinear form $B[u,v]$ by the means of the fractional-linear transformation $S=(I-B)(I+B)^{-1}$. The following proposition can be found in [7].

Proposition 3.5.

(1) *Let B be a non-negative self-adjoint operator and let $S=(I-B)(I+B)^{-1}$ be its Cayley transform. Then*

$$\begin{aligned}\mathcal{D}[B]&=\operatorname{Ran}((I+S)^{1/2}),\\ B[u,v]&=-(u,v)+2\left((I+S)^{-1/2}u,(I+S)^{-1/2}v\right),\quad u,v\in\mathcal{D}[B].\end{aligned}$$

(2) *Let $\dot B$ be a closed densely defined non-negative symmetric operator and let B be its non-negative self-adjoint extension. If $\dot S=(I-\dot B)(I+\dot B)^{-1}$, $S=(I-B)(I+B)^{-1}$, then*

$$\mathcal{D}[B]=\operatorname{Ran}(I+S_\mu)^{1/2}\dotplus\operatorname{Ran}(S-S_\mu)^{1/2}.\tag{3.10}$$

We note that $\operatorname{Ran}(B^{1/2})=\operatorname{Ran}((I-S)^{1/2})$. Now let S_μ and S_M be the rigid and the soft extensions of $\dot S$. Then the Friedrichs and Kreĭn–von Neumann extensions of $\dot B$ are given by

$$B_F=(I-S_\mu)(I+S_\mu)^{-1},\ B_K=(I-S_M)(I+S_M)^{-1}.$$

4. Non-negative self-adjoint bi-extensions

4.1. Disjointness and tranversality of non-negative self-adjoint extensions

Proposition 4.1. *Let $\dot{A}$ be a non-negative closed densely defined operator. Then the following statements hold true for a non-negative self-adjoint extensions A of $\dot{A}$:*

$$A \text{ is disjoint with } A_F \iff \mathcal{D}[A]\cap\mathcal{H}_+ \text{ is dense in } \mathcal{H}_+,$$
$$A \text{ is transversal with } A_F \iff \mathcal{D}[A]\supset\mathcal{H}_+.$$

Proof. Using equality (3.3) in the form

$$\mathcal{D}[A]=\mathcal{D}[\dot{A}]\dot{+}(\mathfrak{N}_{-1}\cap\mathcal{D}[A])$$

and the relation $\mathrm{Dom}(A_F)=\mathcal{D}[\dot{A}]\cap\mathrm{Dom}(\dot{A}^*)$, we get that

$$\mathcal{D}[A]\cap\mathcal{H}_+=\mathrm{Dom}(A_F)\dot{+}(\mathfrak{N}_{-1}\cap\mathcal{D}[A]), \tag{4.1}$$

where $\mathfrak{N}_\lambda$ is the defect subspace of $\dot{A}$. Taking into account the equality

$$\mathcal{H}_+=\mathrm{Dom}(A_F)\dot{+}\mathfrak{N}_{-1},$$

we get that $\mathcal{D}[A]\cap\mathcal{H}_+$ is dense in $\mathcal{H}_+$ if and only if $\mathfrak{N}_{-1}\cap\mathcal{D}[A]$ is dense in $\mathfrak{N}_{-1}$ and

$$\mathcal{D}[A]\cap\mathcal{H}_+=\mathcal{H}_+\iff\mathfrak{N}_{-1}\subset\mathcal{D}[A].$$

Put

$$\dot{S}=(I-\dot{A})(I+\dot{A}),\quad S_\mu=(I-A_F)(I+A_F),\quad S=(I-A)(I+A).$$

Then

$$S-S_\mu=(A+I)^{-1}-(A_F+I)^{-1}. \tag{4.2}$$

Now the equality (see (3.10))

$$\mathcal{D}[A]=\mathcal{D}[\dot{A}]\dot{+}\mathrm{Ran}(S-S_\mu)^{1/2} \tag{4.3}$$

implies the validity of the statements in the proposition. □

From (4.2) and (4.3) we get the following equalities

$$\mathcal{D}[A]\cap\mathcal{H}_+=\mathrm{Dom}(A_F)\dot{+}\mathrm{Ran}(S-S_\mu)^{1/2}=\mathrm{Dom}(A)\dot{+}\mathrm{Ran}(S-S_\mu)^{1/2}.$$

Notice that the equivalence

$$A_F \text{ and } A_K \text{ are transversal} \iff \mathrm{Dom}(\dot{A}^*)\subseteq\mathcal{D}[A_K]$$

has been shown in [25] (see also [11]). The next statement provides one more criteria for A_F and A_K to be transversal.

Proposition 4.2.

$$A_F \text{ and } A_K \text{ are transversal} \iff \sup_{f\in\mathrm{Dom}(\dot{A})}\frac{\|(I+\dot{A}\dot{A}^*)^{-1/2}\dot{A}f\|^2}{(\dot{A}f,f)}<\infty. \tag{4.4}$$

Proof. Let A be a non-negative self-adjoint extension of $\dot{A}$. Since $A^{1/2}$ is closed in $\mathcal{H}$, the closed graph theorem yields that

$$\mathcal{H}_+ \subset \mathcal{D}[A] = \mathrm{Dom}(A^{1/2}) \iff A^{1/2}\upharpoonright \mathcal{H}_+ \in [\mathcal{H}_+, \mathcal{H}],$$

i.e., there exists a number $c > 0$ such that

$$||A^{1/2}u||^2 = A[u] \le c||u||_+^2 \quad \text{for all} \quad u \in \mathcal{H}_+.$$

Take $A = A_K$. Then for $u \in \mathcal{D}[A_K] \cap \mathcal{H}_+$

$$||A_K^{1/2}u||^2 = \sup_{f\in\mathrm{Dom}(\dot{A})} \frac{|(\dot{A}f,u)|^2}{(\dot{A}f,f)} = \sup_{f\in\mathrm{Dom}(\dot{A})} \frac{|(\mathcal{R}\dot{A}f,u)_+|^2}{(\dot{A}f,f)}.$$

Hence

$$|(\mathcal{R}\dot{A}f,u)_+|^2 \le ||A_K^{1/2}u||^2\,(\dot{A}f,f).$$

Then

$$\begin{aligned}
\mathcal{H}_+ \subset \mathcal{D}[A_K] &\iff |(\mathcal{R}\dot{A}f,u)_+|^2 \le c||u||_+^2\,(\dot{A}f,f),\ \forall u \in \mathcal{H}_+,\ \forall f \in \mathrm{Dom}(\dot{A})\\
&\iff ||\mathcal{R}\dot{A}f||_+^2 = \sup_{u\in\mathcal{H}_+} \frac{|(\mathcal{R}\dot{A}f,u)_+|^2}{||u||_+^2} \le c\,(\dot{A}f,f),\ \forall f \in \mathrm{Dom}(\dot{A})\\
&\iff \sup_{f\in\mathrm{Dom}(\dot{A})} \frac{||\mathcal{R}\dot{A}f||_+^2}{(\dot{A}f,f)} < \infty.
\end{aligned}$$

Since

$$||\mathcal{R}g||_+^2 = ||(I+\dot{A}\dot{A}^*)^{-1/2}g||^2,\ g \in \mathcal{H},$$

we arrive at (4.4). □

Notice that due to Theorem 3.4 condition

$$\sup_{f\in\mathrm{Dom}(\dot{A})} \frac{||(I+\dot{A}\dot{A}^*)^{-1/2}\dot{A}f||^2}{(\dot{A}f,f)} < \infty$$

means that the operator $\mathcal{R}\dot{A}$ admits (+)-bounded (+)-self-adjoint non-negative extensions. It is not difficult to show that

$$(I+\dot{A}\dot{A}^*)^{-1/2}\dot{A}f = \dot{A}(I+\dot{A}^*\dot{A})^{-1/2}f, \quad f \in \mathrm{Dom}(\dot{A}).$$

This relation implies that if $\dot{A}$ is positively definite, then A_F and A_K are transversal. Indeed,

$$||(I+\dot{A}\dot{A}^*)^{-1/2}\dot{A}f||^2 = ||\dot{A}(I+\dot{A}^*\dot{A})^{-1/2}f||^2 \le C||f||^2 \le m(\dot{A}f,f),\ f \in \mathrm{Dom}(\dot{A}).$$

Hence,

$$\sup_{f\in\mathrm{Dom}(\dot{A})} \frac{||(I+\dot{A}\dot{A}^*)^{-1/2}\dot{A}f||^2}{(\dot{A}f,f)} < \infty.$$

4.2. Non-negative self-adjoint bi-extensions

Existence. Let $\mathfrak{H}_+ \subset \mathfrak{H} \subset \mathfrak{H}_-$ be a rigged Hilbert space. If $\mathcal{T}$ is a non-negative, densely defined in $\mathfrak{H}_+$ and closed sesquilinear form in $\mathfrak{H}_+$, then there exists a non-negative generalized self-adjoint operator $\mathbb{T}$ acting from $\mathrm{Dom}(\mathbb{T})$ into $\mathcal{H}_-$ *associated* with the form $\mathcal{T}$ in the following sense

$$(\mathbb{T}u, v) = \mathcal{T}[u, v] \text{ for all } u \in \mathrm{Dom}(\mathbb{T}) \text{ and all } v \in \mathrm{Dom}(\mathcal{T}). \tag{4.5}$$

Actually, due to the First Representation Theorem, there is a (+)-non-negative self-adjoint operator $\mathfrak{T}$ associated with the form $\mathcal{T}$ in $\mathfrak{H}_+$, i.e.,

$$(\mathfrak{T}u, v)_+ = \mathcal{T}[u, v] \text{ for all } u \in \mathrm{Dom}(\mathfrak{T}) \text{ and all } v \in \mathrm{Dom}(\mathcal{T}).$$

If $\mathcal{J} \in [\mathfrak{H}_-, \mathfrak{H}_+]$ is the Riesz–Berezansky operator, then $\mathbb{T} = \mathcal{J}^{-1}\mathfrak{T}$ satisfies (4.5). If a non-negative form is defined on $\mathfrak{H}_+$ and is bounded in $\mathfrak{H}_+$, then, clearly, the associated non-negative self-adjoint operator belongs to $[\mathfrak{H}_+, \mathfrak{H}_-]$.

If $\mathfrak{T} : \mathfrak{H}_+ \supseteq \mathrm{Dom}(\mathfrak{T}) \to \mathfrak{H}_-$ is a non-negative generalized self-adjoint operator in the rigged Hilbert space $\mathfrak{H}_+ \subset \mathfrak{H} \subset \mathfrak{H}_-$, i.e., $(\mathfrak{T}f, f) \geq 0$ for all $f \in \mathrm{Dom}(\mathfrak{T})$ and $\mathfrak{T} = \mathfrak{T}^*$, then the sesquilinear form

$$\mathcal{T}_{\mathfrak{T}}[f, g] = (\mathfrak{T}f, g), \quad \mathrm{Dom}(\mathcal{T}_{\mathfrak{T}}) = \mathrm{Dom}(\mathfrak{T})$$

is closable in $\mathfrak{H}_+$. We will denote by $\mathfrak{T}[\cdot, \cdot]$ its closure and by $\mathcal{D}[\mathfrak{T}]$ its domain.

Now we consider a closed non-negative symmetric densely defined operator $\dot{A}$. Let $\mathcal{H}_+ \subset \mathcal{H} \subset \mathcal{H}_-$ be the rigged Hilbert space, where $\mathcal{H}_+ = \mathrm{Dom}(\dot{A}^*)$ and (+)-inner product is defined by (1.1). We are going to study non-negative self-adjoint bi-extensions of the operator $\dot{A}$. Clearly, the operator

$$\dot{B} = \mathcal{R}\dot{A}$$

is non-densely defined in $\mathcal{H}_+$, (+)-bounded and (+)-non-negative. Each non-negative (+)-self-adjoint extension B of $\dot{B}$ in $\mathcal{H}_+$, which is an operator, determines a non-negative self-adjoint bi-extension of $\dot{A}$ by the formula $\mathbb{A} = \mathcal{R}^{-1}B$. Since

$$||\dot{B}\varphi||_+ = ||\mathcal{R}\dot{A}\varphi||_+ = ||(I + \dot{A}\dot{A}^*)^{-1}\dot{A}\varphi||_+ = ||(I + \dot{A}\dot{A}^*)^{-1/2}\dot{A}\varphi||, \ \varphi \in \dot{A},$$

and $(\dot{B}\varphi, \varphi)_+ = (\dot{A}\varphi, \varphi)$, we can use the Ando and Nishio theorem (see [1]) about positively closable symmetric operator and get the following statement.

Proposition 4.3. *A non-negative densely defined closed symmetric operator $\dot{A}$ admits non-negative self-adjoint bi-extension if and only if from*

$$\lim_{n\to\infty} (I + \dot{A}\dot{A}^*)^{-1/2}\dot{A}\varphi_n = g \quad \textit{and} \quad \lim_{n\to\infty} (\dot{A}\varphi_n, \varphi_n) = 0$$

follows $g = 0$, where $\{\varphi_n\} \subset \mathrm{Dom}(\dot{A})$.

Theorem 4.4. *Let $\dot{A}$ be a non-negative closed densely defined operator. The following conditions are equivalent:*

(i) *$\dot{A}$ admits a non-negative self-adjoint bi-extension,*
(ii) *$\dot{A}$ admits t-self-adjoint bi-extension with quasi-kernel A_K,*
(iii) *the Friedrichs and Kreĭn–von Neumann extensions of $\dot{A}$ are disjoint.*

Proof. Clearly (ii)⇒(i). Let us show that (iii)⇒(ii). Suppose that the Friedrichs extension A_F and the Kreĭn–von Neumann extension A_K of the operator $\dot{A}$ are disjoint. Then $\mathrm{Dom}(A_F) + \mathrm{Dom}(A_K)$ is (+)-dense in $\mathcal{H}_+$ or coincides with $\mathcal{H}_+$ (when A_F and A_K are transversal). Then it follows that $\mathcal{D}[A_K] \cap \mathcal{H}_+$ is (+)-dense in $\mathcal{H}_+$ or coincides with $\mathcal{H}_+$. Clearly, the sesquilinear form

$$A_K[u, v], \quad u, v \in \mathcal{D}[A_K] \cap \mathcal{H}_+,$$

is closed in $\mathcal{H}_+$. Because it is at least (+)-densely defined in $\mathcal{H}_+$, there is an associated self-adjoint non-negative operator $\mathbb{A}_K \colon \mathcal{H}_+ \supseteq \mathrm{Dom}(\mathbb{A}_K) \to \mathcal{H}_-$, i.e.,

$$(\mathbb{A}_K u, v) = A_K[u, v] \text{ for all } u \in \mathrm{Dom}(\mathbb{A}_K) \quad \text{and for all} \quad v \in \mathcal{D}[A_K] \cap \mathcal{H}_+.$$

Because $(A_K u, v) = A_K[u, v]$ for all $u \in \mathrm{Dom}(A_K)$ and all $v \in \mathcal{D}[A_K]$, we get that $\mathbb{A}_K \supset A_K$, i.e., the quasi-kernel of $\mathbb{A}_K$ is A_K and therefore, $\mathbb{A}_K$ is t-self-adjoint bi-extension of $\dot{A}$.

Let us prove (i)⇒(iii). Suppose that $\dot{A}$ admits non-negative self-adjoint bi-extensions. Then the Kreĭn–von Neumann extension B_K of the operator $\dot{B} = \mathcal{R}\dot{A}$ in $\mathcal{H}_+$ is an operator. Due to the formula (3.1) the domain $\mathcal{D}[B_K]$ is at least dense in $\mathcal{H}_+$. On the other hand since

$$\frac{|(\dot{B}f, u)_+|^2}{(\dot{B}f, f)_+} = \frac{|(\dot{A}f, u)|^2}{(\dot{A}f, f)},$$

from (3.1) we get

$$\mathcal{D}[B_K] = \mathcal{D}[A_K] \cap \mathcal{H}_+$$

and $B_K[u] = A_K[u]$ for all $u \in \mathcal{D}[A_K] \cap \mathcal{H}_+$. It follows from (4.1) that

$$\mathcal{D}[A_K] \cap \mathcal{H}_+ = \mathrm{Dom}(A_F) \dot{+} (\mathfrak{N}_{-1} \cap \mathcal{D}[A_K]).$$

Therefore, the density of $\mathcal{D}[A_K] \cap \mathcal{H}_+$ implies the density of $\mathfrak{N}_{-1} \cap \mathcal{D}[A_K]$ in $\mathfrak{N}_{-1}$. Equality (3.10) yields that

$$\overline{\mathrm{Ran}}\left((A_K + I)^{-1} - (A_F + I)^{-1}\right) = \mathfrak{N}_{-1},$$

i.e., A_F and A_K are at least disjoint. □

Theorem 4.5.

1) *Let A be a non-negative self-adjoint extension of $\dot{A}$. Then there exists a t-self-adjoint bi-extension $\mathbb{A}$ of $\dot{A}$ with quasi-kernel A if and only if A is disjoint with A_F.*
2) *If a non-negative self-adjoint extension A of $\dot{A}$ is disjoint with A_F, then t-self-adjoint bi-extension $\mathbb{A}$ with quasi-kernel A and generated by A_F is associated with the sesquilinear form $A[u, v]$, $u, v \in \mathcal{D}[A] \cap \mathcal{H}_+$.*

Proof. The form $A[u, v]$ defined on $\mathcal{D}[A] \cap \mathcal{H}_+$ is closed in $\mathcal{H}_+$. By Proposition 4.1 A is disjoint with A_F if and only if the linear manifold $\mathcal{D}[A] \cap \mathcal{H}_+$ is dense in $\mathcal{H}_+$ in which case the non-negative sesquilinear form $A[u, v]$, $u, v \in \mathcal{D}[A] \cap \mathcal{H}_+$ is closed in $\mathcal{H}_+$. The latter implies the existence of a non-negative self-adjoint operator $\mathbb{A} : \mathcal{H}_+ \supseteq \mathrm{Dom}(\mathbb{A}) \to \mathcal{H}_-$ associated with $A[u, v]$, $u, v \in \mathcal{D}[A] \cap \mathcal{H}_+$.

Since $(Au, v) = A[u, v]$ for all $u \in \mathrm{Dom}(A)$ and all $v \in \mathcal{D}[A]$, we get that $\mathbb{A} \supset A$, i.e., the quasi-kernel of $\mathbb{A}$ is A and therefore, A is t-self-adjoint bi-extension of $\dot{A}$. Further we use the following equality (see [6])

$$A[\varphi, u] = (\varphi, \dot{A}^* u), \ \varphi \in \mathcal{D}[\dot{A}], \ u \in \mathcal{D}[A] \cap \mathcal{H}_+.$$

Using (2.7) we get for all $\varphi \in \mathrm{Dom}(A_F)$ and all $u \in \mathcal{D}[A] \cap \mathcal{H}_+$:

$$A[\varphi, u] = (\varphi, \dot{A}^* u) = ((\dot{A}^*)^* \varphi, u) = ((\dot{A}^* - \mathcal{R}^{-1} \dot{A}^* P^+_{\mathfrak{M}}) \varphi, u).$$

Hence, $\mathrm{Dom}(A_F) \subset \mathrm{Dom}(\mathbb{A})$ and

$$\mathbb{A} \varphi = (\dot{A}^* - \mathcal{R}^{-1} \dot{A}^* P^+_{\mathfrak{M}}) \varphi, \ \varphi \in \mathrm{Dom}(A_F).$$

Since $\mathrm{Dom}(A) \subset \mathrm{Dom}(\mathbb{A})$ and $\mathbb{A}$ is a t-self-adjoint bi-extension of $\dot{A}$ with quasi-kernel A, we get

$$\mathrm{Dom}(\mathbb{A}) = \mathrm{Dom}(A) \dotplus \mathrm{Dom}(A_F).$$

Taking into account (2.3), we conclude that $\mathbb{A}$ is generated by A_F. □

The following statement is an immediate consequence of Theorems 4.4 and 4.5.

Corollary 4.6 ([7]). *The operator $\dot{A}$ admits non-negative self-adjoint bi-extensions in $[\mathcal{H}_+, \mathcal{H}_-]$ if and only if A_K and A_F are transversal.*

It was announced in [26] that the transversality condition in Corollary 4.6 is necessary (and sufficient for the case of finite deficiency indices) for the existence of non-negative self-adjoint bi-extensions in $[\mathcal{H}_+, \mathcal{H}_-]$.

Denote by $\mathcal{P}(\dot{A})$ the set of all non-negative self-adjoint bi-extensions of $\dot{A}$. As has been proved in Theorem 4.4 the set $\mathcal{P}(\dot{A})$ is nonempty if and only if A_F and A_K are disjoint in which case the set $\mathcal{P}(\dot{A})$ contains the operator $\mathbb{A}_K$ with the following properties:

1. the operator $\mathbb{A}_K$ is associated with the closed form $A_K[u, v]$, $u, v \in \mathcal{D}[A_K] \cap \mathcal{H}_+$, i.e.,
$$\begin{aligned} \mathcal{D}[\mathbb{A}_K] &= \mathcal{D}[A_K] \cap \mathcal{H}_+, \\ \mathbb{A}_K[u, v] &= A_K[u, v], \quad u \in \mathrm{Dom}(\mathbb{A}_K), \ v \in \mathcal{D}[A_K] \cap \mathcal{H}_+, \end{aligned}$$
2. the operator $\mathbb{A}_K$ is a t-self-adjoint bi-extension of $\dot{A}$ with quasi-kernel A_K and generated by A_F,
3. $\mathcal{P}(\dot{A}) \ni \mathbb{A} \Rightarrow \mathcal{D}[\mathbb{A}] \subseteq \mathcal{D}[\mathbb{A}_K]$, $\mathbb{A}[u] \geq \mathbb{A}_K[u] = A_K[u]$, $u \in \mathcal{D}[\mathbb{A}]$.

Thus, $\mathbb{A}_K$ is the *minimal element* of $\mathcal{P}(\dot{A})$ and is an analog of Kreĭn–von Neumann extension. The minimality property is a consequence of Theorem 3.2. Notice that if A_K and A_F are transversal and the deficiency number of $\dot{A}$ is infinite, then the set $\mathcal{P}(\dot{A})$ contains $+ \to -$ bounded and unbounded operators.

Let A_1 be a non-negative self-adjoint extension of $\dot{A}$. Let $\mathcal{P}(A_1)$ be the set of all non-negative t-self-adjoint bi-extension of $\dot{A}$ with quasi-kernel A_1. According to Theorem 4.5 the set $\mathcal{P}(A_1) \neq \emptyset$ if and only if A_1 is disjoint with A_F. Using Theorem 3.1, and the equalities

$$\mathcal{D}[A_1] = \left\{ f \in \mathcal{H} : \sup_{h \in \mathrm{Dom}(A_1)} \frac{|(A_1 h, f)|^2}{(A_1 h, h)} < \infty \right\},$$

$$\frac{|(\mathcal{R}A_1 h, f)_+|^2}{(\mathcal{R}A_1 h, h)_+} = \frac{|(A_1 h, f)|^2}{(A_1 h, h)}, \ f \in \mathcal{H}_+,$$

we get: if A_1 and A_F are disjoint, then the operator $\mathbb{A}_{1K} : \mathcal{H}_+ \supseteq \mathrm{Dom}(\mathbb{A}_{1K}) \to \mathcal{H}_-$, associated with closed in $\mathcal{H}_+$ non-negative form $A_1[u,v]$, $u, v \in \mathcal{D}[A_1] \cap \mathcal{H}_+$, is the minimal element of the set $\mathcal{P}(A_1)$ in the sense of quadratic forms. According to Theorem 4.5 this operator is generated by A_F. It is an analog of the Kreĭn–von Neumann type extension of A_1 in the rigged Hilbert space $\mathcal{H}_+ \subset \mathcal{H} \subset \mathcal{H}_-$. The operator $\mathbb{A}_K$ is the minimal element of the sets $\mathcal{P}(A_K)$ and $\mathcal{P}(\dot{A})$. The next theorem parameterizes the set $\mathcal{P}(A_1)$.

Theorem 4.7. *Let $\dot{A}$ be a non-negative closed symmetric operator with disjoint non-negative self-adjoint extensions A_F and A_K. Suppose $\mathbb{A}$ is a t-self-adjoint bi-extension of $\dot{A}$ with quasi-kernel A_1 and generated by A_0. Then $\mathbb{A}$ is non-negative if and only if*

$$A_0 \geq A_1 \geq 0.$$

Proof. We will use (2.5)

$$(\mathbb{A}f, f) = (A_1 f_1, f_1) + (A_0 f_0, f_0) + 2\mathrm{Re}\,(A_1 f_1, f_0),$$

for $f = f_1 + f_0$, $f_1 \in \mathrm{Dom}(A_1)$, $f_0 \in \mathrm{Dom}(A_0)$. It follows that $A_1 \geq 0$ and $A_0 \geq 0$. Replacing f_1 by λf_1 and f_0 by μf_0 we have

$$|\lambda|^2 (A_1 f_1, f_1) + |\mu|^2 (A_0 f_0, f_0) + \lambda\bar{\mu}(A_1 f_1, f_0) + \mu\bar{\lambda}(f_0, A_1 f_1) \geq 0$$

for all $\lambda, \mu \in \mathbb{C}$. Thus, the 2×2 matrix

$$\begin{pmatrix} (A_1 f_1, f_1) & (A_1 f_1, f_0) \\ (f_0, A_1 f_1) & (A_0 f_0, f_0) \end{pmatrix}$$

is non-negative. Hence

$$|(A_1 f_1, f_0)|^2 \leq (A_1 f_1, f_1)(A_0 f_0, f_0)$$

and

$$\sup_{f_1 \in \mathrm{Dom}(A_1)} \frac{|(A_1 f_1, f_0)|^2}{(A_1 f_1, f_1)} \leq (A_0 f_0, f_0). \tag{4.6}$$

This means that

$$f_0 \in \mathcal{D}[A_1] \quad \text{and} \quad A_1[f_1] \leq (A_0 f_0, f_0) \ (= A_0[f_0]) \,.$$

If $\{f_2^{(n)}\}_{n=1}^{\infty} \subset \mathrm{Dom}(A_0)$ and A_0-converges to $\varphi_0 \in \mathcal{D}[A_0]$, then (4.6) yields

$$\sup_{f_1 \in \mathrm{Dom}(A_1)} \frac{|(A_1 f_1, \varphi_0)|^2}{(A_1 f_1, f_1)} \leq A_0[\varphi_0].$$

Thus

$$\mathcal{D}[A_0] \subset \mathcal{D}[A_1] \quad \text{and} \quad A_1[\varphi_0] \leq A_0[\varphi_0] \quad \text{for all} \quad \varphi_0 \in \mathcal{D}[A_0],$$

i.e., $A_1 \leq A_0$.

Conversely. Suppose $0 \leq A_1 \leq A_0$. Then for an arbitrary $f_1 \in \mathrm{Dom}(A_1)$, $f_0 \in \mathrm{Dom}(A_0)$ we get

$$\begin{aligned}(\mathbb{A}(f_1+f_0), f_1+f_0) &= (A_1 f_1, f_1) + (A_0 f_0, f_0) + 2\mathrm{Re}\,(A_1 f_1, f_0)\\ &= \|A_1^{1/2} f_1\|^2 + \|A_0^{1/2} f_0\|^2 + 2\mathrm{Re}\,(A_1^{1/2} f_1, A_1^{1/2} f_0)\\ &= \|A_1^{1/2}(f_1+f_0)\|^2 + \|A_0^{1/2} f_0\|^2 - \|A_1^{1/2} f_0\|^2 \geq 0.\end{aligned}$$

This proves the theorem. □

Let A_1 and A_0 be two non-negative self-adjoint extensions of $\dot{A}$. Consider a form defined on $\mathrm{Dom}(A_1) \times \mathrm{Dom}(A_0)$ as follows

$$\mathcal{B}(f_1, f_0) = (A_1 f_1, f_1) + (A_0 f_0, f_0) + 2\mathrm{Re}\,(A_1 f_1, f_0), \tag{4.7}$$

where $f_l \in \mathrm{Dom}(A_l)$, $(l = 0, 1)$. Let

$$\phi_l = \frac{1}{2}(I + A_l) f_l, \qquad S_l \phi_l = \frac{1}{2}(I - A_l) f_l,$$

be the Cayley transform of A_l for $l = 0, 1$. Then

$$f_l = (I + S_l)\phi_l, \qquad A_l f_l = (I - S_l)\phi_l, \quad (l = 0, 1). \tag{4.8}$$

Substituting (4.8) into (4.7) we obtain a form defined on $\mathcal{H} \times \mathcal{H}$

$$\tilde{\mathcal{B}}(\phi_1, \phi_0) = \|\phi_1 + \phi_0\|^2 - \|S_1\phi_1 + S_0\phi_0\|^2 - 2\,\mathrm{Re}\,((S_1 - S_0)\phi_1, \phi_0)\,. \tag{4.9}$$

Let us set

$$F = \frac{1}{2}(S_1 - S_0), \quad G = \frac{1}{2}(S_1 + S_0), \quad u = \frac{1}{2}(\phi_1 + \phi_2), \quad v = \frac{1}{2}(\phi_1 - \phi_0). \tag{4.10}$$

Then

$$\tilde{\mathcal{B}}(\phi_1, \phi_0) = 4H(u, v) := \|u\|^2 + (Fv, v) - (Fu, u) - \|Fv + Gu\|^2. \tag{4.11}$$

Moreover, $F \pm G$ are contractive operators. From the above reasoning we conclude that non-negativity of the form $\mathcal{B}(f_1, f_0)$ on $\mathrm{Dom}(A_1) \times \mathrm{Dom}(A_0)$ is equivalent to non-negativity of the form $H(u, v)$ on $\mathcal{H} \times \mathcal{H}$. The next statement is established in [3], see also [7].

Proposition 4.8. *The form $H(u, v)$ in (4.11) is non-negative for all $u, v \in \mathcal{H}$ if and only if operator F defined in (4.10) is non-negative.*

Proposition 4.8 can be used for another proof of Theorem 4.7 (see [3]).

Let A_1 and A_0 be two disjoint non-negative self-adjoint extensions of $\dot{A}$. We say that A_1 and A_0 form an *admissible pair* $\langle A_1, A_0 \rangle$ if

$$A_0 \geq A_1 \iff (A_1 + I)^{-1} \geq (A_0 + I)^{-1}.$$

If $S_j = (I - A_j))(I + A_j)^{-1}$, $j = 1, 2$, then the pair $\langle A_1, A_0\rangle$ is admissible if and only if $\ker(S_1 - S_0) = \mathrm{Dom}(\dot{S})$ and $S_1 \geq S_0$. Let X_j, $j = 0, 1$ be self-adjoint contractions in $\mathfrak{N}$ such that

$$S_j = \frac{1}{2}(S_M + S_\mu) + \frac{1}{2}(S_M - S_\mu)^{1/2} X_j (S_M - S_\mu)^{1/2}.$$

Then it follows from (3.5) that the pair of non-negative self-adjoint extensions $A_j = (I - S_j)(I + S_j)^{-1}$, $j = 0, 1$ is admissible if and only if

$$\ker(X_1 - X_0) \cap \mathrm{Ran}((S_M - S_\mu)^{1/2}) = \{0\} \quad \text{and} \quad X_1 - X_0 \geq 0.$$

Associated closed forms. The next statement describes $\mathbb{A}[u, v]$ (the closure of the form $(\mathbb{A}f, f)$), where $\mathbb{A}$ is a non-negative t-self-adjoint bi-extension of $\dot{A}$ with quasi-kernel A_1 and generated by A_0 (compare with Theorem 3.3).

Theorem 4.9. *Let $\langle A_1, A_0\rangle$ be an admissible pair and let $\mathbb{A}$ be a non-negative t-self-adjoint bi-extension of $\dot{A}$ with quasi-kernel A_1 and generated by A_0. Let $\mathbb{A}[\cdot,\cdot]$ be the closure of the form $(\mathbb{A}f, g)$, $f, g \in \mathrm{Dom}(\dot{A})$. Then*

$$\begin{aligned}
\mathcal{D}[\mathbb{A}] &= \mathrm{Dom}(A_1)\dot{+}\mathrm{Ran}\left((S_1 - S_0)^{1/2}\right) = \mathrm{Dom}(A_0)\dot{+}\mathrm{Ran}\left((S_1 - S_0)^{1/2}\right),\\
\mathbb{A}[u] &= ||h||^2 - ||S_1 h + w||^2 + 2\mathrm{Re}\,(h, w) + 2||(S_1 - S_0)^{-1/2} w||^2\\
&= A_1[u] + ||(S_1 - S_0)^{-1/2} w||^2 - ||(S_1 - S_\mu)^{-1/2} w||^2,\\
u &= (I + S_1)h + w,
\end{aligned} \tag{4.12}$$

where $S_l = (I - A_l)(I + A_l)$, $l = 0, 1$, $h \in \mathcal{H}$, $w \in \mathrm{Ran}\left((S_1 - S_0)^{1/2}\right)$.

Proof. Let $f = f_1 + f_0$, $f_1 \in \mathrm{Dom}(A_1)$, $f_0 \in \mathrm{Dom}(A_0)$. Then

$$(\mathbb{A}f, f) = (A_1 f_1, f_1) + (A_0 f_0, f_0) + 2\mathrm{Re}\,(A_1 f_1, f_0).$$

Due to (4.9)

$$(\mathbb{A}f, f) = \tilde{\mathcal{B}}(\phi_1, \phi_0) = \|\phi_1 + \phi_0\|^2 - \|S_1\phi_1 + S_0\phi_0\|^2 - 2\,\mathrm{Re}\,((S_1 - S_0)\phi_1, \phi_0),$$

where

$$\phi_l = \frac{1}{2}(I + A_l) f_l,\ S_l\phi_l = \frac{1}{2}(I - A_l) f_l,\ f_l = (I + S_l)\phi_l,\ A_l f_l = (I - S_l)\phi_l,\ l = 0, 1.$$

Represent $f = f_1 + f_0 = (I + S_1)\phi_1 + (I + S_0)\phi_0$ in the form

$$f = (I + S_1)(\phi_1 + \phi_0) - (S_1 - S_0)\phi_0.$$

Then

$$\begin{aligned}
(\mathbb{A}f, f) = \|\phi_1 + \phi_0\|^2 &- \|S_1(\phi_1 + \phi_0) - (S_1 - S_0)\phi_0\|^2\\
&- 2\,\mathrm{Re}\,(\phi_1 + \phi_0, (S_1 - S_0)\phi_0) + 2||(S_1 - S_0)^{1/2}\phi_0||^2.
\end{aligned} \tag{4.13}$$

Suppose that

$$\lim_{n\to\infty} f^{(n)} = u \text{ in } \mathcal{H}_+, \text{ and } \lim_{n\to\infty} (\mathbb{A}(f^{(n)} - f^{(m)}), f^{(n)} - f^{(m)}) = 0.$$

We have

$$f^{(n)} = (I + S_1)(\phi_1^{(n)} + \phi_0^{(n)}) - (S_1 - S_0)\phi_0^{(n)}.$$

Due to the direct decomposition

$$\mathcal{H}_+ = \mathrm{Dom}(A_1)\dot{+}\mathfrak{N}_{-1}$$

and inclusions $\{(I+S_1)(\phi_1^{(n)}+\phi_0^{(n)})\} \subset \mathrm{Dom}(A_1)$, $\{(S_1-S_0)\phi_0^{(n)}\} \subset \mathfrak{N}_{-1}$, we get that the sequences $\{(I+S_1)(\phi_1^{(n)}+\phi_0^{(n)})\}$ and $\{(S_1-S_0)\phi_0^{(n)}\}$ converge in $\mathcal{H}_+$. By definition $||w||_+^2 = 2||w||^2$, $\forall w \in \mathfrak{N}_{-1}$. Hence $\{(S_1 - S_0)\phi_0^{(n)}\}$ converges in $\mathcal{H}$. On the other hand convergence of $\{(I + S_1)(\phi_1^{(n)} + \phi_0^{(n)})\}$ in $\mathcal{H}_+$ yields convergence of $\{\phi_1^{(n)} + \phi_0^{(n)}\}$ in $\mathcal{H}$. Let

$$\begin{aligned} h &= \lim_{n\to\infty} (\phi_1^{(n)} + \phi_0^{(n)}) \text{ in } \mathcal{H}, \\ \mathrm{Dom}(\mathcal{A}_1) \ni y &= \lim_{n\to\infty} (I + S_1)(\phi_1^{(n)} + \phi_0^{(n)}) \text{ in } \mathcal{H}_+, \\ w' &= \lim_{n\to\infty} (S_1 - S_0)\phi_0^{(n)}. \end{aligned}$$

From $\lim_{n\to\infty} (\mathbb{A}(f^{(n)} - f^{(m)}), f^{(n)} - f^{(m)}) = 0$ and (4.13) we obtain that the sequence $\{(S_1 - S_0)^{1/2}\phi_0^{(n)}\}$ converges in $\mathcal{H}$. Let

$$g = \lim_{n\to\infty} (S_1 - S_0)^{1/2}\phi_0^{(n)}.$$

Then $w' = (S_1 - S_0)^{1/2}g$. Set $w = -w'$. Thus

$$u = y + w,$$

where $y = (I + S_1)h \in \mathrm{Dom}(A_1)$, $w \in \mathrm{Ran}\left((S_1 - S_0)^{1/2}\right)$. We get that

$$\begin{aligned} \lim_{n\to\infty} (\mathbb{A}f^{(n)}, f^{(n)}) &= ||h||^2 - ||S_1h - (S_1 - S_0)^{1/2}g||^2 \\ &\quad - 2\mathrm{Re}\,(h, (S_1 - S_0)^{1/2}g) + 2||g||^2 \\ &= ||h||^2 - ||S_1h + w||^2 + 2\mathrm{Re}\,(h, w) + 2||(S_1 - S_0)^{-1/2}w||^2. \end{aligned}$$

Now let us prove that the quadratic form

$$\begin{aligned} \eta(u) &= ||h||^2 - ||S_1h + w||^2 + 2\mathrm{Re}\,(h, w) + 2||(S_1 - S_0)^{-1/2}w||^2, \\ u &= (I + S_1)h + w,\ h \in \mathcal{H}, w \in \mathrm{Ran}(S_1 - S_0)^{1/2} \end{aligned}$$

is non-negative and closed in $\mathcal{H}_+$ as defined on

$$\mathrm{Dom}(\eta) = \mathrm{Dom}(A_1)\dot{+}\mathrm{Ran}\left((S_1 - S_0)^{1/2}\right).$$

Notice that the equality $S_1 - S_0 = 2(A_1 + I)^{-1} - 2(A_0 + I)^{-1}$ yields

$$\mathrm{Dom}(A_1)\dot{+}\mathrm{Ran}\left((S_1 - S_0)^{1/2}\right) = \mathrm{Dom}(A_0)\dot{+}\mathrm{Ran}\left((S_1 - S_0)^{1/2}\right).$$

First we calculate $A_1[u]$ for $u \in \mathrm{Dom}(A_1)\dot{+}(\mathcal{D}[A_1] \cap \mathfrak{N}_{-1})$. Let us represent u as

$$u = (I + S_1)h + (I + S_1)^{1/2}\omega,$$

where $h \in \mathcal{H}$, $\omega \in \Omega = \{g \in \mathcal{H} : (I+S_1)^{1/2}\omega \in \mathfrak{N}_{-1}\}$. Recall that by (3.7) and (3.10) we have

$$\mathrm{Ran}((S_1 - S_\mu)^{1/2}) = (I+S_1)^{1/2}\Omega = \mathcal{D}[A_1] \cap \mathfrak{N}_{-1}.$$

Using (3.5) we obtain

$$\begin{aligned} A_1[u] =& -||u||^2 + 2||I+S_1)^{-1/2}u||^2 \\ =& -||(I+S_1)h + (I+S_1)^{1/2}\omega||^2 + 2||(I+S_1)^{1/2}h + \omega||^2 \\ =& -||I+S_1)h||^2 - ||(I+S_1)^{1/2}\omega||^2 - 2\mathrm{Re}\,((I+S_1)h, (I+S_1)^{1/2}\omega) \\ &+ 2||(I+S_1)^{1/2}h||^2 + 2||\omega||^2 + 4\mathrm{Re}\,((I+S_1)^{1/2}h, \omega) \\ =& \, ||h||^2 - ||S_1 h||^2 - ||(I+S_1)^{1/2}\omega||^2 - 2\mathrm{Re}\,(S_1 h, (I+S_1)^{1/2}\omega) \\ &+ 2\mathrm{Re}\,(h, (I+S_1)^{1/2}\omega) + 2||\omega||^2 \\ =& \, ||h||^2 - ||S_1 h + (I+S_1)^{1/2}\omega||^2 + 2||\omega||^2 + 2\mathrm{Re}\,(h, (I+S_1)^{1/2}\omega). \end{aligned}$$

Denoting $w = (I+S_1)^{1/2}\omega$ and using the equality (see (3.7)) $||(S_1 - S_\mu)^{-1/2}w|| = ||(I+S_1)^{-1/2}w||^2$, we arrive at the equality

$$A_1[u] = ||h||^2 - ||S_1 h + w||^2 + 2\mathrm{Re}\,(h, w) + 2||(S_1 - S_\mu)^{-1/2}w||^2 \geq 0.$$

Furthermore, since $S_1 - S_\mu \geq S_1 - S_0$, we get that

$$\mathrm{Ran}((S_1 - S_\mu)^{1/2}) \supset \mathrm{Ran}((S_1 - S_0)^{1/2})$$

and $||(S_1 - S_0)^{-1/2}w||^2 \geq ||(S_1 - S_\mu)^{-1/2}w||^2$ for all $w \in \mathrm{Ran}((S_1 - S_0)^{1/2})$. So,

$$\begin{aligned} \eta(u) = A_1[u] + ||(S_1 - S_0)^{-1/2}w||^2 - ||(S_1 - S_\mu)^{-1/2}w||^2 \geq 0, \\ u \in \mathrm{Dom}(A_1)\dot{+}\mathrm{Ran}((S_1 - S_0)^{1/2}) \geq 0. \end{aligned}$$

In addition, one can easily see that the right-hand side of (4.12) is closed on $\mathrm{Dom}(A_1)\dot{+}\mathrm{Ran}((S_1 - S_0)^{1/2})$ in $\mathcal{H}_+$. Now we can conclude that (4.12) is valid. □

Define for $\mathbb{A} \in \mathcal{P}(\dot{A})$ the "*dual*" quadratic form

$$\mathbb{A}'[u] = 2\mathrm{Re}\,(\dot{A}^* u, u) - \mathbb{A}[u], \; u \in \mathcal{D}[\mathbb{A}]$$

and let

$$A'_K[u] = 2\mathrm{Re}\,(\dot{A}^* u, u) - A_K[u], \; u \in \mathcal{D}[A_K] \cap \mathcal{H}_+. \tag{4.14}$$

Recall that a linear operator T in a Hilbert space $\mathfrak{H}$ is called *accretive* [20] if $\mathrm{Re}\,(Tf, f) \geq 0$ for all $f \in \mathrm{Dom}(T)$ and *maximal accretive* (m-accretive) if it is accretive and has no accretive extensions in $\mathfrak{H}$. The following statements are equivalent [27]:

(i) the operator T is m-accretive;
(ii) the operator T is accretive and its resolvent set contains points from the left half-plane;
(iii) the operators T and T^* are accretive.

Theorem 4.10. *If A_F and A_K are disjoint, then each non-negative self-adjoint bi-extension $\mathbb{A}$ of $\dot{A}$ possess the properties*

$$\mathcal{D}[\mathbb{A}] \subseteq \mathcal{D}[A_K], \quad \mathbb{A}[u] \geq A_K[u], \quad \mathbb{A}'[u] \leq A'_K[u], \quad u \in \mathcal{D}[\mathbb{A}]. \tag{4.15}$$

In addition, if T is quasi-selfadjoint accretive extension of $\dot{A}$ ($\dot{A} \subset T \subset \dot{A}^*$), then

$$A_K[u] \leq \mathrm{Re}\,(Tu, u) \leq A'_K[u], \quad u \in \mathrm{Dom}(T). \tag{4.16}$$

Proof. As it follows from the proofs of Theorems 4.4 and 4.5 in $\mathcal{H}_+$ the Kreĭn–von Neumann extension of the operator $\dot{B} = \mathcal{R}\dot{A}$ coincides with the Kreĭn–von Neumann extension of the operator $\dot{B}' = \mathcal{R}A_K$. Therefore, using the minimality of A_K among all non-negative self-adjoint extensions of $\dot{A}$ we arrive at (4.15).

It is established in [4] that for each quasi-self-adjoint accretive extension T of $\dot{A}$ one has

$$\mathrm{Dom}(T) \subset \mathcal{D}[A_K], \quad A_K[u] \leq \mathrm{Re}\,(Tu, u), \quad u \in \mathrm{Dom}(T).$$

Using the above and (4.14) we get (4.16). □

Explicit expressions for non-negative t-self-adjoint bi-extensions. Evidently, the linear manifold $\mathrm{Dom}(A_F)$ is a subspace in $\mathcal{H}_+$. Let $\mathfrak{N}_F$ be the orthogonal complement to $\mathrm{Dom}(\dot{A})$ in $\mathrm{Dom}(A_F)$ with respect to the inner product $(\cdot,\cdot)_+$ and let $\mathfrak{M}_F = \mathcal{H}_+ \ominus \mathrm{Dom}(A_F)$. Then $\mathfrak{M}_F = A_F \mathfrak{N}_F$. Thus we have the (+)-orthogonal decomposition

$$\mathcal{H}_+ = \mathrm{Dom}(\dot{A}) \oplus \mathfrak{N}_F \oplus \mathfrak{M}_F.$$

Let

$$\mathfrak{N}_0 = \mathrm{Ran}(A_F^{1/2}) \cap \mathfrak{N}_F.$$

Clearly, $A_F^{-\frac{1}{2}}(\mathfrak{N}_0) \subset \mathrm{Dom}(A_F)$. The following equalities take place

$$\dot{A}^* A_F e = -e, \; e \in \mathfrak{N}_F,$$
$$A_F \dot{A}^* g = -g, \; g \in \mathfrak{M}_F.$$

Theorem 4.11 ([11]). *The condition $\mathfrak{N}_0 = \{0\}$ is necessary and sufficient for the uniqueness of non-negative self-adjoint extension of $\dot{A}$. Suppose $\mathfrak{N}_0 \neq \{0\}$. Then the formulas*

$$\begin{aligned} \mathrm{Dom}(\widetilde{A}) &= \mathrm{Dom}(\dot{A}) \oplus (I + A_F \widetilde{U})\mathrm{Dom}(\widetilde{U}), \\ \widetilde{A}(x + h + A_F \widetilde{U} h) &= A_F(x + h) - \widetilde{U} h, \quad x \in \mathrm{Dom}(\dot{A}), \; h \in \mathrm{Dom}(\widetilde{U}) \end{aligned} \tag{4.17}$$

give a one-to-one correspondence between all non-negative self-adjoint extensions $\widetilde{A}$ of $\dot{A}$ and all (+)-self-adjoint operators $\widetilde{U}$ in $\mathfrak{N}_F$ satisfying the condition

$$0 \leq \widetilde{U} \leq W_0^{-1}$$

where W_0^{-1} determines the operator inverse with respect to the (+)-non-negative self-adjoint relation W_0 in $\mathfrak{N}_F$ associated with the (+)-closed in $\mathfrak{N}_F$ non-negative form

$$\omega_0[x, y] = (A_F^{[-1/2]} x, A_F^{[-1/2]} y)_+ = (A_F^{1/2} x, A_F^{1/2} y) + (A_F^{[-1/2]} x, A_F^{[-1/2]} y), \; x, y \in \mathfrak{N}_0.$$

Here $A_F^{[-1/2]}$ *is the Moore–Penrose pseudo-inverse. Operator* $\widetilde{A}$ *coincides with the Kreĭn–von Neumann non-negative self-adjoint extension* A_K *if and only if* $\widetilde{U} = W_0^{-1}$.

Moreover,

- *the extensions* A_F *and* A_K *are disjoint* $\Longleftrightarrow$ $\mathfrak{N}_0$ *is dense in* $\mathfrak{N}_F$,
- *the extensions* A_F *and* A_K *are transversal* $\Longleftrightarrow$ $\mathfrak{N}_0 = \mathfrak{N}_F$.

The associated with $\widetilde{A}$ *closed form is given by the following equalities:*

$$\mathcal{D}[\widetilde{A}] = \mathcal{D}[\dot{A}]\dot{+}A_F\mathcal{R}(\widetilde{U}^{1/2}), \tag{4.18}$$

$$\widetilde{A}[\varphi + A_F h] = ||A_F^{1/2}\varphi - A_F^{[-1/2]}h||^2 + \widetilde{U}^{-1}[h] - w_0[h], \quad \varphi \in \mathcal{D}[A],\ h \in \mathcal{R}(\widetilde{U}^{1/2}).$$

Let A_1 and A_0 be two non-negative self-adjoint extensions. From (4.17) and (4.18) it follows that A_1 and A_0, determined by parameters U_1 and U_0, respectively, then

- A_1 and A_0 are disjoint if and only if $\mathfrak{N}_0$ is dense in $\mathfrak{N}_F$ and $\ker(U_1 - U_0) = \{0\}$,
- $A_1 \le A_0$ if and only if $U_1 \ge U_0$,
- $A_1 \le A_0$ and A_1 and A_0 are transversal if and only if $\mathfrak{N}_0 = \mathfrak{N}_F$, $\mathrm{Ran}(U_1) = \mathfrak{N}_F$, $U_1 \ge U_0$, and $\mathrm{Ran}(I - U_1^{-1}U_0) = \mathfrak{N}_F$.

Denote by $P^+_{\mathfrak{N}_F}$, $P^+_{\mathfrak{M}_F}$ the orthogonal projection in $\mathcal{H}_+$ onto $\mathfrak{N}_F$ and $\mathfrak{M}_F = A_F\mathfrak{N}_F$. Notice that

$$\mathfrak{M} = \mathfrak{N}_i \oplus \mathfrak{N}_{-i} = \mathfrak{N}_F \oplus \mathfrak{M}_F.$$

Recall that each self-adjoint bi-extensions of $\dot{A}$ is of the form (2.1), where $\mathcal{S}$ is a (+)-self-adjoint operator in $\mathfrak{M}$.

Theorem 4.12. *Suppose* A_K *and* A_F *are disjoint. Then*

1. *the operator* $\mathbb{A}_K$ *is of the form*
$$\mathbb{A}_K = \dot{A}^* - \mathcal{R}^{-1}A_F(P^+_{\mathfrak{N}_F} + W_0\dot{A}^*P^+_{\mathfrak{M}_F}); \tag{4.19}$$
2. *the operator* $\mathbb{A} = A^* + \mathcal{R}^{-1}(\mathcal{S} - \dot{A}^*/2)P^+_{\mathfrak{M}}$ *belongs to* $\mathcal{P}(\dot{A})$ *if and only if*
$$\mathcal{S} \ge \mathcal{S}_K = -A_F W_0 \dot{A}^* P^+_{\mathfrak{M}_F} + \frac{1}{2}\left(\dot{A}^* P^+_{\mathfrak{M}_F} - A_F P^+_{\mathfrak{N}_F}\right)$$
in the sense of quadratic forms;
3. *if* A_1 *is a non-negative self-adjoint extension of* $\dot{A}$ *disjoint with* A_F *and if* $A_0 \ge A_1$, *then the non-negative t-self-adjoint bi-extension of* $\dot{A}$ *with quasi-kernel* A_1 *and generated by* A_0 *is of the form*
$$\mathbb{A} = \dot{A}^* + \mathcal{R}^{-1}\left(\mathcal{S} - \frac{1}{2}\dot{A}^*\right)P^+_{\mathfrak{M}},$$

where $\mathcal{S}$ is a (+)-self-adjoint operator in $\mathfrak{M}$ given by

$$\begin{cases} \mathrm{Dom}(\mathcal{S}) = (I + A_F U_1)\mathrm{Dom}(U_1) \dot{+} (I + A_F U_0)\mathrm{Dom}(U_0) \\ \mathcal{S}(I + A_F U_1)e = \dfrac{1}{2}(A_F - U_1)e, \ e \in \mathrm{Dom}(U_1) \\ \mathcal{S}(I + A_F U_0)g = \dfrac{1}{2}(-A_F + U_0)g, \ g \in \mathrm{Dom}(U_0) \end{cases}, \tag{4.20}$$

and U_1, U_0 determine A_1 and A_0 in formulas (4.17). *In particular, if $A_0 = A_F$, then*

$$\mathcal{S} = -A_F U_1^{-1} \dot{A}^* P^+_{\mathfrak{M}_F} + \frac{1}{2}\left(\dot{A}^* P^+_{\mathfrak{M}_F} - A_F P^+_{\mathfrak{N}_F}\right). \tag{4.21}$$

Proof. From (4.17) we get the equality

$$\mathrm{Dom}(\widetilde{A}) \ominus \mathrm{Dom}(\dot{A}) = (I + A_F \widetilde{U})\mathrm{Dom}(\widetilde{U})$$

for an arbitrary non-negative self-adjoint extension $\widetilde{A}$ of $\dot{A}$. Then equalities (2.4) yield (4.20). When $A_0 = A_F$, we have $U_0 = 0$. This gives the equality

$$f = (I + A_F U_1)(-U_1^{-1} \dot{A}^*) P^+_{\mathfrak{M}_F} f + (P^+_{\mathfrak{N}_F} + U_1^{-1} \dot{A}^* P^+_{\mathfrak{M}_F}) f.$$

Then by virtue of (4.20) we obtain (4.21). The case $A_1 = A_K$ holds true if and only if $U_1 = W_0^{-1}$ and leads to

$$\mathcal{S}_K = -A_F W_0 \dot{A}^* P^+_{\mathfrak{M}_F} + \frac{1}{2}\left(\dot{A}^* P^+_{\mathfrak{M}_F} - A_F P^+_{\mathfrak{N}_F}\right).$$

Then applying (2.1) we get (4.19). Statement (2.) follows from the fact that $\mathbb{A}_K$ is the minimal element of $\mathcal{P}(\dot{A})$. □

5. Extremal non-negative self-adjoint bi-extensions

Let $\dot{S}$ be a symmetric contraction defined in subspace $\mathrm{Dom}(S)$. We call a sc-extension S of $\dot{S}$ *extremal* if

$$\inf_{g_S \in \mathrm{Dom}(\dot{S})} \|(I - S^2)^{1/2}(g - g_S)\| = 0, \quad \forall g \in \mathcal{H}.$$

We can also offer an equivalent definition of an extremal sc-extension. Let $\mathfrak{N} = \mathcal{H} \ominus \mathrm{Dom}(S)$. We call a sc-extension S of $\dot{S}$ *extremal* if $(I - S^2)_{\mathfrak{N}} = 0$, where $(I - S^2)_{\mathfrak{N}}$ is the Kreĭn shorted operator (see (3.8), (3.9)). The following equality was proved in [10]

$$(I - S^2)_{\mathfrak{N}} = (S_M - S_\mu)^{1/2}(I - X^2)(S_M - S_\mu)^{1/2}, \tag{5.1}$$

where X is corresponding to S (via formula (3.5)) contraction in $\overline{\mathrm{Ran}(S_M - S_\mu)}$. Formula (5.1) implies that S is extremal if and only if X is self-adjoint and unitary, i.e., $X = X^*$ and $X^2 = I$.

Now let $\dot{A}$ be a non-negative closed densely defined symmetric operator. Recall (see Section 3) that a non-negative self-adjoint extension A of $\dot{A}$ is *extremal* [3] if

$$\inf_{\varphi \in \mathrm{Dom}(\dot{A})} (A(h-\varphi), h-\varphi) = 0, \quad \forall h \in \mathrm{Dom}(A).$$

If

$$\dot{S} = (I-\dot{A})(I+\dot{A})^{-1}, \qquad S = (I-A)(I+A)^{-1}, \tag{5.2}$$

then $(Ah,h) = ((I-S^2)g,g)$ where $g = (I+S)^{-1}h$. This yields

$$\inf_{\varphi \in \mathrm{Dom}(\dot{A})} (A(h-\varphi), h-\varphi) = \inf_{g_S \in \mathrm{Dom}(\dot{S})} \|(I-S^2)^{1/2}(g-g_S)\|^2,$$

where $\mathrm{Dom}(\dot{S}) = (I+\dot{A})\mathrm{Dom}(\dot{A})$. Therefore, A is extremal non-negative self-adjoint extension of $\dot{A}$ if and only if S is extremal sc-extension of symmetric contraction $\dot{S}$. The Friedrichs and Kreĭn–von Neumann extensions are extremal.

Let $\mathbb{A}$ be a non-negative self-adjoint bi-extension of the symmetric operator $\dot{A}$. We call the operator $\mathbb{A}$ an *extremal bi-extension* if

$$\inf_{\varphi \in \mathrm{Dom}(\dot{A})} (\mathbb{A}(f-\varphi), f-\varphi) = 0, \quad \forall f \in \mathrm{Dom}(\mathbb{A}).$$

In what follows we assume that the operators A_F and A_K are disjoint.

Theorem 5.1. *A t-self-adjoint bi-extension $\mathbb{A}$ is extremal if and only if it is generated by an admissible pair $\langle A_1, A_0\rangle$ of extremal non-negative self-adjoint extensions of $\dot{A}$.*

Proof. Let A_1 and A_0 be the quasi-kernels of $\mathbb{A}$ and $\mathbb{A}'$, respectively. Let also $\mathbb{A}$ be an extremal self-adjoint bi-extension. It follows from (2.5) then that

$$(\mathbb{A}f_k, f_k) = (A_k f_k, f_k), \quad \forall f_k \in \mathrm{Dom}(A_k),\ k = 0,1.$$

Since $\mathbb{A}$ extends A_1 and is generated by A_0, it follows from (2.5) that

$$(\mathbb{A}f, f) = (A_1 f_1, f_1) + (A_0 f_0, f_0) + 2\mathrm{Re}\,(A_1 f_1, f_0) = \mathcal{B}(f_1, f_0),$$

where $f \in \mathrm{Dom}(\mathbb{A})$, $f = f_1 + f_0$, $f_k \in \mathrm{Dom}(A_k)$, $k = 0,1$. Applying (4.10) and (4.11) we get

$$\begin{aligned}\inf_{\varphi \in \mathrm{Dom}(\dot{A})} (\mathbb{A}(f-\varphi), f-\varphi) &= \inf_{h_S \in \mathrm{Dom}(\dot{S})} H(x-h_S, y) \\ &= \inf_{h_S \in \mathrm{Dom}(\dot{S})} \left(\|x-h_S\|^2 - (x,Fx) + (y,Fy) - \|Fy + G(x-h_s)\|^2\right).\end{aligned} \tag{5.3}$$

Since $\inf_{f_A \in \mathrm{Dom}(\dot{A})} (\mathbb{A}(f_k - f_A), f_k - f_A) = 0$ for all $f_k \in \mathrm{Dom}(A_k)$, $k = 0,1$, the operators A_1 and A_0 are extremal non-negative self-adjoint extensions of $\dot{A}$.

Hence, the extremality of $\mathbb{A}$ implies that the non-negative self-adjoint extensions A_1 and A_0 are also extremal. Since $\mathbb{A}$ is a non-negative self-adjoint bi-extension, then the pair $\langle A_1, A_0\rangle$ is an admissible extremal pair.

Conversely, let us assume that $\langle A_1, A_0\rangle$ is an admissible pair of extremal non-negative self-adjoint extensions of $\dot{A}$. We are going to prove that the corresponding non-negative self-adjoint bi-extension $\mathbb{A}$ with quasi-kernel A_1 and generated by A_0 is extremal. The corresponding (via (5.2)) to A_1 and A_0 sc-extensions S_1 and S_0 are extremal. Also, the fact that $\langle A_1, A_0\rangle$ is an admissible pair, implies that $S_1 - S_0 \geq 0$.

Let

$$S_k = \frac{1}{2}(S_M + S_\mu) + \frac{1}{2}(S_M - S_\mu)^{1/2} X_k (S_M - S_\mu)^{1/2}, \quad k = 0, 1,$$

where X_k, $k = 0, 1$ are self-adjoint contractions in $\mathfrak{N}$. Since S_k, $k = 0, 1$ are extremal sc-extensions, then X_k, $k = 0, 1$, are self-adjoint unitary operators and hence $P_k = (I + X_k)/2$, $k = 0, 1$, are orthogonal projections. Also, $X_1 - X_0 \geq 0$ implies that $P_1 - P_0 \geq 0$ and $\operatorname{Ran}(P_1) \supset \operatorname{Ran}(P_0)$. Since $X_k = 2P_k - I$, $k = 0, 1$, then

$$G = \frac{1}{2}(S_M + S_\mu) + \frac{1}{2}(S_M - S_\mu)^{1/2}(P_1 + P_0 - I)(S_M - S_\mu)^{1/2},$$

and

$$F = (S_M - S_\mu)^{1/2}(P_1 - P_0)(S_M - S_\mu)^{1/2}.$$

Since $I - (P_1 + P_0 - I)^2 = P_1 - P_0$, then (5.1) implies that $(I - G^2)\upharpoonright\mathfrak{N} = F$. Consequently, applying the definition of the operator $(I - G^2)\upharpoonright\mathfrak{N}$ we obtain

$$F = (I - G^2)^{1/2} P_G (I - G^2)^{1/2},$$

where P_G is an orthoprojection onto the subspace

$$\mathcal{H}_G = ((I - G^2)^{1/2})^{-1}\{\mathfrak{N}\} \cap \overline{\operatorname{Ran}}\,((I - G^2)^{1/2}).$$

Therefore,

$$\begin{aligned}
H(x-h_s,y) &= \|x-h_S\|^2 - (x,Fx) + (y,Fy) - \|Fy + G(x-h_s)\|^2\\
&= \|(I-G^2)^{1/2}(x-h_s)\|^2 - \|P_G(I-G^2)^{1/2}x\|^2 + \|P_G(I-G^2)^{1/2}y\|^2\\
&\quad - \|(I-G^2)^{1/2}P_G(I-G^2)^{1/2}y\|^2 - 2\mathrm{Re}((I-G^2)^{1/2}P_G(I-G^2)^{1/2}y, G(x-h_s))\\
&= \|(I-G^2)^{1/2}(x-h_s)\|^2 - \|P_G(I-G^2)^{1/2}x\|^2 + \|GP_G(I-G^2)^{1/2}y\|^2\\
&\quad - 2\mathrm{Re}\Big(GP_G(I-G^2)^{1/2}y, (I-G^2)^{1/2}(x-h_s)\Big)\\
&= \|(I-G^2)^{1/2}(x-h_s) - GP_G(I-G^2)^{1/2}y\|^2 - \|P_G(I-G^2)^{1/2}x\|^2.
\end{aligned}$$

Thus, since $(I - G^2)^{1/2}\operatorname{Dom}(\dot{S}) \perp \mathcal{H}_G$, then

$$\begin{aligned}
\inf_{h_S\in\operatorname{Dom}(\dot{S})} H(x - h_S, y) &= \|P_G(I - G^2)^{1/2}x - P_G G P_G (I - G^2)^{1/2}y\|^2\\
&\quad - \|P_G(I - G^2)^{1/2}x\|^2, \quad \forall x, y \in \mathcal{H}.
\end{aligned} \tag{5.4}$$

Since A_1 and A_0 are extremal non-negative self-adjoint extensions, then the definition of the functional H and (4.11) imply

$$\inf_{h_S\in\operatorname{Dom}(\dot{S})} H(x - h_S, x) = 0, \qquad \inf_{h_S\in\operatorname{Dom}(\dot{S})} H(x - h_S, -x) = 0,$$

for all $x \in \mathcal{H}$. Relation (5.4) yields

$$\|P_G(I-G^2)^{1/2}x - P_G G P_G(I-G^2)^{1/2}x\|^2 - \|P_G(I-G^2)^{1/2}x\|^2 = 0,$$

and

$$\|P_G(I-G^2)^{1/2}x + P_G G P_G(I-G^2)^{1/2}x\|^2 - \|P_G(I-G^2)^{1/2}x\|^2 = 0,$$

for all $x \in \mathcal{H}$. Thus, $P_G G P_G(I-G^2)^{1/2}x = 0$ for all $x \in \mathcal{H}$. Applying (5.4) again we get

$$\inf_{h_S \in \mathrm{Dom}(\dot{S})} H(x - h_S, y) = 0, \qquad \forall x, y \in \mathcal{H}.$$

Now we can use (5.3) to confirm that

$$\inf_{f_A \in \mathrm{Dom}(\dot{A})} (\mathbb{A}(f - f_A), f - f_A) = 0,$$

which means that $\mathbb{A}$ is an extremal non-negative self-adjoint bi-extension. □

Recall that the non-negative self-adjoint bi-extension $\mathbb{A}_K$ is associated with the closed in $\mathcal{H}_+$ form $A_K[u,v]$, $u, v \in \mathcal{D}[A_K] \cap \mathcal{H}_+$. The quasi-kernel of $\mathbb{A}_K$ is the Kreĭn–von Neumann extension A_K and $\mathbb{A}_K$ is generated by A_F. Clearly, $\mathbb{A}_K$ is extremal non-negative self-adjoint bi-extension of $\dot{A}$.

Theorem 5.2.

(1) *Let A_F and A_K be transversal. Then the operator $\mathbb{A}_K$ is the unique extremal non-negative t-self-adjoint bi-extension.*

(2) *Let A_F and A_K be disjoint but not transversal. Then except $\mathbb{A}_K$ there exist infinitely many extremal non-negative t-self-adjoint bi-extensions.*

Proof. (1) Suppose that A_F and A_K are transversal. Let also $\mathbb{A}$ be an extremal t-self-adjoint bi-extension with the quasi-kernel A_1 and A_0 be the quasi-kernel of $\mathbb{A}'$. According to Theorem 5.1 for $S_k = (I - A_k)(I + A_k)^{-1}$, $k = 0, 1$ the following relations hold

$$S_k = S_\mu + (S_M - S_\mu)^{1/2} P_k (S_M - S_\mu)^{1/2}, \qquad k = 0, 1, \tag{5.5}$$

where P_k, $k = 0, 1$, are orthoprojections in $\mathfrak{N}$. Since A_1 and A_0 are disjoint, we have $\ker((S_1 - S_0){\upharpoonright}\mathfrak{N} = \{0\}$. But

$$\ker((S_1 - S_0){\upharpoonright}\mathfrak{N} = \ker((S_M - S_\mu)^{1/2}(P_1 - P_0)(S_M - S_\mu)^{1/2}{\upharpoonright}\mathfrak{N}).$$

Since $P_1 - P_0 \geq 0$, then $Q = P_1 - P_0$ is an orthoprojection. Also, $\mathrm{Ran}(S_M - S_\mu) = \mathfrak{N}$ implies $\ker(P_1 - P_0) = \{0\}$ or equivalently $P_1 - P_0 = I$. The latter yields $P_1 = I$ and $P_0 = 0$. Consequently, $S_1 = S_M$, $S_0 = S_\mu$ and the quasi-kernels of $\mathbb{A}$ and $\mathbb{A}'$ coincide with A_F and A_K.

(2) Let A_F and A_K be disjoint but not transversal. Then $\mathrm{Ran}((S_M - S_\mu)^{1/2}) \neq \mathfrak{N}$ and $\ker((S_M - S_\mu)^{1/2}) = \{0\}$. We chose a subspace $\mathfrak{L} \subset \mathfrak{N}$ in a way that $\mathfrak{L} \cap \mathrm{Ran}(S_M - S_\mu)^{1/2} = \{0\}$. Let $\mathfrak{N}_1$ be such that $\{0\} \subseteq \mathfrak{N}_1 \subseteq \mathfrak{L}$. Let also P_1 be an orthogonal projection operator on $\mathfrak{N} \ominus \mathfrak{N}_1$, Q an orthoprojection on $\mathfrak{N} \ominus \mathfrak{L}$, and $P_0 = P_1 - Q$. Then $P_1 - P_0 = Q \geq 0$ and $\ker(P_1) \cap \mathrm{Ran}(S_M - S_\mu)^{1/2} = \{0\}$. Let S_k, $k = 0, 1$, be defined by (5.5). Hence, S_1 and S_0 are extremal sc-extensions and

A_k, $k = 0, 1$ are extremal non-negative self-adjoint extensions of $\dot{A}$ and $\langle A_1, A_0\rangle$ is an admissible pair. Therefore, according to Theorem 5.1, if $\mathbb{A} \supset A_1$ and $\mathbb{A}$ is generated by A_0, then $\mathbb{A}$ is extremal t-self-adjoint bi-extension of $\dot{A}$. It follows from the construction of $\mathbb{A}$ that there is infinite number of these bi-extensions. □

6. Boundary triplets and self-adjoint bi-extensions

Let $\dot{A}$ be a closed densely defined symmetric operator in $\mathcal{H}$ with equal deficiency numbers.

Definition 6.1 (**[21]**). The triplet $\Pi = \{\mathcal{N}, \Gamma_1, \Gamma_0\}$ is called a *boundary triplet* for $\dot{A}^*$ if $\mathcal{N}$ is a Hilbert space and Γ_0, Γ_1 are bounded linear operators from the Hilbert space $\mathcal{H}_+ = \mathrm{Dom}(\dot{A}^*)$ (with the inner product (1.1)) into $\mathcal{N}$ such that the mapping

$$\Gamma := \langle \Gamma_0, \Gamma_1\rangle : \mathcal{H}_+ \to \mathcal{N} \oplus \mathcal{N},$$

is surjective and the abstract Green identity

$$\left(\dot{A}^* f, g\right) - \left(f, \dot{A}^* g\right) = (\Gamma_1 f, \Gamma_0 g)_{\mathcal{N}} - (\Gamma_0 f, \Gamma_1 g)_{\mathcal{N}},$$

holds for all f, $g \in \mathcal{H}_+$.

It follows from Definition 6.1 (see [17], [18]) that the operators

$$\mathrm{Dom}(A_k) := \ker \Gamma_k, \quad A_k := \dot{A}^* \restriction \mathrm{Dom}(A_k), \quad (k = 0, 1),$$

are self-adjoint extensions of $\dot{A}$. Moreover, they are transversal, i.e.,

$$\mathrm{Dom}(\dot{A}^*) = \mathrm{Dom}(A_0) + \mathrm{Dom}(A_1).$$

Notice that if $\Pi = \{\mathcal{N}, \Gamma_1, \Gamma_0\}$ is a boundary triplet for $\dot{A}^*$, then $\Pi' = \{\mathcal{N}, -\Gamma_0, \Gamma_1\}$ is the boundary triplet for $\dot{A}^*$ too.

We are going to provide connections between self-adjoint bi-extensions and boundary triplets [7]. The proposition below follows from Definition 6.1.

Theorem 6.2. *Let $\dot{A}$ be a closed densely defined symmetric operator with equal deficiency indices in the Hilbert space $\mathcal{H}$. Suppose $\mathcal{N}$ is a Hilbert space, $\Gamma_0, \Gamma_1 \in [\mathcal{H}_+, \mathcal{N}]$, and the operator $\langle \Gamma_0, \Gamma_1\rangle \in [\mathcal{H}_+, \mathcal{N} \oplus \mathcal{N}]$ is surjective. Then the following statements are equivalent.*

(i) $\Pi = \{\mathcal{N}, \Gamma_1, \Gamma_0\}$ *is the boundary triplet for* $\dot{A}^*$*;*

(ii) *the sesquilinear form*

$$w(f, g) := (\dot{A}^* f, g) - (\Gamma_1 f, \Gamma_0 g)_{\mathcal{N}}, \quad f, g \in \mathcal{H}_+ = \mathrm{Dom}(\dot{A}^*) \tag{6.1}$$

is Hermitian, i.e., $w(f, g) = \overline{w(g, f)}$*;*

(iii) *the sesquilinear form*

$$w'(f, g) := (\dot{A}^* f, g) + (\Gamma_0 f, \Gamma_1 g)_{\mathcal{N}}, \quad f, g \in \mathcal{H}_+ = \mathrm{Dom}(\dot{A}^*) \tag{6.2}$$

is Hermitian,

If $\mathcal{H}_+ \subset \mathcal{H} \subset \mathcal{H}_-$ is a rigged Hilbert space, $\mathcal{N}$ is a Hilbert space, and $\Gamma \in [\mathcal{H}_+, \mathcal{N}]$, then by $\Gamma^\times$ we will denote the adjoint operator from $[\mathcal{N}, \mathcal{H}_-]$, i.e., $(\Gamma h, g)_\mathcal{N} = (h, \Gamma^\times g)$ for all $h \in \mathcal{H}_+$ and all $g \in \mathcal{N}$.

The following theorem [7] sets up the connection between boundary triplets and t-self-adjoint bi-extensions.

Theorem 6.3. *Let $\dot{A}$ be a closed densely defined symmetric operator with equal deficiency numbers in the Hilbert space $\mathcal{H}$. Consider the rigged Hilbert space $\mathcal{H}_+ \subset \mathcal{H} \subset \mathcal{H}_-$ generated by $\dot{A}$.*

1. *Let $\Pi = \{\mathcal{N}, \Gamma_1, \Gamma_0\}$ for $\dot{A}^*$ be a boundary triplet for $\dot{A}^*$. Define operators $\mathbb{A}$ and $\mathbb{A}'$*
$$\mathbb{A} := \dot{A}^* - \Gamma_0^\times \Gamma_1, \quad \mathbb{A}' := \dot{A}^* + \Gamma_1^\times \Gamma_0,$$
where $\Gamma_0^\times$ and $\Gamma_1^\times \in [\mathcal{N}, \mathcal{H}_-]$ are the adjoint operators to Γ_0 and Γ_1, respectively. Then $\mathbb{A}$ and $\mathbb{A}'$ belong to $[\mathcal{H}_+, \mathcal{H}_-]$ and are t-self-adjoint bi-extensions of $\dot{A}$. Moreover,
$$\mathbb{A} \supset A_1, \quad \mathbb{A}' \supset A_0.$$
2. *If $\mathbb{A}$ is a t-self-adjoint bi-extension of $\dot{A}$ with quasi-kernel A_1 and generated by A_0, then there exists a boundary triplet $\Pi = \{\mathcal{N}, \Gamma_1, \Gamma_0\}$ for $\dot{A}^*$ such that $\dot{A}^* \restriction \ker \Gamma_1 = A_1$ and $\mathbb{A} = \dot{A}^* - \Gamma_0^\times \Gamma_1$.*

It is shown in the proof of Theorem 6.3 that the form $w(f,g)$ in (6.1) corresponds to $\mathbb{A}$, the boundary triplet $\Pi = \{\mathcal{N}, \Gamma_1, \Gamma_0\}$, and $w(f,g) = (\mathbb{A}f, g)$. Similarly, $w'(f,g) = (\mathbb{A}'f, g)$, where $w'(f,g)$ is defined in (6.2), and the boundary triplet is $\Pi' = \{\mathcal{N}, -\Gamma_0, \Gamma_1\}$.

Definition 6.4 ([3]). Suppose that $\dot{A}$ is a non-negative symmetric operator. A boundary triplet $\Pi = \{\mathcal{N}, \Gamma_1, \Gamma_0\}$ is called *non-negative* if
$$w(f,f) = (\dot{A}^* f, f) - (\Gamma_1 f, \Gamma_0 f)_\mathcal{N} \geq 0 \text{ for all } f \in \mathcal{H}_+.$$

The operator $\mathbb{A} = \dot{A}^* - \Gamma_0^\times \Gamma_1$ corresponding to the boundary triplet $\Pi = \{\mathcal{N}, \Gamma_1, \Gamma_0\}$ is [3] a t-self-adjoint non-negative bi-extension of $\dot{A}$ and belongs to $[\mathcal{H}_+, \mathcal{H}_+]$. If $\dot{A}$ is a positive-definite operator, then for the positive-definite self-adjoint extension A we have $\mathcal{H}_+ = \operatorname{Dom}(\dot{A}^*) = \operatorname{Dom}(A) \dotplus \ker(\dot{A}^*)$. Consequently, A_F and A_K are transversal. Let P be a projection in $\mathcal{H}_+$ onto $\operatorname{Dom}(A)$ parallel to $\ker(\dot{A}^*)$, $\Pi = \{\mathcal{N}, \Gamma_K, \Gamma\}$ be a boundary triplet such that $\ker(\Gamma_K) = \operatorname{Dom}(A_K)$, Then
$$(\dot{A}^* f, f) - (\Gamma_K f, \Gamma f)_\mathcal{N} = (APf, Pf), \; f \in \mathcal{H}_+,$$
i.e., $\{\mathcal{N}, \Gamma_K, \Gamma\}$ is a positive boundary triplet. The latter equality has been assumed as the definition of a positive boundary triplet (the space of boundary values) in the case of a positive-definite operator $\dot{A}$ in [22].

It was shown in [3] that a positive boundary triplet exists if and only if A_F and A_K are transversal. The following theorem naturally follows from the preceding discussion.

Theorem 6.5. *Let $\dot{A}$ be a closed densely defined non-negative symmetric operator such that A_F and A_K are transversal. Then*

1. *to every non-negative boundary triplet $\Pi = \{\mathcal{N}, \Gamma_1, \Gamma_0\}$ there corresponds a non-negative t-self-adjoint bi-extension $\mathbb{A} = \dot{A}^* - \Gamma_0^\times \Gamma_1$;*
2. *to every non-negative t-self-adjoint bi-extension $\mathbb{A}$ there corresponds (up to equivalence[1]) a non-negative boundary triplet.*

Let $\Pi = \{\mathcal{N}, \Gamma_F, \Gamma_K\}$ be a non-negative boundary triplet such that $\mathrm{Dom}(A_K) = \ker \Gamma_K$, and $\mathrm{Dom}(A_F) = \ker \Gamma_F$. In [3] this boundary triplet is called *basic.* It is not hard to see that the corresponding to the basic boundary triplet non-negative t-self-adjoint bi-extension

$$\mathbb{A}_0 = \dot{A}^* - \Gamma_F^\times \Gamma_K \tag{6.3}$$

is such that the quasi-kernel of $\mathbb{A}_0$ is A_K. At the same time, A_F is the quasi-kernel of the bi-extension $\mathbb{A}_0' = \dot{A}^* + \Gamma_K^\times \Gamma_F$. It follows that $\mathbb{A}_0 = \mathbb{A}_K$ is the minimal element of $\mathcal{P}(\dot{A})$. The following theorem is established in [3].

Theorem 6.6. *Let $\Pi = \{\mathcal{N}, \Gamma_F, \Gamma_K\}$ be a basic boundary triplet. Then a boundary triplet $\widetilde{\Pi} = \left\{\widetilde{\mathcal{N}}, \widetilde{\Gamma}_1, \widetilde{\Gamma}_0\right\}$ is non-negative if and only if*

$$\widetilde{\Gamma}_1 = X(\Gamma_K - B_1\Gamma_F), \qquad \widetilde{\Gamma}_0 = X^{*-1}[(I + B_2B_1)\Gamma_F - B_2\Gamma_K],$$

where B_1, B_2 are non-negative bounded operators in $\mathcal{H}$ and X is a linear homeomorphism from $\mathcal{H}$ onto $\widetilde{\mathcal{H}}$.

Theorem 6.6 essentially provides us with another way to describe all non-negative t-self-adjoint bi-extensions in $[\mathcal{H}_+, \mathcal{H}_-]$. Namely, if $\Pi = \{\mathcal{N}, \Gamma_F, \Gamma_K\}$ is a basic non-negative boundary triplet, then the formula

$$\mathbb{A} = \dot{A}^* - [\Gamma_F^\times (I + B_1B_2) - \Gamma_K^\times B_2](\Gamma_K - B_1\Gamma_F), \tag{6.4}$$

where B_1, B_2 are non-negative bounded operators in $\mathcal{H}$, gives that description. Formulas (6.3) and (6.4) yield the following expression for quadratic forms

$$(\mathbb{A}f, f) = (\mathbb{A}_0 f, f) + b(f, f), \quad f \in \mathcal{H}_+,$$

where

$$\begin{aligned} b(f,f) &= (B_1\Gamma_F f, \Gamma_F f) + (B_2\Gamma_K f, \Gamma_K f) + (B_1\Gamma_F f, B_2B_1\Gamma_F f) \\ &\quad - 2\mathrm{Re}\,(B_1\Gamma_K f, B_2\Gamma_K f) \\ &= ||B_1^{1/2}\Gamma_F f||^2_{\mathcal{N}} + ||B_2^{1/2}(B_1\Gamma_F - \Gamma_K)f||^2_{\mathcal{N}}. \end{aligned}$$

For the corresponding dual self-adjoint bi-extension

$$\mathbb{A}' = \dot{A}^* + (\Gamma_K^\times - \Gamma_F^\times B_1)((I + B_2B_1)\Gamma_F - B_2\Gamma_K),$$

---

[1]Two boundary triplets $\{\mathcal{N}, \Gamma_1, \Gamma_0\}$ and $\left\{\widetilde{\mathcal{N}}, \widetilde{\Gamma}_1, \widetilde{\Gamma}_0\right\}$ are called *equivalent* [3] if $\ker \Gamma_k = \ker \widetilde{\Gamma}_k$, $k = 0, 1$.

we have

$$(\mathbb{A}'f, f) = (\mathbb{A}'_0 f, f) - b(f, f), \qquad \forall f \in \mathcal{H}_+.$$

Set

$$\mathcal{N} = \mathfrak{N}_F,\ \Gamma_0 = -\dot{A}^* P^+_{\mathfrak{M}_F},\ \Gamma_1 = P^+_{\mathfrak{N}_F}.$$

One can easily check that $\{\mathcal{N}, \Gamma_1, \Gamma_0\}$ is a boundary triplet for $\dot{A}^*$. Clearly

$$\ker(\Gamma_0) = \operatorname{Dom}(A_F).$$

Calculating $\Gamma_0^\times$ and $\Gamma_1^\times$ one obtains

$$\Gamma_0^\times = \mathcal{R}^{-1} A_F P^+_{\mathfrak{N}_F},\ \Gamma_1^\times = \mathcal{R}^{-1} P^+_{\mathfrak{N}_F}.$$

Using Theorem 4.11 we get that the domains of all non-negative self-adjoint extensions $\widetilde{A}$ of $\dot{A}$ takes the form

$$\operatorname{Dom}(\widetilde{A}) = \{v \in \operatorname{Dom}(\dot{A}^*) : \Gamma_0 v = \widetilde{U}\Gamma_1 v\},$$

where $\widetilde{U}$ is an arbitrary (+)-self-adjoint and non-negative operator in $\mathfrak{N}_F$, satisfying $0 \le \widetilde{U} \le W_0^{-1}$, and

$$\operatorname{Dom}(A_K) = \{v \in \operatorname{Dom}(\dot{A}^*) : \Gamma_0 v = W_0^{-1}\Gamma_1 v\}.$$

Now suppose that A_F and A_K are disjoint (transversal). Then W_0 is a densely defined (everywhere defined) in $\mathfrak{N}_F$ and (+)-self-adjoint and we can rewrite $\operatorname{Dom}(A_K)$ as

$$\operatorname{Dom}(A_K) = \ker(\Gamma_1 - W_0\Gamma_0).$$

The operator

$$\mathbb{A}_K = \dot{A}^* - \Gamma_0^\times(\Gamma_1 - W_0\Gamma_0)$$

is t-self-adjoint bi-extension with quasi-kernel A_K and generated by A_F. This is the minimal element of the set $\mathcal{P}(\dot{A})$. Then we get the explicit expressions for $\mathbb{A}_K$ and $\mathbb{A}'_K$ (cf. (4.19)):

$$\begin{aligned}
\mathbb{A}_K &= \dot{A}^* - \mathcal{R}^{-1} A_F (P^+_{\mathfrak{N}_F} + W_0 \dot{A}^* P^+_{\mathfrak{M}_F}),\\
\mathbb{A}'_K &= \dot{A}^* - \mathcal{R}^{-1} (P^+_{\mathfrak{N}_F} - A_F W_0 P^+_{\mathfrak{N}_F}) \dot{A}^* P^+_{\mathfrak{M}_F}.
\end{aligned}$$

If A_F and A_K are transversal, then we set

$$\Gamma_F = \Gamma_0 = -\dot{A}^* P^+_{\mathfrak{M}_F},\ \Gamma_K = \Gamma_1 - W_0\Gamma_0 = P^+_{\mathfrak{N}_F} + W_0 \dot{A}^* P^+_{\mathfrak{M}_F}.$$

Consequently, we obtain that $\{\mathfrak{N}_F, \Gamma_K, \Gamma_F\}$ is a basic boundary triplet for $\dot{A}^*$. Applying (6.4) we get a complete description of the set of all t-self-adjoint non-negative bi-extensions of $\dot{A}$ in $[\mathcal{H}_+, \mathcal{H}_-]$ given by the following formula

$$\mathbb{A} = \dot{A}^* - \mathcal{R}^{-1}\left[A_F(I + (W_0 + B_1)B_2) - B_2\right]\left[P^+_{\mathfrak{N}_F} + (W_0 + B_1)A^* P^+_{\mathfrak{M}_F}\right],$$

where B_1 and B_2 are an arbitrary (+)-bounded and non-negative self-adjoint operators in $\mathfrak{N}_F$.

References

[1] Ando, T., Nishio, K.: Positive self-adjoint extensions of positive symmetric operators. Tohóku Math. J., **22**, 65–75 (1970)

[2] Arlinskiĭ, Yu.M.: Regular (*)-extensions of quasi-Hermitian operators in rigged Hilbert spaces. (Russian), Izv. Akad. Nauk Armyan. SSR, Ser. Mat., **14**, No. 4, 297–312 (1979)

[3] Arlinskiĭ, Yu.M.: Positive spaces of boundary values and sectorial extensions of non-negative operator. (Russian) Ukr. Mat. J. **40**, No. 1, 8–14 (1988)

[4] Arlinskiĭ, Yu.M.: On proper accretive extensions of positive linear relations, Ukr. Math. Journ. 47, no. 6, 723–730 (1995)

[5] Arlinskiĭ, Yu.M.: Maximal sectorial extensions and associated with them closed forms. (Russian) Ukr. Mat. J. **48**, No. 6, 723–739 (1996)

[6] Arlinskiĭ, Yu.M.: Extremal extensions of sectorial Linear relations. Matematychnii Studii, **7**, No. 1, 81–96 (1997)

[7] Arlinskiĭ, Yu.M., Belyi, S., Tsekanovskiĭ, E.R. Conservative realizations of Herglotz–Nevanlinna functions. Operator Theory: Advances and Applications, Vol. 217, Birkhäuser Verlag, 2011.

[8] Arlinskiĭ, Yu.M., Belyi, S., Tsekanovskiĭ, E.R.: Accretive (∗)-extensions and Realization Problems. Operator Theory: Advances and Applications, Vol. 221, 71–108, (2012)

[9] Arlinskiĭ, Yu.M., Kaplan, V.L.: On selfajoint biextensions in rigged Hilbert spaces. (Russian), Functional Analysis, Ulyanovsk, Spektral Theory, No. 29, 11–19 (1989)

[10] Arlinskiĭ, Yu.M., Tsekanovskiĭ, E.R.: Quasi-self-adjoint contractive extensions of Hermitian contractions, Teor. Funkts., Funkts. Anal. Prilozhen, 50 (1988), 9–16 (Russian). English translation in J. Math. Sci. 49, No. 6, 1241–1247 (1990)

[11] Arlinskiĭ, Yu.M., Tsekanovskiĭ, E.R.: The von Neumann problem for non-negative symmetric operators. Integral Equations Operator Theory, **51**, No. 3, 319–356 (2005)

[12] Arlinskiĭ, Yu.M., Tsekanovskiĭ, E.R.: Kreĭn's research on semi-bounded operators, its contemporary developments, and applications, Operator Theory: Advances and Applications, **190**, 65–112 (2009)

[13] Belyi, S.V., Tsekanovskiĭ, E.R.: Stieltjes like functions and inverse problems for systems with Schrödinger operator. Operators and Matrices, vol. 2, No. 2, 265–296 (2008)

[14] Belyi, S.V., Tsekanovskiĭ, E.R.: Inverse Stieltjes like functions and inverse problems for systems with Schrödinger operator. Operator Theory: Advances and Applications, vol. 197, 21–49 (2009)

[15] Berezansky, Yu.M.: Expansions in eigenfunction of selfadjoint operators, Amer. Math. Soc. Providence, 1968.

[16] Coddington, E.A., de Snoo, H.S.V.: Positive selfadjoint extensions of positive symmetric subspaces. Math. Z., **159**, 203–214 (1978)

[17] Derkach, V.A., Malamud, M.M.: Generalized resolvents and the boundary value problems for Hermitian operators with gaps. J. Funct. Anal., **95**, No. 1, 1–95 (1991)

[18] Derkach, V.A., Malamud, M.M.: The extension theory of Hermitian operators and the moment problem. J. of Math. Sci., 73, No. 2, 141–242 (1995)

[19] Hassi, S., Malamud, M.M., de Snoo, H.S.V.: On Kreĭn's extension theory of non-negative operators, Math. Nachr., 274/275, 40–73 (2004)

[20] Kato, T.: Perturbation theory for linear operators. Springer-Verlag, (1966)

[21] Kochubeĭ, A.N.: On extensions of symmetric operators and symmetric binary relations. (Russian), Math. Zametki, **17**, No. 1, 41–48 (1975)

[22] Kochubeĭ, A.N.: Extensions of a positive-definite symmetric operator, Dokl. Akad. Nauk Ukr. SSR, 2 Ser. A No. 3, 168–171 (1979)

[23] Kreĭn, M.G.: The theory of self-adjoint extensions of semibounded Hermitian transformations and its applications. I, (Russian) Mat. Sbornik **20**, No. 3, 431–495 (1947)

[24] Kreĭn, M.G.: The theory of self-adjoint extensions of semibounded Hermitian transformations and its applications, II, (Russian) Mat. Sbornik **21**, No. 3, 365–404 (1947)

[25] Malamud, M.M.: On some classes of Hermitian operators with gaps. (Russian) Ukrainian Mat. J., **44**, No. 2, 215–234 (1992)

[26] Okunskiĭ, M.D., Tsekanovskiĭ, E.R.: On the theory of generalized self-adjoint extensions of semibounded operators. (Russian) Funkcional. Anal. i Prilozen., **7**, No. 3, 92–93 (1973)

[27] Phillips, R.: On dissipative operators, in "Lectures in Differential Equations", vol. II, Van Nostrand-Reinhold, New York, 65–113 (1965)

[28] Rofe-Beketov, F.S.: Numerical range of a linear relation and maximal relations. Theory of Functions, Functional Anal. and Appl., **44**, 103–112 (1985) (Russian)

[29] Shtraus, A.V.: On extensions of semibounded operators, Dokl. Akad. Nauk SSSR. 211, no. 3. 543–546 (1973)

[30] Shtraus, A.V.: On the theory of extremal extensions of a bounded positive operator, Funktional. Anal. Ul'yanovsk. Gos. Ped. Inst., no. 18, 115–126 (1982)

[31] Tsekanovskiĭ, E.R., Šmuljan, Yu.L.: The theory of bi-extensions of operators on rigged Hilbert spaces. Unbounded operator colligations and characteristic functions. Russ. Math. Surv., **32**, 73–131 (1977)

Yury Arlinskiĭ
Department of Mathematics
East Ukrainian National University
Kvartal Molodyozhny, 20-A
91034 Lugansk, Ukraine
e-mail: `yma@snu.edu.ua`

Sergey Belyi
Department of Mathematics
Troy University
Troy, AL 36082, USA
e-mail: `sbelyi@troy.edu`

[19] Hassi, S., Malamud, M., de Snoo, H.S.V.: On Krein's extension theory of nonnegative operators. Math. Nachr. 274/275, 40–73 (2004)

[20] Kato, T.: Perturbation theory for linear operators. Springer-Verlag (1966)

[21] Kochubei, A.N.: On extensions of symmetric operators and symmetric binary relations. Mat. Zametki 17, No. 1, 41–48 (1975)

Operator Theory:
Advances and Applications, Vol. 236, 43–51

Matrices with Bidiagonal Decomposition, Accurate Computations and Corner Cutting Algorithms

A. Barreras and J.M. Peña

Abstract. Some important classes of matrices admit a factorization known as bidiagonal decomposition. Bidiagonal decompositions can provide natural parameters to perform computations with high relative accuracy. We prove that corner cutting algorithms provide bidiagonal decompositions with high relative accuracy.

Mathematics Subject Classification (2010). Primary 15A23; Secondary 65D17.

Keywords. Bidiagonal decomposition, totally positive matrices, high relative accuracy, corner cutting algorithm.

1. Introduction

Some classes of matrices important in applications can be characterized by their bidiagonal decomposition, which uses bidiagonal matrices with unit diagonal and a diagonal matrix. In Section 2 we recall the uniqueness of this decomposition and that its entries can be used as natural parameters to derive algorithms with high relative accuracy.

In Computer Aided Geometric Design, the most important family of algorithms is formed by the corner cutting algorithms. In Section 3 we prove that a corner cutting algorithm allows us to obtain the bidiagonal decomposition of the associated matrix with high relative accuracy. Therefore many computations with this matrix can be performed with high relative accuracy, such as the inversion, the calculation of singular values and the calculation of eigenvalues.

Research Partially Supported by the Spanish Research GrantMTM2009-07315, by Gobierno de Aragón and Fondo Social Europeo.

2. Bidiagonal decompositions and accurate computations

A matrix is totally positive (TP) if all its minors are non-negative. TP matrices are also called totally non-negative matrices. This class of matrices have applications in many fields: see the surveys [1, 5], the classical book [10] and the recent books [6, 16].

Some classes of matrices, including the TP matrices, can be expressed as a product of bidiagonal matrices of the form

$$L^{(k)} = \begin{pmatrix} 1 & & & & & & \\ 0 & 1 & & & & & \\ & \ddots & \ddots & & & & \\ & & 0 & 1 & & & \\ & & & l_{n-k}^{(k)} & 1 & & \\ & & & & \ddots & \ddots & \\ & & & & & l_{n-1}^{(k)} & 1 \end{pmatrix}, \quad U^{(k)} = \begin{pmatrix} 1 & 0 & & & & & \\ & \ddots & \ddots & & & & \\ & & 1 & 0 & & & \\ & & & 1 & u_{n-k}^{(k)} & & \\ & & & & \ddots & \ddots & \\ & & & & & 1 & u_{n-1}^{(k)} \\ & & & & & & 1 \end{pmatrix},$$

where $k = 1, \dots, n-1$.

We now introduce the concept of bidiagonal decomposition of a matrix.

Definition 2.1. Let A be a nonsingular $n \times n$ matrix. Suppose that we can write A as a product of bidiagonal matrices

$$A = L^{(1)} \cdots L^{(n-1)} D U^{(n-1)} \cdots U^{(1)}, \tag{2.1}$$

where $D = \mathrm{diag}(d_1, \dots, d_n)$, and, for $k = 1, \dots, n-1$, $L^{(k)}$ and $U^{(k)}$ are (as above) unit diagonal lower and upper bidiagonal matrices respectively, with off-diagonal entries $l_i^{(k)} := (L^{(k)})_{i+1,i}$ and $u_i^{(k)} := (U^{(k)})_{i,i+1}$, $(i = 1, \dots, n-1)$ satisfying

1. $d_i \neq 0$ for all i,
2. $l_i^{(k)} = u_i^{(k)} = 0$ for $i < n-k$,
3. $l_i^{(k)} = 0 \Rightarrow l_{i+s}^{(k-s)} = 0$ for $s = 1, \dots, k-1$ and
   $u_i^{(k)} = 0 \Rightarrow u_{i+s}^{(k-s)} = 0$ for $s = 1, \dots, k-1$.

Then we denote (2.1) by $\mathcal{BD}(A)$, a bidiagonal decomposition of A satisfying the conditions of this definition.

Recently, matrices with the bidiagonal decomposition presented in Definition 2.1 have been studied (see [11, 12, 13, 14]). Although in general, a bidiagonal decomposition of a matrix is not unique, the following result of [2] uses Theorem 2.2 of [8] to ensure that a bidiagonal factorization as in Definition 2.1 is unique.

Proposition 2.2. *Let A be a nonsingular matrix. If a $\mathcal{BD}(A)$ exists, then it is unique.*

In fact, by Theorem 2.2 of [8], the entries $l_i^{(k)}, u_i^{(k)}$ of the bidiagonal factorization in the previous proposition coincide with the multipliers of an elimination procedure called Neville elimination. This procedure is used in several papers related with TP matrices (cf. [7, 8]).

The following result is a consequence of Theorem 4.2 of [8] and describes the unique bidiagonal decomposition of a nonsingular TP matrix.

Theorem 2.3. *A nonsingular $n \times n$ matrix A is TP if and only if there exists a (unique) $\mathcal{BD}(A)$ such that*

1. $d_i > 0$ *for all* i,
2. $l_i^{(k)} \geq 0$, $u_i^{(k)} \geq 0$ *for* $1 \leq k \leq n-1$ *and* $n-k \leq i \leq n-1$.

Let us recall that an algorithm can be performed with high relative accuracy if it does not include subtractions (except of the initial data), that is, if it only includes products, divisions, sums (of numbers of the same sign) and subtractions of the initial data (cf. [3, 4, 15]).

The nontrivial entries of the matrices in $\mathcal{BD}(A)$ (see (2.1)) have been considered natural parameters associated with A in many recent references ([11, 12, 14]). In many cases, we can assume that we know them with high relative accuracy. If we assume it for a nonsingular totally positive matrix A, then algorithms with high relative accuracy can be applied (see [11, 12, 13, 14]) to compute the singular values of A, the eigenvalues, the inverse or solving certain linear systems $Ax = b$.

Let us now introduce a class of matrices with bidiagonal decomposition that will generalize the class of nonsingular TP matrices. Let us denote a *signature* by ε, which is a vector $\varepsilon = (\varepsilon_1, \ldots, \varepsilon_m)$ with $\varepsilon_j \in \{\pm 1\}$ for $j = 1, \ldots, m$.

Definition 2.4. Given a signature $\varepsilon = (\varepsilon_1, \ldots, \varepsilon_{n-1})$ and a nonsingular $n \times n$ matrix A, we say that A has a signed bidiagonal decomposition with signature ε if there exists a $\mathcal{BD}(A)$ (unique by Proposition 2.2) such that

1. $d_i > 0$ for all i,
2. $l_i^{(k)} \varepsilon_i \geq 0$, $u_i^{(k)} \varepsilon_i \geq 0$ for $1 \leq k \leq n-1$ and $n-k \leq i \leq n-1$.

We say that A has a *signed bidiagonal decomposition* if it has a signed bidiagonal decomposition with some signature ε. By Theorem 2.3, we know that a nonsingular TP matrix has a signed bidiagonal decomposition with signature $\varepsilon = (1, \ldots, 1)$.

If we assume that we know with high relative accuracy the entries of the bidiagonal factors for a signed bidiagonal matrix A, then we can find algorithms with high relative accuracy to perform many computations with A (see [2]), including the inversion and the calculation of singular values and eigenvalues.

3. Corner cutting algorithms and accurate computations

Corner cutting algorithms form the most important family of algorithms in Computer Aided Geometric Design (C.A.G.D.) due to their numerical stability properties as well as to their nice geometric interpretations. They are associated with the factorization of nonsingular stochastic totally positive matrices. Let us recall that a non-negative matrix is called *stochastic* if the entries of each row sum up to 1. Let us now introduce the main definition.

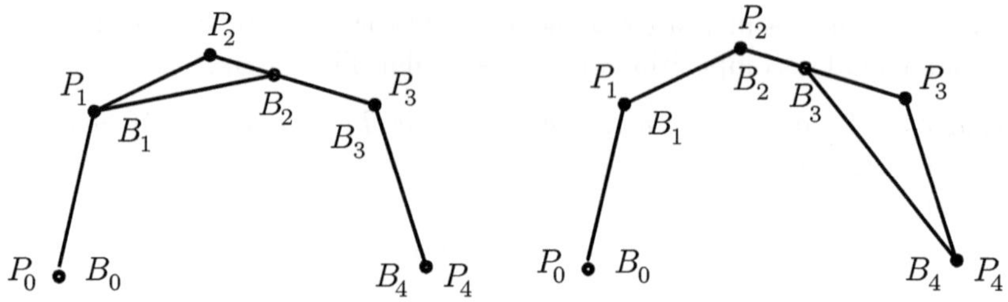

FIGURE 1. Elementary corner cuttings (3.1) and (3.2).

An *elementary corner cutting* is a transformation which maps any polygon $P_0 \cdots P_n$ into another polygon $B_0 \cdots B_n$ defined by one of the following ways

$$\begin{aligned} B_j &= P_j, \quad j \neq i, \\ B_i &= (1-\lambda)P_i + \lambda P_{i+1}, \end{aligned} \tag{3.1}$$

for some $i \in \{0, \dots, n-1\}, 0 \leq \lambda < 1$, or

$$\begin{aligned} B_j &= P_j, \quad j \neq i, \\ B_i &= (1-\lambda)P_i + \lambda P_{i-1}, \end{aligned} \tag{3.2}$$

for some $i \in \{1, \dots, n\}, \quad 0 \leq \lambda < 1$. (See Figure 1).

Clearly, the matrix form of the elementary corner cutting (3.1) is

$$(B_0, \dots, B_n)^T = U(\lambda_i)(P_0, \dots, P_n)^T,$$

where $U(\lambda_i)$ is the *bidiagonal*, nonsingular and upper triangular matrix

$$U(\lambda_i) = \begin{pmatrix} 1 & 0 & & & & & & \\ & 1 & 0 & & & & & \\ & & \ddots & \ddots & & & & \\ & & & 1-\lambda_i & \lambda_i & & & \\ & & & & \ddots & \ddots & & \\ & & & & & 1 & 0 \\ & & & & & & 1 \end{pmatrix}.$$

Analogously, a lower triangular matrix can be used for the elementary corner cutting (3.2).

A *corner cutting algorithm* is any composition of elementary corner cuttings (see [9, 17]). An elementary corner cutting is defined by a bidiagonal, nonsingular, totally positive and stochastic matrix, which is upper (resp., lower) triangular in the case (3.1) (resp., (3.2)). So a corner cutting algorithm is defined by a product of matrices that are simultaneously bidiagonal, nonsingular, totally positive and stochastic. In fact, any upper (resp., lower) triangular bidiagonal, nonsingular, totally positive and stochastic matrix determines a corner cutting algorithm by the factorization

$$\begin{pmatrix} 1-\lambda_0 & \lambda_0 & & & & & \\ & 1-\lambda_1 & \lambda_1 & & & & \\ & & \ddots & \ddots & & & \\ & & & 1-\lambda_i & \lambda_i & & \\ & & & & \ddots & \ddots & \\ & & & & & 1-\lambda_{n-1} & \lambda_{n-1} \\ & & & & & & 1 \end{pmatrix} = U(\lambda_{n-1})U(\lambda_{n-2})\cdots U(\lambda_0)$$

(an analogous factorization can be performed for the lower triangular case).

A corner cutting algorithm described by a nonsingular, totally positive and stochastic matrix can be expressed as a product of bidiagonal, nonsingular, totally positive and stochastic matrices, as shown by the following result, which corresponds to Theorem 4.5 of [8] and summarizes this fact.

Theorem 3.1. *A nonsingular $n \times n$ matrix A is stochastic and totally positive if and only if it can be factorized in the form*

$$A = F_{n-1}F_{n-2}\cdots F_1G_1\cdots G_{n-2}G_{n-1}, \tag{3.3}$$

with

$$F_i = \begin{pmatrix} 1 & & & & & & \\ 0 & 1 & & & & & \\ & \ddots & \ddots & & & & \\ & & 0 & 1 & & & \\ & & & \alpha_{i+1,1} & 1-\alpha_{i+1,1} & & \\ & & & & \ddots & \ddots & \\ & & & & & \alpha_{n,n-i} & 1-\alpha_{n,n-i} \end{pmatrix}$$

and

$$G_i = \begin{pmatrix} 1 & 0 & & & & & \\ & \ddots & \ddots & & & & \\ & & 1 & 0 & & & \\ & & & 1-\alpha_{1,i+1} & \alpha_{1,i+1} & & \\ & & & & \ddots & \ddots & \\ & & & & & 1-\alpha_{n-i,n} & \alpha_{n-i,n} \\ & & & & & & 1 \end{pmatrix},$$

where, $\forall(i,j)$, $0 \le \alpha_{i,j} < 1$ satisfies

$$\begin{aligned} \alpha_{ij} = 0 \Rightarrow \alpha_{hj} = 0 \quad \forall h > i \quad \textit{if} \quad i > j, \\ \alpha_{ij} = 0 \Rightarrow \alpha_{ik} = 0 \quad \forall k > j \quad \textit{if} \quad i < j. \end{aligned} \tag{3.4}$$

Under these conditions, the factorization is unique.

The previous condition for the α_{ij}'s is only used to guarantee the uniqueness of the factorization, analogously to the conditions of zero entries in Definition 2.1.

Many crucial algorithms for the design of curves, such as evaluation, subdivision, degree elevation or knot insertion algorithms, are corner cutting algorithms.

The following result proves that, from a corner cutting algorithm, we can construct with high relative accuracy and $\mathcal{O}(n^2)$ elementary operations the bidiagonal decomposition $\mathcal{BD}(A)$ of the associated $n \times n$ matrix A. Therefore, by the results and methods commented in the previous sections, many computations with A can be performed with high relative accuracy, including the inversion and the calculation of singular values and eigenvalues.

Theorem 3.2. *Let us consider a corner cutting algorithm associated to the matrix factorization* (3.3) *of a nonsingular totally positive matrix A. If we know the entries of F_i, G_i $i = 1, \ldots, n-1$ with high relative accuracy, then we can compute a bidiagonal decomposition $\mathcal{BD}(A)$ with high relative accuracy with $\frac{5}{2}n^2 - \frac{5}{2}n + 3$ products and $\frac{1}{2}n^2 - \frac{3}{2}n + 1$ quotients.*

Proof. We are going to prove that

$$\mathcal{BD}(A) = \hat{F}_{n-1} \cdots \hat{F}_1 D \hat{G}_1 \cdots \hat{G}_{n-1}, \tag{3.5}$$

can be computed from the entries of (3.3) without subtractions.

First, let us show that we can write

$$F_{n-1} \cdots F_1 = \hat{F}_{n-1} \cdots \hat{F}_1 D_1^{(F)} \tag{3.6}$$

with $D_i^{(F)} = \mathrm{diag}(\mathrm{F}_{n\text{-}1}) \cdots \mathrm{diag}(\mathrm{F}_i)$, denoting by $\mathrm{diag}(F_j)$ the diagonal matrix whose diagonal entries coincide with those of F_j.

We have that $F_{n-1} = \hat{F}_{n-1} D_{n-1}^{(F)}$, where

$$\hat{F}_{n-1} = \begin{pmatrix} 1 & & & \\ & \ddots & & \\ & & 1 & \\ & & \alpha_{n1} & 1 \end{pmatrix}.$$

Besides, it can be checked that, for all $i = n-1, \ldots, 2$, $D_i^{(F)} F_{i-1} = \hat{F}_{i-1} D_{i-1}^{(F)}$, where

$$\hat{F}_i = \begin{pmatrix} 1 & & & & & & \\ & \ddots & & & & & \\ & & 1 & & & & \\ & & \alpha_{i+1,1} & 1 & & & \\ & & & \frac{\alpha_{i+2,2}(1-\alpha_{i+2,1})}{1-\alpha_{i+1,1}} & 1 & & \\ & & & & \ddots & \ddots & \\ & & & & & 1 & \\ & & & & & \frac{(1-\alpha_{n1})\cdots(1-\alpha_{n,n-i-1})\alpha_{n,n-i}}{(1-\alpha_{n-1,1})\cdots(1-\alpha_{n-1,n-i-1})} & 1 \end{pmatrix},$$

which has the entry $\alpha_{i+1,1}$ in place $(i+1,i)$. Thus, we conclude that $F_{n-1}\cdots F_1 = \hat{F}_{n-1}\cdots\hat{F}_1 D_1^{(F)}$.

Analogously, we can deduce that $G_1\cdots G_{n-1} = D_1^{(G)}\hat{G}_1\cdots\hat{G}_{n-1}$, with $D_1^{(G)}$ a diagonal matrix and $\hat{G}_j$ a bidiagonal upper triangular matrix with unit diagonal. Then, we have that

$$\begin{aligned} A &= F_{n-1}\cdots F_1 G_1\cdots G_{n-1} \\ &= \hat{F}_{n-1}\cdots\hat{F}_1 D_1^{(F)} D_1^{(G)}\hat{G}_1\cdots\hat{G}_{n-1} \\ &= \hat{F}_{n-1}\cdots\hat{F}_1 D\hat{G}_1\cdots\hat{G}_{n-1} \\ &= \mathcal{BD}(A), \end{aligned}$$

where $D := D_1^{(F)} D_1^{(G)}$.

If we denote by $f_j^{(i)}$ the $(j,j-1)$ entry of the matrix $\hat{F}_i$, it can be checked that it is possible to compute the $f_j^{(i)}$ entry from $f_j^{(i+1)}$ by using the following formula for $j=2,\ldots,n$ and $i=1,\ldots,j-2$,

$$f_j^{(i)} = \frac{\alpha_{j,j-i}(1-\alpha_{j,j-i-1})}{\alpha_{j,j-i-1}(1-\alpha_{j-1,j-i-1})} f_j^{(i+1)}. \tag{3.7}$$

An analogous formula exists for the entries of $\hat{G}_i$. By conditions (3.4) and the previous formula, it can be checked that the zero patterns of Definition 2.1 are satisfied for $\hat{F}_i$, $\hat{G}_i$ and D, and so, we have the $\mathcal{BD}(A)$.

Besides, as we can compute the entries of $\hat{F}_i$, $\hat{G}_i$ and D from the entries of F_j and G_j $(j=i+1,\ldots,n)$ without subtractions, we have computed the factorization with high relative accuracy.

Taking into account (3.7) and the definition of the matrix D, we can conclude that to compute the bidiagonal decomposition $\mathcal{BD}(A)$ are necessary $n(n+2)+\sum_{j=n}^{2}\sum_{i=j-2}^{1} 3 = \frac{5}{2}n^2 - \frac{5}{2}n + 3$ products and $\sum_{j=n}^{2}\sum_{i=j-2}^{1} 1 = \frac{1}{2}n^2 - \frac{3}{2}n + 1$ quotients. □

We finish with the algorithm corresponding to the proof of the previous result to compute $\mathcal{BD}(A)$ from the factorization (3.3). Let us recall that we have denoted by $f_j^{(i)}$ the $(j,j-1)$ entry of $\hat{F}_i$ (resp. by $g_j^{(i)}$ the $(j-i,j)$ entry of $\hat{G}_i$). Let us also initialize the matrix $D := \mathrm{diag}(G_1(1,1),\ldots,G_n(n,n))$.

Algorithm to compute the bidiagonal factorization of a corner cutting algorithm
Input: nontrivial entries of matrices F_i and G_i of (3.3).
Output: matrices $\hat{F}_i$, D and $\hat{G}_i$ of (3.5).

```
Compute the matrices F̂_{n-1} and Ĝ_{n-1}
for i = n - 2 to 1
    f_{i+1}^{(i)} = α_{i+1,1}
    g_{i+1}^{(i)} = α_{1,i+1}(1-α_{1,i+2}) / (1-α_{1,i+1})
endfor
```

for $j = n$ to 2
    for $i = j - 2$ to 2
        $f_j^{(i)} = \frac{\alpha_{j,j-i}(1-\alpha_{j,j-i-1})}{\alpha_{j,j-i-1}(1-\alpha_{j-1,j-i-1})} f_j^{(i+1)}$
        $g_j^{(i)} = \frac{\alpha_{j-i,j}(1-\alpha_{j-i,j+1})}{\alpha_{j-i-1,j}(1-\alpha_{j-i,j})} g_j^{(i+1)}$
    endfor
    for $k = 1$ to $j - 1$
        $D(j,j) = D(j,j) * F_k(j,j) * G_k(j,j)$
    endfor
endfor

References

[1] T. Ando, *Totally positive matrices.* Linear Algebra Appl. **90** (1987), 165–219.

[2] A. Barreras, J.M. Peña, *Accurate computations of matrices with bidiagonal decomposition using methods for totally positive matrices.* Numerical Linear Algebra with Applications **20** (2013), 413–424.

[3] J. Demmel, P. Koev, *Accurate SVDs of weakly diagonally dominant M-matrices.* Numerische Mathematik **98** (2004), 99–104.

[4] J. Demmel, P. Koev, *The accurate and efficient solution of a totally positive generalized Vandermonde linear system.* SIAM Journal on Matrix Analysis and Applications **27** (2005), 142–152.

[5] S.M. Fallat, *Bidiagonal factorizations of totally nonnegative matrices.* Amer. Math. Monthly **108** (2001), 697–712.

[6] S.M. Fallat, C.R. Johnson, *Totally Nonnegative Matrices.* Princeton University Press, Princeton, NJ, 2011.

[7] M. Gasca, J.M. Peña, *Total positivity and Neville elimination.* Linear Algebra and its Applications **165** (1992), 25–44.

[8] M. Gasca, J.M. Peña, *On factorizations of totally positive matrices.* In: *Total Positivity and Its Applications* (M. Gasca and C.A. Micchelli, eds.), Kluver Academic Publishers: Dordrecht, The Netherlands, 1996; 109–130.

[9] T.N.T. Goodman, C.A. Micchelli, *Corner cutting algorithms for the Bézier representation of free form curves.* Linear Algebra Appl. **99** (1988), 225–252.

[10] S. Karlin, *Total Positivity, Vol. I.* Stanford University Press, Stanford, 1968.

[11] P. Koev, *Accurate eigenvalues and SVDs of totally nonnegative matrices.* SIAM Journal on Matrix Analysis and Applications **27** (2005), 1–23.

[12] P. Koev, *Accurate computations with totally nonnegative matrices.* SIAM Journal on Matrix Analysis and Applications **29** (2007), 731–751.

[13] A. Marco, J.J. Martínez, *A fast and accurate algorithm for solving Bernstein–Vandermonde linear systems.* Linear Algebra and its Applications **422** (2007), 616–628.

[14] A. Marco, J.J. Martínez, *Accurate computations with Said–Ball–Vandermonde matrices.* Linear Algebra and its Applications **432** (2010), 2894–2908.

[15] J.M. Peña, *LDU decompositions with L and U well conditioned.* Electronic Transactions on Numerical Analysis **18** (2004), 198–208.

[16] A. Pinkus, *Totally Positive Matrices.* Cambridge Tracts in Mathematics: Cambridge University Press, 2010.

[17] H. Praustzsch, C.A. Michelli, *Computing curves invariant under halving.* Computer-Aided Geom. Design **4** (1987), 133–140.

A. Barreras and J.M. Peña
Dept. Matemática Aplicada/IUMA
Universidad de Zaragoza
e-mail: albarrer@unizar.es
jmpena@unizar.es

Operator Theory:
Advances and Applications, Vol. 236, 53–93

Boundary-value Problems for Higher-order Elliptic Equations in Non-smooth Domains

Ariel Barton and Svitlana Mayboroda

Abstract. This paper presents a survey of recent results, methods, and open problems in the theory of higher-order elliptic boundary value problems on Lipschitz and more general non-smooth domains. The main topics include the maximum principle and pointwise estimates on solutions in arbitrary domains, analogues of the Wiener test governing continuity of solutions and their derivatives at a boundary point, and well-posedness of boundary value problems in domains with Lipschitz boundaries.

Mathematics Subject Classification (2010). Primary 35-02; Secondary 35B60, 35B65, 35J40, 35J55.

Keywords. biharmonic equation, polyharmonic equation, higher-order equation, Lipschitz domain, general domains, Dirichlet problem, regularity problem, Neumann problem, Wiener criterion, maximum principle.

1. Introduction

The last three decades have witnessed a surge of activity on boundary value problems on Lipschitz domains. The Dirichlet, Neumann, and regularity problems for the Laplacian are now well understood for data in L^p, Sobolev, and Besov spaces. More generally, well-posedness in L^p has been established for divergence form elliptic equations with non-smooth coefficients $-\operatorname{div} A\nabla$ and, at least in the context of real symmetric matrices, the optimal conditions on A needed for solvability of the Dirichlet problem in L^p are known. We direct the interested reader to Kenig's 1994 CBMS book [Ken94] for an excellent review of these matters and to [KKPT00, KR09, Rul07, AAA$^+$11, DPP07, DR10, AAH08, AAM10, AA11, AR11, HKMP12] for recent results.

Svitlana Mayboroda is partially supported by the Alfred P. Sloan Fellowship, the NSF CAREER Award DMS 1056004, and the NSF Materials Research Science and Engineering Center Seed Grant.

Unfortunately, this beautiful and powerful theory has mostly been restricted to the case of second-order operators. Higher-order elliptic boundary problems, while having abundant applications in physics and engineering, have mostly been out of reach of the methods devised to study the second-order case. The present survey is devoted to major recent results in this subject, new techniques, and principal open problems.

The prototypical example of a higher-order elliptic operator is the bilaplacian $\Delta^2 = \Delta(\Delta)$ or, more generally, the polyharmonic operator Δ^m, $m \geq 2$. The biharmonic problem in a domain $\Omega \subset \mathbb{R}^n$ with Dirichlet boundary data consists, roughly speaking, of finding a function u such that for given f, g, h,

$$\Delta^2 u = h \text{ in } \Omega, \quad u\big|_{\partial\Omega} = f, \quad \partial_\nu u\big|_{\partial\Omega} = g,$$

subject to the appropriate estimates on u in terms of the data. To make it precise, as usual, one needs to properly interpret restriction of solution to the boundary $u\big|_{\partial\Omega}$ and its normal derivative $\partial_\nu u\big|_{\partial\Omega}$, as well as specify the desired estimates. The biharmonic equation arises in numerous problems of structural engineering. It models the displacements of a thin plate clamped near its boundary, the stresses in an elastic body, the stream function in creeping flow of a viscous incompressible fluid, to mention just a few applications (see, e.g., [Mel03]).

The primary goal of this survey is to address the biharmonic problem and more general higher-order partial differential equations in domains with non-smooth boundaries, specifically, in the class of Lipschitz domains. However, the analysis of such delicate questions as well-posedness in Lipschitz domains requires preliminary understanding of fundamental properties of the solutions, such as boundedness, continuity, and regularity near a boundary point. For the Laplacian, these properties of solutions in general domains are described by the maximum principle and by the 1924 Wiener criterion; for the bilaplacian, they turn out to be highly nontrivial and partially open to date. For the purposes of the introduction, let us mention just a few highlights and outline the paper.

In Section 3, we discuss the maximum principle for higher-order elliptic equations. Loosely, one expects that for a solution u to the equation $Lu = 0$ in Ω, where L is a differential operator of order $2m$, there holds

$$\max_{|\alpha| \leq m-1} \|\partial^\alpha u\|_{L^\infty(\Omega)} \leq C \max_{|\beta| \leq m-1} \|\partial^\beta u\|_{L^\infty(\partial\Omega)},$$

with the usual convention that the zeroth-order derivative of u is simply u itself. For the Laplacian ($m = 1$), this formula is a slightly weakened formulation of the maximum principle. In striking contrast with the case of harmonic functions, the maximum principle for an elliptic operator of order $2m \geq 4$ may fail, even in a Lipschitz domain. To be precise, in general, the derivatives of order $(m-1)$ of a solution to an elliptic equation of order $2m$ need not be bounded. We discuss relevant counterexamples, known positive results, as well as a more general question of pointwise bounds on solutions and their derivatives in arbitrary domains, e.g., whether u (rather than $\nabla^{m-1} u$) is necessarily bounded in Ω.

Section 4 is devoted to continuity of solutions to higher-order equations and their derivatives near the boundary of the domain. Specifically, if for some operator the aforementioned boundedness of the $(m-1)$-st derivatives holds, one next would need to identify conditions assuring their continuity near the boundary. For instance, in the particular case of the bilaplacian, the gradient of a solution is bounded in an arbitrary three-dimensional domain, and one would like to study the continuity of the gradient near a boundary point. As is well known, for second-order equations, necessary and sufficient conditions for continuity of the solutions have been provided by the celebrated Wiener criterion. Analogues of the Wiener test for higher-order PDEs are known only for some operators, and in a restricted range of dimensions. We shall discuss these results, testing conditions, and the associated capacities, as well as similarities and differences with their second-order antecedents.

Finally, Sections 5 and 6 are devoted to boundary-value problems in Lipschitz domains. The simplest example is the Dirichlet problem for the bilaplacian,

$$\Delta^2 u = 0 \quad \text{in } \Omega, \quad u\big|_{\partial\Omega} = f \in W_1^p(\partial\Omega), \quad \partial_\nu u\big|_{\partial\Omega} = g \in L^p(\partial\Omega), \tag{1.1}$$

in which case the expected sharp estimate on the solution is

$$\|N(\nabla u)\|_{L^p(\partial\Omega)} \leq C\|\nabla_\tau f\|_{L^p(\partial\Omega)} + C\|g\|_{L^p(\partial\Omega)}, \tag{1.2}$$

where N denotes the non-tangential maximal function and $W_1^p(\partial\Omega)$ is the Sobolev space of functions with one tangential derivative in L^p (cf. Section 2 for precise definitions). In Sections 5.1–5.6 we discuss (1.1) and (1.2), and more general higher-order homogeneous Dirichlet and regularity boundary value problems with constant coefficients, with boundary data in L^p. Section 5.7 describes the specific case of convex domains. The Neumann problem for the bilaplacian is addressed in Section 5.8. In Section 5.9, we discuss inhomogeneous boundary value problems with data in Besov and Sobolev spaces, which, in a sense, are intermediate between those with Dirichlet and regularity data. Finally, in Section 6, we discuss boundary-value problems with variable coefficients.

The sharp range of p, such that the aforementioned biharmonic (or any other higher-order) Dirichlet problem with data in L^p is well posed in Lipschitz domains, is not yet known in high dimensions. However, over the recent years numerous advances have been made in this direction and some new methods have emerged. For instance, *several* different layer potential constructions have proven to be useful (again, note the difference with the second-order case when the relevant layer potentials are essentially uniquely defined by the boundary problem), as well as recently discovered equivalence of well-posedness to certain reverse Hölder estimates on the non-tangential maximal function. It is interesting to point out that the main local estimates which played a role in recent well-posedness results actually come from the techniques developed in connection with the Wiener test discussed above. Thus, in the higher-order case the two issues are intimately intertwined; this was one of the reasons for the particular choice of topics in the present survey.

The Neumann problem and the variable coefficient case are even more puzzling. The details will be presented in the body of the paper. Here, let us just point out that in both cases even the proper statement of the "natural" boundary problem presents a challenge. For instance, in the higher-order case the choice of Neumann data is not unique. Depending on peculiarities of the Neumann operator, one can be led to well-posed and ill-posed problems even for the bilaplacian, and more general operators give rise to new issues related to the coercivity of the underlying form. However, despite the aforementioned challenges, the first well-posedness results have recently been obtained and will be discussed below.

To conclude this introduction, we refer the reader to the excellent expository paper [Maz99b] by Vladimir Maz'ya on the topic of the Wiener criterion and pointwise estimates. This paper largely inspired the corresponding sections of the present manuscript, and its exposition of the related historical material extends and complements that in Sections 3 and 4. Our main goal here, however, was to discuss the most recent achievements (some of which appeared after the aforementioned survey was written) and their role in the well-posedness results on Lipschitz domains which constitute the main topic of the present review. We also would like to mention that this paper does not touch upon the methods and results of the part of elliptic theory studying the behavior of solutions in the domains with isolated singularities, conical points, cuspidal points, etc. Here, we have intentionally concentrated on the case of Lipschitz domains, which can display accumulating singularities – a feature drastically affecting both the available techniques and the actual properties of solutions.

2. Definitions

As we pointed out in the introduction, the prototypical higher-order elliptic equation is the biharmonic equation $\Delta^2 u = 0$, or, more generally, the polyharmonic equation $\Delta^m u = 0$ for some integer $m \geq 2$. It naturally arises in numerous applications in physics and in engineering, and in mathematics it is a basic model for a higher-order partial differential equation. These operators may be generalized to constant-coefficient differential operators of order $2m$, or to variable-coefficient operators in either divergence or nondivergence form.

Let us discuss the details. To start, a general *constant coefficient* elliptic operator is defined as follows.

Definition 2.1. Let L be an operator acting on functions $u : \mathbb{R}^n \mapsto \mathbb{C}^\ell$. Suppose that we may write

$$(Lu)_j = \sum_{k=1}^{\ell} \sum_{|\alpha|=|\beta|=m} \partial^\alpha a_{\alpha\beta}^{jk} \partial^\beta u_k \tag{2.2}$$

for some coefficients $a_{\alpha\beta}^{jk}$ defined for all $1 \leq j, k \leq \ell$ and all multiindices α, β of length n with $|\alpha| = |\beta| = m$. Then we say that L is a *differential operator of order* $2m$.

Suppose the coefficients $a^{jk}_{\alpha\beta}$ are constant and satisfy the Legendre–Hadamard ellipticity condition

$$\operatorname{Re} \sum_{j,k=1}^{\ell} \sum_{|\alpha|=|\beta|=m} a^{jk}_{\alpha\beta} \xi^\alpha \xi^\beta \zeta_j \bar{\zeta}_k \geq \lambda |\xi|^{2m} |\zeta|^2 \tag{2.3}$$

for all $\xi \in \mathbb{R}^n$ and all $\zeta \in \mathbb{C}^\ell$, where $\lambda > 0$ is a real constant. Then we say that L is an *elliptic operator of order* $2m$.

If $\ell = 1$ we say that L is a *scalar operator* and refer to the equation $Lu = 0$ as an elliptic equation; if $\ell > 1$ we refer to $Lu = 0$ as an elliptic system. If $a^{jk} = a^{kj}$, then we say the operator L is *symmetric*. If $a^{jk}_{\alpha\beta}$ is real for all α, β, j, and k, we say that L has real coefficients.

Here if α is a multiindex of length n, then $\partial^\alpha = \partial^{\alpha_1}_{x_1} \ldots \partial^{\alpha_n}_{x_n}$.

Now let us discuss the case of variable coefficients. A divergence-form higher-order elliptic operator is given by

$$(Lu)_j(X) = \sum_{k=1}^{\ell} \sum_{|\alpha|=|\beta|=m} \partial^\alpha (a^{jk}_{\alpha\beta}(X) \partial^\beta u_k(X)). \tag{2.4}$$

This form affords a notion of weak solution; we say that $Lu = h$ weakly if

$$\sum_{j=1}^{\ell} \int_\Omega \varphi_j \, h_j = \sum_{|\alpha|=|\beta|=m} \sum_{j,k=1}^{\ell} (-1)^m \int_\Omega \partial^\alpha \varphi_j \, a^{jk}_{\alpha\beta} \, \partial^\beta u_k \tag{2.5}$$

for any function $\varphi : \Omega \mapsto \mathbb{C}^\ell$ smooth and compactly supported.

If the coefficients $a^{jk}_{\alpha\beta} : \mathbb{R}^n \to \mathbb{C}$ are sufficiently smooth, we may rewrite (2.4) in nondivergence form

$$(Lu)_j(X) = \sum_{k=1}^{\ell} \sum_{|\alpha|\leq 2m} a^{jk}_\alpha(X) \partial^\alpha u_k(X). \tag{2.6}$$

This form is particularly convenient when we allow equations with lower-order terms (note their appearance in (2.6)).

A simple criterion for ellipticity of the operators L of (2.6) is the condition that (2.3) holds with $a^{jk}_{\alpha\beta}$ replaced by $a^{jk}_\alpha(X)$ for any $X \in \mathbb{R}^n$, that is, that

$$\operatorname{Re} \sum_{j,k=1}^{\ell} \sum_{|\alpha|=2m} a^{jk}_\alpha(X) \xi^\alpha \zeta_j \bar{\zeta}_k \geq \lambda |\xi|^{2m} |\zeta|^2 \tag{2.7}$$

for any fixed $X \in \mathbb{R}^n$ and for all $\xi \in \mathbb{R}^n$, $\zeta \in \mathbb{C}^2$. This means in particular that ellipticity is only a property of the highest-order terms of (2.6); the value of a^{jk}_α, for $|\alpha| < m$, is not considered.

For divergence-form operators, some known results use a weaker notion of ellipticity, namely that $\langle \varphi, L\varphi \rangle \geq \lambda \|\nabla^m \varphi\|^2_{L^2}$ for all smooth compactly supported functions φ; this notion is written out in full in Formula (6.5) below.

Finally, let us mention that throughout we let C and ε denote positive constants whose value may change from line to line. We let $\fint$ denote the average integral, that is, $\fint_E f\,d\mu = \frac{1}{\mu(E)}\int_E f\,d\mu$. The only measures we will consider are the Lebesgue measure dX (on $\mathbb{R}^n$ or on domains in $\mathbb{R}^n$) or the surface measure $d\sigma$ (on the boundaries of domains).

3. The maximum principle and pointwise estimates on solutions

The maximum principle for harmonic functions is one of the fundamental results in the theory of elliptic equations. It holds in arbitrary domains and guarantees that every solution to the Dirichlet problem for the Laplace equation, with bounded data, is bounded. Moreover, it remains valid for all second-order divergence-form elliptic equations with real coefficients.

In the case of equations of higher order, the maximum principle has been established only in relatively nice domains. It was proven to hold for operators with smooth coefficients in smooth domains of dimension two in [Mir48] and [Mir58], and of arbitrary dimension in [Agm60]. In the early 1990s, it was extended to three-dimensional domains diffeomorphic to a polyhedron ([KMR01, MR91]) or having a Lipschitz boundary ([PV93, PV95b]). However, in general domains, no direct analog of the maximum principle exists (see Problem 4.3, p. 275, in Nečas's book [Neč67]). The increase of the order leads to the failure of the methods which work for second-order equations, and the properties of the solutions themselves become more involved.

To be more specific, the following theorem was proved by Agmon.

Theorem 3.1 ([Agm60, Theorem 1]). *Let $m \geq 1$ be an integer. Suppose that Ω is domain with C^{2m} boundary. Let*

$$L = \sum_{|\alpha|\leq 2m} a_\alpha(X)\partial^\alpha$$

be a scalar operator of order $2m$, where $a_\alpha \in C^{|\alpha|}(\overline{\Omega})$. Suppose that L is elliptic in the sense of (2.7). *Suppose further that solutions to the Dirichlet problem for L are unique.*

Then, for every $u \in C^{m-1}(\overline{\Omega}) \cap C^{2m}(\Omega)$ that satisfies $Lu = 0$ in Ω, we have

$$\max_{|\alpha|\leq m-1} \|\partial^\alpha u\|_{L^\infty(\Omega)} \leq C \max_{|\beta|\leq m-1} \|\partial^\beta u\|_{L^\infty(\partial\Omega)}. \tag{3.2}$$

We remark that the requirement that the Dirichlet problem have unique solutions is not automatically satisfied for elliptic equations with lower-order terms; for example, if λ is an eigenvalue of the Laplacian then solutions to the Dirichlet problem for $\Delta u - \lambda u$ are not unique.

Equation (3.2) is called the *Agmon–Miranda maximum principle.* In [Šul75], Šul'ce generalized this to systems of the form (2.6), elliptic in the sense of (2.7), that satisfy a positivity condition (strong enough to imply Agmon's requirement that solutions to the Dirichlet problem be unique).

Thus the Agmon–Miranda maximum principle holds for sufficiently smooth operators and domains. Moreover, for some operators, the maximum principle is valid even in domains with Lipschitz boundary, provided the dimension is small enough. We postpone a more detailed discussion of the Lipschitz case to Section 5.6; here we simply state the main results. In [PV93] and [PV95b], Pipher and Verchota showed that the maximum principle holds for the biharmonic operator Δ^2, and more generally for the polyharmonic operator Δ^m, in bounded Lipschitz domains in $\mathbb{R}^2$ or $\mathbb{R}^3$. In [Ver96, Section 8], Verchota extended this to symmetric, strongly elliptic systems with real constant coefficients in three-dimensional Lipschitz domains.

For Laplace's equation and more general second-order elliptic operators, the maximum principle continues to hold in *arbitrary* bounded domains. In contrast, the maximum principle for higher-order operators in rough domains generally *fails.*

In [MNP83], Maz'ya, Nazarov and Plamenevskii studied the Dirichlet problem (with zero boundary data) for constant-coefficient elliptic systems in cones. Counterexamples to (3.2) for systems of order $2m$ in dimension $n \geq 2m+1$ immediately follow from their results. (See [MNP83, Formulas (1.3), (1.18) and (1.28)].) Furthermore, Pipher and Verchota constructed counterexamples to (3.2) for the biharmonic operator Δ^2 in dimension $n = 4$ in [PV92, Section 10], and for the polyharmonic equation $\Delta^m u = 0$ in dimension n, $4 \leq n < 2m+1$, in [PV95b, Theorem 2.1]. Independently Maz'ya and Rossmann showed that (3.2) fails in the exterior of a sufficiently thin cone in dimension n, $n \geq 4$, where L is any constant-coefficient elliptic scalar operator of order $2m \geq 4$ (without lower-order terms). See [MR92, Theorem 8 and Remark 3].

Moreover, with the exception of [MR92, Theorem 8], the aforementioned counterexamples actually provide a stronger negative result than simply the failure of the maximum principle: they show that the left-hand side of (3.2) may be infinite even if the data of the elliptic problem is as nice as possible, that is, smooth and compactly supported.

The counterexamples, however, pertain to high dimensions and do not indicate, e.g., the behavior of the derivatives of order $(m-1)$ of a solution to an elliptic equation of order $2m$ in the lower-dimensional case.

Recently in [MM09b], the second author of the present paper together with Maz'ya have considered this question for the inhomogeneous Dirichlet problem

$$\Delta^2 u = h \text{ in } \Omega, \quad u \in \mathring{W}_2^2(\Omega). \tag{3.3}$$

Here Ω is a bounded domain in $\mathbb{R}^3$ or $\mathbb{R}^2$, the Sobolev space $\mathring{W}_2^2(\Omega)$ is a completion of $C_0^\infty(\Omega)$ in the norm $\|u\|_{\mathring{W}_2^2(\Omega)} = \|\nabla^2 u\|_{L^2(\Omega)}$, and h is a reasonably nice function (e.g., $C_0^\infty(\Omega)$). We remark that if Ω is an arbitrary domain, defining $\nabla u\big|_{\partial\Omega}$ is a delicate matter, and so considering the Dirichlet problem with homogeneous boundary data is somewhat more appropriate.

Motivated by (3.2), the authors showed that if u solves (3.3), then $\nabla u \in L^\infty(\Omega)$, under no restrictions on Ω other than its dimension. Moreover, they proved the following bounds on the Green function.

Let Ω be an arbitrary bounded domain in $\mathbb{R}^3$ and let G be the Green function for the biharmonic equation. Then

$$|\nabla_X \nabla_Y G(X,Y)| \leq C|X-Y|^{-1}, \qquad X,Y \in \Omega, \tag{3.4}$$

$$|\nabla_X G(X,Y)| \leq C \quad \text{and} \quad |\nabla_Y G(X,Y)| \leq C, \qquad X,Y \in \Omega, \tag{3.5}$$

where C is an absolute constant.

The boundedness of the gradient of a solution to the biharmonic equation in a three-dimensional domain is a sharp property in the sense that the function u satisfying (3.3) generally does not exhibit more regularity. For example, let Ω be the three-dimensional punctured unit ball $B(0,1) \setminus \{0\}$, where $B(X,r) = \{Y \in \mathbb{R}^3 : |X-Y| < r\}$, and consider a function $\eta \in C_0^\infty(B(0,1/2))$ such that $\eta = 1$ on $B(0,1/4)$. Let

$$u(X) := \eta(X)|X|, \qquad X \in B_1 \setminus \{0\}. \tag{3.6}$$

Obviously, $u \in \mathring{W}_2^2(\Omega)$ and $\Delta^2 u \in C_0^\infty(\Omega)$. While ∇u is bounded, it is not continuous at the origin. Therefore, the *continuity* of the gradient *does not hold* in general and must depend on some delicate properties of the domain. These questions will be addressed in Section 4 in the framework of the Wiener criterion.

In the absence of boundedness of the gradient ∇u of a harmonic function, or the higher-order derivatives $\nabla^{m-1}u$ of a solution to a higher-order equation, we may instead consider boundedness of a solution itself. Let

$$\Delta^m u = h \text{ in } \Omega, \quad u \in \mathring{W}_m^2(\Omega), \tag{3.7}$$

and $h \in C_0^\infty(\Omega)$. Observe that if $\Omega \subset \mathbb{R}^n$ for $n \leq 2m-1$, then every $u \in \mathring{W}_m^2(\Omega)$ is Hölder continuous on $\overline{\Omega}$ and so must necessarily be bounded.

In [Maz99b, Section 10], Maz'ya showed that the Green function $G_m(X,Y)$ for Δ^m in an arbitrary bounded domain $\Omega \subset \mathbb{R}^n$ satisfies

$$|G_m(X,Y)| \leq C(2m) \log \frac{C \operatorname{diam} \Omega}{\min(|X-Y|, \operatorname{dist}(Y, \partial\Omega))} \tag{3.8}$$

in dimension $n = 2m$, and satisfies

$$|G_m(X,Y)| \leq \frac{C(n)}{|X-Y|^{n-2m}} \tag{3.9}$$

if $n = 2m+1$ or $n = 2m+2$. If $m = 2$, then (3.9) also holds in dimension $n = 7 = 2m+3$ (cf. [Maz79]). Whether (3.9) holds in dimension $n \geq 8$ (for $m = 2$) or $n \geq 2m+3$ (for $m > 2$) is an open problem; see [Maz99b, Problem 2].

If (3.9) holds, then solutions to (3.7) satisfy

$$\|u\|_{L^\infty(\Omega)} \leq C(m,n,p) \operatorname{diam}(\Omega)^{2m-n/p} \|h\|_{L^p(\partial\Omega)}$$

provided $p > n/2m$ (see, e.g., [Maz99b, Section 2]).

Thus, if $\Omega \subset \mathbb{R}^n$ is bounded for $n \leq 2m+2$, $n \neq 2m$, and if u satisfies (3.7) for a reasonably nice function h, then $u \in L^\infty(\Omega)$. This result also holds if $\Omega \subset \mathbb{R}^7$ and $m = 2$.

As in the case of the Green function estimates, if $\Omega \subset \mathbb{R}^n$ is bounded and $n \geq 2m+3$, or if $m = 2$ and $n \geq 8$, then the question of whether solutions u to (3.7) are bounded is open. In particular, it is not known whether solutions u to

$$\Delta^2 u = h \text{ in } \Omega, \quad u \in \mathring{W}_2^2(\Omega)$$

are bounded if $\Omega \subset \mathbb{R}^n$ for $n \geq 8$. However, there exists another fourth-order operator whose solutions are *not* bounded in higher-dimensional domains. In [MN86], Maz'ya and Nazarov showed that if $n \geq 8$ and if $a > 0$ is large enough, then there exists an open cone $K \subset \mathbb{R}^n$ and a function $h \in C_0^\infty(\overline{K} \setminus \{0\})$ such that the solution u to

$$\Delta^2 u + a\partial_n^4 u = h \text{ in } K, \quad u \in \mathring{W}_2^2(K) \tag{3.10}$$

is unbounded near the origin.

To conclude our discussion of Green's functions, we mention two results from [MM11]; these results are restricted to relatively well-behaved domains. In [MM11], D. Mitrea and I. Mitrea showed that, if Ω is a bounded Lipschitz domain in $\mathbb{R}^3$, and G denotes the Green function for the bilaplacian Δ^2, then the estimates

$$\nabla^2 G(X,\,\cdot\,) \in L^3(\Omega), \quad \operatorname{dist}(\,\cdot\,,\partial\Omega)^{-\alpha}\nabla G(X,\,\cdot\,) \in L^{3/\alpha,\infty}$$

hold, uniformly in $X \in \Omega$, for all $0 < \alpha \leq 1$.

Moreover, they considered more general elliptic systems. Suppose that L is an arbitrary elliptic operator of order $2m$ with constant coefficients, as defined by Definition 2.1, and that G denotes the Green function for L. Suppose that $\Omega \subset \mathbb{R}^n$, for $n > m$, is a Lipschitz domain, and that the unit outward normal ν to Ω lies in the Sarason space $VMO(\partial\Omega)$ of functions of vanishing mean oscillations on $\partial\Omega$. Then the estimates

$$\nabla^m G(X,\,\cdot\,) \in L^{\frac{n}{n-m},\infty}(\Omega), \tag{3.11}$$
$$\operatorname{dist}(\,\cdot\,,\partial\Omega)^{-\alpha}\nabla^{m-1} G(X,\,\cdot\,) \in L^{\frac{n}{n-m-1+\alpha},\infty}(\Omega)$$

hold, uniformly in $X \in \Omega$, for any $0 \leq \alpha \leq 1$.

4. The Wiener test

In this section, we discuss conditions that ensure that solutions (or appropriate gradients of solutions) are continuous up to the boundary. These conditions parallel the famous result of Wiener, who in 1924 formulated a criterion that ensured continuity of *harmonic* functions at boundary points [Wie24]. Wiener's criterion has been extended to a variety of second-order elliptic and parabolic equations ([LSW63, FJK82, FGL89, DMM86, MZ97, AH96, TW02, Lab02, EG82]; see also the review papers [Maz97, Ada97]). However, as with the maximum principle, extending this criterion to higher-order elliptic equations is a subtle matter, and many open questions remain.

We begin by stating the classical Wiener criterion for the Laplacian. If $\Omega \subset \mathbb{R}^n$ is a domain and $Q \in \partial\Omega$, then Q is called *regular* for the Laplacian if every solution u to

$$\Delta u = h \text{ in } \Omega, \quad u \in \mathring{W}_1^2(\Omega)$$

for $h \in C_0^\infty(\Omega)$ satisfies $\lim_{X\to Q} u(X) = 0$. According to Wiener's theorem [Wie24], the boundary point $Q \in \partial\Omega$ is regular if and only if the equation

$$\int_0^1 \operatorname{cap}_2(\overline{B(Q,s)} \setminus \Omega) s^{1-n}\, ds = \infty \tag{4.1}$$

holds, where

$$\operatorname{cap}_2(K) = \inf\Big\{ \|u\|^2_{L^2(\mathbb{R}^n)} + \|\nabla u\|^2_{L^2(\mathbb{R}^n)} : u \in C_0^\infty(\mathbb{R}^n),\ u \geq 1 \text{ on } K \Big\}.$$

For example, suppose Ω satisfies the exterior cone condition at Q. That is, suppose there is some open cone K with vertex at Q and some $\varepsilon > 0$ such that $K \cap B(Q,\varepsilon) \subset \Omega^C$. It is elementary to show that $\operatorname{cap}_2(\overline{B(Q,s)} \setminus \Omega) \geq C(K)s^{n-2}$ for all $0 < s < \varepsilon$, and so (4.1) holds and Q is regular. Regularity of such points was known prior to Wiener (see [Poi90], [Zar09], and [Leb13]) and provided inspiration for the formulation of the Wiener test.

By [LSW63], if $L = -\operatorname{div} A\nabla$ is a second-order divergence-form operator, where the matrix $A(X)$ is bounded, measurable, real, symmetric and elliptic, then $Q \in \partial\Omega$ is regular for L if and only if Q and Ω satisfy (4.1). In other words, $Q \in \partial\Omega$ is regular for the Laplacian if and only if it is regular for all such operators. Similar results hold for some other classes of second-order equations; see, for example, [FJK82], [DMM86], or [EG82].

One would like to consider the Wiener criterion for higher-order elliptic equations, and that immediately gives rise to the question of natural generalization of the concept of a regular point. The Wiener criterion for the second-order PDEs ensures, in particular, that weak $\mathring{W}_1^2$ solutions are *classical.* That is, the solution approaches its boundary values in the pointwise sense (continuously). From that point of view, one would extend the concept of regularity of a boundary point as continuity of derivatives of order $m-1$ of the solution to an equation of order $2m$ up to the boundary. On the other hand, as we discussed in the previous section, even the boundedness of solutions cannot be guaranteed in general, and thus, in lower dimensions the study of the continuity up to the boundary for solutions themselves is also very natural. We begin with the latter question, as it is better understood.

Let us first define a regular point for an arbitrary differential operator L of order $2m$ analogously to the case of the Laplacian, by requiring that every solution u to

$$Lu = h \text{ in } \Omega, \quad u \in \mathring{W}_m^2(\Omega) \tag{4.2}$$

for $h \in C_0^\infty(\Omega)$ satisfy $\lim_{X\to Q} u(X) = 0$. Note that by the Sobolev embedding theorem, if $\Omega \subset \mathbb{R}^n$ for $n \leq 2m-1$, then every $u \in \mathring{W}_m^2(\Omega)$ is Hölder continuous on $\overline{\Omega}$ and so satisfies $\lim_{X \to Q} u(X) = 0$ at every point $Q \in \partial\Omega$. Thus, we are only interested in continuity of the solutions at the boundary when $n \geq 2m$.

In this context, the appropriate concept of capacity is the potential-theoretic Riesz capacity of order $2m$, given by

$$\operatorname{cap}_{2m}(K) = \inf\Big\{ \sum_{0 \leq |\alpha| \leq m} \|\partial^\alpha u\|^2_{L^2(\mathbb{R}^n)} : u \in C_0^\infty(\mathbb{R}^n),\ u \geq 1 \text{ on } K \Big\}. \tag{4.3}$$

The following is known. If $m \geq 3$, and if $\Omega \subset \mathbb{R}^n$ for $n = 2m$, $2m+1$ or $2m+2$, or if $m = 2$ and $n = 4$, 5, 6 or 7, then $Q \in \partial\Omega$ is regular for Δ^m if and only if

$$\int_0^1 \operatorname{cap}_{2m}(\overline{B(Q,s)} \setminus \Omega) s^{2m-n-1}\, ds = \infty. \tag{4.4}$$

The biharmonic case was treated in [Maz77] and [Maz79], and the polyharmonic case for $m \geq 3$ in [MD83] and [Maz99a].

Let us briefly discuss the method of the proof in order to explain the restrictions on the dimension. Let L be an arbitrary elliptic operator, and let F be the fundamental solution for L in $\mathbb{R}^n$ with pole at Q. We say that L is positive with weight F if, for all $u \in C_0^\infty(\mathbb{R}^n \setminus \{Q\})$, we have that

$$\int_{\mathbb{R}^n} Lu(X) \cdot u(X)\, F(X)\, dX \geq c \sum_{k=1}^m \int_{\mathbb{R}^n} |\nabla^k u(X)|^2 |X|^{2k-n}\, dX. \tag{4.5}$$

The biharmonic operator is positive with weight F in dimension n if $4 \leq n \leq 7$, and the polyharmonic operator Δ^m, $m \geq 3$, is positive with weight F in dimension $2m \leq n \leq 2m+2$. (The Laplacian Δ is positive with weight F in any dimension.) The biharmonic operator Δ^2 is not positive with weight F in dimensions $n \geq 8$, and Δ^m is not positive with weight F in dimension $n \geq 2m+3$. See [Maz99a, Propositions 1 and 2].

The proof of the Wiener criterion for the polyharmonic operator required positivity with weight F. In fact, it turns out that positivity with weight F suffices to provide a Wiener criterion for an *arbitrary* scalar elliptic operator with constant coefficients.

Theorem 4.6 ([Maz02, Theorems 1 and 2]). *Suppose $\Omega \subset \mathbb{R}^n$ and that L is a scalar elliptic operator of order $2m$ with constant real coefficients, as defined by Definition* 2.1.

If $n = 2m$, then $Q \in \partial\Omega$ is regular for L if and only if (4.4) *holds.*

If $n \geq 2m+1$, and if the condition (4.5) *holds, then again $Q \in \partial\Omega$ is regular for L if and only if* (4.4) *holds.*

This theorem is also valid for certain variable-coefficient operators in divergence form; see the remark at the end of [Maz99a, Section 5].

Similar results have been proven for some second-order elliptic *systems*. In particular, for the Lamé system $Lu = \Delta u + \alpha \operatorname{grad} \operatorname{div} u$, $\alpha > -1$, positivity with weight F and Wiener criterion have been established for a range of α close to zero, that is, when the underlying operator is close to the Laplacian ([LM10]). It was also shown that positivity with weight F may in general fail for the Lamé system. Since the present review is restricted to the higher-order operators, we shall not elaborate on this point and instead refer the reader to [LM10] for more detailed discussion.

In the absence of the positivity condition (4.5), the situation is much more involved. Let us point out first that the condition (4.5) is *not* necessary for regularity of a boundary point, that is, the continuity of the solutions. There exist fourth-order elliptic operators that are not positive with weight F whose solutions exhibit nice behavior near the boundary; there exist other such operators whose solutions exhibit very bad behavior near the boundary.

Specifically, recall that (4.5) fails for $L = \Delta^2$ in dimension $n \geq 8$. Nonetheless, solutions to $\Delta^2 u = h$ are often well behaved near the boundary. By [MP81], the vertex of a cone is regular for the bilaplacian in any dimension. Furthermore, if the capacity condition (4.4) holds with $m = 2$, then by [Maz02, Section 10], any solution u to

$$\Delta^2 u = h \text{ in } \Omega, \quad u \in \mathring{W}_2^2(\Omega)$$

for $h \in C_0^\infty(\Omega)$ satisfies $\lim_{X\to Q} u(X) = 0$ provided the limit is taken along a *nontangential* direction.

Conversely, if $n \geq 8$ and $L = \Delta^2 + a\partial_n^4$, then by [MN86], there exists a cone K and a function $h \in C_0^\infty(\overline{K} \setminus \{0\})$ such that the solution u to (3.10) is not only discontinuous but *unbounded* near the vertex of the cone. We remark that a careful examination of the proof in [MN86] implies that solutions to (3.10) are unbounded even along some nontangential directions.

Thus, conical points in dimension eight are regular for the bilaplacian and irregular for the operator $\Delta^2 + a\partial_n^4$. Hence, a relevant Wiener condition *must* use different capacities for these two operators. This is a striking contrast with the second-order case, where the same capacity condition implies regularity for all divergence-form operators, even with variable coefficients.

This concludes the discussion of regularity in terms of continuity of the solution. We now turn to regularity in terms of continuity of the $(m-1)$-st derivatives. Unfortunately, much less is known in this case. The first such result has recently appeared in [MM09a]. It pertains to the biharmonic equation in dimension three.

We say that $Q \in \partial\Omega$ is 1-regular for the operator Δ^2 if every solution u to

$$\Delta^2 u = h \text{ in } \Omega, \quad u \in \mathring{W}_2^2(\Omega) \tag{4.7}$$

for $h \in C_0^\infty(\Omega)$ satisfies $\lim_{X\to Q} \nabla u(X) = 0$. In [MM09a] the second author of this paper and Maz'ya proved that in a three-dimensional domain the following holds. If

$$\int_0^c \inf_{P\in\Pi_1} \operatorname{cap}_P(\overline{B(0,as) \setminus B(0,s)} \setminus \Omega)\, ds = \infty \tag{4.8}$$

for some $a \geq 4$ and some $c > 0$, then 0 is 1-regular. Conversely, if $0 \in \partial\Omega$ is 1-regular for Δ^2 then for every $c > 0$ and every $a \geq 8$,

$$\inf_{P\in\Pi_1} \int_0^c \operatorname{cap}_P(\overline{B(0,as) \setminus B(0,s)} \setminus \Omega)\, ds = \infty. \tag{4.9}$$

Here

$$\mathrm{cap}_P(K) = \inf\{\|\Delta u\|^2_{L^2(\mathbb{R}^3)} : u \in \mathring{W}_2^2(\mathbb{R}^3 \setminus \{0\}),\ u = P \text{ in a neighborhood of } K\}$$

and Π_1 is the space of functions $P(X)$ of the form $P(X) = b_0 + b_1 X_1 + b_2 X_2 + b_3 X_3$ with coefficients $b_k \in \mathbb{R}$ that satisfy $\sqrt{b_0^2 + b_1^2 + b_2^2 + b_3^2} = 1$.

Note that the notion of capacity cap_P is quite different from the classical analogues, and even from the Riesz capacity used in the context of the higher-order elliptic operators before (cf. (4.3)). Its properties, as well as properties of 1-regular and 1-irregular points, can be different from classical analogous as well. For instance, for some domains 1-irregularity turns out to be unstable under affine transformations of coordinates.

The slight discrepancy between the sufficient condition (4.8) and the necessary condition (4.9) is needed, in the sense that (4.8) is not always necessary for 1-regularity. However, in an important particular case, there exists a single simpler condition for 1-regularity. To be precise, let $\Omega \subset \mathbb{R}^3$ be a domain whose boundary is the graph of a function φ, and let ω be its modulus of continuity. If

$$\int_0^1 \frac{t\,dt}{\omega^2(t)} = \infty, \tag{4.10}$$

then every solution to the biharmonic equation (4.7) satisfies $\nabla u \in C(\overline{\Omega})$. Conversely, for every ω such that the integral in (4.10) is convergent, there exists a $C^{0,\omega}$ domain and a solution u of the biharmonic equation such that $\nabla u \notin C(\overline{\Omega})$. In particular, as expected, the gradient of a solution to the biharmonic equation is always bounded in Lipschitz domains and is not necessarily bounded in a Hölder domain. Moreover, one can deduce from (4.10) that the gradient of a solution is always bounded, e.g., in a domain with $\omega(t) \approx t \log^{1/2} t$, which is not Lipschitz, and might fail to be bounded in a domain with $\omega(t) \approx t \log t$. More properties of the new capacity and examples can be found in [MM09a].

5. Boundary value problems in Lipschitz domains for elliptic operators with constant coefficients

The maximum principle (3.2) provides estimates on solutions whose boundary data lies in L^∞. Recall that for second-order partial differential equations with real coefficients, the maximum principle is valid in arbitrary bounded domains. The corresponding sharp estimates for boundary data in L^p, $1 < p < \infty$, are much more delicate. They are *not* valid in arbitrary domains, even for harmonic functions, and they depend in a delicate way on the geometry of the boundary. At present, boundary-value problems for the Laplacian and for general real symmetric elliptic operators of the second order are fairly well understood on Lipschitz domains. See, in particular, [Ken94].

We consider biharmonic functions and more general higher-order elliptic equations. The question of estimates on biharmonic functions with data in L^p

was raised by Rivière in the 1970s ([CFS79]), and later Kenig redirected it towards Lipschitz domains in [Ken90, Ken94]. The sharp range of well-posedness in L^p, even for biharmonic functions, remains an open problem (see [Ken94, Problem 3.2.30]). In this section we shall review the current state of the art in the subject, the main techniques that have been successfully implemented, and their limitations in the higher-order case.

Most of the results we will discuss are valid in Lipschitz domains, defined as follows.

Definition 5.1. A domain $\Omega \subset \mathbb{R}^n$ is called a *Lipschitz domain* if, for every $Q \in \partial\Omega$, there is a number $r > 0$, a Lipschitz function $\varphi : \mathbb{R}^{n-1} \mapsto \mathbb{R}$ with $\|\nabla\varphi\|_{L^\infty} \leq M$, and a rectangular coordinate system for $\mathbb{R}^n$ such that

$$B(Q,r) \cap \Omega = \{(x,s) : x \in \mathbb{R}^{n-1},\ s \in \mathbb{R},\ |(x,s) - Q)| < r, \text{ and } s > \varphi(x)\}.$$

If we may take the functions φ to be C^k (that is, to possess k continuous derivatives), we say that Ω is a C^k *domain.*

The outward normal vector to Ω will be denoted ν. The surface measure will be denoted σ, and the tangential derivative along $\partial\Omega$ will be denoted ∇_τ.

In this paper, we will assume that all domains under consideration have connected boundary. Furthermore, if $\partial\Omega$ is unbounded, we assume that there is a single Lipschitz function φ and coordinate system that satisfies the conditions given above; that is, we assume that Ω is the domain above (in some coordinate system) the graph of a Lipschitz function.

In order to properly state boundary-value problems on Lipschitz domains, we will need the notions of non-tangential convergence and non-tangential maximal function.

In this and subsequent sections we say that $u\big|_{\partial\Omega} = f$ if f is the *nontangential limit* of u, that is, if

$$\lim_{X \to Q,\, X \in \Gamma(Q)} u(X) = f(Q)$$

for almost every $(d\sigma)$ $Q \in \partial\Omega$, where $\Gamma(Q)$ is the *nontangential cone*

$$\Gamma(Q) = \{Y \in \Omega : \operatorname{dist}(Y, \partial\Omega) < (1+a)|X - Y|\}. \tag{5.2}$$

Here $a > 0$ is a positive parameter; the exact value of a is usually irrelevant to applications. The *nontangential maximal function* is given by

$$NF(Q) = \sup\{|F(X)| : X \in \Gamma(Q)\}. \tag{5.3}$$

The normal derivative of u of order m is defined as

$$\partial_\nu^m u(Q) = \sum_{|\alpha|=m} \nu(Q)^\alpha \frac{m!}{\alpha!} \partial^\alpha u(Q),$$

where $\partial^\alpha u(Q)$ is taken in the sense of nontangential limits as usual.

5.1. The Dirichlet problem: definitions, layer potentials, and some well-posedness results

We say that the L^p-Dirichlet problem for the biharmonic operator Δ^2 in a domain Ω is well posed if there exists a constant $C > 0$ such that, for every $f \in W_1^p(\partial\Omega)$ and every $g \in L^p(\partial\Omega)$, there exists a unique function u that satisfies

$$\begin{cases} \Delta^2 u = 0 & \text{in } \Omega, \\ u = f & \text{on } \partial\Omega, \\ \partial_\nu u = g & \text{on } \partial\Omega, \\ \|N(\nabla u)\|_{L^p(\partial\Omega)} \leq C\|g\|_{L^p(\partial\Omega)} + C\|\nabla_\tau f\|_{L^p(\partial\Omega)}. \end{cases} \tag{5.4}$$

The L^p-Dirichlet problem for the polyharmonic operator Δ^m is somewhat more involved, because the notion of boundary data is necessarily more subtle. We say that the L^p-Dirichlet problem for Δ^m in a domain Ω is well posed if there exists a constant $C > 0$ such that, for every $g \in L^p(\partial\Omega)$ and every $\dot{f}$ in the Whitney–Sobolev space $WA_{m-1}^p(\partial\Omega)$, there exists a unique function u that satisfies

$$\begin{cases} \Delta^m u = 0 & \text{in } \Omega, \\ \partial^\alpha u\big|_{\partial\Omega} = f_\alpha & \text{for all } 0 \leq |\alpha| \leq m-2, \\ \partial_\nu^{m-1} u = g & \text{on } \partial\Omega, \\ \|N(\nabla^{m-1} u)\|_{L^p(\partial\Omega)} \leq C\|g\|_{L^p(\partial\Omega} + C \sum_{|\alpha|=m-2} \|\nabla_\tau f_\alpha\|_{L^p(\partial\Omega)}. \end{cases} \tag{5.5}$$

The space $WA_m^p(\partial\Omega)$ is defined as follows.

Definition 5.6. Suppose that $\Omega \subset \mathbb{R}^n$ is a Lipschitz domain, and consider arrays of functions $\dot{f} = \{f_\alpha : |\alpha| \leq m-1\}$ indexed by multiindices α of length n, where $f_\alpha : \partial\Omega \mapsto \mathbb{C}$. We let $WA_m^p(\partial\Omega)$ be the completion of the set of arrays $\dot{\psi} = \{\partial^\alpha \psi : |\alpha| \leq m-1\}$, for $\psi \in C_0^\infty(\mathbb{R}^n)$, under the norm

$$\sum_{|\alpha|\leq m-1} \|\partial^\alpha \psi\|_{L^p(\partial\Omega)} + \sum_{|\alpha|=m-1} \|\nabla_\tau \partial^\alpha \psi\|_{L^p(\partial\Omega)}. \tag{5.7}$$

If we prescribe $\partial^\alpha u = f_\alpha$ on $\partial\Omega$ for some $\dot{f} \in WA_m^p(\partial\Omega)$, then we are prescribing the values of $u, \nabla u, \dots, \nabla^{m-1} u$ on $\partial\Omega$, and requiring that (the prescribed part of) $\nabla^m u\big|_{\partial\Omega}$ lie in $L^p(\partial\Omega)$.

The study of these problems began with biharmonic functions in C^1 domains. In [SS81], Selvaggi and Sisto proved that, if Ω is the domain above the graph of a compactly supported C^1 function φ, with $\|\nabla\varphi\|_{L^\infty}$ small enough, then solutions to the Dirichlet problem exist provided $1 < p < \infty$. Their method used certain biharmonic layer potentials composed with the Riesz transforms.

In [CG83], Cohen and Gosselin proved that, if Ω is a bounded, simply connected C^1 domain contained in the plane $\mathbb{R}^2$, then the L^p-Dirichlet problem is well posed in Ω for any $1 < p < \infty$. In [CG85], they extended this result to the complements of such domains. Their proof used multiple layer potentials introduced by Agmon in [Agm57] in order to solve the Dirichlet problem with continuous boundary data. The general outline of their proof parallelled that of the proof of the corresponding result [FJR78] for Laplace's equation.

As in the case of Laplace's equation, a result in Lipschitz domains soon followed. In [DKV86], Dahlberg, Kenig and Verchota showed that the L^p-Dirichlet problem for the biharmonic equation is well posed in any bounded simply connected Lipschitz domain $\Omega \subset \mathbb{R}^n$, provided $2-\varepsilon < p < 2+\varepsilon$ for some $\varepsilon > 0$ depending on the domain Ω.

In [Ver87], Verchota used the construction of [DKV86] to extend Cohen and Gosselin's results from planar C^1 domains to C^1 domains of arbitrary dimension. Thus, the L^p-Dirichlet problem for the bilaplacian is well posed for $1 < p < \infty$ in C^1 domains.

In [Ver90], Verchota showed that the L^p-Dirichlet problem for the polyharmonic operator Δ^m could be solved for $2-\varepsilon < p < 2+\varepsilon$ in starlike Lipschitz domains by induction on the exponent m. He simultaneously proved results for the L^p-regularity problem in the same range; we will thus delay discussion of his methods to Section 5.3.

All three of the papers [SS81], [CG83] and [DKV86] constructed biharmonic functions as potentials. However, the potentials used differ. [SS81] constructed their solutions as

$$u(X) = \int_{\partial\Omega} \partial_n^2 F(X-Y)\, f(Y)\, d\sigma(Y) + \sum_{i=1}^{n-1} \int_{\partial\Omega} \partial_i \partial_n F(X-Y) R_i g(Y)\, d\sigma(Y)$$

where R_i are the Riesz transforms. Here $F(X)$ is the fundamental solution to the biharmonic equation; thus, u is biharmonic in $\mathbb{R}^n \setminus \partial\Omega$. As in the case of Laplace's equation, well-posedness of the Dirichlet problem follows from the boundedness relation $\|N(\nabla u)\|_{L^p(\partial\Omega)} \le C\|f\|_{L^p(\partial\Omega)} + C\|g\|_{L^p(\partial\Omega)}$ and from invertibility of the mapping $(f,g) \mapsto (u\big|_{\partial\Omega}, \partial_\nu u)$ on $L^p(\partial\Omega) \times L^p(\partial\Omega) \mapsto W_1^p(\partial\Omega) \times L^p(\partial\Omega)$.

The multiple layer potential of [CG83] is an operator of the form

$$\mathcal{L}\dot{f}(P) = \text{p.v.} \int_{\partial\Omega} \mathcal{L}(P,Q) \dot{f}(Q)\, d\sigma(Q) \tag{5.8}$$

where $\mathcal{L}(P,Q)$ is a 3×3 matrix of kernels, also composed of derivatives of the biharmonic equation, and $\dot{f} = (f, f_x, f_y)$ is a "compatible triple" of boundary data, that is, an element of $W^{1,p}(\partial\Omega) \times L^p(\partial\Omega) \times L^p(\partial\Omega)$ that satisfies $\partial_\tau f = f_x \tau_x + f_y \tau_y$. Thus, the input is essentially a function and its gradient, rather than two functions, and the Riesz transforms are not involved.

The method of [DKV86] is to compose two potentials. First, the function $f \in L^2(\partial\Omega)$ is mapped to its Poisson extension v. Next, u is taken to be the solution of the inhomogeneous equation $\Delta u(Y) = (n + 2Y \cdot \nabla) v(Y)$ with $u = 0$ on $\partial\Omega$. If $G(X,Y)$ is the Green function for Δ in Ω and k^Y is the harmonic measure density at Y, we may write the map $f \mapsto u$ as

$$u(X) = \int_\Omega G(X,Y)(n+2Y\cdot\nabla)\int_{\partial\Omega} k^Y(Q)\,f(Q)\,d\sigma(Q)\,dY. \tag{5.9}$$

Since $(n+2Y\cdot\nabla)v(Y)$ is harmonic, u is biharmonic, and so u solves the Dirichlet problem.

5.2. The L^p-Dirichlet problem: the summary of known results on well-posedness and ill-posedness

Recall that by [Ver90], the L^p-Dirichlet problem is well posed in Lipschitz domains provided $2-\varepsilon < p < 2+\varepsilon$. As in the case of Laplace's equation (see [FJL77]), the range $p > 2-\varepsilon$ is sharp. That is, for any $p < 2$ and any integers $m \geq 2$, $n \geq 2$, there exists a bounded Lipschitz domain $\Omega \subset \mathbb{R}^n$ such that the L^p-Dirichlet problem for Δ^m is ill posed in Ω. See [DKV86, Section 5] for the case of the biharmonic operator Δ^2, and the proof of Theorem 2.1 in [PV95b] for the polyharmonic operator Δ^m.

The range $p < 2+\varepsilon$ is not sharp and has been studied extensively. Proving or disproving well-posedness of the L^p-Dirichlet problem for $p > 2$ in general Lipschitz domains has been an open question since [DKV86], and was formally stated as such in [Ken94, Problem 3.2.30]. (Earlier in [CFS79, Question 7], the authors had posed the more general question of what classes of boundary data give existence and uniqueness of solutions.)

In [PV92, Theorem 10.7], Pipher and Verchota constructed Lipschitz domains Ω such that the L^p-Dirichlet problem for Δ^2 was ill posed in Ω, for any given $p > 6$ (in four dimensions) or any given $p > 4$ (in five or more dimensions). Their counterexamples built on the study of solutions near a singular point, in particular upon [MNP83] and [MP81]. In [PV95b], they provided other counterexamples to show that the L^p-Dirichlet problem for Δ^m is ill posed, provided $p > 2(n-1)/(n-3)$ and $4 \leq n < 2m+1$. They remarked that if $n \geq 2m+1$, then ill-posedness follows from the results of [MNP83] provided $p > 2m/(m-1)$.

The endpoint result at $p = \infty$ is the Agmon–Miranda maximum principle (3.2) discussed above. We remark that if $2 < p_0 \leq \infty$, and the L^{p_0}-Dirichlet problem is well posed (or (3.2) holds) then by interpolation, the L^p-Dirichlet problem is well posed for any $2 < p < p_0$.

We shall adopt the following definition (justified by the discussion above).

Definition 5.10. Suppose that $m \geq 2$ and $n \geq 4$. Then $p_{m,n}$ is defined to be the extended real number that satisfies the following properties. If $2 \leq p \leq p_{m,n}$, then the L^p-Dirichlet problem for Δ^m is well posed in any bounded Lipschitz domain $\Omega \subset \mathbb{R}^n$. Conversely, if $p > p_{m,n}$, then there exists a bounded Lipschitz domain $\Omega \subset \mathbb{R}^n$ such that the L^p-Dirichlet problem for Δ^m is ill posed in Ω. Here, well-posedness for $1 < p < \infty$ is meant in the sense of (5.5), and well-posedness for $p = \infty$ is meant in the sense of the maximum principle (see (5.24) below).

As in [DKV86], we expect the range of solvability for any *particular* Lipschitz domain Ω to be $2-\varepsilon < p < p_{m,n}+\varepsilon$ for some ε depending on the Lipschitz character of Ω.

Let us summarize here the results currently known for $p_{m,n}$. More details will follow in Section 5.3.

For any $m \geq 2$, we have that

- If $n = 2$ or $n = 3$, then the L^p-Dirichlet problem for Δ^m is well posed in any Lipschitz domain Ω for any $2 \leq p < \infty$. ([PV92, PV95b])
- If $4 \leq n \leq 2m+1$, then $p_{m,n} = 2(n-1)/(n-3)$. ([She06a, PV95b].)
- If $n = 2m+2$, then $p_{m,n} = 2m/(m-1) = 2(n-2)/(n-4)$. ([She06b, MNP83].)
- If $n \geq 2m+3$, then $2(n-1)/(n-3) \leq p_{m,n} \leq 2m/(m-1)$. ([She06a, MNP83].)

The value of $p_{m,n}$, for $n \geq 2m+3$, is open.

In the special case of biharmonic functions ($m = 2$), more is known.

- $p_{2,4} = 6$, $p_{2,5} = 4$, $p_{2,6} = 4$, and $p_{2,7} = 4$. ([She06a] and [She06b])
- If $n \geq 8$, then
$$2 + \frac{4}{n - \lambda_n} < p_{2,n} \leq 4$$
where
$$\lambda_n = \frac{n + 10 + 2\sqrt{2(n^2 - n + 2)}}{7}.$$
([She06c])
- If Ω is a C^1 or convex domain of arbitrary dimension, then the L^p-Dirichlet problem for Δ^2 is well posed in Ω for any $1 < p < \infty$. ([Ver90, She06c, KS11a].)

We comment on the nature of ill-posedness. The counterexamples of [DKV86] and [PV95b] for $p < 2$ are failures of uniqueness. That is, those counterexamples are nonzero functions u, satisfying $\Delta^m u = 0$ in Ω, such that $\partial_\nu^k u = 0$ on $\partial\Omega$ for $0 \leq k \leq m-1$, and such that $N(\nabla^{m-1}u) \in L^p(\partial\Omega)$.

Observe that if Ω is bounded and $p > 2$, then $L^p(\partial\Omega) \subset L^2(\partial\Omega)$. Because the L^2-Dirichlet problem is well posed, the failure of well-posedness for $p > 2$ can only be a failure of the optimal estimate $N(\nabla^{m-1}u) \in L^p(\partial\Omega)$. That is, if the L^p-Dirichlet problem for Δ^m is ill posed in Ω, then for some Whitney array $\dot{f} \in WA^p_{m-1}(\partial\Omega)$ and some $g \in L^p(\partial\Omega)$, the unique function u that satisfies $\Delta^m u = 0$ in Ω, $\partial^\alpha u = f_\alpha$, $\partial_\nu^{m-1} u = g$ and $N(\nabla^{m-1}u) \in L^2(\partial\Omega)$ does not satisfy $N(\nabla^{m-1}u) \in L^p(\partial\Omega)$.

5.3. The regularity problem and the L^p-Dirichlet problem

In this section we elaborate on some of the methods used to prove the Dirichlet well-posedness results listed above, as well as their historical context. This naturally brings up a consideration of a different boundary value problem, the L^q-regularity problem for higher-order operators.

Recall that for second-order equations the regularity problem corresponds to finding a solution with prescribed tangential gradient along the boundary. In analogy, we say that the L^q-regularity problem for Δ^m is well posed in Ω if there exists a constant $C > 0$ such that, whenever $\dot{f} \in WA^q_m(\partial\Omega)$, there exists a unique function u that satisfies

$$\begin{cases} \Delta^m u = 0 & \text{in } \Omega, \\ \partial^\alpha u\big|_{\partial\Omega} = f_\alpha & \text{for all } 0 \le |\alpha| \le m-1, \\ \|N(\nabla^m u)\|_{L^q(\partial\Omega)} \le C \sum_{|\alpha|=m-1} \|\nabla_\tau f_\alpha\|_{L^q(\partial\Omega)}. \end{cases} \tag{5.11}$$

There is an important endpoint formulation at $q = 1$ for the regularity problem. We say that the H^1-regularity problem is well posed if there exists a constant $C > 0$ such that, whenever $\dot{f}$ lies in the Whitney–Hardy space $H^1_m(\partial\Omega)$, there exists a unique function u that satisfies

$$\begin{cases} \Delta^m u = 0 & \text{in } \Omega, \\ \partial^\alpha u\big|_{\partial\Omega} = f_\alpha & \text{for all } 0 \le |\alpha| \le m-1, \\ \|N(\nabla^m u)\|_{L^1(\partial\Omega)} \le C \sum_{|\alpha|=m-1} \|\nabla_\tau f_\alpha\|_{H^1(\partial\Omega)}. \end{cases}$$

The space $H^1_m(\partial\Omega)$ is defined as follows.

Definition 5.12. We say that $\dot{a} \in WA^q_m(\partial\Omega)$ is a $H^1_m(\partial\Omega)$-L^q atom if $\dot{a}$ is supported in a ball $B(Q,r) \cap \partial\Omega$ and if

$$\sum_{|\alpha|=m-1} \|\nabla_\tau a_\alpha\|_{L^q(\partial\Omega)} \le \sigma(B(Q,r) \cap \partial\Omega)^{1/q-1}.$$

If $\dot{f} \in WA^1_m(\partial\Omega)$ and there are H^1_m-L^2 atoms $\dot{a}_k$ and constants $\lambda_k \in \mathbb{C}$ such that

$$\nabla_\tau f_\alpha = \sum_{k=1}^{\infty} \lambda_k \nabla_\tau (a_k)_\alpha \text{ for all } |\alpha| = m-1$$

and such that $\sum |\lambda_k| < \infty$, we say that $\dot{f} \in H^1_m(\partial\Omega)$, with $\|\dot{f}\|_{H^1_m(\partial\Omega)}$ being the smallest $\sum |\lambda_k|$ among all such representations.

In [Ver90], Verchota proved well-posedness of the L^2-Dirichlet problem and the L^2-regularity problem for the polyharmonic operator Δ^m in any bounded starlike Lipschitz domain by simultaneous induction.

The base case $m = 1$ is valid in all bounded Lipschitz domains by [Dah79] and [JK81b]. The inductive step is to show that well-posedness for the Dirichlet problem for Δ^{m+1} follows from well-posedness of the lower-order problems. In particular, solutions with $\partial^\alpha u = f_\alpha$ may be constructed using the regularity problem for Δ^m, and the boundary term $\partial^m_\nu u = g$, missing from the regularity data, may be attained using the *inhomogeneous* Dirichlet problem for Δ^m. On the other hand, it was shown that the well-posedness for the regularity problem for Δ^{m+1} follows from well-posedness of the lower-order problems and from the Dirichlet problem for Δ^{m+1}, in some sense, by realizing the solution to the regularity problem as an integral of the solution to the Dirichlet problem.

As regards a broader range of p and q, Pipher and Verchota showed in [PV92] that the L^p-Dirichlet and L^q-regularity problems for Δ^2 are well posed in all bounded Lipschitz domains $\Omega \subset \mathbb{R}^3$, provided $2 \leq p < \infty$ and $1 < q \leq 2$. Their method relied on duality. Using potentials similar to those of [DKV86], they constructed solutions to the L^2-Dirichlet problem in domains above Lipschitz graphs. The core of their proof was the invertibility on $L^2(\partial\Omega)$ of a certain potential operator T. They were able to show that the invertibility of its adjoint T^* on $L^2(\partial\Omega)$ implies that the L^2-regularity problem for Δ^2 is well posed. Then, using the atomic decomposition of Hardy spaces, they analyzed the H^1-regularity problem. Applying interpolation and duality for T^* once again, now in the reverse regularity-to-Dirichlet direction, the full range for both regularity and Dirichlet problems was recovered in domains above graphs. Localization arguments then completed the argument in bounded Lipschitz domains.

In four or more dimensions, further progress relied on the following theorem of Shen.

Theorem 5.13 ([She06b]). *Suppose that $\Omega \subset \mathbb{R}^n$ is a Lipschitz domain. The following conditions are equivalent.*

- *The L^p-Dirichlet problem for L is well posed, where L is a symmetric elliptic system of order $2m$ with real constant coefficients.*
- *There exists some constant $C > 0$ and some $p > 2$ such that*

$$\left(\fint_{B(Q,r)\cap\partial\Omega} N(\nabla^{m-1}u)^p\,d\sigma\right)^{1/p} \leq C\left(\fint_{B(Q,2r)\cap\partial\Omega} N(\nabla^{m-1}u)^2\,d\sigma\right)^{1/2} \tag{5.14}$$

holds whenever u is a solution to the L^2-Dirichlet problem for L in Ω, with $\nabla u \equiv 0$ on $B(Q,3r)\cap\partial\Omega$.

For the polyharmonic operator Δ^m, this theorem was essentially proven in [She06a]. Furthermore, the reverse Hölder estimate (5.14) with $p = 2(n-1)/(n-3)$ was shown to follow from well-posedness of the L^2-regularity problem. Thus the L^p-Dirichlet problem is well posed in bounded Lipschitz domains in $\mathbb{R}^n$ for $p = 2(n-1)/(n-3)$. By interpolation, and because reverse Hölder estimates have self-improving properties, well-posedness in the range $2 \leq p \leq 2(n-1)/(n-3)+\varepsilon$ for any particular Lipschtiz domain follows automatically.

Using regularity estimates and square-function estimates, Shen was able to further improve this range of p. He showed that with $p = 2+4/(n-\lambda)$, $0 < \lambda < n$, the reverse Hölder estimate (5.14) is true, provided that

$$\int_{B(Q,r)\cap\Omega} |\nabla^{m-1}u|^2 \leq C\left(\frac{r}{R}\right)^\lambda \int_{B(Q,R)\cap\Omega} |\nabla^{m-1}u|^2 \tag{5.15}$$

holds whenever u is a solution to the L^2-Dirichlet problem in Ω with $N(\nabla^{m-1}u) \in L^2(\partial\Omega)$ and $\nabla^k u\big|_{B(Q,R)\cap\Omega} \equiv 0$ for all $0 \leq k \leq m-1$.

It is illuminating to observe that the estimates arising in connection with the pointwise bounds on the solutions in arbitrary domains (cf. Section 3) and the

Wiener test (cf. Section 4), take essentially the form (5.15). Thus, Theorem 5.13 and its relation to (5.15) provide a direct way to transform results regarding local boundary regularity of solutions, obtained via the methods underlined in Sections 3 and 4, into well-posedness of the L^p-Dirichlet problem.

In particular, consider [Maz02, Lemma 5]. If u is a solution to $\Delta^m u = 0$ in $B(Q,R) \cap \Omega$, where Ω is a Lipschitz domain, then by [Maz02, Lemma 5] there is some constant $\lambda_0 > 0$ such that

$$\sup_{B(Q,r)\cap\Omega} |u|^2 \leq \left(\frac{r}{R}\right)^{\lambda_0} \frac{C}{R^n} \int_{B(Q,R)\cap\Omega} |u(X)|^2\, dX \tag{5.16}$$

provided that r/R is small enough, that u has zero boundary data on $B(Q,R)\cap\partial\Omega$, and where $\Omega \subset \mathbb{R}^n$ has dimension $n = 2m+1$ or $n = 2m+2$, or where $m = 2$ and $n = 7 = 2m+3$. (The bound on dimension comes from the requirement that Δ^m be positive with weight F; see equation (4.5).)

It is not difficult to see (cf., e.g., [She06b, Theorem 2.6]), that (5.16) implies (5.15) for some $\lambda > n-2m+2$, and thus implies well-posedness of the L^p-Dirichlet problem for a certain range of p. This provides an improvement on the results of [She06a] in the case $m = 2$ and $n = 6$ or $n = 7$, and in the case $m \geq 3$ and $n = 2m+2$. Shen has stated this improvement in [She06b, Theorems 1.4 and 1.5]: the L^p-Dirichlet problem for Δ^2 is well posed for $2 \leq p < 4+\varepsilon$ in dimensions $n = 6$ or $n = 7$, and the L^p-Dirichlet problem for Δ^m is well posed if $2 \leq p < 2m/(m-1)+\varepsilon$ in dimension $n = 2m+2$.

The method of weighted integral identities, related to positivity with weight F (cf. (4.5)), can be further finessed in a particular case of the biharmonic equation. [She06c] uses this method (extending the ideas from [Maz79]) to show that if $n \geq 8$, then (5.15) is valid for solutions to Δ^2 with $\lambda = \lambda_n$, where

$$\lambda_n = \frac{n+10+2\sqrt{2(n^2-n+2)}}{7}. \tag{5.17}$$

We now return to the L^q-regularity problem. Recall that in [PV92], Pipher and Verchota showed that if $2 < p < \infty$ and $1/p + 1/q < 1$, then the L^p-Dirichlet problem and the L^q-regularity problem for Δ^2 are both well posed in three-dimensional Lipschitz domains. They proved this by showing that, in the special case of a domain above a Lipschitz graph, there is duality between the L^p-Dirichlet and L^q-regularity problems. Such duality results are common. See [KP93], [She07b], and [KR09] for duality results in the second-order case; although even in that case, duality is not always guaranteed. (See [May10].) Many of the known results concerning the regularity problem for the polyharmonic operator Δ^m are results relating the L^p-Dirichlet problem to the L^q-regularity problem.

In [MM10], I. Mitrea and M. Mitrea showed that if $1 < p < \infty$ and $1/p+1/q = 1$, and if the L^q-regularity problem for Δ^2 and the L^p-regularity problem for Δ were both well posed in a particular bounded Lipschitz domain Ω, then the L^p-Dirichlet problem for Δ^2 was also well posed in Ω. They proved this result (in arbitrary dimensions) using layer potentials and a Green representation formula for biharmonic equations. Observe that the extra requirement of well-posedness for the Laplacian is extremely unfortunate, since in bad domains it essentially restricts consideration to $p < 2 + \varepsilon$ and thus does not shed new light on well-posedness in the general class of Lipschitz domains. As will be discussed below, later Kilty and Shen established an optimal duality result for biharmonic Dirichlet and regularity problems.

Recall that the formula (5.14) provides a necessary and sufficient condition for well-posedness of the L^p-Dirichlet problem. In [KS11b], Kilty and Shen provided a similar condition for the regularity problem. To be precise, they demonstrated that if $q > 2$ and L is a symmetric elliptic system of order $2m$ with real constant coefficients, then the L^q-regularity problem for L is well posed if and only if the estimate

$$\left(\fint_{B(Q,r)\cap\Omega} N(\nabla^m u)^q \, d\sigma\right)^{1/q} \leq C\left(\fint_{B(Q,2r)\cap\Omega} N(\nabla^m u)^2 \, d\sigma\right)^{1/2} \tag{5.18}$$

holds for all points $Q \in \partial\Omega$, all $r > 0$ small enough, and all solutions u to the L^2-regularity problem with $\nabla^k u\big|_{B(Q,3r)\cap\partial\Omega} = 0$ for $0 \leq k \leq m-1$. Observe that (5.18) is identical to (5.14) with p replaced by q and $m-1$ replaced by m.

As a consequence, well-posedness of the L^q-regularity problem in Ω for certain values of q implies well-posedness of the L^p-Dirichlet problem for some values of p. Specifically, arguments using interior regularity and fractional integral estimates (given in [KS11b, Section 5]) show that (5.18) implies (5.14) with $1/p = 1/q - 1/(n-1)$. But recall from [She06b] that (5.14) holds if and only if the L^p-Dirichlet problem for L is well posed in Ω. Thus, if $2 < q < n-1$, and if the L^q-regularity problem for a symmetric elliptic system is well posed in a Lipschitz domain Ω, then the L^p-Dirichlet problem for the same system and domain is also well posed, provided $2 < p < p_0 + \varepsilon$ where $1/p_0 = 1/q - 1/(n-1)$.

For the bilaplacian, a full duality result is known. In [KS11a], Kilty and Shen showed that, if $1 < p < \infty$ and $1/p + 1/q = 1$, then well-posedness of the L^p-Dirichlet problem for Δ^2 in a Lipschitz domain Ω, and well-posedness of the L^q-regularity problem for Δ^2 in Ω, were both equivalent to the bilinear estimate

$$\left|\int_\Omega \Delta u \, \Delta v\right| \leq C\left(\|\nabla_\tau \nabla f\|_{L^p} + |\partial\Omega|^{-1/(n-1)}\|\nabla f\|_{L^p} + |\partial\Omega|^{-2/(n-1)}\|f\|_{L^p}\right) \times \left(\|\nabla g\|_{L^q} + |\partial\Omega|^{-1/(n-1)}\|g\|_{L^q}\right) \tag{5.19}$$

for all $f, g \in C_0^\infty(\mathbb{R}^n)$, where u and v are solutions of the L^2-regularity problem with boundary data $\partial^\alpha u = \partial^\alpha f$ and $\partial^\alpha v = \partial^\alpha g$. Thus, if $\Omega \subset \mathbb{R}^n$ is a bounded Lipschitz domain, and if $1/p + 1/q = 1$, then the L^p-Dirichlet problem is well posed in Ω if and only if the L^q-regularity problem is well posed in Ω.

All in all, we see that the L^p-regularity problem for Δ^2 is well posed in $\Omega \subset \mathbb{R}^n$ if

- Ω is C^1 or convex, and $1 < p < \infty$.
- $n = 2$ or $n = 3$ and $1 < p < 2 + \varepsilon$.
- $n = 4$ and $6/5 - \varepsilon < p < 2 + \varepsilon$.
- $n = 5$, 6 or 7, and $4/3 - \varepsilon < p < 2 + \varepsilon$.
- $n \geq 8$, and $2 - \frac{4}{4+n-\lambda_n} < p < 2 + \varepsilon$, where λ_n is given by (5.17). The above ranges of p are sharp, but this range is still open.

5.4. Higher-order elliptic systems

The polyharmonic operator Δ^m is part of a larger class of elliptic higher-order operators. Some study has been made of boundary-value problems for such operators and systems.

The L^p-Dirichlet problem for a strongly elliptic system L of order $2m$, as defined in Definition 2.1, is well posed in Ω if there exists a constant C such that, for every $\dot{f} \in WA^p_{m-1}(\partial\Omega \mapsto \mathbb{C}^\ell)$ and every $\vec{g} \in L^p(\partial\Omega \mapsto \mathbb{C}^\ell)$, there exists a unique vector-valued function $\vec{u} : \Omega \mapsto \mathbb{C}^\ell$ such that

$$\begin{cases} (L\vec{u})_j = \displaystyle\sum_{k=1}^{\ell} \sum_{|\alpha|=|\beta|=m} \partial^\alpha a^{jk}_{\alpha\beta} \partial^\beta u_k = 0 & \text{in } \Omega \text{ for each } 1 \leq j \leq \ell, \\ \partial^\alpha \vec{u} = f_\alpha & \text{on } \partial\Omega \text{ for } |\alpha| \leq m-2, \\ \partial^{m-1}_\nu \vec{u} = \vec{g} & \text{on } \partial\Omega, \\ \|N(\nabla^{m-1} u)\|_{L^q(\partial\Omega)} \leq C \displaystyle\sum_{|\alpha|=m-2} \|\nabla_\tau f_\alpha\|_{L^q(\partial\Omega)} + C\|\vec{g}\|_{L^p(\partial\Omega)}. \end{cases} \tag{5.20}$$

The L^q-regularity problem is well posed in Ω if there is some constant C such that, for every $\dot{f} \in WA^p_m(\partial\Omega \mapsto \mathbb{C}^\ell)$, there exists a unique $\vec{u}$ such that

$$\begin{cases} (L\vec{u})_j = \displaystyle\sum_{k=1}^{\ell} \sum_{|\alpha|=|\beta|=m} \partial^\alpha a^{jk}_{\alpha\beta} \partial^\beta u_k = 0 & \text{in } \Omega \text{ for each } 1 \leq j \leq \ell, \\ \partial^\alpha \vec{u} = f_\alpha & \text{on } \partial\Omega \text{ for } |\alpha| \leq m-1, \\ \|N(\nabla^m u)\|_{L^q(\partial\Omega)} \leq C \displaystyle\sum_{|\alpha|=m-1} \|\nabla_\tau f_\alpha\|_{L^q(\partial\Omega)}. \end{cases} \tag{5.21}$$

In [PV95a], Pipher and Verchota showed that the L^p-Dirichlet and L^p-regularity problems were well posed for $2 - \varepsilon < p < 2 + \varepsilon$, for any higher-order elliptic partial differential equation with real constant coefficients, in Lipschitz domains of arbitrary dimension. This was extended to symmetric elliptic systems in [Ver96]. A key ingredient of the proof was the boundary Gårding inequality

$$\frac{\lambda}{4} \int_{\partial\Omega} |\nabla^m u|(-\nu_n)\, d\sigma \leq \sum_{j,k=1}^{\ell} \sum_{|\alpha|=|\beta|=m} \int_{\partial\Omega} \partial^\alpha a^{jk}_{\alpha\beta} \partial^\beta u_k(-\nu_n)\, d\sigma + C \int_{\partial\Omega} |\nabla^{m-1}\partial_n u|^2\, d\sigma$$

valid if $u \in C_0^\infty(\mathbb{R}^n)^\ell$, if $L = \partial^\alpha a^{jk}_{\alpha\beta}\partial^\beta$ is a symmetric elliptic system with real constant coefficients, and if Ω is the domain above the graph of a Lipschitz function. We observe that in this case, $(-\nu_n)$ is a positive number bounded from below. Pipher and Verchota then used this Gårding inequality and a Green formula to construct the nontangential maximal estimate. See [PV95b] and [Ver96, Sections 4 and 6].

As in the case of the polyharmonic operator Δ^m, this first result concerned the L^p-Dirichlet problem and L^q-regularity problem only for $2-\varepsilon < p < 2+\varepsilon$ and for $2-\varepsilon < q < 2+\varepsilon$. The polyharmonic operator Δ^m is an elliptic system, and so we cannot in general improve upon the requirement that $2-\varepsilon < p$ for well-posedness of the L^p-Dirichlet problem.

However, we can improve on the requirement $p < 2+\varepsilon$. Recall that Theorem 5.13 from [She06b], and its equivalence to (5.15), were proven in the general case of strongly elliptic systems with real symmetric constant coefficients. As in the case of the polyharmonic operator Δ^m, (5.14) follows from well-posedness of the L^2-regularity problem provided $p = 2(n-1)/(n-3)$, and so if L is such a system, the L^p-Dirichlet problem for L is well posed in Ω provided $2-\varepsilon < p < 2(n-1)/(n-3)+\varepsilon$. This is [She06b, Corollary 1.3]. Again, by the counterexamples of [PV95b], this range cannot be improved if $m \geq 2$ and $4 \leq n \leq 2m+1$; the question of whether this range can be improved for general operators L if $n \geq 2m+2$ is still open.

Little is known concerning the regularity problem in a broader range of p. Recall that (5.18) from [KS11b] was proven in the general case of strongly elliptic systems with real symmetric constant coefficients. Thus, we known that for such systems, well-posedness of the L^q-regularity problem for $2 < q < n-1$ implies well-posedness of the L^p-Dirichlet problem for appropriate p. The question of whether the reverse implication holds, or whether this result can be extended to a broader range of q, is open.

5.5. The area integral

One of major tools in the theory of second-order elliptic differential equations is the Lusin area integral, defined as follows. If w lies in $W^2_{1,loc}(\Omega)$ for some domain $\Omega \subset \mathbb{R}^n$, then the area integral (or square function) of w is defined for $Q \in \partial\Omega$ as

$$Sw(Q) = \left(\int_{\Gamma(Q)} |\nabla w(X)|^2 \operatorname{dist}(X,\partial\Omega)^{2-n} dX\right)^{1/2}.$$

In [Dah80], Dahlberg showed that if u is harmonic in a bounded Lipschitz domain Ω, if $P_0 \in \Omega$ and $u(P_0) = 0$, then for any $0 < p < \infty$,

$$\frac{1}{C}\int_{\partial\Omega} Su^p \, d\sigma \leq \int_{\partial\Omega} (Nu)^p \, d\sigma \leq C \int_{\partial\Omega} (Su)^p \, d\sigma \tag{5.22}$$

for some constants C depending only on p, Ω and P_0. Thus, the Lusin area integral bears deep connections to the L^p-Dirichlet problem. In [DJK84], Dahlberg, Jerison and Kenig generalized this result to solutions to second-order divergence-form elliptic equations with real coefficients for which the L^r-Dirichlet problem is well posed for at least one r.

If L is an operator of order $2m$, then the appropriate estimate is

$$\frac{1}{C}\int_{\partial\Omega} N(\nabla u^{m-1})^p\,d\sigma \le \int_{\partial\Omega} S(\nabla u^{m-1})^p\,d\sigma \le C\int_{\partial\Omega} N(\nabla u^{m-1})^p\,d\sigma. \quad (5.23)$$

Before discussing their validity for particular operators, let us point out that such square-function estimates are very useful in the study of higher-order equations. In [She06b], Shen used (5.23) to prove the equivalence of (5.15) and (5.14), above. In [KS11a], Kilty and Shen used (5.23) to prove that well-posedness of the L^p-Dirichlet problem for Δ^2 implies the bilinear estimate (5.19). The proof of the maximum principle (3.2) in [Ver96, Section 8] (to be discussed in Section 5.6) also exploited (5.23). Estimates on square functions can be used to derive estimates on Besov space norms; see [AP98, Proposition S].

In [PV91], Pipher and Verchota proved that (5.23) (with $m = 2$) holds for solutions u to $\Delta^2 u = 0$, provided Ω is a bounded Lipschitz domain, $0 < p < \infty$, and $\nabla u(P_0) = 0$ for some fixed $P_0 \in \Omega$. Their proof was an adaptation of Dahlberg's proof [Dah80] of the corresponding result for harmonic functions. They used the L^2-theory for the biharmonic operator [DKV86], the representation formula (5.9), and the L^2-theory for harmonic functions to prove good-λ inequalities, which, in turn, imply L^p estimates for $0 < p < \infty$.

In [DKPV97], Dahlberg, Kenig, Pipher and Verchota proved that (5.23) held for solutions u to $Lu = 0$, for a symmetric elliptic system L of order $2m$ with real constant coefficients, provided as usual that Ω is a bounded Lipschitz domain, $0 < p < \infty$, and $\nabla^{m-1}u(P_0) = 0$ for some fixed $P_0 \in \Omega$. The argument is necessarily considerably more involved than the argument of [PV91] or [Dah80]. In particular, the bound $\|S(\nabla^{m-1}u)\|_{L^2(\partial\Omega)} \le C\|N(\nabla^{m-1}u)\|_{L^2(\partial\Omega)}$ was proven in three steps.

The first step was to reduce from the elliptic system L of order $2m$ to the scalar elliptic operator $M = \det L$ of order $2\ell m$, where ℓ is as in formula (2.2). The second step was to reduce to elliptic equations of the form $\sum_{|\alpha|=m} a_\alpha \partial^{2\alpha} u = 0$, where $|a_\alpha| > 0$ for all $|\alpha| = m$. Finally, it was shown that for operators of this form

$$\sum_{|\alpha|=m}\int_{\Omega} a_\alpha\, \partial^\alpha u(X)^2 \operatorname{dist}(X,\partial\Omega)\,dX \le C\int_{\partial\Omega} N(\nabla^{m-1}u)^2\,d\sigma.$$

The passage to $0 < p < \infty$ in (5.23) was done, as usual, using good-λ inequalities. We remark that these arguments used the result of [PV95a] that the L^2-Dirichlet problem is well posed for such operators L in Lipschitz domains.

It is quite interesting that for second-order elliptic systems, the only currently known approach to the square-function estimate (5.22) is this reduction to a higher-order operator.

5.6. The maximum principle in Lipschitz domains

We are now in a position to discuss the maximum principle (3.2) for higher-order equations in Lipschitz domains.

We say that the maximum principle for an operator L of order $2m$ holds in the bounded Lipschitz domain Ω if there exists a constant $C > 0$ such that, whenever $\dot f \in \dot{W}A^\infty_{m-1}(\partial\Omega) \subset \dot{W}A^2_{m-1}(\partial\Omega)$ and $g \in L^\infty(\partial\Omega) \subset L^2(\partial\Omega)$, the solution u to the Dirichlet problem (5.20) with boundary data $\dot f$ and g satisfies

$$\|\nabla^{m-1}u\|_{L^\infty} \leq C\|g\|_{L^\infty(\partial\Omega)} + C \sum_{|\alpha|=m-2} \|\nabla_\tau f_\alpha\|_{L^\infty(\partial\Omega)}. \tag{5.24}$$

The maximum principle (5.24) was proven to hold in three-dimensional Lipschitz domains by Pipher and Verchota in [PV93] (for biharmonic functions), in [PV95b] (for polyharmonic functions), and by Verchota in [Ver96, Section 8] (for solutions to symmetric systems with real constant coefficients). Pipher and Verchota also proved in [PV93] that the maximum principle was valid for biharmonic functions in C^1 domains of arbitrary dimension. In [KS11a, Theorem 1.5], Kilty and Shen observed that the same techinque gives validity of the maximum principle for biharmonic functions in convex domains of arbitrary dimension.

The proof of [PV93] uses the L^2-regularity problem in the domain Ω to construct the Green function $G(X,Y)$ for Δ^2 in Ω. Then if u is biharmonic in Ω with $N(\nabla u) \in L^2(\partial\Omega)$, we have that

$$u(X) = \int_{\partial\Omega} u(Q)\, \partial_\nu \Delta G(X,Q)\, d\sigma(Q) + \int_{\partial\Omega} \partial_\nu u(Q)\, \Delta G(X,Q)\, d\sigma(Q)$$

where all derivatives of G are taken in the second variable Q. If the H^1-regularity problem is well posed in appropriate subdomains of Ω, then $\nabla^2\nabla_X G(X,\,\cdot\,)$ is in $L^1(\partial\Omega)$ with L^1-norm independent of X, and so the second integral is at most $C\|\partial_\nu u\|_{L^\infty(\partial\Omega)}$. By taking Riesz transforms, the normal derivative $\partial_\nu \Delta G(X,Q)$ may be transformed to tangential derivatives $\nabla_\tau \Delta G(X,Q)$; integrating by parts transfers these derivatives to u. The square-function estimate (5.23) implies that the Riesz transforms of $\nabla_X \Delta_Q G(X,Q)$ are bounded on $L^1(\partial\Omega)$. This completes the proof of the maximum principle.

Similar arguments show that the maximum principle is valid for more general operators. See [PV95b] for the polyharmonic operator, or [Ver96, Section 8] for arbitrary symmmetric operators with real constant coefficients.

An important transitional step is the well-posedness of the H^1-regularity problem. It was established in three-dimensional (or C^1) domains in [PV93, Theorem 4.2] and [PV95b, Theorem 1.2] and discussed in [Ver96, Section 7]. In each case, well-posedness was proven by analyzing solutions with atomic data $\dot f$ using a technique from [DK90]. A crucial ingredient in this technique is the well-posedness of the L^p-Dirichlet problem for some $p < (n-1)/(n-2)$; the latter is valid if $n = 3$ by [DKV86], and (for Δ^2) in C^1 and convex domains by [Ver90] and [KS11a], but fails in general Lipschitz domains for $n \geq 4$.

5.7. Biharmonic functions in convex domains

We say that a domain Ω is *convex* if, whenever X, $Y \in \Omega$, the line segment connecting X and Y lies in Ω. Observe that all convex domains are necessarily

Lipschitz domains but the converse does not hold. Moreover, while convex domains are in general no smoother than Lipschitz domains, the extra geometrical structure often allows for considerably stronger results.

Recall that in [MM09a], the second author of this paper and Maz'ya showed that the gradient of a biharmonic function is bounded in a three-dimensional domain. This is a sharp property in dimension three, and in higher-dimensional domains the solutions can be even less regular (cf. Section 3). However, using some intricate linear combination of weighted integrals, the same authors showed in [MM08] that *second* derivatives to biharmonic functions were locally bounded when the domain was convex. To be precise, they showed that if Ω is convex, and $u \in \mathring{W}_2^2(\Omega)$ is a solution to $\Delta^2 u = h$ for some $h \in C_0^\infty(\Omega \setminus B(Q, 10R))$, $R > 0$, $Q \in \partial\Omega$, then

$$\sup_{B(Q,R/5)\cap\Omega} |\nabla^2 u| \leq \frac{C}{R^2} \left(\fint_{\Omega\cap B(Q,5R)\setminus B(Q,R/2)} |u|^2 \right)^{1/2}. \tag{5.25}$$

In particular, not only are all boundary points of convex domains 1-regular, but the gradient ∇u is Lipschitz continuous near such points.

Kilty and Shen noted in [KS11a] that (5.25) implies that (5.18) holds in convex domains for any q; thus, the L^q-regularity problem for the bilaplacian is well posed for any $2 < q < \infty$ in a convex domain. Well-posedness of the L^p-Dirichlet problem for $2 < p < \infty$ has been established by Shen in [She06c]. By the duality result (5.19), again from [KS11a], this implies that both the L^p-Dirichlet and L^q-regularity problems are well posed, for any $1 < p < \infty$ and any $1 < q < \infty$, in a convex domain of arbitrary dimension. They also observed that, by the techniques of [PV93] (discussed in Section 5.6 above), the maximum principle (5.24) is valid in arbitrary convex domains.

It is interesting to note how, once again, the methods and results related to pointwise estimates, the Wiener criterion, and local regularity estimates near the boundary are intertwined with the well-posedness of boundary problems in L^p.

5.8. The Neumann problem for the biharmonic equation

So far we have only discussed the Dirichlet and regularity problems for higher-order operators. Another common and important boundary-value problem that arises in applications is the Neumann problem. Indeed, the principal physical motivation for the inhomogeneous biharmonic equation $\Delta^2 u = h$ is that it describes the equilibrium position of a thin elastic plate subject to a vertical force h. The Dirichlet problem $u\big|_{\partial\Omega} = f$, $\nabla u\big|_{\partial\Omega} = g$ describes an elastic plate whose edges are clamped, that is, held at a fixed position in a fixed orientation. The Neumann problem, on the other hand, corresponds to the case of a free boundary. Guido Sweers has written an excellent short paper [Swe09] discussing the boundary conditions that correspond to these and other physical situations.

More precisely, if a thin two-dimensional plate is subject to a force h and the edges are free to move, then its displacement u satisfies the boundary value problem

$$\begin{cases} \Delta^2 u = h & \text{in } \Omega, \\ \rho\Delta u + (1-\rho)\partial_\nu^2 u = 0 & \text{on } \partial\Omega, \\ \partial_\nu \Delta u + (1-\rho)\partial_{\tau\tau\nu} u = 0 & \text{on } \partial\Omega. \end{cases}$$

Here ρ is a physical constant, called the Poisson ratio. This formulation goes back to Kirchhoff and is well known in the theory of elasticity; see, for example, Section 3.1 and Chapter 8 of the classic engineering text [Nad63]. We remark that by [Nad63, Formula (8-10)],

$$\partial_\nu \Delta u + (1-\rho)\partial_{\tau\tau\nu} u = \partial_\nu \Delta u + (1-\rho)\partial_\tau\left(\partial_{\nu\tau} u\right).$$

This suggests the following homogeneous boundary value problem in a Lipschitz domain Ω of arbitrary dimension. We say that the L^p-Neumann problem is well posed if there exists a constant $C > 0$ such that, for every $f_0 \in L^p(\partial\Omega)$ and $\Lambda_0 \in W^p_{-1}(\partial\Omega)$, there exists a function u such that

$$\begin{cases} \Delta^2 u = 0 & \text{in } \Omega, \\ M_\rho u := \rho\Delta u + (1-\rho)\partial_\nu^2 u = f_0 & \text{on } \partial\Omega, \\ K_\rho u := \partial_\nu \Delta u + (1-\rho)\frac{1}{2}\partial_{\tau_{ij}}\left(\partial_{\nu\tau_{ij}} u\right) = \Lambda_0 & \text{on } \partial\Omega, \\ \|N(\nabla^2 u)\|_{L^p(\partial\Omega)} \leq C\|f_0\|_{W^p_1(\partial\Omega)} + C\|\Lambda_0\|_{W^p_{-1}(\partial\Omega)}. \end{cases} \tag{5.26}$$

Here $\tau_{ij} = \nu_i \mathbf{e}_j - \nu_j \mathbf{e}_i$ is a vector orthogonal to the outward normal ν and lying in the $x_i x_j$-plane.

In addition to the connection to the theory of elasticity, this problem is of interest because it is in some sense adjoint to the Dirichlet problem (5.4). That is, if $\Delta^2 u = \Delta^2 w = 0$ in Ω, then $\int_{\partial\Omega} \partial_\nu w\, M_\rho u - w\, K_\rho u\, d\sigma = \int_{\partial\Omega} \partial_\nu u\, M_\rho w - u\, K_\rho w\, d\sigma$, where M_ρ and K_ρ are as in (5.26); this follows from the more general formula

$$\int_\Omega w\,\Delta^2 u = \int_\Omega (\rho\Delta u\,\Delta w + (1-\rho)\partial_{jk} u\,\partial_{jk} w) + \int_{\partial\Omega} w\, K_\rho u - \partial_\nu w\, M_\rho u\, d\sigma \tag{5.27}$$

valid for arbitrary smooth functions. This formula is analogous to the classical Green identity for the Laplacian

$$\int_\Omega w\,\Delta u = -\int_\Omega \nabla u \cdot \nabla w + \int_{\partial\Omega} w\,\nu\cdot\nabla u\, d\sigma. \tag{5.28}$$

Observe that, contrary to the Laplacian or more general second-order operators, there is a *family* of relevant Neumann data for the biharmonic equation. Moreover, different values (or, rather, ranges) of ρ correspond to different natural physical situations. We refer the reader to [Ver05] for a detailed discussion.

In [CG85], Cohen and Gosselin showed that the L^p-Neumann problem (5.26) was well posed in C^1 domains contained in $\mathbb{R}^2$ for for $1 < p < \infty$, provided in addition that $\rho = -1$. The method of proof was as follows. Recall from (5.8) that Cohen and Gosselin showed that the L^p-Dirichlet problem was well posed by constructing a multiple layer potential $\mathcal{L}\dot{f}$ with boundary values $(I+\mathcal{K})\dot{f}$, and showing that $I+\mathcal{K}$ is invertible. We remark that because Cohen and Gosselin preferred to work with Dirichlet boundary data of the form $(u,\partial_x u,\partial_y u)\big|_{\partial\Omega}$ rather than of the form $(u,\partial_\nu u)\big|_{\partial\Omega}$, the notation of [CG85] is somewhat different from that of the present paper. In the notation of the present paper, the method of proof of [CG85] was to observe that $(I+\mathcal{K})^*\dot{\theta}$ is equivalent to $(K_{-1}v\dot{\theta}, M_{-1}v\dot{\theta})_{\partial\Omega^C}$, where v is another biharmonic layer potential and $(I+\mathcal{K})^*$ is the adjoint to $(I+\mathcal{K})$. Well-posedness of the Neumann problem then follows from invertibility of $I+\mathcal{K}$ on $\partial\Omega^C$.

In [Ver05], Verchota investigated the Neumann problem (5.26) in full generality. He considered Lipschitz domains with compact, connected boundary contained in $\mathbb{R}^n$, $n\geq 2$. He showed that if $-1/(n-1)\leq\rho<1$, then the Neumann problem is well posed provided $2-\varepsilon<p<2+\varepsilon$. That is, the solutions exist, satisfy the desired estimates, and are unique either modulo functions of an appropriate class, or (in the case where Ω is unbounded) when subject to an appropriate growth condition. See [Ver05, Theorems 13.2 and 15.4]. Verchota's proof also used boundedness and invertibility of certain potentials on $L^p(\partial\Omega)$; a crucial step was a coercivity estimate $\|\nabla^2 u\|_{L^2(\partial\Omega)}\leq C\|K_\rho u\|_{W^2_{-1}(\partial\Omega)}+C\|M_\rho u\|_{L^2(\partial\Omega)}$. (This estimate is valid provided u is biharmonic and satisfies some mean-value hypotheses; see [Ver05, Theorem 7.6]).

In [She07a], Shen improved upon Verchota's results by extending the range on p (in bounded simply connected Lipschitz domains) to $2(n-1)/(n+1)-\varepsilon<p<2+\varepsilon$ if $n\geq 4$, and $1<p<2+\varepsilon$ if $n=2$ or $n=3$. This result again was proven by inverting layer potentials. Observe that the L^p-regularity problem is also known to be well posed for p in this range, and (if $n\geq 6$) in a broader range of p; see Section 5.3. The question of the sharp range of p for which the L^p-Neumann problem is well posed is still open.

It turns out that extending the well-posedness results for the Neumann problem beyond the case of the bilaplacian is an excruciatingly difficult problem, even if one considers only fourth-order operators with constant coefficients.

The solutions to (5.26) in [CG85], [Ver05] and [She07a] were constructed using layer potentials. It is possible to construct layer potential operators, and to prove their boundedness, for a fairly general class of higher-order operators. However, the problems arise at a much more fundamental level.

In analogy to (5.27) and (5.28), one can write

$$\int_\Omega w\,Lu = A[u,w]+\int_{\partial\Omega} w\,K_A u-\partial_\nu w\,M_A u\,d\sigma, \tag{5.29}$$

where $A[u,w]=\sum_{|\alpha|=|\beta|=2}a_{\alpha\beta}\int_\Omega D^\beta u\,D^\alpha w$ is an energy form associated to the operator $L=\sum_{|\alpha|=|\beta|=2}a_{\alpha\beta}D^\alpha D^\beta$. Note that in the context of fourth-order operators, the pair $(w,\partial_\nu w)$ constitutes the Dirichlet data for w on the boundary, and so one can say that the operators $K_A u$ and $M_A u$ define the Neumann data for u. One immediately faces the problem that the same higher-order operator L

can be rewritten in many different ways and gives rise to different energy forms. The corresponding Neumann data will be different. (This is the reason why there is a family of Neumann data for the biharmonic operator.)

Furthermore, whatever the choice of the form, in order to establish well-posedness of the Neumann problem, one needs to be able to estimate all second derivatives of a solution on the boundary in terms of the Neumann data. In the analogous second-order case, such an estimate is provided by the Rellich identity, which shows that the tangential derivatives are equivalent to the normal derivative in L^2 for solutions of elliptic PDEs. In the higher-order scenario, such a result calls for certain coercivity estimates which are still rather poorly understood. We refer the reader to [Ver10] for a detailed discussion of related results and problems.

5.9. Inhomogeneous problems for the biharmonic equation and other classes of boundary data

In [AP98], Adolfsson and Pipher investigated the inhomogeneous Dirichlet problem for the biharmonic equation with data in Besov and Sobolev spaces. While resting on the results for homogeneous boundary value problems discussed in Sections 5.1 and 5.3, such a framework presents a completely new setting, allowing for the inhomogeneous problem and for consideration of the classes of boundary data which are, in some sense, intermediate between the Dirichlet and the regularity problems.

They showed that if $f \in WA^p_{1+s}(\partial\Omega)$ and $h \in L^p_{s+1/p-3}(\Omega)$, then there exists a unique function u that satisfies

$$\begin{cases} \Delta^2 u = h & \text{in } \Omega, \\ \operatorname{Tr} \partial^\alpha u = f_\alpha, & \text{for } 0 \leq |\alpha| \leq 1 \end{cases} \tag{5.30}$$

subject to the estimate

$$\|u\|_{L^p_{s+1/p+1}(\Omega)} \leq C\|h\|_{L^p_{s+1/p-3}(\Omega)} + C\|\dot{f}\|_{WA^p_{1+s}(\partial\Omega)} \tag{5.31}$$

provided $2-\varepsilon < p < 2+\varepsilon$ and $0 < s < 1$. Here $\operatorname{Tr} w$ denotes the trace of w in the sense of Sobolev spaces; that these may be extended to functions $u \in L^p_{s+1+1/p}$, $s > 0$, was proven in [AP98, Theorem 1.12].

In Lipschitz domains contained in $\mathbb{R}^3$, they proved these results for a broader range of p and s, namely for $0 < s < 1$ and for

$$\max\left(1, \frac{2}{s+1+\varepsilon}\right) < p < \begin{cases} \infty, & s < \varepsilon, \\ \frac{2}{s-\varepsilon}, & \varepsilon \leq s < 1. \end{cases} \tag{5.32}$$

Finally, in C^1 domains, they proved these results for any p and s with $1 < p < \infty$ and $0 < s < 1$.

In [MMW11], I. Mitrea, M. Mitrea and Wright extended the three-dimensional results to $p = \infty$ (for $0 < s < \varepsilon$) or $2/(s+1+\varepsilon) < p \leq 1$ (for $1-\varepsilon < s < 1$). They also extended these results to data h and $\dot{f}$ in more general Besov or Triebel–Lizorkin spaces.

Let us define the function spaces appearing above. $L^p_\alpha(\mathbb{R}^n)$ is defined to be $\{g : (I-\Delta)^{\alpha/2} g \in L^p(\mathbb{R}^n)\}$; we say $g \in L^p_\alpha(\Omega)$ if $g = h\big|_\Omega$ for some $h \in L^p_\alpha(\mathbb{R}^n)$. If k is a non-negative integer, then $L^p_k = W^p_k$. If m is an integer and $0 < s < 1$, then the Whitney–Besov space WA^p_{m-1+s} is defined analogously to WA^p_m (see Definition 5.6), except that we take the completion with respect to the Whitney–Besov norm

$$\sum_{|\alpha|\leq m-1} \|\partial^\alpha \psi\|_{L^p(\partial\Omega)} + \sum_{|\alpha|=m-1} \|\partial^\alpha \psi\|_{B^{p,p}_s(\partial\Omega)} \tag{5.33}$$

rather than the Whitney–Sobolev norm

$$\sum_{|\alpha|\leq m-1} \|\partial^\alpha \psi\|_{L^p(\partial\Omega)} + \sum_{|\alpha|=m-1} \|\nabla_\tau \partial^\alpha \psi\|_{L^p(\partial\Omega)}.$$

The general problem (5.30) was first reduced to the case $h = 0$ (that is, to a homogeneous problem) by means of trace/extension theorems, that is, subtracting $w(X) = \int_{\mathbb{R}^n} F(X,Y)\,\tilde{h}(Y)\,dY$, and showing that if $h \in L^p_{s+1/p-3}(\Omega)$ then $(\operatorname{Tr} w, \operatorname{Tr} \nabla w) \in WA^p_{1+s}(\partial\Omega)$. Next, the well-posedness of Dirichlet and regularity problems discussed in Sections 5.1 and 5.3 provide the endpoint cases $s = 0$ and $s = 1$, respectively. The core of the matter is to show that, if u is biharmonic, k is an integer and $0 \leq \alpha \leq 1$, then $u \in L^p_{k+\alpha}(\Omega)$ if and only if

$$\int_\Omega |\nabla^{k+1} u(X)|^p \operatorname{dist}(X,\partial\Omega)^{p-p\alpha} + |\nabla^k u(X)|^p + |u(X)|^p \, dX < \infty, \tag{5.34}$$

(cf. [AP98, Proposition S]). With this at hand, one can use square-function estimates to justify the aforementioned endpoint results. Indeed, observe that for $p = 2$ the first integral on the left-hand side of (5.34) is exactly the L^2 norm of $S(\nabla^k u)$. The latter, by [PV91] (discussed in Section 5.5), is equivalent to the L^2 norm of the corresponding non-tangential maximal function, connecting the estimate (5.31) to the nontangential estimates in the Dirichlet problem (5.4) and the regularity problem 5.11. Finally, one can build an interpolation-type scheme to pass to well-posedness in intermediate Besov and Sobolev spaces.

6. Boundary value problems with variable coefficients

In this section we discuss divergence-form operators with variable coefficients. At the moment, well-posedness results for such operators are restricted in two serious ways. First, the coefficients cannot oscillate too much. Secondly, the boundary problems treated fall *strictly* between the range of L^p-Dirichlet and L^p-regularity, in the sense of Section 5.9. That is, the L^p-Dirichlet, regularity, and Neumann problems on Lipschitz domains with the usual sharp estimates in terms of the non-tangential maximal function for these divergence-form operators seem to be completely open.

To be more precise, recall from the discussion in Section 5.9 that the classical Dirichlet and regularity problems, with boundary data in L^p, can be viewed as the

$s = 0, 1$ endpoints of the boundary problem studied in [AP98] and [MMW11]

$$\Delta^2 u = h \text{ in } \Omega, \quad \partial^\alpha u\big|_{\partial\Omega} = f_\alpha \text{ for all } |\alpha| \leq 1$$

with $\dot{f}$ lying in an *intermediate* smoothness space $WA^p_{1+s}(\partial\Omega)$, $0 \leq s \leq 1$. In the context of divergence-form higher-order operators with variable coefficients, essentially the known results pertain *only* to boundary data of intermediate smoothness.

We now establish some terminology. A divergence-form operator L, acting on $W^2_{m,loc}(\Omega \mapsto \mathbb{C}^\ell)$, may be defined weakly via (2.5); we say that $Lu = h$ if

$$\sum_{j=1}^{\ell} \int_\Omega \varphi_j \, h_j = (-1)^m \sum_{j,k=1}^{\ell} \sum_{|\alpha|=|\beta|=m} \int_\Omega \partial^\alpha \varphi_j(X) \, a^{jk}_{\alpha\beta}(X) \, \partial^\beta u_k(X) \, dX \tag{6.1}$$

for all φ smooth and compactly supported in Ω.

In [Agr07], Agranovich investigated the inhomogeneous Dirichlet problem, in Lipschitz domains, for such operators L that are elliptic (in the sense of (2.7)) and whose coefficients $a^{jk}_{\alpha\beta}$ are Lipschitz continuous in Ω.

He showed that if $h \in L^p_{-m-1+1/p+s}(\Omega)$ and $\dot{f} \in WA^p_{m-1+s}(\partial\Omega)$, for some $0 < s < 1$, and if $|p-2|$ is small enough, then the Dirichlet problem

$$\begin{cases} Lu = h & \text{in } \Omega, \\ \operatorname{Tr} \partial^\alpha u = f_\alpha & \text{for all } 0 \leq |\alpha| \leq m-1 \end{cases} \tag{6.2}$$

has a unique solution u that satisfies the estimate

$$\|u\|_{L^p_{m-1+s+1/p}(\Omega)} \leq C\|h\|_{L^p_{-m-1+1/p+s}(\partial\Omega)} + C\|\dot{f}\|_{WA^p_{m-1+s}(\partial\Omega)}. \tag{6.3}$$

Agranovich also considered the Neumann problem for such operators. As we discussed in Section 5.8, defining Neumann problem is a delicate matter. In the context of zero boundary data, the situation is a little simpler as one can take a formal functional analytic point of view and avoid to some extent the discussion of estimates at the boundary. First, observe that if the test function φ does not have zero boundary data, then formula (6.1) becomes

$$\sum_{j=1}^{\ell} \int_\Omega (Lu)_j \, \varphi_j = (-1)^m \sum_{j,k=1}^{\ell} \sum_{|\alpha|=|\beta|=m} \int_\Omega \partial^\alpha \varphi_j(X) \, a^{jk}_{\alpha\beta}(X) \, \partial^\beta u_k(X) \, dX$$
$$+ \sum_{i=0}^{m-1} \int_{\partial\Omega} B_{m-1-i} u \, \partial^i_\nu \varphi \, d\sigma$$

where $B_i u$ is an appropriate linear combination of the functions $\partial^\alpha u$ where $|\alpha| = m+i$. The expressions $B_i u$ may then be regarded as the Neumann data for u. Notice that if L is a fourth-order constant-coefficient operator, then $B_0 = -M_A$ and $B_1 = K_A$, where K_A, M_A are given by (5.29).

We say that u solves the Neumann problem for L, with homogeneous boundary data, if (6.1) is valid for all test functions φ compactly supported in $\mathbb{R}^n$ (but not necessarily in Ω.) Agranovich showed that, if $h \in \mathring{L}^p_{-m-1+1/p+s}(\Omega)$, then there exists a unique function $u \in L^p_{m-1+1/p+s}(\Omega)$ that solves this Neumann problem with homogeneous boundary data, under the same conditions on p, s, L as for his results for the Dirichlet problem. He also provided some brief discussion (see [Agr07, Section 5.2]) of the conditions needed to resolve the Neumann problem with inhomogeneous boundary data. Here $h \in \mathring{L}^p_\alpha(\Omega)$ if $h = g\big|_\Omega$ for some $g \in L^p_\alpha(\mathbb{R}^n)$ that in addition is supported in $\bar{\Omega}$.

In [MMS10], Maz'ya, M. Mitrea and Shaposhnikova considered the Dirichlet problem, again with boundary data in intermediate Besov spaces, for much rougher coefficients. They showed that if $f \in WA^p_{m-1+s}$, for some $0 < s < 1$ and some $1 < p < \infty$, if h lies in an appropriate space, and if L is a divergence-form operator of order $2m$ (as defined by (2.5)), then under some conditions, there is a unique function u that satisfies (6.2) subject to the estimate

$$\sum_{|\alpha|\leq m} \int_\Omega |\partial^\alpha u(X)|^p \operatorname{dist}(X, \partial\Omega)^{p-ps-1}\, dX < \infty. \tag{6.4}$$

See [MMS10, Theorem 8.1]. The inhomogeneous data h is required to lie in the space $V^p_{-m,1-s-1/p}(\Omega)$, the dual space to $V^q_{m,s+1/p-1}(\Omega)$, where

$$\|w\|_{V^p_{m,a}} = \left(\sum_{|\alpha|\leq m} \int_\Omega |\partial^\alpha u(X)|^p \operatorname{dist}(X, \partial\Omega)^{pa+p|\alpha|-pm}\, dX\right)^{1/p}.$$

The conditions are that Ω be a Lipschitz domain whose normal vector ν lies in $VMO(\partial\Omega)$, and that the coefficients $a^{ij}_{\alpha\beta}$ lie in $VMO(\mathbb{R}^n)$. Recall that this condition on Ω has also arisen in [MM11] (it ensures the validity of formula (3.11)). The ellipticity condition they required was that the coefficients be bounded and that $\langle \varphi, L\varphi\rangle \geq \lambda \|\nabla^m \varphi\|^2_{L^2}$ for all smooth compactly supported functions φ, that is, that

$$\operatorname{Re} \sum_{|\alpha|=|\beta|=m} \sum_{j,k=1}^{\ell} \int_\Omega a^{jk}_{\alpha\beta}(X)\partial^\beta \varphi_k(X)\partial^\alpha \overline{\varphi_j}(X)\, dX \geq \lambda \sum_{|\alpha|=m} \sum_{k=1}^{\ell} \int_\Omega |\partial^\alpha \varphi_k|^2 \tag{6.5}$$

for all functions $\varphi \in C^\infty_0(\Omega \mapsto \mathbb{C}^\ell)$. This is a weaker requirement than condition (2.7).

In fact, [MMS10] provides a more intricate result, allowing one to deduce a well-posedness range of s and p, given information about the oscillation of the coefficients $a^{jk}_{\alpha\beta}$ and the normal to the domain ν. In the extreme case, when the oscillations for both are vanishing, the allowable range expands to $0 < s < 1$, $1 < p < \infty$, as stated above.

We comment on the estimate (6.4). First, by [AP98, Propositon S] (listed above as formula (5.34)), if u is biharmonic then the estimate (6.4) is equivalent to the estimate (6.3) of [Agr07]. Second, by (5.23), if the coefficients $a^{jk}_{\alpha\beta}$ are constant, one can draw connections between (6.4) for $s = 0, 1$ and the nontangential maximal estimates of the Dirichlet or regularity problems (5.20) or (5.21). However, as we pointed out earlier, this endpoint case, corresponding to the true L^p-Dirichlet and regularity problems, has not been achieved.

6.1. The Kato problem and the Riesz transforms

An important topic in elliptic theory, which formally stands somewhat apart from the well-posedness issues, is the Kato problem and the properties of the Riesz transform. In the framework of elliptic boundary problems, the related results can be viewed as the estimates for the solutions with data in L^p for certain operators in block form.

Suppose that L is a variable-coefficient operator in divergence form, that is, an operator defined by (2.5). Suppose that L satisfies the ellipticity estimate (6.5), and the bounds

$$\left|\sum_{|\alpha|=|\beta|=m}\sum_{j,k=1}^{\ell}\int_{\mathbb{R}^n} a^{jk}_{\alpha\beta}\partial^\beta f_k\,\partial^\alpha g_j\right| \leq C\|\nabla^m f\|_{L^2(\mathbb{R}^n)}\|\nabla^m g\|_{L^2(\mathbb{R}^n)}. \tag{6.6}$$

(This is a weaker condition than the assumption of [MMS10] that $a^{ij}_{\alpha\beta}$ be bounded pointwise.) Auscher, Hofmann, McIntosh and Tchamitchian [AHMT01] proved that under these conditions, the Kato estimate

$$\frac{1}{C}\|\nabla^m f\|_{L^2(\mathbb{R}^n)} \leq \|\sqrt{L}f\|_{L^2(\mathbb{R}^n)} \leq C\|\nabla^m f\|_{L^2(\mathbb{R}^n)} \tag{6.7}$$

is valid for some constant C. They also proved similar results for operators with lower-order terms.

It was later observed in [Aus04] that by the methods of [AT98], if $1 \leq n \leq 2m$, then the bound on the Riesz transform $\nabla^m L^{-1/2}$ in L^p (that is, the first inequality in (6.7)) extends to the range $1 < p < 2+\varepsilon$, and the reverse Riesz transform bound (that is, the second inequality in (6.7)) extends to the range $1 < p < \infty$. This also holds if the Schwartz kernel $W_t(X,Y)$ of the operator e^{-tL} satisfies certain pointwise bounds (e.g., if the coefficients of A are real).

In the case where $n > 2m$, the inequality $\|\nabla^m L^{-1/2} f\|_{L^p(\mathbb{R}^n)} \leq C\|f\|_{L^p(\mathbb{R}^n)}$ holds for $2n/(n+2m) - \varepsilon < p \leq 2$; see [BK04, Aus04]. The reverse inequality holds for $\max(2n/(n+4m)-\varepsilon, 1) < p < 2$ by [Aus04, Theorem 18], and for $2 < p < 2n/(n-2m)+\varepsilon$ by duality (see [Aus07, Section 7.2]).

Going further, let us consider the second-order divergence-form operator $\mathbb{L} = -\operatorname{div}\mathbb{A}\nabla$ in $\mathbb{R}^{n+1}$, where $\mathbb{A}$ is an $(n+1)\times(n+1)$ matrix in block form; that is, $\mathbb{A}_{j,n+1} = \mathbb{A}_{n+1,j} = 0$ for $1 \leq j \leq n$, and $\mathbb{A}_{n+1,n+1} = 1$. It is fairly easy to see that one can formally realize the solution to $\mathbb{L}u = 0$ in $\mathbb{R}^{n+1}_+$, $u\big|_{\mathbb{R}^n} = f$, as the Poisson semigroup $u(x,t) = e^{-t\sqrt{L}}f(x)$, $(x,t) \in \mathbb{R}^{n+1}_+$. Then (6.7) essentially provides an analogue of the Rellich identity-type estimate for the block operator $\mathbb{L}$, that is, the L^2-equivalence between normal and tangential derivatives of the solution on the boundary

$$\|\partial_t u(\,\cdot\,,0)\|_{L^2(\mathbb{R}^n)} \approx \|\nabla_x u(\,\cdot\,,0)\|_{L^2(\mathbb{R}^n)}.$$

As we discussed in Section 5, such a Rellich identity-type estimate is, in a sense, a core result needed to approach Neumann and regularity problems, and for second-order equations it was formally shown that it translates into familiar well-posedness results with the sharp non-tangential maximal function bounds. ([May10]; see also [AAA+11].)

Following the same line of reasoning, one can build a higher-order "block-type" operator $\mathbb{L}$, for which the Kato estimate (6.7) would imply a certain comparison between normal and tangential derivatives on the boundary

$$\|\partial_t^m u(\,\cdot\,,0)\|_{L^2(\mathbb{R}^n)} \approx \|\nabla_x^m u(\,\cdot\,,0)\|_{L^2(\mathbb{R}^n)}.$$

It remains to be seen whether these bounds lead to standard well-posedness results. However, we would like to emphasize that such a result would be restricted to very special, block-type, operators.

References

[AA11] P. Auscher and A. Axelsson, *Weighted maximal regularity estimates and solvability of non-smooth elliptic systems I*, Invent. Math. **184** (2011), no. 1, 47–115.

[AAA+11] M.A. Alfonseca, P. Auscher, A. Axelsson, S. Hofmann, and S. Kim, *Analyticity of layer potentials and L^2 solvability of boundary value problems for divergence form elliptic equations with complex L^∞ coefficients*, Adv. Math. **226** (2011), no. 5, 4533–4606.

[AAH08] P. Auscher, A. Axelsson, and S. Hofmann, *Functional calculus of Dirac operators and complex perturbations of Neumann and Dirichlet problems*, J. Funct. Anal. **255** (2008), no. 2, 374–448.

[AAM10] P. Auscher, A. Axelsson, and A. McIntosh, *Solvability of elliptic systems with square integrable boundary data*, Ark. Mat. **48** (2010), no. 2, 253–287.

[Ada97] D.R. Adams, *Potential and capacity before and after Wiener*, Proceedings of the Norbert Wiener Centenary Congress, 1994 (East Lansing, MI, 1994) (Providence, RI), Proc. Sympos. Appl. Math., vol. 52, Amer. Math. Soc., 1997, pp. 63–83.

[Agm57] S. Agmon, *Multiple layer potentials and the Dirichlet problem for higher order elliptic equations in the plane. I*, Comm. Pure Appl. Math **10** (1957), 179–239.

[Agm60] ———, *Maximum theorems for solutions of higher order elliptic equations*, Bull. Amer. Math. Soc. **66** (1960), 77–80.

[Agr07] M.S. Agranovich, *On the theory of Dirichlet and Neumann problems for linear strongly elliptic systems with Lipschitz domains*, Funktsional. Anal. i Prilozhen. **41** (2007), no. 4, 1–21, 96. English translation: Funct. Anal. Appl. **41** (2007), no. 4, 247–263.

[AH96] D.R. Adams and L.I. Hedberg, *Function spaces and potential theory*, Grundlehren der Mathematischen Wissenschaften [Fundamental Principles of Mathematical Sciences], vol. 314, Springer-Verlag, Berlin, 1996.

[AHMT01] P. Auscher, S. Hofmann, A. McIntosh, and P. Tchamitchian, *The Kato square root problem for higher order elliptic operators and systems on* $\mathbb{R}^n$, J. Evol. Equ. **1** (2001), no. 4, 361–385.

[AP98] V. Adolfsson and J. Pipher, *The inhomogeneous Dirichlet problem for* Δ^2 *in Lipschitz domains*, J. Funct. Anal. **159** (1998), no. 1, 137–190.

[AR11] P. Auscher and A. Rosén, *Weighted maximal regularity estimates and solvability of non-smooth elliptic systems II*, ArXiv e-prints (2011).

[AT98] P. Auscher and P. Tchamitchian, *Square root problem for divergence operators and related topics*, Astérisque (1998), no. 249, viii+172.

[Aus04] P. Auscher, *On* L^p *estimates for square roots of second order elliptic operators on* $\mathbb{R}^n$, Publ. Mat. **48** (2004), no. 1, 159–186.

[Aus07] ———, *On necessary and sufficient conditions for* L^p*-estimates of Riesz transforms associated to elliptic operators on* $\mathbb{R}^n$ *and related estimates*, Mem. Amer. Math. Soc. **186** (2007), no. 871, xviii+75.

[BK04] S. Blunck and P.C. Kunstmann, *Weak type* (p,p) *estimates for Riesz transforms*, Math. Z. **247** (2004), no. 1, 137–148.

[CFS79] A. Calderón, E.B. Fabes, and Y. Sagher, *Nestor M. Rivière: June* 10, 1940–*January* 3, 1978, Harmonic analysis in Euclidean spaces (Proc. Sympos. Pure Math., Williams Coll., Williamstown, Mass., 1978), Part 1, Proc. Sympos. Pure Math., XXXV, Part, Amer. Math. Soc., Providence, R.I., 1979, pp. vii–xvii (1 plate).

[CG83] J. Cohen and J. Gosselin, *The Dirichlet problem for the biharmonic equation in a* C^1 *domain in the plane*, Indiana Univ. Math. J. **32** (1983), no. 5, 635–685.

[CG85] ———, *Adjoint boundary value problems for the biharmonic equation on* C^1 *domains in the plane*, Ark. Mat. **23** (1985), no. 2, 217–240.

[Dah79] B.E.J. Dahlberg, *On the Poisson integral for Lipschitz and* C^1*-domains*, Studia Math. **66** (1979), no. 1, 13–24.

[Dah80] ———, *Weighted norm inequalities for the Lusin area integral and the nontangential maximal functions for functions harmonic in a Lipschitz domain*, Studia Math. **67** (1980), no. 3, 297–314.

[DJK84] B.E.J. Dahlberg, D.S. Jerison, and C.E. Kenig, *Area integral estimates for elliptic differential operators with nonsmooth coefficients*, Ark. Mat. **22** (1984), no. 1, 97–108.

[DK90] B.E. J. Dahlberg and C.E. Kenig, L^p *estimates for the three-dimensional systems of elastostatics on Lipschitz domains*, Analysis and partial differential equations, Lecture Notes in Pure and Appl. Math., vol. 122, Dekker, New York, 1990, pp. 621–634.

[DKPV97] B.E.J. Dahlberg, C.E. Kenig, J. Pipher, and G.C. Verchota, *Area integral estimates for higher order elliptic equations and systems*, Ann. Inst. Fourier (Grenoble) **47** (1997), no. 5, 1425–1461.

[DKV86] B.E.J. Dahlberg, C.E. Kenig, and G.C. Verchota, *The Dirichlet problem for the biharmonic equation in a Lipschitz domain*, Ann. Inst. Fourier (Grenoble) **36** (1986), no. 3, 109–135.

[DMM86] G. Dal Maso and U. Mosco, *Wiener criteria and energy decay for relaxed Dirichlet problems*, Arch. Rational Mech. Anal. **95** (1986), no. 4, 345–387.

[DPP07] M. Dindoš, S. Petermichl, and J. Pipher, *The L^p Dirichlet problem for second order elliptic operators and a p-adapted square function*, J. Funct. Anal. **249** (2007), no. 2, 372–392.

[DR10] M. Dindoš and D.J. Rule, *Elliptic equations in the plane satisfying a Carleson measure condition*, Rev. Mat. Iberoam. **26** (2010), no. 3, 1013–1034.

[EG82] L.C. Evans and R.F. Gariepy, *Wiener's criterion for the heat equation*, Arch. Rational Mech. Anal. **78** (1982), no. 4, 293–314.

[FGL89] E.B. Fabes, N. Garofalo, and E. Lanconelli, *Wiener's criterion for divergence form parabolic operators with C^1-Dini continuous coefficients*, Duke Math. J. **59** (1989), no. 1, 191–232.

[FJK82] E.B. Fabes, D.S. Jerison, and C.E. Kenig, *The Wiener test for degenerate elliptic equations*, Ann. Inst. Fourier (Grenoble) **32** (1982), no. 3, vi, 151–182.

[FJL77] E.B. Fabes, M. Jodeit, Jr., and J.E. Lewis, *Double layer potentials for domains with corners and edges*, Indiana Univ. Math. J. **26** (1977), no. 1, 95–114.

[FJR78] E.B. Fabes, M. Jodeit, Jr., and N.M. Rivière, *Potential techniques for boundary value problems on C^1-domains*, Acta Math. **141** (1978), no. 3-4, 165–186.

[HKMP12] S. Hofmann, C.E. Kenig, S. Mayboroda, and J. Pipher, *Square function/non-tangential maximal function estimates and the Dirichlet problem for non-symmetric elliptic operators*, ArXiv e-prints (2012).

[JK81a] D.S. Jerison and C.E. Kenig, *The Dirichlet problem in nonsmooth domains*, Ann. of Math. (2) **113** (1981), no. 2, 367–382.

[JK81b] ———, *The Neumann problem on Lipschitz domains*, Bull. Amer. Math. Soc. (N.S.) **4** (1981), no. 2, 203–207.

[Ken90] C.E. Kenig, *Progress on two problems posed by Rivière*, Harmonic analysis and partial differential equations (Boca Raton, FL, 1988), Contemp. Math., vol. 107, Amer. Math. Soc., Providence, RI, 1990, pp. 101–107.

[Ken94] ———, *Harmonic analysis techniques for second order elliptic boundary value problems*, CBMS Regional Conference Series in Mathematics, vol. 83, Published for the Conference Board of the Mathematical Sciences, Washington, DC, 1994.

[KKPT00] C.E. Kenig, H. Koch, J. Pipher, and T. Toro, *A new approach to absolute continuity of elliptic measure, with applications to non-symmetric equations*, Adv. Math. **153** (2000), no. 2, 231–298.

[KMR01] V.A. Kozlov, V.G. Maz'ya, and J. Rossmann, *Spectral problems associated with corner singularities of solutions to elliptic equations*, Mathematical Surveys and Monographs, vol. 85, American Mathematical Society, Providence, RI, 2001.

[KP93] C.E. Kenig and J. Pipher, *The Neumann problem for elliptic equations with nonsmooth coefficients*, Invent. Math. **113** (1993), no. 3, 447–509.

[KR09] C.E. Kenig and D.J. Rule, *The regularity and Neumann problem for non-symmetric elliptic operators*, Trans. Amer. Math. Soc. **361** (2009), no. 1, 125–160.

[KS11a] J. Kilty and Z. Shen, *A bilinear estimate for biharmonic functions in Lipschitz domains*, Math. Ann. **349** (2011), no. 2, 367–394.

[KS11b] ———, *The L^p regularity problem on Lipschitz domains*, Trans. Amer. Math. Soc. **363** (2011), no. 3, 1241–1264.

[Lab02] D.A. Labutin, *Potential estimates for a class of fully nonlinear elliptic equations*, Duke Math. J. **111** (2002), no. 1, 1–49.

[Leb13] H. Lebesgue, *Sur des cas d'impossibilité du problème de Dirichlet ordinaire*, C.R. des Séances de la Société Mathématique de France **17** (1913).

[LM10] G. Luo and V.G. Maz'ya, *Wiener type regularity of a boundary point for the 3D Lamé system*, Potential Anal. **32** (2010), no. 2, 133–151.

[LSW63] W. Littman, G. Stampacchia, and H.F. Weinberger, *Regular points for elliptic equations with discontinuous coefficients*, Ann. Scuola Norm. Sup. Pisa (3) **17** (1963), 43–77.

[May10] S. Mayboroda, *The connections between Dirichlet, regularity and Neumann problems for second order elliptic operators with complex bounded measurable coefficients*, Adv. Math. **225** (2010), no. 4, 1786–1819.

[Maz77] V.G. Maz'ya, *On the behavior near the boundary of solutions of the Dirichlet problem for the biharmonic operator*, Dokl. Akad. Nauk SSSR **18** (1977), 15–19. English translation: Soviet Math. Dokl. **18** (1977), no. 4, 1152–1155.

[Maz79] ———, *Behaviour of solutions to the Dirichlet problem for the biharmonic operator at a boundary point*, Equadiff IV (Proc. Czechoslovak Conf. Differential Equations and their Applications, Prague, 1977), Lecture Notes in Math., vol. 703, Springer, Berlin, 1979, pp. 250–262.

[Maz97] ———, *Unsolved problems connected with the Wiener criterion*, The Legacy of Norbert Wiener: A Centennial Symposium (Cambridge, MA, 1994), Proc. Sympos. Pure Math., vol. 60, Amer. Math. Soc., Providence, RI, 1997, pp. 199–208.

[Maz99a] ———, *On the Wiener type regularity of a boundary point for the polyharmonic operator*, Appl. Anal. **71** (1999), no. 1-4, 149–165.

[Maz99b] ———, *On Wiener's type regularity of a boundary point for higher order elliptic equations*, Nonlinear analysis, function spaces and applications, Vol. 6 (Prague, 1998), Acad. Sci. Czech Repub., Prague, 1999, pp. 119–155.

[Maz02] ———, *The Wiener test for higher order elliptic equations*, Duke Math. J. **115** (2002), no. 3, 479–512.

[MD83] V.G. Maz'ya and T. Donchev, *Regularity in the sense of Wiener of a boundary point for a polyharmonic operator*, C. R. Acad. Bulgare Sci. **36** (1983), no. 2, 177–179.

[Mel03] V.V. Meleshko, *Selected topics in the theory of two-dimensional biharmonic problem*, Appl. Mech. Rev. **56** (2003), no. 1, 33–85.

[Mir48] C. Miranda, *Formule di maggiorazione e teorema di esistenza per le funzioni biarmoniche de due variabili*, Giorn. Mat. Battaglini (4) **2(78)** (1948), 97–118.

[Mir58] ______, *Teorema del massimo modulo e teorema di esistenza e di unicità per il problema di Dirichlet relativo alle equazioni ellittiche in due variabili*, Ann. Mat. Pura Appl. (4) **46** (1958), 265–311.

[Mit10] I. Mitrea, *Mapping properties of layer potentials associated with higher-order elliptic operators in Lipschitz domains*, Topics in operator theory. Volume 2. Systems and mathematical physics, Oper. Theory Adv. Appl., vol. 203, Birkhäuser Verlag, Basel, 2010, pp. 363–407.

[MM08] S. Mayboroda and V.G. Maz'ya, *Boundedness of the Hessian of a biharmonic function in a convex domain*, Comm. Partial Differential Equations **33** (2008), no. 7-9, 1439–1454.

[MM09a] ______, *Boundedness of the gradient of a solution and Wiener test of order one for the biharmonic equation*, Invent. Math. **175** (2009), no. 2, 287–334.

[MM09b] ______, *Pointwise estimates for the polyharmonic Green function in general domains*, Analysis, partial differential equations and applications, Oper. Theory Adv. Appl., vol. 193, Birkhäuser Verlag, Basel, 2009, pp. 143–158.

[MM10] I. Mitrea and M. Mitrea, *On the Dirichlet and regularity problems for the bi-Laplacian in Lipschitz domains*, Integral methods in science and engineering. Vol. 1, Birkhäuser Boston Inc., Boston, MA, 2010, pp. 245–254.

[MM11] D. Mitrea and I. Mitrea, *On the regularity of Green functions in Lipschitz domains*, Comm. Partial Differential Equations **36** (2011), no. 2, 304–327.

[MMS10] V.G. Maz'ya, M. Mitrea, and T. Shaposhnikova, *The Dirichlet problem in Lipschitz domains for higher order elliptic systems with rough coefficients*, J. Anal. Math. **110** (2010), 167–239.

[MMW11] I. Mitrea, M. Mitrea, and M. Wright, *Optimal estimates for the inhomogeneous problem for the bi-laplacian in three-dimensional Lipschitz domains*, J. Math. Sci. **172** (2011), no. 1, 24–134.

[MN86] V.G. Maz'ya and S.A. Nazarov, *The apex of a cone can be irregular in Wiener's sense for a fourth-order elliptic equation*, Mat. Zametki **39** (1986), no. 1, 24–28, 156. English translation: Math. Notes **39** (1986), no. 1-2, 14–16.

[MNP83] V.G. Maz'ya, S.A. Nazarov, and B.A. Plamenevskiĭ, *Singularities of solutions of the Dirichlet problem in the exterior of a thin cone*, Mat. Sb. (N.S.) **122 (164)** (1983), no. 4, 435–457. English translation: Math. USSR-Sb. **50** (1985), no. 2, 415–437.

[MP81] V.G. Maz'ya and B.A. Plamenevskiĭ, *On the maximum principle for the biharmonic equation in a domain with conical points*, Izv. Vyssh. Uchebn. Zaved. Mat. (1981), no. 2, 52–59. English translation: Soviet Math. (Iz. VUZ) **25** (1981), no. 2, 61–70.

[MR91] V.G. Maz'ya and J. Rossmann, *On the Agmon-Miranda maximum principle for solutions of elliptic equations in polyhedral and polygonal domains*, Ann. Global Anal. Geom. **9** (1991), no. 3, 253–303.

[MR92] ______, *On the Agmon-Miranda maximum principle for solutions of strongly elliptic equations in domains of* $\mathbf{R}^n$ *with conical points*, Ann. Global Anal. Geom. **10** (1992), no. 2, 125–150.

[MZ97] J. Malý and W.P. Ziemer, *Fine regularity of solutions of elliptic partial differential equations*, Mathematical Surveys and Monographs, vol. 51, American Mathematical Society, Providence, RI, 1997.

[Nad63] A. Nadai, *Theory of Flow and Fracture of Solids*, vol. II, McGraw-Hill, 1963.

[Neč67] J. Nečas, *Les méthodes directes en théorie des équations elliptiques*, Masson et Cie, Éditeurs, Paris, 1967.

[Poi90] H. Poincaré, *Sur les Equations aux Dérivées Partielles de la Physique Mathématique*, Amer. J. Math. **12** (1890), no. 3, 211–294.

[PV91] J. Pipher and G.C. Verchota, *Area integral estimates for the biharmonic operator in Lipschitz domains*, Trans. Amer. Math. Soc. **327** (1991), no. 2, 903–917.

[PV92] ———, *The Dirichlet problem in L^p for the biharmonic equation on Lipschitz domains*, Amer. J. Math. **114** (1992), no. 5, 923–972.

[PV93] ———, *A maximum principle for biharmonic functions in Lipschitz and C^1 domains*, Comment. Math. Helv. **68** (1993), no. 3, 385–414.

[PV95a] ———, *Dilation invariant estimates and the boundary Gårding inequality for higher order elliptic operators*, Ann. of Math. (2) **142** (1995), no. 1, 1–38.

[PV95b] ———, *Maximum principles for the polyharmonic equation on Lipschitz domains*, Potential Anal. **4** (1995), no. 6, 615–636.

[Rul07] D.J. Rule, *Non-symmetric elliptic operators on bounded Lipschitz domains in the plane*, Electron. J. Differential Equations (2007), No. 144, 8.

[She06a] Z. Shen, *The L^p Dirichlet problem for elliptic systems on Lipschitz domains*, Math. Res. Lett. **13** (2006), no. 1, 143–159.

[She06b] ———, *Necessary and sufficient conditions for the solvability of the L^p Dirichlet problem on Lipschitz domains*, Math. Ann. **336** (2006), no. 3, 697–725.

[She06c] ———, *On estimates of biharmonic functions on Lipschitz and convex domains*, J. Geom. Anal. **16** (2006), no. 4, 721–734.

[She07a] ———, *The L^p boundary value problems on Lipschitz domains*, Adv. Math. **216** (2007), no. 1, 212–254.

[She07b] ———, *A relationship between the Dirichlet and regularity problems for elliptic equations*, Math. Res. Lett. **14** (2007), no. 2, 205–213.

[SS81] R. Selvaggi and I. Sisto, *An existence theorem for the Dirichlet problem with respect to the operator Δ^2 in certain domains of class C^1*, Boll. Un. Mat. Ital. B (5) **18** (1981), no. 2, 473–483.

[Šul75] B.V. Šul'ce, *A priori estimates in uniform norms for strongly elliptic systems*, Sibirsk. Mat. Ž. **16** (1975), 384–394, 422, English translation: Siberian Math. J. 16 (1975), no. 2, 297–305.

[Swe09] G. Sweers, *A survey on boundary conditions for the biharmonic*, Complex Var. Elliptic Equ. **54** (2009), no. 2, 79–93.

[TW02] N.S. Trudinger and X.J. Wang, *On the weak continuity of elliptic operators and applications to potential theory*, Amer. J. Math. **124** (2002), no. 2, 369–410.

[Ver87] G.C. Verchota, *The Dirichlet problem for the biharmonic equation in C^1 domains*, Indiana Univ. Math. J. **36** (1987), no. 4, 867–895.

[Ver90] ______, *The Dirichlet problem for the polyharmonic equation in Lipschitz domains*, Indiana Univ. Math. J. **39** (1990), no. 3, 671–702.

[Ver96] ______, *Potentials for the Dirichlet problem in Lipschitz domains*, Potential theory – ICPT 94 (Kouty, 1994), de Gruyter, Berlin, 1996, pp. 167–187.

[Ver05] ______, *The biharmonic Neumann problem in Lipschitz domains*, Acta Math. **194** (2005), no. 2, 217–279.

[Ver10] ______, *Boundary coerciveness and the Neumann problem for 4th order linear partial differential operators*, Around the research of Vladimir Maz'ya. II, Int. Math. Ser. (N. Y.), vol. 12, Springer, New York, 2010, pp. 365–378.

[Wie24] N. Wiener, *The Dirichlet problem*, J. Math. Phys. **3** (1924), 127–146.

[Zar09] S.C. Zaremba, *Sur le principe du minimum*, Bull. Acad. Sci. Cracovie (1909), 197–264.

Ariel Barton and Svitlana Mayboroda
Department of Mathematics
University of Minnesota
Minneapolis, Minnesota 55455, USA
e-mail: `abarton@math.umn.edu`
`svitlana@math.umn.edu`

Operator Theory:
Advances and Applications, Vol. 236, 95–102

Additive Maps Preserving the Inner Local Spectral Radius

M. Bendaoud and M. Sarih

Abstract. Let X be a complex Banach space and let $\mathcal{L}(X)$ be the algebra of all bounded linear operators on X. We characterize additive continuous maps from $\mathcal{L}(X)$ onto itself which preserve the inner local spectral radius at a nonzero fixed vector.

Mathematics Subject Classification (2010). Primary 47B49; Secondary 47A10, 47A53.

Keywords. Local spectrum; inner local spectral radius; additive preservers.

1. Introduction

Throughout this paper, X and Y will denote infinite-dimensional complex Banach spaces and $\mathcal{L}(X)$ and $\mathcal{L}(Y)$ will denote the algebras of all bounded linear operators on X and Y with unit I, respectively. For $T \in \mathcal{L}(X)$ we will denote by $\sigma(T)$, $\sigma_{ap}(T)$, and $\sigma_{su}(T) := \{\lambda \in \mathbb{C} : \lambda - T \text{ is not surjective}\}$ the spectrum, the approximate point spectrum, and the surjectivity spectrum of T; respectively. The local resolvent set of $T \in \mathcal{L}(X)$ at a point $x \in X$, $\rho_T(x)$, is the set of all $\lambda \in \mathbb{C}$ for which there exists an open neighborhood U_λ of λ in $\mathbb{C}$ and an X-valued analytic function on U_λ such that $(\mu - T)f(\mu) = x$ for all $\mu \in U_\lambda$. Its complement denoted by $\sigma_T(x)$ is called the local spectrum of T at x. We denote as usual the spectral radius of T by $r(T) := \max\{|\lambda| : \lambda \in \sigma(T)\}$ which coincides, by Gelfand's formula for the spectral radius, with the limit of the convergent sequence $(\|T^n\|^{\frac{1}{n}})_n$. The lower-boundedness spectral radius $\ell(T)$ and the surjectivity spectral radius $\omega(T)$ of T are given by

$$\ell(T) = \sup\{\varepsilon \geq 0 : \lambda - T \text{ is bounded below for } |\lambda| < \varepsilon\},$$
$$\omega(T) = \sup\{\varepsilon \geq 0 : \lambda - T \text{ is surjective for } |\lambda| < \varepsilon\}.$$

The authors thank the support and the hospitality of the organizers of the "22nd International Workshop on Operator Theory and its Applications, Sevilla, Spain, July 3–9, 2011", where the main result of this paper was announced.

These quantities are quite useful for the localization of the approximate point (surjectivity) spectrum and the spectrum; see for instance [1] and [9]. In [10], E. Makai and J. Zemánek proved, in fact, that $\ell(T)$ (resp. $\omega(T)$) is nothing but the minimum modulus of $\sigma_{ap}(T)$ (resp. $\sigma_{su}(T)$) that coincides with the limit $\lim_{n\to\infty} m(T^n)^{\frac{1}{n}}$ (resp. $\lim_{n\to\infty} q(T^n)^{\frac{1}{n}}$). Here $m(T) := \inf\{\|Tx\| : x \in X, \|x\| \leq 1\}$ (resp. $q(T) := \sup\{\varepsilon \geq 0; \varepsilon B(0,1) \subseteq T(B(0,1))\}$) is the so-called minimum (resp. surjectivity) modulus of T; where $B(0,1)$ denotes the closed unit ball of X. In the same paper a counter-example was given showing that $\ell(T)$ and $\omega(T)$ are not determined by the spectrum of T. The inner local spectral radius of T at a point $x \in X$, $\iota_T(x)$, is defined by

$$\iota_T(x) := \sup\{\varepsilon \geq 0 : x \in \mathcal{X}_T(\mathbb{C} \setminus D(0,\varepsilon))\},$$

where $D(0,\varepsilon)$ denotes the open disc of radius ε centered at 0 and $\mathcal{X}_T(\mathbb{C} \setminus D(0,\varepsilon))$ is the so-called local spectral subspace of T associated with $\mathbb{C} \setminus D(0,\varepsilon)$, that is, the set of all $x \in X$ for which there is an X-valued analytic function f on $D(0,\varepsilon)$ such that $(\lambda - T)f(\lambda) = x$ for all $\lambda \in D(0,\varepsilon)$. The local spectral radius of T at x is given by

$$r_T(x) := \limsup_{n\to+\infty} \|T^n x\|^{\frac{1}{n}}.$$

The inner local (resp. local) spectral radius of T at x coincides with the minimum (resp. maximum) modulus of $\sigma_T(x)$ provided that T has the single-valued extension property; see [9] and [11]. Recall that T is said to have the single-valued extension property if for every open set U of $\mathbb{C}$, the equation $(T-\lambda)\phi(\lambda) = 0$, $(\lambda \in U)$, has no nontrivial analytic solution on U. For more details and basic facts concerning the spectral quantities $\ell(T)$, $\omega(T)$, and $\iota_T(x)$ we refer the reader to [1, 9, 10], and [11].

We will say that an additive map $\phi : \mathcal{L}(X) \to \mathcal{L}(X)$ compresses the local spectrum at a fixed nonzero vector $e \in X$ if $\sigma_{\phi(T)}(e) \subseteq \sigma_T(e)$ holds for all $T \in \mathcal{L}(X)$ and preserves the local spectrum (resp. local spectral radius) at e if the reverse set-inclusion holds too (resp. $r_{\phi(T)}(e) = r_T(e)$ for all $T \in \mathcal{L}(X)$).

In [7], Bračič and Müller characterized continuous surjective linear maps from $\mathcal{L}(X)$ into itself that preserve the local spectrum and the local spectral radius at a nonzero fixed vector in X. In [4], the authors treated the problem of characterizing locally spectrally bounded linear maps on the algebra $\mathcal{L}(\mathcal{H})$ of all bounded linear operators on a complex Hilbert space $\mathcal{H}$, and they described continuous linear maps from $\mathcal{L}(\mathcal{H})$ onto itself that compress the local spectrum at a fixed nonzero vector in $\mathcal{H}$. The surjective continuous additive mappings ϕ on $\mathcal{L}(X)$ which are local spectrum compressing or local spectral radius preserving at a nonzero vector were characterized in [5].

In this paper, we first collect in the next section some results concerning additive maps from $\mathcal{L}(X)$ onto $\mathcal{L}(Y)$ that preserve the lower-boundedness (surjectivity) of operators in both directions and the ones that preserve the lower-boundedness (surjectivity) spectral radius of operators. This allows us to characterize in the last section additive maps from $\mathcal{L}(X)$ onto $\mathcal{L}(Y)$ that preserve the inner local spectral radius at a fixed nonzero vector. It should be pointed out that our proofs use some arguments which are influenced by ideas from Bračič and Müller [7].

2. Preliminaries

We first fix some notation and terminology. The duality between the Banach spaces X and its dual, X^*, will be denoted by $\langle .,. \rangle$. For $x \in X$ and $f \in X^*$, as usual we denote by $x \otimes f$ the rank at most one operator on X given by $z \mapsto \langle z, f \rangle x$. For $T \in \mathcal{L}(X)$ we will denote by $\ker(T)$, $\operatorname{range}(T)$, and T^* the null space, the range, and the adjoint of T; respectively. The operator T is said to be semi-Fredholm if $\operatorname{range}(T)$ is closed and $\dim(\ker(T))$ or $\dim(X/\operatorname{range}(T))$ is finite, and is said to be semi-invertible if it is left or right invertible. An additive mapping $A : X \to Y$ is called semilinear if $A(\lambda x) = \tau(\lambda)A(x)$ holds for all scalars $\lambda \in \mathbb{C}$ and vectors $x \in X$, where τ is a ring automorphism of $\mathbb{C}$. It is called conjugate linear if $A(\lambda x) = \overline{\lambda}A(x)$ holds for all scalars $\lambda \in \mathbb{C}$ and vectors $x \in X$.

Recall that an additive map $\phi : \mathcal{L}(X) \to \mathcal{L}(Y)$ is called unital if $\phi(I) = I$, and is said to preserve the lower-boundedness of operators in both directions provided that $\phi(T)$ is bounded below if and only if T is. The additive maps preserving the surjectivity in both directions are defined in a similar way.

The following elementary lemmas, inspired by [3], are on the straightforward side. We include them for the sake of completeness.

Lemma 2.1. *Let $\phi : \mathcal{L}(X) \to \mathcal{L}(Y)$ be a surjective additive map. If ϕ either preserves lower-boundedness or surjectivity of operators in both directions, then either*

(i) *there exist invertible bounded both linear or both conjugate linear operators $A : X \to Y$ and $B : Y \to X$ such that $\phi(T) = ATB$ for all $T \in \mathcal{L}(X)$, or*
(ii) *there exist invertible bounded both linear or both conjugate linear operators $A : X^* \to Y$ and $B : Y \to X^*$ such that $\phi(T) = AT^*B$ for all $T \in \mathcal{L}(X)$.*

The last case occurs only if X and Y are reflexive.

Proof. Assume that ϕ preserves the lower-boundedness of operators in both directions. It is easy to check that T is lower bounded if and only if T is not left topological divisor of zero; i.e., there is no sequence $(S_n)_{n\geq 1} \subseteq \mathcal{L}(X)$ satisfying $\|S_n\| = 1$ and $TS_n \to 0$ as $n \to \infty$. So, by using the same approach as in [8, Theorem 3.1] one can see that ϕ is injective and either

(a) there exist semilinear bijective maps $C : X \to Y$ and $D : X^* \to Y^*$ such that $\phi(x \otimes f) = Cx \otimes Df$ for all $x \in X$ and all $f \in X^*$, or
(b) there exist semilinear bijective maps $C : X^* \to Y$ and $D : X \to Y^*$ such that $\phi(x \otimes f) = Cf \otimes Dx$ for all $x \in X$ and all $f \in X^*$.

Now, let us show that $\phi(I)$ is invertible. Note that $\phi(I)$ is injective with closed range, and let us show by way of contradiction that $\phi(I)$ is surjective. So, assume that there exists a nonzero element $y_0 \in Y \setminus \operatorname{range}(\phi(I))$. We claim that the operator $\phi(I) - y_0 \otimes g$ is injective with closed range for all $g \in Y^*$. Indeed, the operator $\phi(I)$ is semi-Fredholm since it is bounded below. Thus the operator $\phi(I) - y_0 \otimes g$ is semi-Fredholm for every $g \in Y^*$. On the other hand, $\phi(I) - y_0 \otimes g$ is injective because $\phi(I)$ is injective and $y_0 \notin \operatorname{range} \phi(I)$. This yields the claim. So, if the case (a) occurs we can find an element $x_0 \in X$ and a linear functional

$f_0 \in X^*$ such that $Ax_0 = y_0$ and $\langle x_0, f_0 \rangle = 1$. Thus, we have $I - x_0 \otimes f_0$ as well as $\phi(I - x_0 \otimes f_0) = \phi(I) - Ax_0 \otimes Cf_0$ is bounded below; which contradicts $\sigma(x_0 \otimes f_0) = \{0, 1\}$. By similarity, in the case when (b) occurs we get a contradiction, too. Hence $\phi(I)$ is invertible. Set

$$\chi(T) = \phi(I)^{-1}\phi(T), \quad (T \in \mathcal{L}(X)).$$

The map χ is a unital surjective additive map preserving lower-boundedness of operators in both directions, and so by applying [8, Corollary 3.5] the map ϕ takes one of the desired forms.

The case when ϕ preserves the surjectivity of operators in both directions is treated in [3]; and the proof is therefore complete. □

Let us recall the following useful facts that will be often used in the sequel. For $T \in \mathcal{L}(X)$ it is straightforward that $\ell(T) > 0$ (resp. $\omega(T) > 0$) if and only if T is bounded below (resp. surjective), that is equivalent in the Hilbert space setting that T is left (resp. right) invertible. Notice that $\sigma_{ap}(T) = \sigma_{su}(T^*)$ and $\sigma_{su}(T) = \sigma_{ap}(T^*)$, and so $\ell(T) = \omega(T^*)$ and $\omega(T) = \ell(T^*)$; see [9] and [10].

We will say that an additive map $\phi : \mathcal{L}(X) \to \mathcal{L}(Y)$ preserves the lower-boundedness spectral radius if $\ell(\phi(T)) = \ell(T)$ for all $T \in \mathcal{L}(X)$. The additive maps preserving the surjectivity spectral radius are defined analogously.

Lemma 2.2. *Let $\varphi : \mathcal{L}(X) \to \mathcal{L}(Y)$ be a surjective additive map. If φ either preserves the lower-boundedness radius or surjectivity radius, then there exists a scalar $c \in \mathbb{C}$ of modulus one and either*

(i) *there exists an invertible bounded linear or conjugate linear operator $A : X \to Y$ such that $\varphi(T) = cATA^{-1}$ for all $T \in \mathcal{L}(X)$, or*
(ii) *there exists an invertible bounded linear or conjugate linear operator $A : X^* \to Y$ such that $\varphi(T) = cAT^*A^{-1}$ for all $T \in \mathcal{L}(X)$.*

The last case occurs only if X and Y are reflexive.

Proof. Note that, if ϕ preserves the spectral radius $\ell(.)$ (resp. $\omega(.)$) then ϕ preserves the lower-boundedness (resp. surjectivity) of operators in both directions; and thus by Lemma 2.1 either

(a) there exist invertible bounded both linear or both conjugate linear operators $A : X \to Y$ and $B : Y \to X$ such that $\phi(T) = ATB$ for all $T \in \mathcal{L}(X)$, or
(b) there exist invertible bounded both linear or both conjugate linear operators $A : X^* \to Y$ and $B : Y \to X^*$ such that $\phi(T) = AT^*B$ for all $T \in \mathcal{L}(X)$.

To complete the proof it suffices to show that AB is a multiple of the unit by a unimodular scalar.

Assume that ϕ preserves the lower-boundedness radius. First, we claim that

$$\ell(RQ) = \ell(Q) \tag{2.1}$$

for all $Q \in \mathcal{L}(Y)$, where $R := B^{-1}A^{-1}$. Indeed, if the case (a) occurs we have

$$\ell(R\phi(T)) = \ell(T) = \ell(\phi(T))$$

for all $T \in \mathcal{L}(X)$, and the surjectivity of ϕ yields the claim. If the case (b) occurs we have

$$\ell(R\phi(T)) = \ell(T^*) = \omega(T)$$

for all $T \in \mathcal{L}(X)$. Particulary we have

$$\begin{aligned} T \text{ is surjective } &\Leftrightarrow \phi(T) \text{ bounded below} \\ &\Leftrightarrow T \text{ is bounded below} \end{aligned}$$

for all $T \in \mathcal{L}(X)$. From this we infer that $\sigma_{ap}(T) = \sigma_{su}(T)$, and so $\ell(R\phi(T)) = \ell(T) = \ell(\phi(T))$ for all $T \in \mathcal{L}(X)$. Again the surjectivity of ϕ yields the claim in this case, too. Next, assume by way of contradiction that R and I are linearly independent. So, we can find a nonzero element $y_0 \in Y$ such that y_0 and Ry_0 are linearly independent, and let W be a topological complement of the linear subspace spanned by $\{y_0, Ry_0\}$ in Y. Fix a nonzero complex number α for which $|\alpha| < 1$, and define linearly the operator $Q_0 \in \mathcal{L}(Y)$ by

$$Q_0 y := \begin{cases} \alpha^{-1} R y_0 & \text{if } y = y_0 \\ \alpha y_0 & \text{if } y = Ry_0 \\ y & \text{if } y \in W \end{cases}$$

It easy to check that $\ell(Q_0) = 1$, and that $RQ_0(Ry_0) = \alpha R y_0$. These show that

$$\ell(RQ_0) \leq |\alpha| < 1 = \ell(Q_0),$$

and lead to a contradiction; see (2.1). Thus AB as well as R is a multiple of the unit with a scalar $c \in \mathbb{C}$, and $| c |= \ell(AB) = \ell(I) = 1$.

By similarity, if ϕ preserves the surjectivity radius we have $\omega(QR) = \omega(Q)$ for all $Q \in \mathcal{L}(Y)$; and so $\ell(R^*Q^*) = \ell(Q^*)$ for all $Q \in \mathcal{L}(Y)$. Thus by what has been shown above, we have R as well as R^* is a multiple of the unit by a scalar of modulus one. The proof is therefore complete. □

In the finite-dimensional case, from the fact that a matrix T in the algebra $M_n(\mathbb{C})$ of all complex $n \times n$ matrices is invertible if and only if it is semi-invertible, one can see that

$$\ell(T) = \omega(T),$$

for all $T \in M_n(\mathbb{C})$.

The following characterizes additive maps from $M_n(\mathbb{C})$ onto itself that preserve the lower-boundedness or surjectivity spectral radius of matrices.

Proposition 2.3. *Let $\phi : M_n(\mathbb{C}) \to M_n(\mathbb{C})$ be a surjective additive map. The following are equivalent:*

(i) $\ell(\phi(T)) = \ell(T)$ *for every* $T \in M_n(\mathbb{C})$.

(ii) $\omega(\phi(T)) = \omega(T)$ *for every* $T \in M_n(\mathbb{C})$.

(iii) *There exist a scalar $c \in \mathbb{C}$ of modulus one and an invertible matrix A in $M_n(\mathbb{C})$ such that either $\phi(T) = cATA^{-1}$, $\phi(T) = cAT^{tr}A^{-1}$, $\phi(T) = cAT^*A^{-1}$, or $\phi(T) = cA(T^{tr})^*A^{-1}$; for every $T \in M_n(\mathbb{C})$. Here T^{tr} denotes the transpose of the matrix T.*

Proof. As the sufficiency condition is obvious, we only need to prove the necessity. So assume that ϕ preserves either the lower boundedness or surjectivity spectral radius of matrices, and note that, in this case, ϕ is a bijective map preserving invertibility in both directions. So, using the same approach as in [2, Theorem 4.1] one can see that ϕ takes one of the desired forms; and the necessity condition is established. □

3. Main result and proof

We will say that an additive map $\phi : \mathcal{L}(X) \to \mathcal{L}(X)$ preserves the inner local spectral radius at a fixed nonzero vector $x \in X$ if $\iota_{\phi(T)}(x) = \iota_T(x)$ for all $T \in \mathcal{L}(X)$.

The following is the main result of this paper. It characterizes additive maps from $\mathcal{L}(X)$ onto itself that preserve the inner local spectral radius at a fixed nonzero vector and extends [6, Theorem 2.1] from linear case to additive case. Its proof use some arguments which are influenced by ideas from Bračič and Müller [7].

Theorem 3.1. *Let e be a fixed nonzero vector in X. An additive continuous map ϕ from $\mathcal{L}(X)$ onto itself preserves the inner local spectral radius at e if and only if there exist a scalar c of modulus one and a linear or conjugate linear bijective bounded operator $A : X \to X$ such that $Ae = e$, and $\phi(T) = cATA^{-1}$ for all $T \in \mathcal{L}(X)$.*

The proof of this theorem uses some auxiliary lemmas. The first is quoted from Bračič and Müller [7, Lemma 2.2].

Lemma 3.2. *Let e be a fixed nonzero vector in X, and let $T \in \mathcal{L}(X)$. If $\lambda \in \sigma_{su}(T)$, then for every $\varepsilon > 0$, there exists $T' \in \mathcal{L}(X)$ such that $\|T - T'\| < \varepsilon$ and $\lambda \in \sigma_{T'}(e)$.*

Proof. See [7, Lemma 2.2]. □

Lemma 3.3. *Let e be a fixed nonzero vector in X. For a linear or conjugate linear bijective bounded operator $A : X \to X$, the map $\phi : T \in \mathcal{L}(X) \mapsto ATA^{-1} \in \mathcal{L}(X)$ preserves the inner local spectrum at e if and only if $Ae = \lambda e$ for some $\lambda \in \mathbb{C}$.*

Proof. We shall only deal with the case when A is conjugate linear, because the linear case follows analogously. First, we claim that for every $T \in \mathcal{L}(X)$ and $\varepsilon > 0$ we have $Ae \in \mathcal{X}_{ATA^{-1}}(\mathbb{C} \setminus D(0, \varepsilon))$ whenever $e \in \mathcal{X}_T(\mathbb{C} \setminus D(0, \varepsilon))$. Indeed, assume that $e \in \mathcal{X}_T(\mathbb{C} \setminus D(0, \varepsilon))$ and let f be a X-valued analytic function on $D(0, \varepsilon)$ such that $(\mu - T)f(\mu) = x$ for all $\mu \in D(0, \varepsilon)$. We have

$$(\mu^\eta - ATA^{-1})Af(\mu) = Ae$$

for all $\mu \in D(0, \varepsilon)$; where $\eta : \mathbb{C} \to \mathbb{C}$ is the complex conjugation. Set

$$\widetilde{f}(\mu^\eta) := Af(\mu), \quad (\mu \in D(0, \varepsilon)),$$

and note that the map $\widetilde{f}$ is an analytic function on $D(0, \varepsilon)^\eta = D(0, \varepsilon)$ since

$$\lim_{h \to 0} \frac{\widetilde{f}(\mu^\eta + h) - \widetilde{f}(\mu^\eta)}{h} = \lim_{h \to 0} A(\frac{f(\mu + h^\eta) - f(\mu)}{h^\eta}) = Af'(\mu)$$

for all $\mu \in D(0,\varepsilon)$, where $f'(\mu)$ is the derivative of f at μ. This shows that $Ae \in \mathcal{X}_T(\mathbb{C}\setminus D(0,\varepsilon))$ and yields the claim. When Ae and e are linearly dependent, the reverse implication can be obtained by similarity, and thus, we in fact have $e \in \mathcal{X}_T(\mathbb{C}\setminus D(0,\varepsilon))$ if and only if $Ae \in \mathcal{X}_{ATA^{-1}}(\mathbb{C}\setminus D(0,\varepsilon))$ for all $\varepsilon > 0$; which show that $\iota_{ATA^{-1}}(e) = \iota_T(e)$ and ϕ preserves the inner local spectrum at e in this case.

Conversely, assume that ϕ preserves the inner local spectrum at e, but Ae and e are linearly independent. Let $f \in X^*$ be a linear functional such that $\langle e, f\rangle = 1$ and $\langle A^{-1}e, f\rangle = 0$. Set $T =: e \otimes f$ and note that $\iota_{ATA^{-1}}(e) = 0$ and $\iota_T(e) = 1$; which leads to a contradiction and completes the proof. □

We have now collected all the necessary ingredients and are therefore in a position to prove our main result.

Proof of Theorem 3.1. As the sufficiency condition is a consequence of the above Lemma, we only need to prove the necessity. So, assume that ϕ preserves the inner local spectral radius at e. We claim that ϕ preserves the spectral radius function $\omega(.)$. For this, let $T \in \mathcal{L}(X)$ and let $\lambda \in \sigma_{su}(\phi(T))$ satisfy $|\lambda| = \omega(\phi(T))$. The Lemma 3.2 ensures that for each integer $n \geq 1$ there exists an operator T'_n in $\mathcal{L}(X)$ such that $\|T'_n - \phi(T)\| < n^{-1}$ and $\lambda \in \sigma_{T'_n}(e)$. Since ϕ is continuous and surjective, by the Banach open mapping theorem there exists $\eta > 0$ such that $\eta B(0,1) \subseteq \phi(B(0,1))$, where $B(0,1)$ denotes the open unit ball of $\mathcal{L}(X)$. Therefore, for each n there exists $T_n \in \mathcal{L}(X)$ such that $\phi(T_n) = T'_n$ and $\|T_n - T\| \leq \eta^{-1}\|T'_n - \phi(T))\| \leq \eta^{-1}n^{-1}$. Thus $T_n \to T$ and $\lambda \in \sigma_{\phi(T_n)}(e)$ for all $n \geq 1$. On the other hand, again by the Banach open mapping theorem and by applying [12, Propositions 6.9 and 9.9] to the set of all surjective operators on X one can see that the surjectivity spectrum is an upper semi-continuity function. Thus, the spectral function $\omega(.)$ is upper semi-continuous and so

$$\omega(T) \leq \liminf_{n\to\infty} \omega(T_n) \leq \liminf_{n\to\infty} \iota_{T_n}(e) = \liminf_{n\to\infty} \iota_{\phi(T_n)}(e) \leq |\lambda| = \omega(\phi(T)).$$

To establish the reverse inequality, pick an arbitrary $T \in \mathcal{L}(X)$ and $\lambda \in \sigma_{su}(T)$ such that $|\lambda| = \omega(T)$. By Lemma 3.2 there exists a sequence of operators (T_n) in $\mathcal{L}(X)$ converging to T such that $\lambda \in \sigma_{T_n}(e)$ for all n, and consequently we have

$$\omega(\phi(T)) \leq \liminf_{n\to\infty} \omega(\phi(T_n)) \leq \liminf_{n\to\infty} \iota_{\phi(T_n)}(e) = \liminf_{n\to\infty} \iota_{T_n}(e) \leq |\lambda| = \omega(T).$$

From this, we infer that $\omega(\phi(T)) = \omega(T)$ for all $T \in \mathcal{L}(X)$, and so by Theorem 2.2, there exists a scalar c of modulus one and either there exists a linear or conjugate linear invertible bounded operator $A : X \to X$ such that $\phi(T) = cATA^{-1}$ for all $T \in \mathcal{L}(X)$, or there exists a linear or conjugate linear invertible bounded operator $A : X^* \to X$ such that $\phi(T) = cAT^*A^{-1}$ for all $T \in \mathcal{L}(X)$. By the same argument given in the end of the proof of Lemma 3.3, one can see that when A is defined from X^* into X we can find an operator $T \in \mathcal{L}(X)$ such that $\iota_T(e) = 1$ and $\iota_{AT^*A^{-1}}(e) = 0$; which shows that the second form is excluded, and consequently ϕ takes only the first one with $Ae = \lambda e$ for some nonzero $\lambda \in \mathbb{C}$. Dividing A by λ or its complex conjugate $\overline{\lambda}$ if necessary, we may assume that $Ae = e$, and thus the necessity condition is established. □

References

[1] P. Aiena, *Fredholm and local spectral theory, with applications to Multipliers*, Kluwer Academic Publishers, 2004.

[2] M. Bendaoud, *Additive maps preserving the minimum and surjectivity moduli of operators*, Arch. Math. **92**(2009), 257–265.

[3] M. Bendaoud and M. Sarih, *Additive maps preserving the ascent and descent of operators*, Linear Algabra Appl. **431**(2009), 1740–1744.

[4] M. Bendaoud and M. Sarih, *Locally spectrally bounded linear maps*, Mathematica Bohemica, 136, No. **1**(2011), 81–89.

[5] M. Bendaoud and M. Sarih, *Additive local spectrum compressors*, Linear Algabra Appl. **435** (6) (2011), 1473–1478.

[6] M. Bendaoud and M. Sarih, *Surjective linear maps preserving certain spectral radii*, Rocky Mountain Journal of Mathematics, **41** (3) (2011), 727–735.

[7] J. Bračič and V. Müller, *Local spectrum and local spectral radius at a fixed vector*, Studia Math. **194** (2009), 155–162.

[8] J. Cui and J. Hou, *Additive maps on standard operator algebras preserving invertibilities or zero divisors*, Linear Algebra Appl. **359**(2003), 219–233.

[9] K.B. Laursen and M.M. Neumann, *An introduction to local spectral theory*, Oxford University, Press, New York, 2000.

[10] E. Makai and J. Zemánek, *The surjectivity radius, packing numbers and boundedness below of linear operators*, Integ. Equ. Oper. Theory **6**(1983), 372–384.

[11] T.L. Miller, V.G. Miller and M. Neumann, *Local spectral properties of weighted shifts*, J. Oper. Theory **51**(2004), 71–88.

[12] V. Müller, *Spectral theory of linear operators and spectral systems in Banach algebras*, Second Edition, Operator Theory: Advances and applications, vol. **139**, Birkhäuser 2007.

M. Bendaoud
Université Moulay Ismail
Ecole Nationale Supérieure d'Arts et Métiers
Département de Mathématiques et Informatique
BP 4024 Béni M'hamed, Marjane II
Meknès, Morocco
e-mail: m.bendaoud@ensam-umi.ac.ma

M. Sarih
Université Moulay Ismail
Faculté des Sciences
Département de mathématiques
BP 11201, Zitoune
Meknès, Morocco
e-mail: sarih@fs-umi.ac.ma

Operator Theory:
Advances and Applications, Vol. 236, 103–115

On some Generalized Riemann Boundary Value Problems with Shift on the Real Line

L.F. Campos, A.B. Lebre and J.S. Rodríguez

Dedicated to Professor V.G. Kravchenko on the occasion of his seventieth birthday.

Abstract. On the real line we consider singular integral operators with a linear Carleman shift and complex conjugation, acting in $\widetilde{L}_2(\mathbb{R})$, the real space of all Lebesgue measurable complex value functions on $\mathbb{R}$ with $p = 2$ power. We show that the original singular integral operator with shift and conjugation is, after extension, equivalent to a singular integral operator without shift and with a 4×4 matrix coefficients. By exploiting the properties of the factorization of the symbol of this last operator, it is possible to describe the solution of a generalized Riemann boundary value problem with a Carleman shift.

Mathematics Subject Classification (2010). Primary: 47G10; Secondary: 47A68.

Keywords. Generalized Riemann boundary value problem, singular integral operators, Carleman shift, factorization.

1. Introduction

Let $\widetilde{L}_2(\mathbb{R})$ denote the real space of all Lebesgue measurable complex-valued functions on $\mathbb{R}$ with $p = 2$ power. We consider the generalized Riemann boundary value problem:

Find $\widetilde{L}_2(\mathbb{R})$ functions $\varphi_+(z)$ and $\varphi_-(z)$ analytic in $\operatorname{Im} z > 0$ and $\operatorname{Im} z < 0$, respectively, satisfying the conditions

$$\varphi_+ = a_{00}\varphi_- + a_{10}\varphi_-(\alpha) + a_{01}\overline{\varphi_-} + a_{11}\overline{\varphi_-}(\alpha), \quad \varphi_-(\infty) = 0, \tag{1.1}$$

imposed on their boundary values on $\mathbb{R}$, where

$$\alpha(t) = -t + h, \quad h \in \mathbb{R}, \tag{1.2}$$

is a Carleman shift on the real line, and $a_{ik} \in \widetilde{L}_\infty(\mathbb{R}), i, k = 0, 1$.

Research supported by Centro de Análise Funcional e Aplicações do Instituto Superior Técnico (CEAF), which is financed by FCT (Portugal).

It is clear that α is a backward Carleman shift of order $n = 2$ which has on $\mathbb{R}$ one fixed point $t_0 = \frac{h}{2}$.

It is well known (see for instance, [5]) and [8]) that generalized Riemann boundary value problems (BVP) found applications in physics, mechanics and geometry of surfaces. Actually, already in the early 1959 was mentioned by I. Vekua in his book [10], that the problem of rigidity of a closed surface which consist of two glued pieces, leads to the solvability of generalized Riemann BVP (1.1) with a non-Carleman shift. The above-mentioned BVP was studied in the papers [6] and [7], where the authors constructed an estimate for the number of linearly independent solutions.

The paper is organized as follows. In Section 2 we derive some basic properties and equalities that will be used in the rest of the paper. The Section 3 contains the main results of our work. There it is shown that the solution of the considered generalized Riemann boundary value problem with a Carleman shift and conjugation is, after extension, equivalent to description of the kernel of singular integral operator with 4×4 matrix coefficients. In Section 4 we state some results about the generalized factorization of a matrix-function related to the BVP under consideration.

2. Preliminaries

Associated with a Carleman shift (1.2) we consider the shift operator $U : \widetilde{L}_2(\mathbb{R}) \to \widetilde{L}_2(\mathbb{R})$, defined by

$$U\varphi = \varphi \circ \alpha := \varphi(\alpha). \tag{2.1}$$

By $C : \widetilde{L}_2(\mathbb{R}) \longrightarrow \widetilde{L}_2(\mathbb{R})$ we denote a complex conjugation operator

$$C\varphi = \overline{\varphi}. \tag{2.2}$$

The operators U and C possess a number of properties, namely:

$$\text{(i) } U^2 = I, \text{ (ii) } C^2 = I, \text{ (iii) } CU = UC, \tag{2.3}$$

$$\text{(iv) } P_\pm U = UP_\mp, \text{ (v) } P_\pm C = CP_\mp, \tag{2.4}$$

where $P_\pm$ are the pair of complementary projection operators,

$$P_\pm = \frac{1}{2}(I \pm S),$$

generated by S, the singular operator with Cauchy kernel

$$S\varphi(t) = \frac{1}{\pi i} \int_{\mathbb{R}} \frac{1}{\tau - t} \varphi(\tau)\, d\tau,$$

where the integral is understood in the sense of principal value.

The homographic transformation

$$z = \theta(t) = \frac{t - i}{t + i} \tag{2.5}$$

is a homeomorphism of $\dot{\mathbb{R}}$ onto $\mathbb{T}$ (the unit circle) with inverse

$$\theta_{-1}(z) = i\frac{1+z}{1-z}.$$

Later the following factorization on $\mathbb{R}$ will be useful,

$$\theta(\alpha) = \theta_+\theta^{-1}\theta_-, \tag{2.6}$$

where the outer factors θ_+ and θ_- are given by

$$\theta_+ = \frac{t+i-h}{t+i}, \quad \theta_- = \frac{t-i}{t-i-h},$$

and satisfy the following relations:

$$\theta_\pm(\alpha) = \theta_\mp, \quad \overline{\theta_\pm} = \theta_\mp^{-1}. \tag{2.7}$$

3. Singular integral operators with linear-fractional Carleman shift and conjugation

It is clear that the generalized Riemann BVP (1.1) is equivalent to the problem of the description of the kernel of the paired operator

$$T_A = P_+ - AP_- \tag{3.1}$$

where $A : \widetilde{L}_2(\mathbb{R}) \longrightarrow \widetilde{L}_2(\mathbb{R})$ is the functional operator with conjugation of the form

$$A = a_{00}I + a_{10}U + a_{01}C + a_{11}UC. \tag{3.2}$$

To find the kernel of operator T_A we use similar ideas to those of [4]. In fact, as we shall see, the study of operator (3.1) can be reduced to the study of a singular integral operator without shift.

For $\Phi \in \widetilde{L}_2^4(\mathbb{R})$ put $\Phi = (\Phi_1, \ldots, \Phi_4)$. Let

$$\mathfrak{E} = \left\{\Phi \in \widetilde{L}_2^4(\mathbb{R}) : \Phi_1 = \cdots = \Phi_4\right\}$$

and define

$$\pi : \widetilde{L}_2(\mathbb{R}) \to \mathfrak{E} \tag{3.3}$$

as the map that associates to the function $\varphi \in \widetilde{L}_2(\mathbb{R})$ the vector function $\Phi \in \mathfrak{E}$ with $\Phi_j = \varphi, j = 1, \ldots, 4$. Further, consider the subspace $\widetilde{\mathfrak{E}}$ with $\widetilde{\mathfrak{E}} = \operatorname{im}\Delta$, where $\Delta : \mathfrak{E} \to \widetilde{\mathfrak{E}}$ denotes the invertible operator

$$\Delta = \operatorname{diag}(I, U, C, UC)\,|_{\mathfrak{E}}\,. \tag{3.4}$$

Proposition 3.1. *The subspace $\widetilde{\mathfrak{E}}$ is characterized by the equalities*

$$U\Psi = \begin{pmatrix} \mathcal{E} & 0 \\ 0 & \mathcal{E} \end{pmatrix}\Psi, \quad C\Psi = \begin{pmatrix} 0 & \mathcal{I}_2 \\ \mathcal{I}_2 & 0 \end{pmatrix}\Psi, \quad \Psi \in \widetilde{\mathfrak{E}}, \tag{3.5}$$

where $\mathcal{E}$ is the 2×2 constant matrix

$$\mathcal{E} = \begin{pmatrix} 0 & 1 \\ 1 & 0 \end{pmatrix}, \tag{3.6}$$

$\mathcal{I}_2$ is the 2×2 identity matrix and we suppose that the vector-valued operators U and C act componentwise.

Let us introduce the operator $T : \mathfrak{E} \to \widetilde{L}_2^4(\mathbb{R})$ defined by

$$T = \operatorname{diag}\,(T_A, \dots, T_A)\,|_{\mathfrak{E}}\,.$$

The main advantage of considering this extension of the operator T_A is the fact that operator T is similar to a singular integral operator without shift and matrix coefficients, as stated in the next theorem.

Theorem 3.2. *The following operator relation holds*

$$T = \Delta^{-1} S_{\mathcal{A},\mathcal{B}} \Delta, \tag{3.7}$$

where $S_{\mathcal{A},\mathcal{B}} : \widetilde{L}_2^4(\mathbb{R}) \to \widetilde{L}_2^4(\mathbb{R})$ is a singular integral operator without either shift or conjugation and with 4×4 matrix coefficients,

$$S_{\mathcal{A},\mathcal{B}} = \mathcal{A}\,P_+ + \mathcal{B}\,P_-,$$

and the matrix-functions $\mathcal{A}$ and $\mathcal{B}$ satisfy the following relations:

$$\mathcal{A} = \begin{pmatrix} \mathcal{E} & 0 \\ 0 & \mathcal{E} \end{pmatrix} \mathcal{B}(\alpha) \begin{pmatrix} \mathcal{E} & 0 \\ 0 & \mathcal{E} \end{pmatrix}, \tag{3.8}$$

$$\mathcal{A} = \begin{pmatrix} 0 & \mathcal{I}_2 \\ \mathcal{I}_2 & 0 \end{pmatrix} \overline{\mathcal{B}} \begin{pmatrix} 0 & \mathcal{I}_2 \\ \mathcal{I}_2 & 0 \end{pmatrix}. \tag{3.9}$$

Proof. Let $\varphi \in \widetilde{L}_2(\mathbb{R})$, set $\Phi = \pi\varphi$, $\Psi = \Delta\Phi$. Then, making use of (2.3)–(2.4), we successively obtain

$$\begin{aligned}
T_A &= P_+ - a_{10}P_+U - a_{01}P_+C - a_{00}P_- - a_{11}P_-UC,\\
UT_A &= -a_{00}(\alpha)P_+U - a_{11}(\alpha)P_+C - a_{10}(\alpha)P_- + P_-U - a_{01}(\alpha)P_-UC,\\
CT_A &= -\overline{a_{11}}P_+U - \overline{a_{00}}P_+C - \overline{a_{01}}P_- + P_-C - \overline{a_{10}}P_-UC,\\
UCT_A &= -\overline{a_{01}}(\alpha)P_+U - \overline{a_{10}}(\alpha)P_+C + P_+UC - \overline{a_{11}}(\alpha)P_- - \overline{a_{00}}(\alpha)P_-UC.
\end{aligned}$$

If $\mathcal{A}$ and $\mathcal{B}$ are the matrix-functions

$$\mathcal{A} = \begin{pmatrix} 1 & -a_{10} & -a_{01} & 0 \\ 0 & -a_{00}(\alpha) & -a_{11}(\alpha) & 0 \\ 0 & -\overline{a_{11}} & -\overline{a_{00}} & 0 \\ 0 & -\overline{a_{01}}(\alpha) & -\overline{a_{10}}(\alpha) & 1 \end{pmatrix} \tag{3.10}$$

and

$$\mathcal{B} = \begin{pmatrix} -a_{00} & 0 & 0 & -a_{11} \\ -a_{10}(\alpha) & 1 & 0 & -a_{01}(\alpha) \\ -\overline{a_{01}} & 0 & 1 & -\overline{a_{10}} \\ -\overline{a_{11}}(\alpha) & 0 & 0 & -\overline{a_{00}}(\alpha) \end{pmatrix}, \tag{3.11}$$

then we have

$$\Delta T\Phi = (\mathcal{A}P_+ + \mathcal{B}P_-)\Psi = S_{\mathcal{A},\mathcal{B}}\Delta\Phi,$$

which completes the proof of (3.7), since $\Phi = \pi\varphi$ is arbitrary.

The other relations stated in the theorem are easily verified by direct computation. □

Next we need to recall some results from the general theory of singular integral operators without shift (see [1] and [9]). We start with the notion of factorization, the main tool for studying singular integral operators with matrix coefficients.

Definition 3.3. A matrix function $\mathcal{C} \in L_\infty^{n\times n}(\mathbb{R})$ is said to admit a generalized (left) factorization in $L_2^n(\mathbb{R})$ if the following two conditions hold:

(i) It can be given the form

$$\mathcal{C} = \mathcal{C}_+ \Lambda \mathcal{C}_-$$

where

$$(t+i)^{-1}\mathcal{C}_+^{\pm 1} \in \widehat{L}_2^+(\mathbb{R}), (t-i)^{-1}\mathcal{C}_-^{\pm 1} \in \widehat{L}_2^-(\mathbb{R}),$$

$$\Lambda = \operatorname{diag}\{\theta^{\kappa_1}, \dots, \theta^{\kappa_n}\},$$

the integers $\kappa_j, j = 1, \dots, n$, with $\kappa_1 \geq \dots \geq \kappa_n$ are the partial indices of the factorization, θ is the function (2.5), $\widehat{L}_2^\pm$ are the spaces of the Fourier transforms of the $L_2(\mathbb{R})$ functions which vanish on $\mathbb{R}^-$ (plus sign) or on $\mathbb{R}^+$ (minus sign).

(ii) The linear operator D acting according to the rule $D\varphi = \mathcal{C}_+ P_- \mathcal{C}_+^{-1}\varphi$ is bounded in $L_2^n(\mathbb{R})$.

The numbers κ_j, $j = 1, \dots, n$, taken in non-deceasing order as in the above definition and then called partial indices of $\mathcal{C}$, are uniquely determined by this matrix-function and sometimes it is convenient to consider their number $\ell \leq n$ which are pairwise distinct and we will write $\Lambda = \operatorname{diag}\{t^{\kappa_1}\mathcal{I}_{d_1}, \dots, t^{\kappa_\ell}\mathcal{I}_{d_\ell}\}$, where d_i is the multiplicity of the partial index κ_i, $i = 1, \dots, \ell$.

Let $\mathcal{A}, \mathcal{B} \in L_\infty(\mathbb{R})^{n\times n}$ and consider in $L_2(\mathbb{R})^n$ the operator

$$S_{\mathcal{A},\mathcal{B}} = \mathcal{A}\, P_+ + \mathcal{B}\, P_-.$$

Theorem 3.4 ([9]). *The operator $S_{\mathcal{A},\mathcal{B}}$ is a Fredholm operator in $L_2^n(\mathbb{R})$ if and only if the following conditions hold:*

(i) $\mathcal{A}, \mathcal{B} \in \mathcal{G}L_\infty^{n\times n}(\mathbb{R})$ *(the group of invertible elements in $L_\infty^{n\times n}(\mathbb{R})$),*

(ii) $\mathcal{C} = \mathcal{A}^{-1}\,\mathcal{B}$ *admits a generalized factorization in $L_2^n(\mathbb{R})$.*

Moreover, in case $S_{\mathcal{A},\mathcal{B}}$ is a Fredholm operator in $L_2^n(\mathbb{R})$ its kernel can be derived from a factorization of $\mathcal{C} = \mathcal{A}^{-1}\,\mathcal{B}$, say $\mathcal{C} = \mathcal{C}_+\,\Lambda\,\mathcal{C}_-$, as follows:

$$\ker S_{\mathcal{A},\mathcal{B}} = \left\{\varphi \in L_2^n(\mathbb{R}) : \varphi = \left(\mathcal{C}_+ - \mathcal{C}_-^{-1}\Lambda^{-1}\right) r_+ v, v \in \mathfrak{P}\right\}, \tag{3.12}$$

where $r_+(t) = (t+i)^{-1}, t \in \mathbb{R}$,

$$\mathfrak{P} = \left\{v \in L_p^n(\mathbb{T}) : v_i = p_i(\theta), p_i \in P^{\kappa_i - 1} \text{ if } \kappa_i \in \mathbb{N} \text{ or } p_i \equiv 0 \text{ if } \kappa_i \leq 0\right\}, \tag{3.13}$$

and $P^{\kappa-1}$ denotes the space of all polynomials with degree at most $\kappa - 1 \in \mathbb{N}_0$.

Proposition 3.5. *Let $\mathcal{A}, \mathcal{B} \in \mathcal{G}L^{4\times 4}_{\infty}(\mathbb{R})$ be the matrix-functions* (3.10) *and* (3.11), *respectively. If*

$$\mathcal{C} = \mathcal{A}^{-1}\mathcal{B}, \tag{3.14}$$

admits a factorization in $L^4_2(\mathbb{R})$, say $\mathcal{C} = \mathcal{C}_+ \Lambda \mathcal{C}_-$ with $\Lambda = \operatorname{diag}\{\theta^{\kappa_1}\mathcal{I}_{d_1}, \ldots, \theta^{\kappa_\ell}\mathcal{I}_{d_\ell}\}$, where $\kappa_1 > \cdots > \kappa_\ell$ are the pairwise distinct partial indices of the factorization, then the outer factors $\mathcal{C}_+$ and $\mathcal{C}_-$ fulfill the following identities:

$$\mathcal{C}_+ = \begin{pmatrix} \mathcal{E} & 0 \\ 0 & \mathcal{E} \end{pmatrix} \mathcal{C}_-^{-1}(\alpha)\, \Lambda_+^{-1}\mathcal{H}_1, \tag{3.15}$$

$$\mathcal{C}_- = \Lambda^{-1}\mathcal{H}_1^{-1}\Lambda\,\Lambda_-^{-1}\mathcal{C}_+^{-1}(\alpha) \begin{pmatrix} \mathcal{E} & 0 \\ 0 & \mathcal{E} \end{pmatrix}, \tag{3.16}$$

and

$$\mathcal{C}_+ = \begin{pmatrix} 0 & \mathcal{I}_2 \\ \mathcal{I}_2 & 0 \end{pmatrix} \left(\overline{\mathcal{C}_-}\right)^{-1} \mathcal{H}_2, \tag{3.17}$$

$$\mathcal{C}_- = \Lambda^{-1}\mathcal{H}_2^{-1}\Lambda \left(\overline{\mathcal{C}_+}\right)^{-1} \begin{pmatrix} 0 & \mathcal{I}_2 \\ \mathcal{I}_2 & 0 \end{pmatrix}, \tag{3.18}$$

where $\Lambda_\pm = \operatorname{diag}\{\theta_\pm^{\kappa_1}\mathcal{I}_{d_1}, \ldots, \theta_\pm^{\kappa_\ell}\mathcal{I}_{d_\ell}\}$, $\theta_\pm$ are the outer factors of the factorization (2.6), *$\mathcal{E}$ denotes the matrix* (3.6), *$\mathcal{I}_2$ is a 2×2 identity matrix and $\mathcal{H}_i, i = 1, 2$ is a rational matrix-function, with an upper triangular block structure*

$$\mathcal{H}_i = \begin{pmatrix} \mathcal{H}_{i1} & * & * \\ 0 & \ddots & * \\ 0 & 0 & \mathcal{H}_{il} \end{pmatrix}, \tag{3.19}$$

where each block $\mathcal{H}_{ij}$, $i = 1, 2$, $j = 1, \ldots \ell$, in the main diagonal is a non-singular constant matrix of order d_j, the multiplicity of the partial index κ_j, and the possible nonzero entries, marked with $*$ above, have the form

$$h_{ms} = p_{ms}(\theta), \quad m < s,\ m, s = 1, \ldots, 4,$$

with p_{ms} polynomials of degree at most $\kappa_m - \kappa_s$.

Proof. Taking into account that the matrix-functions (3.10) and (3.11) satisfy (3.8) and (3.9), we can conclude that the matrix-function (3.14) enjoy the following equalities:

$$\mathcal{C} = \begin{pmatrix} \mathcal{E} & 0 \\ 0 & \mathcal{E} \end{pmatrix} \mathcal{C}^{-1}(\alpha) \begin{pmatrix} \mathcal{E} & 0 \\ 0 & \mathcal{E} \end{pmatrix} \tag{3.20}$$

and

$$\mathcal{C} = \begin{pmatrix} 0 & \mathcal{I}_2 \\ \mathcal{I}_2 & 0 \end{pmatrix} \left(\overline{\mathcal{C}}\right)^{-1} \begin{pmatrix} 0 & \mathcal{I}_2 \\ \mathcal{I}_2 & 0 \end{pmatrix}. \tag{3.21}$$

Since $\Lambda(\alpha) = \Lambda_-\Lambda^{-1}\Lambda_+$, we have

$$\mathcal{C}(\alpha) = \mathcal{C}_+(\alpha)\Lambda_-\Lambda^{-1}\Lambda_+\mathcal{C}_-(\alpha).$$

Thus, according to the last equality and (3.20), we have

$$\mathcal{C} = \left[\begin{pmatrix} \mathcal{E} & 0 \\ 0 & \mathcal{E} \end{pmatrix} \mathcal{C}_-^{-1}(\alpha)\, \Lambda_+^{-1}\right] \Lambda \left[\Lambda_-^{-1}\mathcal{C}_+^{-1}(\alpha) \begin{pmatrix} \mathcal{E} & 0 \\ 0 & \mathcal{E} \end{pmatrix}\right].$$

So, we obtain another factorization of $\mathcal{C}$ in $L_2(\mathbb{R})$. It is very well known (see, for instance, [1] and [9]) that exist a matrix-function $\mathcal{H}_1$ with upper block triangular structure (3.19), such that

$$\mathcal{C}_+ = \begin{pmatrix} \mathcal{E} & 0 \\ 0 & \mathcal{E} \end{pmatrix} \mathcal{C}_-^{-1}(\alpha)\,\Lambda_-^{-1}\mathcal{H}_1, \quad \mathcal{C}_- = \Lambda^{-1}\mathcal{H}_1^{-1}\Lambda\Lambda_+^{-1}\mathcal{C}_+^{-1}(\alpha) \begin{pmatrix} \mathcal{E} & 0 \\ 0 & \mathcal{E} \end{pmatrix}.$$

On the other hand, as $\overline{\theta}(t) = \theta^{-1}(t), t \in \mathbb{R}$, then

$$\left(\overline{\mathcal{C}}\right)^{-1} = \left(\overline{\mathcal{C}_-}\right)^{-1} \Lambda \left(\overline{\mathcal{C}_+}\right)^{-1}. \tag{3.22}$$

Thus, using the last equality and (3.21), we have

$$\mathcal{C} = \left[\begin{pmatrix} 0 & \mathcal{I}_2 \\ \mathcal{I}_2 & 0 \end{pmatrix} \left(\overline{\mathcal{C}_-}\right)^{-1}\right] \Lambda \left[\left(\overline{\mathcal{C}_+}\right)^{-1} \begin{pmatrix} 0 & \mathcal{I}_2 \\ \mathcal{I}_2 & 0 \end{pmatrix}\right].$$

Similarly, we get yet another factorization of $\mathcal{C}$ in $L_2^4(\mathbb{R})$ and, therefore, there exists a matrix-function $\mathcal{H}_2$ such that

$$\mathcal{C}_+ = \begin{pmatrix} 0 & \mathcal{I}_2 \\ \mathcal{I}_2 & 0 \end{pmatrix} \left(\overline{\mathcal{C}_-}\right)^{-1} \mathcal{H}_2, \quad \mathcal{C}_- = \Lambda^{-1}\mathcal{H}_2^{-1}\Lambda \left(\overline{\mathcal{C}_+}\right)^{-1} \begin{pmatrix} 0 & \mathcal{I}_2 \\ \mathcal{I}_2 & 0 \end{pmatrix},$$

where $\mathcal{H}_2$ has the structure (3.19). □

Proposition 3.6. *Under the conditions of the previous proposition, the matrix $\mathcal{H}_1$ satisfies the equality*

$$\mathcal{H}_1(\alpha) = \Lambda_-\Lambda^{-1}\mathcal{H}_1^{-1}\Lambda\Lambda_-^{-1}, \tag{3.23}$$

and for each block in the main diagonal $\mathcal{H}_{1j}$, $j = 1, \dots \ell$, there holds

$$\mathcal{H}_{1j}^2 = \mathcal{I}_{d_j}. \tag{3.24}$$

Proof. From (3.15) we have

$$\mathcal{H}_1(\alpha) = \Lambda_+(\alpha)\,\mathcal{C}_- \begin{pmatrix} \mathcal{E} & 0 \\ 0 & \mathcal{E} \end{pmatrix} \mathcal{C}_+(\alpha).$$

Using (3.16) and $\Lambda_+(\alpha) = \Lambda_-$ we obtain (3.23), with the result

$$\Lambda_-\Lambda^{-1}\mathcal{H}_1\Lambda\Lambda_-^{-1}\mathcal{H}_1(\alpha) = \mathcal{I}_4. \tag{3.25}$$

Then, taking into account the $\mathcal{H}_1$ block structure, the previous equality can be written in the form:

$$\begin{pmatrix} \mathcal{H}_{11}^2 & * & * \\ 0 & \ddots & * \\ 0 & 0 & \mathcal{H}_{1\ell}^2 \end{pmatrix} = \mathcal{I}_4,$$

from which it follows that $\mathcal{H}_{1j}^2 = \mathcal{I}_{d_j}$, $j = 1 \dots, \ell$. □

Proposition 3.7. *Under the conditions of the Proposition 3.5, the matrix $\mathcal{H}_2$ satisfies the equality*

$$\overline{\mathcal{H}_2} = \Lambda^{-1}\mathcal{H}_2^{-1}\Lambda,$$

and for each block in the main diagonal $\mathcal{H}_{2j}$, $j = 1, \dots \ell$, there holds

$$\overline{\mathcal{H}_{2j}}\mathcal{H}_{2j} = \mathcal{I}_{d_j}. \tag{3.26}$$

Proof. From (3.17), we have

$$\overline{\mathcal{C}_+} = \begin{pmatrix} 0 & \mathcal{I}_2 \\ \mathcal{I}_2 & 0 \end{pmatrix} \mathcal{C}_-^{-1}\overline{\mathcal{H}_2}$$

and, from (3.18), it follows that

$$\overline{\mathcal{C}_+} = \begin{pmatrix} 0 & \mathcal{I}_2 \\ \mathcal{I}_2 & 0 \end{pmatrix} \mathcal{C}_-^{-1}\Lambda^{-1}\mathcal{H}_2^{-1}\Lambda.$$

Hence, $\overline{\mathcal{H}_2} = \Lambda^{-1}\mathcal{H}_2^{-1}\Lambda$ and, using the block structure of $\mathcal{H}_2$, we may conclude that

$$\begin{pmatrix} \overline{\mathcal{H}_{21}}\mathcal{H}_{21} & * & * \\ 0 & \ddots & * \\ 0 & 0 & \overline{\mathcal{H}_{2\ell}}\mathcal{H}_{2\ell} \end{pmatrix} = \mathcal{I}_4,$$

from which the second assertion follows. □

Now we can state the main result of the paper. To its formulation it is convenient to introduce the following notation. For $\kappa \in \mathbb{N}$ let $\widetilde{P}^{\kappa-1}$ denote the linear space of all polynomials with degree at most $\kappa - 1$, over the field of real numbers; if $\kappa \in \mathbb{Z} \setminus \mathbb{N}$, set $\widetilde{P}^{\kappa-1} = \{0\}$. For $\kappa_1, \dots, \kappa_4 \in \mathbb{Z}$ set

$$\widetilde{\mathfrak{P}} = \{v \in \widetilde{L}_2^4(\mathbb{R}) : v_i = p_i(\theta),\ p_i \in \widetilde{P}^{\kappa_i - 1},\ i = 1, \dots, 4\}.$$

Theorem 3.8. *Let T_A be the singular integral operator* (3.1). *If the matrix-functions* (3.10) *and* (3.11) *are invertible in $L_\infty^{4\times 4}(\mathbb{R})$ and $\mathcal{C} = \mathcal{A}^{-1}\mathcal{B}$ admits a generalized factorization in $L_2^4(\mathbb{R})$, say $\mathcal{C} = \mathcal{C}_+\Lambda\,\mathcal{C}_-$ with $\Lambda = \operatorname{diag}\{\theta^{\kappa_1}, \dots, \theta^{\kappa_4}\}$, then*

$$\dim\ker T_A = \dim\left(\ker Q_1 \mid_{\widetilde{\mathfrak{P}}} \cap \ker Q_2 \mid_{\widetilde{\mathfrak{P}}}\right),$$

$$\ker T_A = \left\{\pi^{-1}(\Phi) : \Phi = \Delta^{-1}\left(\mathcal{C}_+ - \mathcal{C}_-^{-1}\Lambda^{-1}\right) r_+ v\right\},$$

where $r_+(t) = (t+i)^{-1}$, $v \in \ker Q_1 \mid_{\widetilde{\mathfrak{P}}} \cap \ker Q_2 \mid_{\widetilde{\mathfrak{P}}}$, and Q_i, $i = 1, 2$, are the linear operators

$$\begin{aligned} Q_1 &= \frac{1}{2}(I - \theta\theta_-^{-1}\mathcal{H}_1^{-1}\Lambda\Lambda_-^{-1}U), \\ Q_2 &= \frac{1}{2}(I + \theta^{-1}\mathcal{H}_2^{-1}\Lambda C), \end{aligned}$$

Δ is the invertible operator (3.4), *π is the map* (3.3) *and $\mathcal{H}_i$, $i = 1, 2$, is the rational matrix-function* (3.19).

Proof. If $\varphi \in \ker T_A$, then, taking into account (3.7), Theorem 3.4 and (3.12), we have

$$\Psi = \Delta\pi(\varphi) = \left(\mathcal{C}_+ - \mathcal{C}_-^{-1}\Lambda^{-1}\right) r_+ v,$$

$r_+ = (t+i)^{-1}$ and $v \in \widetilde{\mathfrak{P}}$. It is clear that

$$\Psi_+ = P_+\Psi = \mathcal{C}_+ r_+ v, \quad \Psi_- = P_-\Psi = -\mathcal{C}_-^{-1}\Lambda^{-1} r_+ v.$$

Since $\Psi \in \widetilde{\mathfrak{E}}$, it must satisfy the characterization conditions (3.5), whence

$$P_\pm U\Psi = \begin{pmatrix} \mathcal{E} & 0 \\ 0 & \mathcal{E} \end{pmatrix} \Psi_\pm, \quad P_\pm C\Psi = \begin{pmatrix} 0 & \mathcal{I} \\ \mathcal{I} & 0 \end{pmatrix} \Psi_\pm,$$

and, from $P_\pm U = UP_\mp$, $CP_\pm = P_\mp C$ (see 2.3), we get

$$U\Psi_\pm = \begin{pmatrix} \mathcal{E} & 0 \\ 0 & \mathcal{E} \end{pmatrix} \Psi_\mp, \quad C\Psi_\pm = \begin{pmatrix} 0 & \mathcal{I} \\ \mathcal{I} & 0 \end{pmatrix} \Psi_\mp.$$

The last two equalities mean that

$$\mathcal{C}_+(\alpha) r_+(\alpha) U v = -\begin{pmatrix} \mathcal{E} & 0 \\ 0 & \mathcal{E} \end{pmatrix} \mathcal{C}_-^{-1}\Lambda^{-1} r_+ v,$$

$$\mathcal{C}_-^{-1}(\alpha)\Lambda^{-1}(\alpha) r_+(\alpha) U v = -\begin{pmatrix} \mathcal{E} & 0 \\ 0 & \mathcal{E} \end{pmatrix} \mathcal{C}_+ r_+ v,$$

$$\overline{\mathcal{C}_+} C r_+ v = -\begin{pmatrix} 0 & \mathcal{I} \\ \mathcal{I} & 0 \end{pmatrix} \mathcal{C}_-^{-1}\Lambda^{-1} r_+ v,$$

$$(\overline{\mathcal{C}_-})^{-1}(\overline{\Lambda})^{-1} C r_+ v = -\begin{pmatrix} 0 & \mathcal{I} \\ \mathcal{I} & 0 \end{pmatrix} \mathcal{C}_+ r_+ v.$$

Since $r_+(\alpha) = -r_+\theta\theta_-^{-1}$ and using (3.15) and (3.16) we conclude that the first two previous equalities imply that

$$\theta\theta_-^{-1}\mathcal{H}_1^{-1}\Lambda\Lambda_-^{-1} U v = v.$$

Using (3.17), (3.18) and $\overline{r_+} r_+^{-1} = \theta^{-1}, \theta = (\overline{\theta})^{-1}$ we conclude that the last two previous equalities imply that

$$\theta^{-1}\mathcal{H}_2^{-1}\Lambda C v = -v,$$

it follows that $Q_i v = 0$, $i = 1, 2$.

Clearly the above reasoning can be reversed and, thus, if $v \in \ker Q_i|_{v \in \widetilde{\mathfrak{P}}}$, then the function $\varphi = \pi^{-1}\Delta^{-1}(\mathcal{C}_+ - \mathcal{C}_-^{-1}\Lambda^{-1}) r_+ v$ belongs to $\ker T_A$, which completes the proof. □

4. About the factorization of the matrix-function $\mathcal{C}$

In the last section was shown that the solution of the generalized BVP (1.1) is closely related to the factorization of the matrix-function $\mathcal{C}$ in (3.14). In this section we study this factorization problem, which still open in general. However, in some particular cases it is possible to derive some result concerning the factorization of $\mathcal{C}$.

Let $\mathcal{K}$ be the constant matrix-function

$$\mathcal{K} = \begin{pmatrix} 1 & 0 & 0 & 0 \\ 0 & 0 & 0 & 1 \\ 0 & 0 & 1 & 0 \\ 0 & 1 & 0 & 0 \end{pmatrix},$$

then (3.10) can be written as the following product:

$$\mathcal{A} = \mathcal{K}\begin{pmatrix} \mathcal{I}_2 & \mathcal{A}_{12} \\ 0 & \mathcal{A}_{22} \end{pmatrix}\mathcal{K}, \tag{4.27}$$

where

$$\mathcal{A}_{12} = -\begin{pmatrix} a_{01} & a_{10} \\ \overline{a_{10}}(\alpha) & \overline{a_{01}}(\alpha) \end{pmatrix}, \quad \mathcal{A}_{22} = -\begin{pmatrix} \overline{a_{00}} & \overline{a_{11}} \\ a_{11}(\alpha) & a_{00}(\alpha) \end{pmatrix}.$$

Thus the matrix-function (3.14) admits the following representation:

$$\mathcal{C} = \mathcal{K}\begin{pmatrix} \mathcal{I}_2 & -\mathcal{A}_{12} \\ 0 & \mathcal{I}_2 \end{pmatrix}\begin{pmatrix} \overline{\mathcal{A}_{22}} & 0 \\ 0 & \mathcal{A}_{22}^{-1} \end{pmatrix}\begin{pmatrix} \mathcal{I}_2 & 0 \\ \overline{\mathcal{A}_{12}} & \mathcal{I}_2 \end{pmatrix}\mathcal{K}. \tag{4.28}$$

If $\mathcal{A}_{22}^{-1}$ admits a generalized factorization in $L_2^2(\mathbb{R})$, say

$$\mathcal{A}_{22}^{-1} = \mathcal{D}_+\mathcal{L}\mathcal{D}_-, \tag{4.29}$$

introducing the notation

$$\mathcal{A}_{12} = \mathcal{F}_+ + \mathcal{F}_-, \quad \mathcal{F}_\pm = P_\pm\mathcal{A}_{12},$$

and

$$\mathcal{M} = \overline{\mathcal{D}_-}\mathcal{F}_-\mathcal{D}_+,$$

it is possible to rewrite (4.28) as follows,

$$\mathcal{C} = \mathcal{R}_+\begin{pmatrix} \mathcal{I}_2 & -\mathcal{M} \\ 0 & \mathcal{I}_2 \end{pmatrix}\begin{pmatrix} \overline{\mathcal{L}} & 0 \\ 0 & \mathcal{L} \end{pmatrix}\begin{pmatrix} \mathcal{I}_2 & 0 \\ \overline{\mathcal{M}} & \mathcal{I}_2 \end{pmatrix}\mathcal{R}_-,$$

where

$$\mathcal{R}_+ = \mathcal{K}\begin{pmatrix} \mathcal{I}_2 & -\mathcal{F}_+ \\ 0 & \mathcal{I}_2 \end{pmatrix}\begin{pmatrix} (\overline{\mathcal{D}_-})^{-1} & 0 \\ 0 & \mathcal{D}_+ \end{pmatrix},$$

$$\mathcal{R}_- = \begin{pmatrix} (\overline{\mathcal{D}_+})^{-1} & 0 \\ 0 & \mathcal{D}_- \end{pmatrix}\begin{pmatrix} \mathcal{I}_2 & 0 \\ \overline{\mathcal{F}_+} & \mathcal{I}_2 \end{pmatrix}\mathcal{K}.$$

Finally, if

$$\mathcal{M} = \mathcal{M}_+ + \mathcal{M}_-, \quad \mathcal{M}_\pm = P_\pm\mathcal{M} \tag{4.30}$$

then

$$\mathcal{C} = \mathcal{G}_+\begin{pmatrix} \mathcal{I}_2 & -\mathcal{M}_- \\ 0 & \mathcal{I}_2 \end{pmatrix}\begin{pmatrix} \overline{\mathcal{L}} & 0 \\ 0 & \mathcal{L} \end{pmatrix}\begin{pmatrix} \mathcal{I}_2 & 0 \\ \overline{\mathcal{M}_-} & \mathcal{I}_2 \end{pmatrix}\mathcal{G}_-,$$

with

$$\mathcal{G}_+ = \mathcal{R}_+\begin{pmatrix} \mathcal{I}_2 & -\mathcal{M}_+ \\ 0 & \mathcal{I}_2 \end{pmatrix}, \quad \mathcal{G}_- = \begin{pmatrix} \mathcal{I}_2 & 0 \\ \overline{\mathcal{M}_+} & \mathcal{I}_2 \end{pmatrix}\mathcal{R}_-.$$

Now we are in conditions to state some results about the factorization of the matrix function $\mathcal{C}$ in (3.14).

Proposition 4.1. *Let $\mathcal{A}_{12}$ and $\mathcal{A}_{22}$ be the blocks in the equality* (4.27).

If $\mathcal{F}_- = P_-\mathcal{A}_{12} \equiv 0$ and (4.29) *is a generalized factorization of $\mathcal{A}_{22}^{-1}$ in $L_2^2(\mathbb{R})$, then*

$$\mathcal{C} = \mathcal{G}_+ \begin{pmatrix} \mathcal{L} & 0 \\ 0 & \mathcal{L} \end{pmatrix} \mathcal{G}_-,$$

where $\mathcal{L} = \operatorname{diag}(\theta^{\lambda_1}, \theta^{\lambda_2})$, $\lambda_1 \geq \lambda_2$ and the partial indices of $\mathcal{C}$ are the integers

$$\kappa_1 = \kappa_2 = \lambda_1, \quad \kappa_3 = \kappa_4 = \lambda_2.$$

Proposition 4.2. *If the matrix-function $\mathcal{A}_{22}^{-1}$ admits a generalized factorization* (4.29) *with both partial indices equal to λ, then*

$$\mathcal{C} = \theta^\lambda \mathcal{G}_+ \mathcal{N} \mathcal{G}_-,$$

where

$$\mathcal{N} = \begin{pmatrix} \mathcal{I}_2 & -\mathcal{M}_- \\ 0 & \mathcal{I}_2 \end{pmatrix} \begin{pmatrix} \mathcal{I}_2 & 0 \\ \overline{\mathcal{M}_-} & \mathcal{I}_2 \end{pmatrix},$$

and $\mathcal{M}_-$ is the matrix-function (4.30), *which additionally must be symmetric.*

The partial indices of $\mathcal{C}$ are given by

$$\kappa_1 = \mu_1 + \lambda, \quad \kappa_2 = \mu_2 + \lambda, \quad \kappa_3 = -\mu_2 + \lambda, \quad \kappa_4 = -\mu_1 + \lambda,$$

where $\mu_1, \mu_2, -\mu_2, -\mu_1$ are the partial indices of $\mathcal{N}$.

Proof. It is clear that we can pass from a factorization of $\mathcal{C}$ to a factorization of $\mathcal{N}$ and vice versa. Thus, the factorization problem under consideration can be reduced to the factorization of the Hermitian matrix-function

$$\begin{pmatrix} 0 & \mathcal{I}_2 \\ \mathcal{I}_2 & 0 \end{pmatrix} \mathcal{N} \begin{pmatrix} 0 & -\mathcal{I}_2 \\ \mathcal{I}_2 & 0 \end{pmatrix}.$$

According to [9], p. 258, the partial indices of $\mathcal{N}$, $\mu_j, j = 1, \ldots, 4$, are related by the equalities

$$\mu_1 + \mu_4 = 0, \quad \mu_2 + \mu_3 = 0.$$

□

For deeper results about the factorization of the matrix-function $\mathcal{N}$, the reader can consult [2, 9].

Above we mentioned some cases for which a factorization of $\mathcal{C}$ can be derived from a known factorization of $\mathcal{A}_{22}^{-1}$. The factorization problem for $\mathcal{A}_{22}^{-1}$ can also be handled, again not with full generality but in particular cases. Below we mention one of them, that was already considered by two of the authors together with V.G. Kravchenko in a former publication [3]. If the block $\mathcal{A}_{22}$ is an invertible matrix function, then $\mathcal{A}_{22}^{-1}$ can be written in the form:

$$\mathcal{A}_{22}^{-1} = \begin{pmatrix} f & g \\ \overline{g}(\alpha) & \overline{f}(\alpha) \end{pmatrix}.$$

We assume additionally that the functions f and g are connected by the relation:

$$f - g = \overline{f}(\alpha) - \overline{g}(\alpha).$$

Put $h = -g(f-g)^{-1}$. Then $\mathcal{A}_{22}^{-1} = (f-g)\mathcal{H}$ with

$$\mathcal{H} = \begin{pmatrix} 1-h & -h \\ -\overline{h}(\alpha) & 1-\overline{h}(\alpha) \end{pmatrix}. \tag{4.31}$$

So we are left with the factorization problem for $\mathcal{H}$. Concerning this problem here we just give the statement of a proposition. The proof of it and the explicit formulas for the factors can be obtained following the lines of the proof of Theorem 3.6 in [3].

Proposition 4.3. *Let $\mathcal{H}$ be the matrix function* (4.31). *Further, let $\mathrm{d} = \mathrm{d}_+\theta^\kappa \mathrm{d}_-$ be a factorization of $\mathrm{d} = \det \mathcal{H}$, let $\mathrm{c} = (a - \overline{a}(\alpha))/2$, decompose it, $\mathrm{c} = \mathrm{c}_+ + \mathrm{c}_-$ with $\mathrm{c}_\pm = P_\pm \mathrm{c}$, and set $\mathrm{e}_\pm = 2P_\pm(\mathrm{c}_+\mathrm{d}_-^{-1})$. Then, the factorization of $\mathcal{H}$ is equivalent to the factorization of*

$$\Lambda\mathcal{E}_+ = \begin{pmatrix} \theta^\kappa & 0 \\ 0 & 1 \end{pmatrix} \begin{pmatrix} 1 & 0 \\ e_+ & 1 \end{pmatrix}.$$

The sense of equivalence of factorization mentioned here means that from a factorization of $\mathcal{H}$ we can obtain a factorization of $\Lambda\mathcal{E}_+$, vice versa and, moreover, the partial indices of the two matrix functions coincide.

References

[1] K. Clancey and I. Gohberg. *Factorization of Matrix Functions and Singular Integral Operators*, vol. 3 of *Operator Theory: Advances and Applications*, Birkhäuser, 1981.

[2] A.C. Conceição, V.G. Kravchenko and J.C. Pereira. *Factorization algorithm for some special non-rational matrix functions*, *Operator Theory: Advances and Applications* 202, 87–109, Birkhäuser, 2012.

[3] V.G. Kravchenko, A.B. Lebre, and J.S. Rodríguez. *Factorization of singular integral operators with a Carleman shift and spectral problems*, *J. Integral Equations Appl.* 13(4): 339–383, 2001.

[4] V.G. Kravchenko, A.B. Lebre, and J.S. Rodríguez. *The kernel of singular integral operator with a finite group of liner-fractional shifts*, *Operator Theory* 20, Theta: 143–156, 2006.

[5] V.G. Kravchenko and G.S. Litvinchuk. *Introduction to the Theory of Singular Integral Operators with shift*, Kluwer Academic Publishers, 1994.

[6] V.G. Kravchenko, R.C. Marreiros and J.S. Rodríguez. *On an estimate for the number of solutions of the generalized Riemann boundary value problem with shift*, Conference on Differential and Difference Equations and Applications, 605–614, Hindawi Publishing Corporation, 2006.

[7] V.G. Kravchenko, R.C. Marreiros and J.S. Rodríguez. *An estimate for the number of solutions of a homogeneous generalized Riemann boundary value problem with shift*, *Operator Theory: Advances and Applications* 220, 163–178, Birkhäuser, 2012.

[8] G.S. Litvinchuk. *Solvability Theory of Boundary Value Problems and Singular Integral Equations with Shift*, vol. 523 of *Mathematics and its Applications*, Kluwer Academic Publishers, 2000.

[9] G.S. Litvinchuk and I.M. Spitkovskii. *Factorization of Measurable Matrix Functions*, vol. 25 of *Operator Theory: Advances and Applications*, Birkhäuser, 1987.

[10] I.N. Vekua. *Generalized Analytics Functions*, Nauka, 1988.

L.F. Campos
Escola Secundária de Loulé
Avenida Laginha Serafim
P-8100-740 Loulé, Portugal
e-mail: linacampos1@hotmail.com

A.B. Lebre
Departamento de Matemática
Instituto Superior Técnico
Av. Rovisco Pais
P-1049-001 Lisboa, Portugal
e-mail: alebre@math.ist.utl.pt

J.S. Rodríguez
Departamento de Matemática, F.C.T.
Universidade do Algarve
Campus de Gambelas
P-8000-810 Faro, Portugal
e-mail: jsanchez@ualg.pt

Operator Theory:
Advances and Applications, Vol. 236, 117–131

Generalized Extremal Vectors and Some New Properties

Gilles Cassier and Jérôme Verliat

Abstract. Extremal vectors were introduced by S. Ansari and P. Enflo in [2], this method produced new and more constructive proofs of existence of invariant subspaces. In this paper, our purpose is to introduce generalized extremal vectors and to study their properties. We firstly check that general properties of extremal vectors also hold for generalized extremal vectors. We give a new useful characterization of generalized extremal vectors. We show that there exist relationships between these vectors and the famous Moore–Penrose pseudo-inverse showing their intrinsic nature. Applications to weighted shift operators are given. In particular, we discuss for quasinilpotent backward weighted shifts the following question: Can the Ansari–Enflo method be used in order to obtain all hyper-invariant subspaces?

Mathematics Subject Classification (2010). Primary 47A15, 47B37; Secondary 47A50, 47S30.

Keywords. Extremal vectors, Moore–Penrose pseudo-inverse, weighted shifts, hyper-invariant subspaces.

1. Introduction

Given any operator T acting on a separable infinite-dimensional complex Hilbert space $\mathcal{H}$, it is quite natural to estimate its surjectivity default. This purpose, in the particular case of operators with dense range, is the starting point of Ansari–Enflo theory [2].

The method consists to obtain the best approximate solution of the equation $Ty = x$. Applying to the powers of T, this process allows Ansari and Enflo to produce hyper-invariant subspace results. Afterward, many authors took up the theory and made important investigations in the invariant subspace problem for some classes of operators (see for instance [5]).

Consequently, the basic point of the theory, that is extremal vectors, deserves to be clarified. The aim of this paper is to extend the notion of extremal vector to any arbitrary operator and to study these generalized extremal vectors. We give a new characterization, which is useful for concrete calculus. We notice a link between Ansari–Enflo theory and the Moore–Penrose pseudo-inverse which emphasize that this method is constructive in other areas than the existence of hyper-invariant subspaces. Finally, we study operator weighted shifts and we discuss the Ansari–Enflo method for unilateral weighted shifts.

2. Generalization of extremal vectors definition for arbitrary operators

Lemma 1. *Let T be any operator of $\mathscr{B}(\mathcal{H})$. Let x be a vector of $\mathcal{H}\backslash\mathcal{N}(T^*)$ and ε be a scalar in $]\,\mathrm{d}(x;\mathcal{R}(T));\|x\|[$. Thus there exists a unique vector y with the smallest norm such that*

$$\|Ty - x\| \leqslant \varepsilon.$$

Proof. The set $\mathcal{F} = \{z \in \mathcal{H}; \|Tz - x\| \leqslant \varepsilon\}$ is a convex closed subspace of $\mathcal{H}$. Moreover, it is not empty: indeed, suppose $\mathcal{F} = \varnothing$. Then $\forall z \in \mathcal{H}$, $\|Tz - x\| > \varepsilon$. It implies that $\mathrm{d}(x;\mathcal{R}(T)) \geq \varepsilon$, which is absurd. Consequently, the projection theorem in Hilbert spaces claims the existence and unicity of y. □

Definition 2. Such a vector y is called extremal (or minimal) vector associated with $(T;x;\varepsilon)$. It will be denoted by $y_{T,x,\varepsilon}$ in the sequel of the paper.

We can notice that the Banach structure cannot assure the unicity of y: consider $\mathcal{H} = \ell^\infty(\mathbb{R}^2)$, $T = \mathrm{Id}$ and $x = (1;0)$; then the set $\mathcal{F}$ in the previous proof is the segment $[a;b]$ where $a = \left(\frac{1}{2};-\frac{1}{2}\right)$ and $b = \left(\frac{1}{2};\frac{1}{2}\right)$.

To find the previous vector y amounts to a minimization problem on the ball centered in x and with radius ε. Actually, we can only consider vectors of $(\mathcal{N}(T))^\perp$ such that their image by T belongs to the sphere corresponding to that minimization.

Remark 3. Let T be any operator of $\mathscr{B}(\mathcal{H})$. Let x be a vector of $\mathcal{H}\backslash\mathcal{N}(T^*)$ and ε be a scalar in $]\,\mathrm{d}(x;\mathcal{R}(T));\|x\|[$. The extremal vector associated with $(T;x;\varepsilon)$ is the vector y in $(\mathcal{N}(T))^\perp$ with the smallest norm such that

$$\|Ty - x\| = \varepsilon.$$

Proof. Let $y = y_{T,x,\varepsilon}$. Suppose that $\|Ty - x\| < \varepsilon$. On the one hand, we can write $\alpha = \varepsilon - \|Ty - x\| > 0$. On the other hand, there exists a real number $\delta > 0$ such that $\mathcal{B}(y,\delta] \subset \mathcal{F}$. Indeed, every vector z belonging to $\mathcal{B}(y,\delta]$ satisfies: $\|Tz - x\| \leqslant \|T(z-y)\| + \|Ty - x\| \leqslant \|T\|\delta + \varepsilon - \alpha \leq \varepsilon$ for an appropriate choice of $\delta > 0$ provided that $\delta \leqslant \frac{\alpha}{\|T\|}$ holds. Consequently, $\left(1 - \frac{\delta}{2}\right)y$ has a smaller norm than y, which is impossible.

Write $y = y' + y''$ with respect to the direct sum $\mathcal{H} = \mathcal{N}(T) \oplus (\mathcal{N}(T))^\perp$. Then $Ty = Ty''$. By $\|Ty'' - x\| = \|Ty - x\| = \varepsilon$, $\|y''\| \leq \|y\|$ and the unicity of y, we can affirm that $y = y''$ which means $y \in (\mathcal{N}(T))^\perp$. The proof is complete. □

The collinearity between the two vectors $T^*(Ty_{T,x,\varepsilon} - x)$ and $y_{T,x,\varepsilon}$ (see [2]) can be easily extended for generalized extremal vectors.

Lemma 4. *Let T be any operator of $\mathscr{B}(\mathcal{H})$. Let x be a vector of $\mathcal{H}\backslash\mathcal{N}(T^*)$ and ε be a scalar in $]\,\mathrm{d}(x;\mathcal{R}(T));\|x\|[$. There exists a positive number c such that*

$$T^*(Ty_{T,x,\varepsilon} - x) = -cy_{T,x,\varepsilon}. \tag{1}$$

Given any $(T;x;\varepsilon) \in \mathscr{B}(\mathcal{H})\times\mathcal{H}\times\mathbb{R}$ such that $x \in \mathcal{H}\backslash\mathcal{N}(T^*)$ and $\mathrm{d}(x;\mathcal{R}(T)) < \varepsilon < \|x\|$, the previous scalar c is well defined and unique.

Definition 5. The previous scalar c is called collinearity coefficient associated with $(T;x;\varepsilon)$. It will be denoted by $c_{T,x,\varepsilon}$ in the sequel.

For now, we propose a new reduction of the minimization place. We are only satisfied with the minimization on a kind of cap of the sphere $\partial\mathcal{B}(x;\varepsilon)$. A similar proof as the one exposed in [11] allows us to state the following result.

Proposition 6. *Let $T \in \mathscr{B}(\mathcal{H})$, $x \in \mathcal{H}\backslash\mathcal{N}(T^*)$ and $\varepsilon \in]\,\mathrm{d}(x;\mathcal{R}(T));\|x\|[$. The minimal vector $y_{T,x,\varepsilon}$ is the vector of smallest norm such that $Ty_{T,x,\varepsilon}$ belongs to*

$$\mathcal{V}_{x,\varepsilon} = \partial\mathcal{B}(x;\varepsilon) \cap \mathcal{B}\big(0;\sqrt{\|x\|^2 - \varepsilon^2}\big[,$$

where $\partial\mathcal{B}(a;r)$ and $\mathcal{B}\big(a;r\big[$ denotes respectively the sphere and the open ball centered at a with radius r.

We notice the next result, which gives an optimal bound for $c_{T,x,\varepsilon}$.

Proposition 7. *Let $(T;x;\varepsilon) \in \mathscr{B}(\mathcal{H}) \times \mathcal{H} \times \mathbb{R}$ such that $x \in \mathcal{H}\backslash\mathcal{N}(T^*)$ and $\mathrm{d}(x;\mathcal{R}(T)) < \varepsilon < \|x\|$. Then we have*

$$c_{T,x,\varepsilon} \leqslant \frac{\varepsilon\|T\|^2}{\|x\| - \varepsilon}.$$

Moreover this inequality is sharp.

Proof. Using Lemma 4, $c_{T,x,\varepsilon}$ satisfies

$$\langle T^*(Ty_{T,x,\varepsilon} - x)|T^*Ty_{T,x,\varepsilon}\rangle = -c_{T,x,\varepsilon}\langle y_{T,x,\varepsilon}|T^*Ty_{T,x,\varepsilon}\rangle = -c_{T,x,\varepsilon}\|Ty_{T,x,\varepsilon}\|^2$$

and we deduce

$$c_{T,x,\varepsilon} = \frac{|\langle T^*(Ty_{T,x,\varepsilon} - x)|T^*Ty_{T,x,\varepsilon}\rangle|}{\|Ty_{T,x,\varepsilon}\|^2}.$$

However, according to Cauchy–Schwarz inequality, we get

$$\begin{aligned}|\langle T^*(Ty_{T,x,\varepsilon} - x)|T^*Ty_{T,x,\varepsilon}\rangle| &\leqslant \|T^*(Ty_{T,x,\varepsilon} - x)\|\|T^*Ty_{T,x,\varepsilon}\| \\ &\leqslant \|T^*\|\|Ty_{T,x,\varepsilon} - x\|\|T^*\|\|Ty_{T,x,\varepsilon}\| \leqslant \varepsilon\|T\|^2\|Ty_{T,x,\varepsilon}\|.\end{aligned}$$

Since $y_{T,x,\varepsilon} \in \mathcal{V}_{x,\varepsilon}$, we have $\|Ty_{T,x,\varepsilon}\|^2 \geqslant (\|x\| - \varepsilon)^2$. It implies $1/\|Ty_{T,x,\varepsilon}\| \leqslant 1/(\|x\| - \varepsilon)$ and $c_{T,x,\varepsilon} \leqslant \varepsilon\|T\|^2\|Ty_{T,x,\varepsilon\|} \leqslant \varepsilon\|T\|^2/(\|x\| - \varepsilon)$.

Let us consider the classical bilateral shift S acting on $\ell^2(\mathbb{Z})$ and $x = (\delta_{0,n})_{n\in\mathbb{Z}}$ (where $\delta_{i,j}$ denotes the Kronecker symbol). Using Proposition 21, we see that $c_{T,x,\varepsilon} = \varepsilon/(1 - \varepsilon) = \varepsilon\|S\|^2/(\|x\| - \varepsilon)$. It clearly shows that this inequality is sharp. □

Corollary 8. *Let* $(T;x;\varepsilon) \in \mathscr{B}(\mathcal{H}) \times \mathcal{H} \times \mathbb{R}$ *such that* $x \in \mathcal{H}\backslash\overline{(\bigcup_{n\geqslant 1}\mathcal{N}(T^{*n}))}$ *and* ε *is a real number in the open interval* $]\lim_{n\longrightarrow\infty} \mathrm{d}(x;\mathcal{R}(T^n)); \|x\|[$. *Then we have*

$$\overline{\lim}\ (c_{T^n,x,\varepsilon})^{\frac{1}{n}} \leqslant r(T)^2,$$

where $r(T)$ *denotes the spectral radius of* T. *In particular, when* T *is a quasinilpotent operator, we get* $\lim_{n\to+\infty}(c_{T^n,x,\varepsilon})^{1/n} = 0$.

Proof. From Proposition 7, we obtain

$$\overline{\lim}\ (c_{T^n,x,\varepsilon})^{\frac{1}{n}} \leqslant \lim \frac{\varepsilon^{\frac{1}{n}}\|T^n\|^{\frac{2}{n}}}{(\|x\|-\varepsilon)^{\frac{1}{n}}} = r(T)^2.$$ □

Recall that an operator C is said to be stable if $\lim C^n x = 0$ for every $x \in \mathcal{H}$.

Theorem 9. *Let* T *be an operator whose adjoint is stable,* $x \in \mathcal{H}\backslash\overline{(\bigcup_{n\geqslant 1}\mathcal{N}(T^{*n}))}$ *and* ε *is a real number in the open interval* $]\lim_{n\longrightarrow\infty} \mathrm{d}(x;\mathcal{R}(T^n)); \|x\|[$. *Then we have*

$$\lim_{n\longrightarrow+\infty} c_{T^n,x,\varepsilon} = 0$$

Proof. Firstly, by using the Banach–Steinhaus theorem (see for instance [3]), we observe that T^*, and hence T is necessarily power bounded and we set $M = \sup\{\|T^n\|; n \geqslant 0\}$. Suppose the contrary, that is $c_n := c_{T^n,x,\varepsilon}$ does not converge to 0. From Proposition 7, we easily see that the sequence $(c_n)_{n\geqslant 0}$ is bounded, then we can consider a subsequence $c_{\varphi(n)}$ converging to a positive real number c. Let n_0 be such that $c_{\varphi(n)} \geqslant c/2$ for any $n \geqslant n_0$. For $n \geqslant n_0$, we have

$$\begin{aligned}\|c_{\varphi(n)}(c_{\varphi(n)}I + T^{\varphi(n)}T^{*\varphi(n)})^{-1}x - x\| &= \|(c_{\varphi(n)}I + T^{\varphi(n)}T^{*\varphi(n)})^{-1}T^{\varphi(n)}T^{*\varphi(n)}x\| \\ &\leqslant M\|(c_{\varphi(n)}I + T^{\varphi(n)}T^{*\varphi(n)})^{-1}\|\|T^{*\varphi(n)}x\| \\ &\leqslant \frac{2M}{c}\|T^{*\varphi(n)}x\| \longrightarrow 0.\end{aligned}$$

Then, by using Corollary 12, we deduce that

$$\varepsilon = \lim_{n\longrightarrow+\infty} \|c_{\varphi(n)}(c_{\varphi(n)}I + T^{\varphi(n)}T^{*\varphi(n)})^{-1}x\| = \|x\|$$

which is a contradiction. □

3. A characterization of generalized extremal vectors

Instead of using the minimal condition associated with extremal vectors in order to find them (which leads sometimes to some difficult computations), the following criterion shows that it suffices to solve a system of two equations which can be very useful for concrete operators.

Theorem 10. *Let T be a dense range operator acting on a Hilbert space $\mathcal{H}$, $x \in \mathcal{H} \backslash \mathcal{N}(T^*)$ and $\varepsilon \in]\, \mathrm{d}(x;\mathcal{R}(T)); \|x\|[$. Then the pair $(y_{T,x,\varepsilon}, c_{T,x,\varepsilon})$ is the unique solution $(y,c) \in \mathcal{H} \times \mathbb{R}_+^*$ of the following system of equations*

$$\begin{cases} \|Ty - x\| = \varepsilon \,; \\ (T^*T + cI)y = T^*x. \end{cases}$$

Proof. For any positive real number c, set:

$$T_c = T(T^*T + cI)^{-1}T^* - I.$$

Since the operator $T^*T + cI$ is strictly positive, the operator T_c is clearly well defined. A key of the proof is to write T_c in a different form. Firstly, notice that $T_c = \frac{1}{c}T\big(I + \frac{1}{c}T^*T\big)^{-1}T^* - I$ for every positive c. Then, for any $c > 1$ the operator $\big(I + \frac{1}{c}T^*T\big)^{-1}$ can be developed into an infinite series and hence:

$$\begin{aligned} T_c &= \frac{1}{c}T\Big(\sum_{k=0}^{+\infty} \frac{(-1)^k}{c^k}\big(T^*T\big)^k\Big)T^* - I \\ &= \Big(\sum_{k=0}^{+\infty} \frac{(-1)^k}{c^{k+1}} T\big(T^*T\big)^k T^*\Big) - I \\ &= -\Big(\sum_{k=0}^{+\infty} \frac{(-1)^{k+1}}{c^{k+1}} \big(TT^*\big)^{k+1}\Big) - I \\ &= -\Big(\sum_{k=1}^{+\infty} \frac{(-1)^k}{c^k} \big(TT^*\big)^k\Big) - I. \end{aligned}$$

We recognize the development of $-c(TT^* + cI)^{-1}$, which coincides with T_c as soon as $c > 1$. From the uniqueness of analytic continuation we then get

$$\forall c > 0, \quad T_c = -c(TT^* + cI)^{-1}. \tag{2}$$

Set $\mathscr{F} = \{c > 0 \,;\, \|T_c x\| = \varepsilon\}$. Observe that it is enough to prove that $\mathscr{F}$ is reduced to the single set $\{c_{T,x,\varepsilon}\}$. We know that $T_c x = T y_{T,x,\varepsilon} - x$ and that we necessarily have $\|T y_{T,x,\varepsilon} - x\| = \varepsilon$. Hence it implies $c_{T,x,\varepsilon} \in \mathscr{F}$. Let us introduce the function F defined on $\mathbb{R}_+^*$ by: $\forall c > 0, \quad F(c) = \|T_c x\|^2$. On the one hand, for any $c > 0$ we have

$$\begin{aligned} F(c) &= c^2 \|(TT^* + cI)^{-1}x\|^2 \\ &= c^2 \langle (TT^* + cI)^{-2}x | x \rangle \end{aligned}$$

since $(TT^* + cI)$ is a positive operator. On the other hand, using the functional calculus for the positive operator TT^* we obtain

$$F(c) = \int_0^{\|T\|^2} \frac{c^2}{(t+c)^2}\, \mathrm{d}E_{x,x}(t),$$

where $E_{x,x}$ is the scalar spectral measure associated with the pair (x, x).

Fix the positive parameter t, then we easily see that the function $f : c \mapsto \frac{c^2}{(t+c)^2}$ is strictly increasing on $\mathbb{R}_+^*$. Thus the function F is increasing. Assume that there exists $0 < c_1 < c_2$ such that $F(c_1) = F(c_2)$, we deduce that we have necessarily

$$\frac{c_1^2}{(t+c_1)^2} = \frac{c_2^2}{(t+c_2)^2}$$

almost everywhere with respect to the measure $E_{x,x}$. It implies that $E_{x,x} \in \mathbb{R}_+^* \delta_0$ where δ_0 is the Dirac measure at 0, hence we have $x \in \mathcal{N}(T^*)$, a contradiction. Therefore, the function F is one to one and $\mathscr{F}$ contains at most one point. Consequently, $\mathscr{F}$ is reduced to $c_{T,x,\varepsilon}$, and the result follows. □

Remark 11. In particular, it seems that the characterization given in Theorem 10 is sometime useful for studying concrete operators because the calculus of the inverse of an operator is not required.

Besides, the following result states that we can completely determined $c_{T,x,\varepsilon}$ by a single useful equation.

Corollary 12. *Let T be an dense range operator acting on a Hilbert space $\mathcal{H}$, $x \in \mathcal{H}\backslash\mathcal{N}(T^*)$ and $\varepsilon \in]\,\mathrm{d}(x;\mathcal{R}(T)); \|x\|[$. Then $c_{T,x,\varepsilon}$ is uniquely determined by the following single relation*

$$\varepsilon = c_{T,x,\varepsilon}\|(TT^* + c_{T,x,\varepsilon}I)^{-1}x\|.$$

Proof. Combining the two equations in the system given by Theorem 10, we easily obtain

$$\varepsilon = \|(T(T^*T + cI)^{-1}T^* - I)x\|.$$

Then, we get the result by using (2). □

Remark 13. Notice that the end of the proof of Theorem 10 shows that $c_{T,x,\varepsilon}$ is the unique solution of the equation $\varepsilon = c\|(TT^* + cI)^{-1}x\|$.

4. Regularities for extremal vectors

In this section, we observe regularities of four maps that we define as follow. Either $(T;x) \in \mathscr{B}(\mathcal{H}) \times \mathcal{H}$ is fixed such that $x \in \mathcal{H}\backslash\mathcal{N}(T^*)$, and we define the maps

$$y_{T,x} : \begin{array}{ccc}]\,\mathrm{d}(x;\mathcal{R}(T)); \|x\|[& \longrightarrow & \mathcal{H} \\ \varepsilon & \longmapsto & y_{T,x,\varepsilon} \end{array} \qquad c_{T,x} : \begin{array}{ccc}]\,\mathrm{d}(x;\mathcal{R}(T)); \|x\|[& \longrightarrow & \mathcal{H} \\ \varepsilon & \longmapsto & c_{T,x,\varepsilon} \end{array}$$

or $(T;\varepsilon) \in \mathscr{B}(\mathcal{H}) \times \mathbb{R}_+^*$ is fixed, and we define the maps

$$y_{T,\varepsilon} : \begin{array}{ccc} \mathcal{K} & \longrightarrow & \mathcal{H} \\ x & \longmapsto & y_{T,x,\varepsilon} \end{array} \qquad c_{T,\varepsilon} : \begin{array}{ccc} \mathcal{K} & \longrightarrow & \mathcal{H} \\ x & \longmapsto & c_{T,x,\varepsilon} \end{array}$$

where $\mathcal{K} = \{x \in \mathcal{H}\backslash(\mathcal{N}(T)^\perp \cup \mathcal{B}(0;\varepsilon]); \mathrm{d}(x;\mathcal{R}(T)) < \varepsilon\}$.

A same proof as the one exposed in [2] allows us to establish analyticity of the two first maps.

Proposition 14. *Let $(T;x) \in \mathscr{B}(\mathcal{H}) \times \mathcal{H}$ such that $x \in \mathcal{H}\backslash\mathcal{N}(T^*)$. Then the two maps $y_{T,x}$ and $c_{T,x}$ are analytic over* $]\, \mathrm{d}(x;\mathcal{R}(T)); \|x\|\,[$.

Remark 15. With the same hypotheses, we can easily notice that the map $\varepsilon \mapsto \|y_{T,x,\varepsilon}\|$ is decreasing over $]\, \mathrm{d}(x;\mathcal{R}(T)); \|x\|\,[$.

Our next aim is to study the local behavior of $y_{T,x}$ and $c_{T,x}$ around the bounds of the set $]\, \mathrm{d}(x;\mathcal{R}(T)); \|x\|\,[$. We have chosen to emphasize the study of the particular bound $\mathrm{d}(x;\mathcal{R}(T))$ where we recognize a strong link between generalized minimal vectors and the famous notion of Moore–Penrose pseudo-inverse. For that reason, this study is kept for Section 5. For now, let us expose the behaviour around $\|x\|$.

Proposition 16. *Let $(T;x) \in \mathscr{B}(\mathcal{H}) \times \mathcal{H}$ such that $x \in \mathcal{H}\backslash\mathcal{N}(T^*)$. When ε tends to $\|x\|$,*

- *$y_{T,x}$ strongly converges to 0;*
- *$c_{T,x}$ tends to $+\infty$.*

Proof. Let us prove that $y_{T,x}$ weakly converges to 0 when ε tends to $\|x\|$.

Proposition 6 allows us to claim that for any $\varepsilon \in]\, \mathrm{d}(x;\mathcal{R}(T)); \|x\|\,[$, we get $Ty_{T,x} \in \mathcal{V}_{x,\varepsilon}$, and then

$$\|Ty_{T,x}\| < \sqrt{\|x\|^2 - \varepsilon^2}.$$

It follows that $\|Ty_{T,x}\| \to 0$, therefore $Ty_{T,x}$ strongly converges to 0 when ε tends to $\|x\|$. Moreover, using Remark 15, $\varepsilon \mapsto \|y_{T,x}\|$ is decreasing and non-negative, which means that $\|y_{T,x}\|$ is convergent when ε tends to $\|x\|$. All the more, $y_{T,x}$ is bounded when ε is closed to $\|x\|$. Consequently, we can build a sequence $(\varepsilon_k)_{k\in\mathbb{N}}$ which tends to $\|x\|$ and such that $(y_{T,x,\varepsilon_k})_{k\in\mathbb{N}}$ is weakly convergent to some vector $b \in \mathcal{H}$. We set $y_k = y_{T,x,\varepsilon_k}$ in the sequel of the proof. We deduce that Ty_k weakly converges to Tb. But $Ty_{T,x}$ strongly converges to 0 when ε tends to $\|x\|$, so Ty_k strongly converges to 0. It implies $Tb = 0$, that is $b \in \mathcal{N}(T)$.

On an other hand, we can write

$$\forall k \in \mathbb{N}, \quad y_k \in (\mathcal{N}(T))^{\perp}.$$

Indeed, for any non-negative integer k, we decompose y_k with respect to the direct sum $\mathcal{H} = \mathcal{N}(T) \oplus (\mathcal{N}(T))^{\perp}$:

$$y_k = y_k' + y_k'' \quad \text{with } (y_k', y_k'') \in (\mathcal{N}(T)) \times (\mathcal{N}(T))^{\perp}.$$

Then we obtain $Ty_k = Ty_k''$ and $\|Ty_k'' - x\| = \varepsilon$. But the inequality $\|y_k''\| \leqslant \|y_k\|$ is obvious. The unicity of minimal vector y_k allows us to affirm that $y_k'' = y_k$, that is $y_k \in (\mathcal{N}(T))^{\perp}$. Because the vector subspace $(\mathcal{N}(T))^{\perp}$ is closed, the weak limit b of the sequence (y_k) belongs to $(\mathcal{N}(T))^{\perp}$. As a result, we get

$$b \in \mathcal{N}(T) \cap (\mathcal{N}(T))^{\perp} = \{0\}.$$

Hence 0 is the only weak limit point of $y_{T,x}$, so that $y_{T,x}$ is weakly convergent to 0 when ε tends to $\|x\|$.

For now, we proceed by absurdum to show that $c_{T,x}$ tends to $+\infty$ when ε tends to $\|x\|$. Suppose that $c_{T,x}$ does not tend to $+\infty$. Therefore, there exists a sequence $(\varepsilon_k)_{k\in\mathbb{N}}$ that tends to $\|x\|$ and such that (c_{T,x,ε_k}) is bounded. Set $c_k = c_{T,x,\varepsilon_k}$ and $y_k = y_{T,x,\varepsilon_k}$. Up to a subsequence, we may assume that (c_k) is convergent to some non-negative number c. On the one hand, (T^*T+c_kI) converges in $\mathscr{B}(\mathcal{H})$ to T^*T+cI. On the other hand, using the previous argument, we can say that (y_k) weakly converges to 0. According to Theorem 10, we have

$$\forall k \in \mathbb{N}, \quad (T^*T + c_kI)y_k = T^*x,$$

so $T^*x = 0$, that is $x \in \mathcal{N}(T^*)$. But the equalities

$$\mathcal{N}(T^*) \subset \overline{\mathcal{N}(T^*)} = (\mathcal{R}(T))^{\perp} = \left(\overline{\mathcal{R}(T)}\right)^{\perp} = \{0\},$$

imply that $x = 0$, which is a contradiction. Consequently, $c_{T,x}$ tends to $+\infty$.

According to Theorem 10, the equality $y_{T,x} = (T^*T + c_{T,x}I)^{-1}T^*x$ holds. Since $c_{T,x} \to +\infty$, $(T^*T + c_{T,x}I)^{-1}$ converges in $\mathscr{B}(\mathcal{H})$ to 0. We deduce that $y_{T,x}$ strongly converges to 0, and the proof is ended. □

Finally, we focus on $y_{T,\varepsilon}$ et $c_{T,\varepsilon}$, defined at the beginning of the section. We obtain the continuity of these two maps.

Theorem 17. *Let $(T;\varepsilon) \in \mathscr{B}(\mathcal{H}) \times \mathbb{R}_+^*$. Then the two maps $y_{T,\varepsilon}$ and $c_{T,\varepsilon}$ are continuous on the open set $\mathcal{K} = \left\{x \in \mathcal{H}\backslash(\mathcal{N}(T)^{\perp} \cup \mathcal{B}(0;\varepsilon]); \mathrm{d}(x;\mathcal{R}(T)) < \varepsilon\right\}$.*

Proof. Using a similar proof as in [11], $y_{T,\varepsilon}$ is obviously continuous on $\mathcal{K}$.

Let $(x_n)_{n\in\mathbb{N}}$ be a sequence of vectors in $\mathcal{K}$ such that (x_n) converges to some vector $x \in \mathcal{K}$. Let $\rho \in]0; \|x\| - \varepsilon[$. There exists a non-negative integer N such that $\forall n \geqslant N$, $\|x_n\| - \varepsilon \geqslant \rho$ and $\mathrm{d}(x_n;\mathcal{R}(T)) < \varepsilon$. Thus $c_n = c_{T,x_n,\varepsilon}$ and $y_n = y_{T,x_n,\varepsilon}$ are well defined from the rank N. We also set $c = c_{T,x,\varepsilon}$ and $y = y_{T,x,\varepsilon}$. Let us prove that the sequence $(c_n)_{n\geqslant N}$ converges to c. According to Proposition 7, we get

$$\forall n \geqslant N, \quad 0 < c_n \leqslant \frac{\varepsilon\|T\|^2}{\|x_n\| - \varepsilon} \leqslant \frac{\varepsilon\|T\|^2}{\rho}.$$

Hence $(c_n)_{n\geqslant N}$ is bounded. Let c' be any limit point of $(c_n)_{n\geqslant N}$. There exists a subsequence $(c_{\varphi(n)})$ that converges c'. Moreover, since $y_{T,\varepsilon}$ is continuous at x, $y_{\varphi(n)} \to y$. From Theorem 10, the equality

$$(T^*T + c_{\varphi(n)}I)y_{\varphi(n)} = T^*x_{\varphi(n)}$$

holds for any $n \in \mathbb{N}$ which leads to $(T^*T + c'I)y = T^*x$. The uniqueness of the solution of the system

$$\begin{cases} \|Ty - x\| = \varepsilon; \\ (T^*T + cI)y = T^*x, \end{cases}$$

allows us to affirm that $(c';y) = (c;y)$ and obviously $c' = c$. Thus c is the only limit point of the bounded sequence $(c_n)_{n\geqslant N}$. It means that $(c_n)_{n\geqslant N}$ converges to c. □

5. Relationships between generalized extremal vectors and Moore–Penrose pseudo-inverse

The extremal vectors were introduced by Ansari and Enflo in order to supply to the default of surjectivity for a densely range operator. It would be interesting to test the notion of generalized extremal vectors in the following way: Let $T \in \mathscr{B}(\mathcal{H})$ and $x \in \overline{\mathcal{R}(T)}$, what is the behavior of $y_{T,x,\varepsilon}$ when ε goes to 0? An interesting case will be the case where $x \in \mathcal{R}(T)$.

Theorem 18. *Let $T \in \mathscr{B}(\mathcal{H})$ and $x \in \overline{\mathcal{R}(T)}\backslash\{0\}$. The following assertions are equivalent:*

(i) $x \in \mathcal{R}(T)$;
(ii) y_ε *weakly converges in* $\mathcal{H}$ *when* ε *goes to* 0;
(iii) y_ε *is norm-convergent to the image of* x *by the Moore–Penrose pseudo-inverse of* T, *when* ε *goes to* 0.

Moreover, when $x \notin \mathcal{R}(T)$, we have $\|y_\varepsilon\| \underset{\varepsilon\to 0}{\longrightarrow} +\infty$; in particular, $\varepsilon \mapsto y_\varepsilon$ is divergent when ε tends to 0.

Proof. Let us prove that (i) implies (iii). Assume $x \in \mathcal{R}(T)$. Set

$$\rho = \inf\{\|u\|; Tu = x\}$$

and consider a minimizing sequence (b_n), that is $Tb_n = x$ and $\|b_n\|$ decreases to ρ. Since the sequence (b_n) is bounded, we can find a subsequence $(b_{\varphi(n)})$ which weakly converges to some vector $b \in \mathcal{H}$. Therefore, it derives $Tb = x$ and $\|b\| = \rho$. Now suppose that there is another vector c satisfying $Tc = x$ and $\|c\| = \rho$. Obviously we have $T(ta + (1-t)c) = x$, and using triangular inequality and the minimality of ρ, we obtain $\|ta + (1-t)c\| = \rho$. The strict convexity of $\mathcal{H}$ ensures that $b = c$. We have thus shown that

$$\exists! b \in H; Tb = x \text{ and } \rho = \|b\|. \tag{3}$$

From $\|Tb - x\| = 0$ and the minimality condition on y_ε, it follows that

$$\|y_\varepsilon\| \leqslant \rho = \|b\|$$

for any $\varepsilon > 0$.

We will now prove that $\|y_\varepsilon - b\| \to 0$ when ε goes to 0. Suppose the contrary, then there exist a positive number δ and a sequence $\varepsilon_n \downarrow 0$ such that

$$\|y_{\varepsilon_n} - b\| \geqslant \delta.$$

We can extract a subsequence $(y_{\varepsilon_{\varphi(n)}})$ which converges weakly to some vector b'. Observe that $\|b'\| \leqslant \limsup(\|y_{\varepsilon_n}\|) \leqslant \rho = \|b\|$. Since $\|Ty_{\varepsilon_{\varphi(n)}} - x\| \to 0$, we derive that $Tb' = x$ and we have necessarily $\rho = \|b'\|$. Therefore, from (3) we get $b' = b$. As the function $\varepsilon \to \|y_\varepsilon\|$ is decreasing, we see that the sequence $(\|y_{\varepsilon_{\varphi(n)}}\|)$ is increasing and bounded above by $\rho = \|b\|$, hence convergent to some $l \leqslant \rho$. Considering a subsequence $(y_{\varepsilon_{\psi\circ\varphi(n)}})$ weakly convergent to some b'', from (3) we deduce as before that $b'' = b$ and $\rho = \|b''\| \leqslant l$, hence $l = \rho$. Finally, we have proved that the subsequence $\left(y_{\varepsilon_{\varphi(n)}}\right)$ is weakly convergent to b and that the sequence $\left(\left\|y_{\varepsilon_{\varphi(n)}}\right\|\right)$ converges to $\|b\|$. In view of the Kadec–Klee property of $\mathcal{H}$, we obtain that the subsequence $\left(y_{\varepsilon_{\varphi(n)}}\right)$ converges in norm to b, which is a contradiction.

The implication (iii) $\Longrightarrow$ (ii) is clearly true. Next, we prove that (ii) implies (i). Assume that y_ε converges weakly to some vector a when ε tends to 0. On the one hand Ty_ε is weakly convergent to Ta and on the other hand the inequality $\|Ty_\varepsilon - x\| \leqslant \varepsilon$ implies $Ty_\varepsilon \to x$. Therefore $x = Ta \in \mathcal{R}(T)$.

It remains to prove the last assertion. Assume the contrary, then there exists a bounded sequence (y_{ε_n}) for which $\varepsilon_n \downarrow 0$. Then, we can consider a subsequence $\left(y_{\varepsilon_{\varphi(n)}}\right)$ which is weakly convergent to a point c. As before, we derive that $x = Tc \in \mathcal{R}(T)$ which is absurd. It ends the proof of Theorem 18. □

Remark 19. The notion of extremal vector was extended by G. Androulakis in the setting of super-reflexive Banach spaces [1]. In the previous proposition, when T is densely defined we can replace $\mathcal{H}$ by any super-reflexive Banach space which is strictly convex and has the Kadec–Klee property.

Theorem 18 allows us to compute the limit of $c_{T,x}$ when $\varepsilon \longrightarrow 0$.

Corollary 20. *Let $T \in \mathscr{B}(\mathcal{H})$ and $x \in \overline{\mathcal{R}(T)}\backslash\{0\}$. Then, we have* $\lim\limits_{\varepsilon\to 0} c_{T,x,\varepsilon} = 0$.

Proof. Since $Ty_{T,x,\varepsilon} \xrightarrow[\varepsilon\to 0]{} x$, using Theorem 10, we get

$$c_{T,x,\varepsilon} y_{T,x,\varepsilon} = T^*(x - Ty_{T,x,\varepsilon}) \xrightarrow[\varepsilon\to 0]{} T^*(0) = 0.$$

We derive that $c_{T,x,\varepsilon}\|y_{T,x,\varepsilon}\| \xrightarrow[\varepsilon\to 0]{} 0$ and $c_{T,x,\varepsilon} = \underset{\varepsilon\to 0}{o}\left(1/\|y_\varepsilon\|\right)$. In the two following cases, applying Proposition 18, we obtain:

- If $x_0 \notin \mathcal{R}(T)$, we have $\|y_{T,x,\varepsilon}\| \xrightarrow[\varepsilon\to 0]{} +\infty$, hence $1/\|y_{T,x,\varepsilon}\| \xrightarrow[\varepsilon\to 0]{} 0$ and $c_{T,x,\varepsilon} \xrightarrow[\varepsilon\to 0]{} 0$.
- If $x_0 \in \mathcal{R}(T)$, then the function $\varepsilon \mapsto y_{T,x,\varepsilon}$ converges to a non null vector, thus the function $\varepsilon \mapsto 1/\|y_{T,x,\varepsilon}\|$ is bounded. Therefore, we deduce immediately that $c_{T,x,\varepsilon} \xrightarrow[\varepsilon\to 0]{} 0$. □

6. Applications to weighted shifts

Let $(E_n)_{n\in\mathbb{Z}}$ be a family of Hilbert spaces and $\mathcal{H} = \oplus_{n\in\mathbb{Z}} E_n$ be the direct sum of this family of Hilbert spaces. We consider a bounded sequence of operators $(W_n)_{n\in\mathbb{Z}}$ such that $W_n \in B(E_n, E_{n+1})$ for any $n \in \mathbb{Z}$ and we define the operator weighted shift S acting on H by setting

$$\begin{aligned} &S(\dots, x_{-2}, x_{-1}, \boxed{x_0}, x_1, x_2, \dots) \\ &\quad = (\dots, W_{-2}x_{-2}, \boxed{W_{-1}x_{-1}}, W_0x_0, W_1x_1, W_2x_2, \dots) \end{aligned}$$

where the components, which are contained in boxes, belong to E_0.

Proposition 21. *Let $x = (x_k)_{k\in\mathbb{Z}}$ be in $\mathcal{H}\backslash\mathcal{N}(T^{*n})$ where $n \in \mathbb{N}^*$. Let*

$$\varepsilon \in]\, \mathrm{d}(x; \mathcal{R}(T^n)); \|x\|[,$$

then the generalized extremal vector associated with (T^n, x, ε) (resp. (T^{*n}, x, ε)) is

$$y_{T^n,x,\varepsilon} = \sum_{k\in\mathbb{Z}} (W_k^* \dots W_{k+n-1}^* W_{k+n-1} \dots W_k + c_n \operatorname{Id}_{E_k})^{-1} W_k^* \dots W_{k+n-1}^* x_{k+n}$$

(resp.

$$y_{T^{*n},x,\varepsilon} = \sum_{k\in\mathbb{Z}} (W_{k-1} \dots W_{k-n} W_{k-n}^* \dots W_{k-1}^* + \tilde{c}_n \operatorname{Id}_{E_k})^{-1} W_{k-1} \dots W_{k-n} x_{k-n})$$

where c_n (resp. $\tilde{c}_n$) is the unique positive real number satisfying the following equality

$$\varepsilon^2 = \sum_{k\in\mathbb{Z}} c_n^2 \|(W_{k-1} \dots W_{k-n} W_{k-n}^* \dots W_{k-1}^* + c_n \operatorname{Id}_{E_k})^{-1} x_k\|^2$$

$$(\textit{resp. } \varepsilon^2 = \sum_{k\in\mathbb{Z}} \tilde{c}_n^2 \|(W_k^* \dots W_{k+n-1}^* W_{k+n-1} \dots W_k + \tilde{c}_n \operatorname{Id}_{E_k})^{-1} x_k\|^2).$$

Proof. Let $x \in \oplus_{n=-q}^{q} E_n$, according to Theorem 10, we have to solve the system of equations

$$\begin{cases} (T^{*n}T^n + c_n I)y = T^{*n}x; \\ \|T^n y - x\| = \varepsilon. \end{cases} \tag{4}$$

The first equation leads to

$$\sum_{k\in\mathbb{Z}} (W_k^* \dots W_{k+n-1}^* W_{k+n-1} \dots W_k + c_n \operatorname{Id}_{E_k}) y_k = \sum_{k\in\mathbb{Z}} W_k^* \dots W_{k+n-1}^* x_{k+n}$$

where both sums are finite. Let $k \in \mathbb{Z}$, taking the projection onto the subspace E_k, we see that $(W_k^* \dots W_{k+n-1}^* W_{k+n-1} \dots W_k + c_n \operatorname{Id}_{E_k}) y_k = W_k^* \dots W_{k+n-1}^* x_{k+n}$, hence

$$y_k = (W_k^* \dots W_{k+n-1}^* W_{k+n-1} \dots W_k + c_n \operatorname{Id}_{E_k})^{-1} W_k^* \dots W_{k+n-1}^* x_{k+n}.$$

By a similar computation than in the proof of Theorem 10 we get

$$\begin{aligned} &W_{k-1} \dots W_{k-n} (W_{k-n}^* \dots W_{k-1}^* W_{k-1} \dots W_{k-n} + c_n \operatorname{Id}_{E_{k-n}})^{-1} \\ &\quad \times W_{k-n}^* \dots W_{k-1}^* - \operatorname{Id}_{E_k} \\ &= -c_n (W_{k-1} \dots W_{k-n} W_{k-n}^* \dots W_{k-1}^* + c_n \operatorname{Id}_{E_k})^{-1}, \end{aligned}$$

and taking into account the second equation in system (4), we obtain

$$\begin{aligned} \varepsilon^2 &= \sum_{k\in\mathbb{Z}} \|W_{k-1} \dots W_{k-n} (W_{k-n}^* \dots W_{k-1}^* W_{k-1} \dots W_{k-n} + c_n \operatorname{Id}_{E_k})^{-1} \\ &\quad \times W_{k-n}^* \dots W_{k-1}^* x_k - x_{k+n}\|^2 \\ &= \sum_{n\in\mathbb{Z}} c_n^2 \|(W_{k-1} \dots W_{k-n} W_{k-n}^* \dots W_{k-1}^* + c_n \operatorname{Id}_{E_k})^{-1} x_k\|^2. \end{aligned}$$

Let $x = \sum_{n\in\mathbb{Z}} x_n \in \mathcal{H}\backslash\mathcal{N}(T^*)$ with respect to the direct sum decomposition $\mathcal{H} = \oplus_{n\in\mathbb{Z}} E_n$, we clearly have $x = \lim_{q\longrightarrow+\infty} \sum_{n=-q}^{q} x_n$. Using the continuity of the two maps $y_{T^n,\varepsilon}$ and $c_{T^n,\varepsilon}$ (Theorem 17) we easily deduce the result in the general case. In the same manner, we find formulas associated with $y_{T^{*n},x,\varepsilon}$ and $\tilde{c}_n$. □

Remark 22. Taking $E_n = 0$ when $n \leqslant -1$ in Proposition 21, we get the characterization of the generalized extremal vector associated with a vector $\mathcal{H}\backslash\mathcal{N}(T^{*n})$ for T^n, where T is a forward or backward unilateral operator weighted shift.

The following proposition is concerned with the weak closure of the sequence $(T^n y_{T^n,x,\varepsilon})_{n\geqslant 1}$.

Proposition 23. *Let $T \in B(\mathcal{H})$, $x \in \mathcal{H}\backslash\overline{(\bigcup_{n\geqslant 1}\mathcal{N}(T^{*n}))}$ and*

$$\varepsilon \in]\lim_{n\longrightarrow\infty} \mathrm{d}(x;\mathcal{R}(T^n)); \|x\|[.$$

Then, the null vector is not a weak limit point of the sequence $(T^n y_{T^n,x,\varepsilon})_{n\geqslant 1}$.

Proof. For convenience we set $y^{(n)} := y_{T^n,x,\varepsilon}$ for the generalized extremal vector associated with (T^n, x, ε). We proceed by absurdum and suppose that there exists a subsequence $(T^{\varphi(n)} y^{\varphi(n)})_{n\geqslant 1}$ converging weakly to the null vector 0. Then, we have

$$\|x\| \leqslant \underline{\lim} \|T^{\varphi(n)} y^{\varphi(n)} - x\| = \varepsilon,$$

which is a contradiction. □

Now we focus on classical unilateral weighted shifts. Firstly, we study the variations of the sequence $(c_{T^n,x,\varepsilon})$ where T is the adjoint of some classical weighted shifts.

Proposition 24. *We denote by T a backward weighted shift acting on $\ell^2(\mathbb{N})$. We suppose that the non-negative weight $(w_n)_{n\in\mathbb{N}}$ of T^* tends to 0. Let us fix $x \in \mathcal{H}\backslash\{0\}$ and $\varepsilon \in]0; \|x\|[$. Then there exists some positive integer N such that $(c_{T^n,x,\varepsilon})_{n\geq N}$ is decreasing to 0.*

Proof. Set $c_n = c_{T^n,x,\varepsilon}$ and $y^{(n)} = y_{T^n,x,\varepsilon}$ for any $n \in \mathbb{N}^*$: $y^{(n)} = (y_k^{(n)})_{k\in\mathbb{N}} \in \ell^2(\mathbb{N})$. Using Proposition 21, we can see that

$$\varepsilon^2 = \sum_{i=0}^{+\infty} \frac{c_n^2}{(w_i^2 \cdots w_{i+n-1}^2 + c_n)^2} |x_i|^2 = \sum_{i=0}^{+\infty} \frac{c_{n+1}^2}{(w_i^2 \cdots w_{i+n}^2 + c_{n+1})^2} |x_i|^2.$$

It follows that

$$\begin{aligned} 0 &= \sum_{i=0}^{+\infty} \left[\frac{c_{n+1}^2}{(w_i^2\cdots w_{i+n}^2+c_{n+1})^2} - \frac{c_n^2}{(w_i^2\cdots w_{i+n-1}^2+c_n)^2}\right] |x_i|^2 \\ &= \sum_{i=0}^{+\infty} \left[\frac{c_{n+1}}{w_i^2\cdots w_{i+n}^2+c_{n+1}} - \frac{c_n}{w_i^2\cdots w_{i+n-1}^2+c_n}\right]\left[\frac{c_{n+1}}{w_i^2\cdots w_{i+n}^2+c_{n+1}} + \frac{c_n}{w_i^2\cdots w_{i+n-1}^2+c_n}\right] |x_i|^2. \end{aligned}$$

Let us denote by S the last sum. The sign of I, defined by

$$I = \frac{c_{n+1}}{w_i^2 \cdots w_{i+n}^2 + c_{n+1}} - \frac{c_n}{w_i^2 \cdots w_{i+n-1}^2 + c_n}$$
$$= \frac{w_i^2 \cdots w_{i+n-1}^2 [c_{n+1} - w_{i+n}^2 c_n]}{(w_i^2 \cdots w_{i+n}^2 + c_{n+1})(w_i^2 \cdots w_{i+n-1}^2 + c_n)},$$

is the same as the one of $c_{n+1} - w_{i+n}^2 c_n$. Since (w_n) tends to 0 when n tends to $+\infty$, there exists a positive integer N such that for every $n \geq N$, we have: $\forall i \in \mathbb{N}$, $w_{i+n} < 1$. Thus, $\forall n \geqslant N$, $c_{n+1} \leqslant c_n$. Indeed, if we suppose that $c_{n+1} > c_n$, then $\forall i \in \mathbb{N}$, $c_{n+1} > w_{i+n}^2 c_n$ and we deduce that $S > 0$, which is a contradiction.
As a direct consequence of Theorem 9, we obtain that (c_n) converges to 0. □

A crucial point in the Ansari–Enflo method is to obtain norm convergence for a subsequence of $(\{T^{\varphi(n)} y_{T^{\varphi(n)},x,\varepsilon}\})_{n\geqslant 1}$. In the following proposition, we see that it is always the case when T a bounded backward weighted shift acting on $\ell^2(\mathbb{N})$.

Proposition 25. *We denote by T a bounded backward weighted shift acting on $\ell^2(\mathbb{N})$. Let us fix $x \in \mathcal{H}\backslash\{0\}$ and $\varepsilon \in]0; \|x\|[$. Then the weak closure and the norm closure of the set $\{T^n y_{T^n,x,\varepsilon}\}$ coincide and do not contain 0. More precisely, every subsequence $\{T^{\varphi(n)} y_{T^{\varphi(n)},x,\varepsilon}\}$ which is weakly convergent is also norm convergent.*

Proof. Let $(T^{\varphi(n)} y_{T^{\varphi(n)},x,\varepsilon}\})_{n\geqslant 1}$ be a weakly convergent subsequence and denote by y its limit. Then, we have

$$\lim_{n\longrightarrow+\infty} \frac{w_k^2 \cdots w_{k+\varphi(n)-1}^2}{w_k^2 \cdots w_{k+\varphi(n)-1}^2 + c_{\varphi(n)}} x_k = y_k$$

for any $k \in \mathbb{N}$. Therefore, we get

$$y_k - x_k = \lim_{n\longrightarrow+\infty} \frac{c_{\varphi(n)}}{w_k^2 \cdots w_{k+\varphi(n)-1}^2 + c_{\varphi(n)}} x_k,$$

and applying the Lebesgue theorem, with respect to the discrete measure $\mu = \sum_{k=0}^{+\infty} |x_k|^2 \delta_k$ and the bounded sequence of functions $(f_n)_{n\in\mathbb{N}}$ defined by setting $f_n(x) = c_{\varphi(n)}^2 / [w_{[x]}^2 \cdots w_{[x]+\varphi(n)-1}^2]^2$, we obtain

$$\varepsilon^2 = \|T^{\varphi(n)} y_{T^{\varphi(n)},x,\varepsilon} - x\|^2 = \sum_{k=0}^{+\infty} \frac{c_{\varphi(n)}^2}{(w_k^2 \cdots w_{k+\varphi(n)-1}^2 + c_{\varphi(n)})^2} |x_k|^2 \longrightarrow \|y - x\|^2.$$

Then, using the Kadec–Klee property, we see that the subsequence

$$(T^{\varphi(n)} y_{T^{\varphi(n)},x,\varepsilon}\})_{n\geqslant 1}$$

is necessarily norm convergent. Finally, the fact that 0 is not a weak limit point follows directly from Proposition 23. □

Recall that a operator A is said to be unicellular if its lattice of invariant subspaces is totally ordered by inclusion.

Theorem 26. *We denote by T a quasinilpotent backward weighted shift acting on $\ell^2(\mathbb{N})$. Then all its invariant subspaces can be obtained by using the Ansari–Enflo method if and only if T is unicellular.*

Proof. We can extract a subsequence $(y^{(\varphi(n))})_{n\geqslant 0}$ such that the subsequences

$$(T^{\varphi(n)}y^{(\varphi(n))})_{n\geqslant 0} \quad \text{and} \quad (T^{\varphi(n)-1}y^{(\varphi(n)-1)})_{n\geqslant 0}$$

converge weakly respectively to y and z. From Proposition 25, we derive that both subsequences strongly converge and have non-zero limit points. Since T is quasinilpotent, using Lemma 1 in [2] we may suppose that $\|y_{\varphi(n)-1}\|/\|y_{\varphi(n)}\|$ converges to 0. Then, following the method of [2], we can see that the subspace $E = \{Rz : RT = TR\}$ is a non-trivial hyper-invariant subspace of T. Since we have

$$T^{\varphi(n)-1}y^{(\varphi(n)-1)} = \sum_{k=0}^{+\infty} \frac{w_k^2 \cdots w_{k+\varphi(n)-2}^2}{w_k^2 \cdots w_{k+\varphi(n)-2}^2 + c_{\varphi(n)-1}} x_k e_k, \tag{5}$$

we that deduce that $|z_k| \leqslant |x_k|$ and that z_k/x_k is non-negative when $x_k \neq 0$. There is at least one component z_k which is different from x_k. Suppose not, then from Proposition 3 it follows that

$$\|x\|^2 = \|z\|^2 \leqslant \underline{\lim}\|T^{\varphi(n)-1}y^{(\varphi(n)-1)}\|^2 \leqslant \|x\|^2 - \varepsilon^2$$

which is a contradiction. Let p be the first integer such that $z_p \neq x_p$. Using Equation 5, we easily obtain

$$\frac{c_{\varphi(n)-1}}{w_p^2 \cdots w_{p+\varphi(n)-2}^2} \longrightarrow \frac{x_p}{z_p} - 1,$$

where we have made the convention that $x_p/z_p - 1 = +\infty$ if $z_p = 0$. Let $q > p$ and suppose that $z_q \neq 0$. On the one hand we get, as before

$$\frac{c_{\varphi(n)-1}}{w_q^2 \cdots w_{q+\varphi(n)-2}^2} \longrightarrow \frac{x_q}{z_q} - 1.$$

On the other hand, we see that

$$\begin{aligned}\frac{c_{\varphi(n)-1}}{w_q^2 \cdots w_{q+\varphi(n)-2}^2} &= \frac{w_p^2 \cdots w_{q-1}^2}{w_{p+\varphi(n)-1} \cdots w_{q+\varphi(n)-2}} \frac{c_{\varphi(n)-1}}{w_p^2 \cdots w_{p+\varphi(n)-2}^2} \\ &\longrightarrow (+\infty) \times (\frac{x_p}{z_p} - 1) = +\infty,\end{aligned}$$

which leads to a contradiction. Thus, we necessarily have $z_q = 0$. Consequently, we obviously have $E = \mathcal{H}_p := Span\{e_k : k \leqslant m\}$. Conversely, using previous properties, it is clear that if we take $x = e_p$ we obtain $E = \mathcal{H}_p$. It clearly proves Theorem 26. □

Let T be backward shift with positive weights $(w_n)_{n\geqslant 0}$, we set $\omega_0 := 1$ and $\omega_n := w_0 \cdots w_{n-1}$ when $n \geqslant 1$.

Remark 27. We refer to [6] for several results concerning non-unicellular quasinilpotent operators, and for examples involving unicellular quasinilpotent operators see for instance [7] and [9].

Corollary 28. *We denote by T a backward weighted shift acting on $\ell^2(\mathbb{N})$. Suppose that for every $i \geqslant 0$, there exists a number $N_i \geqslant i$ such that*

$$\sum_{n,m \geqslant N_i} \frac{\omega_{n+m-i}}{\omega_n \omega_m} < +\infty.$$

Then all invariant subspaces of T can be obtained by using the Ansari–Enflo method.

Proof. The fact that T is unicellular is a direct consequence of Theorem 2 in [9]. Notice that T is necessarily quasinilpotent, then it suffices to apply Theorem 26. □

References

[1] G. Androulakis, *A note on the method of minimal vectors*, Trendin Banach spaces and operator theory (Memphis, TN, 2001), 29–36, Contemp. Math., 321, Amer. Math. Soc., Providence, RI, 2003.

[2] S. Ansari, P. Enflo, *Extremal Vectors and Invariant Subspaces*, Trans. Amer. Math. Soc., **350** (1998), 539–558.

[3] B. Beauzamy, *Introduction to Banach spaces and their geometry*, second edition. North-Holland Mathematics Studies, 68, 1985.

[4] J.B. Conway, *A Course in Functional Analysis*, Second Edition, Springer, New York.

[5] C. Foias, I.B. Jung, E. Ko, C. Pearcy, *On quasinilpotent operators. III*, J. Operator Theory **54:2**, 401–414, 2005.

[6] S. Grabiner and M.P. Thomas, *Non-unicellular strictly cyclic quasi-nilpotent shifts on Banach spaces*, J. Operator Theory, 13 (1985), 163–170.

[7] D.A. Herrero, *A unicellular universal quasinilpotent weighted shift*, Proc. American Mathematical Society, Vol. 110, No 3 (1990), 649–652.

[8] K. Hoffman, *Banach Spaces of Analytic Functions*, Prentice Hall Inc., 1962.

[9] M.T. Karaev and M. Gürdal, *Strongly splitting weighted shift operators on Banach spaces and unicellularity*, Operators and Matrices **5** No. 1 (2011), 157–171.

[10] B. Sz.-Nagy, C. Foias, *Harmonic Analysis of Operators on Hilbert Space*, North-Holland, Amsterdam 1970.

[11] J. Verliat, *Regularity Problem for Extremal Vectors*, Extracta Mathematicae, **22** No. 1 (2007), 35–43.

Gilles Cassier and Jérôme Verliat
Institut Camille Jordan, CNRS UMR 5208, Université de Lyon
Université Lyon 1, 43 boulevard du 11 Novembre 1918
F-69622 Villeurbanne, France
e-mail: `cassier@math.univ-lyon1.fr`
`verliat@math.univ-lyon1.fr`

Operator Theory:
Advances and Applications, Vol. 236, 133–157

Weighted Composition Operators from the Analytic Besov Spaces to BMOA

Flavia Colonna and Maria Tjani

Abstract. Let ψ and φ be analytic functions on the open unit disk $\mathbb{D}$ with $\varphi(\mathbb{D}) \subseteq \mathbb{D}$ and let $1 \leq p < \infty$. We characterize the bounded and the compact weighted composition operators $W_{\psi,\varphi}$ from the analytic Besov space B_p into *BMOA* and into *VMOA*. We also show that there are no isometries among the composition operators.

Mathematics Subject Classification (2010). Primary: 47B38, 30H35, 30H30; Secondary: 30H10.

Keywords. Weighted composition operators, Besov space, Dirichlet space, *BMOA*, *VMOA*, Bloch space, Hardy space.

1. Introduction

Let X and Y be Banach spaces of analytic functions on a domain Ω in $\mathbb{C}$, ψ an analytic function on Ω and let φ be an analytic function mapping Ω into itself such that $\psi(f \circ \varphi) \in Y$ for each $f \in X$. The *weighted composition operator with symbols ψ and φ* from X to Y is the operator $W_{\psi,\varphi}$ defined by

$$W_{\psi,\varphi} f = M_\psi C_\varphi f = \psi(f \circ \varphi), \text{ for } f \in X,$$

where M_ψ denotes the multiplication operator with symbol ψ and C_φ denotes the composition operator with symbol φ.

Recently, there has been an increasing interest in the study of the weighted composition operators, since they arise naturally in the study of the isometries of many functional Banach spaces.

In this work, we investigate the bounded and the compact weighted composition operators from the analytic Besov spaces B_p (with $1 \leq p < \infty$) into the space *BMOA* of analytic functions of bounded mean oscillation as well as into its subspace *VMOA* of functions of vanishing mean oscillation. To carry out this study, we make use of different approaches, since research on composition operators on *BMOA*, has produced different criteria for compactness [5], [16], [20], and

[22]. See also [18] for a special case. For general weighted composition operators on *BMOA* and on *VMOA*, boundedness and compactness have been characterized in [12] and new criteria were derived in [6]. The compact composition operators on the Besov spaces have been characterized in [19] for the case $1 < p < \infty$ and in [7] (Proposition 5.3) and independently in [21] for the case $p = 1$ in terms of the norm in B_p of the operator applied to the automorphisms of the open unit disk $\mathbb{D}$.

The compact composition operators from the analytic Besov spaces to *BMOA* were characterized in [18] in terms of the norm in *BMOA* of the operator applied to the automorphisms of $\mathbb{D}$. In fact, by Corollary 5 in [13], $C_\varphi : B_p \to BMOA$ (with $1 < p < \infty$) is a compact operator if and only if $C_\varphi : BMOA \to BMOA$ is a compact operator.

In this paper, we obtain another proof of this result as well as a proof of the equivalence between the compact composition operators on *VMOA* and those from B_p to *VMOA*, which we show also holds in the case $p = 1$. This interesting equivalence prompted us to explore whether a similar equivalence holds for the weighted composition operators between these same spaces and whether the norm of the operator applied to the automorphisms of $\mathbb{D}$ plays a similar role.

In the present work, our main objective is to seek and extend this type of result to the weighted composition operator for establishing both boundedness and compactness. Parallel to this, we derive criteria in the spirit of those used in [12] and [6]. As expected, the presence of a nonconstant multiplicative symbol affects boundedness and compactness by adding at least one extra condition. Thus, the norm of the operator applied to the automorphisms of $\mathbb{D}$ alone is not sufficient to characterize the bounded and the compact weighted composition operators.

After giving the background on the spaces under consideration, in Section 2, we give several characterizations of the bounded weighted composition operators from B_p to *BMOA* and to *VMOA* for $1 \leq p < \infty$ and show that there are no isometries among the composition operators.

In Section 3, we derive a sufficient condition for compactness of the weighted composition operator from B_p to *BMOA* for $1 < p \leq 2$ in terms of a limiting weighted integral condition involving the symbols of the operator over sets of the form

$$\{z \in \mathbb{D} : |z - e^{i\theta}| < h\},$$

where h is a small positive number and $\theta \in \mathbb{R}$. We then obtain characterizations of compactness of the weighted composition operator mapping the Besov space B_1 to *BMOA* under a restriction on the multiplicative symbol in relation to the set of points where the composition symbol approaches the unit circle (which covers the case when the symbol belongs to *VMOA*) and we use this result to characterize the compact weighted composition operators from the Besov space B_p (for $1 < p < \infty$) to *BMOA* without any restriction set on the symbols.

In Section 4, we characterize compactness of the weighted composition operators from the Besov spaces to *VMOA*. A comparison between these conditions, one of which depends on the index p, to the conditions that characterize the compact weighted composition operators on *VMOA* obtained in [12], indicates that the equivalence of compactness for the composition operators between these spaces does not extend to the case of the weighted composition operators.

1.1. Preliminaries

Let A denote the area measure on $\mathbb{D}$ normalized by the condition $A(\mathbb{D}) = 1$ and let $1 < p < \infty$. The *analytic Besov space* B_p is the Banach space consisting of the analytic functions f on $\mathbb{D}$ such that

$$b_p(f)^p := \int_{\mathbb{D}} (1 - |z|^2)^{p-2} |f'(z)|^p \, dA(z) < \infty,$$

with Besov norm

$$\|f\|_{B_p} = |f(0)| + b_p(f).$$

For $p = 2$, B_p is the classical Dirichlet space $\mathcal{D}$. An equivalent norm, called the *Dirichlet norm*, is defined as

$$\|f\|_{\mathcal{D}} = \left(|f(0)|^2 + \int_{\mathbb{D}} |f'(z)|^2 \, dA(z) \right)^{1/2}.$$

The Besov spaces are Möbius invariant and the Dirichlet space is the unique Möbius invariant Hilbert space that is continuously embedded in the Bloch space [2].

The space *BMOA* is the space of functions f on $\mathbb{D}$ which are analytic Poisson integrals of a function of bounded mean oscillation on the unit circle $\partial\mathbb{D}$. Equivalently, *BMOA* may be defined as the set of functions f in the Hardy space H^2 such that

$$\|f\|_* = \sup_{a \in \mathbb{D}} \|f \circ L_a - f(a)\|_{H^2} < \infty,$$

where for $g \in H^2$,

$$\|g\|_{H^2} = \sup_{0<r<1} \left(\int_{\partial\mathbb{D}} |g(re^{i\theta})|^2 \, dm(\theta) \right)^{1/2}.$$

Here m is the one-dimensional Lebesgue measure on $\partial\mathbb{D}$ normalized by the condition $m(\partial\mathbb{D}) = 1$ and

$$L_a(z) = \frac{a - z}{1 - \overline{a}z}, \quad \text{for } z \in \mathbb{D}.$$

The correspondence $f \mapsto \|f\|_*$ is a seminorm, and the norm defined as

$$\|f\|_{BMOA} = |f(0)| + \|f\|_*$$

yields a Banach space structure on *BMOA*.

Another seminorm on *BMOA* equivalent to $\|\cdot\|_*$ is

$$\|f\|_{**} = \sup_{q \in \mathbb{D}} \left(\int_{\mathcal{D}} |f'(z)|^2 \, (1 - |L_q(z)|^2) \, dA(z) \right)^{1/2}, \quad \text{for } f \in BMOA.$$

The space $VMOA$ of vanishing mean oscillation is defined as the subspace of $BMOA$ consisting of the functions f such that

$$\lim_{|a|\to 1} \|f \circ L_a - f(a)\|_{H^2} = 0,$$

or equivalently,

$$\lim_{|q|\to 1} \int_{\mathbb{D}} |f'(z)|^2 \, (1 - |L_q(z)|^2) \, dA(z) = 0.$$

It is well known (see, e.g., [18]) that if $1 < p < q < \infty$, then

$$B_p \subset B_q \subset VMOA \subset BMOA \subset \mathcal{B},$$

where $\mathcal{B}$ is the Bloch space defined as the set of analytic functions f on $\mathbb{D}$ such that

$$\|f\| = \sup_{z\in\mathbb{D}} (1 - |z|^2)|f'(z)| < \infty.$$

In fact, B_p is continuously embedded in B_q and $BMOA$ is continuously embedded in the Bloch space, since for $a \in \mathbb{D}$ and $f \in BMOA$,

$$(1 - |a|^2)|f'(a)| = |(f \circ L_a - f(a))'(0)| \leq \|f \circ L_a - f(a)\|_{H^2}$$

so $\|f\| \leq \|f\|_*$. Note also that if $f \in H^\infty$ and $a \in \mathbb{D}$, then

$$\|f \circ L_a - f(a)\|_{H^2}^2 = \int_{\partial\mathbb{D}} |f(L_a(\zeta))|^2 \, dm(\zeta) - |f(a)|^2 \leq \|f\|_\infty^2,$$

which implies that

$$\|f\|_{BMOA} \leq |f(0)| + \|f\|_\infty \leq 2\|f\|_\infty. \tag{1}$$

The analytic Besov space B_1 is defined as the set consisting of the functions f on $\mathbb{D}$ that admit the representation

$$f(z) = \sum_{n=1}^{\infty} a_n L_{\lambda_n}(z), \;\; z \in \mathbb{D},$$

where $\{a_n\} \in \ell^1$ and $\lambda_n \in \mathbb{D}$ for $n \in \mathbb{N}$. The norm in B_1 is defined as

$$\|f\|_{B_1} = \inf\left\{ \sum_{n=1}^{\infty} |a_n| : f(z) = \sum_{n=1}^{\infty} a_n L_{\lambda_n}(z), \; z \in \mathbb{D} \right\}.$$

It is evident that B_1 is a Möbius invariant subset of H^∞. In fact, as noted in [1], the functions in B_1 can be extended continuously to the closed unit disk. Moreover, as is the case for the Besov spaces B_p for $1 < p < \infty$, B_1 is a subset of the little Bloch space defined as the subspace $\mathcal{B}_0$ of $\mathcal{B}$ consisting of those $f \in \mathcal{B}$ such that

$$\lim_{|z|\to 1} (1 - |z|^2)|f'(z)| = 0.$$

(See [23]).

The space B_1 was extensively studied in [3], where it was shown that B_1 is the smallest Möbius invariant space. For this reason, the Besov space B_1 is also known as *the minimal Möbius invariant space*. Furthermore, in [3] and [23] it was shown that there exists a constant $C > 0$ such that for every $f \in B_1$,

$$C^{-1} \int_{\mathbb{D}} |f''(z)| dA(z) \leq \|f - f(0) - f'(0)z\|_{B_1} \leq C \int_{\mathbb{D}} |f''(z)| dA(z).$$

For $f \in B_1$, let

$$b(f) = \int_{\mathbb{D}} |f''(z)| dA(z).$$

The quantity $\|f\|_{B_1,*} := \max\{|f(0)|, |f'(0)|, b(f)\}$ defines a non-Möbius-invariant norm on B_1 (see, e.g., [4]) equivalent to $\|\cdot\|_{B_1}$.

The Besov spaces B_p are contained in the Hardy space H^2. In fact, for $1 < p < \infty$, using the Möbius invariance of B_p, one can show (see (1.7) in [18]) that for some $C > 0$,

$$\|f\|_*^2 \leq C \|f\|_{B_p}^p \quad \text{for } f \in B_p. \tag{2}$$

From the inclusion $B_1 \subset H^\infty \cap \mathcal{B}_0$, we establish a relationship between the Bloch seminorm and the norms in H^∞ and B_1. For $f = \sum_{k=1}^\infty a_k L_{\lambda_k} \in B_1$, with $\{a_k\} \in \ell^1$ and $\lambda_k \in \mathbb{D}$, $k \in \mathbb{N}$,

$$\|f\| \leq \|f\|_\infty \leq \sum_{k=1}^\infty |a_k| \|L_{\lambda_k}\|_\infty = \sum_{k=1}^\infty |a_k|.$$

Taking the infimum over all such sequences $\{a_k\}$ in the above representation of f, we obtain

$$\|f\| \leq \|f\|_\infty \leq \|f\|_{B_1}.$$

For $1 < p < \infty$, B_1 is continuously embedded in B_p. Indeed, observe that for $a \in \mathbb{D}$, by the conformal invariance of the Besov seminorm,

$$\|L_a\|_{B_p} = |a| + b_p(L_a) = |a| + \left(\int_{\mathbb{D}} (1 - |z|^2)^{p-2} \, dA(z) \right)^{1/p} < 1 + \frac{1}{(p-1)^{1/p}}. \tag{3}$$

Therefore, for $f \in B_1$ represented as above, we have

$$\|f\|_{B_p} \leq \sum_{k=1}^\infty |a_k| \|L_{\lambda_k}\|_{B_p} < \left(1 + \frac{1}{(p-1)^{1/p}}\right) \sum_{k=1}^\infty |a_k|.$$

Thus, taking the infimum over all such sequences $\{a_k\}$ in the above representation of f, we obtain

$$\|f\|_{B_p} \leq \left(1 + \frac{1}{(p-1)^{1/p}}\right) \|f\|_{B_1}.$$

Another noteworthy property of the Besov spaces is that the polynomials are dense in B_p $(1 \leq p < \infty)$.

We recommend to the interested reader [3], [8], [9], and [24] for an in-depth study on the spaces BMO, $BMOA$, and the analytic Besov spaces.

Throughout this paper we shall adopt the convention of denoting by C a positive constant which may differ from one occurrence to the next.

2. Boundedness

We begin the section by recalling some useful results that shall be used later.

For ψ, φ analytic functions on $\mathbb{D}$ with $\varphi(\mathbb{D}) \subseteq \mathbb{D}$ and $a \in \mathbb{D}$, define

$$\alpha(\psi, \varphi, a) = |\psi(a)| \|L_{\varphi(a)} \circ \varphi \circ L_a\|_{H^2}.$$

Lemma 2.1 ([12], Lemma 3.4). *For $\psi \in BMOA$ and φ an analytic self-map of $\mathbb{D}$, there exists $C \geq 1$ such that*

$$\|(\psi \circ L_a - \psi(a))(f \circ \varphi \circ L_a - f(\varphi(a)))\|_{H^2}^2 \leq C\|\psi\|_* \|\psi \circ L_a - \psi(a)\|_{H^2}$$

and

$$|\psi(a)| \, \|f \circ \varphi \circ L_a - f(\varphi(a))\|_{H^2} \leq C\, \alpha(\psi, \varphi, a)\|f\|_*$$

for all $f \in BMOA$ and all $a \in \mathbb{D}$.

Lemma 2.2 ([23]). *For $z \in \mathbb{D}$ and $t > -1$,*

$$\int_{\mathbb{D}} \frac{(1 - |w|^2)^t}{|1 - \overline{w}z|^{t+2}} \, dA(w) \asymp \log \frac{2}{1 - |z|^2}, \quad \text{as } |z| \to 1.$$

Lemma 2.3 ([6], Corollary 2.1). *Let φ be an analytic self-map of $\mathbb{D}$. If $\psi \in BMOA$, then the sequence $\{\|\psi\varphi^n\|_{BMOA}\}$ is bounded if and only if $\sup_{a \in \mathbb{D}} \alpha(\psi, \varphi, a) < \infty$.*

Suppose $1 \leq p < \infty$. In the following theorem, we characterize the bounded weighted composition operators from B_p to $BMOA$.

Let ψ be an analytic function on $\mathbb{D}$ and let φ be an analytic self-map of $\mathbb{D}$. For $a \in \mathbb{D}$, define

$$\gamma(\psi, \varphi, p, a) = \left(\log \frac{2}{1 - |\varphi(a)|^2}\right)^{1-1/p} \|\psi \circ L_a - \psi(a)\|_{H^2}.$$

Theorem 2.1. *Let ψ be an analytic function on $\mathbb{D}$, φ an analytic self-map of $\mathbb{D}$, and $1 \leq p < \infty$. The following statements are equivalent.*

(a) *The operator $W_{\psi,\varphi} : B_p \to BMOA$ is bounded.*

(b) $\alpha_{\psi,\varphi} := \sup_{a \in \mathbb{D}} \alpha(\psi, \varphi, a) < \infty$, *and* $\gamma_{\psi,\varphi,p} =: \sup_{a \in \mathbb{D}} \gamma(\psi, \varphi, p, a) < \infty$.

(c) $\sup_{n \in \mathbb{N}} \|\psi\varphi^n\|_{BMOA} < \infty$, *and* $\gamma_{\psi,\varphi,p} < \infty$.

(d) $\sup_{q \in \mathbb{D}} \|\psi(L_q \circ \varphi)\|_{BMOA} < \infty$, *and* $\gamma_{\psi,\varphi,p} < \infty$.

Note that for $p = 1$, the condition $\gamma_{\psi,\varphi,p} < \infty$ means $\psi \in BMOA$.

Proof. (a) $\Rightarrow$ (b) Assume $W_{\psi,\varphi} : B_p \to BMOA$ is bounded. Then $\psi = W_{\psi,\varphi}1 \in BMOA$. Suppose first $p = 1$. Then for $b \in \mathbb{D}$, $\|L_b\|_{B_1} \leq 1$. Thus, by the triangle inequality and the boundedness of $L_{\varphi(a)} \circ \varphi \circ L_a$, we obtain

$$\begin{aligned}\alpha(\psi, \varphi, a) &= |\psi(a)| \|L_{\varphi(a)} \circ \varphi \circ L_a\|_{H^2} \\ &\le \|(\psi \circ L_a - \psi(a))(L_{\varphi(a)} \circ \varphi \circ L_a)\|_{H^2} \\ &\quad + \|(\psi \circ L_a)(L_{\varphi(a)} \circ \varphi \circ L_a)\|_{H^2} \\ &\le \|\psi\|_* + \|W_{\psi,\varphi} L_{\varphi(a)}\|_* \\ &\le \|\psi\|_* + \|W_{\psi,\varphi}\| \|L_{\varphi(a)}\|_{B_1}, \\ &\le \|\psi\|_* + \|W_{\psi,\varphi}\|.\end{aligned} \tag{4}$$

Therefore, $\alpha_{\psi,\varphi}$ is finite, which, by the observation before the proof, proves (b) in the case $p = 1$.

Next suppose $1 < p < \infty$ and fix $a \in \mathbb{D}$. By the triangle inequality and the boundedness of $L_{\varphi(a)} \circ \varphi \circ L_a$ and of the operator $W_{\psi,\varphi}$, recalling that $\psi \in BMOA$, as done in (4) and using (3), we obtain

$$\begin{aligned}\alpha(\psi, \varphi, a) &\le \|\psi\|_* + \|W_{\psi,\varphi} L_{\varphi(a)}\|_* \\ &\le \|\psi\|_* + \left(1 + \frac{1}{(p-1)^{1/p}}\right) \|W_{\psi,\varphi}\|,\end{aligned} \tag{5}$$

which is finite and independent of a. Therefore, $\alpha_{\psi,\varphi}$ is finite.

Next, we show that $\gamma_{\psi,\varphi,p} < \infty$. For $a \in \mathbb{D}$, define

$$f_a(z) = \left(\log \frac{2}{1 - |a|^2}\right)^{-1/p} \log \frac{2}{1 - \overline{a}z}, \quad \text{for } z \in \mathbb{D}. \tag{6}$$

By Lemma 2.2, we have that

$$b_p(f_a) = \left(\log \frac{2}{1 - |a|^2}\right)^{-1/p} \left(\int_{\mathbb{D}} (1 - |z|^2)^{p-2} \left|\frac{\overline{a}}{1 - \overline{a}z}\right|^p dA(z)\right)^{1/p}$$

is bounded by a constant independent of a. Thus,

$$f_a \in B_p \quad \text{and} \quad M = \sup_{a \in \mathbb{D}} \|f_a\|_{B_p} < \infty.$$

Since $W_{\psi,\varphi}$ is bounded, it follows that for each $a \in \mathbb{D}$,

$$\|(\psi(f_{\varphi(a)} \circ \varphi)) \circ L_a) - \psi(a) f_{\varphi(a)}(\varphi(a))\|_{H^2} \le \|W_{\psi,\varphi} f_{\varphi(a)}\|_{BMOA} \le M \|W_{\psi,\varphi}\|.$$

Noting that $f_{\varphi(a)}(\varphi(a)) = \left(\log \frac{2}{1-|\varphi(a)|^2}\right)^{1-1/p}$, we obtain

$$\begin{aligned}\gamma(\psi, \varphi, p, a) &= f_{\varphi(a)}(\varphi(a)) \|\psi \circ L_a - \psi(a)\|_{H^2} \\ &\le \|(\psi(f_{\varphi(a)} \circ \varphi)) \circ L_a - \psi(a) f_{\varphi(a)}(\varphi(a))\|_{H^2} \\ &\quad + \|(\psi \circ L_a)(f_{\varphi(a)} \circ \varphi \circ L_a - f_{\varphi(a)}(\varphi(a)))\|_{H^2} \\ &\le M \|W_{\psi,\varphi}\| + \|(\psi \circ L_a - \psi(a))(f_{\varphi(a)} \circ \varphi \circ L_a - f_{\varphi(a)}(\varphi(a)))\|_{H^2} \\ &\quad + |\psi(a)| \|f_{\varphi(a)} \circ \varphi \circ L_a - f_{\varphi(a)}(\varphi(a))\|_{H^2} \\ &= M \|W_{\psi,\varphi}\| + I_a + II_a,\end{aligned} \tag{7}$$

where I_a and II_a are the second and the third term of the above sum. Recalling that $B_p \subset BMOA$, by Lemma 2.1, we obtain

$$I_a \leq C\|\psi\|_*^{1/2}\|\psi\circ L_a - \psi(a)\|_{H^2}^{1/2} \leq C\|\psi\|_*,$$

so $\sup_{a\in\mathbb{D}} I_a < \infty$.

On the other hand, by Lemma 2.1 and (2), we have

$$II_a \leq C\,\alpha(\psi,\varphi,a)\|f_{\varphi(a)}\|_* \leq C\,\alpha(\psi,\varphi,a)\|f_{\varphi(a)}\|_{B_p}^{p/2},$$

which is bounded by a constant independent of a. Therefore, by (7), taking the supremum over all $a \in \mathbb{D}$, it follows that $\gamma_{\psi,\varphi,p}$ is finite.

(b) $\Rightarrow$ (a) Suppose (b) holds. Note that for $a \in \mathbb{D}$,

$$\|\psi\circ L_a - \psi(a)\|_{H^2} = \left(\log\frac{2}{1-|\varphi(a)|^2}\right)^{\frac{1}{p}-1}\gamma(\psi,\varphi,p,a) \leq (\log 2)^{\frac{1}{p}-1}\gamma_{\psi,\varphi,p},$$

so $\psi \in BMOA$.

Let $f \in B_1$, $f(z) = \sum a_n L_{\lambda_n}(z)$ for $z \in \mathbb{D}$, with $\{a_n\} \in \ell_1$ and $\lambda_n \in \mathbb{D}$, $n \in \mathbb{N}$. Then $|f(z)| \leq \sum |a_n|$, so taking the infimum over all above atomic decompositions of f, we have

$$|f(z)| \leq \|f\|_{B_1}. \tag{8}$$

Next, let $f \in B_p$ with $1 < p < \infty$. Then by Theorem 9 in [24], we have

$$|f(z)| \leq C\|f\|_{B_p}\left(\log\frac{2}{1-|z|^2}\right)^{1-1/p}. \tag{9}$$

By (8), condition (9) holds also for $p = 1$.

Now suppose $\|f\|_{B_p} \leq 1$. For $a \in \mathbb{D}$, by the triangle inequality, we have

$$\begin{aligned}\|W_{\psi,\varphi}f\circ L_a - W_{\psi,\varphi}f(a)\|_{H^2} &\leq \|(\psi\circ L_a - \psi(a))f(\varphi(a))\|_{H^2}\\ &\quad+ \|(\psi\circ L_a)(f\circ\varphi\circ L_a - f(\varphi(a)))\|_{H^2}\\ &= III_a + IV_a,\end{aligned} \tag{10}$$

where, III_a and IV_a are the first and second term, respectively, of the above sum. Using (9), we have

$$\sup_{a\in\mathbb{D}} III_a \leq \sup_{a\in\mathbb{D}} C\left(\log\frac{2}{1-|\varphi(a)|^2}\right)^{1-1/p}\|\psi\circ L_a - \psi(a)\|_{H^2} \leq C\,\gamma_{\psi,\varphi,p}, \tag{11}$$

which is finite by the hypothesis. On the other hand, by the triangle inequality, Lemma 2.1, and (2),

$$\begin{aligned}IV_a &= \|(\psi\circ L_a)(f\circ\varphi\circ L_a - f(\varphi(a)))\|_{H^2}\\ &\leq \|(\psi\circ L_a - \psi(a))(f\circ\varphi\circ L_a - f(\varphi(a)))\|_{H^2}\\ &\quad+ |\psi(a)|\|f\circ\varphi\circ L_a - f(\varphi(a))\|_{H^2}\\ &\leq C\|\psi\|_*^{1/2}\|\psi\circ L_a - \psi(a)\|_{H^2}^{1/2} + C\,\alpha(\psi,\varphi,a)\|f\|_* &(12)\\ &\leq C\,\|\psi\|_* + C\,\alpha_{\psi,\varphi}. &(13)\end{aligned}$$

From (10), (11) and (13), we obtain

$$\|W_{\psi,\varphi}f\|_* \leq C(\gamma_{\psi,\varphi,p} + \|\psi\|_* + \alpha_{\psi,\varphi}).$$

Finally, using (9), we have

$$\begin{aligned}\|W_{\psi,\varphi}f\|_{BMOA} &\leq |\psi(0)f(\varphi(0))| + C(\gamma_{\psi,\varphi,p} + \|\psi\|_* + \alpha_{\psi,\varphi}) \\ &\leq C\left(|\psi(0)|\left(\log\frac{2}{1-|\varphi(0)|^2}\right)^{1-1/p} + \gamma_{\psi,\varphi,p} + \|\psi\|_* + \alpha_{\psi,\varphi}\right).\end{aligned}$$

Therefore $W_{\psi,\varphi}$ is bounded.

The equivalence of (b) and (c) follows from Lemma 2.3.

(a) $\Rightarrow$ (d) Since $\{\|L_q\|_{B_p} : q \in \mathbb{D}\}$ is bounded, the boundedness of $\{\|\psi(L_q \circ \varphi)\|_{BMOA} : q \in \mathbb{D}\}$ follows immediately from the boundedness of $W_{\psi,\varphi}$. The finiteness of $\gamma_{\psi,\varphi,p}$ follows from the equivalence of (a) and (b) or (c).

(d) $\Rightarrow$ (a) By the equivalence of (a) and (b), it suffices to show that $\alpha_{\psi,\varphi} < \infty$. Fix $a \in \mathbb{D}$ and set $M = \sup_{q\in\mathbb{D}} \|\psi(L_q \circ \varphi)\|_{BMOA}$. As noted above, since $\gamma_{\psi,\varphi,p} < \infty$, $\psi \in BMOA$. Then, by (4) and (5),

$$\alpha(\psi,\varphi,a) \leq \|\psi\|_* + \|\psi(L_{\varphi(a)} \circ \varphi)\|_* \leq \|\psi\|_* + M,$$

completing the proof. □

Remark 2.1. For $1 \leq p < \infty$, the boundedness of the sequence $\{\|\psi\varphi^n\|_{BMOA}\}$ in Theorem 2.1, which clearly holds if the symbol ψ is bounded, cannot be shown using the boundedness of $W_{\psi,\varphi}$, because the sequence $\{p_n\}$ defined by $p_n(z) = z^n$, $z \in \mathbb{D}$, is not bounded in B_p, as it will be shown in the proof of Theorem 2.2.

From Theorem 2.1 of [6] and Theorem 2.1 for the case when $p = 1$, we deduce the following result.

Corollary 2.1. *Let ψ and φ be analytic on $\mathbb{D}$ with $\varphi(\mathbb{D}) \subseteq \mathbb{D}$. The following statements are equivalent:*

(a) *The operator $W_{\psi,\varphi} : B_1 \to BMOA$ is bounded.*
(b) $\sup_{a\in\mathbb{D}} \alpha(\psi,\varphi,a) < \infty$, *and* $\psi \in BMOA$.
(c) $\sup_{n\in\mathbb{N}\cup\{0\}} \|\psi\varphi^n\|_{BMOA} < \infty$.
(d) $\sup_{q\in\mathbb{D}} \|\psi(L_q \circ \varphi)\|_{BMOA} < \infty$, *and* $\psi \in BMOA$.
(e) *The operator $W_{\psi,\varphi} : H^\infty \to BMOA$ is bounded.*

For $1 < p < \infty$, define p-$LMOA$ to be the space of analytic functions f of p-logarithmic mean oscillation, that is, satisfying the condition

$$\|f\|_{L,p} := \sup_{a\in\mathbb{D}} \left(\log\frac{2}{1-|a|^2}\right)^{1-1/p} \|f \circ L_a - f(a)\|_{H^2} < \infty.$$

Clearly, the space $LMOA$ of functions of logarithmic mean oscillation, defined as the set of functions f such that

$$\sup_{a\in\mathbb{D}} \log \frac{2}{1-|a|^2} \|f \circ L_a - f(a)\|_{H^2} < \infty,$$

is contained in p-$LMOA$, which in turn is properly contained in $VMOA$. It is well known (see [10], [17], and [14]) that a multiplication operator M_ψ is bounded on $BMOA$ if and only if $\psi \in H^\infty \cap LMOA$.

In the degenerate case of the multiplication operator, we deduce the following result.

Corollary 2.2. *Let ψ be analytic on $\mathbb{D}$. Then*

(a) *$M_\psi : B_1 \to BMOA$ is bounded if and only if $\psi \in H^\infty$.*
(b) *For $1 < p < \infty$, $M_\psi : B_p \to BMOA$ is bounded if and only if $\psi \in H^\infty \cap$ $p-LMOA$.*

Condition (a) follows from the fact that for φ equal to the identity, $\alpha_{\psi,\varphi} = \sup_{a\in\mathbb{D}} |\psi(a)|$, so $\alpha_{\psi,\varphi} < \infty$ means that $\psi \in H^\infty$.

We turn our attention to the search of the isometries among the composition operators from the Besov spaces to $BMOA$. We now show that, although there is a very rich supply of isometries among the composition operators on $BMOA$ [11], there are none from the Besov spaces to $BMOA$.

Theorem 2.2. *For $1 \le p < \infty$, there are no isometries among the composition operators from B_p to $BMOA$.*

Proof. Suppose $C_\varphi : B_p \to BMOA$ is an isometry. For $n \in \mathbb{N}$ and $z \in \mathbb{D}$, let $p_n(z) = z^n$. Then

$$\|\varphi^n\|_{BMOA} = \|C_\varphi p_n\|_{BMOA} = \|p_n\|_{B_p}.$$

Since, by (1),

$$\|\varphi^n\|_{BMOA} \le 2\|\varphi^n\|_\infty \le 2,$$

to reach a contradiction it suffices to show that $\|p_n\|_{B_p} > 2$ for n sufficiently large.

This is clear for $p = 1$ since, by Corollary 3.4 of [1],

$$\|p_n\|_{B_1} = \frac{n+1}{2}\left(1 + \frac{2}{n-1}\right)^{\frac{n-1}{2}}, \quad n \ge 2.$$

Let us next consider the case $1 < p < \infty$.

By the formula

$$\|p_n\|_{B_p}^p = 2n^p \int_0^1 r^{np-p+1}(1-r^2)^{p-2}\, dr,$$

making the change of variable $r^2 = t$, we obtain

$$\|p_n\|_{B_p}^p = n^p \int_0^1 t^{(np-p)/2}(1-t)^{p-2}\, dt = n^p B\left(\frac{np-p}{2} + 1, p-1\right), \tag{14}$$

where $B(x,y) = \int_0^1 t^{x-1}(1-t)^{p-1}\,dt$ denotes the Beta function. Recalling that, by Stirling's formula, for x large and y fixed,

$$B(x,y) \approx x^{-y},$$

from (14), we obtain

$$\|p_n\|_{B_p}^p \approx n^p \left(\frac{np-p}{2}+1\right)^{-p+1} \approx n^p n^{-p+1} = n.$$

Thus, $\|p_n\|_{B_p}^p \to \infty$ as $n \to \infty$, proving the result. □

We end the section by characterizing the bounded weighted composition operators from the Besov space B_p to *VMOA*.

Theorem 2.3. *Let $1 \le p < \infty$, ψ an analytic function on $\mathbb{D}$ and let φ be an analytic self-map of $\mathbb{D}$. Then the following statements are equivalent:*

(a) *$W_{\psi,\varphi} : B_p \to VMOA$ is bounded.*

(b) *$W_{\psi,\varphi} : B_p \to BMOA$ is bounded, $\psi \in VMOA$ and*

$$\lim_{|a|\to 1} |\psi(a)|\|\varphi \circ L_a - \varphi(a)\|_{H^2} = 0. \tag{15}$$

(c) *$W_{\psi,\varphi} : B_p \to BMOA$ is bounded, $\psi \in VMOA$ and*

$$\lim_{|q|\to 1} \int_{\mathbb{D}} |\psi(z)|^2 |\varphi'(z)|^2 (1-|L_q(z)|^2)\,dA(z) = 0. \tag{16}$$

Proof. (a) ⇒ (b) Assume $W_{\psi,\varphi} : B_p \to VMOA$ is bounded. Then $W_{\psi,\varphi}$ is clearly bounded as an operator from B_p to *BMOA*, $\psi = W_{\psi,\varphi}1 \in VMOA$, and $W_{\psi,\varphi}\mathrm{id} = \psi\varphi \in VMOA$, where id denotes the identity function on $\mathbb{D}$. Thus, as shown in the proof of Proposition 4.1 of [12], for $a \in \mathbb{D}$,

$$\begin{aligned} |\psi(a)|\|\varphi \circ L_a - \varphi(a)\|_{H^2} &\le \|(\psi\varphi)\circ L_a - \psi(a)\varphi(a)\|_{H^2} \\ &\quad + \|(\psi \circ L_a - \psi(a))(\varphi \circ L_a)\|_{H^2} \\ &\le \|(\psi\varphi)\circ L_a - \psi(a)\varphi(a)\|_{H^2} + \|\psi \circ L_a - \psi(a)\|_{H^2}, \end{aligned}$$

which approaches 0 as $|a| \to 1$.

(b) ⇒ (a) Suppose $W_{\psi,\varphi}$ is bounded as an operator from B_p to *BMOA*, $\psi \in VMOA$ and (15) holds. To prove (a) it suffices to show that for each $f \in B_p$, $W_{\psi,\varphi}f \in VMOA$. Since the polynomials are dense in B_p, we only need to prove this for the polynomials. This can be easily proved using the argument of the proof of Proposition 4.1 of [12].

(a) ⇒ (c) Suppose (a) holds. Arguing as in (a) ⇒ (b), the functions ψ and $\psi\varphi$ are in *VMOA*. Since for $z \in \mathbb{D}$,

$$|\psi(z)\varphi'(z)|^2 \le 2(|(\psi\varphi)'(z)|^2 + |\psi'(z)\varphi(z)|^2) \le 2(|(\psi\varphi)'(z)|^2 + |\psi'(z)|^2),$$

then, for $q \in \mathbb{D}$, we have

$$\int_{\mathbb{D}} |\psi(z)|^2|\varphi'(z)|^2(1-|L_q(z)|^2)\,dA(z)$$
$$\leq 2\int_{\mathbb{D}} |(\psi\varphi)'(z)|^2(1-|L_q(z)|^2)\,dA(z) + 2\int_{\mathbb{D}} |\psi'(z)|^2(1-|L_q(z)|^2)\,dA(z) \to 0$$

as $|q| \to 1$.

(c) ⇒ (a) Assume $W_{\psi,\varphi} : B_p \to BMOA$ is bounded, $\psi \in VMOA$ and (16) holds. To prove that $W_{\psi,\varphi}$ is bounded as an operator mapping B_p into $VMOA$, it suffices to show that $W_{\psi,\varphi}f \in VMOA$ for any polynomial f. Let $f(z) = \sum_{n=0}^{N} a_n z^n$, $z \in \mathbb{D}$, $N \in \mathbb{N}$. Then, for $q \in \mathbb{D}$, we have

$$\begin{aligned}\int_{\mathbb{D}} &|\psi'(z)|^2|f(\varphi(z))|^2(1-|L_q(z)|^2)\,dA(z)\\ &\leq \|f\|_\infty^2 \int_{\mathbb{D}} |\psi'(z)|^2(1-|L_q(z)|^2)\,dA(z) \to 0\end{aligned} \tag{17}$$

as $|q| \to 1$. Moreover, noting that $|f'(\varphi(z))| \leq \sum_{n=1}^{N} n|a_n| = C$, by (16) we also have

$$\begin{aligned}\int_{\mathbb{D}} &|\psi(z)|^2|(f\circ\varphi)'(z)|^2(1-|L_q(z)|^2)\,dA(z)\\ &\leq C^2 \int_{\mathbb{D}} |\psi(z)|^2|\varphi'(z)|^2(1-|L_q(z)|^2)\,dA(z) \to 0\end{aligned} \tag{18}$$

as $|q| \to 1$.

From (17) and (18) it follows that

$$\int_{\mathbb{D}} |(\psi(f\circ\varphi))'(z)|^2(1-|L_q(z)|^2)\,dA(z) \to 0$$

as $|q| \to 1$, completing the proof. □

For the composition operator, from part (ii) of Corollary 4.2 of [12], and as a special case of Theorem 2.3 of [6], we deduce the following result.

Corollary 2.3. *Let $1 \leq p < \infty$ and φ an analytic self-map of $\mathbb{D}$. Then the following statements are equivalent.*

(a) *$C_\varphi : B_p \to VMOA$ is bounded.*
(b) *$C_\varphi : VMOA \to VMOA$ is bounded.*
(c) *$C_\varphi : H^\infty \to VMOA$ is bounded.*
(d) *$\varphi \in VMOA$.*

The equivalence of (b) and (d) was noted in [3], Theorem 12.

For the multiplication operator, note that by part (a) of Corollary 2.2, the boundedness of $M_\psi : B_1 \to BMOA$ requires $\psi \in H^\infty$. Moreover, if ψ is bounded and φ is the identity, then condition (15) is automatically satisfied. Thus, we obtain the following result.

Corollary 2.4. *Let ψ be analytic on $\mathbb{D}$. Then, $M_\psi : B_1 \to VMOA$ is bounded if and only if $\psi \in H^\infty \cap VMOA$.*

In the case $1 < p < \infty$, by part (b) of Corollary 2.2, the boundedness of $M_\psi : B_p \to BMOA$ requires $\psi \in H^\infty \cap p-LMOA$, and in particular, $\psi \in VMOA$. Moreover, if ψ is bounded and φ is the identity, then condition (15) is automatically satisfied. Therefore, we obtain the following result.

Corollary 2.5. *Let $1 < p < \infty$ and ψ be analytic on $\mathbb{D}$. Then, the following propositions are equivalent:*

(a) $M_\psi : B_p \to VMOA$ *is bounded.*
(b) $M_\psi : B_p \to BMOA$ *is bounded.*
(c) $\psi \in H^\infty \cap p-LMOA$.

3. Compact weighted composition operators from B_p to $BMOA$

The following compactness criterion can be easily proved using Lemma 3.7 of [19].

Lemma 3.1. *Let $1 \leq p < \infty$. A weighted composition operator $W_{\psi,\varphi}$ from B_p to BMOA is compact if and only if, for every bounded sequence $\{f_n\}$ in B_p converging to 0 uniformly on compact subsets of $\mathbb{D}$, the sequence $\{\|W_{\psi,\varphi} f_n\|_{BMOA}\}$ approaches 0 as $n \to \infty$.*

Notation 3.1. For $\theta \in [0, 2\pi)$ and $h \in (0, 1)$, define

$$S(h, \theta) = \{z \in \mathbb{D} : |z - e^{i\theta}| < h\}.$$

The following result is an immediate corollary of Proposition 3.4 in [19].

Proposition 3.1. *Let $\{\mu_q : q \in I\}$ be a collection of positive measures on $\mathbb{D}$. Then, for $1 < p < \infty$, the following statements are equivalent:*

(a) $\displaystyle\lim_{h\to 0} \sup_{\theta\in[0,2\pi), q\in I} \frac{\mu_q(S(h,\theta))}{h^p} = 0.$

(b) $\displaystyle\lim_{|a|\to 1} \sup_{q\in I} \int_{\mathbb{D}} |L_a'(w)|^p \, d\mu_q(w) = 0.$

We now provide a sufficient condition for compactness of the weighted composition operator from B_p to $BMOA$ for $1 < p \leq 2$.

Theorem 3.1. *Let ψ be an analytic function on $\mathbb{D}$, φ be an analytic self-map of $\mathbb{D}$, and $1 < p \leq 2$. If $\lim_{|a|\to 1} \|\psi(L_a \circ \varphi) - a\psi\|_{**} = 0$ and, for some fixed $t > 0$,*

$$\lim_{h\to 0} \sup_{q\in\mathbb{D}, \theta\in[0,2\pi)} \frac{\log \frac{1}{h}}{h^{2-t}} \int_{\varphi^{-1}(S(h,\theta))} \frac{|\psi'(w)|^2(1 - |L_q(w)|^2)}{(1 - |\varphi(w)|)^t} \, dA(w) = 0,$$

then $W_{\psi,\varphi} : B_p \to BMOA$ is a compact operator.

Proof. Suppose that $\{f_n\}$ is a bounded sequence in $\mathcal{D}$ converging to 0 uniformly on compact subsets of $\mathbb{D}$. Fix $\varepsilon > 0$ and let $\delta > 0$ be such that for any $\theta \in [0, 2\pi]$, $q \in \mathbb{D}$ and any $h \in (0, \delta)$,

$$\int_{\varphi^{-1}(S(h,\theta))} \frac{|\psi'(w)|^2\,(1-|L_q(w)|^2)}{(1-|\varphi(w)|)^t}\,dA(w) < \varepsilon\,\frac{h^{2-t}}{\log\frac{1}{h}}\,. \tag{19}$$

For $w \in \mathbb{D}$, the mean value property and Jensen's inequality yield,

$$|f_n(\varphi(w))|^2 \leq \frac{4}{(1-|\varphi(w)|)^2}\int_{|\varphi(w)-z|<\frac{1-|\varphi(w)|}{2}} |f_n(z)|^2\,dA(z). \tag{20}$$

Multiply (20) by $|\psi'(w)|^2(1-|L_q(w)|^2)$, integrate and use Fubini's theorem to get

$$\begin{aligned} I &= \int_{\mathbb{D}} |f_n(\varphi(w))|^2|\psi'(w)|^2(1-|L_q(w)|^2)\,dA(w) \qquad (21)\\ &\leq \int_{\mathbb{D}} \frac{4}{(1-|\varphi(w)|)^2}\int_{|\varphi(w)-z|<\frac{1-|\varphi(w)|}{2}} |f_n(z)|^2\,dA(z)\,|\psi'(w)|^2(1-|L_q(w)|^2)\,dA(w)\\ &= \int_{\mathbb{D}} |f_n(z)|^2\int_{\mathbb{D}} \frac{4\,\chi_{\{z:|\varphi(w)-z|<\frac{1-|\varphi(w)|}{2}\}}(z)}{(1-|\varphi(w)|)^{2-t}}\,\frac{|\psi'(w)|^2(1-|L_q(w)|^2)}{(1-|\varphi(w)|)^t}\,dA(w)\,dA(z). \end{aligned}$$

Note that, if $|\varphi(w)-z| < \frac{1-|\varphi(w)|}{2}$ and $z = |z|e^{i\theta} \in \mathbb{D}$, then

$$\begin{aligned} 1-|\varphi(w)| &\leq |\varphi(w)-e^{i\theta}| \leq |\varphi(w)-z| + |z-e^{i\theta}|\\ &= |\varphi(w)-z| + ||z|-1| < \frac{1-|\varphi(w)|}{2} + 1 - |z|. \end{aligned}$$

Hence $\frac{1-|\varphi(w)|}{2} \leq 1-|z|$ and so $|\varphi(w)-e^{i\theta}| < 2(1-|z|)$, or $\varphi(w) \in S(2(1-|z|),\theta)$. Furthermore, $1-|z| \leq 1-|\varphi(w)| + |\varphi(w)-z| < \frac{3}{2}(1-|\varphi(w)|)$. So

$$\frac{1}{2}\,\frac{1}{1-|z|} \leq \frac{1}{1-|\varphi(w)|} \leq \frac{3}{2}\,\frac{1}{1-|z|}.$$

Therefore (21) yields

$$\begin{aligned} I &\leq C\int_{\mathbb{D}} \frac{|f_n(z)|^2}{(1-|z|)^{2-t}}\int_{\varphi^{-1}(S(2(1-|z|),\theta))} \frac{|\psi'(w)|^2(1-|L_q(w)|^2)}{(1-|\varphi(w)|)^t}\,dA(w)\,dA(z)\\ &= C\int_{|z|>1-\frac{\delta}{2}} \frac{|f_n(z)|^2}{(1-|z|)^{2-t}}\left(\int_{\varphi^{-1}(S(2(1-|z|),\theta))} \frac{|\psi'(w)|^2(1-|L_q(w)|^2)}{(1-|\varphi(w)|)^t}\,dA(w)\right)dA(z)\\ &\quad + C\int_{|z|\leq 1-\frac{\delta}{2}} \frac{|f_n(z)|^2}{(1-|z|)^{2-t}}\left(\int_{\varphi^{-1}(S(2(1-|z|),\theta))} \frac{|\psi'(w)|^2(1-|L_q(w)|^2)}{(1-|\varphi(w)|)^t}\,dA(w)\right)dA(z)\\ &= C(I_1+I_2), \qquad (22) \end{aligned}$$

for any $0 < \delta < 1$.

By (19) and (22), and since each $f_n \in \mathcal{D}$,

$$I_1 \leq C\,\varepsilon\int_{|z|>1-\frac{\delta}{2}} \frac{|f_n(z)|^2}{\log\frac{2}{1-|z|}}\,dA(z) \leq C\,\varepsilon.$$

Since $f_n \to 0$ uniformly on compact sets, there exists $N \in \mathbb{N}$ such that if $n \geq N$, then $|f_n(z)| \leq \varepsilon$ for all z with $|z| \leq 1 - \frac{\delta}{2}$. Therefore

$$I_2 \leq C \, \|\psi\|_{**}^2 \, \varepsilon.$$

Above we proved that

$$\lim_{n\to\infty} \sup_{q\in\mathbb{D}} I = 0. \tag{23}$$

Given any sequence $\{a_n\}$ in $\mathbb{D}$ with $|a_n| \to 1$ as $n \to \infty$, we have that $\lim_{n\to\infty} \|\psi L_{a_n} \circ \varphi - a_n\psi\|_{**} = 0$, or

$$\begin{aligned}\lim_{n\to\infty} \sup_{q\in\mathbb{D}} \int_{\mathbb{D}} &|\psi'(z)L_{a_n}(\varphi(z)) - a_n\psi'(z) + \psi(z)L'_{a_n}(\varphi(z))\varphi'(z)|^2 \\ &\times (1-|L_q(z)|^2)\, dA(w) = 0\,.\end{aligned} \tag{24}$$

Since $\{L_{a_n} - a_n\}$ is a bounded sequence in $\mathcal{D}$ that converges to 0 uniformly on compact subsets of $\mathbb{D}$, we may use an argument similar to the proof of (23) to get

$$\lim_{n\to\infty} \sup_{q\in\mathbb{D}} \int_{\mathbb{D}} |\psi'(z)|^2 \, |L_{a_n}(\varphi(z)) - a_n|^2 \, (1-|L_q(z)|^2)\, dA(z) = 0\,.$$

Thus by (24) we conclude that

$$\lim_{n\to\infty} \sup_{q\in\mathbb{D}} \int_{\mathbb{D}} |\psi(z)|^2 \, |L'_{a_n}(\varphi(z))|^2 \, |\varphi'(z)|^2 \, (1-|L_q(z)|^2) dA(z) = 0.$$

Hence

$$\lim_{|a|\to 1} \sup_{q\in\mathbb{D}} \int_{\mathbb{D}} |\psi(z)|^2 \, |L'_a(\varphi(z))|^2 \, |\varphi'(z)|^2 \, (1-|L_q(z)|^2) dA(z) = 0\,.$$

Thus by making a non-univalent change of variables as done in [15, page 186], we see that

$$\lim_{|a|\to 1} \sup_{q\in\mathbb{D}} \int_{\mathbb{D}} |L'_a(w)|^2 \, N(w,q,\varphi,\psi)\, dA(w) = 0,$$

where

$$N(w,q,\varphi,\psi) = \sum_{\varphi(z)=w} |\psi(z)|^2(1-|L_q(z)|^2)\,.$$

Therefore, by Proposition 3.1,

$$\lim_{h\to 0} \sup_{\theta\in[0,2\pi),\ q\in\mathbb{D}} \frac{1}{h^2} \int_{S(h,\theta)} N(w,q,\varphi,\psi)\, dA(w) = 0\,.$$

Fix $\varepsilon > 0$ and let $\delta > 0$ be such that for any $\theta \in [0, 2\pi)$ and any $q \in \mathbb{D}$, if $h < \delta$ then

$$\int_{S(h,\theta)} N(w,q,\varphi,\psi)\, dA(w) < \varepsilon h^2\,.$$

The mean value property and Jensen's inequality yield

$$|f_n'(w)|^2 \leq \frac{4}{(1-|w|)^2} \int_{|w-z|<\frac{1-|w|}{2}} |f_n'(z)|^2 \, dA(z) \, .$$

Thus, fixing $q \in \mathbb{D}$, we obtain

$$\begin{aligned} II &= \int_{\mathbb{D}} |f_n'(w)|^2 \, N(w,q,\varphi,\psi) \, dA(w) \\ &\leq \int_{\mathbb{D}} \frac{4}{(1-|w|)^2} \left(\int_{|w-z|<\frac{1-|w|}{2}} |f_n'(z)|^2 \, dA(z) N(w,q,\varphi,\psi) \right) \, dA(w). \end{aligned} \tag{25}$$

Note that if $|w-z| < \frac{1-|w|}{2}$, then $w \in S(2(1-|z|),\theta)$ and

$$\frac{1}{2} \, \frac{1}{1-|z|} \leq \frac{1}{1-|w|} \leq \frac{3}{2} \, \frac{1}{1-|z|} \, ,$$

where $z = |z|e^{i\theta}$. Therefore by (25) and Fubini's theorem, we have

$$II \leq C \int_{\mathbb{D}} \frac{|f_n'(z)|^2}{(1-|z|)^2} \left(\int_{S(2(1-|z|,\theta))} N(w,q,\varphi,\psi) \, dA(w) \right) \, dA(z) \, . \tag{26}$$

Next split the first integral in (26) into two pieces, one over the set $\{z \in \mathbb{D} : |z| > 1 - \frac{\delta}{2}\}$ and the other over the complementary set. Then,

$$\begin{aligned} &\int_{|z|>1-\frac{\delta}{2}} \frac{|f_n'(z)|^2}{(1-|z|)^2} \left(\int_{S(2(1-|z|),\theta)} N(w,q,\varphi,\psi) \, dA(w) \right) \, dA(z) \\ &\quad \leq C \, \varepsilon \int_{|z|>1-\frac{\delta}{2}} \frac{|f_n'(z)|^2}{(1-|z|)^2} (1-|z|)^2 \, dA(z) \leq C \, \varepsilon \|f_n\|_{\mathcal{D}}^2 < C \, \varepsilon, \end{aligned} \tag{27}$$

and

$$\begin{aligned} &\int_{|z|\leq 1-\frac{\delta}{2}} \frac{|f_n'(z)|^2}{(1-|z|)^2} \left(\int_{S(2(1-|z|),\theta)} N(w,q,\varphi,\psi) \, dA(w) \right) \, dA(z) \\ &\quad \leq C \left(\sup_{q\in\mathbb{D}} \int_{\mathbb{D}} N(w,q,\varphi,\psi) \, dA(w) \right) \int_{|z|\leq 1-\frac{\delta}{2}} |f_n'(z)|^2 \, dA(z) \quad \leq C \, \varepsilon \end{aligned} \tag{28}$$

for n large enough, since $f_n' \to 0$ uniformly on $\{z \in \mathbb{D} : |z| \leq 1 - \frac{\delta}{2}\}$.

Thus (26), (27), and (28) yield

$$\sup_{q\in\mathbb{D}} \int_{\mathbb{D}} |f_n'(w)|^2 \, N(w,q,\varphi,\psi) \, dA(w) < C \, \varepsilon$$

for n large enough, or

$$\sup_{q\in\mathbb{D}} \int_{\mathbb{D}} |\psi(z)|^2 |f_n'(\varphi(z))|^2 |\varphi'(z)|^2 \, (1-|L_q(z)|^2) \, dA(z) < C \, \varepsilon \, .$$

This shows that

$$\lim_{n\to\infty} \sup_{q\in\mathbb{D}} II = 0 \, . \tag{29}$$

Note that

$$\|W_{\psi,\varphi}f_n\|_{**}^2 = \sup_{q\in\mathbb{D}} \int_{\mathbb{D}} |\psi'(z)f_n(\varphi(z)) + \psi(z)f_n'(\varphi(z))\varphi'(z)|^2 \left(1 - |L_q(z)|^2\right) dA(z)\,.$$

Thus, by (23) and (29), it follows that $\lim_{n\to\infty} \|W_{\psi,\varphi}f_n\|_{**} = 0$, and Lemma 3.1 shows that the operator $W_{\psi,\varphi} : \mathcal{D} \to BMOA$ is compact. Therefore, since B_p is continuously embedded in $\mathcal{D}$ if $1 < p \le 2$, $W_{\psi,\varphi} : B_p \to BMOA$ is also a compact operator. □

We obtain compactness criteria for the weighted composition operator from B_1 to $BMOA$ under a restriction on the symbol ψ. In particular, under this restriction, we obtain that $W_{\psi,\varphi}$ is compact as an operator from B_1 to $BMOA$ if and only if it is compact as an operator from H^∞ to $BMOA$.

For an analytic self-map φ of $\mathbb{D}$, for $t, R \in (0,1)$, and for $a \in \mathbb{D}$, define

$$\tilde{E}(\varphi, a, t) = \{\zeta \in \partial\mathbb{D} : |\varphi(L_a(\zeta))| > t\} \quad \text{and} \quad \Omega_R = \{a \in \mathbb{D} : |\varphi(a)| \le R\}.$$

Theorem 3.2. *Let ψ be analytic on $\mathbb{D}$, φ an analytic self-map of $\mathbb{D}$ such that $W_{\psi,\varphi} : B_1 \to BMOA$ is bounded, and suppose*

$$\lim_{|\varphi(a)|\to 1} \|\psi \circ L_a - \psi(a)\|_{H^2} = 0.$$

Then the following propositions are equivalent:

(a) *$W_{\psi,\varphi} : B_1 \to BMOA$ is compact.*

(b) $\lim_{n\to\infty} \|\psi\varphi^n\|_{BMOA} = 0$.

(c) $\lim_{|\varphi(a)|\to 1} \alpha(\psi, \varphi, a) = 0$ *and for all $R \in (0,1)$,*

$$\lim_{t\to 1} \sup_{a\in\Omega_R} \int_{\tilde{E}(\varphi,a,t)} |\psi \circ L_a(\zeta)|^2 \, dm(\zeta) = 0. \tag{30}$$

(d) *$W_{\psi,\varphi} : H^\infty \to BMOA$ is compact.*

Proof. Note that by Corollary 2.1, $W_{\psi,\varphi}$ is bounded as an operator from B_1 to $BMOA$ if and only if it is bounded as an operator from H^∞ to $BMOA$.

The equivalence of (b), (c) and (d) was shown in [6]. Thus, it suffices to show that (a) implies (c) and (d) implies (a).

(a) $\Longrightarrow$ (c) Suppose $W_{\psi,\varphi} : B_1 \to BMOA$ is compact. Proceeding as in the proof of (b) implies (c) of Theorem 2.2 in [6], let $\{a_n\}$ be a sequence in $\mathbb{D}$ such that $|\varphi(a_n)| \to 1$ and let $\{h_n\}$ be the bounded sequence in B_1 defined by $h_n = L_{\varphi(a_n)} - \varphi(a_n)$. By the compactness of $W_{\psi,\varphi}$, we see that $\|W_{\psi,\varphi}h_n\|_* \to 0$ as $n \to \infty$, and hence, as shown there, $\alpha(\psi, \varphi, a_n) \to 0$ as $n \to \infty$. Moreover, since the sequence $\{p_n\}$ defined as $p_n(z) = \frac{1}{n}z^n$ is bounded in B_1 and converges to 0 uniformly on compact subsets of $\mathbb{D}$, by Lemma 3.1, it follows that $\|\frac{1}{n}\psi\varphi^n\|_{BMOA} \to 0$ as $n \to \infty$. Condition (30) now follows in a similar fashion as in the proof of Proposition 2.1 of [6].

(d) $\Rightarrow$ (a) Since B_1 is continuously embedded in H^∞, if $W_{\psi,\varphi} : H^\infty \to BMOA$ is a compact operator, then so is $W_{\psi,\varphi} : B_1 \to BMOA$. □

We now turn our attention to the case $1 < p < \infty$. Unlike the case $p = 1$, we obtain full compactness criteria without setting any restriction on the symbols.

Theorem 3.3. *Let $1 < p < \infty$, ψ an analytic function on $\mathbb{D}$ and φ an analytic self-map of $\mathbb{D}$. If $W_{\psi,\varphi} : B_p \to BMOA$ is bounded, then the following statements are equivalent:*

(a) *$W_{\psi,\varphi} : B_p \to BMOA$ is compact.*
(b) *$\lim_{|\varphi(a)|\to 1} \alpha(\psi, \varphi, a) = 0$, $\lim_{|\varphi(a)|\to 1} \gamma(\psi, \varphi, p, a) = 0$, and condition (30) holds for all $R \in (0,1)$.*
(c) *$\lim_{n\to\infty} \|\psi\varphi^n\|_{BMOA} = 0$ and $\lim_{|\varphi(a)|\to 1} \gamma(\psi, \varphi, p, a) = 0$.*

Proof. (a) $\Rightarrow$ (b) Suppose $W_{\psi,\varphi} : B_p \to BMOA$ is compact. Since B_1 is continuously embedded in B_p, $W_{\psi,\varphi}$ is also compact as an operator from B_1 to $BMOA$. By Theorem 2.1, the boundedness of $W_{\psi,\varphi} : B_p \to BMOA$ implies that $\gamma_{\psi,\varphi,p}$ is finite. Let $\{a_n\}$ be a sequence in $\mathbb{D}$ such that $|\varphi(a_n)| \to 1$ as $n \to \infty$. Since $\left(\log \frac{2}{1-|\varphi(a_n)|^2}\right)^{1-1/p} \to \infty$ as $n \to \infty$, the quantity $\|\psi \circ L_{a_n} - \psi(a_n)\|_{H^2}$ must converge to 0. Thus, by Theorem 3.2, we deduce that $\lim_{|\varphi(a)|\to 1} \alpha(\psi, \varphi, a) = 0$ and (30) holds for each $R \in (0,1)$.

Next, we wish to show that $\lim_{n\to\infty} \gamma(\psi, \varphi, p, a_n) = 0$.

For $n \in \mathbb{N}$, let $f_{\varphi(a_n)}$ be defined as in (6). Then, $\{f_{\varphi(a_n)}\}$ is bounded in B_p and converges to 0 uniformly on compact subsets of $\mathbb{D}$. Thus, by Lemma 3.1, $\|W_{\psi,\varphi} f_{\varphi(a_n)}\|_{BMOA} \to 0$ as $n \to \infty$. In particular, for each $n \in \mathbb{N}$, as shown in the proof of Theorem 2.1, we have

$$
\begin{aligned}
\gamma(\psi, \varphi, p, a_n) &= f_{\varphi(a_n)}(\varphi(a_n))\|\psi \circ L_{a_n} - \psi(a_n)\|_{H^2} \\
&\le \|(\psi(f_{\varphi(a_n)} \circ \varphi)) \circ L_{a_n} - \psi(a_n) f_{\varphi(a_n)}(\varphi(a_n))\|_{H^2} \\
&\quad + \|(\psi \circ L_{a_n})(f_{\varphi(a_n)} \circ \varphi \circ L_{a_n} - f_{\varphi(a_n)}(\varphi(a_n)))\|_{H^2} \\
&\le \|W_{\psi,\varphi} f_{\varphi(a_n)}\|_{BMOA} \\
&\quad + \|(\psi \circ L_{a_n} - \psi(a_n))(f_{\varphi(a_n)} \circ \varphi \circ L_{a_n} - f_{\varphi(a_n)}(\varphi(a_n)))\|_{H^2} \\
&\quad + |\psi(a_n)| \|f_{\varphi(a_n)} \circ \varphi \circ L_{a_n} - f_{\varphi(a_n)}(\varphi(a_n))\|_{H^2} \\
&\le \|W_{\psi,\varphi} f_{\varphi(a_n)}\|_{BMOA} + C\|\psi\|_*^{1/2} \|\psi \circ L_{a_n} - \psi(a_n)\|_{H^2}^{1/2} \\
&\quad + C\,\alpha(\psi, \varphi, a_n)\|f_{\varphi(a_n)}\|_{B_p}^{p/2} \to 0,
\end{aligned}
$$

as $n \to \infty$, due to the boundedness of the sequence $\{\|f_{\varphi(a_n)}\|_{B_p}\}$. Therefore (b) holds.

(b) $\Rightarrow$ (a) Suppose (b) holds. The proof of the compactness of $W_{\psi,\varphi} : B_p \to BMOA$ is modeled after the proof of Theorem 3.1 in [12]. Let $\{f_n\}$ be a bounded sequence in B_p converging to 0 uniformly on compact subsets of $\mathbb{D}$. By Lemma 3.1, it suffices to show that $\|W_{\psi,\varphi} f_n\|_{BMOA} \to 0$ as $n \to \infty$. Fix $\varepsilon \in (0,1]$. By the hypothesis, there exist $R \in \left(\frac{2}{3}, 1\right)$ and $t \in [\frac{1}{2}, 1)$ such that $\alpha(\psi, \varphi, a) < \varepsilon$, $\gamma(\psi, \varphi, p, a) < \varepsilon$

for $|\varphi(a)| > R$, and

$$\sup_{a\in\Omega_R}\int_{\tilde{E}(\varphi,a,t)} |(\psi\circ L_a)(\zeta)|^2\,dm(\zeta) < \varepsilon^4.$$

For $a \in \mathbb{D}$, with $|\varphi(a)| > R$, we have

$$\begin{aligned}
&\|(\psi\circ L_a)(f_n\circ\varphi\circ L_a) - \psi(a)f_n(\varphi(a))\|_{H^2}\\
&\quad\le \|(\psi\circ L_a - \psi(a))f_n(\varphi(a))\|_{H^2}\\
&\qquad + \|(\psi\circ L_a - \psi(a))(f_n\circ\varphi\circ L_a - f_n(\varphi(a)))\|_{H^2}\\
&\qquad + |\psi(a)|\,\|f_n\circ\varphi\circ L_a - f_n(\varphi(a))\|_{H^2}\\
&\quad = I + II + III.
\end{aligned}\tag{31}$$

Then, by (9) and the boundedness of the sequence $\{f_n\}$ in B_p, for $|\varphi(a)| > R$, we obtain

$$I \le C\,\|f_n\|_{B_p}\left(\log\frac{2}{1-|\varphi(a)|^2}\right)^{1-1/p}\|\psi\circ L_a - \psi(a)\|_{H^2} \le C\,\gamma(\psi,\varphi,p,a) < C\,\varepsilon.$$

On the other hand, by Lemma 2.1 and (2), we have

$$\begin{aligned}
II &\le C\,\|\psi\|_*^{1/2}\|\psi\circ L_a - \psi(a)\|_{H^2}^{1/2} < C\,\varepsilon^{1/2}, \quad\text{and}\\
III &\le C\,\|f_n\|_*\alpha(\psi,\varphi,a) \le C\,\|f_n\|_{B_p}^{p/2}\alpha(\psi,\varphi,a) < C\,\varepsilon.
\end{aligned}$$

Therefore, from (31), it follows that for $|\varphi(a)| > R$,

$$\|(\psi\circ L_a)(f_n\circ\varphi\circ L_a) - \psi(a)f_n(\varphi(a))\|_{H^2} < C\,\varepsilon^{1/2}.$$

Next, assume $|\varphi(a)| \le R$. Then

$$\begin{aligned}
&\|(\psi\circ L_a)(f_n\circ\varphi\circ L_a) - \psi(a)f_n(\varphi(a))\|_{H^2}\\
&\quad\le \|(\psi\circ L_a - \psi(a))f_n(\varphi(a))\|_{H^2} + \|(\psi\circ L_a)F_{n,a}\|_{H^2} = IV + V,
\end{aligned}$$

where $F_{n,a} := f_n\circ\varphi\circ L_a - f_n(\varphi(a))$. By the uniform convergence of $\{f_n\}$ to 0 on compact subsets of $\mathbb{D}$,

$$IV \le \|\psi\|_*\max_{|z|\le R}|f_n(z)| < \varepsilon,$$

for all n sufficiently large. On the other hand,

$$\begin{aligned}
V \le &\sup_{a\in\Omega_R}\int_{\partial\mathbb{D}\setminus E(\varphi,a,t)} |(\psi\circ L_a)(\zeta)F_{n,a}(\zeta)|^2\,dm(\zeta)\\
&+\sup_{a\in\Omega_R}\int_{E(\varphi,a,t)} |(\psi\circ L_a)(\zeta)F_{n,a}(\zeta)|^2\,dm(\zeta),
\end{aligned}$$

where $E(\varphi,a,t) := \{\zeta\in\partial\mathbb{D} : |(L_{\varphi(a)}\circ\varphi\circ L_a)(\zeta)| > t\}$. In [12] it was observed that for all $\zeta\in\partial\mathbb{D}$ such that $|(\varphi\circ L_a)(\zeta)| < 1$ and $|\varphi(a)| \le R$,

$$\frac{1-R}{1+R} \le \frac{1-|(L_{\varphi(a)}\circ\varphi\circ L_a)(\zeta)|^2}{1-|(\varphi\circ L_a)(\zeta)|^2} \le \frac{1+R}{1-R}.$$

In particular, it is easy to see that condition (30) is equivalent to

$$\lim_{t\to 1}\sup_{a\in\Omega_R}\int_{E(\varphi,a,t)}|(\psi\circ L_a)(\zeta)|^2\,dm(\zeta)=0. \tag{32}$$

Let us now obtain an upper estimate for V. On page 37 in [12], it was shown that letting $G_{n,a}=f_n\circ L_{\varphi(a)}-f_n(\varphi(a))$ and $\lambda_a=L_{\varphi(a)}\circ\varphi\circ L_a$, one has $G_{n,a}(0)=0$ and $F_{n,a}=G_{n,a}\circ\lambda_a$. Furthermore, if $|\varphi(a)|\le R$ and $\zeta\in\partial\mathbb{D}\backslash E(\varphi,a,t)$, then

$$\begin{aligned}|F_{n,a}(\zeta)|=|G_{n,a}(\lambda_a(\zeta))|&\le 2|\lambda_a(\zeta)|\max_{|w|\le t}|G_{n,a}(w)|\\ &\le 2\Big(\max_{|w|\le t}|f_n(L_{\varphi(a)}(w))|+|f_n(\varphi(a))|\Big)<\varepsilon,\end{aligned}$$

for all n sufficiently large. Moreover, it was also shown that

$$\begin{aligned}\|(\psi\circ L_a)\lambda_a\|_{H^2}&\le\|(\psi\circ L_a-\psi(a))\lambda_a\|_{H^2}+\|\psi(a)\lambda_a\|_{H^2}\\ &\le\|\psi\|_*\|\lambda_a\|_\infty+\alpha_{\psi,\varphi}<\infty.\end{aligned}$$

Therefore

$$\sup_{a\in\Omega_R}\int_{\partial\mathbb{D}\backslash E(\varphi,a,t)}|(\psi\circ L_a)(\zeta)F_{n,a}(\zeta)|^2\,dm(\zeta)\le C\,\varepsilon^2 \tag{33}$$

for all n sufficiently large, where C is a positive constant depending only on ψ and $\alpha_{\psi,\varphi}$.

Using Hölder's inequality, we have

$$\begin{aligned}&\sup_{a\in\Omega_R}\int_{E(\varphi,a,t)}|(\psi\circ L_a)(\zeta)F_{n,a}(\zeta)|^2\,dm(\zeta)\\ &\quad\le\sup_{a\in\Omega_R}\left(\int_{E(\varphi,a,t)}|(\psi\circ L_a)(\zeta)|^2\,dm(\zeta)\right)^{1/2} \qquad (34)\\ &\quad\times\left(\int_{E(\varphi,a,t)}|(\psi\circ L_a)(\zeta)|^2|F_{n,a}(\zeta)|^4\,dm(\zeta)\right)^{1/2}.\end{aligned}$$

On the other hand, again using Hölder's inequality, we obtain

$$\int_{E(\varphi,a,t)}|(\psi\circ L_a)(\zeta)|^2|F_{n,a}(\zeta)|^4\,dm(\zeta)\le\|(\psi\circ L_a)F_{n,a}\|_{H^4}\|F_{n,a}\|_{H^4}.$$

Since, as shown on page 38 in [12], $\|(\psi\circ L_a)F_{n,a}\|_{H^4}$ and $\|F_{n,a}\|_{H^4}$ are bounded by constants independent of a, from (32), (33) and (34) we obtain

$$V<C\,\varepsilon^2,$$

for some positive constant C. Hence $\|W_{\psi,\varphi}f_n\|_*\to 0$ as $n\to\infty$.

Since $|\psi(0)f_n(\varphi(0))|\to 0$, it follows that $\|W_{\psi,\varphi}f_n\|_{BMOA}\to 0$ as $n\to\infty$, as desired. Finally, note that the equivalence of (b) and (c) follows at once from Theorem 3.2. □

By Theorem 2 in [13], C_φ is a compact operator on $BMOA$ if and only if

$$\lim_{|\varphi(a)|\to 1}\|L_\varphi(a)\circ\varphi\circ L_a\|_{H^2}=0.$$

Therefore, from Theorem 3.3 we obtain the following result.

Corollary 3.1.
(a) *Let φ be an analytic self-map of $\mathbb{D}$. Then $C_\varphi : B_p \to BMOA$ is a compact operator if and only if $C_\varphi : BMOA \to BMOA$ is a compact operator.*
(b) *The only compact multiplication operator from B_p to BMOA has symbol identically* 0.

We end the section with the following:

Open Question. Does $W_{\psi,\varphi} : B_1 \to BMOA$ compact imply that

$$\lim_{|\varphi(a)|\to 1} \|\psi \circ L_a - \psi(a)\|_{H^2} = 0?$$

4. Compact weighted composition operators from B_p to VMOA

We begin the section with two lemmas that will be needed to prove the main result of this section.

Lemma 4.1. *Assume $1 < p < \infty$, ψ is analytic on $\mathbb{D}$ and φ is an analytic self-map of $\mathbb{D}$. Then, $\lim_{|a|\to 1} \gamma(\psi, \varphi, p, a) = 0$ if and only if $\psi \in VMOA$ and $\lim_{|\varphi(a)|\to 1} \gamma(\psi, \varphi, p, a) = 0$.*

Proof. Suppose $\lim_{|a|\to 1} \gamma(\psi, \varphi, p, a) = 0$. Then, for $a \in \mathbb{D}$, we have

$$(\log 2)^{1-1/p} \|\psi \circ L_a - \psi(a)\|_{H^2} \le \gamma(\psi, \varphi, p, a) \to 0$$

as $|a| \to 1$. Thus, $\psi \in VMOA$. Moreover, if $|\varphi(a)| \to 1$, then $|a| \to 1$, so $\gamma(\psi, \varphi, p, a) \to 0$.

Conversely, suppose $\psi \in VMOA$ and $\gamma(\psi, \varphi, p, a) \to 0$ as $|\varphi(a)| \to 1$. Then, for every $\varepsilon > 0$, there exists $r \in (0, 1)$ such that $\gamma(\psi, \varphi, p, a) < \varepsilon$ for $r < |\varphi(a)| < 1$. Since $\psi \in VMOA$, there exists a $\delta \in (0, 1)$, such that $\|\psi \circ L_a - \psi(a)\|_{H^2} < \varepsilon$ for $\delta < |a| < 1$.

Therefore, if $\delta < |a| < 1$ and $r < |\varphi(a)| < 1$, then $\gamma(\psi, \varphi, p, a) < \varepsilon$. On the other hand, if $|\varphi(a)| \le r$ and $\delta < |a| < 1$, then

$$\gamma(\psi, \varphi, p, a) \le \left(\log \frac{2}{1 - r^2}\right)^{1-1/p} \|\psi \circ L_a - \psi(a)\|_{H^2} < \left(\log \frac{2}{1 - r^2}\right)^{1-1/p} \varepsilon.$$

The result follows at once. □

Lemma 4.2. *Let ψ be analytic on $\mathbb{D}$ and φ is an analytic self-map of $\mathbb{D}$. Then $\lim_{|a|\to 1} \alpha(\psi, \varphi, a) = 0$ if and only if $\lim_{|a|\to 1} |\psi(a)| \|\varphi \circ L_a - \varphi(a)\|_{H^2} = 0$ and $\lim_{|\varphi(a)|\to 1} \alpha(\psi, \varphi, a) = 0$.*

Proof. Suppose $\lim_{|a|\to 1} \alpha(\psi, \varphi, a) = 0$. Then, by Lemma 2.1 applied to the identity function, for $a \in \mathbb{D}$, we have

$$|\psi(a)| \|\varphi \circ L_a - \varphi(a)\|_{H^2} \le C\, \alpha(\psi, \varphi, a) \to 0$$

as $|a| \to 1$. Furthermore, if $|\varphi(a)| \to 1$, then $|a| \to 1$, so $\alpha(\psi, \varphi, a) \to 0$.

Conversely, suppose $\lim_{|a|\to 1}|\psi(a)|\|\varphi\circ L_a-\varphi(a)\|_{H^2}=0$ and $\alpha(\psi,\varphi,a)\to 0$ as $|\varphi(a)|\to 1$. Then, for every $\varepsilon>0$, there exists $R\in(0,1)$ such that $\alpha(\psi,\varphi,a)<\varepsilon$ for $R<|\varphi(a)|<1$. Also, there exists a number $\delta\in(0,1)$, such that $|\psi(a)|\|\varphi\circ L_a-\varphi(a)\|_{H^2}<\varepsilon$ for $\delta<|a|<1$.

Therefore, if $\delta<|a|<1$ and $R<|\varphi(a)|<1$, then $\alpha(\psi,\varphi,a)<\varepsilon$. On the other hand, if $|\varphi(a)|\le R$ and $\delta<|a|<1$, then

$$\begin{aligned}\alpha(\psi,\varphi,a)&=|\psi(a)|\left(\lim_{r\to 1}\frac{1}{2\pi}\int_0^{2\pi}\left|\frac{\varphi(a)-\varphi(L_a(re^{i\theta}))}{1-\overline{\varphi(a)}\varphi(L_a(re^{i\theta}))}\right|^2 d\theta\right)^{1/2}\\&\le|\psi(a)|\left(\lim_{r\to 1}\frac{1}{2\pi}\int_0^{2\pi}\frac{\left|\varphi(a)-\varphi(L_a(re^{i\theta}))\right|^2}{(1-|\varphi(a)|)^2}\,d\theta\right)^{1/2}\\&\le\frac{1}{1-R}|\psi(a)|\|\varphi\circ L_a-\varphi(a)\|_{H^2}<\frac{1}{1-R}\,\varepsilon.\end{aligned}$$

Thus $\alpha(\psi,\varphi,a)\to 0$ as $|a|\to 1$, as desired. □

Theorem 4.1. *Suppose $1\le p<\infty$, ψ is analytic on $\mathbb{D}$, and φ is an analytic self-map of $\mathbb{D}$. If $W_{\psi,\varphi}:B_p\to VMOA$ is bounded, then $W_{\psi,\varphi}:B_p\to VMOA$ is compact if and only if*

$$\lim_{|a|\to 1}\gamma(\psi,\varphi,p,a)=0\quad\textit{and}\quad\lim_{|a|\to 1}\alpha(\psi,\varphi,a)=0.\tag{35}$$

Proof. Suppose $W_{\psi,\varphi}:B_p\to VMOA$ is compact. Then $\psi=W_{\psi,\varphi}1\in VMOA$, $\psi\varphi=W_{\psi,\varphi}id\in VMOA$, and $W_{\psi,\varphi}:B_p\to BMOA$ is compact, so for $1<p<\infty$, by Theorem 3.3 and Lemma 4.1, it follows that $\lim_{|a|\to 1}\gamma(\psi,\varphi,p,a)=0$. Since $\gamma(\psi,\varphi,1,a)=\|\psi\circ L_a-\psi(a)\|_{H^2}$ and $\psi\in VMOA$, this limit is 0 also for $p=1$.

Next observe that since $W_{\psi,\varphi}:B_p\to VMOA$ is bounded, by part (b) of Theorem 2.3,

$$\lim_{|a|\to 1}|\psi(a)|\|\varphi\circ L_a-\varphi(a)\|_{H^2}=0.$$

By Theorem 3.2 (in the case $p=1$), Theorem 3.3 (for $1<p<\infty$), and Lemma 4.2, it follows that $\lim_{|a|\to 1}\alpha(\psi,\varphi,a)=0$.

Conversely, suppose (35) holds. To prove that $W_{\psi,\varphi}$ is compact, we shall adapt the proof of Theorem 4.3 in [12] to our setting. Let $\{r_n\}$ be a sequence in $(0,1)$ such that $r_n\to 1$ as $n\to\infty$ and, for $n\in\mathbb{N}$, define the operator K_n on B_p by

$$K_nf(z)=f(r_nz),\quad\text{for } z\in\mathbb{D}.$$

Fix $n\in\mathbb{N}$. We now prove that $K_n:B_p\to B_p$ is compact. By Lemma 2.11 of [19], it suffices to show that if $\{f_k\}$ is a bounded sequence in B_p converging to 0 uniformly on compact subsets of $\mathbb{D}$, then $\|K_nf_k\|_{B_p}\to 0$ as $k\to\infty$. Since the sequence $\{f_k'\}$ converges uniformly on the disk centered at 0 of radius r_n, it follows that $b_p(K_nf_k)\to 0$ as $k\to\infty$. On the other hand $K_nf_k(0)=f_k(0)\to 0$ as $k\to\infty$. Therefore, $\|K_nf_k\|_{B_p}\to 0$ as $k\to\infty$, proving the compactness of K_n.

Since $W_{\psi,\varphi} : B_p \to VMOA$ is bounded, the operator $W_{\psi,\varphi}K_n : B_p \to VMOA$ is also compact.

By Theorem 5.3.3 of [23] for $n = 2$ and $1 \le p < \infty$, if $f \in B_p$, then

$$\|f\|_{B_p} \approx |f(0)| + |f'(0)| + \left(\int_{\mathbb{D}} |f''(z)|^p(1-|z|^2)^{2p-2}\, dA(z)\right)^{1/p},$$

which, applied to the function $K_n f$ (with $f \in B_p$), yields

$$\begin{aligned}\|K_n f\|_{B_p} &\approx |f(0)| + r_n|f'(0)| + \left(\int_{\mathbb{D}} r_n^{2p}|f''(r_n z)|^p(1-|z|^2)^{2p-2}\, dA(z)\right)^{1/p}\\ &\le |f(0)| + |f'(0)| + r_n^{2-2/p}\left(\int_{\mathbb{D}} |f''(r_n z)|^p(1-|r_n z|^2)^{2p-2}\, dA(r_n z)\right)^{1/p}\\ &\le C\,\|f\|_{B_p}. \end{aligned}\tag{36}$$

For $f \in B_p$, define $S_n f = f - K_n f$. Then, by (36), S_n is bounded on B_p and $\|S_n f\|_{B_p} \le \|f\|_{B_p} + \|K_n f\|_{B_p} \le C\,\|f\|_{B_p}$. Hence, as an operator on B_p, $\|S_n\| \le C$.

Regarding $W_{\psi,\varphi}$ as an operator from B_p to $VMOA$, we may now estimate its essential norm by

$$\|W_{\psi,\varphi}\|_e \le \|W_{\psi,\varphi} - W_{\psi,\varphi}K_n\| = \sup_{\|f\|_{B_p}\le 1} \|W_{\psi,\varphi}S_n f\|_{BMOA}.$$

Fix $r \in [\frac{1}{2}, 1)$. Applying Lemma 4.6 of [12], for each $n \in \mathbb{N}$, we have

$$\begin{aligned}\|W_{\psi,\varphi}S_n f\|_{BMOA} \le C\Big((1-r)^{-5/2} &\max_{|z|\le r} |(W_{\psi,\varphi}S_n f)(z)|\\ &+ \sup_{|a|\ge r} \|(W_{\psi,\varphi}S_n f)\circ L_a - \psi(a)(S_n f)(\varphi(a))\|_{H^2}\Big).\end{aligned}$$

Thus, taking the supremum over all $f \in B_p$ such that $\|f\|_{B_p} \le 1$, we obtain

$$\begin{aligned}\|W_{\psi,\varphi}\|_e \le C\Big((1-r)^{-5/2} &\sup_{\|f\|_{B_p}\le 1}\max_{|z|\le r} |(W_{\psi,\varphi}S_n f)(z)|\\ &+ \sup_{\|f\|_{B_p}\le 1}\sup_{|a|\ge r} \|(W_{\psi,\varphi}S_n f)\circ L_a - \psi(a)(S_n f)(\varphi(a))\|_{H^2}\Big).\end{aligned}$$

As shown in [12], $\displaystyle\sup_{\|f\|_{B_p}\le 1}\max_{|z|\le r} |(W_{\psi,\varphi}S_n f)(z)| \to 0$ as $n \to \infty$.

Since $\|S_n\|_{B_p\to B_p} \le C$, we obtain

$$\|W_{\psi,\varphi}\|_e \le C \sup_{g\in B_p, \|g\|_{B_p}\le C}\ \sup_{|a|\ge r} \|(W_{\psi,\varphi}g)\circ L_a - \psi(a)g(\varphi(a))\|_{H^2}. \tag{37}$$

Using (10), (11), and (12), for $\|g\|_{B_p} \le C$, we obtain

$$\|(W_{\psi,\varphi}g)\circ L_a - \psi(a)g(\varphi(a))\|_{H^2} \le C(\gamma(\psi,\varphi,p,a) + \|\psi\circ L_a - \psi(a)\|_{H^2}^{1/2} + \alpha(\psi,\varphi,a)).$$

From (37), taking the supremum over all $g \in B_p$ with $\|g\|_{B_p} \le C$, we conclude that

$$\|W_{\psi,\varphi}\|_e \le C \sup_{|a|\ge r} (\gamma(\psi,\varphi,p,a) + \|\psi\circ L_a - \psi(a)\|_{H^2}^{1/2} + \alpha(\psi,\varphi,a)).$$

Since $\psi \in VMOA$, letting $r \to 1$, we deduce that $\|W_{\psi,\varphi}\|_e = 0$. Hence $W_{\psi,\varphi} : B_p \to VMOA$ is compact. □

Remark 4.1. If $1 < p < q < \infty$ then B_p is continuously embedded in B_q, therefore if $C_\varphi : B_q \to BMOA$ is compact, then $C_\varphi : B_p \to BMOA$ is also compact.

From Theorem 4.1, Corollary 2.2, Remark 4.5 of [12], Theorem 2.4 of [6], Remark 4.1, and using the fact that B_1 is continuously embedded in H^∞, we obtain the following result.

Corollary 4.1. *For an analytic self-map φ of $\mathbb{D}$ and for $1 \le p < \infty$, the following statements are equivalent:*

(a) $C_\varphi : B_p \to VMOA$ *is compact.*
(b) $C_\varphi : VMOA \to VMOA$ *is compact.*
(c) $C_\varphi : H^\infty \to VMOA$ *is compact.*
(d) $\varphi \in VMOA$ *and* $\lim_{|a|\to 1} \|L_{\varphi(a)} \circ \varphi \circ L_a\|_{H^2} = 0$.
(e) $\varphi \in VMOA$ *and* $\lim_{n\to\infty} \|\varphi^n\|_{BMOA} = 0$.

Acknowledgment

We wish to express our gratitude to the referee for showing us how to extend the sufficiency in Theorem 4.1 to the case $1 < p < 2$ and for his/her suggestions on how to improve the manuscript.

References

[1] J. Arazy and S.D. Fisher, *Some aspects of the minimal, Möbius-invariant space of analytic functions on the unit disc*, Interpolation spaces and allied topics in analysis (Lund, 1983), 2444, Lecture Notes in Math., **1070**, Springer, Berlin, 1984.

[2] J. Arazy and S.D. Fisher, *The uniqueness of the Dirichlet space among Möbius-invariant Hilbert spaces*, Illinois J. Math. **29** (3) (1985), 449–462.

[3] J. Arazy, S.D. Fisher, and J. Peetre, *Möbius invariant function spaces*, J. Reine Angew. Math. **363** (1985), 110–145.

[4] O. Blasco, *Composition operators on the minimal space invariant under Möbius transformations*, Complex and harmonic analysis, 157–166, DEStech Publ. Inc., Lancaster, PA, 2007.

[5] P.S. Bourdon, J. Cima and A.L. Matheson, *Compact composition operators on BMOA*, Trans. Amer. Math. Soc. **351** (1999), 2183–2196.

[6] F. Colonna, *Weighted composition operators between H^∞ and BMOA*, Bull. Korean Math. Soc. **50** (1) (2013), 185–200.

[7] R.L. Donaway, *Norm and Essential Norm Estimates of Composition Operators on Besov Type Spaces*, Doctoral Dissertation, University of Virginia, 1999.

[8] J.B. Garnett, *Bounded Analytic Functions*, Pure and Applied Math., **96**, Academic Press, Orlando FL, 1981.

[9] D. Girela, *Analytic functions of bounded mean oscillation*, Complex Function Spaces (Mekrijärvi, 1999), 61–170, Univ. Joensuu Dept. Math. Rep. Ser. No. **4** (2001).

[10] S. Janson, *On functions with conditions on the mean oscillation*, Ark. Math. **14** (1976), 189–196.

[11] J. Laitila, *Isometric composition operators on BMOA*, Math. Nachr. **283** (2010), 1646–1653.

[12] J. Laitila, *Weighted composition operators on BMOA*, Comput. Methods Funct. Theory **9**, no. 1 (2009), 27–46.

[13] J. Laitila, P. Nieminen, E. Saksman, and H. Tylli, *Compact and weakly compact composition operators on BMOA*, Compl. Anal. Oper. Theory **7** (1) (2013), 163–181.

[14] J.M. Ortega and J. Fàbrega, *Pointwise multipliers and corona type decompositions in BMOA*, Ann. Inst. Fourier (Grenoble) **46** (1976), 111–137.

[15] J.H. Shapiro, *Composition Operators and Classical Function Theory*, Springer-Verlag, New York, 1993.

[16] W. Smith, *Compactness of composition operators on BMOA*, Proc. Amer. Math. Soc. **127** (1999), 2715–2725.

[17] D.A. Stegenga, *Bounded Toeplitz operators on H^1 and applications of the duality between H^1 and the functions of bounded mean oscillation*, Amer. J. Math. **98** (1976), 573–589.

[18] M. Tjani, *Compact composition operators on some Möbius invariant Banach spaces*, Doctoral Dissertation, Michigan State University, 1996.

[19] M. Tjani, *Compact composition operators on Besov spaces*, Trans. Amer. Math. Soc. **355** (2003), 4683–4698.

[20] H. Wulan, *Compactness of composition operators on BMOA and VMOA*, Sci. China **50** (2007), 997–1004.

[21] H. Wulan and C. Xiong, Composition operators on the minimal Möbius invariant space, *Hilbert Spaces of Analytic Functions, CRM proceedings and lecture notes*, Volume 51, 2010.

[22] H. Wulan, D. Zheng, and K. Zhu, *Compact composition operators on BMOA and the Bloch space*, Proc. Amer. Math. Soc. **137** (2009), 3861–3868.

[23] K. Zhu, *Operator Theory in Function Spaces*, Marcel Dekker, New York, 1990.

[24] K. Zhu, *Analytic Besov spaces*, J. Math. Anal. Appl. **157** (1991), 318–336.

Flavia Colonna
Department of Mathematical Sciences
George Mason University
Fairfax, VA 22030, USA
e-mail: `fcolonna@gmu.edu`

Maria Tjani
Department of Mathematical Sciences
University of Arkansas
Fayetteville, AR 72701, USA
e-mail: `mtjani@uark.edu`

Operator Theory:
Advances and Applications, Vol. 236, 159–170

Some Remarks on Essentially Normal Submodules

Ronald G. Douglas and Kai Wang

Abstract. Given a $*$-homomorphism $\sigma : C(M) \to \mathscr{L}(\mathcal{H})$ on a Hilbert space $\mathcal{H}$ for a compact metric space M, a projection P onto a subspace $\mathcal{P}$ in $\mathcal{H}$ is said to be essentially normal relative to σ if $[\sigma(\varphi), P] \in \mathcal{K}$ for $\varphi \in C(M)$, where $\mathcal{K}$ is the ideal of compact operators on $\mathcal{H}$. In this note we consider two notions of span for essentially normal projections P and Q, and investigate when they are also essentially normal. First, we show the representation theorem for two projections, and relate these results to Arveson's conjecture for the closure of homogenous polynomial ideals on the Drury–Arveson space. Finally, we consider the relation between the relative position of two essentially normal projections and the K homology elements defined for them.

Mathematics Subject Classification (2010). Primary 47A13; Secondary 46E22, 46H25, 47A53.

Keywords. Arveson conjecture, essentially normal submodules.

1. Introduction

Spurred by a question of Arveson [1, 2, 3, 4], several researchers have been considering when certain submodules of various Hilbert modules of holomorphic functions on the unit ball in $\mathbb{C}^n$ are essentially normal. In particular, Guo and the second author showed in [17] that the closure of a principal homogenous polynomial ideal in the Drury–Arveson space in $\mathbb{B}^n$ is essentially normal. More recently, the authors have shown in [13] that the closure of all principal polynomial ideals in the Bergman module on the unit ball are essentially normal. Other results have been obtained by Arveson [5], Douglas [10, 11], the first author and Sarkar [12], Eschmeier [14], Kennedy [19], and Shalit [20].

The second author was supported by NSFC 11271075, the Department of Mathematics at Texas A&M University and Laboratory of Mathematics for Nonlinear Science at Fudan University.

The Arveson conjecture concerns the closure of an arbitrary homogeneous polynomial ideal which, in general, is not singly generated. For the case of $n = 1$, one knows that a pure hyponormal operator submodule is essentially normal if it is finitely generated. The basis on which this result depends is the Berger–Shaw Theorem [8].

For ideals that are not principal, or singly generated, the results in the several variable case are few. Guo in [15] firstly proved Arveson's conjecture in case of the dimension $n = 2$. Guo and the second author established in [16, 17] essential normality when $n = 3$ or the dimension of the zero variety of the homogeneous ideal is one or less, the opposite extreme, more or less, of the case of principal ideals. There is also a result of Shalit [20] which holds for ideals having a "very nice" basis relative to the norm. More recently, Kennedy [19] extended that result in another direction, considering when the linear span of the closures of polynomial ideals is closed. He gives some examples, but it would appear that not all non-principal ideals are covered by this result. One should note that these results when the linear span of two essentially normal submodules is closed is implicit in the work of Arveson [5, Theorem 4.4].

In this note, we explore a more general version of the question of when the linear span of two essentially normal submodules is also essentially normal. We show that this result contains one aspect of the results of Shalit and Kennedy.

Our work does not depend on the special nature of the submodules; that is, we do not assume any connection with any underlying algebraic structure, only the fact that the linear span is closed.

There is more than one sense of the span of two submodules relevant in this context: the first is the obvious one defined to be the closure of the linear span of two submodules $\mathcal{P}$ and $\mathcal{Q}$, while the second one considers the span modulo the ideal of compact operators. If P and Q denote the orthogonal projections onto $\mathcal{P}$ and $\mathcal{Q}$, respectively, then we will show that this notion makes sense if 0 is an isolated point in the essential spectrum of $P + Q$.

We consider the first notion in Section 2, and the results obtained are based on the structure theorem for two projections. The latter notion of the essential span is taken up in Section 3. We apply these results to the context of Arveson's conjecture and raise some questions. In particular, we assume that there is a $*$-homomorphism σ of $C(M)$ for some compact metric space M and the projections essentially commute with the range of σ. Finally, in Section 4 we observe that an essentially normal projection determines an element of the odd K-homology group for some compact subset of M and consider the relation of the K-homology elements defined by two essentially normal submodules and their sum.

2. Refinement of the two projection representation

Our results in this section are based on refinements of the structure theorem for two projections [6, 18].

Let P and Q be two projections on the Hilbert space $\mathcal{H}$. Then there exist operators $S_1 : \mathcal{P} \to \mathcal{P}, S_2 : \mathcal{P}^\perp \to \mathcal{P}^\perp$ and $X : \mathcal{P}^\perp \to \mathcal{P}$, where $\mathcal{P} = \operatorname{ran} P$ with $0_\mathcal{P} \leq S_1 \leq I_\mathcal{P}, 0_{\mathcal{P}^\perp} \leq S_2 \leq I_{\mathcal{P}^\perp}$ and $\|X\| \leq 1$, such that

$$P = \begin{pmatrix} I_\mathcal{P} & 0 \\ 0 & 0_{\mathcal{P}^\perp} \end{pmatrix} \text{ and } Q = \begin{pmatrix} S_1 & X \\ X^* & S_2 \end{pmatrix}.$$

Moreover, if we set $\mathcal{P} = \mathcal{P}_1 \oplus \mathcal{P}_2 \oplus \mathcal{P}_3$ and $\mathcal{P}^\perp = \mathcal{Q}_1 \oplus \mathcal{Q}_2 \oplus \mathcal{Q}_3$, where $\mathcal{P}_1 = \{x \in \mathcal{P} : S_1 x = 0\}, \mathcal{P}_2 = \{x \in \mathcal{P} : S_1 x = x\}, \mathcal{P}_3 = \mathcal{P} \ominus (\mathcal{P}_1 \oplus \mathcal{P}_2)$, $\mathcal{Q}_2 = \{x \in \mathcal{P}^\perp : S_2 x = x\}$, $\mathcal{Q}_3 = \{x \in \mathcal{P}^\perp : S_2 x = 0\}$, and $\mathcal{Q}_1 = \mathcal{P}^\perp \ominus (\mathcal{Q}_2 \oplus \mathcal{Q}_3)$, then we have

$$S_1 = \begin{pmatrix} 0_{\mathcal{P}_1} & 0 & 0 \\ 0 & I_{\mathcal{P}_2} & 0 \\ 0 & 0 & S_1' \end{pmatrix} \in \mathscr{L}(\mathcal{P}_1 \oplus \mathcal{P}_2 \oplus \mathcal{P}_3) \text{ with } S_1' \in \mathscr{L}(\mathcal{P}_3),$$

$$S_2 = \begin{pmatrix} S_2' & 0 & 0 \\ 0 & I_{\mathcal{Q}_2} & 0 \\ 0 & 0 & 0_{\mathcal{Q}_3} \end{pmatrix} \in \mathscr{L}(\mathcal{Q}_1 \oplus \mathcal{Q}_2 \oplus \mathcal{Q}_3) \text{ with } S_2' \in \mathscr{L}(\mathcal{Q}_1), \text{ and}$$

$$X = \begin{pmatrix} 0 & 0 & 0 \\ 0 & 0 & 0 \\ X' & 0 & 0 \end{pmatrix} \in \mathscr{L}(\mathcal{Q}_1 \oplus \mathcal{Q}_2 \oplus \mathcal{Q}_3, \mathcal{P}_1 \oplus \mathcal{P}_2 \oplus \mathcal{P}_3) \text{ with } X' \in \mathscr{L}(\mathcal{Q}_1, \mathcal{P}_3).$$

These results are all straightforward.

Further, using matrix computations and the fact that $Q^2 = Q = Q^*$, one shows that there exists an isometry V from $\mathcal{Q}_1$ onto $\mathcal{P}_3$ such that $V^* S_1' V = I_{\mathcal{Q}_1} - S_2'$. We refer the reader to [18] for a detailed argument. Therefore, we have derived the standard model for two projections.

Theorem 2.1. *Two projections P and Q on a Hilbert space $\mathcal{H}$ are determined by*

(1) *a decomposition $\mathcal{H} = \mathcal{H}_0 \oplus \mathcal{H}_1 \oplus \mathcal{H}' \oplus \mathcal{H}' \oplus \mathcal{H}_2 \oplus \mathcal{H}_3$, and*

(2) *a positive contraction $S \in \mathcal{L}(\mathcal{H}')$ with $\{0, 1\}$ not in its point spectrum.*

In this case, one has

$$P = \begin{pmatrix} I_{\mathcal{H}_0} & 0 & 0 & 0 & 0 & 0 \\ 0 & I_{\mathcal{H}_1} & 0 & 0 & 0 & 0 \\ 0 & 0 & I_{\mathcal{H}'} & 0 & 0 & 0 \\ 0 & 0 & 0 & 0_{\mathcal{H}'} & 0 & 0 \\ 0 & 0 & 0 & 0 & 0_{\mathcal{H}_2} & 0 \\ 0 & 0 & 0 & 0 & 0 & 0_{\mathcal{H}_3} \end{pmatrix}$$

and

$$Q = \begin{pmatrix} 0_{\mathcal{H}_0} & 0 & 0 & 0 & 0 & 0 \\ 0 & I_{\mathcal{H}_1} & 0 & 0 & 0 & 0 \\ 0 & 0 & S & X & 0 & 0 \\ 0 & 0 & X & I_{\mathcal{H}'} - S & 0 & 0 \\ 0 & 0 & 0 & 0 & I_{\mathcal{H}_2} & 0 \\ 0 & 0 & 0 & 0 & 0 & 0_{\mathcal{H}_3} \end{pmatrix},$$

where $X = \sqrt{S(I_{\mathcal{H}'} - S)} \in \mathcal{L}(\mathcal{H}')$.

Proof. Again, the representation results follow from standard matrix computations. □

The following question happens frequently in many concrete problems in operator theory.

Question 2.2. When is $\mathcal{P} + \mathcal{Q}$ closed in $\mathcal{H}$, where $\mathcal{P} = \operatorname{ran} P$ and $\mathcal{Q} = \operatorname{ran} Q$?

Note we have:

$$\mathcal{P} = \left\{ \begin{pmatrix} x_0 \\ x_1 \\ x' \\ 0 \\ 0 \\ 0 \end{pmatrix} : x_0 \in \mathcal{H}_0, x_1 \in \mathcal{H}_1, x' \in \mathcal{H}' \right\}$$

and

$$\mathcal{Q} = \left\{ \begin{pmatrix} 0 \\ x_1 \\ Sx' + Xy' \\ Xx' + (I_{\mathcal{H}'} - S)y' \\ x_2 \\ 0 \end{pmatrix} : x_1 \in \mathcal{H}_1, x', y' \in \mathcal{H}', x_2 \in \mathcal{H}_2 \right\}.$$

Therefore,

$$\mathcal{P} + \mathcal{Q} = \left\{ \begin{pmatrix} x_0 \\ x_1 \\ x' \\ z \\ x_2 \\ 0 \end{pmatrix} : \begin{array}{l} x_0 \in \mathcal{H}_0, x_1 \in \mathcal{H}_1, x' \in \mathcal{H}', \\ \qquad z \in \operatorname{ran} X + \operatorname{ran}(I_{\mathcal{H}'} - S), x_2 \in \mathcal{H}_2 \end{array} \right\}.$$

This implies that $\mathcal{P} + \mathcal{Q}$ is closed if and only if $\operatorname{ran} X + \operatorname{ran}(I_{\mathcal{H}'} - S)$ is closed. Since $X = \sqrt{S(I_{\mathcal{H}'} - S)}$, we have that $\operatorname{ran} X \subseteq \operatorname{ran}(I_{\mathcal{H}'} - S)^{\frac{1}{2}}$. Moreover, by the spectral theorem for the positive contraction S, one sees that $\sqrt{S} + \sqrt{I_{\mathcal{H}'} - S}$ is invertible on $\mathcal{H}'$. This implies that $\operatorname{ran} X + \operatorname{ran}(I_{\mathcal{H}'} - S) \supseteq \operatorname{ran}(I_{\mathcal{H}'} - S)^{\frac{1}{2}}$. Therefore, $\operatorname{ran} X + \operatorname{ran}(I_{\mathcal{H}'} - S)$ is closed if and only if $\operatorname{ran}(I_{\mathcal{H}'} - S)^{\frac{1}{2}}$ is closed. Since 1 is not in the point spectrum of S, it follows from the spectral theorem that $\operatorname{ran}(I_{\mathcal{H}'} - S)^{\frac{1}{2}}$ is closed if and only if 1 is not in the spectrum of S. Hence we have the following result.

Theorem 2.3. *For two projections P and Q on the Hilbert space $\mathcal{H}$ with $\mathcal{P} = \operatorname{ran} P$ and $\mathcal{Q} = \operatorname{ran} Q$, the linear span $\mathcal{P} + \mathcal{Q}$ is closed if and only if $1 \notin \sigma(S)$, or equivalently, $\sigma(PQP) \cap (\varepsilon, 1) = \phi$ for some $0 < \varepsilon < 1$, where S is the same as in*

Theorem 2.1. *Moreover, in the case that $\mathcal{R} = \mathcal{P} + \mathcal{Q}$ is closed, the projection R onto $\mathcal{R}$ has the form*

$$R = \begin{pmatrix} I_{\mathcal{H}_0} & 0 & 0 & 0 & 0 & 0 \\ 0 & I_{\mathcal{H}_1} & 0 & 0 & 0 & 0 \\ 0 & 0 & I_{\mathcal{H}'} & 0 & 0 & 0 \\ 0 & 0 & 0 & I_{\mathcal{H}'} & 0 & 0 \\ 0 & 0 & 0 & 0 & I_{\mathcal{H}_2} & 0 \\ 0 & 0 & 0 & 0 & 0 & 0_{\mathcal{H}_3} \end{pmatrix}$$

Proof. Only the last statement remains to be proved and that follows from the fact that $1 \notin \sigma(S)$ if and only if $0 \notin \sigma(I_{\mathcal{H}'} - S)$, which implies $\operatorname{ran}(I_{\mathcal{H}'} - S) = \mathcal{H}'$. □

A nearly immediate consequence of the representation theorem is the following characterization of the C^*-algebra $\mathcal{A}(P, Q, I)$ generated by projections P, Q and the identity operator on the Hilbert space $\mathcal{H}$. This is usually attributed to Dixmier [9].

Theorem 2.4. *Let P and Q be the projections onto the subspaces $\mathcal{P}$ and $\mathcal{Q}$ of the Hilbert space $\mathcal{H}$, respectively. If $\mathcal{P} \cap \mathcal{Q} = \mathcal{P}^\perp \cap \mathcal{Q} = \mathcal{P} \cap \mathcal{Q}^\perp = \mathcal{P}^\perp \cap \mathcal{Q}^\perp = \{0\}$, then $\mathcal{A}(P, Q, I)$ is $*$-algebraically isomorphic to a $*$-subalgebra $\mathcal{C}$ of $M_2(C(M))$, where $M = \sigma(PQP)$, $M_2(C(M))$ denotes the algebra of two by two matrices with entries in $C(M)$ and*

$$\mathcal{C} = \left\{ \begin{pmatrix} \phi_{11} & \phi_{12} \\ \phi_{21} & \phi_{22} \end{pmatrix} \in M_2(C(M)) : \phi_{12}(i) = \phi_{21}(i) = 0, \textit{ if } i = 0, 1 \textit{ and } i \in M \right\}.$$

Proof. Applying the spectral theorem to the operators $I_{\mathcal{P}}$ and S, one obtains the correspondence from which the result follows:

$$P = \begin{pmatrix} I_{\mathcal{P}} & 0 \\ 0 & 0 \end{pmatrix} \sim \begin{pmatrix} 1 & 0 \\ 0 & 0 \end{pmatrix} \in M_2(C(M))$$

and

$$Q = \begin{pmatrix} S & \sqrt{S(1-S)} \\ \sqrt{S(1-S)} & 1-S \end{pmatrix}$$
$$\sim \begin{pmatrix} \chi & \sqrt{\chi(1-\chi)} \\ \sqrt{\chi(1-\chi)} & 1-\chi \end{pmatrix} \in M_2(C(M)),$$

where 1 and χ denote the functions on M defined by $1(x) = 1$ and $\chi(x) = x$ for $x \in M$. The fact that the functions ϕ_{12} and ϕ_{21} in the definition of $\mathcal{C}$ vanish at $0, 1 \in M$ follows from the fact that the function $\sqrt{\chi(1-\chi)}$ does. □

We now use the characterization of the C^*-algebra generated by two projections to get our first result on the essential normality of the projection onto the linear span when it is closed.

Theorem 2.5. *For two projections P and Q on the Hilbert space $\mathcal{H}$, if $\mathcal{R} = \operatorname{ran} P + \operatorname{ran} Q$ is closed, then the C^*-algebra $\mathcal{A}(P, Q, I_{\mathcal{H}})$ generated by P, Q and the identity operator $I_{\mathcal{H}}$ contains the projection R onto the subspace $\mathcal{R}$.*

Proof. Using a direct matrix computation, one sees that the operator $P + (I - P)Q(I - P)$, which is in the C^*-algebra $\mathcal{A}(P, Q, I)$, has the form

$$P + (I - P)Q(I - P) = \begin{pmatrix} I_{\mathcal{H}_0} & 0 & 0 & 0 & 0 & 0 \\ 0 & I_{\mathcal{H}_1} & 0 & 0 & 0 & 0 \\ 0 & 0 & I_{\mathcal{H}'} & 0 & 0 & 0 \\ 0 & 0 & 0 & I_{\mathcal{H}'} - S & 0 & 0 \\ 0 & 0 & 0 & 0 & I_{\mathcal{H}_2} & 0 \\ 0 & 0 & 0 & 0 & 0 & 0_{\mathcal{H}_3} \end{pmatrix}.$$

This implies that $\sigma(P + (I - P)Q(I - P)) \subseteq \{0\} \cup [\varepsilon, 1]$ for some $0 < \varepsilon < 1$. Since $[\varepsilon, 1] \cap \sigma(P + (I - P)Q(I - P))$ is an open and closed subset of the spectrum $\sigma(P + (I - P)Q(I - P))$, it follows from the spectral theorem that the spectral projection $\mathbb{1}_{[\varepsilon,1]}(P + (I - P)Q(I - P))$ is in $\mathcal{A}(P, Q, I)$, which leads to the desired result since $\mathbb{1}_{[\varepsilon,1]}(P + (I - P)Q(I - P)) = R$. □

We now relate the representation result to a question in the context of Arveson's conjecture. We will provide a more precise statement in Section 4.

Theorem 2.6. *Suppose $\sigma : C(M) \to \mathcal{L}(\mathcal{H})$ is a $*$-homomorphism for some compact metric space M, and P, Q are projections on the Hilbert space $\mathcal{H}$ such that the commutators $[\sigma(\varphi), P] \in \mathcal{K}$ and $[\sigma(\varphi), Q] \in \mathcal{K}$ for $\varphi \in C(M)$, where $\mathcal{K}$ denotes the ideal of compact operators on $\mathcal{H}$. If $\mathcal{R} = \operatorname{ran} P + \operatorname{ran} Q$ is closed and R is the projection onto $\mathcal{R}$, then $[\sigma(\varphi), R] \in \mathcal{K}$ for $\varphi \in C(M)$.*

Proof. Using an elementary C^*-algebra argument, one shows that $[\sigma(\varphi), T] \in \mathcal{K}$ for any operator $T \in \mathcal{A}(P, Q, I)$. Combining this fact with Theorem 2.5, one obtains the desired result. □

Corollary 2.7. *With the same hypotheses, the projection $\widetilde{R}$ onto $\mathcal{P} \cap \mathcal{Q}$ essentially commutes with the range of σ.*

Proof. This is an immediate consequence of Theorem 1 in [11] and the exact sequence

$$0 \to \widetilde{R} \xrightarrow{i} \mathcal{P} \oplus \mathcal{Q} \xrightarrow{j} \mathcal{R} \to 0,$$

where $i(r) = (r, -r)$, and $j(p, q) = p + q$ for $r \in \widetilde{R}, p \in \mathcal{P}, q \in \mathcal{Q}$. □

Remark 2.8. In both the theorem and corollary, $C(M)$ can be replaced by any C^*-subalgebra of $\mathscr{L}(\mathcal{H})$.

Remark 2.9. These results are related to a theorem of Arveson [5, Theorem 4.4] and the more recent work of Kennedy [19] in which essential normality is replaced by p-essential normality, where the commutators are assumed to be in the Schatten p-class for $1 \leq p < \infty$. If one examines the proof of Theorem 2.5 more closely, the preceding arguments can be refined to obtain analogous results for p-essential normality. Basically, this is true because the functional calculus which yields the spectral projection $\mathbb{1}_{[\varepsilon,1]}(P + (I - P)Q(I - P))$ can be approximated on a neighborhood of the spectrum with analytic functions.

We can extend these results somewhat using the following reduction which is essentially algebraic.

Theorem 2.10. *Suppose $\mathcal{P}$ and $\mathcal{Q}$ are subspaces of the Hilbert space $\mathcal{H}$ and $\mathcal{R}^\sharp$ is a subspace of $\mathcal{P} \cap \mathcal{Q}$. Then $\mathcal{P} + \mathcal{Q}$ is closed if and only if $\mathcal{P}/\mathcal{R}^\sharp + \mathcal{Q}/\mathcal{R}^\sharp$ is closed in $\mathcal{H}/\mathcal{R}^\sharp$.*

Proof. It follows from the fact that $\mathcal{P}/\mathcal{R}^\sharp + \mathcal{Q}/\mathcal{R}^\sharp = (\mathcal{P} + \mathcal{Q})/\mathcal{R}^\sharp$ and the fact that for any linear manifold $\mathcal{L}$ containing $\mathcal{R}^\sharp$, $\mathcal{L}/\mathcal{R}^\sharp$ is closed if and only if $\mathcal{L}$ is closed. □

Corollary 2.11. *With the same hypotheses, the closeness of $\mathcal{P}/\mathcal{R}^\sharp + \mathcal{Q}/\mathcal{R}^\sharp$ is equivalent to the closeness of $\mathcal{P}/(\mathcal{P} \cap \mathcal{Q}) + \mathcal{Q}/(\mathcal{P} \cap \mathcal{Q})$.*

Proof. Both of these statements are equivalent to $\mathcal{P} + \mathcal{Q}$ being closed in $\mathcal{H}$. □

3. Essential span of subspaces

The following question and corresponding result are important for considering the notion of essential span in this section.

Question 3.1. When do two projections P and Q on a Hilbert space $\mathcal{H}$ almost commute; that is, when is $[P, Q] \in \mathcal{K}(\mathcal{H})$?

Using the representation theorem for P and Q above, we see that $[P, Q] \in \mathcal{K}$ if and only if $X \in \mathcal{K}$ and so we have the following result.

Theorem 3.2. *For projections P and Q onto subspaces $\mathcal{P}$ and $\mathcal{Q}$, respectively, on a Hilbert space $\mathcal{H}$, $[P, Q] \in \mathcal{K}$ if and only if $\sigma_e(S) \subset \{0, 1\}$. Moreover, $PQ \in \mathcal{K}$ if and only if $S \in \mathcal{K}$ and* $\dim \mathcal{P} \cap \mathcal{Q} < \infty$ *in the representation appearing in Theorem* 2.1.

Proof. The proof follows from a matrix calculation in the above representation theorem which shows that $[P, Q] \in \mathcal{K}$ if and only if $X = \sqrt{S(I_{\mathcal{H}'} - S)}$ is compact. For $PQ \in \mathcal{K}$, it is necessary and sufficient for S and $I_{\mathcal{H}_1}$ to be compact. □

If P and Q are projections on the Hilbert space $\mathcal{H}$, then another notion of the span of the ranges of P and Q is relevant when considering questions of essential normality, which involves the images of P and Q in the Calkin algebra. If 0 is an isolated point in the essential spectrum, $\sigma_e(P + Q)$, of $P + Q$, or $0 \notin \sigma_e(P + Q)$, then the image in the Calkin algebra of the spectral projection, $P \bigvee_e Q$, for $[\varepsilon, \infty]$, where $(0, \varepsilon) \cap \sigma_e(P + Q) = \phi$, can be thought of as the "essential span" of $\operatorname{ran} P$ and $\operatorname{ran} Q$. (Note that the image of this spectral projection in the Calkin algebra does not depend on ε whenever $(0, \varepsilon) \cap \sigma_e(P + Q) = \phi$.) One result related to this notion is the following.

Theorem 3.3. *If $[P, Q] \in \mathcal{K}$, then 0 is isolated in $\sigma_e(P + Q)$. Moreover, if P and Q almost commute with a C^*-algebra $\mathfrak{A}$, then so does any projection on $\mathcal{H}$ with the image $P \bigvee_e Q$ in the Calkin algebra.*

Proof. Considering the standard model for two projection in Section 2, one sees that $[P,Q] \in \mathcal{K}$ implies that X is compact and $\sigma_e(P+Q) \subseteq \{0,1,2\}$. This implies that $[\varepsilon, 2]$ is an open and closed subset of $\sigma(P+Q)$ for some $0 < \varepsilon < 1$ and hence $\mathbb{1}_{[\varepsilon,2]}(P+Q)$ is in $\mathcal{A}(P,Q,I)$, where $\mathcal{A}(P,Q,I)$ is the C^*-algebra generated by P, Q and the identity operator I. Therefore, its image in the Calkin algebra, $P \bigvee_e Q$, commutes with the image of $\mathfrak{A}$, which completes the proof. □

One thing one needs to be clear on is that the image of $P \bigvee Q$ in the Calkin algebra and $P \bigvee_e Q$ are not necessarily the same. Consider, for example, the subspaces $\mathcal{P} = \overline{\text{span}}\{e_n \oplus 0 : n \in \mathbb{N}\}$ in $\ell^2 \oplus \ell^2$ and $\mathcal{Q} = \overline{\text{span}}\{e_n \oplus \frac{1}{n} e_n : n \in \mathbb{N}\}$. These subspaces have the images of $\pi(P \bigvee Q)$ and $P \bigvee_e Q$ in the Calkin algebra which are the images of the projections onto $\ell^2 \oplus \ell^2$ and $\ell^2 \oplus (0)$, respectively. Note that in this case $\mathcal{P} + \mathcal{Q}$ is not closed, which is the key as the following result shows.

Theorem 3.4. *Let P, Q be the projections onto the subspaces $\mathcal{P}$ and $\mathcal{Q}$ of a Hilbert space $\mathcal{H}$, respectively. Then $\mathcal{P} + \mathcal{Q}$ is closed if and only if $\pi(P \bigvee Q) = P \bigvee_e Q$.*

Proof. We first suppose that $\mathcal{P} + \mathcal{Q}$ is closed. Using the notation in Theorem 2.1, we have that

$$P + Q = \begin{pmatrix} I_{\mathcal{H}_0} & 0 & 0 & 0 & 0 & 0 \\ 0 & 2I_{\mathcal{H}_1} & 0 & 0 & 0 & 0 \\ 0 & 0 & I_{\mathcal{H}'} + S & \sqrt{S(I_{\mathcal{H}'} - S)} & 0 & 0 \\ 0 & 0 & \sqrt{S(I_{\mathcal{H}'} - S)} & I_{\mathcal{H}'} - S & 0 & 0 \\ 0 & 0 & 0 & 0 & I_{\mathcal{H}_2} & 0 \\ 0 & 0 & 0 & 0 & 0 & 0_{\mathcal{H}_3} \end{pmatrix}.$$

By Theorem 2.3 we know that $1 \notin \sigma(S)$ when $\mathcal{P} + \mathcal{Q}$ is closed. Applying the spectral theorem to the operator S, one obtains that 0 is isolated in $\sigma(P+Q)$. This implies that the notion $P \bigvee_e Q$ makes sense and, in fact, it is the image in the Calkin algebra of the projection onto $\text{ran}(P+Q)$. Moreover, by the above representation of $P+Q$, one sees that

$$\text{ran}(P+Q) = \mathcal{H}_0 \oplus \mathcal{H}_1 \oplus \mathcal{H}' \oplus \mathcal{H}' \oplus \mathcal{H}_2 = \mathcal{P} + \mathcal{Q}.$$

It follows that $P \bigvee_e Q$ is the image of the projection onto $\mathcal{P} + \mathcal{Q}$.

On the other hand, in case that $\pi(P \bigvee Q) = P \bigvee_e Q$, there exists $0 < \varepsilon < 1$ such that $(0, \varepsilon) \cap \sigma_e(P+Q) = \phi$ and $P \bigvee Q - \mathbb{1}_{[\varepsilon,\infty]}(P+Q)$ is a finite-dimensional projection. Applying the spectral theorem for S to the matrix representation of $P + Q$, one sees that the spectral projection of S for $(1-\varepsilon, 1)$ is also a finite-dimensional projection. Combing this fact with that 1 is not in the point spectrum of S, we have that $1 \notin \sigma(S)$, which leads to the desired result using Theorem 2.3. □

While it seems inconceivable that $[p] + [q]$ is always closed for polynomials p and q in $\mathbb{C}[z_1, \dots, z_n]$; here $[\cdot]$ denotes the closure in the Hardy, Bergman or Drury–Arveson modules on the unit ball, it seems quite possible that the projections onto $[p]$ and $[q]$ always almost commute. One thing making the answering of such a question difficult is the fact that $[p] \cap [q]$ is always large containing $[pq]$. One possible way to circumvent this problem might be to consider the quotient modules $[p]^\perp$ and $[q]^\perp$. We'll have something more to say about them in the next section.

Another possibility to handle the fact that $[p] \cap [q]$ is large might be to use Theorem 2.10 to reduce the matter to $[p]/([p] \cap [q])$ and $[q]/([p] \cap [q])$. In this case, $[p]/([p] \cap [q])$ and $[q]/([p] \cap [q])$ are semi-invariant modules. We will obtain a result using this approach in the following section.

4. Locality of essentially normal projections

Let M be a compact metric space and $\sigma : C(M) \to \mathscr{L}(\mathcal{H})$ be a $*$-homomorphism for a Hilbert space $\mathcal{H}$. We say that a projection P on $\mathcal{H}$ is essentially normal relative to σ if $[\sigma(\varphi), P] \in \mathcal{K}$ for any $\varphi \in C(M)$. This implies that the map $\sigma_P : \varphi \to \pi(P\sigma(\varphi)P) \in \mathscr{Q}(\mathcal{H})$ into the Calkin algebra $\mathscr{Q}(\mathcal{H}) = \mathscr{L}(\mathcal{H})/\mathcal{K}(\mathcal{H})$ is a $*$-homomorphism. Hence, there exists a compact subset M_P of M such that the following diagram commutes:

$$\begin{array}{ccc} C(M) & \xrightarrow{\sigma} & \mathscr{L}(\mathcal{H}) \\ \downarrow & & \downarrow \\ C(M_P) & \xrightarrow{\hat{\sigma}_P} & \mathscr{Q}(P\mathcal{H}) \end{array} .$$

Here the vertical arrow on the left is defined by restriction; that is, $\varphi \to \varphi|_{M_P}$, and the one on the right is the compression to $\operatorname{ran} P$ followed by the map onto the Calkin algebra. Therefore, using [7], one knows that (σ, P) defines an element $[\sigma, P] \in K_1(M_P)$. An interesting question concerns the relation of elements $[\sigma, P]$ and $[\sigma, Q]$ for two essentially normal projections P and Q relative to σ.

Now this relationship can't be too simple. In particular, consider the representation τ of $C(clos\mathbb{B}^n)$ in $L^2(\mathbb{B}^n)$ and the projection P onto the Bergman space $L^2_a(\mathbb{B}^n)$. For $p \in \mathbb{C}[z_1, \dots, z_n]$, one knows [13] that the projection Q_p of $L^2(\mathbb{B}^n)$ onto the closure $[p]$ of the ideal (p) in $\mathbb{C}[z_1, \dots, z_n]$ generated by p is essentially normal; that is, $[\tau(\varphi), Q_p] \in \mathcal{K}$ for $\varphi \in C(clos\mathbb{B}^n)$. Further, we have that $R_p = P - Q_p$ is also essentially normal and $M_{[\tau, R_p]} \subseteq Z(p) \cap \partial\mathbb{B}^n$, where $Z(p)$ is the zero variety of the polynomial p. It follows that the image of $[\tau, R_p] \in K_1(\partial\mathbb{B}^n)$ is zero since $Z(p) \cap \partial\mathbb{B}^n$ is a proper subset of $\partial\mathbb{B}^n$. Therefore, one has $[\tau, P] = [\tau, Q_p]$ for every polynomial $p \in \mathbb{C}[z_1, \dots, z_n]$. Hence, there is a great variety of essentially normal projections defining the same element in $K_1(\partial\mathbb{B}^n)$.

However, we do have a result for what happens at the opposite extreme.

Theorem 4.1. *Suppose that P and Q are essentially normal projections on the Hilbert space $\mathcal{H}$ for the $*$-homomorphism $\sigma : C(M) \to \mathscr{L}(\mathcal{H})$ for some compact space M. If $M_P \cap M_Q = \phi$, then $PQ \in \mathcal{K}$.*

Proof. By the assumption that P and Q are essentially normal relative to σ, one sees that the operator PQ almost intertwines the two representations $\sigma|_{C(M_P)}$ and $\sigma|_{C(M_Q)}$; that is, one has that $P\sigma(\varphi)P(PQ) - (PQ)Q\sigma(\varphi)Q \in \mathcal{K}$ for $\varphi \in C(M)$. Thus, in the Calkin algebra, if $\varphi \in C(M)$ satisfies $\varphi|_{M_P} \equiv 1$ and $\varphi|_{M_Q} \equiv 0$, we obtain that $\pi(P\sigma(\phi)P)\pi(PQ) = \pi(PQ)\pi(Q\sigma(\phi)Q)$. But, $\pi(Q\sigma(\varphi)Q) = 0$ and $\pi(P\sigma(\varphi)P) = \pi(P)$, this means that $\pi(PQ) = 0$, which implies $PQ \in \mathcal{K}$ and completes the proof. □

We can use this theorem to obtain a partial result concerning the relation of the projections onto $[p]$ and $[q]$ for $p, q \in \mathbb{C}[z_1, \ldots, z_n]$.

Corollary 4.2. *For two polynomials $p, q \in \mathbb{C}[z_1, \ldots, z_n]$, let P and Q be the projections onto the submodules $\mathcal{P} = [p], \mathcal{Q} = [q]$ on $L^2_a(\mathbb{B}^n)$, respectively. If p, q satisfy $Z(p) \cap Z(q) \cap \partial\mathbb{B}^n = \phi$, then we have that $[P, Q] \in \mathcal{K}$.*

Proof. Note that $I - P, I - Q$ are the projections onto the quotient modules $\mathcal{P}^\perp$ and $\mathcal{Q}^\perp$, respectively. Using the notation in the above, by [13] we know that $M_{I-P} \subseteq Z(p) \cap \partial\mathbb{B}^n$ and $M_{I-Q} \subseteq Z(q) \cap \partial\mathbb{B}^n$. It follows from the hypothesis that $M_{I-P} \cap M_{I-Q} = \phi$. By Theorem 4.1 we have that $(I-P)(I-Q)$ and $(I-Q)(I-P)$ are compact. Therefore, one sees that $[P, Q] \in \mathcal{K}$, which completes the proof. □

We can extend this result using Theorem 2.10 as follows.

Corollary 4.3. *For polynomials $p, q, r \in \mathbb{C}[z_1, \ldots, z_n]$ with $Z(p) \cap Z(q) \cap \partial\mathbb{B}^n = \phi$, let $\mathcal{P} = [pr]$ and $\mathcal{Q} = [qr]$ be the submodules in $L^2_a(\mathbb{B}^n)$. Then one has $[P, Q] \in \mathcal{K}$, where P, Q are the projections onto $\mathcal{P}$ and $\mathcal{Q}$, respectively.*

Proof. One can generalize the argument in [13] to show that

$$M_{[pr]^\perp/[r]^\perp} \subseteq Z(p) \cap \partial\mathbb{B}^n \textit{ and } M_{[qr]^\perp/[r]^\perp} \subseteq Z(q) \cap \partial\mathbb{B}^n.$$

By Theorem 4.1, this implies that $(R - P)(R - Q)$ and $(R - Q)(R - P)$ are compact, where R is the projection onto the submodule $[r]$. This means that $[P, Q] = [R - P, R - Q] \in \mathcal{K}$, which completes the proof. □

Another example of the application of the notion of the locality of essentially normal projection is the following result which is more or less the opposite situation of the previous theorem.

Theorem 4.4. *Assume that $\sigma : C(M) \to \mathcal{L}(\mathcal{H})$ is a $*$-homomorphism on the Hilbert space $\mathcal{H}$ for a compact metric space M, and P and Q are two essentially normal projections such that $\mathcal{P} \cap \mathcal{Q}^\perp = \mathcal{P}^\perp \cap \mathcal{Q} = \{0\}$, where $\mathcal{P} = \operatorname{ran} P$ and $\mathcal{Q} = \operatorname{ran} Q$. If $\mathcal{P} + \mathcal{Q}^\perp$ is closed, then $M_P = M_Q$ and $[\hat{\sigma}_P] = [\hat{\sigma}_Q] \in K_1(M_P)$.*

Proof. In the representation theorem for P, Q, the spaces $\mathcal{H}_0$ and $\mathcal{H}_2$ are $\{0\}$ by assumption and we can write $\mathcal{P} = \mathcal{H}_1 \oplus \mathcal{P}'$ and $\mathcal{Q} = \mathcal{H}_1 \oplus \mathcal{Q}'$ corresponding to $P' = P - I_{\mathcal{H}_1}$ and $Q' = Q - I_{\mathcal{H}_1}$. As in the proof of Theorem 4.1, the image $\pi(P'Q')$ of $P'Q'$ in the Calkin algebra intertwines the operators $\pi(P'\sigma(\varphi)P')$ and $\pi(Q'\sigma(\varphi)Q')$. Using Theorem 2.3 and the assumption $\mathcal{P} + \mathcal{Q}^\perp$ is closed, we have $0 \notin \sigma(S) = \sigma(P'Q'P')$. Combining this with the fact $\ker P'Q' = \mathcal{P}'^\perp \cap \mathcal{Q}' = \{0\}$, one sees that $P'Q' : \mathcal{Q}' \to \mathcal{P}'$ is invertible. Therefore, using the polar decomposition in the Calkin algebra, one sees that $M_P = M_Q$ and that the K_1 elements are equal. □

There would seem to be a limit to what can be concluded about the K_1 element. If $k \in K_1(M_P)$ for some essentially normal projection P on the Hilbert space $\mathcal{H}$ with a $*$-homomorphism $\sigma : C(M) \to \mathscr{L}(\mathcal{H})$, then there exists an essentially normal projection $Q \leq P$ such that $[\sigma, Q] = k \in K_1(M_P)$.

References

[1] W.B. Arveson, *Subalgebras of C^*-algebras III: Multivariate operator theory*, Acta Math., **181**(1998), 159–228.

[2] W.B. Arveson, *The curvature invariant of a Hilbert module over* $\mathbb{C}[z_1, \dots, z_d]$, J. reine angew. Math., **522**(2000), 173–236.

[3] W.B. Arveson, *The Dirac operator of a commuting d-tuple*, J. Funct. Anal., **189** (2002), 53–79.

[4] W.B. Arveson, *p-summable commutators in dimension d*, J. Oper. Theory, **54**(2005), 101–117.

[5] W.B. Arveson, *Quotients of standard Hilbert modules*, Trans. Amer. Math. Soc., **359**(2007), 6027–6055.

[6] A. Brown, The unitary equivalence of binormal operators, Amer. J. Math., **76**(1954), 414–434.

[7] L.G. Brown, R.G. Douglas and P.A. Fillmore, Extensions of C^*-algebras and K-homology, Ann. Math., **105** (1977), 265–324.

[8] C. Berger and B. Shaw, *Selfcommutators of multicyclic hyponormal operators are always trace class*, Bull. Amer. Math. Soc., **79**(1973), 1193–1199.

[9] J. Dixmier, *Position relative de deux variétés linéaires fermées dans un espace de Hilbert*, Rev. Sci., **86**(1948), 387–399.

[10] R.G. Douglas, *A new kind of index theorem*, Analysis, Geometry and Topology of Elliptic Operators (Roskilde, Denmark, 2005), 369–382, World Sci. Publishing, 2006.

[11] R.G. Douglas, *Essentially reductive Hilbert modules*, J. Oper. Theory, **55**(2006), 117–133.

[12] R.G. Douglas and J. Sarkar, *Essentially reductive weighted shift Hilbert modules*, J. Oper. Theory, **65**(2011), 101–133.

[13] R.G. Douglas and K. Wang, *A harmonic analysis approach to essential normality of principal submodules,* J. Funct. Anal., **261**(2011), 3155–3180.

[14] J. Eschmeier, *Essential normality of homogeneous submodules*, Int. Eq. Oper. Ther., **69**(2011), 171–182.

[15] K. Guo, *Defect operators for submodules of H_d^2*, J. reine angew. Math., **573**(2004), 181–209.

[16] K. Guo and K. Wang, *Essentially normal Hilbert modules and K-homology II: Quasi-homogeneous Hilbert modules over the two dimensional unit ball*, J. Ramanujan Math. Soc., **22** (2007), 259–281.

[17] K. Guo and K. Wang, *Essentially normal Hilbert modules and K-homology*, Math. Ann., **340**(2008), 907–934.

[18] P.R. Halmos, *Two subspaces*, Trans. Amer. Math. Soc., **144**(1969), 381–389.

[19] M.Kennedy, *Essential normality and the decomposability of homogeneous submodules*, Trans. Amer. Math. Soc. (to appear), arxiv:1202.1797.

[20] O. Shalit, *Stable polynomial division and essential normality of graded Hilbert modules*, J. London Math. Soc., **83**(2011), 273–289.

Ronald G. Douglas
Department of Mathematics
Texas A&M University
College Station, TX 77843, USA
e-mail: `rdouglas@math.tamu.edu`

Kai Wang
School of Mathematical Sciences
Fudan University
Shanghai, 200433, P. R. China
e-mail: `kwang@fudan.edu.cn`

Operator Theory:
Advances and Applications, Vol. 236, 171–179

Which Weighted Composition Operators are Complex Symmetric?

Stephan Ramon Garcia and Christopher Hammond

Abstract. Recent work by several authors has revealed the existence of many unexpected classes of normal weighted composition operators. On the other hand, it is known that every normal operator is a complex symmetric operator. We therefore undertake the study of complex symmetric weighted composition operators, identifying several new classes of such operators.

Mathematics Subject Classification (2010). 47B33, 47B32, 47B99.

Keywords. Complex symmetric operator, conjugation, composition operator, weighted composition operator, Hermitian operator, normal operator, self-map, Koenigs eigenfunction, disk automorphism, involution.

1. Introduction

In 2010, C. Cowen and E. Ko obtained an explicit characterization and spectral description of all Hermitian weighted composition operators on the classical Hardy space H^2 [5]. This work was later extended to certain weighted Hardy spaces by C. Cowen, G. Gunatillake, and E. Ko [4]. Along similar lines, P. Bourdon and S. Narayan have recently studied normal weighted composition operators on H^2 [1]. Taken together, these articles have established the existence of several unexpected families of normal weighted composition operators.

It turns out that normal operators are the simplest examples of complex symmetric operators. We say that a bounded operator T on a complex Hilbert space $\mathcal{H}$ is *complex symmetric* if there exists a *conjugation* (i.e., a conjugate-linear, isometric involution) J such that $T = JT^*J$. The general study of such operators was undertaken by the first author, M. Putinar, and W. Wogen, in various combinations, in [7–10]. A number of other authors have also made significant contributions [3, 11–14, 18–21].

First author partially supported by National Science Foundation Grant DMS-1001614.

We consider here the problem of describing all complex symmetric weighted composition operators. Among other results, we produce a class of complex symmetric weighted composition operators which includes the Hermitian examples obtained in [4, 5] as special cases. We also raise a number of open questions which we hope will spur further research.

2. Observations and results

In what follows, we let $H^2(\beta)$ denote the weighted Hardy space which corresponds to the weight sequence $\{\beta(n)\}_{n=0}^{\infty}$ [6, Sect. 2.1]. For each w in the open unit disk $\mathbb{D}$ and every integer $n \geq 0$, we let $K_w^{(n)}$ denote the unique function in $H^2(\beta)$ which satisfies $\langle f, K_w^{(n)} \rangle = f^{(n)}(w)$ for every f in $H^2(\beta)$. For convenience, we often choose to write K_w in place of $K_w^{(0)}$. If $\varphi : \mathbb{D} \to \mathbb{D}$ is analytic, then the *composition operator* $C_\varphi : H^2(\beta) \to H^2(\beta)$ is defined by setting

$$C_\varphi(f) = f \circ \varphi.$$

Given another analytic function $\psi : \mathbb{D} \to \mathbb{C}$, we define the *weighted composition operator* $W_{\varphi,\psi}$ by setting

$$W_{\varphi,\psi}(f) = \psi \cdot (f \circ \varphi).$$

Assuming that $W_{\varphi,\psi}$ is bounded, one has the useful formulae

$$W_{\varphi,\psi}^*(K_w) = \overline{\psi(w)} K_{\varphi(w)}, \tag{1}$$

$$W_{\varphi,\psi}^*(K_w^{(1)}) = \overline{\psi(w)}\,\overline{\varphi'(w)} K_{\varphi(w)}^{(1)} + \overline{\psi'(w)} K_{\varphi(w)}. \tag{2}$$

2.1. Composition operators

One initially expects few unweighted composition operators to be complex symmetric. In fact, the only obvious candidates which come to mind are the normal composition operators. These are precisely the operators $C_\varphi : H^2(\beta) \to H^2(\beta)$ where $\varphi(z) = az$ and $|a| \leq 1$ [6, Thm. 8.2]. One might initially suspect that these are the *only* complex symmetric composition operators. This naïve conjecture proves to be false, however, as there exist at least two other basic families of complex symmetric composition operators.

Proposition 2.1. *If φ is either* (i) *constant, or* (ii) *an involutive disk automorphism, then $C_\varphi : H^2(\beta) \to H^2(\beta)$ is a complex symmetric operator.*

The preceding follows immediately from the fact that an operator which is algebraic of degree two is complex symmetric [10, Thm. 2]. In what follows, we work only with nonconstant symbols φ. It turns out that (ii) prompts an elementary question whose answer has so far eluded us.

Question 1. Let φ be an involutive disk automorphism. Find an explicit conjugation $J : H^2(\beta) \to H^2(\beta)$ such that $C_\varphi = JC_\varphi^* J$.

Naturally, one is also interested in determining whether there are any additional classes of complex symmetric composition operators.

Question 2. Characterize all complex symmetric composition operators C_φ on the classical Hardy space H^2 or, more generally, on weighted Hardy spaces $H^2(\beta)$.

In the negative direction, we have the following results.

Proposition 2.2. *If $C_\varphi : H^2(\beta) \to H^2(\beta)$ is a hyponormal composition operator which is complex symmetric, then $\varphi(z) = az$ where $|a| \le 1$.*

Proof. Suppose C_φ is hyponormal; that is, $\|C_\varphi f\| \ge \|C_\varphi^* f\|$ for all f in $H^2(\beta)$. If C_φ is J-symmetric, then it follows that

$$\|C_\varphi^* f\| = \|JC_\varphi J f\| = \|C_\varphi J f\| \ge \|C_\varphi^* J f\| = \|JC_\varphi f\| = \|C_\varphi f\|.$$

Thus $\|C_\varphi f\| = \|C_\varphi^* f\|$ for all f in H^2 whence C_φ is normal. By [6, Thm. 8.2] we conclude that $\varphi(z) = az$ where $|a| \le 1$. □

Proposition 2.3. *Suppose that $C_\varphi : H^2(\beta) \to H^2(\beta)$ is J-symmetric. If $J(1)$ is a constant multiple of a kernel function K_w, then $\varphi(w) = w$. The converse holds whenever φ is not an automorphism.*

Proof. If $J(1) = \gamma K_w$ for some constant $\gamma \neq 0$ and C_φ is J-symmetric, then

$$\gamma K_w = J(1) = JC_\varphi(1) = C_\varphi^* J(1) = C_\varphi^*(\gamma K_w) = \gamma K_{\varphi(w)},$$

from which we conclude that $\varphi(w) = w$. On the other hand, suppose that $\varphi(w) = w$. Since $C_\varphi^*(K_w) = K_{\varphi(w)} = K_w$, we see that

$$C_\varphi J(K_w) = JC_\varphi^*(K_w) = J(K_w).$$

As long as φ is not an automorphism, the only eigenvectors for C_φ corresponding to the eigenvalue 1 are the constant functions [17, p. 90]. Therefore $J(K_w)$ must be a constant function, which means that $J(1)$ must be a scalar multiple of K_w. □

In light of the preceding, we see that if J is a conjugation on $H^2(\beta)$ such that $J(1)$ is not a constant multiple of a kernel function, then there does not exist a J-symmetric composition operator C_φ on $H^2(\beta)$ whose symbol fixes a point in $\mathbb{D}$. If $J(1)$ is a constant multiple of 1, then we can say even more about φ. The following is inspired by an unpublished result of P. Bourdon and D. Szajda [6, Ex. 8.1.2].

Proposition 2.4. *Suppose that $J : H^2(\beta) \to H^2(\beta)$ is a conjugation, $J(1)$ is a constant multiple of 1, and $J(z)$ is a constant multiple of z^m for some $m \ge 1$. If C_φ is J-symmetric, then $\varphi(z) = az$ for some $|a| \le 1$.*

Proof. Since $1 = \beta(0)K_0$, it follows from Proposition 2.3 that $\varphi(0) = 0$, whence

$$C_\varphi^*(K_0^{(1)}) = \overline{\varphi'(0)} K_{\varphi(0)}^{(1)} = \overline{\varphi'(0)} K_0^{(1)}$$

by (2). Thus $z = \beta(1)K_0^{(1)}$ is an eigenvector for C_φ^* corresponding to the eigenvalue $\overline{\varphi'(0)}$. Since C_φ is J-symmetric, z^m must be an eigenvector for C_φ corresponding to the eigenvalue $\varphi'(0)$. Observe that $C_\varphi(z^m) = \varphi^m$, which means that $\varphi(z)^m = \varphi'(0)z^m$. Consequently $\varphi(z) = az$, where $|a| \le 1$. □

2.2. Weighted composition operators

Although our list of complex symmetric composition operators is somewhat sparse, there are a variety of *weighted* composition operators which are known to be complex symmetric. Indeed, the study of Hermitian, normal, and unitary weighted composition operators has been the focus of intense research [1, 4, 5]. The following is a generalization of [1, Lem. 2, Prop. 3], where the same conclusion is obtained under the assumption that $W_{\varphi,\psi}$ is normal.

Proposition 2.5. *If $W_{\varphi,\psi} : H^2(\beta) \to H^2(\beta)$ is complex symmetric, then either ψ is identically zero or ψ is nonvanishing on $\mathbb{D}$. Moreover, if φ is not a constant function and ψ is not identically zero, then φ is univalent.*

Proof. Suppose that $W_{\varphi,\psi}$ is complex symmetric and that ψ does not vanish identically. Since $\ker W_{\varphi,\psi} = \{0\}$, we conclude that $\ker W^*_{\varphi,\psi} = \{0\}$ by [7, Prop. 1]. If $\psi(w) = 0$ for some w in $\mathbb{D}$, then $W^*_{\varphi,\psi}(K_w) = 0$ by (1). Since this contradicts the fact that $\ker W^*_{\varphi,\psi}$ is trivial, we conclude that ψ is nonvanishing on $\mathbb{D}$. Now suppose that there are points w_1 and w_2 in $\mathbb{D}$ such that $\varphi(w_1) = \varphi(w_2)$. It follows that

$$W^*_{\varphi,\psi}\big(\overline{\psi(w_2)}K_{w_1} - \overline{\psi(w_1)}K_{w_2}\big) = \overline{\psi(w_2)}\,\overline{\psi(w_1)}K_{\varphi(w_1)} - \overline{\psi(w_1)}\,\overline{\psi(w_2)}K_{\varphi(w_2)} = 0.$$

Since any distinct pair of reproducing kernel functions is linearly independent, we conclude that $w_1 = w_2$. In other words, φ is univalent. □

The following result provides a severe restriction on the spectrum of a complex symmetric weighted composition operator whose symbol has a fixed point in $\mathbb{D}$.

Proposition 2.6. *Suppose that $W_{\varphi,\psi} : H^2(\beta) \to H^2(\beta)$ is a complex symmetric operator. If $\varphi(w_0) = w_0$ for some w_0 in $\mathbb{D}$, then $\psi(w_0)\,\varphi'(w_0)^n$ is an eigenvalue of $W_{\varphi,\psi}$ for every integer $n \geq 0$.*

Proof. Since $W_{\varphi,\psi}$ is complex symmetric, by [7, Prop. 1] it suffices to prove that

$$\overline{\psi(w_0)}\,\overline{\varphi'(w_0)^n} \tag{3}$$

is an eigenvalue for $W^*_{\varphi,\psi}$. Let us first assume that $\varphi'(w_0)$ is not a root of unity. We claim that for each $n \geq 0$, the function $K^{(n)}_{w_0}$ can be written in the form

$$v_n + \alpha_{n-1}v_{n-1} + \alpha_{n-2}v_{n-2} + \cdots + \alpha_0 v_0,$$

where v_j is an eigenvector for $W_{\varphi,\psi}$ corresponding to the eigenvalue $\overline{\psi(w_0)}\,\overline{\varphi'(w_0)^j}$. We prove this assertion by induction. Note that

$$W^*_{\varphi,\psi}(K_{w_0}) = \overline{\psi(w_0)}K_{\varphi(w_0)} = \overline{\psi(w_0)}K_{w_0},$$

so the claim holds when $n = 0$. Suppose then that the claim holds for all $n \leq k$ and consider the kernel function $K^{(k+1)}_{w_0}$. Now recall that $W^*_{\varphi,\psi}(K^{(k+1)}_{w_0})$ equals $\overline{\psi(w_0)}\,\overline{\varphi'(w_0)^{k+1}}K^{(k+1)}_{\varphi(w_0)}$ plus a linear combination of kernel functions $K^{(j)}_{w_0}$ with

$j \leq k$. Our induction hypothesis implies that each of these kernel functions is a linear combination of eigenvectors v_j. Therefore we may write

$$W^*_{\varphi,\psi}(K^{(k+1)}_{w_0}) = \overline{\psi(w_0)}\,\overline{\varphi'(w_0)^{k+1}} K^{(k+1)}_{w_0} + \beta_k v_k + \beta_{k-1} v_{k-1} + \cdots + \beta_0 v_0$$

for some constants $\beta_0, \beta_1, \ldots, \beta_k$. Observe that the function

$$v_{k+1} = K^{(k+1)}_{w_0} + \sum_{j=0}^{k} \frac{\beta_j}{\overline{\psi(w_0)}\big(\overline{\varphi'(w_0)^{k+1}} - \overline{\varphi'(w_0)^j}\big)} v_j$$

is an eigenvector for $W^*_{\varphi,\psi}$ corresponding to the eigenvalue $\overline{\psi(w_0)}\,\overline{\varphi'(w_0)^{k+1}}$. Consequently our claim holds for all n. In other words, every term (3) is an eigenvalue for $W^*_{\varphi,\psi}$. If $\varphi'(w_0)$ is an mth root of unity, then a similar argument shows that

$$K^{(n)}_{w_0} = v_n + \alpha_{n-1} v_{n-1} + \alpha_{n-2} v_{n-2} + \cdots + \alpha_0 v_0$$

whenever $0 \leq n \leq m-1$. Hence (3) is an eigenvalue for $W^*_{\varphi,\psi}$ when $n \leq m-1$ and hence for all n. In either case, every number (3) is an eigenvalue for $W^*_{\varphi,\psi}$, which means that $\psi(w_0)\,\varphi'(w_0)^n$ is an eigenvalue for $W_{\varphi,\psi}$. □

Example 1. Fix $a \in \mathbb{D}\backslash\{0\}$ and let

$$\varphi = \frac{a - z}{1 - \overline{a}z}.$$

Since φ is an involutive automorphism, the composition operator $C_\varphi : H^2(\beta) \to H^2(\beta)$ is complex symmetric by Proposition 2.1. Moreover, observe that the spectrum $\sigma(C_\varphi)$ of C_φ is precisely $\{-1, 1\}$. On the other hand, Proposition 2.6 implies that $\varphi'(w_0)^n$ belongs to $\sigma(C_\varphi)$ whenever w_0 is a fixed point of w_0. However, the only fixed point of φ which lies inside of $\mathbb{D}$ is

$$w_0 = \frac{1 - \sqrt{1 - |a|^2}}{\overline{a}},$$

which happens to satisfy $\varphi'(w_0) = -1$, in accordance with Proposition 2.6.

2.3. Koenigs eigenfunctions

For any nonconstant non-automorphism $\varphi : \mathbb{D} \to \mathbb{D}$ which has a fixed point w_0 in $\mathbb{D}$ and for which $\varphi'(w_0) \neq 0$, there is an analytic $\kappa : \mathbb{D} \to \mathbb{C}$ such that $\kappa \circ \varphi = \varphi'(w_0)\kappa$. This function, called the *Koenigs eigenfunction* for φ, is unique up to scalar multiplication [6, p. 62, p. 93]. Furthermore, κ^n (or any constant multiple thereof) is the only analytic function for which $\kappa^n \circ \varphi = \varphi'(w_0)^n \kappa^n$. Proposition 2.6, together with the details of its proof, yields the following result pertaining to unweighted composition operators.

Proposition 2.7. *Let $\varphi : \mathbb{D} \to \mathbb{D}$ be an analytic selfmap which is not an automorphism and suppose that $\varphi(w_0) = w_0$ and $\varphi'(w_0) \neq 0$ for some w_0 in $\mathbb{D}$. If $C_\varphi : H^2(\beta) \to H^2(\beta)$ is complex symmetric, then every power κ^n of the Koenigs eigenfunction for φ belongs to $H^2(\beta)$.*

It is not difficult to construct a univalent map $\varphi : \mathbb{D} \to \mathbb{D}$ in such a way that one can readily determine whether its Koenigs eigenfunction belongs to $H^2(\beta)$ [17, pp. 93-94]. Let $\kappa : \mathbb{D} \to \mathbb{C}$ be a univalent function that vanishes at some point w_0 and consider the region $\kappa(\mathbb{D})$. Suppose that $\lambda\kappa(\mathbb{D}) \subseteq \kappa(\mathbb{D})$ for some complex λ with $|\lambda| < 1$. Define the map φ by $\varphi(z) = \kappa^{-1}(\lambda\kappa(z))$. Then, by construction, φ is a univalent self-map of $\mathbb{D}$ that fixes w_0 and whose Koenigs eigenfunction is κ. Hence, by starting with a κ that belongs to $H^2(\beta)$, we construct a φ whose Koenigs function belongs to $H^2(\beta)$. Similarly, if we take κ such that κ^n does not belong to $H^2(\beta)$ for some n, we obtain a map whose corresponding composition operator is not complex symmetric by Proposition 2.7. For example, consider any such λ and take $\kappa(z) = 2z/(1-z)$, which does not belong to the Hardy space H^2. From this we obtain the map $\varphi(z) = (\lambda z)/(1 + (\lambda - 1)z)$, which induces a composition operator $C_\varphi : H^2 \to H^2$ which is not complex symmetric.

Much work has been done to determine the conditions under which a Koenigs eigenfunction κ belongs to the Hardy space H^2. In this context, Proposition 2.7 is equivalent to saying that κ belongs to H^p for every $0 < p < \infty$. The following proposition follows directly from [16, Thm. 2.2].

Proposition 2.8. *Suppose that $\varphi : \mathbb{D} \to \mathbb{D}$ is not an automorphism and that φ has a fixed point w_0 in $\mathbb{D}$ such that $\varphi'(w_0) \neq 0$. If $C_\varphi : H^2 \to H^2$ is complex symmetric, then the essential spectral radius of C_φ is 0. In other words, C_φ must be a Riesz composition operator.*

A good deal of work has been done to study Riesz composition operators on H^2. Bourdon and Shapiro's paper [2] serves as an excellent starting point.

Suppose that φ is not an automorphism, $\varphi(w_0) = w_0$, $\varphi'(w_0) \neq 0$, and that C_φ is J-symmetric. As we have already observed, $J(1)$ must be a constant multiple of K_{w_0}. Let κ denote the Koenigs eigenfunction for φ, normalized so that $\|\kappa\| = 1$. We also know that $J(\kappa)$ equals a constant multiple of $K_{w_0}^{(1)}$. In particular, taking into account the norms of these functions, we can write

$$J(1) = \frac{\gamma\,\beta(0)\,K_{w_0}}{\|K_{w_0}\|}, \qquad J(\kappa) = \frac{\delta\,K_{w_0}^{(1)}}{\|K_{w_0}^{(1)}\|},$$

where $|\gamma| = |\delta| = 1$. Since $\langle \kappa, 1\rangle = \langle J(1), J(\kappa)\rangle$, we see that

$$|\kappa(0)| = \frac{|K_{w_0}^{(1)}(w_0)|}{\|K_{w_0}\|\,\|K_{w_0}^{(1)}\|}.$$

If $w_0 = 0$, then this tells us nothing. If $w_0 \neq 0$, however, it places a major restriction upon the function κ. In essence, most functions in $H^2(\beta)$ cannot be Koenigs eigenfunctions for complex symmetric composition operators.

2.4. An instructive example

We conclude this note by producing a class of complex symmetric weighted composition operators which includes the Hermitian examples obtained in [4, 5] as special cases. For each $\kappa \geq 1$, let $H^2(\beta_\kappa)$ denote the weighted Hardy space whose reproducing kernel is $K_w(z) = (1-\overline{w}z)^{-\kappa}$. We now explicitly characterize all weighted composition operators on $H^2(\beta_\kappa)$ which are J-symmetric with respect to the conjugation

$$[Jf](z) = \overline{f(\overline{z})} \tag{4}$$

on $H^2(\beta_\kappa)$. For the sake of convenience, we sometimes write $\widetilde{f} := Jf$.

Proposition 2.9. *A weighted composition operator $W_{\varphi,\psi} : H^2(\beta_\kappa) \to H^2(\beta_\kappa)$ is J-symmetric with respect to the conjugation (4) if and only if*

$$\psi(z) = \frac{b}{(1-a_0 z)^\kappa}, \qquad \varphi(z) = a_0 + \frac{a_1 z}{1-a_0 z}, \tag{5}$$

where a_0 and a_1 are constants such that φ maps $\mathbb{D}$ into $\mathbb{D}$. Moreover, such an operator is normal if and only if either,

(i) $b = 0$,

(ii) $b \neq 0$ *and* $\operatorname{Im} a_0\overline{a_1} = (1-|a_0|^2)\operatorname{Im} a_0$.

Moreover, $W_{\varphi,\psi}$ is Hermitian if and only if a_0, a_1, and b each belong to $\mathbb{R}$.

Proof. To streamline our notation, we let $W := W_{\varphi,\psi}$. A simple computation now confirms that if ψ and φ are given by (5), then $WJK_w = JW^*K_w$ for all w in $\mathbb{D}$, implying that $W = JW^*J$. On the other hand, if $W = JW^*J$, then $WJK_w = JW^*K_w$ for all w in $\mathbb{D}$. Since $JK_w = K_{\overline{w}}$, this implies that

$$\psi(z)K_{\overline{w}}(\varphi(z)) = \psi(w)K_{\overline{\varphi(w)}}(z) \tag{6}$$

holds for all z, w in $\mathbb{D}$. Setting $w = 0$ in the preceding we find that

$$\psi(z) = \frac{\psi(0)}{(1-\varphi(0)z)^\kappa}.$$

Thus ψ is of the form (5) with $b = \psi(0)$ and $a_0 = \varphi(0)$. From (6) it follows that

$$\frac{1-\varphi(w)z}{1-a_0 z} = \frac{1-\varphi(z)w}{1-a_0 w}.$$

Writing $\varphi(z) = a_0 + z\xi(z)$ where ξ is analytic on $\mathbb{D}$, we see that

$$(1-a_0 z)\xi(z) = (1-a_0 w)\xi(w)$$

for all z, w in $\mathbb{D}$. Thus the function $(1-a_0 z)\xi(z)$ is constant. Letting $\xi(0) = a_1$, we conclude that φ has the form (5).

Suppose that ψ and φ are given by (5) and note that W is normal if and only if $JWW^*K_w = WW^*JK_w$ for all w in $\mathbb{D}$. The preceding condition is equivalent to asserting that

$$\frac{\psi(w)\widetilde{\psi}(z)}{1-\varphi(w)\widetilde{\varphi}(z)} = \frac{\widetilde{\psi}(w)\psi(z)}{1-\widetilde{\varphi}(w)\varphi(z)}$$

holds for all z, w in $\mathbb{D}$. Taking the reciprocal of the preceding and simplifying, we see that equality holds for all z, w if and only if either $b = 0$ or $b \neq 0$ and $\operatorname{Im} a_0\overline{a_1} = (1-|a_0|^2)\operatorname{Im} a_0$.

We also note that $W = W^*$ if and only if $WJK_w = JWK_w$, which yields

$$\psi(z)K_{\overline{w}}(\varphi(z)) = \widetilde{\psi}(z)K_{\overline{w}}(\widetilde{\varphi}(z)).$$

Setting $w = 0$ in the preceding yields $\psi(z) = \widetilde{\psi}(z)$ so that a_0 and b are real. This implies that $\varphi(z) = \widetilde{\varphi}(z)$ whence a_1 is also real. Conversely, it is easy to see that if a_0, a_1, and b are real, then W is Hermitian. □

It follows from the preceding that if a_0, a_1, b are chosen so that φ maps $\mathbb{D}$ into $\mathbb{D}$ and so that (i) and (ii) both fail to hold, then the operator $W_{\varphi,\psi} : H^2(\beta_\kappa) \to H^2(\beta_\kappa)$ will be complex symmetric and non-normal. Moreover, the operators produced by Proposition 2.9 include the Hermitian examples considered in [4, 5].

Question 3. The detailed spectral structure of *Hermitian* weighted composition operators $W_{\varphi,\psi} : H^2(\beta_\kappa) \to H^2(\beta_\kappa)$ with ψ and φ given by (5) is studied in [4, 5]. What is the corresponding spectral theory for the non-normal weighted composition operators arising from Proposition 2.9?

Note added in proof. Recent work by S. Waleed Noor [15] appears to have answered Question 1 in the context of the weighted Hardy spaces on the unit ball $\mathbb{B}_n$.

References

[1] Paul S. Bourdon and Sivaram K. Narayan. Normal weighted composition operators on the Hardy space $H^2(\mathbb{U})$. *J. Math. Anal. Appl.*, 367(1):278–286, 2010.

[2] Paul S. Bourdon and Joel H. Shapiro. Riesz composition operators. *Pacific J. Math.*, 181(2):231–246, 1997.

[3] Nicolas Chevrot, Emmanuel Fricain, and Dan Timotin. The characteristic function of a complex symmetric contraction. *Proc. Amer. Math. Soc.*, 135(9):2877–2886 (electronic), 2007.

[4] Carl C. Cowen, Gajath Gunatillake, and Eungil Ko. Hermitian weighted composition operators and Bergman extremal functions. *Complex Anal. Oper. Theory*, 7(1):69–99, 2013.

[5] Carl C. Cowen and Eungil Ko. Hermitian weighted composition operators on H^2. *Trans. Amer. Math. Soc.*, 362(11):5771–5801, 2010.

[6] Carl C. Cowen and Barbara D. MacCluer. *Composition operators on spaces of analytic functions.* Studies in Advanced Mathematics. CRC Press, Boca Raton, FL, 1995.

[7] Stephan Ramon Garcia and Mihai Putinar. Complex symmetric operators and applications. *Trans. Amer. Math. Soc.*, 358(3):1285–1315 (electronic), 2006.

[8] Stephan Ramon Garcia and Mihai Putinar. Complex symmetric operators and applications. II. *Trans. Amer. Math. Soc.*, 359(8):3913–3931 (electronic), 2007.

[9] Stephan Ramon Garcia and Warren R. Wogen. Complex symmetric partial isometries. *J. Funct. Anal.*, 257(4):1251–1260, 2009.

[10] Stephan Ramon Garcia and Warren R. Wogen. Some new classes of complex symmetric operators. *Trans. Amer. Math. Soc.*, 362(11):6065–6077, 2010.

[11] T.M. Gilbreath and Warren R. Wogen. Remarks on the structure of complex symmetric operators. *Integral Equations Operator Theory*, 59(4):585–590, 2007.

[12] Sungeun Jung, Eungil Ko, and Ji Eun Lee. On scalar extensions and spectral decompositions of complex symmetric operators. *J. Math. Anal. Appl.*, 384(2):252–260, 2011.

[13] S. Jung, E. Ko, M. Lee, and J. Lee. On local spectral properties of complex symmetric operators. *J. Math. Anal. Appl.*, 379:325–333, 2011.

[14] Chun Guang Li, Sen Zhu, and Ting Ting Zhou. Foguel operators with complex symmetry. Preprint.

[15] S. Waleed Noor, Complex symmetry of composition operators induced by involutive ball automorphisms. Preprint.

[16] Pietro Poggi-Corradini. The Hardy class of Koenigs maps. *Michigan Math. J.*, 44(3):495–507, 1997.

[17] Joel H. Shapiro. *Composition operators and classical function theory.* Universitext: Tracts in Mathematics. Springer-Verlag, New York, 1993.

[18] James E. Tener. Unitary equivalence to a complex symmetric matrix: an algorithm. *J. Math. Anal. Appl.*, 341(1):640–648, 2008.

[19] Xiao Huan Wang and Zong Sheng Gao. Some equivalence properties of complex symmetric operators. *Math. Pract. Theory*, 40(8):233–236, 2010.

[20] Sergey M. Zagorodnyuk. On a J-polar decomposition of a bounded operator and matrix representations of J-symmetric, J-skew-symmetric operators. *Banach J. Math. Anal.*, 4(2):11–36, 2010.

[21] Sen Zhu, Chun Guang Li, and You Qing Ji. The class of complex symmetric operators is not norm closed. *Proc. Amer. Math. Soc.*, 140(5):1705–1708, 2012.

Stephan Ramon Garcia
Department of Mathematics
Pomona College
Claremont, California 91711, USA
e-mail: Stephan.Garcia@pomona.edu
URL: http://pages.pomona.edu/~sg064747

Christopher Hammond
Department of Mathematics
Connecticut College
270 Mohegan Avenue
New London, CT 06320, USA
e-mail: cnham@conncoll.edu
URL: http://math.conncoll.edu/faculty/chammond/

Operator Theory:
Advances and Applications, Vol. 236, 181–192

C^*-algebras Generated by Truncated Toeplitz Operators

Stephan Ramon Garcia, William T. Ross and Warren R. Wogen

Dedicated to the memory of William Arveson.

Abstract. We obtain an analogue of Coburn's description of the Toeplitz algebra in the setting of truncated Toeplitz operators. As a byproduct, we provide several examples of complex symmetric operators which are not unitarily equivalent to truncated Toeplitz operators having continuous symbols.

Mathematics Subject Classification (2010). 46Lxx, 47A05, 47B35, 47B99.

Keywords. C^*-algebra, Toeplitz algebra, Toeplitz operator, model space, truncated Toeplitz operator, compact operator, commutator ideal.

1. Introduction

In the following, we let $\mathcal{H}$ denote a separable complex Hilbert space and $\mathcal{B}(\mathcal{H})$ denote the set of all bounded linear operators on $\mathcal{H}$. For each $\mathscr{X} \subseteq \mathcal{B}(\mathcal{H})$, let $C^*(\mathscr{X})$ denote the unital C^*-algebra generated by $\mathscr{X}$. Since we are frequently interested in the case where $\mathscr{X} = \{A\}$ is a singleton, we often write $C^*(A)$ in place of $C^*(\{A\})$ in order to simplify our notation.

Recall that the *commutator ideal* $\mathscr{C}(C^*(\mathscr{X}))$ of $C^*(\mathscr{X})$ is the smallest norm closed two-sided ideal which contains the *commutators* $[A, B] := AB - BA$, where A and B range over all elements of $C^*(\mathscr{X})$. Since the quotient algebra

$$C^*(\mathscr{X})/\mathscr{C}(C^*(\mathscr{X}))$$

is an abelian C^*-algebra, it is isometrically $*$-isomorphic to $C(Y)$, the set of all continuous functions on some compact Hausdorff space Y [12, Thm. 1.2.1]. If we

The first named author was partially supported by National Science Foundation Grant DMS-1001614.

agree to denote isometric a $*$-isomorphism by $\cong$, then we may write

$$\frac{C^*(\mathscr{X})}{\mathscr{C}(C^*(\mathscr{X}))} \cong C(Y). \tag{1}$$

This yields the short exact sequence

$$0 \longrightarrow \mathscr{C}(C^*(\mathscr{X})) \xrightarrow{\iota} C^*(\mathscr{X}) \xrightarrow{\pi} C(Y) \longrightarrow 0, \tag{2}$$

where $\iota : \mathscr{C}(C^*(\mathscr{X})) \to C^*(\mathscr{X})$ is the inclusion map and $\pi : C^*(\mathscr{X}) \to C(Y)$ is the composition of the quotient map with the map which implements (1).

The *Toeplitz algebra* $C^*(T_z)$, where T_z denotes the unilateral shift on the classical Hardy space H^2, has been extensively studied since the seminal work of Coburn in the late 1960s [10, 11]. Indeed, the Toeplitz algebra is now one of the standard examples discussed in many well-known texts (e.g., [3, Sect. 4.3], [13, Ch. V.1], [14, Ch. 7]). In this setting, we have $\mathscr{C}(C^*(T_z)) = \mathscr{K}$, the ideal of compact operators on H^2, and $Y = \mathbb{T}$ (the unit circle), so that the short exact sequence (2) takes the form

$$0 \longrightarrow \mathscr{K} \xrightarrow{\iota} C^*(T_z) \xrightarrow{\pi} C(\mathbb{T}) \longrightarrow 0. \tag{3}$$

In other words, $C^*(T_z)$ is an *extension* of $\mathscr{K}$ by $C(\mathbb{T})$. In fact, one can prove that

$$C^*(T_z) = \{T_\varphi + K : \varphi \in C(\mathbb{T}), K \in \mathscr{K}\}$$

and that each element of $C^*(T_z)$ enjoys a unique decomposition of the form $T_\varphi + K$ [3, Thm. 4.3.2]. Indeed, it is well known that the only compact Toeplitz operator is the zero operator [3, Cor. 1, p. 109]. We also note that the surjective map $\pi : C^*(T_z) \to C(\mathbb{T})$ in (3) is given by $\pi(T_\varphi + K) = \varphi$.

The preceding results have spawned numerous generalizations and variants over the years. For instance, one can consider C^*-algebras generated by matrix-valued Toeplitz operators or by Toeplitz operators which act upon other Hilbert function spaces (e.g., the Bergman space [4, 25]). As another example, if $\mathscr{X}$ denotes the space of functions on $\mathbb{T}$ which are both piecewise and left continuous, then a fascinating result of Gohberg and Krupnik asserts that $\mathscr{C}(C^*(\mathscr{X})) = \mathscr{K}$ and provides the short exact sequence

$$0 \longrightarrow \mathscr{K} \xrightarrow{\iota} C^*(\mathscr{X}) \xrightarrow{\pi} C(Y) \longrightarrow 0,$$

where Y is the cylinder $\mathbb{T} \times [0, 1]$, endowed with a certain nonstandard topology [21].

Along different lines, we seek here to replace Toeplitz operators with *truncated Toeplitz operators*, a class of operators whose study has been largely motivated by a seminal 2007 paper of Sarason [26]. Let us briefly recall the basic definitions which are required for this endeavor. We refer the reader to Sarason's paper or to the recent survey article [18] for a more thorough introduction.

For each nonconstant inner function u, we consider the *model space*

$$\mathcal{K}_u := H^2 \ominus uH^2,$$

which is simply the orthogonal complement of the standard Beurling-type subspace uH^2 of H^2. Letting P_u denote the orthogonal projection from $L^2 := L^2(\mathbb{T})$ onto $\mathcal{K}_u$, for each φ in $L^\infty(\mathbb{T})$ we define the *truncated Toeplitz operator* $A^u_\varphi : \mathcal{K}_u \to \mathcal{K}_u$ by setting

$$A^u_\varphi f = P_u(\varphi f)$$

for f in $\mathcal{K}_u$. The function φ in the preceding is referred to as the *symbol* of the operator A^u_φ.[1] In particular, let us observe that A^u_φ is simply the compression of the standard Toeplitz operator $T_\varphi : H^2 \to H^2$ to the subspace $\mathcal{K}_u$. Unlike traditional Toeplitz operators, however, the symbol of a truncated Toeplitz is not unique. In fact, $A^u_\varphi = 0$ if and only if φ belongs to $uH^2 + \overline{uH^2}$ [26, Thm. 3.1].

In our work, the *compressed shift* A^u_z plays a distinguished role analogous to that of the unilateral shift T_z in Coburn's theory. In light of this, let us recall that the spectrum $\sigma(A^u_z)$ of A^u_z coincides with the so-called *spectrum*

$$\sigma(u) := \left\{ \lambda \in \mathbb{D}^- : \liminf_{z \to \lambda} |u(z)| = 0 \right\} \tag{4}$$

of the inner function u [26, Lem. 2.5]. In particular, if $u = b_\Lambda s_\mu$, where b_Λ is a Blaschke product with zero sequence $\Lambda = \{\lambda_n\}$ and s_μ is a singular inner function with corresponding singular measure μ, then

$$\sigma(u) = \Lambda^- \cup \operatorname{supp} \mu.$$

With this terminology and notation in hand, we are ready to state our main result, which provides an analogue of Coburn's description of the Toeplitz algebra in the truncated Toeplitz setting.

Theorem 1. *If u is an inner function, then*

(i) $\mathscr{C}(C^*(A^u_z)) = \mathscr{K}$, *the algebra of compact operators on* $\mathcal{K}_u$,
(ii) $C^*(A^u_z)/\mathscr{K}$ *is isometrically* $*$*-isomorphic to* $C(\sigma(u) \cap \mathbb{T})$,
(iii) *For* φ *in* $C(\mathbb{T})$, A^u_φ *is compact if and only if* $\varphi(\sigma(u) \cap \mathbb{T}) = \{0\}$,
(iv) $C^*(A^u_z) = \{A^u_\varphi + K : \varphi \in C(\mathbb{T}), K \in \mathscr{K}\}$,
(v) *For* φ *in* $C(\mathbb{T})$, $\sigma_e(A^u_\varphi) = \varphi(\sigma_e(A^u_z))$,
(vi) *For* φ *in* $C(\mathbb{T})$, $\|A^u_\varphi\|_e = \sup\{|\varphi(\zeta)| : \zeta \in \sigma(u) \cap \mathbb{T}\}$,
(vii) *Every operator in* $C^*(A^u_z)$ *is of the form normal plus compact.*

Moreover,

$$0 \longrightarrow \mathscr{K} \xrightarrow{\iota} C^*(A^u_z) \xrightarrow{\pi} C(\sigma(u) \cap \mathbb{T}) \longrightarrow 0$$

is a short exact sequence and thus $C^*(A^u_z)$ is an extension of the compact operators by $C(\sigma(u) \cap \mathbb{T})$. In particular, the map $\pi : C^*(A^u_z) \to C(\sigma(u) \cap \mathbb{T})$ is given by

$$\pi(A^u_\varphi + K) = \varphi|_{\sigma(u) \cap \mathbb{T}}.$$

[1] It is possible to consider truncated Toeplitz operators with symbols in $L^2(\mathbb{T})$, although we have little need to do so here.

The proof of Theorem 1 is somewhat involved and requires a number of preliminary lemmas. It is therefore deferred until Section 3. However, let us remark now that the same result holds when the hypothesis that f belongs to $C(\mathbb{T})$ is replaced by the weaker assumption that f is in $\mathscr{X}_u$, the class of L^∞ functions which are continuous at each point of $\sigma(u) \cap \mathbb{T}$. In fact, given f in $\mathscr{X}_u$, there exists a g in $C(\mathbb{T})$ so that

$$A_f^u \equiv A_g^u \pmod{\mathscr{K}}.$$

Thus one can replace A_f^u with A_g^u when working modulo the compact operators and adapt the proof of Theorem 1 so that $C(\mathbb{T})$ is replaced by $\mathscr{X}_u$.

Using completely different language and terminology, some aspects of Theorem 1 can be proven by triangularizing the compressed shift A_z^u according to the scheme discussed at length in [24, Lec. V]. For instance, items (vi) and (iii) of the preceding theorem are [1, Cor. 5.1] and [1, Thm. 5.4], respectively (we should also mention related work of Kriete [22, 23]). From an operator algebraic perspective, however, we believe that a different approach is desirable. Our approach is similar in spirit to the original work of Coburn and forms a possible blueprint for variations and extensions (see Section 4). Moreover, our approach does not require the detailed consideration of several special cases (i.e., Blaschke products, singular inner functions with purely atomic spectra, etc.) as does the approach pioneered in [1, 2]. In particular, we are able to avoid the somewhat involved computations and integral transforms encountered in the preceding references.

2. Continuous symbols and the TTO-CSO problem

Recall that a bounded operator T on a Hilbert space $\mathcal{H}$ is called *complex symmetric* if there exists a conjugate-linear, isometric involution J on $\mathcal{H}$ such that $T = JT^*J$. It was first recognized in [16, Prop. 3] that every truncated Toeplitz operator is complex symmetric (see also [15] where this is discussed in great detail). This hidden symmetry turns out to be a crucial ingredient in Sarason's general treatment of truncated Toeplitz operators [26].

A significant amount of evidence is mounting that truncated Toeplitz operators may play a significant role in some sort of model theory for complex symmetric operators. Indeed, a surprising and diverse array of complex symmetric operators can be concretely realized in terms of truncated Toeplitz operators (or direct sums of such operators). The recent articles [7, 8, 19, 27] all deal with various aspects of this problem and a survey of this work can be found in [18, Sect. 9].

It turns out that viewing truncated Toeplitz operators in the C^*-algebraic setting can shed some light on the question of whether every complex symmetric operator can be written in terms of truncated Toeplitz operators (the *TTO-CSO Problem*). Corollaries 2 and 3 below provide examples of complex symmetric operators which are not unitarily equivalent to truncated Toeplitz operators having continuous symbols. To our knowledge, this is the first *negative* evidence relevant to the TTO-CSO Problem which has been obtained.

Corollary 2. *If A is a noncompact operator on a Hilbert space $\mathcal{H}$, then the operator $T : \mathcal{H} \oplus \mathcal{H} \to \mathcal{H} \oplus \mathcal{H}$ defined by*

$$T = \begin{pmatrix} 0 & A \\ 0 & 0 \end{pmatrix}$$

is a complex symmetric operator which is not unitarily equivalent to a truncated Toeplitz operator with continuous symbol.

Proof. Since T is nilpotent of degree two, it is complex symmetric by [20, Thm. 2]. However, T is not of the form normal plus compact since $[T, T^*]$ is noncompact. Thus T cannot belong to $C^*(A_z^u)$ for any u by (vii) of Theorem 1. □

Corollary 3. *If S denotes the unilateral shift, then $T = \bigoplus_{i=1}^{\infty}(S \oplus S^*)$ is a complex symmetric operator which is not unitarily equivalent to a truncated Toeplitz operator with continuous symbol.*

Proof. First note that the operator $S \oplus S^*$ is complex symmetric by [17, Ex. 5] whence T itself is complex symmetric. Since $[S, S^*]$ has rank one, it follows that $[T, T^*]$ is noncompact. Therefore T is not of the form normal plus compact whence T cannot belong to $C^*(A_z^u)$ for any u by (vii) of Theorem 1. □

Unfortunately, the preceding corollary sheds no light on the following apparently simple problem.

Problem 1. Is $S \oplus S^*$ unitarily equivalent to a truncated Toeplitz operator? If so, can the symbol be chosen to be continuous?

3. Proof of Theorem 1

To prove Theorem 1, we first require a few preliminary lemmas. The first lemma is well known and we refer the reader to [24, p. 65] or [6, p. 84] for its proof.

Lemma 1. *Each function in $\mathcal{K}_u$ can be analytically continued across $\mathbb{T} \backslash \sigma(u)$.*

The following description of the spectrum and essential spectrum of the compressed shift can be found in [26, Lem. 2.5], although portions of it date back to the work of Livšic and Moeller [24, Lec. III.1]. The essential spectrum of A_z^u was computed in [1, Cor. 5.1].

Lemma 2. $\sigma(A_z^u) = \sigma(u)$ *and* $\sigma_e(A_z^u) = \sigma(u) \cap \mathbb{T}$.

Although the following must certainly be well known among specialists, we do not recall having seen its proof before in print. We therefore provide a short proof of this important fact.

Lemma 3. *A_z^u is irreducible.*

Proof. Let $\mathcal{M}$ be a nonzero reducing subspace of $\mathcal{K}_u$ for the operator A_z^u. In light of the fact that $\mathcal{M}$ is invariant under the operator $I - A_z^u(A_z^u)^* = k_0 \otimes k_0$ [26, Lem. 2.4], it follows that the nonzero vector k_0 belongs to either $\mathcal{M}$ or $\mathcal{M}^\perp$. Since k_0 is a cyclic vector for A_z^u [26, Lem. 2.3], we conclude that $\mathcal{M} = \mathcal{K}_u$ or $\mathcal{M} = \{0\}$. □

Lemma 4. *If $\varphi \in C(\mathbb{T})$, then A_φ^u is compact if and only if $\varphi|_{\sigma(u)\cap\mathbb{T}} \equiv 0$.*

Proof. ($\Leftarrow$) Suppose that $\varphi|_{\sigma(u)\cap\mathbb{T}} \equiv 0$. Let $\varepsilon > 0$ and pick ψ in $C(\mathbb{T})$ such that ψ vanishes on an open set containing $\sigma(u)\cap\mathbb{T}$ and $\|\varphi-\psi\|_\infty < \varepsilon$. Since $\|A_\varphi^u - A_\psi^u\| \leq \|\varphi - \psi\|_\infty < \varepsilon$, it suffices to show that A_ψ^u is compact. To this end, we prove that if f_n is a sequence in $\mathcal{K}_u$ which tends weakly to zero, then $A_\psi^u f_n \to 0$ in norm.

Let K denote the closure of $\psi^{-1}(\mathbb{C}\backslash\{0\})$ and note that $K \subset \mathbb{T}\backslash\sigma(u)$. By Lemma 1, we know that each f_n has an analytic continuation across K from which it follows that $f_n(\zeta) = \langle f_n, k_\zeta\rangle \to 0$, where

$$k_\zeta(z) = \frac{1 - \overline{u(\zeta)}u(z)}{1 - \overline{\zeta}z}$$

denotes the reproducing kernel corresponding to a point ζ in K [26, p. 495]. Since u is analytic on a neighborhood of the compact set K we obtain

$$|f_n(\zeta)| = |\langle f_n, k_\zeta\rangle| \leq \|f_n\||u'(\zeta)|^{\frac{1}{2}} \leq \sup_n \|f_n\| \sup_{\zeta\in K} |u'(\zeta)|^{\frac{1}{2}} = C < \infty$$

for each ζ in K. By the dominated convergence theorem, it follows that

$$\|A_\psi^u f_n\|^2 = \|P_u(\psi f_n)\| \leq \|\psi f_n\|^2 = \int_K |\psi|^2|f_n|^2 \to 0$$

whence $A_\psi^u f_n$ tends to zero in norm, as desired.

($\Rightarrow$) Suppose that φ belongs to $C(\mathbb{T})$, ξ belongs to $\sigma(u)\cap\mathbb{T}$, and A_φ^u is compact. Let

$$F_\lambda(z) := \frac{1-|\lambda|^2}{1-|u(\lambda)|^2}\left|\frac{1-\overline{u(\lambda)}u(z)}{1-\overline{\lambda}z}\right|^2,$$

which is the absolute value of the normalized reproducing kernel for $\mathcal{K}_u$. Observe that $F_\lambda(z) \geq 0$ and

$$\frac{1}{2\pi}\int_{-\pi}^{\pi} F_\lambda(e^{it})\,dt = 1$$

by definition.

By (4) there is sequence λ_n in $\mathbb{D}$ such that $|u(\lambda_n)| \to 0$. Suppose that $\xi = e^{i\alpha}$ and note that if $|t-\alpha| \geq \delta$, then

$$F_{\lambda_n}(e^{it}) \leq C_\delta \frac{1-|\lambda_n|^2}{1-|u(\lambda_n)|^2} \to 0. \tag{5}$$

This is enough to make the following approximate identity argument go through. Indeed,

$$\left|\varphi(\xi) - \frac{1}{2\pi}\int_{-\pi}^{\pi} \varphi(e^{it})F_{\lambda_n}(e^{it})\,dt\right| \leq \frac{1}{2\pi}\int_{|t-\alpha|\leq\delta} |\varphi(\xi) - \varphi(e^{it})|F_{\lambda_n}(e^{it})\,dt$$
$$+ \frac{1}{2\pi}\int_{|t-\alpha|\geq\delta} |\varphi(\xi) - \varphi(e^{it})|F_{\lambda_n}(e^{it})\,dt.$$

This first integral can be made small by the continuity of φ. Once $\delta > 0$ is fixed, the second term goes to zero by (5). □

Remark 1. We would like to thank the referee for suggesting this elegant normalized kernel function proof of the ($\Rightarrow$) direction of this lemma. Our original argument was somewhat longer.

Lemma 5. *For each $\varphi, \psi \in C(\mathbb{T})$, the semicommutator $A^u_\varphi A^u_\psi - A^u_{\varphi\psi}$ is compact. In particular, the commutator $[A^u_\varphi, A^u_\psi]$ is compact.*

Proof. Let $p(z) = \sum_i p_i z^i$ and $q(z) = \sum_j q_j z^j$ be trigonometric polynomials on $\mathbb{T}$ and note that

$$A^u_p A^u_q - A^u_{pq} = \sum_{i,j} p_i q_j (A^u_{z^i} A^u_{z^j} - A^u_{z^{i+j}}).$$

We claim that the preceding operator is compact. Since all sums involved are finite, it suffices to prove that $A^u_{z^i} A^u_{z^j} - A^u_{z^{i+j}}$ is compact for each pair (i,j) of integers.

If i and j are of the same sign, then $A^u_{z^i} A^u_{z^j} - A^u_{z^{i+j}} = 0$ is trivially compact. If i and j are of different signs, then upon relabeling and taking adjoints, if necessary, it suffices to show that if $n \geq m \geq 0$, then the operator $A^u_{z^n} A^u_{\overline{z}^m} - A^u_{z^{n-m}}$ is compact (the case $n \leq m \leq 0$ being similar). In light of the fact that

$$A^u_{z^n} A^u_{\overline{z}^m} - A^u_{z^{n-m}} = A^u_{z^{n-m}}(A^u_{z^m} A^u_{\overline{z}^m} - I),$$

we need only show that $A^u_{z^m} A^u_{\overline{z}^m} - I$ is compact for each $m \geq 1$. However, since $A^u_z A^u_{\overline{z}} - I$ has rank one [26, Lem. 2.4], this follows immediately from the identity

$$A^u_{z^m} A^u_{\overline{z}^m} - I = \sum_{\ell=0}^{m-1} A^u_{z^\ell}(A^u_z A^u_{\overline{z}} - I)A^u_{\overline{z}^\ell}.$$

Having shown that $A^u_p A^u_q - A^u_{pq}$ is compact for every pair of trigonometric polynomials p and q, the desired result follows since we may uniformly approximate any given φ, ψ in $C(\mathbb{T})$ by their respective Cesàro means. □

Remark 2. For Toeplitz operators, it is known that the semicommutator $T_\varphi T_\psi - T_{\varphi\psi}$ is compact under the assumption that one of the symbols is continuous, while the other belongs to L^∞ [3, Prop. 4.3.1], [13, Cor. V.1.4]. Though not needed for the proof of our main theorem, the same is true for truncated Toeplitz operators. This was kindly pointed out to us by Trieu Le. Here is his proof: For f in L^∞, define the Hankel operator $H_f^u : \mathcal{K}_u \to L^2$ by $H_f^u := (I - P_u)M_f$ and note that $(H_f^u)^* = P_u M_{\overline{f}}(I - P_u)$. For φ, ψ in L^∞ a computation shows that

$$A_{\varphi\psi}^u - A_\varphi^u A_\psi^u = (H_{\overline{\varphi}}^u)^* H_\psi^u. \tag{6}$$

If f belongs to L^∞, then setting $\varphi = \overline{f}$ and $\varphi = f$, we have

$$(H_f^u)^* H_f^u = A_{\overline{f}f}^u - A_{\overline{f}}^u A_f^u.$$

For continuous f it follows from the previous Lemma that $(H_f^u)^* H_f^u$ and hence H_f^u is compact whenever f is continuous. From (6) we see that if one of φ or ψ is continuous then $A_{\varphi\psi}^u - A_\varphi^u A_\psi^u$ is compact.

Proof of Theorem 1. Before proceeding further, let us remark that statement (iii) has already been proven (see Lemma 4). We first claim that

$$C^*(A_z^u) = C^*(\{A_\varphi^u : \varphi \in C(\mathbb{T})\}), \tag{7}$$

noting that the containment $\subseteq$ in the preceding holds trivially. Since $(A_z^u)^* = A_{\overline{z}}^u$, it follows that A_p^u belongs to $C^*(A_z^u)$ for any trigonometric polynomial p. We may then uniformly approximate any given φ in $C(\mathbb{T})$ by its Cesàro means to see that A_φ^u belongs to $C^*(A_z^u)$. This establishes the containment $\supseteq$ in (7).

We next prove statement (i) of Theorem 1, which states that the commutator ideal $\mathscr{C}(C^*(A_z^u))$ of $C^*(A_z^u)$ is precisely $\mathscr{K}$, the set of all compact operators on the model space $\mathcal{K}_u$:

$$\mathscr{C}(C^*(A_z^u)) = \mathscr{K}. \tag{8}$$

The containment $\mathscr{C}(C^*(A_z^u)) \subseteq \mathscr{K}$ follows easily from (7) and Lemma 5. On the other hand, Lemma 3 tells us that A_z^u is irreducible, whence the algebra $C^*(A_z^u)$ itself is irreducible. Since $[A_z^u, A_{\overline{z}}^u] \neq 0$ is compact, it follows that $C^*(A_z^u) \cap \mathscr{K} \neq \{0\}$. By [12, Cor. 3.16.8], we conclude that $\mathscr{K} \subseteq \mathscr{C}(C^*(A_z^u))$, which establishes (8).

We now claim that

$$C^*(A_z^u) = \{A_\varphi^u + K : \varphi \in C(\mathbb{T}), K \in \mathscr{K}\}, \tag{9}$$

which is statement (iv) of Theorem 1. The containment $\subseteq$ in the preceding holds because the right-hand side of (9) is a C^*-algebra which contains A_z^u (mimic the first portion of the proof of [3, Thm. 4.3.2] to see this). On the other hand, the containment $\supseteq$ in (9) follows because $C^*(A_z^u)$ contains $\mathscr{K}$ by (8) and contains every operator of the form A_φ^u with φ in $C(\mathbb{T})$ by (7).

The map $\gamma : C(\mathbb{T}) \to C^*(A_z^u)/\mathscr{K}$ defined by

$$\gamma(\varphi) = A_\varphi^u + \mathscr{K}$$

is a homomorphism by Lemma 5 and hence $\gamma(C(\mathbb{T}))$ is a dense subalgebra of $C^*(A_z^u)/\mathscr{K}$ by (7). In light of Lemma 4, we see that

$$\ker\gamma = \{\varphi \in C(\mathbb{T}) : \varphi|_{\sigma(u)\cap\mathbb{T}} \equiv 0\}, \tag{10}$$

whence the map

$$\widetilde{\gamma} : C(\mathbb{T})/\ker\gamma \to C^*(A_z^u)/\mathscr{K} \tag{11}$$

defined by

$$\widetilde{\gamma}(\varphi + \ker \gamma) = A_\varphi^u + \mathscr{K}$$

is an injective $*$-homomorphism. By [13, Thm. I.5.5], it follows that $\widetilde{\gamma}$ is an isometric $*$-isomorphism. Since

$$C(\mathbb{T})/\ker \gamma \cong C(\sigma(u) \cap \mathbb{T}) \tag{12}$$

by (10), it follows that

$$\sigma_e(A_\varphi^u) = \sigma_{C(\sigma(u)\cap\mathbb{T})}(\varphi) = \varphi(\sigma(u) \cap \mathbb{T}) = \varphi(\sigma_e(A_z^u)),$$

where $\sigma_{C(\sigma(u)\cap\mathbb{T})}(\varphi)$ denotes the spectrum of φ as an element of the Banach algebra $C(\sigma(u) \cap \mathbb{T})$. This yields statement (v). We also note that putting (11) and (12) together shows that $C^*(A_z^u)/\mathscr{K}$ is isometrically $*$-isomorphic to $C(\sigma(u)\cap\mathbb{T})$, which is statement (ii).

We now need only justify statement (vii). To this end, recall that a seminal result of Clark [9] asserts that for each α in $\mathbb{T}$, the operator

$$U_\alpha := A_z^u + \frac{\alpha}{1 - \overline{u(0)}\alpha} k_0 \otimes C k_0 \tag{13}$$

on $\mathcal{K}_u$ is a cyclic unitary operator and, moreover, that every unitary, rank-one perturbation of A_z^u is of the form (13). A complete exposition of this important result can be found in the text [5]. Since

$$U_\alpha \equiv A_z^u \pmod{\mathscr{K}},$$

it follows that

$$\varphi(U_\alpha) \equiv A_\varphi^u \pmod{\mathscr{K}} \tag{14}$$

for every φ in $C(\mathbb{T})$. This is because the norm on $\mathcal{B}(\mathcal{K}_u)$ dominates the quotient norm on $\mathcal{B}(\mathcal{K}_u)/\mathscr{K}$ and since any φ in $C(\mathbb{T})$ can be uniformly approximated by trigonometric polynomials. Since $\mathscr{K} \subseteq C^*(A_z^u)$, it follows that

$$C^*(U_\alpha) + \mathscr{K} = C^*(A_z^u),$$

which yields the desired result. □

4. Piecewise continuous symbols

Having obtained a truncated Toeplitz analogue of Coburn's work, it is of interest to see if one can also obtain a truncated Toeplitz version of Gohberg and Krupnik's results concerning Toeplitz operators with piecewise continuous symbols [21]. Although we have not yet been able to complete this work, we have obtained a few partial results which are worth mentioning.

Let $PC := PC(\mathbb{T})$ denote the $*$-algebra of piecewise continuous functions on $\mathbb{T}$. To get started, we make the simplifying assumption that u is inner and that

$$\sigma(u) \cap \mathbb{T} = \{1\}.$$

For instance, u could be a singular inner function with a single atom at 1 or a Blaschke product whose zeros accumulate only at 1. Let

$$\mathscr{A}^u_{PC} = \{A^u_\varphi : \varphi \in PC\}$$

denote the set of all truncated Toeplitz operators on $\mathcal{K}_u$ having symbols in PC. The following lemma identifies the commutator ideal of $C^*(\mathscr{A}^u_{PC})$.

Lemma 6. $\mathscr{C}(C^*(\mathscr{A}^u_{PC})) = \mathscr{K}$.

Proof. Let

$$\chi(e^{i\theta}) := 1 - \frac{\theta}{2\pi}, \qquad 0 \leq \theta < 2\pi, \tag{15}$$

and notice that χ belongs to PC and satisfies

$$\chi_+(1) := \lim_{\theta \to 0} \chi(e^{i\theta}) = 1, \qquad \chi_-(1) := \lim_{\theta \to 2\pi} \chi(e^{i\theta}) = 0.$$

If φ is any function in PC, then it follows that

$$\varphi - \varphi_+(1)\chi - \varphi_-(1)(1-\chi)$$

is continuous at 1 and assumes the value zero there. By the remarks following Theorem 1 in the introduction, we see that

$$A^u_\varphi \equiv \alpha A^u_\chi + \beta I \pmod{\mathscr{K}}, \tag{16}$$

where $\alpha = \varphi_+(1) - \varphi_-(1)$ and $\beta = \varphi_-(1)$. In light of (16) it follows that

$$[A_\varphi, A_\psi] \equiv 0 \pmod{\mathscr{K}}$$

for any φ, ψ in PC whence $\mathscr{C}(C^*(\mathscr{A}^u_{PC})) \subseteq \mathscr{K}$. Since A^u_z belongs to $\mathscr{A}^u_{PC}$, we conclude that $\mathscr{C}(C^*(\mathscr{A}^u_{PC}))$ contains the nonzero commutator $[A^u_z, A^u_{\overline{z}}]$ whence $C^*(\mathscr{A}^u_{PC})$ is irreducible by Lemma 3. Moreover, By [12, Cor. 3.16.8] we conclude that $\mathscr{K} \subseteq \mathscr{C}(C^*(\mathscr{A}^u_{PC}))$ which concludes the proof. □

Lemma 7. $C^*(\mathscr{A}^u_{PC}) = C^*(A^u_\chi) + \mathscr{K}$.

Proof. The containment $\supseteq$ is clear from (16) since $C^*(\mathscr{A}^u_{PC}))$ contains $\mathscr{K}$. Conversely, the containment $\subseteq$ follows immediately from (16). □

From the discussion above and [13, Cor. I.5.6] we know that

$$\frac{C^*(\mathscr{A}^u_{PC})}{\mathscr{C}(C^*(\mathscr{A}^u_{PC}))} = \frac{C^*(A^u_\chi) + \mathscr{K}}{\mathscr{K}} \cong \frac{C^*(A^u_\chi)}{C^*(A^u_\chi) \cap \mathscr{K}}.$$

is a commutative C^*-algebra. Unfortunately, we are unable to identify the algebra $C^*(A^u_\chi)$ in a more concrete manner. This highlights the important fact that truncated Toeplitz operators such as A^u_χ, whose symbols are neither analytic nor coanalytic, are difficult to deal with.

Problem 2. Suppose that $\sigma(u) = \{1\}$. Give a concrete description of $C^*(A^u_\chi)$ where χ denotes the piecewise continuous function (15).

Problem 3. Provide an analogue of the Gohberg–Krupnik result for $\mathscr{A}^u_{PC}$. In other words, give a description of $C^*(\mathscr{A}^u_{PC})$ analogous to that of Theorem 1.

References

[1] P.R. Ahern and D.N. Clark, *On functions orthogonal to invariant subspaces*, Acta Math. **124** (1970), 191–204.

[2] ———, *On star-invariant subspaces*, Bull. Amer. Math. Soc. **76** (1970), 629–632.

[3] William Arveson, *A short course on spectral theory*, Graduate Texts in Mathematics, vol. 209, Springer-Verlag, New York, 2002.

[4] S. Axler, J.B. Conway, and G. McDonald, *Toeplitz operators on Bergman spaces*, Canad. J. Math. **34** (1982), no. 2, 466–483.

[5] J.A. Cima, A.L. Matheson, and W.T. Ross, *The Cauchy transform*, Mathematical Surveys and Monographs, vol. 125, American Mathematical Society, Providence, RI, 2006.

[6] J.A. Cima and W.T. Ross, *The backward shift on the Hardy space*, Mathematical Surveys and Monographs, vol. 79, American Mathematical Society, Providence, RI, 2000.

[7] J.A. Cima, W.T. Ross, and W.R. Wogen, *Truncated Toeplitz operators on finite dimensional spaces*, Oper. Matrices **2** (2008), no. 3, 357–369.

[8] Joseph A. Cima, Stephan Ramon Garcia, William T. Ross, and Warren R. Wogen, *Truncated Toeplitz operators: spatial isomorphism, unitary equivalence, and similarity*, Indiana Univ. Math. J. **59** (2010), no. 2, 595–620.

[9] D.N. Clark, *One dimensional perturbations of restricted shifts*, J. Analyse Math. **25** (1972), 169–191.

[10] L.A. Coburn, *The C^*-algebra generated by an isometry*, Bull. Amer. Math. Soc. **73** (1967), 722–726.

[11] ———, *The C^*-algebra generated by an isometry. II*, Trans. Amer. Math. Soc. **137** (1969), 211–217.

[12] J.B. Conway, *A course in operator theory*, Graduate Studies in Mathematics, vol. 21, American Mathematical Society, Providence, RI, 2000.

[13] Kenneth R. Davidson, *C^*-algebras by example*, Fields Institute Monographs, vol. 6, American Mathematical Society, Providence, RI, 1996.

[14] Ronald G. Douglas, *Banach algebra techniques in operator theory*, second ed., Graduate Texts in Mathematics, vol. 179, Springer-Verlag, New York, 1998.

[15] S.R. Garcia, *Conjugation and Clark operators*, Recent advances in operator-related function theory, Contemp. Math., vol. 393, Amer. Math. Soc., Providence, RI, 2006, pp. 67–111.

[16] S.R. Garcia and M. Putinar, *Complex symmetric operators and applications*, Trans. Amer. Math. Soc. **358** (2006), no. 3, 1285–1315 (electronic).

[17] ———, *Complex symmetric operators and applications. II*, Trans. Amer. Math. Soc. **359** (2007), no. 8, 3913–3931 (electronic).

[18] S.R. Garcia and W.T. Ross, *Recent progress on truncated Toeplitz operators*, Fields Institute Proceedings **65** (2013) 275–319.

[19] Stephan Ramon Garcia, Daniel E. Poore, and William T. Ross, *Unitary equivalence to a truncated Toeplitz operator: analytic symbols*, Proc. Amer. Math. Soc. **140** (2012), no. 4, 1281–1295.

[20] Stephan Ramon Garcia and Warren R. Wogen, *Some new classes of complex symmetric operators*, Trans. Amer. Math. Soc. **362** (2010), no. 11, 6065–6077.

[21] I.C. Gohberg and N.Ja. Krupnik, *The algebra generated by the Toeplitz matrices*, Funkcional. Anal. i Priložen. **3** (1969), no. 2, 46–56.

[22] T.L. Kriete, III, *A generalized Paley-Wiener theorem*, J. Math. Anal. Appl. **36** (1971), 529–555.

[23] Thomas L. Kriete, *Fourier transforms and chains of inner functions*, Duke Math. J. **40** (1973), 131–143.

[24] N. Nikolski, *Treatise on the shift operator*, Springer-Verlag, Berlin, 1986.

[25] W.T. Ross, *Analytic continuation in Bergman spaces and the compression of certain Toeplitz operators*, Indiana Univ. Math. J. **40** (1991), no. 4, 1363–1386.

[26] D. Sarason, *Algebraic properties of truncated Toeplitz operators*, Oper. Matrices **1** (2007), no. 4, 491–526.

[27] E. Strouse, D. Timotin, and M. Zarrabi, *Unitary equivalence to truncated Toeplitz operators*, `http://arxiv.org/abs/1011.6055`.

Stephan Ramon Garcia
Department of Mathematics
Pomona College
Claremont, California 91711, USA
e-mail: `Stephan.Garcia@pomona.edu`
URL: `http://pages.pomona.edu/~sg064747`

William T. Ross
Department of Mathematics and Computer Science
University of Richmond
Richmond, Virginia 23173, USA
e-mail: `wross@richmond.edu`
URL: `http://facultystaff.richmond.edu/~wross`

Warren R. Wogen
Department of Mathematics
University of North Carolina
Chapel Hill, North Carolina 27599, USA
e-mail: `wrw@email.unc.edu`
URL: `http://www.math.unc.edu/Faculty/wrw/`

Operator Theory:
Advances and Applications, Vol. 236, 193–203

On Some Vector Differential Operators of Infinite Order

Sergey Gefter and Tetiana Stulova

Abstract. In the paper some classes of vector differential operators of infinite order are studied and their use for constructing the entire solutions of implicit linear differential equations in a Banach space is considered. In addition, the integral representations of the Cauchy type for vector differential operators of infinite order are obtained.

Mathematics Subject Classification (2010). Primary 47B38; Secondary 46E40.

Keywords. Differential operator of infinite order, Banach space, implicit linear differential equation, spaces of entire functions.

1. Introduction

Let E be a complex Banach space and $\varphi(z) = \sum_{n=0}^{\infty} C_n z^n$ be a formal power series, for which the coefficients are bounded linear operators on E. In the paper the applicability of the differential operator of infinite order

$$\varphi\left(\frac{d}{dz}\right) g = \sum_{n=0}^{\infty} C_n g^{(n)}(z)$$

to two spaces of entire E-valued functions is considered, namely in the space of all entire vector-functions and in the space of all entire vector-functions of zero exponential type (see Theorems 2.1 and 2.5). Moreover, we obtained the Cauchy type integral representation for this operator (see Corollary 4.3 and Theorem 4.4). We apply these results for studying the well-posedness of some implicit inhomogeneous linear differential equations in a Banach space in the above-mentioned spaces of entire functions (Corollaries 2.2, 2.4, 4.5). Note that the cases for which there does not exist any nontrivial scalar analogue, occur herein (see Corollary 2.2). Also note that some differential equations in which the corresponding operator is not densely defined are considered (see Corollary 2.4 and Examples 3.1, 3.6). In the scalar case differential operators of infinite order were studied in different points of view in works of Valiron, Polya, Muggli, Sikkema, Korobeinik, Leont'ev, Dickson and many other mathematicians (see, for example [1], [2]). For implicit linear differential equations in a Banach space we refer to [3]–[5] and to references that were therein. The entire and holomorphic solutions of explicit and implicit linear differential equations in a complex Banach space are considered in [6]–[9] and other works. In [10] holomorphic solutions of linear differential equations in a Banach space over a non-Archimedean field are examined. The equation on the semi-axis with an operator which has a nondense domain was studied by P. Sobolevsky, Ju. Silchenko, Da Prato, and E. Sinestrati ([16]–[18]).

2. On applicability of the vector differential operator of infinite order

For a complex Banach space E we denote by $\mathcal{H}(\mathbb{C}, E)$ the space of all entire E-valued functions. We endow $\mathcal{H}(\mathbb{C}, E)$ with the topology of uniform convergence on compact sets.

Theorem 2.1. *Let* $\varphi(z) = \sum_{n=0}^{\infty} C_n z^n$ *be an entire operator-valued function. Then* $\varphi\left(\frac{d}{dz}\right)$ *is a continuous operator on* $\mathcal{H}(\mathbb{C}, E)$ *if and only if* φ *is of exponential type.*

Proof. Let φ be of exponential type. Then there are such $\gamma_1 > 0$ and $M > 0$, that $\|C_n\| \leq \gamma_1 \frac{M^n}{n!}$ for all $n = 0, 1, 2, \ldots$ (see [11], Appendix B). Let now $g(z) = \sum_{m=0}^{\infty} \alpha_m z^m$, $R > 0$, $0 < \varepsilon < \frac{1}{M+R}$ and $|z| \leq R$. Then $\|\alpha_m\| \leq \gamma_2 \cdot \varepsilon^m$ for some $\gamma_2 > 0$ and for all m. Hence, $\left\|g^{(n)}(z)\right\| \leq \gamma_2 \varepsilon^n \sum_{m=0}^{\infty} \frac{(m+n)!}{m!} (\varepsilon |z|)^m \leq \frac{\gamma_2 n! \varepsilon^n}{(1-\varepsilon R)^{n+1}}$ and $\left\|C_n g^{(n)}(z)\right\| \leq \frac{\gamma_1 \gamma_2}{1-\varepsilon R} \left(\frac{M\varepsilon}{1-\varepsilon R}\right)^n$, that is the series $\sum_{n=0}^{\infty} C_n g^{(n)}(z)$ converges uniformly in the disk $|z| \leq R$. As $\mathcal{H}(\mathbb{C}, E)$ is a Fréchet space, then the continuity $\varphi\left(\frac{d}{dz}\right)$ follows from the Banach–Steinhaus theorem. Let us prove the reverse statement. Let $g_0(z) = \sum_{n=0}^{\infty} \beta_n z^n$, $\beta_n \in \mathbb{C}$ be an arbitrary entire function, $\xi \in E$ and $g(z) = g_0(z)\xi$. Since the series $\sum_{n=0}^{\infty} C_n g^{(n)}(0)$ converges, then $\|n! C_n \xi\| \cdot |\beta_n| \leq M$ for some $M > 0$ and $n = 0, 1, 2 \ldots$. Hence $\varlimsup_{n\to\infty} \left(\sqrt[n]{n! \|C_n \xi\|}\, \sqrt[n]{|\beta_n|}\right) \leq 1$. As the function $g_0(z)$ is an arbitrary one, then $\lim_{n\to\infty} \left(\sqrt[n]{n! \|C_n \xi\|}\, \sqrt[n]{|\beta_n|}\right) = 0$ and the sequence $\sqrt[n]{n! \|C_n \xi\|}$ is bounded. Using reasoning in a proof of the Banach–Steinhaus theorem we obtain boundedness of the sequence $\sqrt[n]{n! \|C_n\|}$. Thus $\varphi(z)$ is an entire function of exponential type. □

The following corollary is of interest in the vector case only.

Corollary 2.2. *Let $T : E \to E$ be a bounded quasinilpotent operator, and let $g(z)$ be an arbitrary E-valued entire function. Assume that the Fredholm resolvent $(1 - zT)^{-1}$ is of exponential type. Then the differential equation*

$$Tw' + g(z) = w, \tag{2.1}$$

has a unique entire solution $w(z) = \sum_{n=0}^{\infty} T^n g^{(n)}(z)$ and this solution continuously depends on g in the topology of the space $\mathcal{H}(\mathbb{C}, E)$.

Proof. In the considering case $\varphi(z) = \sum_{n=0}^{\infty} T^n z^n$ and $\varphi(\frac{d}{dz})g = \sum_{n=0}^{\infty} T^n g^{(n)}$. It is easy to check that $w(z)$ is this solution of Equation (2.1). Let us prove the uniqueness of this solution. Let $w_0(z) = \sum_{n=0}^{\infty} a_n z^n$ be an entire solution of the homogeneous equation $Tw' = w$. Then it is not difficult to verify that $a_0 = n!T^n a_n$ (see [8], Lemma 1). Therefore $\sqrt[n]{\|a_0\|} \leq \sqrt[n]{\|a_n\|} \cdot \sqrt[n]{n!\,\|T^n\|} \to 0$ since the sequence $\sqrt[n]{n!\|T^n\|}$ is bounded. Hence $a_0 = 0$. As the function ${w_0}^{(k)}(z)$ satisfies the homogeneous equation as well, we have that $a_k = 0,\ k \in \mathbb{N}$, that is $w_0 = 0$. □

Remark 2.3. The existence and uniqueness of an entire solution of Equation (2.1) were proved in ([8], Theorem 2.6) using a different technique.

Corollary 2.4. *Let $A : D(A) \to E$ be a closed invertible operator on E, $T = A^{-1}$, and let $f(z)$ be an arbitrary E-valued entire function. Assume that the Fredholm resolvent $(1 - zT)^{-1}$ is an entire function of exponential type. Then the differential equation*

$$w' = Aw + f(z), \tag{2.2}$$

has the unique entire solution $w(z) = -\sum_{n=0}^{\infty} A^{-(n+1)} f^{(n)}(z)$ and this solution continuously depends on f in the topology of the space $\mathcal{H}(\mathbb{C}, E)$.

Proof. Let $g(z) = -A^{-1}f(z)$. Then $g(z)$ is an entire function and Equation (2.2) is equivalent to (2.1). □

Now consider the space $\mathcal{H}_0(\mathbb{C}, E)$ of entire E-valued functions of zero exponential type. That is, if $f \in \mathcal{H}_0(\mathbb{C}, E)$, then it satisfies the following condition:

$$\forall \varepsilon > 0\ \exists \beta_\varepsilon > 0\ \forall z \in \mathbb{C} : \|f(z)\| \leq \beta_\varepsilon e^{\varepsilon|z|}.$$

On $\mathcal{H}_0(\mathbb{C}, E)$ we will consider a natural topology of projective limit of Banach spaces $\mathcal{H}_\sigma(\mathbb{C}, E)$, $\sigma > 0$, where

$$\mathcal{H}_\sigma(\mathbb{C}, E) = \left\{ g \in \mathcal{H}(\mathbb{C}, E) : \sup_{z\in\mathbb{C}} \|g(z)\|\, e^{-\sigma|z|} < +\infty \right\}.$$

Theorem 2.5. *Let $\varphi(z) = \sum_{n=0}^{\infty} C_n z^n$ be a holomorphic operator-valued function, $g \in \mathcal{H}_0(\mathbb{C}, E)$ and $w(z) = \sum_{n=0}^{\infty} C_n g^{(n)}(z)$. Then the last series converges uniformly* *in every disk, $w(z)$ is of zero exponential type and $\varphi\left(\frac{d}{dz}\right)$ is a continuous operator on $\mathcal{H}_0(\mathbb{C}, E)$. Here the zero exponential type of $g(z)$ is a necessary condition: if $g(z)$ is an E-valued entire function and the series $\sum_{n=0}^{\infty} C_n g^{(n)}(0)$ converges for all power series $\varphi(z) = \sum_{n=0}^{\infty} C_n z^n$ with positive radius of convergence, then $g(z)$ is of zero exponential type.*

Proof. Let R be a radius of convergence for the series $\varphi(z)$. If $\frac{1}{M} < R$, then $\|C_n\| \leq \gamma \cdot M^n$ for some $\gamma > 0$ and for all $n = 0, 1, 2 \ldots$. Let $0 < \varepsilon < \frac{1}{M}$. As $g(z)$ is a function of zero exponential type, then it is easy to verify that there is a such $\beta > 0$, for which $\left\|g^{(n)}(z)\right\| \leq \beta \varepsilon^n e^{\varepsilon|z|}$, $z \in \mathbb{C}$, $n = 0, 1, 2 \ldots$. Show that the series $\sum_{n=0}^{\infty} C_n g^{(n)}(z)$ converges uniformly in any disk and its sum is of zero exponential type. If $r > 0$ and $|z| \leq r$, then $\left\|C_n g^{(n)}(z)\right\| \leq \|C_n\| \cdot \left\|g^{(n)}(z)\right\| \leq \beta \cdot \gamma M^n \varepsilon^n e^{\varepsilon r} = \beta\gamma(\varepsilon M)^n e^{\varepsilon r}$ and $\sum_{n=0}^{\infty} (\varepsilon \|C_n\|)^n < +\infty$. Hence, the series $\sum_{n=0}^{\infty} C_n g^{(n)}(z)$ converges uniformly in the disk $|z| \leq r$. Moreover

$$\|g(z)\| \leq \sum_{n=0}^{\infty} \left\|C_n g^{(n)}(z)\right\| \leq \beta\gamma e^{\varepsilon|z|} \sum_{n=0}^{\infty} (\varepsilon M)^n = \frac{\beta\gamma}{1 - \varepsilon M} \cdot e^{\varepsilon|z|}.$$

Let now $\sigma > 0$. Using the Cauchy integral formula and the Stirling formula one can show that $\left\|g^{(n)}\right\|_\sigma \leq 2\sqrt{\pi n}\sigma^n \|g\|_\sigma$. Thus the operator of differentiation D is bounded on $\mathcal{H}_\sigma(\mathbb{C}, E)$ and $\|D^n\| \leq 2\sqrt{\pi n}\sigma^n$. Therefore the series $\sum_{n=0}^{\infty} C_n g^{(n)}$ converges in $\mathcal{H}_\sigma(\mathbb{C}, E)$ for all $\sigma < R$. Hence this series converges in the space $\mathcal{H}_0(\mathbb{C}, E)$. We obtain that $\varphi(\frac{d}{dz})$ is continuous since $\mathcal{H}_0(\mathbb{C}, E)$ is a Fréchet space. The necessity that $g(z)$ is an entire function of zero exponential type follows from the consideration of the class of the series $\varphi_r(z) = \sum_{n=0}^{\infty} \frac{z^n}{r^n} I$, $r > 0$, where I is the identity operator. □

Corollary 2.6. *Let E_1 and E_2 be Banach spaces, $Q : E_1 \to E_2$ be an arbitrary bounded linear operator, and let $A : D(A) \to E_2$ be a closed invertible linear operator with $D(A) \subset E_1$. Consider the following implicit differential equation*

$$Qw' = Aw + f(z), \tag{2.3}$$

where f is an E_2-valued entire function. If f is of zero exponential type then Equation (2.3) has the unique entire solution of zero exponential type

$$w(z) = -\sum_{n=0}^{\infty} \left(A^{-1}Q\right)^n A^{-1} f^{(n)}(z)$$

and this solution continuously depends on f in the topology of the space $\mathcal{H}_0(\mathbb{C}, E_2)$.

Proof. Let $T = A^{-1}Q$ and $g(z) = -A^{-1}f(z)$. Then g is of zero exponential type and Equation (2.3) is equivalent to Equation (2.1). Now the proof is similar to the proof of Corollary 2.2. □

3. Examples

Let us give some examples.

Example 3.1. Let $E = C[0,1]$, $A = \frac{d}{dx}$ and $D(A) = \{u \in C^1[0,1] : u(0) = 0\}$. Then $D(A)$ is not dense in E, the operator A is invertible, $\left(A^{-1}h\right)(x) = \int\limits_0^x h(y)\,dy$ and $\left(A^{-(n+1)}h\right)(x) = \frac{1}{n!}\int\limits_0^x (x-y)^n h(y)\,dy$. Hence, if $T = A^{-1}$, then T is a bounded quasinilpotent operator and its Fredholm resolvent is of exponential type. By transition to real axes Equation (2.2) has the form

$$\begin{cases} \frac{\partial w}{\partial t} = \frac{\partial w}{\partial x} + f(t,x), & t \in \mathbb{R},\ x \in (0,1) \\ w(t,0) = 0 \end{cases} \tag{3.1}$$

If in the second variable f can be extended to an entire function, then in this class of functions Problem (3.1) has the unique solution

$$w(t,x) = -\sum_{n=0}^{\infty} \frac{1}{n!} \int\limits_0^x (x-y)^n \frac{\partial^n f}{\partial t^n}(t,y)\,dy = -\int\limits_0^x f(t+x-y,y)\,dy.$$

It is important to note that Problem (3.1) has only a zero solution for the homogeneous equation even in the class of continuously differentiable functions. Specifically, A is not a Hille–Yosida operator (see [12], Section 3.5)

Example 3.2. Now let us show that when T is quasinilpotent, but its Fredholm resolvent is not of exponential type, then Equation (2.1) can have no smooth solution onto $[0,t_0], t_0 > 0$ at all.

Let E be a Hilbert space with an orthonormalized basis $\{e_n\}_{n=0}^{\infty}$, T be the weighted shift operator such that $Te_n = \frac{1}{\sqrt{n+1}}e_{n+1}$, and $g(z) = e^{z^2}e_0$. It is easy to check that $\|T^n\| = \frac{1}{\sqrt{n!}}$. Therefore T is quasinilpotent but $\sqrt[n]{n!\|T^n\|} \to +\infty$. Hence the Fredholm resolvent $(1-zT)^{-1}$ is not of exponential type. If $w(t) = \sum\limits_{n=0}^{\infty} w_n(t)\,e_n$ is a solution of Equation (2.1) on the real axes then

$$\begin{cases} e^{t^2} = w_0(t) \\ \frac{1}{\sqrt{n+1}} w_n'(t) = w_{n+1}(t), & n \geq 0. \end{cases}$$

Hence $w_n(t) = \frac{1}{\sqrt{n!}}\left(e^{t^2}\right)^{(n)}$ and $w_{2n}(0) = \frac{\sqrt{(2n)!}}{n!}$. Therefore $\sum\limits_{n=0}^{\infty} |w_n(0)|^2 = +\infty$, i.e., Equation (2.1) has no smooth solutions on $[0,t_0], t_0 > 0$.

Let us consider some examples of differential operators of infinite order in the space of entire functions of zero exponential type. In our opinion the interesting examples even appear in the finite-dimensional case.

Example 3.3. Let $E = \mathbb{C}$, and $a \in \mathbb{C}$, $a \neq 0$. Consider the differential equation $w' = aw + f(z)$. If $f(z)$ is an entire function of zero exponential type, then this equation has a unique entire solution of zero exponential type $w(z) = -\sum\limits_{n=0}^{\infty} \frac{1}{a^{n+1}} f^{(n)}(z)$ and this solution continuously depends on f in the topology of the space $\mathcal{H}_0(\mathbb{C}, \mathbb{C})$.

Example 3.4. Consider the following equation of forced oscillations $\ddot{x} + \omega^2 x = f(t)$, where $\omega > 0$ and $f(t)$ is the trace on the real axes of an entire function of zero exponential type. This equation has a unique solution $x(t) = \sum\limits_{k=0}^{\infty} \frac{(-1)^k}{\omega^{2k+2}} f^{(2k)}(t)$, which can be extended to an entire function of zero exponential type.

Example 3.5. Let E be a Hilbert space, A be a closed normal operator on E with discrete spectrum and $0 \notin \sigma(A)$. Let $\{e_k\}$ be an orthonormalized eigenbasis for A, $A\, e_k = \lambda_k\, e_k$, where $\lambda_k \to \infty$. If $f : \mathbb{C} \to E$, $f(z) = \sum\limits_k f_k(z)\, e_k$ is an entire function of zero exponential type, then Equation (2.2) has the following unique entire solution of zero exponential type $w(z) = -\sum\limits_{n=0}^{\infty} \left(\sum\limits_k \lambda_k^{-(n+1)} f_k^{(n)}(z)\, e_k \right)$.

Example 3.6. Let $E = C[0,\, 1]$, $A = \frac{d^2}{dx^2}$, and

$$D(A) = \left\{ u \in C^2[0,\, 1] : u(0) = u(1) = 0 \right\}.$$

Then the operator A is invertible and $\left(A^{-1}h\right)(x) = \int\limits_0^1 G(x,\, y)\; h(y) dy$, where G is the Green function of the corresponding boundary problem. Moreover,

$$\left(A^{-(n+1)}h\right)(x) = \int\limits_0^1 G_{n+1}(x,\, y)\; h(y)\, dy,$$

where $G_1(x,\, y) = G(x,y)$ and $G_{n+1}(x,\, y) = \int\limits_0^1 G_n(x,\, s)\; G(s,\, y)\, ds$. Now on real axes Equation (2.2) has the form of the heat equation with zero boundary conditions:

$$\begin{cases} \frac{\partial w}{\partial t} = \frac{\partial^2 w}{\partial x^2} + f(t,x), & t \in \mathbb{R},\ x \in (0,\, 1) \\ w(t,\, 0) = w(t,\, 1) = 0 \end{cases} \tag{3.2}$$

If $f(t,x) = \sum\limits_{n=0}^{\infty} c_n(x)\, t^n$, where $c_n \in C[0,1]$ and $\lim\limits_{n\to\infty} \sqrt[n]{n!\, \|c_n\|} = 0$, then the problem (2.2) has the solution $w(t,x) = -\sum\limits_{n=0}^{\infty} \int\limits_0^1 G_{n+1}(x,y) \frac{\partial^n f}{\partial t^n}(t,y)\, dy$.

4. Integral representations of Cauchy type for a vector differential operator of infinite order

Let $\varphi(z) = \sum_{n=0}^{\infty} C_n z^n$ be a formal power series with operator coefficients and $g(z)$ be an entire vector function. Using the following Cauchy integral formula for the nth derivative

$$g^{(n)}(z) = \frac{n!}{2\pi i} \oint_{|\zeta|=|z|+r} \frac{g(\zeta)}{(\zeta - z)^{n+1}} d\zeta, r > 0,$$

we make such a formal transformation of the formal differential operator $\varphi(\frac{d}{dz})$:

$$\begin{aligned}(\varphi\left(\frac{d}{dz}\right) g)(z) &= \sum_{n=0}^{\infty} C_n g^{(n)}(z) = \frac{1}{2\pi i} \sum_{n=0}^{\infty} n! C_n \oint_{|\zeta|=|z|+r} \frac{g(\zeta)}{(\zeta - z)^{n+1}} d\zeta \\ &= \frac{1}{2\pi i} \oint_{|\zeta|=|z|+r} \left(\sum_{n=0}^{\infty} \frac{n! C_n}{(\zeta - z)^{n+1}} \right) g(\zeta) d\zeta.\end{aligned}$$

Let us give the meaning of this formal transformation in the considered two cases in the Section 2. Namely, $\varphi(z)$ is an entire function of exponential type (see, Theorem 2.1) and $\varphi(z)$ has positive radius of convergence (see Theorem 2.5).

Proposition 4.1. *Let the conditions of Theorem* 2.1 *be fulfilled and let* $\Phi(\zeta) = \sum_{n=0}^{\infty} \frac{n! C_n}{\zeta^{n+1}}$ *be the Borel transform of* $\varphi(z)$. *Then*

$$(\varphi\left(\frac{d}{dz}\right) g)(z) = \frac{1}{2\pi i} \oint_{|\zeta|=|z|+r} \Phi(\zeta - z) g(\zeta) d\zeta, \tag{4.1}$$

where r *is greater than the exponential type of* $\varphi(z)$.

Proof. Let the exponential type of $\varphi(z)$ is equal to σ. Then $\Phi(\zeta)$ is holomorphic out of the circle $|\zeta| = \sigma$. If $r > \sigma$ and $|\zeta| = |z| + r$, then $|\zeta - z| \geq ||\zeta| - |z|| = r > \sigma$. Therefore for any $z \in \mathbb{C}$ the function $\Phi(\zeta - z) g(\zeta)$ is holomorphic out of the circle $|\zeta| = |z| + \sigma$. Having changed the variables $s = \zeta - z$ on the right-hand side of the equality (4.1) we have

$$\begin{aligned}\frac{1}{2\pi i} \oint_{|\zeta|=|z|+r} \Phi(\zeta - z) g(\zeta) d\zeta &= \frac{1}{2\pi i} \oint_{|s+z|=|z|+r} \Phi(s) g(s + z) ds \\ &= \frac{1}{2\pi i} \oint_{|s|=r} \Phi(s) g(s + z) ds.\end{aligned}$$

The last obtained integral is the classical convolution integral representation for the differential operator of infinite order (in the scalar case see, for example, [13], Ch. 5, §7, formula (77)). □

Remark 4.2. The closed path in the integral representation (4.1) depends on z. A simpler form for the representation of the Cauchy type can be obtained by using the the following notion of a formal integral in the space of formal Laurent series (see, [14], [15]).

Let V be an arbitrary complex vector space and $V\left[\left[\zeta, \frac{1}{\zeta}\right]\right]$ be the space of all formal Laurent series with coefficient of V. For $f(\zeta) = \sum_{n=-\infty}^{\infty} b_n \zeta^n \in V\left[\left[\zeta, \frac{1}{\zeta}\right]\right]$ we set

$$\oint f(\zeta) d\zeta = 2\pi i\, b_{-1}. \tag{4.2}$$

Now from Proposition 4.1 we obtain

Corollary 4.3. *Let φ, Φ, g be the same as in Proposition 4.1. Then for any given $z \in \mathbb{C}$ the following inclusion*

$$\Phi(\zeta - z) g(\zeta) \in E\left[\left[\zeta, \frac{1}{\zeta}\right]\right]$$

is fulfilled and

$$w(z) = \frac{1}{2\pi i} \oint \Phi(\zeta - z) g(\zeta) d\zeta,$$

where the integral is considered in the sense of (4.2).

If the power series $\varphi(z) = \sum_{n=0}^{\infty} C_n z^n$ is not an entire function of exponential type, then its formal Laplace–Borel transform $\Phi(\zeta) = \sum_{n=0}^{\infty} \frac{n! C_n}{\zeta^{n+1}}$ diverges for all ζ. In this case we will consider $\Phi(\zeta - z)$ as the following power series in power $\frac{1}{\zeta}$:

$$\Phi(\zeta - z) = \sum_{n=0}^{\infty} \frac{n! C_n}{(\zeta - z)^{n+1}} \overset{\text{def}}{=} \sum_{n=0}^{\infty} \frac{1}{\zeta^{n+1}} \frac{n! C_n}{(1 - \frac{z}{\zeta})^{n+1}} = \sum_{n=0}^{\infty} \left(\sum_{k=0}^{\infty} \frac{z^k}{\zeta^k} \right)^{n+1} \frac{n! C_n}{\zeta^{n+1}}.$$

We have that $\Phi(\zeta - z) \in B(E)[[z]][[\zeta, \frac{1}{\zeta}]]$, i.e., $\Phi(\zeta - z)$ is a formal Laurent series in ζ with formal power series over $B(E)$ as coefficients, where $B(E)$ is the space of all bounded linear operator on E.

Theorem 4.4. *Let the conditions of Theorem 2.5 be fulfilled. Then the product $\Phi(\zeta - z) g(\zeta)$ is well defined as an element of the space $E[[z]]\left[\left[\zeta, \frac{1}{\zeta}\right]\right]$ and*

$$w(z) = \frac{1}{2\pi i} \oint \Phi(\zeta - z) g(\zeta) d\zeta. \tag{4.3}$$

Proof. First we need to show that $\Phi(\zeta-z)g(\zeta)$ exists as an element of $E[[z]][[\zeta,\frac{1}{\zeta}]]$. We have

$$\begin{aligned}\Phi(\zeta-z) &= \sum_{n=0}^{\infty} n!\frac{C_n}{\zeta^{n+1}}\left(1+\frac{z}{\zeta}+\frac{z^2}{\zeta^2}+\cdots\right)^{n+1}\\ &= \sum_{n=0}^{\infty} n!C_n\sum_{j=n}^{\infty} C_j^n\frac{z^{j-n}}{\zeta^{j+1}} = \sum_{j=0}^{\infty}\left(\sum_{n=0}^{j} n!C_j^n C_n z^{-n}\right)\frac{z^j}{\zeta^{j+1}}\\ &= \sum_{j=0}^{\infty}\left(\sum_{n=0}^{j}\frac{j!}{(j-n)!z^n}C_n\right)\frac{z^j}{\zeta^{j+1}}.\end{aligned}$$

If $g(\zeta)=\sum\limits_{k=0}^{\infty}\alpha_k\zeta^k$, then the product $\Phi(\zeta-z)g(\zeta)$ can be formally rewritten as follows:

$$\begin{aligned}\Phi(\zeta-z)g(\zeta) &= \left(\sum_{j=0}^{\infty}\left(\sum_{n=0}^{j}\frac{j!}{(j-n)!z^n}C_n\right)\frac{z^j}{\zeta^{j+1}}\right)\cdot\left(\sum_{k=0}^{\infty}\alpha_k\zeta^k\right)\\ &= \sum_{n=0}^{\infty}\left(\sum_{k=0}^{\infty}\alpha_k\sum_{m=0}^{n+k}\frac{(n+k)!z^{n+k-m}}{(n+k-m)!}C_m\right)\frac{1}{\zeta^{n+1}}\\ &\quad+\sum_{n=0}^{\infty}\left(\sum_{k=0}^{\infty}\alpha_{k+n+1}\sum_{m=0}^{k}\frac{k!z^{k-m}}{(k-m)!}C_m\right)\zeta^n.\end{aligned}$$

One can use estimations for the function $g(z)$ to show that the coefficients at each term ζ^n, $n\in\mathbb{Z}$ are convergent series. We check this only for the coefficient at the term $\frac{1}{\zeta}$, which we are mostly interested in. Write out separately this coefficient:

$$\begin{aligned}\sum_{k=0}^{\infty}\alpha_k\left(\sum_{m=0}^{k}\frac{k!C_m}{(k-m)!}\right)z^{k-m} &= \sum_{m=0}^{\infty}C_m\sum_{k=m}^{\infty}\alpha_k\frac{k!}{(k-m)!}z^{k-m}\\ &= \sum_{m=0}^{\infty}C_m g^{(m)}(z) = w(z),\end{aligned}$$

and

$$\sum_{k=0}^{\infty}\alpha_k\Big(\sum_{m=0}^{k}\frac{k!C_m}{(k-m)!}\Big)z^{k-m} = \sum_{k=0}^{\infty}\left(\sum_{m=0}^{\infty}\alpha_{k+m}\frac{(k+m)!}{k!}C_m\right)z^k.$$

To validate this formal transformations above it is sufficient to show

$$\sum_{k=0}^{\infty}\sum_{m=0}^{\infty}||C_m|||\alpha_{k+m}|\frac{(k+m)!}{k!}|z|^k<+\infty,$$

for all $z\in\mathbb{C}$. Let R be a radius of convergence of the series $\varphi(z)$ and $0<\sigma<r<R$. Then $||C_m||\le\frac{M_1}{r^m}$ and $|\alpha_{k+m}|\le\frac{M_2\sigma^{k+m}}{(k+m)!}$ for certain $M_1,M_2>0$ and all m,k.

Therefore,

$$\sum_{k=0}^{\infty}\sum_{m=0}^{\infty}\frac{||C_m|||\alpha_{k+m}|(k+m)!}{k!}|z|^k \le \sum_{k=0}^{\infty}\sum_{m=0}^{\infty}\frac{M_1M_2}{r^m}\cdot\frac{\sigma^{k+m}}{k!}|z|^k$$

$$\le M_1M_2\sum_{m=0}^{\infty}(\frac{\sigma}{r})^m\sum_{k=0}^{\infty}\frac{\sigma^k}{k!}|z|^k = \frac{M_1M_2}{1-\frac{\sigma}{r}}e^{\sigma|z|}, z\in\mathbb{C}. \tag{4.4}$$

Thus, $\sum\limits_{k=0}^{\infty}\left(\sum\limits_{m=0}^{\infty}\alpha_{k+m}\frac{(k+m)!}{k!}C_m\right)z^k$ is an entire function, and we obtain

$$\frac{1}{2\pi i}\oint\Phi(\zeta-z)g(\zeta)d\zeta = w(z).$$

Finally, point out that the evaluation (4.4) reveals again that $w(z)$ is an entire function of zero exponential type. The theorem is proved. □

From Corollary 2.6 and Theorem 4.4 we obtain the following integral representation of Cauchy type for the solution of Equation (2.3).

Corollary 4.5. *Let all conditions of Corollary* 2.6 *be fulfilled and let*

$$\Psi(\zeta) = \sum_{n=0}^{\infty}\frac{n!(A^{-1}Q)^nA^{-1}}{\zeta^{n+1}}$$

be the formal Laplace–Borel transform of the resolvent $R(z) = (zQ-A)^{-1}$ *of the pencil* $L(z) = zQ - A$. *The unique solution of zero exponential type of Equation* (2.3) *can be represented in the following integral form*

$$w(z) = \frac{1}{2\pi i}\oint\Psi(\zeta-z)f(\zeta)d\zeta,$$

where the integral is considered in the sense of (4.2).

References

[1] P.C. Sikkema, *Differential operators and differential equations of infinite order with constant coefficients.* Groningen, 1953.

[2] Yu.F. Korobeinik, *Shift operators on families of numbers.* Rostov-on-Don (Russian), 1983.

[3] A. Favini, and A. Yagi, *Degenerate differential equations in Banach spaces.* New York-Basel-Hong Kong: Marcel Dekker, Inc., 1999, 313.

[4] L.A. Vlasenko, *Evolutionary patterns with implicit and singular differential equations.* Systemnye Technology, Dnepropetrovsk (Russian), 2006.

[5] A.G. Rutkas, *Spectral methods for studying degenerate differential-operator equations. I.* Journal of Mathematical Sciences, 144: 4 (2007), 4246–4262.

[6] Ju. Dalec'kii, and M. Kreĭn, *Stability of differential equations in Banach space.* Amer. Math. Soc., Providence, R.I., 1974.

[7] W. Balser, A. Duval, and S. Malek, *Summability of formal solutions for abstract Cauchy problems and related convolution equations.* Ulmer Seminare über Funktionalanalysis und Differentialgleichungen, **11** (2007), 29–44.

[8] S. Gefter, and T. Stulova, *On Holomorphic Solutions of Some Implicit Linear Differential Equation in a Banach Space.* Operator Theory: Advances and Applications, Birkhäuser Verlag, Basel, Switzerland, **191** (2009), 331–340.

[9] M. Gorbachuk, and V. Gorbachuk, *On the well-posed solvability in some classes of entire functions of the Cauchy problem for differential equations in a Banach space.* Methods Funct. Anal. Topology, 11:**2** (2005), 113–125.

[10] V.I. Gorbachuk, and V.M. Gorbachuk, *On holomorphic solutions of some inhomogeneous linear differential equations in a Banach space over a non-Archimedean field.* p-Adic Numbers, Ultrametric Analysis and Applications, 2:**2** (2010), 114–121.

[11] W. Balser, *Formal Power series and Linear Systems of Meromorphic Ordinary Differential Equations.* Springer-Verlag New York Inc., 2000.

[12] W. Arendt, C.J.K. Batty, M. Hieber, F. Neubrander, *Vector-valued Laplace Transforms and Cauchy Problems.* Monographs in Mathematics, 96:**XI**, 2001.

[13] Gelfond, A.O. *Calculus of finite differences.* International Monographs on Advanced Mathematics and Physics. Hindustan Publishing Corp., Delhi, 1971, vi+451.

[14] G.P. Egorychev. *Integral representation and the computation of combinatorial sums.* Transl. of Math. Monographs, **59**, AMS, 1984.

[15] I. Frenkel, J. Lepovsky and A. Meurman. *Vertex operator algebras and the monster.* Academic Press, New York, 1988.

[16] Yu.T. Sil'chenko, P.E. Sobolevskii. *Solvability of the Cauchy problem for an evolution equation in a Banach space with a non-densely given operator coefficient which generates a semigroup with a singularity (Russian).* Siberian. Math. J., 27:**4** (1986), 544–553.

[17] Yu. T.Sil'chenko. *Differential equations with non-densely defined operator coefficients, generating semigroups with singularities.* Nonlinear Anal., Ser. A: Theory Methods, 36:**3** (1999), 345–352.

[18] G. Da Prato, E. Sinestrati. *Differential operators with non dense domain.* Annali della Scuola Normale Superiore di Pisa, **14** (1987), 285–344.

Sergey Gefter and Tetiana Stulova
V.N. Karazin Kharkiv National University
Department of Mechanics and Mathematics
4 sq. Svoboda
61022 Kharkiv, Ukraine
e-mail: `gefter@univer.kharkov.ua`
`stutestella@rambler.ru`

Operator Theory:
Advances and Applications, Vol. 236, 205–230

Wiener–Hopf Type Operators and Their Generalized Determinants

James F. Glazebrook

Abstract. We recall some results on generalized determinants which support a theory of operator τ-functions in the context of their predeterminants which are operators valued in a Banach–Lie group that are derived from the transition maps of certain Banach bundles. Related to this study is a class of Banach–Lie algebras known as L*-algebras from which several results are obtained in relationship to tau functions. We survey the applicability of this theory to that of Schlesinger systems associated with (operator) equations of Fuschsian type and discuss how meromorphic connections may play a role here.

Mathematics Subject Classification (2010). Primary: 47A05, 47B10, 53B10; Secondary: 58B99, 58B25.

Keywords. Toeplitz operator, Fredholm operator, Banach algebra, L*-algebra, integral operator, tau function, Kac–Moody algebra, Schlesinger system, meromorphic connection.

1. Introduction

This contribution is in part based on my talk at IWOTA 2011 in Seville (Spain). At first glances it has the look of a survey bringing together some known results under one roof, but then it unfolds to a more general perspective, and eventually suggests some new directions via applications.

After delving into the background to generalized determinants and operator-valued meromorphic functions, it seemed fitting for these Proceedings to acknowledge the important work in this area that was accomplished by Professor Gohberg along with several of his coworkers towards the development of some foundational concepts which enter into part of the survey here (as realized in [28, 29, 30, 31], for instance). We expect that the fruits of his profound mathematical insight will continue to influence many research projects in the years to come.

The scene is set by recalling some earlier work regarding the existence of determinants in the Banach algebra category, along with a class of operators belonging to a Banach–Lie group for which the concept of determinant can be defined. This is developed in the context of transition maps of a bundle theory over a class of infinite-dimensional manifolds, as was the case in [23]. The resulting operators, in a certain sense, can be viewed as *generalized Wiener–Hopf operators*, and these turn out to be a shade more general than the meaning that can be found in the current literature (see, e.g., [16, 35]). It is mainly this class of operators that encapsulates several of the 'predeterminants' that are probed into.

One principal theme deals with a particular class of generalized determinant operators giving rise to an assortment of τ*-functions*, the background to which is discussed in §4.1. This subject is motivated from several sources such as [36, 37, 59, 63] in the Grassmannian setting which includes flows on invariant subspaces [35], and the close relationship with the Painlevé equations (see Appendix B). Familiar examples arise from Toeplitz and Fredholm determinants in the case where the algebra is $\mathcal{L}(H)$ (H a Hilbert space), in which case, studying the various classes of integral operators and their corresponding determinants seems to be relevant here. In this respect it is worth mentioning several ideas that were previously introduced in [22, 23, 24, 25] connecting to the theory of integrable systems (for instance, involving Lax Pairs and the KP-Hierarchy) with operator theory. Further, we will recall, from, e.g., [5], the class of Banach–Lie algebras known as (simple) L*-algebras, which along with Kac–Moody algebras can be interwoven into this study. Some new observations in this direction are obtained in the form of Propositions 6.1, 6.2, and 6.3.

There is already a significant amount of work that links the τ-function theory to *Schlesinger systems* in the framework of the Riemann–Hilbert problem and isomonodromic deformations (see [3, 9, 11, 36, 37] and references therein). The present approach, as taken in an infinite-dimensional vector bundle (with connection) setting, suggests something more general since we introduce and apply certain operator-valued mechanisms. Of interest are (closed) 1-forms of the type $d \ln \tau$. Partial motivation for doing this is suggested by the work of Katsnelson and Volok [39] who considered this problem from the point of view of matrix-operator differential equations of Fuchsian type along with their associated Schlesinger systems. The instrumentation of generalized determinants and meromorphic operator-valued functions is one such example, and here some attention is paid to the idea of an *operator meromorphic connection* besides suggesting several examples where this can be realized (§7.2).

2. Background to the geometry

2.1. A principal bundle and its transition map

We will start by outlining a general construction from which a large class of interesting and well-studied operators can be obtained directly from the transition functions of an infinite-dimensional bundle theory. A class of these operators will in fact produce the 'predeterminants' for several types of operator-valued functions of a determinant type that we keep in mind.

Let A be a (complex, associative) unital Banach algebra with group of units $G(A)$ and space of idempotents $P(A)$ (in some cases A may be semisimple, and this is assumed if needs be). For a given $p \in P(A)$, we denote by $\Lambda = \mathrm{Sim}(p, A)$ the similarity orbit of p under the inner automorphic action of $G(A)$. There exists a natural map [23, §5]

$$\pi_\Lambda : \Lambda \longrightarrow \mathrm{Gr}(p, A), \tag{2.1}$$

where $\mathrm{Gr}(p, A)$ is an associated Grassmannian of closed subspaces $W = \mathrm{Im}(p)$ (see [21, §6]). In the following we shall be considering certain Banach–Lie subgroups (subalgebras) of $G(A)$ (respectively, of $\mathfrak{g}(A)$).

Remark 2.1. For the standard theory of Banach algebras and associated classes of linear operators we refer to [18, 29]. For the general theory of Banach–Lie groups (algebras) and the infinite-dimensional manifolds modeled on these (such as $\mathrm{Gr}(p, A)$ above) reference [5](cf. [20]) provides a comprehensive account from the operator algebra perspective including many references to the related work of other authors, while much of the development of the relationships between the Banach manifolds $\mathrm{Gr}(p, A)$ and Λ appeared in [21, 23]. As far as parts of this preliminary section is concerned there is a significant amount of related work that has been progressively developed in [5, 6, 7] pertaining to the theory of holomorphic vector bundles over infinite-dimensional flag manifolds.

The detailed framework outlined in [23, §5 and §6] produces a principal $G(pAp)$-bundle with connection

$$(Q', \omega_{Q'}) \longrightarrow \Lambda, \tag{2.2}$$

and an associated vector bundle (with Koszul connection) $(\gamma'_\Lambda, \nabla'_\Lambda) \longrightarrow \Lambda$, whose structure group is $G(pAp)$. This associated vector bundle is constructed via the usual means (cf. [43, Chap. 37]). There is the transition map

$$t_\Lambda : Q' \times_\Lambda Q' \longrightarrow G(pAp), \tag{2.3}$$

which for a pair of sections α, β, is given by $\alpha t_\Lambda(\alpha, \beta) = \beta$. For now we take $A = \mathcal{L}(E)$ where E is a (complex) Banach space, and then observe that the operator $\mathsf{T}_{(\alpha,\beta)} = t_\Lambda(\alpha, \beta) \in G(pAp)$ belongs to some class (to be made more precise later).

Remark 2.2. It will be instructive to point out that $\omega_{Q'}$ in (2.2) is constructed via the properties of a $\mathrm{Lie}(G(pAp))$-valued connection map $\mathcal{V} : TQ' \longrightarrow TQ'$ (see [23, §5.2]) to be used within the meromorphic context later in §7.2.

2.2. Reduction of the structure group

Suppose $B \subseteq A$ is a Banach subalgebra and $G \subseteq G(pBp) \subseteq G(pAp)$ is a Banach–Lie subgroup. Then in the standard way, granting the existence of a cross section $\Lambda \longrightarrow Q'/G$, we obtain from (2.2), a (reduced) principal G-bundle $Q \longrightarrow \Lambda$. The means of doing this is formally the same as seen in, e.g., [41, Propositions 5.5, 5.6] and [43, p. 381]. We also assume, that under appropriate conditions (cf. [43, p. 381] and [41, Theorem 7.1]), that the connection $\omega_{Q'}$ has been reduced accordingly. This can be achieved by commencing from a Banach–Lie group homomorphism $\psi : G \longrightarrow G(pAp)$, and a principal bundle homomorphism

$$\Psi : (Q, G, \Lambda) \longrightarrow (Q', G(pAp), \Lambda), \tag{2.4}$$

such that $\Psi : Q \longrightarrow Q'$ is smooth, and $\Psi(u \cdot g) = \Psi(u) \cdot \psi(g)$ (see [43, p. 381]). Hence we obtain a commutative diagram

$$\begin{array}{ccc} Q & \xrightarrow{\Psi} & Q' \\ \downarrow & & \downarrow \\ \Lambda & \xrightarrow{\widetilde{\Psi}} & \Lambda \end{array} \tag{2.5}$$

A pull-back connection ω_Q on Q is then obtained as $\omega_Q = \Psi^* \omega_{Q'}$, and so leads to a principal G-bundle with connection $(Q, \omega_Q) \longrightarrow \Lambda$, along with its related objects. In particular, these likewise include a G-valued transition map

$$t_\Lambda : Q \times_\Lambda Q \longrightarrow G, \tag{2.6}$$

and an associated vector bundle (with Koszul connection) $(\gamma, \nabla_\gamma) \longrightarrow \Lambda$, whose structure group is G.

3. On generalized determinants in the Banach algebra setting

3.1. Two approaches for generalized determinants

Proceeding with $A = \mathcal{L}(E)$, let us recall the notion of the *socle* of A, denoted soc(A). This consists of the sum of all minimal left ideals (or right ideals) if they exist, or else it is zero. For the situation in question we follow [2, §2] and take soc(A) to be generated by the minimal projections of A, that is, elements $p \in P(A)$ such that $pAp = \mathbb{C}p$ (meaning that the restriction of elements of A to $\mathrm{Im}(p)$ is the identity on $\mathrm{Im}(p)$). Also, given $\mathcal{L}(E)$ contains finite rank operators, we have $\mathrm{soc}(A) \neq 0$. For $a \in A$, the spectral rank of a, is given by $\sup_{x \in A} \#(\mathrm{spec}(xa) - \{0\})$.

The *maximal finite rank elements* are those elements of A such that we have rank $a = \#(\mathrm{spec}(a) - \{0\})$, and these elements admit spectral representations of the form $a = \lambda_1 p_1 + \cdots + \lambda_n p_n$ (for some n, and where the p_i are minimal projections).

For arbitrary $a \in \mathrm{soc}(A)$, we have rank $a = \sum m(\lambda_i, a)$ (sum over non-zero elements) where the multiplicity $m(\lambda_i, a)$ is the rank of the Riesz projection for λ_i (see, e.g., [18]). Taking λ (below) as a sum over spec(A), we have the trace and determinant well-defined for $a \in \mathrm{soc}(A)$, as given by [2, §2]:

$$\begin{aligned} \mathrm{Tr}\ a &= \sum\nolimits_\lambda \lambda\ m(\lambda, a), \\ \mathrm{Det}(1 + a) &= \Pi_\lambda (1 + \lambda)^{m(\lambda, a)}. \end{aligned} \tag{3.1}$$

We refer to [2] (cf. [55]) for further consequences of these constructions.

Another approach for the case $A = \mathcal{L}(E)$ [28] involves taking the subalgebra $\mathcal{F}(E) \subset \mathcal{L}(E)$ of *operators of finite rank* and then define $\mathrm{Det}(1 + F)$ on certain normed subalgebras of E. This starts by considering certain (Banach) subalgebras $B \subset \mathcal{L}(E)$ which are embedded continuously in $\mathcal{L}(E)$, meaning that there is a norm $\|\cdot\|_B$ on B such that:

i) $\|F\|_{\mathcal{L}(E)} \leq C\|F\|_B$, for all $F \in B$, where $C = \text{const.}$, and also assume that,
ii) $\|SF\|_B \leq \|S\|_B\|F\|_B$, for all $S, F \in B$.

If i) and ii) hold then B is called *an embedded subalgebra of* $\mathcal{L}(E)$. If also we have $\mathcal{F}_B = \mathcal{F}(E) \cap B$ dense in B with respect to $\|\cdot\|_B$, then B is said to have *the approximation property.* This property assists the continuous extension of trace and determinant from $\mathcal{F}_B$ to B. If then $B \subset \mathcal{L}(E)$ is an embedded subalgebra with the approximation property, then $\mathrm{Det}(1 + F) : \mathcal{F}_B \longrightarrow \mathbb{C}$ admits a continuous extension in the B-norm from $\mathcal{F}_B$ to B (see [28, Theorem 2.1] which includes related results), and for $F \in \mathcal{F}(E)$ with $|z|$ sufficiently small [28, Theorem 3.3]:

$$\mathrm{Det}(1 + zF) = \exp\left(\sum_{n=1}^{\infty} \frac{(-1)^{n+1}}{n} \mathrm{Tr}(F^n) z^n\right). \tag{3.2}$$

If $E = H$ is a Hilbert space and F is a trace class operator, then there is the usual Fredholm determinant given by

$$\mathrm{Det}(1 + F) = \sum_{n=0}^{\infty} \mathrm{Tr}\ \Lambda^n(F), \tag{3.3}$$

(see, e.g., [28, 64, 67]).

3.2. Admissible elements

From [24, §6] we have the principal $G(pAp)$-bundle (*the Stiefel bundle*) denoted $V(p, A) \longrightarrow \mathrm{Gr}(p, A)$ for which an element $v \in V(p, A)$ is manifestly a framing for the Banach algebra A, or simply a basis for its underlying (Banach) vector space. Let us then say that $v \in V(p, A)$ is *admissible* when $\mathrm{Det}(v)$ is defined in the context of a suitable generalized determinant. A particular instance concerns that of 'admissible bases' relative to polarized Hilbert spaces (modules) that can be used to produce an important class of determinant line bundles [57, §7.7] and [63, §3] (cf. [22, 23] and see §5.1 below). Accordingly, when we speak of 'determinants' we will take for granted the existence of the corresponding admissible elements as they were defined in this general criteria.

3.3. Determinants of generalized Wiener–Hopf operators

Again taking $A = \mathcal{L}(E)$ and $B \subset A$ a Banach subalgebra, let $E_1 \subset E$ be a closed subspace, and let $p : E \longrightarrow E_1$ be the projection onto E_1. For some $L \in B$, one could in principle define a *generalized Wiener–Hopf operator* T simply in terms of the relationship $\mathsf{T}_p(L) = pLp$ (see Example 3.1 below). Thus for an appropriate choice of sections, the operator $\mathsf{T}_{(\alpha,\beta)} = t_\Lambda(\alpha, \beta) \in G$ seen above, would then be such an example. Accordingly, if $\mathsf{T} \in \mathrm{soc}(A)$, or if $\mathsf{T} \in B$ where B has the approximation property, then $\mathrm{Tr}(\mathsf{T})$ and $\mathrm{Det}(1 \pm \mathsf{T})$ are well defined as we have seen in §3.1.

Along with applications, much of the work that had been available concerns taking $A = \mathcal{L}(H)$, and so let us recall some particular examples.

Example 3.1. Let us take $B \subset \mathcal{L}(H)$ to be a maximal abelian von Neumann algebra of operators and $H_1 \subset H$ a proper closed subspace of H such that any non-zero vector in H_1 or $H_1^{\perp}$ is separating for B. In [19, §2] an operator $L \in B$ is decreed to be a *generalized Laurent operator*, in which case the triple (H, B, H_1) is called a *Riesz system* for which $\mathsf{T}_p = pLp$ is a *generalized Toeplitz operator* (cf. [16, 51]). This generalizes the well-known case where $H = L^2(S^1, \mathbb{C})$ and $H_1 = H^2(S^1, \mathbb{C})$ is the Hardy space consisting of those Fourier coefficients that vanish on $\mathbb{Z}^-$. Here $p : L^2(S^1, \mathbb{C}) \longrightarrow H^2(S^1, \mathbb{C})$ is the Riesz projection and $\mathsf{T}_p = pm(f)p$ is then the classical Toeplitz operator, where $m(f) \in B$ is multiplication by an essentially bounded function. The term generalized Laurent operator is fitted to such a Riesz system. For the more usual notion of this class of operators see, e.g., [29, §3.1,§16.1]. A further study of the theory of Toeplitz operators relative to bounded domains in $\mathbb{C}^n$ is treated in [69, Chap. 4].

4. Generalized determinants and the τ-function

4.1. Background to the τ-function

For the benefit of readers we provide a short background to the nature of the τ-function besides motivating its introduction into the operator theory context. Originally the function seemed to have played a significant role in the theory of *the Painlevé transcendents* and Hamiltonian systems (see, e.g., [1, Chap. 7] and [14, 54]), whereas in classical Sturm–Liouville theory, the logarithmic derivative of the τ-function differentiates to a 'potential'. Let us exemplify this latter case now, and postpone a short discussion of the Painlevé equations to Appendix B.

Example 4.1. For the Sturm–Liouville (SL) theory, the classical setting considers the Hilbert space $H = L^2([a, b], r(x)dx)$ endowed with the usual inner product $\langle f, g\rangle = \int_a^b f(x)\overline{g(x)}r(x)\, dx$. As seen in, e.g., [29, §6.5], for the appropriate differentiable (real-valued) functions p, q and $r(\neq 0)$, we have the SL-operator L as given by

$$Ly(x) = \frac{1}{r(x)}\left(-\frac{d}{dx}\left[p(x)y'(x)\right] + q(x)y(x)\right). \tag{4.1}$$

The SL-equation $Ly = \lambda y$, along with boundary conditions, is a well-mined eigenvalue problem. Typically, L is a self-adjoint (unbounded) operator with ordered real eigenvalues $\lambda_1 < \lambda_2 < \cdots$, with associated orthonormal eigenfunctions y_1, $y_2, \ldots$ Just as any good student knows, there are many classes of second-order (linear) ODEs that can be expressed in SL-form and numerous applications of the SL-equation to mathematical physics. It is here that we discover 'potentials' $q(x)$

as given by

$$q(x) = -\left(\frac{d}{dx}\right)^2 \left[\ln \tau\right] = -\frac{d}{dx}\left[\frac{\tau'}{\tau}\right]. \tag{4.2}$$

Of interest is the type of functions such as τ and how they may arise. For instance, on commencing from the SL-equation, the work in [49] leads to aspects of such a τ-function relative to differential rings and 'vessels' in a fashion applicable to systems theory.

The main thrust of the τ-function theory has arisen in the theory of integrable systems, particulary in the theory of nonlinear waves (such as in the KP-hierarchy), the inverse scattering transform method, statistical physics and in a number of related areas enjoying deep mathematical connections, as seen in [1, 3, 17, 36, 37, 59, 60, 63, 66]. For instance, there are certain τ-functions for which the expression $\omega = \partial_x \ln \tau$ (where ∂_x denotes a spatial derivative) provides a 'Kähler potential' in the theory of self-dual Einstein gravity [56]. The point being that it is mainly through this work that the τ-function is realizable as a (generalized) determinant, though often in a seemingly formal sense.

Example 4.2. The formal way of defining a τ-function commences with symmetric functions in n-variables indexed by partitions (cf. characters of irreducible representations of $\mathrm{GL}(n,\mathbb{R})$). Consider a partition $\nu = \nu_1 + \nu_2 + \cdots + \nu_n$, with ν_k non-negative integers, $\nu_1 \geq \nu_2 \geq \cdots \geq \nu_n$. Alternating polynomials are given by

$$\begin{aligned} A_{(\nu_1,\ldots,\nu_n)}(x_1,\ldots,x_n) &= \det\left[x_i^{\nu_j}\right] \\ &= \sum_{\sigma\in S_n} \epsilon(\sigma)\, x_{\sigma(1)}^{\nu_1}\cdots x_{\sigma(n)}^{\nu_n}, \end{aligned} \tag{4.3}$$

leading to the following expression for the *Schur function*

$$\mathsf{s}_\nu(x) := \frac{A_{(\nu_1+n-1,\nu_2+n-2,\ldots,\nu_n)}(x_1,x_2,\ldots,x_n)}{\Delta_{\mathrm{vm}}}, \tag{4.4}$$

where $\Delta_{\mathrm{vm}} = \prod_{1\leq j<k\leq n}(x_j - x_k)$ is a Vandermonde determinant.

For a large class of wave functions the formal τ-function is given by a linear combination of Schur functions relative to a partition

$$\tau(\mathbf{t}) = \sum_{\nu} c_\nu\, \mathsf{s}_\nu(\mathbf{t}), \tag{4.5}$$

where the constants c_ν depend on 'embedding coordinates' [63, §8].

Many significant developments, somewhat in the context of Example 4.2, blossomed forth in the 1980's (see for instance [36, 37, 59, 60]). The essential ideas were later studied in a more unified geometric-analytic context in the ground-breaking paper of Segal and Wilson [63]. In this latter setting the τ-functions, such as those to be defined in (5.5) below (see also Example 4.3), are associated to points W in a Grassmannian of the type $\mathrm{Gr}(p, A)$ (here A denotes the 'restricted'

Banach *-algebra to be outlined in 5.2) and are denoted τ_W-functions. These τ_W-functions are shown in [63, §8] to be closely related to the formal τ-function of Example 4.2 and the basic construction of these will be seen in §5.3.

We have in mind a short survey showing how the τ-function arises from a generalized (e.g., Fredholm or Toeplitz) determinant along with several algebraic, and analytic ramifications (see, e.g., [9, 11]). Also significant is the construction of the associated predeterminant operators in §5.1. A further aspect of the τ-function is its role in the theory of Schlesinger systems as associated with differential equations of Fuchsian type [39, 40, 42, 46], where the operator-theoretic setting is briefly discussed in §7. The classical theta functions are in fact closely tied to τ-functions via exponential multiplication [63, Theorem 9.11] (see also [58]). Hence the acclaimed *Fay trisecant identity* [26], which is ubiquitous in the theory of integrable systems, is expressible in terms of τ-functions as seen in, e.g., [58, §10](cf. [50, §2] in the setting of the KP-hierarchy). Perhaps more in line with this present work is a matrix-operator account of this subject as treated in [4, §4] which touches on 'moduli' questions, as does [8, §5] which also establishes such a trisecant identity in terms of nonabelian theta functions.

Example 4.3. This involves briefly introducing loop groups 'LU' [57, Chap. 6]. Take $H = L^2(S^1, \mathbb{C})$ and a polarization $H = H_+ \oplus H_-$ (with $H_+ \cap H_- = \{0\}$). Following [48, §2.3], for $g \in \Gamma^+ \subset \mathrm{LU}(1)$, we have a Toeplitz operator $\mathsf{T}_g : H_- \longrightarrow H_+$ given by $v \mapsto (gv)_-$. Note that $\mathsf{T}_{g_1 g_2} \neq \mathsf{T}_{g_2 g_1}$, in general, but $\mathsf{T}_{g_+ g g_-} = \mathsf{T}_{g_- g g_+}$ when $g_\pm \in \mathrm{GL}(H_\pm)$. Taking $K_+ = f^{-1}H_+$, for some $f \in \mathrm{GL}(H)$, leads to another polarization (K_+, K_-), and a Toeplitz operator given by $\mathsf{T}'_g = \mathsf{T}_f^{-1}\mathsf{T}_{fg}$. Taking $k \in \mathrm{GL}_+(H)$ and $h \in \mathrm{GL}_-(H)$, then there exists a *Toeplitz-τ-function* given by

$$\tau(h^{-1}H_+, H_+, kH_-, H_-) = \mathrm{Det}(\mathsf{T}_h^{-1}\mathsf{T}_{hk}\mathsf{T}_k^{-1}). \tag{4.6}$$

Example 4.4. References [11, 12](cf. [67]) also reveal a large class of Fredholm determinants, within the representation theory of the infinite-dimensional unitary group, actually to be τ-functions of various integrable systems associated to a particular Painlevé type [1]. Typical of this approach is to commence with T taken to be an integral operator, and on restricting its kernel $K_{\mathsf{T}}(x, y)$ to an interval $\mathcal{J} = \bigcup_{j=1}^m (a_{2j-1}, a_{2j}) \subset \mathbb{R}$, to take the Fredholm determinant $\mathrm{Det}(1 - \lambda\mathsf{T}|_{\mathcal{J}})$, for a suitable $\lambda \in \mathbb{C}$ [11, 67]. In all cases, the kernel $K_{\mathsf{T}}(x, y)$ is explicit, though in [11, §2] the authors implement a continuous, so-called ${}_2F_1$-kernel based on the Gauss hypergeometric function. This ${}_2F_1$-kernel, $K_{\mathsf{T}}(x, y)|_{\mathcal{J}}$ is seen to be of trace class and, in particular, $\mathrm{Det}(1 - \lambda\mathsf{T}|_{\mathcal{J}})$ is a τ-function for the Schlesinger equation (see, e.g., [11, §11] and [36, §5]; also see the assortment of examples in §7).

Example 4.5. In [9] a moment functional $\mathfrak{M} : \mathbb{C}[z, z^{-1}] \longrightarrow \mathbb{C}$, is considered along with a corresponding Lax operator $Q_{\mathcal{I}}(\mathfrak{M})$ depending on an index set $\mathcal{I}$ (here $Q_{ij} = \mathfrak{M}(r_i x p_j)$). The shifted Toeplitz determinant yields a τ-function such that

$$d \ln \tau = \sum_{\rho=1}^{n} \frac{1}{\rho}\, \mathrm{Tr}_n(Q_{\mathcal{I}})\; dt_\rho. \tag{4.7}$$

The Hankel determinant of a semiclassical moment functional on the space of polynomials can be identified with the isomonodromic τ-functions of [36] and this together with the above Toeplitz determinant (of equivalent order) with respect to $\mathfrak{M}$ above can be related by sequences of Schlesinger transformations (cf. §7).

5. The predeterminant $\mathcal{T}$-function and its τ-function

5.1. The predeterminant $\mathcal{T}$-function

We return to the general setting of §2.1, with A a unital Banach algebra and then specialize. Firstly, we recall from §3.2 the principal $G(pAp)$-Stiefel bundle $V(p, A) \longrightarrow \mathrm{Gr}(p, A)$, endowed with a *canonical section* denoted S_p, and proceed to extract part of a construction in [23, §8] to which we refer for complete details. Recalling the map π_Λ in (2.1), and for $p \in P(A)$, we set

$$W_p = \pi_\Lambda^{-1}(p + pA\hat{p}), \tag{5.1}$$

(where $\hat{p} = 1 - p$), and then consider the subset

$$W_p^0 = \{r \in W_p : \phi_p(r) := rp + \hat{r}\hat{p} \in G(A)\}. \tag{5.2}$$

Also recalling $\Lambda = \mathrm{Sim}(p, A)$, the development of [23, §8.1] entailed defining two sections α, β of the principal $G(pAp)$-bundle $Q' \longrightarrow \Lambda$ having the following properties:

i) With respect to π_Λ in (2.1), $\alpha_p = \pi_\Lambda^*(S_p)$ is defined over $\pi_\Lambda(W_p) \subset \mathrm{Gr}(p, A)$.
ii) For β_p with $g = \phi_p(r)$ and $r \in W_p^0$, we have $g \in G(A)$. The assignment $rp : p \longrightarrow r$, yields a proper partial isomorphism which projects along $\mathrm{Ker}(r)$. Then we set $\beta_p(r) = (r, rp)$.

Next, on recalling the transition map t_Λ in (2.3), we set

$$\mathcal{T}(r) = t_\Lambda(\alpha_p(r), \beta_p(r)) \in G(pAp). \tag{5.3}$$

In [23, §8] we called the left operator of (5.3) a $\mathcal{T}$*-function* which can be viewed as a type of generalized Wiener–Hopf operator T as described above. Note that this essentially algebraic construction would work equally well for any of the reduced subgroups G in §2.1, and hence for the principal bundles $Q \longrightarrow \Lambda$ described in §2.2.

5.2. The restricted Banach algebra A

To see how this was used in [23], we first of all take a (separable) Hilbert module $H_{\mathcal{A}}$ over a unital C*-algebra $\mathcal{A}$, and a polarization $H_{\mathcal{A}} = \mathsf{H}_+ \oplus \mathsf{H}_-$ (with $\mathsf{H}_+ \cap \mathsf{H}_- = \{0\}$). Let J be an $\mathcal{A}$-module map satisfying $J^2 = 1$. The 'restricted' Banach algebra $A = \mathcal{L}_J(H_{\mathcal{A}})$ is a Banach *-algebra under the Hilbert–Schmidt modified norm $\|T\|_J = \|T\| + \|[J, T]\|_2$ (here we have used the generalization of the Schatten p-classes to Hilbert modules following [65] and cf. [28]).

If $\mathfrak{P}$ denotes the space of polarizations on $H_{\mathcal{A}}$, then by [23, Theorem 4.1], there exists an analytic diffeomorphism $\varphi : \mathfrak{P} \longrightarrow \Lambda$. Instrumental in [23, §8] and [70, §3] was to take the above sections $\alpha_p(r)$ and $\beta_p(r)$, when viewed as sections of the (universal) vector bundle $\gamma_\Lambda \longrightarrow \Lambda$ in §2.2, to be *covariantly constant* with respect to the connection ∇_Λ. The $\mathcal{T}$-function (operator) is more general than the Zelikin $\mathfrak{T}$-function which for the case $\mathcal{A} \cong \mathbb{C}$ utilizes a cross-ratio coordinate system on $\mathfrak{P}$ [70, §3]. Also, Det $\mathcal{T}$ is definable in terms of 'admissible elements (bases)' for a sufficiently large class of Banach algebras in the setting of §3.

5.3. The τ_W-function

Suppose we take a pair of polarizations $(\mathsf{H}_+, \mathsf{H}_-), (\mathsf{K}_+, \mathsf{K}_-) \in \mathfrak{P}$ to be such that H_+ is the graph of a linear map $S : \mathsf{K}_+ \longrightarrow \mathsf{K}_-$, and H_- is the graph of a linear map $T : \mathsf{K}_- \longrightarrow \mathsf{K}_+$. Then on $H_{\mathcal{A}}$ consider the identity map $\mathsf{H}_+ \oplus \mathsf{H}_- \longrightarrow \mathsf{K}_+ \oplus \mathsf{K}_-$, as represented by the block form

$$\begin{bmatrix} a & b \\ c & d \end{bmatrix} \tag{5.4}$$

where $a : \mathsf{H}_+ \longrightarrow \mathsf{K}_+$, $d : \mathsf{H}_- \longrightarrow \mathsf{K}_-$ are zero-index Fredholm operators, and $b : \mathsf{H}_- \longrightarrow \mathsf{K}_+$, $c : \mathsf{H}_+ \longrightarrow \mathsf{K}_-$, belong to $\mathcal{K}(H_{\mathcal{A}})$ (the compact operators), such that $S = ca^{-1}$ and $T = bd^{-1}$. When b, c are taken to be Hilbert–Schmidt operators, then ST is of trace-class, and the operator $(1 - ST)$ can be expressed as a $\mathfrak{T}$-function as above. We denote this by $\mathfrak{T}(\mathsf{H}_+, \mathsf{H}_-, \mathsf{K}_+, \mathsf{K}_-)$ (it is essentially a transition map in the $\mathcal{T}$-function setting).

Following [63, §2] and [57, §7.1] (cf. [24, A2]) points $W \in \mathrm{Gr}(p, A)$ represent closed subspaces W of $H_{\mathcal{A}}$ such that i) the orthogonal projection $p_+ : W \longrightarrow \mathsf{H}_+$ is in $\mathrm{Fred}(H_{\mathcal{A}})$, and ii) the orthogonal projection $p_- : W \longrightarrow \mathsf{H}_-$ is in $\mathcal{L}_2(\mathsf{H}_+, \mathsf{H}_-)$ (Hilbert–Schmidt operators). Relative to such points $W \in \mathrm{Gr}(p, A)$, the corresponding τ_W-function is constructed in [63, §3] and is seen to be of the form

$$\begin{aligned} \tau_W(\mathsf{H}_+, \mathsf{H}_-, \mathsf{K}_+, \mathsf{K}_-) &= \mathrm{Det}\ \mathfrak{T}(\mathsf{H}_+, \mathsf{H}_-, \mathsf{K}_+, \mathsf{K}_-) \\ &= \mathrm{Det}(1 - ca^{-1}bd^{-1}) \in \mathbb{C} \otimes 1_{\mathcal{A}}, \end{aligned} \tag{5.5}$$

(cf. [23, 48, 70]). This particular τ-function frequently appears in the Lax equation method in nonlinear wave theory (for instance, *solitons*) and the general setting for the latter within the KP-hierarchy (see, e.g., [1, 3, 17, 63, 60]).

Remark 5.1. In several examples above and in those in the sequel, the main class of operators surveyed will often turn out to be integral operators. One way of seeing this proceeds as follows. Suppose (Ω, μ) is some σ-finite measure space. Let $H = L^2(\Omega, \mu)$ (be separable), $A = \mathcal{L}(H)$ and let $\mathcal{L}_2(H)$ denote the Hilbert–Schmidt operators on H. Then projections $p \in P(A) \cap \mathcal{L}_2(H)$ may be realizable as integral operators, and likewise for elements $q \in U(p, A)$ in the unitary orbit of p (see, e.g., [32, §15] and related results therein). For instance, let $\Omega \subset\subset \mathbb{C}^n$ be a bounded domain. A familiar example is the Bergman projection $p_B : L^2(\Omega) \longrightarrow \mathrm{Hol}(\Omega) \cap L^2(\Omega)$, orthogonally projecting $L^2(\Omega)$ onto its holomorphic subspace. In terms of its associated kernel function $K_\Omega(z, w)$, it is given by $p_B f(z) = \int_\Omega K_\Omega(z, w) f(w)\, dw$. This projection p_B may be viewed as a generalized Calderón–Zygmund operator with respect to certain local pseudometrics [45]. Other familiar and partially related examples are presented in [69, Chap. 4].

6. Simple L*-algebras and Kac–Moody algebras

6.1. Simple L*-algebras

Henceforth, unless otherwise stated, we now restrict in this section to $A = \mathcal{L}(H)$, where H is a separable Hilbert space, and proceed to consider a class of involutive Banach–Lie algebras called *L*-algebras* (see [5, Chap. 7] and [34, Chap. II, III] following the earlier work of [61, 62]).

An *L*-algebra* is a Lie algebra $\mathfrak{g}$ (over $\mathbb{R}$ or $\mathbb{C}$) whose underlying vector space is a Hilbert space, together with a map $x \mapsto x^*$ that satisfies $\langle [x,y], z\rangle = \langle y, [x^*, z]\rangle$, for all $x, y, z \in \mathfrak{g}$. In particular, the Hilbert–Schmidt operators $\mathcal{L}_2(H)$ form a complex simple L*-algebra (see [5, Theorem 7.18] and [34, II.5]) in which the $*$-map specifies an adjoint representation, while noting that by Remark 5.1 above, this latter observation subsequently provides a potential supply of integral operators on H. We denote this L*-algebra by $\mathfrak{g}_{\mathbf{A}}$, which is a simple Lie algebra of type $\mathbf{A}$ in the Cartan classification. In fact, all simple Lie algebras of type $\mathbf{A}$ are isomorphic, up to some multiple of the inner product, to an L*-subalgebra of $\mathfrak{g}_{\mathbf{A}}$.

There are also the simple L*-algebras of Cartan type $\mathbf{B}, \mathbf{C}$ and $\mathbf{D}$ (see [5, 34, 52, 61, 62] for a comprehensive treatment of this infinite-dimensional structure theory). Because of the frequent instrumentation of the Hilbert–Schmidt operators, we will restrict matters here to $\mathfrak{g}_{\mathbf{A}}$, and note that $\mathfrak{g}_{\mathbf{A}}$ as a simple L*-algebra can be approximated by taking the limit of a strictly increasing sequence $\{\mathfrak{g}_n\}$ of simple finite-dimensional Lie algebras of Cartan type $\mathbf{A}$ [61, §3.2]. Specifically, we shall take $G \subseteq G(pBp) \subseteq G(pAp)$ to be a Banach–Lie group whose Banach–Lie algebra (over $\mathbb{C}$) is the simple L*-algebra $\mathfrak{g} = \mathfrak{g}_{\mathbf{A}}$.

A *Cartan subalgebra* $\mathfrak{h}$ of $\mathfrak{g}$ is defined as a maximal self-adjoint abelian (closed) subalgebra. With respect to $\mathfrak{h}$, a Cartan decomposition of $\mathfrak{g}$ can be formed (see below). Let Δ denote the set of non-zero roots of $\mathfrak{g}$ relative to $\mathfrak{h}$. In the form of a lemma we collect together several basic results about the structural theory of the L*-algebra $\mathfrak{g}$ to be used later.

Lemma 6.1. *Let $\mathfrak{g} = \mathfrak{g}_{\mathbf{A}}$ be the (complex) simple L*-algebra as above. Then we have the properties:*

(1) *There exist simple closed $\mathfrak{g}$-ideals $\mathfrak{g}_k$, indexed by some set $\mathcal{J}$, leading to a Hilbert space direct sum*
$$\mathfrak{g} = \bigoplus_{k \in \mathcal{J}} \mathfrak{g}_k. \tag{6.1}$$

(2) *Given a Cartan subalgebra $\mathfrak{h} \subset \mathfrak{g}$, there exists a Cartan (Hilbert space) decomposition*
$$\mathfrak{g} = \mathfrak{h} \oplus \bigoplus_{\nu \in \Delta} V_\nu, \tag{6.2}$$
where the root spaces V_ν are one-dimensional.

(3) *We have $\mathfrak{g} = \mathrm{cl}(\bigcup \mathfrak{g}_n)$, where each $\mathfrak{g}_n$ is a finite-dimensional simple Lie algebra (of Cartan classification $\mathbf{A}$) appearing in a strictly increasing sequence $\cdots \subset \mathfrak{g}_n \subset \mathfrak{g}_{n+1} \subset \cdots$*

Proof. Part (1) is stated in [61, §1.2, Theorem 1]. The existence of the Cartan decomposition in Part (2) follows from [61, §2.2, Theorem 2] (cf. [62]) and the one-dimensionality of the root spaces V_ν follows from [61, §2.3]. As for Part (3), this follows from [61, §3.2, (iv)]. □

Remark 6.1. The corresponding results for the other types in the Cartan classification are essentially the same. A treatment of these results and the Cartan decomposition also appear in [34, Proposition 11, II.22] (cf. [5, Chap. 7] for certain extensions of the theory) and more generally in [13, Theorem 1] for semisimple Banach–Lie algebras of compact operators.

Having introduced this class of L*-algebras, we recall the setting of the principal G-bundles $Q \longrightarrow \Lambda$ in §2.2, where G is the corresponding Lie group of the L*-algebra $\mathfrak{g}$ and the operator $\mathcal{T}$ arises via the transition map t_Λ in (2.6). This leads to some further observations concerning generalized determinants and τ-functions in particular. The first follows by a straightforward restriction argument.

Proposition 6.1. *For each operator $\mathcal{T} \in G$ of the (predetermiant) type in* (5.3) *there exists a family of generalized determinants defined relative to the closed ideals of $\mathfrak{g}$.*

Proof. For each $\mathcal{T} \in G$ in (5.3) we can assign a corresponding Hilbert–Schmidt operator $\mathsf{T} \in \mathfrak{g}$. Recall from Lemma 6.1(1) we have a Hilbert space direct sum

$$\mathfrak{g} = \mathfrak{g}_1 \oplus \mathfrak{g}_2 \oplus \cdots \oplus \mathfrak{g}_k \oplus \cdots \tag{6.3}$$

in terms of closed $\mathfrak{g}$-ideals. We then consider restrictions $\mathsf{T}|\mathfrak{g}_k$, which for each k, define a family of generalized determinants by taking the Fredholm determinant $\mathrm{Det}(1 - \lambda\mathsf{T}|\mathfrak{g}_k)$ (for some suitable λ). □

It is clear that the means behind Proposition 6.1 would apply by restricting to a Hilbert space summand in any decomposition of $\mathfrak{g}$. The next observation is in the context of Example 4.4 and follows from the development of ideas in [11].

Proposition 6.2. *Let $G \subseteq U(pAp)$ be a simple Banach–Lie subgroup with associated (simple) L^*-algebra which is a real form of $\mathfrak{g} = \mathfrak{g}_{\mathbf{A}}$ as above. Then for each $\mathcal{T} \in G$ in* (5.3) *we obtain a family of τ-functions via a smooth integrable kernel* K *defined on a subset $\mathfrak{J} \subset \mathbb{R}$.*

Proof. Without loss of generality we assume that $G \subseteq U(pAp)$ is isomorphic to a unitary (Banach–Lie) subgroup of the infinite-dimensional unitary group $U(\infty)$. The idea is that starting from the operator $\mathcal{T} \in G \subseteq U(\infty)$, we arrive at a integrable kernel defined with respect to $\mathfrak{J}$. Firstly, the results of [12, §1-§3] show that G can be decomposed into irreducible parts via probability measures on the space of all irreducible representations. Specifically, if $\chi : G \longrightarrow \mathbb{C}$ denotes a character of G, then there exists a spectral measure μ_χ on parameter set Ω so that

$$\chi = \int_\Omega \chi_w \mu_\chi \, dw, \tag{6.4}$$

as established in [11, Theorem 1.1]. Following from this, the construction of the kernel K is given in [11, §5]. Taking a projection $p(\mu_\chi)$ onto $\mathfrak{J} = (s, \infty)$ to give a restriction, (see Example 4.4) the Fredholm determinant $\mathcal{D}(s) = \mathrm{Det}(1 - \lambda \mathsf{K}|_{(s,\infty)})$ leads to a family of τ-functions (see [11, §7]). □

Remark 6.2. In [11, 12] the distribution $\mathcal{D}(s)$ and the corresponding τ-function above lead to classes of differential equations of the various Painlevé type (see Appendix B. In [12, §10] the τ-functions arising there are interpreted as correlation functions that are manifestly determinants of certain kernels. As we pointed out in §4.1, there is a considerable amount of background material to this topic in the framework of integrable systems and statistical physics, and so for now we are obliged to refer the reader to works such as [1, 11, 12, 15, 36, 37, 67] in order to get a more complete picture.

6.2. A link with affine Kac–Moody algebras

In Appendix §A.1 we recall how the Cartan matrix of any semisimple Lie algebra induces an associated (*affine*) *Kac–Moody algebra* [38, Chap. 7]. This concept we will adopt in view of the simple L*-algebras discussed previously. Firstly, a lemma that uses the construction and which will be put to use in §6.3 below.

Lemma 6.2. *Consider the (complex) simple L^*-algebra $\mathfrak{g} = \mathfrak{g}_{\mathbf{A}}$. Then there is an associated (strictly) increasing sequence of affine Kac–Moody algebras $\{\widehat{\mathfrak{g}}_{\kappa_n}\}$ where κ_n is an invariant inner product on each simple finite-dimensional Lie algebra $\mathfrak{g}_n$ as in Lemma* 6.1(3) *above.*

Proof. Recall from Lemma 6.1(3) we have $\mathfrak{g} = \mathrm{cl}(\bigcup \mathfrak{g}_n)$ with each $\mathfrak{g}_n$ a simple finite-dimensional Lie algebra. Let $\mathbb{C}((t)) = \mathbb{C}[t, t^{-1}]$ denote the algebra of Laurent polynomials in t, and let κ_n denote an invariant inner product on $\mathfrak{g}_n$. Following [38, §7.1] (see also [27, §1]), the affine Kac–Moody (Lie) algebra $\widehat{\mathfrak{g}}_{\kappa_n}$ is the one-dimensional central extension of $\mathfrak{g}_n \otimes \mathbb{C}((t))$ with

$$[\phi_1 \otimes f(t), \phi_2 \otimes g(t)] = [\phi_1, \phi_2] \otimes f(t)g(t) - (\kappa_n(\phi_1, \phi_2)\mathrm{Res}\, f dg) K_n, \tag{6.5}$$

where $\phi_1, \phi_2 \in \mathfrak{g}_n$, $f, g \in \mathbb{C}((t))$, and K_n denotes the central element. The fact that the sequence $\{\widehat{\mathfrak{g}}_{\kappa_n}\}$ is increasing, is clear, since because $\mathfrak{g}_n \subset \mathfrak{g}_{n+1}$, we have a naturally induced sequence $\mathfrak{g}_n \otimes \mathbb{C}((t)) \subset \mathfrak{g}_{n+1} \otimes \mathbb{C}((t))$. □

6.3. A representation of the Weyl group and τ-functions

Here we use the root lattice (for roots ν) and refer to [53, 68](see Appendix §A.2). Following [53, 68], we give a short outline of a representation-theoretic definition of the τ-function. Firstly, for each generalized Cartan matrix $C = [c_{ij}]$ of affine type, there exists a representation of the Weyl group $\mathcal{W} = \mathcal{W}(C)$ on a field $\mathbb{C}(\nu; \theta; \tau)$ of rational functions with respect to infinitely many variables ν_i, θ_i, τ_i $(i \in \mathcal{I})$,

$$\varrho : \mathcal{W} \longrightarrow \mathbb{C}(\nu; \theta; \tau). \tag{6.6}$$

This representation is characterized by the action of the generators s_i (see §A.2), such that whenever defined,

$$\begin{aligned} s_i(\nu_j) &= \nu_j - \nu_i c_{ij} \text{ (reflections)}, \ s_i(\theta_j) = \theta_j + \frac{\nu_i}{\theta_i} u_{ij}, \\ s_i(\tau_j) &= \tau_j(\theta_j \ \Pi_{k \in \mathcal{I}} \ \tau_k^{-c_{kj}})^{\delta_{ij}}, \end{aligned} \tag{6.7}$$

where the $u_{ij}(i,j \in \mathcal{I})$ satisfy certain conditions as seen in [53, §2] and [68, §1]. The τ_j are realizable as τ-functions for the variables θ_j, and in this representation the τ_j are seen to correspond to the *fundamental weights* Ξ_j, while the θ-variables correspond to the simple roots ν_i [53, §2]. The motivation for regarding the τ_j as τ-functions is commented upon in Remark 6.3 below.

Proposition 6.3. *Consider the (complex) simple L^*-algebra $\mathfrak{g} = \mathfrak{g}_{\mathbf{A}}$. Then there exists a representation $\mathfrak{g} \longrightarrow \mathbb{C}(\nu;\theta;\tau)$ relative to which a family of Fredholm determinants of an (integral) operator $\mathsf{T} \in \mathfrak{g}$ is assigned to a family of τ-functions of the above type.*

Proof. Firstly, we recall from Lemma 6.1(3) the sequence of finite-dimensional simple Lie algebras $\mathfrak{g}_j$. Commencing from a Hilbert–Schmidt (integral) operator $\mathsf{T} \in \mathfrak{g}$, we proceed to the Fredholm determinant $\mathrm{Det}(1 - \lambda\mathsf{T}|\mathfrak{g}_j)$. The latter as a generalized determinant is assigned to each τ_j, and hence to the corresponding weight Ξ_j.

For each $\mathfrak{g}_j$, we have by Lemma 6.2 an associated increasing sequence of affine Kac–Moody algebras $\{\widehat{\mathfrak{g}}_{\kappa_j}\}$. The affine Weyl group $\mathcal{W}(C_n)$ acts upon the root system of each $\widehat{\mathfrak{g}}_{\kappa_j}$ (for each j) as well as on the rational function field $\mathbb{C}(\nu;\theta;\tau)$ [53, §2]. Now since $\mathbb{C}(\nu;\theta;\tau)$ depends on the sum $\sum_j \Xi_j$ of fundamental weights, we thus arrive at a representation

$$\begin{aligned} \mathfrak{g} &\longrightarrow \mathbb{C}(\nu;\theta;\tau), \\ \mathrm{Det}(1 - \lambda\mathsf{T}|\mathfrak{g}_j) &\mapsto (\Xi_j \leftrightarrow \tau_j), \end{aligned} \tag{6.8}$$

which yields the desired result. □

On referring to Appendix §A.2, we introduce the dual $\mathbb{Z}$-module of the coroot lattice $R^\vee$ denoted in (A.4) by $R^\star = \mathrm{Hom}_{\mathbb{Z}}(R^\vee, \mathbb{Z})$. As shown in [53, §2], any family $\{\phi_w(\lambda)\}$ $(w \in \mathcal{W}, \lambda \in R^\star)$ can be identified with a mapping

$$\begin{aligned} \phi : \mathcal{W} &\longrightarrow \mathrm{Hom}_{\mathbb{Z}}(R^\star, \mathbb{C}(\nu;\theta)^*), \\ w &\mapsto \phi_w, \end{aligned} \tag{6.9}$$

where $\mathbb{C}(\nu;\theta)^*$ denotes the multiplicative group of $\mathbb{C}(\nu;\theta)$ regarded as a $\mathbb{Z}$-module, and for which

$$\phi_{w_1 w_2}(\lambda) = w_1(\phi_{w_2}(\lambda))\phi_{w_1}(w_2 \cdot \lambda), \ \forall w_1, w_2, \in \mathcal{W}, \text{ and } \lambda \in R^\star. \tag{6.10}$$

The map ϕ in (6.9) now viewed as a (linear) map $\phi : \mathbb{C}[\mathcal{W}] \longrightarrow M$ on the group algebra, where $M = \mathrm{Hom}_{\mathbb{Z}}(R^\star, \mathbb{C}(\nu;\theta)^*)$, is shown in [53, §2] to define a Hochschild 1-cocycle of $\mathbb{C}[\mathcal{W}]$ with respect to the natural $\mathcal{W}$-bimodule structure of M (we refer to [44, §1-§4] for the basic theory of Hochschild complexes). Since it can be shown that $w(\tau^\lambda) = \phi_w(\lambda)\tau^{w\cdot\lambda}$ $(w \in \mathcal{W}, \lambda \in R^\star)$, then the cocycle induced by ϕ in (6.10) becomes the coboundary of the 0-cochain

$$\begin{aligned} &\tau \in \mathrm{Hom}_{\mathbb{Z}}(R^\star, \mathbb{C}(\nu;\theta;\tau)^*), \\ &\lambda \mapsto \tau^\lambda, \end{aligned} \tag{6.11}$$

on extending the $\mathcal{W}$-module $\mathbb{C}(\nu;\theta)$ to $\mathbb{C}(\nu;\theta;\tau)$. It is in this way that the τ-functions are seen to trivialize the Hochschild 1-cocycle as defined by these variables.

Remark 6.3. As outlined in [53, 68], triples $(\nu_j, \theta_j, \tau_j)$ can be seen to lead to a family of discrete dynamical systems classified by certain types denoted $\mathbf{A}_\ell^{(1)}, \ldots, \mathbf{D}_\ell^{(1)}$, for various ℓ, whose corresponding affine Weyl groups are realized as Bäcklund transformations of the Painlevé equations of class $P_{\mathrm{II}}, \ldots, P_{\mathrm{VI}}$ (see Appendix B). Here the pairs (ν_j, θ_j) play the role of discrete time dependent variables and the τ_j are τ-functions associated to the corresponding Painlevé types.

7. Applications: Schlesinger systems, isomonodromic transformations and meromorphic connections

7.1. Holomorphic maps to $\Lambda = \mathrm{Sim}(p, A)$

We return now to the case where $A = \mathcal{L}(H_{\mathcal{A}})$ in §5.2, and commence by considering holomorphic maps $\hat{f} : X \longrightarrow \mathrm{Gr}(p, A)$, where X in this section denotes a compact Riemann surface of genus g_X.

Proposition 7.1. *Given a (non-constant) holomorphic map $\hat{f} : X \longrightarrow \mathrm{Gr}(p, A)$, then $\hat{f}$ can be extended to a holomorphic map $f : X \longrightarrow \Lambda$.*

Proof. For each $x \in X$, let us set $\hat{f}(x) = (\mathsf{K}_+)_x$, and then let $(\mathsf{K}_-)_x$ be a closed complemented subspace for $(\mathsf{K}_+)_x$ in $H_{\mathcal{A}}$, so that $(\mathsf{K}_+)_x \cap (\mathsf{K}_-)_x = \{0\}$. Thus we have produced a polarizing pair $((\mathsf{K}_+)_x, (\mathsf{K}_-)_x)$ that depends on $x \in X$, and hence f extends to the space $\mathfrak{P}$ of polarizations in §5.2. Following from [23, Theorem 4.1 (3)], we next make use of the analytic diffeomorphism $\varphi : \mathfrak{P} \xrightarrow{\cong} \Lambda$, and then finally, the desired holomorphic map $f : X \longrightarrow \Lambda$ is taken to be the composition $f = \varphi \circ \hat{f}$. □

The Krichever correspondence which is based on certain holomorphic data as described in [63, §6] provides a prototypical example, namely, a holomorphic embedding $\hat{f} : X \longrightarrow \mathrm{Gr}(p, A)$ (as applied in [24, 25]). The point of extending such a map to Λ is that the subsequent calculations as carried out in [23] tend to be relatively straightforward, and the relevant analytic objects defined over Λ can be pulled-back via f. For instance, recall the setting of §2.1 and consider the possible pull-back vector bundles with connection under $f : X \longrightarrow \Lambda$:

(i) $(f^*\gamma, {}^*\nabla_\gamma) \longrightarrow X$
(ii) $(f^*\mathrm{Det}_\gamma, f^*\nabla_{\mathrm{Det}_\gamma}) \longrightarrow X$

(iii) More generally, consider $(V, \nabla_V) \longrightarrow X$, with structure group $G \subseteq G(pAp)$ and corresponding connection form ω_V as pulled back under f from a vector bundle with connection $(\mathbf{V}, \nabla_{\mathbf{V}}) \longrightarrow \Lambda$, with corresponding connection form denoted $\omega_{\mathbf{V}}$ (typically associated to a principal G-bundle $Q \longrightarrow \Lambda$ as in (2.5) with connection 1-form ω_Q).

Next we take an integral operator T that corresponds to the $G \subseteq G(pAp)$-valued $\mathcal{T}$-function for Q as described in §5.1 for which, via a holomorphic embedding $f : X \longrightarrow \Lambda$, the Fredholm determinant

$$\mathrm{Det}(1 - \lambda \mathsf{T}|_{\mathfrak{J}}), \tag{7.1}$$

is suitably supported on a countable number of points, or on a union of curve segments $\mathfrak{J}$ in the (complex) ζ-plane supporting T in X. In the context of Example 4.4, the determinant (7.1) provides the τ-function of an isomonodromic family of meromorphic covariant derivative operators D_ζ.

Example 7.1. Specifically from [33] (for $g_X = 1$), the τ-functions in question are of the form $\tau = \tau(a_1, \ldots, a_r; b_1, \ldots, b_n)$, where the $a_1, \ldots, a_r$ are 'asymptotic elements' and the $b_1, \ldots, b_n$ are 'pole locations' collectively parametrizing τ for the given pair (r, n). Moreover, as shown in [33, Theorem 2.6], τ is effectively the Segal–Wilson τ_W-function in [63, §3]. In this case there is the 1-form

$$\omega = d \ln \tau = \sum_{i=1}^{r} K_i \, da_i + \sum_{j=1}^{n} H_j \, db_j, \tag{7.2}$$

where the pairs (H_i, K_j) are Poisson-commuting Hamiltonians. In [25, §4.2] it was shown that a class of τ-functions, denoted τ_Λ, are essentially the same as the τ_W, and are linked via pullback. Each of the τ_Λ serves as a logarithmic potential for the curvature of the connection $\nabla_{\mathrm{Det}_\gamma}$ seen above, and so (7.2) can likewise be interpreted for $\omega = d \ln \tau_\Lambda$.

The further significance of relations of the type (7.2) has been pointed out in [46], where ω on the one hand represents the poles of solutions to the Schlesinger equation, and on the other hand, it is the Hamiltonian of this equation with respect to a natural Poisson structure. The task undertaken in [46] was to give a detailed explanation of the equivalence of these two approaches.

7.2. Operator meromorphic connection

Next we consider how a *meromorphic connection* denoted $\omega_{\mathbf{V}}$ can be constructed on a complex vector bundle $\mathbf{V} \longrightarrow \Lambda$, by commencing from a *meromorphic operator* ($\mathfrak{g}$-*valued*) *function.* Operator-valued meromorphic functions as providing suitable potentials, denoted say $B(\zeta)$ with respect to a local coordinate ζ on some domain in X, have been studied to a significant extent in, e.g., [30, 31, 47]. Typically $B(\zeta)$ is of the form

$$B(\zeta) = \sum_{j=-n}^{\infty} (\zeta - \zeta_0)^j B_j, \tag{7.3}$$

satisfying some mild technical condition (such as 'normality' [30, §3]).

Recall from Remark 2.2 the connection map $\mathcal{V} : TQ \longrightarrow TQ$ used in constructing the connection 1-form ω_Q. To say that $\mathcal{V}$ is *meromorphic* means that $\mathcal{V}$ is pointwise an analytic map with a countable number of poles. The resulting connection ω_Q is then said to be *a meromorphic connection* if, in its local representation, it contains the data of the Laurent coefficients as exhibited in (7.3). This property passes over to the induced connection 1-form $\omega_{\mathbf{V}}$ on the associated complex vector bundle $\mathbf{V} \longrightarrow \Lambda$ (see below).

Granted a supply of holomorphic maps from X to $\mathrm{Gr}(p, A)$, and thus to Λ by Proposition 7.1, we give an application in the following. To do this, let us first recall that $\gamma \longrightarrow \Lambda$ is the vector bundle associated to a principal G-bundle $Q \longrightarrow \Lambda$ as in §2.1 (where $G \subseteq G(pAp)$). Let $\mathcal{V} : TQ \longrightarrow TQ$ be a meromorphic connection map on the principal bundle $Q \longrightarrow \Lambda$. In the usual way, this induces a connection map, denoted the same, $\mathcal{V} : T\gamma \longrightarrow T\gamma$ on the associated vector bundle (the formal details for doing this can be seen in, e.g., [43, p. 381]), and hence we obtain a meromorphic connection on the holomorphic vector bundle $\gamma \longrightarrow \nabla_\gamma$.

When 'Det' is well defined under the criteria we had discussed earlier, then we obtain an induced connection map $\mathcal{V}_{Det} : T\mathrm{Det}_\gamma \longrightarrow T\mathrm{Det}_\gamma$ yielding a local (operator) connection meromorphic 1-form $\widehat{\omega}$. Thus as an immediate consequence of the usual pull-back construction, if $f : X \longrightarrow \Lambda$ is a non-constant holomorphic map, and $\widehat{\omega}$ is a meromorphic connection 1-form on Det_γ, then on X we obtain a holomorphic vector bundle $(f^*\mathrm{Det}_\gamma, f^*\widehat{\omega}) \longrightarrow X$ with a meromorphic connection.

Remark 7.1. Following, e.g., [66], one may look at 'free energy pre-potentials' in the context of matrix models and topological conformal field theories where the relevant object to consider is $\ln \tau$. Indeed, as we have pointed out in §4.1, it has been known that $\ln \tau$ is a 'prepotential' (for a local connection 1-form, say) as apparent in [10, 36, 37, 42, 46, 56, 66]. Thus somewhat in the spirit of how one applies monodromy preserving transformations [33, 36, 66], we are led from the matrix case to operator-valued meromorphic functions comprised of Laurent coefficients (such as $B(\zeta)$ above) and hence towards a class of distinguished meromorphic (connection) 1-forms

$$\omega = d \ln \tau = \sum_{n=-\infty}^{\infty} \mathrm{Tr}\; B_n (\zeta - \zeta_0)^n \; dt_n. \tag{7.4}$$

7.3. Examples

Given that a meromorphic 1-form in the operator setting of Λ can be pulled back via a holomorphic map from X to $\mathrm{Gr}(p, A)$, or to Λ using Proposition 7.1, let us see some motivation for doing this in the operator context as provided by the following examples of Schlesinger systems relative to X (cf. [39] for matrix operator coefficients). One should keep in mind the role of the τ-function, and the setting of the determinant line bundles with meromorphic connection $(f^*\mathrm{Det}_\gamma, f^*\widehat{\omega}) \longrightarrow X$, as we have described above.

Example 7.2. Inspired by the setting of the Riemann–Hilbert problem, let us consider an equation of the form

$$\frac{d\Psi}{d\zeta}(\zeta) = B(\zeta)\Psi(\zeta), \tag{7.5}$$

where $\Psi(\zeta)$ is a meromorphic G-valued function, and $B(\zeta)$ in (7.3) has poles $\{b_1, b_2, \ldots, b_n\}$ that are viewed as variables, and $\{b_1, b_2, \ldots, b_n, \infty\}$ branch points for $B(\zeta)$ when viewed as points of X. Granted Ψ can be continued along a closed path γ, away from the branch points within this embedding, we consider a transformation of the type

$$\Psi(\zeta) \longrightarrow \Psi(\zeta)Z_\gamma, \tag{7.6}$$

where $Z_\gamma \in G \subseteq G(pAp)$ is a (constant) invertible operator depending only on the homotopy class $[\gamma]$ of γ. In this way Z_γ induces a monodromy representation of the fundamental group

$$\pi_1(X/\{b_1, b_2, \ldots, b_n\}) \longrightarrow G, \text{ where } [\gamma] \mapsto Z_\gamma. \tag{7.7}$$

We recall that *a Schlesinger transformation* can be regarded as a discrete monodromy preserving transformation of a meromorphic connection matrix that shifts by elements of $\mathbb{Z}$, the eigenvalues of its residues [11, 42]. A modification of this theory to matrix operator coefficients of differential equations of Fuchsian type is studied in [39], and as was mentioned in the Introduction, motivates a broader scope of applications using operator-valued functions.

Example 7.3. For $g_X = 0$ and $G \cong \mathrm{SL}(2, \mathbb{C})$, the corresponding Schlesinger system satisfies [42, §II]

$$\begin{aligned} \frac{\partial B_i}{\partial \zeta_j} &= \frac{[B_i, B_j]}{\zeta_i - \zeta_j}, \text{ for } i \neq j, \\ \frac{\partial B_i}{\partial \zeta_i} &= -\sum_{j \neq i} \frac{[B_i, B_j]}{\zeta_i - \zeta_j}, \end{aligned} \tag{7.8}$$

where the $B_j = B_j(b_1, b_2, \ldots, b_n)$ are taken to be certain meromorphic functions (cf. [36, 37, 39, 46]). Here the role of the meromorphic connection is prominent in [46]. One starts with a holomorphically trivial vector bundle $(\mathbf{V}_0, \nabla_{\mathbf{V}_0}) \longrightarrow \mathbb{C}P^1$, with meromorphic connection having logarithmic poles at points $(b_1^0, \ldots, b_n^0, \infty)$. This can be deformed isomonodromically to a holomorphic vector bundle with meromorphic connection $(\mathbf{V}, \nabla_{\mathbf{V}}) \longrightarrow \mathbb{C}P^1 \times \mathcal{D}$, where $\mathcal{D}$ is certain deformation space, and for which the restriction $(\mathbf{V}, \nabla_{\mathbf{V}})|(z, b_1^0, \ldots, b_n^0, \infty) \cong (\mathbf{V}_0, \nabla_{\mathbf{V}_0})$. The integrability of this isomonodromic extension leads to the Schlesinger equation:

$$dB_i = -\sum_{i \neq j} \frac{[B_i, B_j]\; d(b_i - b_j)}{(b_i - b_j)}. \tag{7.9}$$

A class of meromorphic connections also appears in [10] which considers a similar deformation problem.

Example 7.4. There is a class of wave functions $\Psi = \Psi(\zeta)$, defined relative to a spectral parameter ζ, whose monodromy data is independent of the deformation parameter if and only if Ψ satisfies the deformation equation (cf. (7.5))

$$\partial_\zeta \Psi = dB(\zeta)\Psi, \tag{7.10}$$

where $dB(\zeta) = \sum_i B_i d\zeta_i$ is a 1-form with rational coefficients $B_i = B_i(\zeta)$. Now to any solution of (7.10) there is an associated 1-form

$$\omega = -\sum_k \operatorname{Res}_{\zeta=\zeta_k} \operatorname{Tr}[g^{(k)-1}\partial_\zeta g^{(k)} d\xi^{(k)}]d\zeta, \tag{7.11}$$

for certain analytic functions $g = g(\zeta)$ and diagonal matrices ξ [3, §8.4], where the deformation equation (7.10) implies that $d\omega = 0$. In fact, we have $\omega = d\ln\tau$, and hence for each solution of (7.10) a τ-function can be associated and one that is transformable under an elementary Schlesinger transformation [3, §8.6].

Example 7.5. Following, e.g., [11, §6] and [36, 42], for the case $g_X = 1$, we have an $\mathfrak{sl}(2,\mathbb{C})$-valued 1-form ω given by

$$\omega = \sum_{j=1}^{k} \sum_{\substack{1\le \ell \le n \\ \ell \neq j}} \frac{\operatorname{Tr}(B_j B_k)}{b_j - b_\ell}\, db_\ell. \tag{7.12}$$

In this case $\tau = \tau(b_1, b_2, \ldots, b_k)$ is a τ-function for the system such as appears in (7.8) if $d\ln\tau = \omega$ (see [11, 36]), so τ is defined at least locally. With regards to the Painlevé property, any solution $\{B_j\}_{j=1}^n$ of the Schlesinger equations are analytic functions in $(b_1, b_2, \ldots, b_n)$ that have at most poles in addition to the fixed singularities $b_j = b_\ell$, for some $j \neq \ell$ [11] (see also [40]).

Appendix A. Briefly Kac–Moody–Lie algebras

A.1. The basic definitions

The $\ell \times \ell$ Cartan matrix $C = [c_{ij}]$ $(i, j \in \mathcal{I})$ of any semisimple Lie algebra $\mathfrak{g}_{\mathbb{C}}$ satisfies the conditions

(1) $c_{ij} \in \mathbb{Z}$, for all i, j;
(2) $c_{ii} = 2$, for all i;
(3) $c_{ij} \le 0$ if $i \neq j$;
(4) $c_{ij} = 0$ whenever $c_{ji} = 0$;
(5) The matrix C is positive definite in the sense that all of the principal minors of C are positive.

Conversely, if we have an $\ell \times \ell$ matrix C satisfying (1)–(4), the Cartan structural relations of $\mathfrak{g}_{\mathbb{C}}$ define an abstract complex Lie algebra $\mathfrak{g}'$ called *the Kac–Moody–Lie algebra defined by C*. If in addition C satisfies (5), then $\mathfrak{g}'$ will be finite-dimensional and also semisimple. But if (5) does not hold, then $\mathfrak{g}'$ will be infinite-dimensional. There is a way of modifying this latter case so that much of the finite-dimensional theory can apply directly. We refer to [38, Chap. 1] and [57, §5.3] for details and retain the notation $\mathfrak{g}'$ once this modification has been done.

Let $\mathfrak{L} = \mathbb{C}((t))(= \mathbb{C}[t, t^{-1}])$ denote the algebra of Laurent polynomials in t. Following [38, §7.1], we have the loop algebra $\mathfrak{L}(\mathfrak{g}') = \mathfrak{L} \otimes_{\mathbb{C}} \mathfrak{g}'$. The central extension $\widetilde{\mathfrak{L}}\mathfrak{g}'$ of $\mathfrak{L}\mathfrak{g}'$ satisfies (1)–(4), and (5)$'$: det $C = 0$, and all the proper principal minors of C are positive. Conversely, the Kac–Moody–Lie algebras corresponding to Cartan matrices following (1)–(4) and (5)$'$ are called *affine Kac–Moody–Lie algebras* (see in particular [38, Theorem 7.4]).

Remark A.1. We could assume that C is *locally finite*, meaning that for each $j \in \mathbb{Z}$, we have $c_{ij} = 0$, except for a finite number of the i's.

A.2. The Weyl–Coxeter group

With respect to a collection of roots $\{\nu_j\}$, the *root lattice* $R = R(C)$ and the *co-root lattice* $R^\vee = R^\vee(C)$ are defined by

$$R = \bigoplus_{j \in \mathcal{I}} \mathbb{Z}\nu_j, \text{ and } R^\vee = \bigoplus_{j \in \mathcal{I}} \mathbb{Z}\nu_j^\vee, \tag{A.1}$$

respectively, together with the pairing $\langle , \rangle : R^\vee \times R \longrightarrow \mathbb{Z}$. Then let $\mathcal{W} = \mathcal{W}(C)$ be the *Weyl–Coxeter group* as defined by generators $s_i (i \in \mathcal{I})$ satisfying the relations

$$s_i^2 = 1, \ (s_i s_j)^{m_{ij}} = 1, \ (i, j \in \mathcal{I}, i \neq j), \tag{A.2}$$

where $m_{ij} = 2, 3, 4, 6$ or ∞, according to whether $c_{ij}c_{ji} = 0, 1, 2, 3$ or ≥ 4, respectively. As is well known in the theory of Coxeter groups, the generators s_i act naturally on the root lattice R by reflections

$$s_i(\nu_j) = \nu_j - \nu_i < \nu_i^\vee, \nu_j > = \nu_j - \nu_i c_{ij}, \tag{A.3}$$

(see, e.g., [38, Chap. 6], and in particular [53, §2] in relationship to §6.3). Let

$$R^\star = \mathrm{Hom}_{\mathbb{Z}}(R^\vee, \mathbb{Z}), \tag{A.4}$$

denote the dual $\mathbb{Z}$-module of the coroot lattice $R^\vee$. Taking the dual basis $\{\Lambda_j\}_{j \in \mathcal{I}}$ of $\{\nu_j^\vee\}$ so that $R^\star = \oplus_{j \in \mathcal{I}} \mathbb{Z}\Lambda_j$, there exists a natural $\mathcal{W}$-homomorphism $R \longrightarrow R^\star$, such that $\nu_j \longrightarrow \sum_{i \in \mathcal{I}} \Lambda_i c_{ij}$ [53, §2].

Appendix B. The Painlevé equations

This subject has a rich and illustrious history as finely surveyed in the monograph [15](see also [14]) to which we refer the reader who wishes pursue further details as well as the interesting historical background.

Painlevé studied second-order nonlinear ODEs of the form

$$y'' = F(y, y', z), \ y = y(z), \tag{B.1}$$

where $F(y', y, z)$ is rational in y and y', and analytic in z. The problem (originally posed by E. Picard) was to identify all equations of the form (B.1) for which the solutions have "no movable critical points", i.e., the locations of any branch points or essential singularities do not depend on the constants of integration of

(B.1). Painlevé studied an assortment of 50 'canonical equations', up to Möbius transformations, whose solutions have no movable critical points and then reduced the study to six particular types (P_I–P_{VI}, the solutions to which are often called *the Painlevé transcendents*) [15] and [14, §1]:

P_I $y'' = 6y^2 + z$

P_{II} $y'' = 2y^3 + zy + \alpha$

P_{III} $y'' = \frac{(y')^2}{y} - \frac{y'}{z} + \frac{\alpha y^2+\beta}{z} + \gamma y^3 + \frac{\delta}{y}$

P_{IV} $y'' = \frac{(y')^2}{2y} + \frac{3}{2}y^3 + 4zy^2 + 2(z^2-\alpha)y + \frac{\beta}{y}$

P_V $y'' = \left(\frac{1}{2y} + \frac{1}{y-1}\right)(y')^2 - \frac{y'}{z} + \frac{(y-1)^2}{z^2}\left(\alpha y + \frac{\beta}{y}\right) + \frac{\gamma y}{z} + \frac{\delta y(y+1)}{y-1}$

P_{VI} $y'' = \frac{1}{2}\left(\frac{1}{y} + \frac{1}{y-1} + \frac{1}{y-z}\right)(y')^2 - \left(\frac{1}{z} + \frac{1}{z-1} + \frac{1}{y-z}\right)y'$
$+ \frac{y(y-1)(y-z)}{z^2(z-1)^2}\left\{\alpha + \frac{\beta z}{y^2} + \frac{\gamma(z-1)}{(y-1)^2} + \frac{\delta z(z-1)}{(y-z)^2}\right\}$

where the $\alpha, \dots, \delta$ are constants. The solutions are aptly named 'transcendents' because they cannot be expressed in terms of traditional special functions.

Example B.1. Following [54], we exemplify properties of P_I. Let $y = y(z)$ be a solution of

$$y'' = 6y^2 + z. \tag{B.2}$$

The function y is meromorphic on $\mathbb{C}$, and there is a function $\tau = \tau(z)$ holomorphic on $\mathbb{C}$ such that

$$y(z) = -\left(\frac{d}{dz}\right)^2 \ln \tau = \frac{(\tau')^2 - \tau\tau''}{\tau^2}. \tag{B.3}$$

Setting $\eta = \frac{\tau'}{\tau}$, then η is a solution of a third-order ODE in τ:

$$\eta'' - 4(\eta')^3 - 2z - 2\eta = 0. \tag{B.4}$$

Consider the polynomial in y and μ:

$$H_I(z; y, \mu) = \frac{1}{2}\mu^2 - 2y^3 - zy. \tag{B.5}$$

Equation (B.2) is equivalent to a Hamiltonian system:

$$\begin{cases} y' &= \frac{\partial H}{\partial \mu} \\ \mu' &= -\frac{\partial H}{\partial y} \end{cases} \tag{B.6}$$

In fact, each type of Painlevé equation can be written as a system such as (B.6) [54]. If $(y(z), \mu(z))$ is a solution to (B.6), then

$$\tau(z) = \exp\left[\int H(z)\, dz\right], \tag{B.7}$$

where $H(z) = H_I(z; y(z), \mu(z))$.

Example B.2. A modified version of the Korteweg–de Vries equation is

$$u_t - 6u^2 u_x + u_{xxx} = 0. \tag{B.8}$$

As pointed out in [14], (B.8) can be re-scaled by setting $u(x,t) = (3t)^{-\frac{1}{3}}y(z)$ where $z = x(3t)^{-\frac{1}{3}}$, and this is solvable via an inverse scattering transform, where $y(z)$ satisfies the equation P_{II}: $y'' = 2y^3 + zy + \alpha$.

In applications of the Painlevé equations to integrable systems it is usually the case that for each type PI-PVI there is a companion τ-function playing a significant role, as Example B.1 reveals. Likewise, this special relationship shows up in other areas such as plasma physics, quantum optics, general relativity and quantum gravity (cf. §4.1).

Acknowledgment

I wish to thank colleagues Maurice Dupré and Emma Previato, and several of the participants at IWOTA 2011 for their various comments concerning this topic. My gratitude is extended to the referees and editor who provided useful suggestions towards improving the final version.

References

[1] M. Ablowitz and P. Clarkson, *Solitons, Nonlinear Evolution Equations and Inverse Scattering.* London Math. Soc. Lecture Notes **149**, Cambridge University Press, Cambridge, 1991.

[2] B. Aupetit and H. du T. Mouton, Trace and determinant in Banach algebras. *Studia Math.* **121** (2) (1996), 115–136.

[3] O. Babelon, D. Bernard and M. Talon, *Introduction to Classical Integrable Systems.* Cambridge University Press, Cambridge, 2003.

[4] J. Ball and V. Vinnikov, Zero-pole interpolation for matrix meromorphic functions on a compact Riemann surface and a matrix Fay trisecant identity, *Amer. J. Math.* **121** (1999), no. 4, 841–888.

[5] D. Beltiţă, *Smooth Homogeneous Structures in Operator Theory.* Monographs and Surveys in Pure and Appl. Math. **137**, Chapman and Hall/CRC, Boca Raton Fl. 2006.

[6] D. Beltiţă and J. Galé, Holomorphic geometric models for representations of $C*$-algebras. *J. Functional Anal.* **255** (10) (2008), 2888–2932.

[7] D. Beltiţă and J. Galé, On complex infinite-dimensional Grassmann manifolds. *Complex Anal. Oper. Theory* **3** (2009), 739–758.

[8] D. Ben-Zvi and I. Biswas, Theta functions and Szegő kernels. *Int. Math. Res. Notes* **24** (2003), 1305–1340.

[9] M. Bertola and M. Gekhtman: Biorthogonal Laurent polynomials, Toeplitz determinants, minimal Toda orbits and isomonodromic tau functions. *Constr. Approx.* **26** (2007), no. 3, 383–430.

[10] Y.P. Bibilo and R.R. Gontsov, On the Malgrange isomonodromic deformations of non-resonant meromorphic (2×2) connections. http://arxiv.org/abs/1201.0181v1 [math.CA]

[11] A. Borodin and P. Deift, Fredholm determinants, Jimbo-Miwa-Ueno τ-functions, and representation theory. *Comm. Pure Appl. Math.* **55** (2002), no. 9, 1160–1230.

[12] A. Borodin and G. Olshanski, Harmonic analysis on the infinite-dimensional unitary group and determinantal point processes. *Ann. of Math.*(2) **161** (2005), no. 3, 1319–1422.

[13] A.J. Calderón Martín and F. Forero Piulestán, Roots and root spaces of compact Banach–Lie algebras. *Irish Math. Soc. Bulletin* **49** (2002), 15–22.

[14] P.A. Clarkson, Painlevé equations – nonlinear special functions. Proceedings of the 6th International Symposium on Orthogonal Polynomials, Special Functions and their Applications (Rome 2001), *J. Comput. Appl. Math.* **153**, (2003), no. 1-2, 127–140.

[15] R. Conte and M. Musette, *The Painlevé handbook.* Springer-Verlag, Dordrecht, 2008.

[16] A. Devinatz and M. Shibrot, General Wiener–Hopf operators. *Trans. Amer. Math. Soc.* **145** (1969), 467–494.

[17] L.A. Dickey, Another example of a τ-function. In (J. Harnad and J.E. Marsden, eds.) *Proceedings of the CRM Workshop on Hamiltonian Systems, Transformation Groups and Spectral Transform Methods* (1989*: Montréal, Québec*) (pp. 39–44). Montréal, QC, Canada: Centre de Recherches Mathématiques, 1990.

[18] R.G. Douglas, *Banach Algebra Techniques in Operator Theory* (2nd Edition). Graduate Texts in Mathematics **179**, Springer-Verlag, New York–Berlin, 1998.

[19] R.G. Douglas and C. Pearcy, Spectral theory of generalized Toeplitz operators. *Trans. Amer. Math. Soc.* **115** (1965), 443–444.

[20] M.J. Dupré and J.F. Glazebrook, Infinite dimensional manifold structures on principal bundles. *J. of Lie Theory* **10** (2000), 359–373.

[21] M.J. Dupré and J.F. Glazebrook, The Stiefel bundle of a Banach algebra. *Integral Equations Operator Theory* **41** (3) (2001), 264–287.

[22] M.J. Dupré, J.F. Glazebrook and E. Previato, A Banach algebra version of the Sato Grassmannian and commutative rings of differential operators. *Acta Applicandae Math.* **92** (3) (2006), 241–267.

[23] M.J. Dupré, J.F. Glazebrook and E. Previato, The curvature of universal bundles of Banach algebras. In, 'Proceedings of the International Workshop on Operator Theory and Applications – 2008', eds. J. Ball et al., *Operator Theory: Advances and Applications* **202** (2010), 195–222, Birkhäuser-Verlag, Basel-Boston-Berlin.

[24] M.J. Dupré, J.F. Glazebrook and E. Previato, Differential algebras with Banach-algebra coefficients I: From C*-algebras to the K-theory of the spectral curve. *Complex Anal. Oper. Theory.* (DOI) 10.1007/s11785-012-0221-2

[25] M.J. Dupré, J.F. Glazebrook and E. Previato, Differential algebras with Banach-algebra coefficients II, The operator cross-ratio Tau function and the Schwarzian derivative. *Complex Anal. Oper. Theory.* (DOI) 10.1007/s11785-012-0219-9

[26] J. Fay, *Theta functions on Riemann surfaces.* Lect. Notes in Math. **352**, Springer-Verlag, Berlin 1973.

[27] E. Frenkel, Affine Kac–Moody algebras, integrable systems and their deformations ('Hermann Weyl Prize Lecture'). In (J.-P. Gazeau et al. eds.) *Group* 24*: Physical and mathematical aspects of symmetries – Proceedings of the Colloquium on Group Theoretical Methods in Physics* (24*th:* 2002*: Paris, France*), pp. 21–32, Institute of Physics Conference Series **173**, Bristol, UK, 2003.

[28] I. Gohberg, S. Goldberg and N. Krupnik, Traces and determinants of linear operators. *Operator Theory: Advances and Applications* **116** (2), Birkhäuser-Verlag, Basel-Boston-Berlin, 2000.

[29] I. Gohberg, S. Goldberg and M.A. Kaashoek, *Basic classes of linear operators.* Birkhäuser-Verlag, Basel, 2003.

[30] I. Gohberg and J. Leiterer, Factorization of operator functions with respect to a contour.I. Finitely meromorphic operator functions. (Russian) *Math. Nachr.* **72** (1972), 259–282.

[31] I.C. Gohberg and E.I. Sigal, An operator generalization of the logarithmic residue and Rouché's theorem. (Russian). *Math. Sb.* (*N.S.*) **84**(**126**) (1971), 607–629.

[32] P.R. Halmos and V.S. Sunder, Bounded integral operators on L^2-spaces. Ergebnisse der Mathematik und ihrer Grenzgebiete **96**, Springer-Verlag, Berlin-New York, 1978.

[33] J. Harnad and A.R. Its, Integrable Fredholm operators and dual isomonodromic deformations. *Commun. Math. Phys.* **226** (2002), 497–530.

[34] P. de la Harpe, *Classical Banach–Lie Algebras and Banach–Lie groups of operators in Hilbert space.* Lect. Notes in Math. **285**, Springer-Verlag, Berlin-New York, 1972.

[35] J.W. Helton, *Operator Theory, Analytic Functions, Matrices and Electrical Engineering.* CBMS Reg. Conf. Ser. **68**, Amer. Math. Soc., Providence RI, 1987.

[36] M. Jimbo, T. Miwa and K. Ueno, Monodromy preserving deformation of linear ordinary differential equations with rational coefficients I. General theory and τ-function. *Phys. D* **2** (1981), no. 2, 306–352.

[37] M. Jimbo, T. Miwa, Monodromy preserving deformation of linear ordinary differential equations with rational coefficients II. *Phys. D* **2** (1981), no. 3, 407–448.

[38] V.G. Kac, *Infinite Dimensional Lie Algebras.* 3rd edition, Cambridge University Press, Cambridge-New York, 1990.

[39] V. Katsnelson and D. Volok, Rational solutions of the Schlesinger system and isoprincipal deformations of rational matrix functions. I. 'Current trends in operator theory and its applications', 291–348. *Operator Theory Advances and Applications* **149**, Birkhäuser Verlag, Basel, 2004.

[40] A.V. Kitaev and D.A. Korotkin, On solutions of the Schlesinger equations in terms of Θ-functions. *Internat Math. Res. Notices* **17**, (1998), 877–905.

[41] S. Kobayashi and K. Nomizu, *Foundations of Differential Geometry I.* Interscience Tracts Pure and Appl. Math. **15**, John Wiley & Sons, 1963.

[42] D. Korotkin, N. Manojlović and H. Sambtleben, Schlesinger transformations for elliptic isomonodromic deformations. *J. Math. Phys.* **41** (2000), no. 5, 3125–3141.

[43] A. Kriegl and P.W. Michor, *The Convenient Setting of Global Analysis.* Math. Surveys and Monographs **53**, Amer. Math. Soc., 1997.

[44] J.L. Loday, *Cyclic Homology.* Grund. der math. Wissenschaften **301**, Springer-Verlag, Berlin-New York, 1998.

[45] J.D. McNeal, The Bergman projection as a singular integral operator. *The Journal of Geometric Analysis* **6**, no. 1, (1994), 91–103.

[46] B. Malgrange, Déformations isomonodromiques, forme de Liouville, fonction τ. *Ann. Inst. Fourier* (*Grenoble*) **54**, no. 5, 1371–1392.

[47] A.S. Markus and E.I. Sigal, The multiplicity of the characteristic of an analytic operator function. (Russian) *Mat. Issled.* **5** (1970), no. 3(17), 129–147.

[48] L.J. Mason, M.A. Singer and N.M.J. Woodhouse, Tau functions and the twistor theory of integrable systems. *J. Geom. and Phys.* **32** (2000), 397–430.

[49] A. Melnikov, Finite-dimensional Sturm–Liouville vessels and their tau functions. *Integral Equations Operator Theory* **71** (2011), 455–490.

[50] T. Miwa, On Hirota's difference equations. *Proc. Japan Acad. Ser. A Math. Sci.* **58** (1982), no. 1, 9–12.

[51] T. Nakazi and T. Yamamoto, Generalized Riesz projections and Toeplitz operators. *Math. Inequal. Appl.* **11** (2008), no. 3, 507–528.

[52] K.-H. Neeb, Infinite-dimensional groups and their representations. In (J.P. Anker and B. Orsted, eds.) *Lie Theory-Lie Algebras and Representations*, pp. 213–328. *Prog. in Math.* **228**, Birkhäuser-Verlag, Boston, 2004.

[53] M. Noumi and Y. Yamada, Affine Weyl groups, discrete dynamical systems and Painlevé equations. *Commun. Math. Phys.* **199** (1988), 281–295.

[54] K. Okamoto, On the τ-function of the Painlevé equations. *Physica D: Nonlinear Phenomena* **2**(3), 525–535.

[55] A. Pietsch, *Eigenvalues and s-numbers.* Cambridge Studies in Advanced Mathematics **13**. Cambridge University Press, Cambridge, 1987.

[56] J.F. Plebanski, Some solutions of complex Einstein equations. *J. Math. Phys.* **16** (1975), 2395–2404.

[57] A. Pressley and G. Segal, *Loop Groups and their Representations.* Oxford University Press, Oxford, 1986.

[58] A.K. Raina, Fay's trisecant identity and conformal field theory. *Commun. Math. Phys.* **122** (1989) (4), 625–641.

[59] M. Sato, The KP-hierarchy and infinite dimensional Grassmann manifolds. *Proc. Sympos. Pure Math.* **49** Part 1, pp. 51–66, *Amer. Math. Soc.*, 1989.

[60] M. Sato, T. Miwa and M. Jimbo, Aspects of holonomic quantum fields. Isomonodromic deformations and Ising model. In (D. Iagolnitzer, ed.) 'Complex analysis, microlocal calculus and relativistic quantum field theory (La Houches, 1979)'. *Lect. Notes in Physics* **126** (1980), Springer-Verlag, Berlin-Heidelberg-New York, 429–491.

[61] J.R. Schue, Hilbert space methods in the theory of Lie algebras. *Trans. Amer. Math. Soc.* **95** (1960), no. 1, 69–80.

[62] J.R. Schue, Cartan decompositions for L*-algebras. *Trans. Amer. Math. Soc.* **98** (1961), no. 1, 334–349.

[63] G. Segal and G. Wilson, Loop groups and equations of KdV type. *Publ. Math. IHES* **61** (1985), 5–65.

[64] B. Simon, *Trace Ideals and Their Applications.* London Math. Soc. Lecture Notes **35**, Cambridge University Press, Cambridge, 1979.

[65] J.F. Smith, The p-classes of a Hilbert module. *Proc. Amer. Math. Soc.* **36** (2) (1972), 428–434.

[66] K. Takasaki, Symmetries and tau function of higher-dimensional dispersionless integrable hierarchies. *J. Math. Phys.* **36** (1995), no. 7, 3574–3607.

[67] C.A. Tracy and H. Widom, Fredholm determinants, differential equations and matrix models. *Commun. Math. Phys.* **163** (1994), no. 1, 33–72.

[68] Y. Yamada, Determinant formulas for the τ-functions of the Painlevé equations of type A. *Nagoya Math. J.* **156** (1999), 123–134.

[69] H. Upmeier, *Toeplitz operators and Index Theory in several complex variables. Operator Theory: Advances and Applications* **81**, Birkhäuser Verlag, Basel – Boston – Berlin, 1996.

[70] M.I. Zelikin, Geometry of operator cross ratio. *Math. Sbornik* **197** (1) (2006), 39–54.

James F. Glazebrook
Department of Mathematics and Computer Science
Eastern Illinois University
600 Lincoln Avenue
Charleston, IL 61920–3099, USA

and

Adjunct Faculty
Department of Mathematics
University of Illinois at Urbana–Champaign
Urbana, IL 61801, USA
e-mail: jfglazebrook@eiu.edu

Operator Theory:
Advances and Applications, Vol. 236, 231–242

Tauberian Operators. Properties, Applications and Open Problems

Manuel González

To the memory of Israel Gohberg

Abstract. Tauberian operators have been useful in the study of many different topics of functional analysis. Here we describe some properties and the main applications of tauberian operators, and we point out several concrete problems that remain open.

Mathematics Subject Classification (2010). Primary 46B03, 47A53; Secondary 46B10, 47L20.

Keywords. Tauberian operator, factorization, operator ideal, Fredholm theory, Banach space.

1. Introduction

The concept of tauberian operator was introduced by Kalton and Wilansky [19] as a tool in the investigation of some questions in summability theory from an abstract point of view. We refer to [13, Chapter 1] for an historical account of the work that culminated with this concept.

Tauberian operators have been useful in the study of real interpolation theory of Banach spaces and operators [6], factorization of operators [18, 10], equivalence between the Kreĭn–Milman property and the Radon–Nikodým property [24], weak Calkin algebras of operators [5], embedding of dual separable Banach spaces in spaces with boundedly complete bases [8], refinements of James' characterization of reflexive Banach spaces [21], preservation of isomorphic properties of Banach spaces [20], convergence of bounded martingales [14], construction of hereditarily indecomposable Banach spaces [4], extension to operators of the principle of local reflexivity [7], etc.

Supported in part by MICINN (Spain), Grant MTM2010–20190.

In this paper we survey the main properties and applications of tauberian operators, and we describe some lines of research and some open problems. Since there is an up to date exposition of the theory of tauberian operators [13], we refer for details to this reference, including proper credits of the results, and we emphasize on questions that may deserve further study.

Throughout the paper, we denote by $\mathcal{L}(X, Y)$ the bounded operators between Banach spaces X and Y, $T^* \in \mathcal{L}(Y^*, X^*)$ is the conjugate operator of $T \in \mathcal{L}(X, Y)$, T^{**} is the second conjugate, and in the case $X = Y$ we write $\mathcal{L}(X)$ instead of $T \in \mathcal{L}(X, X)$.

2. The residuum operator

In the study of tauberian operators it is convenient to consider the dual concept of cotauberian operators.

Definition 2.1. An operator $T \in \mathcal{L}(X, Y)$ is called *tauberian* when it satisfies $(T^{**})^{-1}(Y) = X$. The operator T is called *cotauberian* when T^* is tauberian.

We denote by $\mathcal{T}(X, Y)$ and $\mathcal{T}^d(X, Y)$ the subsets of tauberian operators and cotauberian operators in $\mathcal{L}(X, Y)$.

A useful tool is the *residuum operator* T^{co} associated to each $T \in \mathcal{L}(X, Y)$, which is the operator $T^{co} \in \mathcal{L}\big(X^{**}/X, Y^{**}/Y\big)$ defined by

$$T^{co}(x^{**} + X) := T^{**}x^{**} + Y.$$

It is easy to check that T is tauberian if and only if T^{co} is injective, and by means of duality arguments we can prove that T is cotauberian if and only if T^{co} has dense range [13, Section 3.1].

Proposition 2.2 ([3]). *Given $T \in \mathcal{L}(X, Y)$, T^* cotauberian implies T tauberian but the converse implication is not valid.*

Proof. Note that $(T^{**})^{co}$ can be identified with $(T^{co})^{**}$ [13, Proposition 3.1.11]. The implication is a direct consequence of this fact.

To show the failure of the converse implication, we consider $S \in \mathcal{L}(\ell_1)$ given by $S(x_k) := (x_k/k)$, and find an operator $T \in \mathcal{L}(Z)$ such that T^{co} can be identified with S. See [13, Theorem 3.1.18]. □

Question 1. *Find conditions on X and Y under which T tauberian implies T^* cotauberian.*

The operators T for which T^{co} is injective and has closed range were studied by Rosenthal [23], who called them *strongly tauberian operators*. Note that T strongly tauberian implies T^* cotauberian. In some cases each tauberian operator $T \in \mathcal{L}(X, Y)$ is strongly tauberian. For example, when X is $L_1(\mu)$ and, more generally, when X is L-embedded in X^{**} [13, Theorem 6.2.18] and when the reflexive subspaces of X are superreflexive (see [13, Proposition 6.5.3 and Theorem 6.5.16]). However, these conditions are far from necessary.

Proposition 2.3. *In general, neither of the sets $\mathcal{T}(X,Y)$ and $\mathcal{T}^d(X,Y)$ is open in $\mathcal{L}(X,Y)$.*

Proof. Given a non-reflexive Banach space X, neither $\mathcal{T}(\ell_2(X))$ nor $\mathcal{T}^d(\ell_2(X))$ is open in $\mathcal{L}(\ell_2(X))$. See [13, Example 2.1.17]. □

Question 2. *Find conditions on X and Y implying that $\mathcal{T}(X,Y)$ or $\mathcal{T}^d(X,Y)$ is an open subset of $\mathcal{L}(X,Y)$.*

When every $T \in \mathcal{T}(X,Y)$ is strongly tauberian, $\mathcal{T}(X,Y)$ is an open subset of $\mathcal{L}(X,Y)$. Similarly, when $T \in \mathcal{T}^d(X,Y)$ implies T^* strongly tauberian, $\mathcal{T}^d(X,Y)$ is an open subset of $\mathcal{L}(X,Y)$.

Remark 2.4. In the case $X = Y$ it is interesting that $\mathcal{T}(X)$ or $\mathcal{T}^d(X)$ be open because it allows to introduce notions of "essential spectra" and apply techniques of spectral theory in the study of tauberian operators, like it is done in [25].

There are many Banach spaces X that are isomorphic to Z^{**}/Z for some space Z. This is proved in [8, Proposition 1] for X weakly compactly generated (see the definition in [2, page 308]). Note that reflexive spaces and separable spaces are weakly compactly generated.

Question 3. *Given two concrete Banach spaces X and Y, characterize the operators $S \in \mathcal{L}(X^{**}/X, Y^{**}/Y)$ for which there exists $T \in \mathcal{L}(X,Y)$ such that $T^{co} = S$.*

The previous question was studied in [16] where some partial results were obtained. For example, in [16] we can find a Banach space Z with Z^{**}/Z isomorphic to ℓ_2 such that $\{T^{co} : T \in \mathcal{L}(Z)\}$ coincides with the set of $S \in \mathcal{L}(\ell_2)$ which are regular with respect to the natural Banach lattice structure of ℓ_2.

Examples of tauberian and cotauberian operators

The main source of examples of tauberian and cotauberian operators is the following result which is a refinement of the main result in [8].

Theorem 2.5 ([10]). *For every $T \in \mathcal{L}(X,Y)$ there exist Banach spaces E and F, and operators $j \in \mathcal{L}(F,Y)$, $U \in \mathcal{L}(E,F)$ and $k \in \mathcal{L}(X,E)$ such that j is tauberian, U is a bijective isomorphism, k is cotauberian, and $T = jUk$.*

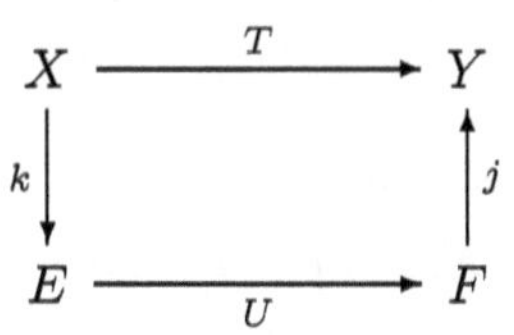

For the proof see [10] or [13, Section 3.2].

Remark 2.6. In the factorization of Theorem 2.5, if $T = jUk$ is the factorization of T, then $T^* = k^*U^*j^*$ and $T^{co} = j^{co}U^{co}k^{co}$ can be identified with the factorizations of T^* and T^{co} [13, Theorem 3.2.8].

3. Perturbative and algebraic characterizations

Recall that an operator $K \in \mathcal{L}(X,Y)$ is *weakly compact* when it takes bounded subsets of X to relatively weakly compact subsets of Y.

It follows from [2, Theorem 17.2] that K is weakly compact if and only if $K^{co} = 0$. As a consequence, $\mathcal{T}(X,Y)$ and $\mathcal{T}^d(X,Y)$ are stable under additive perturbation by weakly compact operators.

Theorem 3.1 (**[15]**). *Let $T \in \mathcal{L}(X,Y)$.*

(a) *T is tauberian if and only if for every compact operator $K \in \mathcal{L}(X,Y)$, the kernel $N(T+K)$ is reflexive;*

(b) *T is cotauberian if and only if for every compact operator $K \in \mathcal{L}(X,Y)$, the cokernel $Y/\overline{R(T+K)}$ is reflexive.*

Recall that $T \in \mathcal{L}(X,Y)$ is said to be *upper semi-Fredholm* if its kernel $N(T)$ is finite dimensional and its range $R(T)$ is closed. The operator T is said to be *lower semi-Fredholm* if its range is finite codimensional in Y, hence closed.

Remark 3.2. Theorem 3.1 was inspired by classical characterizations for the upper and the lower semi-Fredholm operators, in which finite-dimensional spaces replace reflexive spaces. It shows that $\mathcal{T}(X,Y)$ and $\mathcal{T}^d(X,Y)$ allow to develop a generalized Fredholm theory associated to the reflexive Banach spaces.

Part (b) also shows that the cotauberian operators can be defined without reference to conjugate operators.

The following characterization was obtained in [15].

Theorem 3.3. *Let $T \in \mathcal{L}(X,Y)$.*

(a) *T tauberian if and only if for every Banach space Z and every $S \in \mathcal{L}(Z,X)$, TS weakly compact implies S weakly compact;*

(b) *T cotauberian if and only if for every Banach space Z and every $S \in \mathcal{L}(Y,Z)$, ST weakly compact implies S weakly compact.*

Remark 3.4. Theorem 3.3 has a version that characterizes the upper and the lower semi-Fredholm operators, in which compact operators replace weakly compact operators. Moreover, this result inspired the introduction of operator semigroups associated to an operator ideal in [1]. See Definition 3.5 below.

Recall that an *operator ideal* (in the sense of Pietsch [22]) is a subclass $\mathcal{A}$ of the class $\mathcal{L}$ of all operators between Banach spaces such that, given Banach spaces V, X, Y and Z, the *component* $\mathcal{A}(X,Y)$ is a subspace of $\mathcal{L}(X,Y)$ that contains the finite rank operators, and satisfies

$$S \in \mathcal{L}(Y,Z), K \in \mathcal{A}(X,Y), T \in \mathcal{L}(V,X) \Longrightarrow SKT \in \mathcal{A}(V,Z).$$

Examples of operator ideals are the compact operators $\mathcal{K}$ and the weakly compact operators $\mathcal{W}$. For many other examples and an account of the properties of operator ideals, we refer to [22].

Next for every operator ideal $\mathcal{A}$ we define two classes of operators $\mathcal{A}_+$ and $\mathcal{A}_-$ in terms of their components $\mathcal{A}_+(X,Y)$ and $\mathcal{A}_-(X,Y)$.

Definition 3.5 ([1]). Let $\mathcal{A}$ be an operator ideal and $T \in \mathcal{L}(X, Y)$.

(a) $T \in \mathcal{A}_+$ if for every Banach space Z and every $S \in \mathcal{L}(Z, X)$,
$TS \in \mathcal{A}$ implies $S \in \mathcal{A}$;

(b) $T \in \mathcal{A}_-$ if for every Banach space Z and every $S \in \mathcal{L}(Y, Z)$,
$ST \in \mathcal{A}$ implies $S \in \mathcal{A}$.

Note that $\mathcal{A}_+(X, Y)$ and $\mathcal{A}_-(X, Y)$ can be empty when $X \neq Y$. However the identity I_X belongs to both $\mathcal{A}_+(X)$ and $\mathcal{A}_-(X)$.

Remark 3.6. For every operator ideal $\mathcal{A}$, the classes $\mathcal{A}_+$ and $\mathcal{A}_-$ are stable under product: given $S \in \mathcal{L}(Y, Z)$ and $T \in \mathcal{L}(X, Y)$,

$$S, T \in \mathcal{A}_+ \Rightarrow ST \in \mathcal{A}_+ \text{ and } S, T \in \mathcal{A}_- \Rightarrow ST \in \mathcal{A}_-.$$

For this reason $\mathcal{A}_+$ and $\mathcal{A}_-$ are called *operator semigroups* associated to $\mathcal{A}$.

It follows from Theorem 3.3 and Remark 3.4 that $\mathcal{W}_+$ and $\mathcal{W}_-$ are the classes of tauberian and cotauberian operators, and $\mathcal{K}_+$ and $\mathcal{K}_-$ are the classes of upper semi-Fredholm and lower semi-Fredholm operators. The class $\mathcal{A}_+$ has been studied for the operator ideals of Rosenthal operators $\mathcal{R}$, completely continuous operators $\mathcal{C}$, weakly completely continuous operators $\mathcal{WC}$, and unconditionally converging operators $\mathcal{U}$. And denoting $\mathcal{A}^d := \{T : T^* \in \mathcal{A}\}$, the dual operator ideal of $\mathcal{A}$, the class $\mathcal{A}_-$ has been studied for $\mathcal{A}$ one of the operator ideals $\mathcal{R}^d$, $\mathcal{C}^d$, $\mathcal{WC}^d$, and $\mathcal{U}^d$. Note that by the Schauder and Gantmacher theorems, $\mathcal{K}^d = \mathcal{K}$ and $\mathcal{W}^d = \mathcal{W}$. We refer to [13, Sections 3.5 and 6.1] for details.

Recall that $T \in \mathcal{L}(X, Y)$ is a *Banach-Saks* operator if every bounded sequence (x_n) in X has a subsequence (x_{n_k}) for which $\left(m^{-1} \sum_{k=1}^{m} T x_{n_k}\right)$ is convergent. A Banach space X has the *Banach-Saks property* if the identity I_X is Banach-Saks. This is an intermediate property between reflexivity and super-reflexivity.

Question 4. *Study the semigroups $\mathcal{A}_+$ and $\mathcal{A}_-$ for other operator ideals $\mathcal{A}$. In particular, we point out the operator ideal of the* Banach-Saks operators *considered in* [18]. *It could provide new results on the Banach-Saks property, whose study presents notable technical difficulties.*

The semigroups $\mathcal{A}_+$ and $\mathcal{A}_-$ are specially useful when they admit perturbative characterizations similar to Theorem 3.1. In order to be precise, we consider the *space ideal* associated to $\mathcal{A}$, which is the class of Banach spaces defined by $Sp(\mathcal{A}) := \{X : I_X \in \mathcal{A}\}$.

Question 5. *Find examples of semigroups $\mathcal{A}_+$ and $\mathcal{A}_-$ satisfying a* perturbative characterization, *in the sense that for $T \in \mathcal{L}(X, Y)$,*

- *$T \in \mathcal{A}_+$ if and only if for every compact operator $K \in \mathcal{L}(X, Y)$, the kernel $N(T + K)$ belongs to $Sp(\mathcal{A})$;*
- *$T \in \mathcal{A}_-$ if and only if for every compact operator $K \in \mathcal{L}(X, Y)$, the cokernel $Y/\overline{R(T + K)}$ belongs to $Sp(\mathcal{A})$.*

The semigroups Co_+, WCo_+, $\mathcal{R}_+$, $\mathcal{C}_+$, $\mathcal{WC}_+$ and $\mathcal{U}_+$, and the semigroups Co_-, WCo_-, $\mathcal{R}^d_-$, $\mathcal{C}^d_-$, $\mathcal{WC}^d_-$ and $\mathcal{U}^d_-$ satisfy perturbative characterizations [13, Section 3.5]. See [11] and [14] for further examples.

4. Tauberian operators acting on Banach lattices

There are not many results on tauberian operators acting on Banach lattices in which the lattice structure is applied, like in the following one.

Theorem 4.1 (**[12]**). *An operator $T \in \mathcal{L}(L_1(0,1), Y)$ is tauberian if and only if*

$$\inf\{\liminf_{n\to\infty} \|Tf_n\| \,:\, (f_n) \subset L_1(0,1) \text{ normalized and disjoint}\} > 0.$$

It is easy to derive from this characterization that $\mathcal{T}(L_1(0,1), Y)$ is open in $\mathcal{L}(L_1(0,1), Y)$. Another consequence shows that tauberian operators on $L_1(0,1)$ are not far from into isomorphisms:

Corollary 4.2 (**[12]**). *For every tauberian operator $T \in \mathcal{L}(L_1(0,1), Y)$ there exists a finite partition of $(0,1)$ consisting of intervals $\{I_1, \dots, I_n\}$ such that the restriction $T|_{L_1(I_k)}$ is an into isomorphism for each $k = 1, \dots, n$.*

Here $L_1(I_k)$ is the set of functions in $L_1(0,1)$ with support essentially contained in the interval I_k.

The space $L_1(0,1)$ contains many infinite-dimensional reflexive subspaces R, and for these subspaces the quotient map $Q : L_1(0,1) \to L_1(0,1)/R$ is tauberian [13, Theorem 2.1.5]. However, the following question remains open.

Question 6. *Suppose that $T \in \mathcal{L}(L_1(0,1))$ is tauberian. Is T upper semi-Fredholm?*

In [26] we can find a positive answer to this problem for some special operators. In order to state the result, recall that for every $T \in \mathcal{L}(L_1(0,1))$ we can find a family $\{\mu_s : s \in (0,1)\}$ of Borel measures in $(0,1)$ so that $Tf(s) = \int_0^1 f(t)d\mu_s(t)$ a.e. for every $f \in L_1(0,1)$.

Using the fact that every Borel measure μ admits a decomposition with respect to the Lebesgue measure $\mu = \mu^{at} + \mu^{ac} + \mu^{sc}$ in atomic part, absolutely continuous part and singular continuous part, for every operator $T \in \mathcal{L}(L_1(0,1))$ we obtain a decomposition $T = T^{at} + T^{ac} + T^{sc}$. We refer to [13, Section 4.3] for more details.

Theorem 4.3 (**[26]**). *Every tauberian operator $T \in \mathcal{L}(L_1(0,1))$ with $T^{sc} = 0$ is upper semi-Fredholm.*

Question 6 is likely to be difficult. So it could be interesting to begin with the case of multiplier operators. Observe that the set of multipliers depends on the isometric representation of $L_1(0,1)$ that we consider. For simplicity, we consider the space $L_1(\mathbb{T})$ of integrable functions in the unit circle $\mathbb{T}$ endowed with the Lebesgue measure m, and we denote by $M(\mathbb{T})$ the space of Borel measures on $\mathbb{T}$.

The *convolution product* $\mu * \nu$ of two measures $\mu, \nu \in M(\mathbb{T})$ is defined as follows. Given a Borel subset E of $\mathbb{T}$, we consider the set

$$E_{(2)} := \{(e^{is}, e^{it}) \in \mathbb{T} \times \mathbb{T} : e^{i(s+t)} \in E\},$$

and denoting by $\mu \times \nu$ the product of measures we define

$$(\mu * \nu)(E) := (\mu \times \nu)(E_{(2)}).$$

Definition 4.4. Let $\mu \in M(\mathbb{T})$. The *convolution operator* associated to μ is the operator $T_\mu \in \mathcal{L}(L_1(\mathbb{T}))$ is defined by $T_\mu f := \mu * f$.

Recall that $L_1(\mathbb{T})$ can be identified with the subspace of $M(\mathbb{T})$ consisting of the measures that are absolutely continuous with respect to the Lebesgue measure, and $M(\mathbb{T})$ can be identified in a natural way with a closed subspace of $L_1(\mathbb{T})^{**}$. Indeed, $L_1(\mathbb{T})^* = L_\infty(\mathbb{T})$ and for every $\mu \in M(\mathbb{T})$ the expression

$$\langle g, \mu \rangle := \int_{\mathbb{T}} g(e^{it}) d\mu, \quad g \in L_\infty(\mathbb{T}).$$

defines an element $L_\infty(\mathbb{T})^*$.

The convolution operator T_μ admits an extension $\widehat{T_\mu} \in \mathcal{L}(M(\mathbb{T}))$ which is also defined by $\widehat{T_\mu}\nu := \mu * \nu$. Moreover, with the natural identifications, the second conjugate ${T_\mu}^{**}$ is an extension of $\widehat{T_\mu}$.

The proof of the following result is not difficult, and it is left to the interested reader.

Proposition 4.5. *Given* $\mu \in M(\mathbb{T})$, *consider the following assertions:*

(a) T_μ *is upper semi-Fredholm;*
(b) T_μ *is tauberian;*
(c) $\nu \in M(\mathbb{T})$ *and* $\mu * \nu \in L_1(\mathbb{T})$ *imply* $\nu \in L_1(\mathbb{T})$.

Then (a) $\Longrightarrow$ (b) $\Longrightarrow$ (c).

We can now state a special case of Question 6.

Question 7. *Are the converse implications in Proposition* 4.5 *true?*

We refer to [14] for examples of convolution operators in other representations of $L_1(0, 1)$.

Tauberian positive operators

Given Banach lattices E and F, and positive operators S, T in $\mathcal{L}(E, F)$, a classical problem, known as the *domination problem* for compact operators, is to find conditions on E and F so that

$$0 \leq S \leq T \text{ and } T \text{ compact } \Longrightarrow S \text{ compact}.$$

A sufficient condition is that both E^* and F have order continuous norm [2, Theorem 16.20]. Moreover, in the case $E = F$, T compact implies S^3 compact with no conditions on E [2, Theorem 16.13].

Similar results have been obtained for other operator ideals like the weakly compact operators, the completely continuous operators, and the strictly singular operators. We refer to the introduction of [9] for more information.

Next we consider the domination problem for tauberian and cotauberian operators. Observe that the implications are reversed.

Question 8. *Find conditions on Banach lattices E and F so that* $0 \leq S \leq T$ *in* $\mathcal{L}(E, F)$ *and S tauberian (cotauberian)imply T tauberian (cotauberian).*

Remark 4.6. In the case $E = F$, observe that there is a difference with the problem for operator ideals because, given $T \in \mathcal{L}(X)$ and n a positive integer, T^n is tauberian (cotauberian) if and only if T is tauberian (cotauberian).

It is also interesting to consider the domination problem for semigroups associated to other operator ideals $\mathcal{A}$ besides $\mathcal{W}$.

Question 9. *Given an operator ideal $\mathcal{A}$, investigate the problems in Question* 8 *and Remark* 4.6 *for the semigroups $\mathcal{A}_+$ and $\mathcal{A}_-$.*

In the previous question, note that we need the condition $\mathcal{A}$ injective in order to have $T^n \in \mathcal{A}_+$ if and only if $T \in \mathcal{A}_+$, and we need $\mathcal{A}$ surjective in order to have $T^n \in \mathcal{A}_-$ if and only if $T \in \mathcal{A}_-$.

5. Applications

Here we only consider a few of the applications of tauberian operators mentioned in the introduction of the paper. For a more complete account and a detailed description, we refer to [13].

5.1. Factorization of operators

The factorization result obtained in [8] was improved in [18] and [10], and applied to prove that some operator ideals $\mathcal{A}$ satisfy the *factorization property;* i.e., that every $T \in \mathcal{A}(X, Y)$ can be factorized through a Banach space E such that $I_E \in \mathcal{A}$. The weakly compact, the weakly precompact, and the Banach-Saks operators satisfy the factorization property. We refer to [13, Section 5.3] for additional examples. Next result shows that Theorem 2.5 provides a method to find operator ideals satisfying the factorization property.

Proposition 5.1. *Let $\mathcal{A}$ be an operator ideal. Suppose that for every operator T, the operators j and k obtained in Theorem* 2.5 *satisfy $j \in \mathcal{A}_+$ and $k \in \mathcal{A}_-$. Then $\mathcal{A}$ satisfies the factorization property.*

The proof is a consequence of the definitions of $\mathcal{A}_+$ and $\mathcal{A}_-$. Let $T \in \mathcal{L}(X, Y)$ be an operator in $\mathcal{A}$, and let $T = jUk$ be the factorization obtained in Theorem 2.5. Since $j \in \mathcal{A}_+$, $k \in \mathcal{A}_-$ and U is a bijective isomorphism,

$$jUk \in \mathcal{A} \Rightarrow Uk \in \mathcal{A} \Rightarrow U \in \mathcal{A} \Rightarrow I_E \in \mathcal{A}.$$

So $\mathcal{A}$ satisfies the factorization property.

Question 10. *Find operator ideals $\mathcal{A}$ such that for every operator T, the operators j and k obtained in Theorem* 2.5 *satisfy $j \in \mathcal{A}_+$ and $k \in \mathcal{A}_-$.*

Every operator ideal $\mathcal{A}$ has associated the dual operator ideal $\mathcal{A}^d$ and the residuum operator ideal $\mathcal{A}^{co}$ that are defined by

$$\mathcal{A}^d := \{T \in \mathcal{L} : T^* \in \mathcal{A}\} \quad \text{and} \quad \mathcal{A}^{co} := \{T \in \mathcal{L} : T^{co} \in \mathcal{A}\}.$$

The following result is a consequence of Remark 2.6.

Proposition 5.2. *Let $\mathcal{A}$ be an operator ideal such that for every operator T, the operators j and k obtained in Theorem 2.5 satisfy $j \in \mathcal{A}_+$ and $k \in \mathcal{A}_-$. Then the same is true for $\mathcal{A}^d$ and $\mathcal{A}^{co}$.*

5.2. James' characterization of reflexive Banach spaces

A classical result of James establishes that a Banach space X is reflexive if and only if each $f \in X^*$ attains its norm in the closed unit ball B_X; i.e., $f\big(B_X\big)$ is closed for every $f \in X^*$. An improvement was obtained by Neidinger and Rosenthal in [21] as an application on their investigations on the behavior of tauberian operators.

Theorem 5.3 ([21]). *Let $T \in \mathcal{L}(X,Y)$ be a non-zero operator. The following assertions are equivalent:*

(a) *T is tauberian;*
(b) *$T(A)$ is closed for all closed bounded convex subsets A;*
(c) *$T(B_Y)$ is closed for all closed subspaces Y.*

From this characterization of tauberian operators, the following improvement of James' theorem was obtained.

Theorem 5.4 ([21]). *Let X be a non-reflexive Banach space. Then for every non-zero $f \in X^*$ there exists a closed subspace Y of X such that the restriction $f|_Y$ does not attain its norm on B_Y. Moreover, the subspace Y may be chosen of codimension one in X.*

We refer to [13, Section 2.3] for an alternative exposition of the results of Neidinger and Rosenthal.

5.3. Equivalence between the KMP and the RNP

The equivalence between the Radon–Nikodym property (RNP) and the Kreĭn–Milman property (KMP) for Banach spaces is a problem that it is still open, but has produced an intense investigation on the structure of the extreme points of convex sets that has been useful in other topics, like optimization theory. We refer to [13, Section 5.1] for an account of the relevant concepts.

The characterization of tauberian operators in terms of their action on closed bounded convex subsets given in Theorem 5.3 was applied by Schachermayer to obtain a necessary condition for the equivalence of the RNP and the KMP.

Theorem 5.5 ([24]). *Let X be a Banach space for which there exists an injective tauberian operator $T \in \mathcal{L}(X \times X, X)$. Then X has the RNP if and only if it has the KMP.*

This result was formulated in [24] in a weaker version. For the previous formulation we refer to [13, Theorem 5.1.12]. A key in its proof is the fact that a Banach space X has the RNP if and only if the space $\ell_2(X)$ of norm square summable sequences in X has the KMP.

Question 11. *Find conditions on a Banach space X guaranteeing the existence of an injective tauberian operator $T \in \mathcal{L}(X \times X, X)$.*

5.4. Preservation of isomorphic properties of Banach spaces

One of the key points in the main result of [8] is that tauberian operators preserve the non-weak compactness of bounded sets; i.e., given $T \in \mathcal{T}(X,Y)$ and a bounded subset A of X, if $T(A)$ is relatively weakly compact then A is relatively weakly compact. In [20], the preservation of isomorphic properties of Banach spaces by tauberian operators and other related classes of operators was thoroughly investigated. From the preservation of isomorphic properties of bounded sets, we can derive the preservation of properties of the whole space. Next we give a sample of results of this kind.

Theorem 5.6 ([20]). *Let $\mathcal{P}$ be one of the following properties of a Banach space: Kreĭn–Milman property, Radon–Nikodým property, quasi-reflexivity, somewhat reflexivity, weak sequential completeness, containing no copies of ℓ_1, containing no copies of c_0, separability, or separability of the dual space.*

Suppose that Y satisfies $\mathcal{P}$ and there exists an injective tauberian operator $T \in \mathcal{L}(X,Y)$. Then X satisfies $\mathcal{P}$.

We observe that the preservation of properties by cotauberian operators have not received attention.

Question 12. *Determine isomorphic properties $\mathcal{P}$ such that if X satisfies $\mathcal{P}$ and there exists a dense range cotauberian operator $T \in \mathcal{L}(X,Y)$, then Y satisfies $\mathcal{P}$.*

Similarly, given an operator ideal $\mathcal{A}$, the definition of the semigroups $\mathcal{A}_+$ and $\mathcal{A}_-$ can be described as the preservation of an isomorphic property.

Question 13. *For an operator ideal $\mathcal{A}$, determine the isomorphic properties preserved by the semigroups $\mathcal{A}_+$ and $\mathcal{A}_-$.*

The answers to the last question would be useful in the search for examples of operator ideals with the factorization property.

5.5. Construction of hereditarily indecomposable Banach spaces

A Banach space X is said to be *indecomposable* if it cannot be decomposed as the direct sum of two infinite-dimensional closed subspaces; equivalently, if each complemented subspace of X is finite dimensional or finite codimensional. The space X is said to be *hereditarily indecomposable* (H.I. for short) if every closed subspace of X is indecomposable. The existence of infinite-dimensional H.I. Banach spaces was proved by Gowers and Maurey in [17]. These spaces have been useful to give answers to a good number of classical problems in Banach space theory.

In [4], Argyros and Felouzis introduced a factorization whose construction have some similarities with the one given in [8]. In particular, the construction gives a tauberian operator between the intermediate space and the final space, like in Theorem 2.5. However, for compact operators the intermediate space obtained in [4] is H.I. [4, Theorem 8.5], while the one obtained in [8] is hereditarily ℓ_2 (each infinite-dimensional closed subspace contains a subspace isomorphic to ℓ_2) [13, Corollary 3.2.12].

Question 14. *Develop a systematic study of the factorization in [4] parallel to the one that have been done for the factorization in [8].*

The construction in [4] is technically complicated, and it can be described as a conditional version of the construction in [8]. It is very flexible and it may provide a great variety of examples of Banach spaces and operators.

Acknowledgment

Many thanks to the organizers of IWOTA 2011 for their efforts that facilitated a smooth development of the Meeting.

References

[1] P. Aiena, M. González and A. Martínez-Abejón, *Operator semigroups in Banach space theory.* Boll. Unione Mat. Ital. (8) **4** (2001), 157–205.

[2] C.D. Aliprantis and O. Burkinshaw, *Positive operators.* Academic Press, New York, 1985.

[3] T. Álvarez and M. González, *Some examples of tauberian operators.* Proc. Amer. Math. Soc. **111** (1991), 1023–1027.

[4] S.A. Argyros and V. Felouzis, *Interpolating hereditarily indecomposable Banach spaces* J. Amer. Math. Soc. **13** (2000), 243–294.

[5] K. Astala and H.-O. Tylli, *Seminorms related to weak compactness and to tauberian operators.* Math. Proc. Cambridge Phil. Soc. **107** (1990), 365–375.

[6] B. Beauzamy, *Espaces d'interpolation réels: Topologie et Géométrie.* Lecture Notes in Math. 666; Springer, Berlin, 1978.

[7] E. Behrends, *On the principle of local reflexivity.* Studia Math. **100** (1991), 109–128.

[8] W.J. Davis, T. Figiel, W.B. Johnson and A. Pełczyński, *Factoring weakly compact operators.* J. Funct. Anal. **17** (1974), 311–327.

[9] J. Flores and F.L. Hernández, *Domination by positive strictly singular operators.* J. London Math. Soc. **66** (2002), 433–452.

[10] M. González, *Dual results of factorization for operators.* Ann. Acad. Sci. Fennicae Math. **18** (1993), 3–11.

[11] M. González, A. Martínez-Abejón Supertauberian operators and perturbations Arch. Math. 64 1995 423–433

[12] M. González and A. Martínez-Abejón, *Tauberian operators on $L_1(\mu)$-spaces.* Studia Math. **125** (1997), 289–303.

[13] M. González, A. Martínez-Abejón, *Tauberian operators.* Operator Theory: Advances and applications 194; Birkhäuser, Basel, 2010.

[14] M. González, A. Martínez-Abejón and J. Pello, *Conjugate operators that preserve the nonconvergence of bounded martingales.* J. Funct. Anal. **252** (2007), 566–580.

[15] M. González and V.M. Onieva, *Characterizations of tauberian operators and other semigroups of operators.* Proc. Amer. Math. Soc. **108** (1990), 399–405.

[16] M. González, E. Saksman and H.-O. Tylli, *Representing non-weakly compact operators.* Studia Math. **113** (1995), 289–303.

[17] W.T. Gowers and B. Maurey, *The unconditional basic sequence problem.* J. Amer. Math. Soc. **6** (1993), 851–874.

[18] S. Heinrich, *Closed operator ideals and interpolation.* J. Funct. Anal. **35** (1980), 397–411.

[19] N. Kalton and A. Wilansky, *Tauberian operators on Banach spaces.* Proc. Amer. Math. Soc. **57** (1976), 251–255.

[20] R. Neidinger, *Properties of tauberian operators on Banach spaces* Ph. D. Thesis. The University of Texas at Austin, Austin TX, 1984.

[21] R. Neidinger and H.P. Rosenthal, *Norm-attainment of linear functionals on subspaces and characterizations of tauberian operators.* Pacific J. Math. **118** (1985), 215–228.

[22] A. Pietsch, *Operator ideals.* North Holland, Amsterdam, 1980.

[23] H. Rosenthal, *On wide-(s) sequences and their applications to certain classes of operators.* Pacific J. Math. **189** (1999), 311–338.

[24] W. Schachermayer, *For a Banach space isomorphic to its square the Radon–Nikodým property and the Kreĭn–Milman property are equivalent.* Studia Math. **81** (1985), 329–339.

[25] H.-O. Tylli, *A spectral radius problem connected with weak compactness.* Glasgow Math. J. **35** (1993), 85–94.

[26] L.W. Weis, *On the computation of some quantities in the theory of Fredholm operators.* Rend. Circ. Mat. Palermo Suppl. **(2)** Suppl. No. 5, (1984), 109–133.

Manuel González
Departamento de Matemáticas
Facultad de Ciencias
Universidad de Cantabria
E-39071 Santander, Spain
e-mail: manuel.gonzalez@unican.es

Operator Theory:
Advances and Applications, Vol. 236, 243–256

Products of Toeplitz and Hankel Operators on the Hardy Space of the Unit Sphere

Hocine Guediri

Abstract. The aim of this note is to discuss boundedness and compactness of Hankel products and mixed Toeplitz–Hankel products on the Hardy space of the unit sphere in several complex variables. The main adopted tool is an auxiliary pioneering operator involved in an earlier investigation of dual Toeplitz operators on the orthogonal complement of the Hardy space on the unit sphere.

Mathematics Subject Classification (2010). 47B35.

Keywords. Dual Toeplitz operator, Hardy space of the unit sphere, Hankel products, mixed Haplitz products.

1. Introduction

Dual Toeplitz operators on the orthogonal complement of the Bergman space have been introduced and well investigated by Stroethoff and Zheng [11]. The higher-dimensional case of dual Toeplitz operators in both Hardy and Bergman space settings has been studied in [2, 5]; see also the relevant references therein. On the other hand, Toeplitz and Hankel operators in the latter setting have been extensively studied; in this respect we refer to [3, 4, 7, 8, 9, 12, 13].

For our purpose, let $\mathbb{B}_n$, $n > 1$, be the unit ball of $\mathbb{C}^n$ and $\mathbb{S}_n$ be its boundary (the unit sphere). Denote by $L^2(\mathbb{S}_n)$ the Lebesgue space of square integrable functions and by $\mathcal{H}^2(\mathbb{S}_n)$ its Hardy subspace, (for more details see [1, 3, 10, 15]). While on the circle the orthogonal complement of the Hardy space can be characterized by $(H^2)^\perp = \overline{zH^2}$, the matter is much more involved in higher dimensions because

$$L^2(\mathbb{S}_n) \ominus \left\{\mathcal{H}^2(\mathbb{S}_n) + \overline{\mathcal{H}^2(\mathbb{S}_n)}\right\}$$

is large enough to cause capital differences from the one-dimensional case.

This project was supported by King Saud University, Deanship of Scientific Research, College of Science Research Center.

Therefore, in contrast to the case of the circle, (where dual Toeplitz operators are anti-unitarily equivalent to Toeplitz operators in view of the above symmetry between $(H^2)^{\perp}$ and H^2, [11] or [5]), dual Toeplitz operators on the orthogonal complement of the Hardy space on the unit sphere cannot be analogously reduced to Toeplitz operators. Accordingly, they may constitute a worth studying new class of Haplitz-type operators. In [5], such dual Toeplitz operators have been introduced and studied from various points of view.

An interesting auxiliary operator, namely $\mathscr{S}_w$, has been introduced and used in studying products of these dual Toeplitz operators. In particular, commuting dual Toeplitz operators have then been characterized through certain necessary and sufficient conditions on the symbols. Besides, a Brown–Halmos type theorem has been proved; it tells us when exactly the product of two dual Toeplitz operators is again a dual Toeplitz operator. Several consequences of the two latter issues, such as the characterization of zero divisors, have been also inferred. For the sake of completeness, we summarize a few of these results here and refer to [5] for details.

Sarason's problem related to the boundedness of Toeplitz products has been extensively investigated by many authors, we refer to [6, 14] for details. The higher-dimensional case of Toeplitz products and Hankel products has been considered by many authors [2, 7, 8, 9, 12, 13]; see also the relevant references therein.

In the present paper, a more prominent role of the operator $\mathscr{S}_w$ is emphasized. More precisely, making use of this transformation, we discuss necessary conditions ensuring boundedness and compactness of products of Hankel operators $H_f H_g^*$, (equivalently the "dual" semicommutators $\mathcal{S}_{fg} - \mathcal{S}_f \mathcal{S}_g$), and mixed Hankel–Toeplitz products $T_f H_g^*$ and $H_g T_{\overline{f}}$; as well as the commutators $\mathcal{S}_f \mathcal{S}_g - \mathcal{S}_g \mathcal{S}_f$. These represent the main results of this communication.

2. Preliminaries

For $f \in L^\infty(\mathbb{S}_n)$, define the dual Toeplitz operator $\mathcal{S}_f$ as the operator on $\left(\mathcal{H}^2(\mathbb{S}_n)\right)^{\perp}$ defined to be a multiplication followed by a projection as follows:

$$\begin{aligned} \mathcal{S}_f : \ \left(\mathcal{H}^2(\mathbb{S}_n)\right)^{\perp} &\longrightarrow \left(\mathcal{H}^2(\mathbb{S}_n)\right)^{\perp} \\ u &\longrightarrow \mathcal{S}_f(u) := \mathcal{Q}(fu). \end{aligned}$$

Here, $\mathcal{Q}$ is the orthogonal projection from $L^2(\mathcal{S}_n)$ onto $\left(\mathcal{H}^2(\mathbb{S}_n)\right)^{\perp}$ defined by

$$\begin{aligned} \mathcal{Q} : \ L^2(\mathbb{S}_n) &\longrightarrow \left(\mathcal{H}^2(\mathbb{S}_n)\right)^{\perp} \\ g &\longrightarrow \mathcal{Q}(g) := (I - \mathcal{P})(g), \end{aligned}$$

with $\mathcal{P}$ being the customary (Hardy) orthogonal projection from $L^2(\mathbb{S}_n)$ onto the Hardy space $\mathcal{H}^2(\mathbb{S}_n)$. Since the projection $\mathcal{Q}$ has norm 1, then for any $h \in \left(\mathcal{H}^2(\mathbb{S}_n)\right)^{\perp}$, we have

$$||\mathcal{S}_f(h)||_2 = ||\mathcal{Q}(fh)||_2 \leq ||fh||_2 \leq ||f||_\infty ||h||_2.$$

Immediate algebraic properties of dual Toeplitz operators can be easily observed; for instance for $f, g \in L^\infty(\mathbb{S}_n)$, $\alpha, \beta \in \mathbb{C}$, we have

$$\mathcal{S}_f^* = \mathcal{S}_{\overline{f}} \text{ and } \mathcal{S}_{\alpha f+\beta g} = \alpha\mathcal{S}_f + \beta\mathcal{S}_g.$$

Dual Toeplitz operators appear naturally if one observes that under the orthogonal decomposition:

$$L^2(\mathcal{S}_n) = \mathcal{H}^2(\mathbb{S}_n) \oplus \left(\mathcal{H}^2(\mathbb{S}_n)\right)^\perp,$$

the multiplication operator $\mathcal{M}_f$, $f \in L^\infty(\mathbb{S}_n)$, can be represented as follows:

$$\mathcal{M}_f = \begin{pmatrix} T_f & H_{\overline{f}}^* \\ H_f & \mathcal{S}_f \end{pmatrix},$$

where the Toeplitz and Hankel operators are defined respectively by

$$\begin{aligned} T_f : \mathcal{H}^2(\mathbb{S}_n) &\longrightarrow \mathcal{H}^2(\mathbb{S}_n) \\ g &\longrightarrow T_f(g) := \mathcal{P}(fg), \end{aligned}$$

and

$$\begin{aligned} H_f : \mathcal{H}^2(\mathbb{S}_n) &\longrightarrow \left(\mathcal{H}^2(\mathbb{S}_n)\right)^\perp \\ g &\longrightarrow H_f(g) := \mathcal{Q}(fg). \end{aligned}$$

This representation gives rise to *dual Toeplitz operators* on $\left(\mathcal{H}^2(\mathbb{S}_n)\right)^\perp$. At once, we observe the following algebraic relationships connecting them with Haplitz operators, namely: for $f, g \in L^\infty(\mathbb{S}_n)$, the product identity $\mathcal{M}_f\mathcal{M}_g = \mathcal{M}_{fg}$ implies that

$$\begin{pmatrix} T_f & H_{\overline{f}}^* \\ H_f & \mathcal{S}_f \end{pmatrix}\begin{pmatrix} T_g & H_{\overline{g}}^* \\ H_g & \mathcal{S}_g \end{pmatrix} = \begin{pmatrix} T_{fg} & H_{\overline{fg}}^* \\ H_{fg} & \mathcal{S}_{fg} \end{pmatrix}.$$

Hence, we infer that

$$\begin{aligned} T_{fg} &= T_fT_g + H_{\overline{f}}^*H_g. \\ \mathcal{S}_{fg} &= H_fH_{\overline{g}}^* + \mathcal{S}_f\mathcal{S}_g. \\ H_{fg} &= H_fT_g + \mathcal{S}_fH_g. \end{aligned} \tag{2.1}$$

It follows that the commutator $[\,\mathcal{S}_f,\ \mathcal{S}_g] = \ \mathcal{S}_f\mathcal{S}_g - \mathcal{S}_g\mathcal{S}_f$ is given by

$$[\,\mathcal{S}_f,\ \mathcal{S}_g] = H_gH_{\overline{f}}^* - H_fH_{\overline{g}}^*. \tag{2.2}$$

In particular, such identities reduce to the following ones, since the Hankel operator is trivial if the symbol is analytic:

Lemma 2.1. *Let $f \in \mathcal{H}^\infty(\mathbb{S}_n)$, then we have*

i) $H_gT_f = \mathcal{S}_fH_g$.
ii) $T_{\overline{f}}H_g^* = H_g^*\mathcal{S}_{\overline{f}}$.
iii) $\mathcal{S}_{fg} = \mathcal{S}_f\mathcal{S}_g$.
iv) $\mathcal{S}_{g\overline{f}} = \mathcal{S}_g\mathcal{S}_{\overline{f}}$.

A key property which usually proves very useful in establishing more fundamental properties of Toeplitz type operators is the so-called the *spectral inclusion theorem.* It turns out [5] that our dual Toeplitz operators do satisfy such property.

Let us denote by $\mathscr{R}(f)$ the essential range of the essentially bounded function f, and by $\sigma(T)$ the spectrum of an operator T. Then, we have

Proposition 2.2. [5]

1. *If f is in $L^\infty(\mathbb{S}_n)$, then $\mathscr{R}(f) = \sigma(M_f) \subseteq \sigma(\mathcal{S}_f)$.*
2. *Let f be in $L^\infty(\mathbb{S}_n)$. Then, we have $||\mathcal{S}_f|| = ||f||_\infty$.*
3. *If f is in $L^\infty(\mathbb{S}_n)$, then $\mathcal{S}_f = 0$ if and only if $f = 0$.*

3. The auxiliary operator $\mathscr{S}_w$

Let z, w be in $\mathbb{B}_n$, and recall that the Hardy space $\mathcal{H}^2(\mathbb{B}_n)$ is a reproducing kernel Hilbert space with kernel function given by

$$K_w(z) = \frac{1}{(1 - \langle z, w\rangle)^n},$$

while the normalized reproducing kernel is denoted by k_w.

For f and g in $L^2(\mathbb{S}_n)$, consider the rank one operator defined by $(f \otimes g)h = \langle h, g\rangle f, \forall f \in L^2(\mathbb{S}_n)$; and note that $\|f \otimes g\| = \|f\| \, \|g\|$.

The unitary operator $\mathbb{U}_w$ is defined by

$$\mathbb{U}_w f = (f \circ \varphi_w)k_w. \tag{3.1}$$

Observe that $\mathbb{U}_w 1 = k_w$. Also, for a Toeplitz operator, we have

$$\mathbb{U}_w T_f \mathbb{U}_w = T_{f\circ\varphi_w}. \tag{3.2}$$

Further, we know that

$$\langle z, w\rangle^j = \sum_{|m|=j} \frac{j!}{m!} z^m \, \overline{w}^m.$$

Thus, by the binomial rule, we obtain

$$\begin{aligned} K_w^{-1}(z) = (1 - \langle z, w\rangle)^n &= \sum_{j=0}^{n} \frac{n!}{j!(n-j)!} \; (-1)^j \langle z, w\rangle^j \\ &= \sum_{j=0}^{n} \sum_{|m|=j} \frac{(-1)^j n!}{j!(n-j)!} \frac{j!}{m!} \; z^m \overline{w}^m \\ &= \sum_{j=0}^{n} \sum_{|m|=j} \lambda_{j,m} \; z^m \overline{w}^m, \quad \text{with } \lambda_{j,m} = \frac{(-1)^j n!}{(n-j)! m!}. \end{aligned} \tag{3.3}$$

On the other hand, the customary ball automorphism $\varphi_w \in \mathbb{A}ut(\mathbb{B}_n)$ is defined by

$$\varphi_w(z) = \frac{w - P_w(z) - \sqrt{1 - |w|^2} Q_w(z)}{1 - \langle z, w\rangle}, \quad z, w \in \mathbb{B}_n, \tag{3.4}$$

where P_w denotes the orthogonal projection onto the subspace generated by w defined by $P_0 = 0$ and $P_w(z) = \frac{\langle z,w\rangle}{\langle w,w\rangle}$, $w \neq 0$, and Q_w denotes the orthogonal projection onto its orthogonal complement given by $Q_w(z) = z - P_w(z)$. In particular, it satisfies the universal identity:

$$1-|\varphi_w(z)|^2=\frac{\left(1-|w|^2\right)\left(1-|z|^2\right)}{|1-\langle z,w\rangle|^2},\quad z,w\in\mathbb{B}_n. \tag{3.5}$$

Finally, for operators $\mathbf{T}$ and $\mathbf{S}$, we can easily verify that:

$$\mathbf{T}(f\otimes g)\mathbf{S}^*=\mathbf{T}f\otimes\mathbf{S}g. \tag{3.6}$$

Matching all that together, we obtain the following key assertion:

Proposition 3.1. *On the Hardy space of the unit sphere $\mathcal{H}^2(\mathbb{S}_n)$, we have*

$$k_w\otimes k_w=\sum_{j=0}^{n}\sum_{|m|=j}\lambda_{j,m}T_{\varphi_w^m}T_{\overline{\varphi_w}^m},\quad\forall w\in B_n, \tag{3.7}$$

with $\lambda_{j,m}$ as in (3.3) *above.*

Proof. Consider a Hardy function $f\in\mathcal{H}^2(\mathbb{B}_n)$. The invariant volume mean value property tells us that

$$f(\psi(0))=\int_{\mathbb{B}_n}(f\circ\psi)(w)d\nu(w),\ \ \forall f\in L^\infty(\mathbb{B}_n),\forall\psi\in\mathbb{A}ut(\mathbb{B}_n).$$

In particular, for the identity map which is in $\mathbb{A}ut(\mathbb{B}_n)$, we get

$$f(0)=\int_{\mathbb{B}_n}f(w)dA(w),\forall f\in\mathcal{H}^\infty(\mathbb{B}_n).$$

Inserting $K_w(z)K_w^{-1}(z)$ and noticing that $(1\otimes 1)f=f(0)$, we obtain

$$(1\otimes 1)f=\int_{\mathbb{B}_n}K_w^{-1}(z)K_w(z)f(w)dA(w),\forall f\in\mathcal{H}^\infty(\mathbb{B}_n). \tag{3.8}$$

Owing to Formula (3.3), we infer that

$$\begin{aligned}(1\otimes 1)f&=\sum_{j=0}^{n}\sum_{|m|=j}\lambda_{j,m}z^m\int_{\mathbb{B}_n}\overline{w}^m f(w)\overline{K_z(w)}dA(w)\\&=\sum_{j=0}^{n}\sum_{|m|=j}\lambda_{j,m}z^m\left(T_{\overline{w}^m}f\right)(z),\ \ \forall f\in\mathcal{H}^\infty(\mathbb{B}_n).\end{aligned}$$

Therefore, we obtain the following operator identity

$$(1\otimes 1)=\sum_{j=0}^{n}\sum_{|m|=j}\lambda_{j,m}T_{z^m}T_{\overline{w}^m}.$$

Introducing the unitary operator $\mathbb{U}_w$, we get

$$\mathbb{U}_w(1\otimes 1)\mathbb{U}_w=\sum_{j=0}^{n}\sum_{|m|=j}\lambda_{j,m}\left(\mathbb{U}_wT_{z^m}\mathbb{U}_w\right)\left(\mathbb{U}_wT_{\overline{w}^m}\mathbb{U}_w\right).$$

Notice also that by Formulas (3.1) and (3.6), we have

$$\mathbb{U}_w(1 \otimes 1)\mathbb{U}_w = (\mathbb{U}_w 1) \otimes (\mathbb{U}_w 1) = k_w \otimes k_w.$$

Using the latter two equations along with Identity (3.2), we infer that on $\mathcal{H}^2(\mathbb{B}_n)$ we have

$$k_w \otimes k_w = \sum_{j=0}^{n} \sum_{|m|=j} \lambda_{j,m} T_{\varphi_w^m} T_{\overline{\varphi_w}^m}, \ \forall w \in B_n,$$

which is valid on the sphere as well. □

Proposition 3.1 suggests the introduction of the following transformation of operators: for a bounded linear operator T on $\left(\mathcal{H}^2(\mathbb{S}_n)\right)^\perp$ and $w \in \mathbb{B}_n$, define the linear operator $\mathscr{S}_w(T)$ by

$$\mathscr{S}_w(T) = \sum_{j=0}^{n} \sum_{|m|=j} \lambda_{j,m} \ \mathcal{S}_{\varphi_w^m} T \mathcal{S}_{\overline{\varphi_w}^m}. \tag{3.9}$$

This pioneering operator $\mathscr{S}_w$ has an amazing story. It has been discovered first by K. Stroethoff and D. Zheng [11] in the Bergman space setting, where it looks like a two-term perturbation of the identity. In the setting of the Hardy space on the circle, it was adopted by H. Guediri and took the form of a one-term perturbation of the identity. In Lu & Chang [8], in H. Guediri [5] and in the present paper it takes the form of a multi-term perturbation of the identity as in the latter formula (3.9). This phenomenon seems to be very connected to the degree of the denominator, (equivalently to the dimension of the manifold), in the reproducing kernel expression of the underlying space.

The operator $\mathscr{S}_w$ reveals on a characterization of Hardy space dual Toeplitz operators:

Proposition 3.2. *If $\mathcal{S}_f$ is a dual Toeplitz operator on $\left(\mathcal{H}^2(\mathbb{S}_n)\right)^\perp$, then*

$$\mathscr{S}_w(\mathcal{S}_f) = 0, \quad \textit{for all } w \in \mathbb{B}_n.$$

Proof. Fix a $w \in \mathbb{B}_n$ and consider a dual Toeplitz operator $\mathcal{S}_f$ on $\left(\mathcal{H}^2(\mathbb{S}_n)\right)^\perp$, with symbol $f \in L^\infty(\mathbb{S}_n)$. By (3.9), we have

$$\mathscr{S}_w(\mathcal{S}_f) = \sum_{j=0}^{n} \sum_{|m|=j} \lambda_{j,m} \ \mathcal{S}_{\varphi_w^m} \mathcal{S}_f \mathcal{S}_{\overline{\varphi_w}^m} = \sum_{j=0}^{n} \sum_{|m|=j} \lambda_{j,m} \ \mathcal{S}_{|\varphi_w^m|^2 f} = \mathcal{S}_\Psi,$$

with $\Psi = f \sum_{j=0}^{n} \sum_{|m|=j} \lambda_{j,m} |\varphi_w^m|^2$.

Now, applying Formula (3.3) to $\varphi_w(z)$ with $z \in \mathbb{S}_n$, and invoking Identity (3.5), we see that

$$\sum_{j=0}^{n} \sum_{|m|=j} \lambda_{j,m} |\varphi_w^m(z)|^2 = (1 - \langle \varphi_w(z), \varphi_w(z) \rangle)^n = 0.$$

Therefore, we see that $\mathscr{S}_w(\mathcal{S}_f) = 0$. □

Remark 3.3. *Theorem 3.2 can be restated as follows: if T is a dual Toeplitz operator, then* $T \in \bigcap_{w\in\mathbb{D}} \ker \mathscr{S}_w$.

It seems to be interesting to see whether the following complete characterization is valid: given $w \in \mathbb{D}$ (fixed), T is a dual Toeplitz operator if and only if $\mathscr{S}_w(T) = 0$?

The following novel assertion plays a central role in the sequel:

Theorem 3.4. *Let T be a compact operator on $\left(\mathcal{H}^2(\mathbb{S}_n)\right)^{\perp}$, then $\|\mathscr{S}_w(T)\| \longrightarrow 0$ as $|w| \to 1^-$.*

Proof. First, we claim that the operator $\mathscr{S}_w$ admits the following representation:

$$\mathscr{S}_w(T) = \sum_{j=0}^{n-1}\sum_{|m|=j} \lambda_{j,m}^{(n-1)} S_{\varphi_w^m}\left(T - \sum_{i=1}^{n} S_{\varphi_{w,i}} T S_{\overline{\varphi_{w,i}}}\right) S_{\overline{\varphi_w}^m}, \tag{3.10}$$

where $\varphi_{w,i}$ denotes the ith component of φ_w and $\lambda_{j,m}^{(n-1)} = \dfrac{(-1)^j(n-1)!}{(n-1-j)!m!}$.

Indeed, setting $\alpha_i = (\underbrace{0,0,\ldots,1}_{i\text{th component}},0,\ldots,0)$ for $1 \le i \le n$, we have

$$\begin{aligned}
&\sum_{j=0}^{n-1}\sum_{|m|=j} \lambda_{j,m}^{(n-1)} S_{\varphi_w^m}\left(T - \sum_{i=1}^{n} S_{\varphi_{w,i}} T S_{\overline{\varphi_{w,i}}}\right) S_{\overline{\varphi_w}^m}\\
&= \sum_{j=0}^{n-1}\sum_{|m|=j} \lambda_{j,m}^{(n-1)} S_{\varphi_w^m} T S_{\overline{\varphi_w}^m} - \sum_{j=0}^{n-1}\sum_{|m|=j} \lambda_{j,m}^{(n-1)} \sum_{i=1}^{n} S_{\varphi_w^{m+\alpha_i}} T S_{\overline{\varphi_w}^{m+\alpha_i}}\\
&= \sum_{j=0}^{n-1}\sum_{|m|=j} \lambda_{j,m}^{(n-1)} S_{\varphi_w^m} T S_{\overline{\varphi_w}^m} - \sum_{k=1}^{n}\sum_{|p|=k} |p|\lambda_{k-1,p}^{(n-1)} S_{\varphi_w^p} T S_{\overline{\varphi_w}^p}\\
&= T + \sum_{j=1}^{n-1}\sum_{|m|=j} \left(\lambda_{j,m}^{(n-1)} - |m|\lambda_{j-1,m}^{(n-1)}\right) S_{\varphi_w^m} T S_{\overline{\varphi_w}^m} - \sum_{|m|=n} n\lambda_{n-1,m}^{(n-1)} S_{\varphi_w^m} T S_{\overline{\varphi_w}^m}\\
&= T + \sum_{j=1}^{n-1}\sum_{|m|=j} \lambda_{j,m} S_{\varphi_w^m} T S_{\overline{\varphi_w}^m} + \sum_{|m|=n} \lambda_{n,m} S_{\varphi_w^m} T S_{\overline{\varphi_w}^m}\\
&= \sum_{j=0}^{n}\sum_{|m|=j} \lambda_{j,m} S_{\varphi_w^m} T S_{\overline{\varphi_w}^m} = \mathscr{S}_w(T).
\end{aligned}$$

Next, using identity (3.10), we only need to verify that

$$\left\| T - \sum_{i=1}^{n} S_{\varphi_{w,i}} T S_{\overline{\varphi_{w,i}}} \right\| \longrightarrow 0 \quad \text{as} \quad |w| \longrightarrow 1^-. \tag{3.11}$$

Owing to the density of finite rank operators in the set of compact operators, we only need to verify the latter for rank one operators. For let $f, g \in \left(\mathcal{H}^2(\mathbb{S}_n)\right)^{\perp}$;

then one has

$$\left\| f\otimes g-\sum_{i=1}^{n}\mathcal{S}_{\varphi_{w,i}}(f\otimes g)\mathcal{S}_{\overline{\varphi_{w,i}}}\right\| = \left\|\sum_{i=1}^{n}\{(\zeta_i f)\otimes(\zeta_i g)-(\mathcal{S}_{\varphi_{w,i}}f)\otimes(\mathcal{S}_{\varphi_{w,i}}g)\}\right\|$$

$$\leq \sum_{i=1}^{n}\left\{\left\|(\zeta_i f-\mathcal{S}_{\varphi_{w,i}}f)\otimes(\zeta_i g)\right\|+\left\|(\mathcal{S}_{\varphi_{w,i}}f)\otimes(\zeta_i g-\mathcal{S}_{\varphi_{w,i}}g)\right\|\right\}. \tag{3.12}$$

Now, for $z\in\mathbb{S}_n$ and $w\in\mathbb{B}_n$, observe that $w-\varphi_w(z)\longrightarrow 0$ *a.e.* as $|w|\to 1^-$; and thus componentwise, for $i=1,2,3,\ldots,n$, we have $w_i-\varphi_{w,i}(z)\longrightarrow 0$ *a.e.* as $|w|\to 1^-$. Making appeal to the dominated convergence theorem, we infer that for $f\in\left(\mathcal{H}^2(\mathbb{S}_n)\right)^{\perp}$ one has

$$\|w_i f-\varphi_{w,i}f\|_2^2=\int_{\mathbb{S}_n}|w_i f(z)-\varphi_{w,i}(z)f(z)|^2\,d\sigma(z)\longrightarrow 0 \text{ as } |w|\longrightarrow 1^-.$$

Hence, for $i=1,2,3,\ldots,n$, we see that $\|\zeta_i f-\varphi_{w,i}f\|_2\longrightarrow 0$ as $\mathbb{B}_n\ni w\longrightarrow\zeta\in\mathbb{S}_n$. Because of the identity $(I-\mathcal{P})(\zeta_i f(z))=\zeta_i f(z)$, we see that

$$\left\|\zeta_i f-\mathcal{S}_{\varphi_{w,i}}f\right\|_2=\|(I-\mathcal{P})\left(\zeta_i f-\varphi_{w,i}f\right)\|_2\longrightarrow 0 \text{ as } \mathbb{B}_n\ni w\longrightarrow\zeta\in\mathbb{S}_n.$$

The latter together with Inequality (3.12) yield:

$$\left\| f\otimes g-\sum_{i=1}^{n}\mathcal{S}_{\varphi_{w,i}}(f\otimes g)\mathcal{S}_{\overline{\varphi_{w,i}}}\right\|\longrightarrow 0 \text{ as } \mathbb{B}_n\ni w\longrightarrow\zeta\in\mathbb{S}_n.$$ □

4. Products of dual Toeplitz operators

Lemma 2.1 suggests that $\mathcal{S}_f$ and $\mathcal{S}_g$ commute if f and g are both analytic or conjugate analytic. If a non-trivial linear combination of the symbols f and g is constant, they do commute as well. In this section, we are interested to see whether these are the only cases where commutativity holds. The same question in related settings has been considered for instance in [2, 11]. An answer to this question [5] is reported again in the following:

Theorem 4.1. *Suppose that φ,ψ are bounded functions on the unit sphere $\mathbb{S}^n$. Then, the corresponding dual Toeplitz operators commute on $\left(\mathcal{H}^2(\mathbb{S}_n)\right)^{\perp}$, (i.e., $\mathcal{S}_\varphi\mathcal{S}_\psi=\mathcal{S}_\psi\mathcal{S}_\varphi$), if and only if φ and ψ satisfy one of the following conditions:*

1. *They are both analytic on $\mathbb{S}_n$.*
2. *They are both co-analytic on $\mathbb{S}_n$.*
3. *One of them is constant on $\mathbb{S}_n$.*
4. *A non-trivial linear combination of them is constant on $\mathbb{S}_n$.*

Proof. The if part is trivial due to Lemma 2.1. Regarding the only if part, observe that by Proposition 3.1 and parts (i) and (ii) of Lemma 2.1 one has

$$H_f(k_w\otimes k_w)H_{\overline{g}}^*=\sum_{j=0}^{n}\sum_{|m|=j}\lambda_{j,m}\left(\mathcal{S}_{\varphi_w^m}H_f\right)\left(H_{\overline{g}}^*\mathcal{S}_{\overline{\varphi_w}^m}\right)=\mathscr{S}_w(H_fH_{\overline{g}}^*). \tag{4.1}$$

Similarly, we have

$$H_g(k_w \otimes k_w)H_{\overline{f}}^* = \mathscr{S}_w(H_g H_{\overline{f}}^*). \tag{4.2}$$

Combining the two latter identities and owing again to Identity (3.6) as well as to Equation (2.2), we see that

$$(H_g k_w) \otimes (H_{\overline{f}} k_w) - (H_f k_w) \otimes (H_{\overline{g}} k_w) = \mathscr{S}_w([\mathcal{S}_f, \mathcal{S}_g]).$$

By assumption, we get

$$(H_g k_w) \otimes (H_{\overline{f}} k_w) = (H_f k_w) \otimes (H_{\overline{g}} k_w), \forall w \in \mathbb{B}_n.$$

In particular, for $w = 0$ one has $k_0 = 1$; whence $H_g 1 \otimes H_{\overline{f}} 1 = H_f 1 \otimes H_{\overline{g}} 1$, which can be rewritten as

$$\left\langle h, H_{\overline{f}} 1 \right\rangle H_g 1 = \langle h, H_{\overline{g}} 1 \rangle H_f 1, \forall h \in \left(\mathcal{H}^2(\mathbb{B}_n)\right)^{\perp}.$$

At this stage, we distinguish several cases:

1) If $H_g 1 = 0$, then g is analytic. Also we must have either $H_f 1 = 0$ or $H_{\overline{g}} 1 = 0$, which means that either f is analytic, (which corresponds to condition (1)), or g is co-analytic, (in this case g must be constant, which corresponds to condition (3)).

2) If $H_{\overline{g}} 1 = 0$, then g is co-analytic. Also we see that either $H_g 1 = 0$ or $H_{\overline{f}} 1 = 0$. This means that either g is analytic, (which implies that g is constant and corresponds to condition (3)), or f is co-analytic, (which agrees with condition (2)).

3) If both $H_g 1 \neq 0$ and $H_{\overline{g}} 1 \neq 0$, then there exists a complex number $\lambda \neq 0$ such that $H_f 1 = \lambda H_g 1$ and $H_{\overline{f}} 1 = \overline{\lambda} H_{\overline{g}} 1$. That is to say $\mathcal{Q}(f - \lambda g) = \mathcal{Q}(\overline{f} - \overline{\lambda}\overline{g}) = 0$; whence $f - \lambda g$ and $\overline{f} - \overline{\lambda}\overline{g}$ are both analytic. Thus $f - \lambda g$ is constant, which corresponds to condition (4). □

Products of bounded dual Toeplitz operators can be bounded operators in numerous cases. But the crucial question is when does the product of two dual Toeplitz operators produce a dual Toeplitz operator? The answer of this question [5] is given in the following Brown-Halmos type theorem:

Theorem 4.2. *Let f and g be in $L^\infty(\mathbb{S}_n)$. Then, the dual Toeplitz product $\mathcal{S}_f \mathcal{S}_g$ is again a dual Toeplitz operator if and only if one of the following conditions holds:*

1. *f is analytic.*
2. *g is co-analytic.*

In either cases $\mathcal{S}_f \mathcal{S}_g = \mathcal{S}_{fg}$.

Proof. From the elementary properties of dual Toeplitz operators, namely Lemma 2.1, the "if part" is obvious, whereas the "only if part" is less trivial. For, suppose that $\mathcal{S}_f \mathcal{S}_g = \mathcal{S}_h$ for some $h \in L^\infty(\mathbb{S}_n)$. From Identity (2.1), we have

$$0 = \mathcal{S}_h - \mathcal{S}_f \mathcal{S}_g = \mathcal{S}_{h-fg} + H_f H_{\overline{g}}^*.$$

Introducing the operator $\mathscr{S}_w$, we see from Relation (4.1) that

$$\mathscr{S}_w(S_{fg-h}) = \mathscr{S}_w\left(H_f H_{\overline{g}}^*\right) = H_f(k_w \otimes k_w)H_{\overline{g}}^*. \tag{4.3}$$

Since S_{fg-h} is a dual Toeplitz operator, Proposition 3.2 reduces Equation (4.3) to

$$(H_f(k_w)) \otimes (H_{\overline{g}}(k_w)) = 0.$$

In particular, if $w = 0$ one gets $k_0 = 1$; whence we obtain

$$(H_f 1) \otimes (H_{\overline{g}}^* 1) = 0.$$

Since $\|H_f 1 \otimes H_{\overline{g}} 1\| = \|H_f 1\| \, \|H_{\overline{g}} 1\|$, we see that at least one of the two factors vanishes. Therefore, we have two possibilities

- If $H_f 1 = 0$, we see that f is analytic, (which corresponds to condition (1)).
- If $H_{\overline{g}} 1 = 0$, then $\overline{g}$ is analytic, whence g is co-analytic, (which corresponds to (2)).

The additional conclusion of the theorem, then, follows from Lemma 2.1. □

The so-called zero product problem is then a simple corollary of the latter:

Corollary 4.3. *$S_f S_g = 0$ if and only if either $f = 0$ or $g = 0$; i.e., among the class of dual Toeplitz operators on $\left(\mathcal{H}^2(\mathbb{B}_n)\right)^{\perp}$ there are no zero divisors.*

5. Haplitz products

Based on the above concepts, (namely the operator $\mathscr{S}_w$ defined by (3.9), Proposition 3.1 and Theorem 3.4), we now discuss certain characterizations of boundedness and compactness of Hankel products $H_f H_g^*$ as well as of mixed Haplitz products $T_f H_g^*$ and $H_g T_f$ on the sphere. Notice that Toeplitz products on the circle have been studied by Zheng [14], whereas Hankel and mixed Haplitz products have been discussed by Hamada [6]. In case of several complex variables, analog investigations have been done by Zheng [13], Nie [9], Xia [12], Le [7] and Lu & Shang [8].

The following theorem gives a necessary condition for the boundedness of a Hankel product $H_f H_g^*$:

Theorem 5.1. *Let f and g be in $L^2(\mathbb{S}_n)$. If the Hankel product $H_f H_g^*$ is bounded, then*

$$\sup_{w \in \mathbb{B}_n} \|f \circ \varphi_w - \mathcal{P}(f \circ \varphi_w)\|_2 \, \|g \circ \varphi_w - \mathcal{P}(g \circ \varphi_w)\|_2 < \infty. \tag{5.1}$$

Proof. According to Zheng [13], we have

$$\|H_f k_w\|_2 \, \|H_g k_w\|_2 = \|f \circ \varphi_w - \mathcal{P}(f \circ \varphi_w)\|_2 \, \|g \circ \varphi_w - \mathcal{P}(g \circ \varphi_w)\|_2 . \tag{5.2}$$

On the other hand, by the above norm formula of rank one operators and Equation (3.6), we have

$$\|H_f k_w\|_2 \, \|H_g k_w\|_2 = \|(H_f k_w) \otimes (H_g k_w)\| = \left\|H_f(k_w \otimes k_w)H_g^*\right\| . \tag{5.3}$$

So, it suffices that the R.H.S. of the latter is bounded.

Since $\varphi_w \in \mathcal{H}^\infty(\mathbb{S}_n)$, we see by Lemma 2.1 that $H_f T_{\varphi_w} = \mathcal{S}_{\varphi_w} H_f$ and $T_{\overline{\varphi_w}} H_g^* = H_g^* \mathcal{S}_{\overline{\varphi_w}}$. Thus, inserting H_f and H_g^* into Formula (3.7), we see that

$$\begin{aligned} H_f(k_w \otimes k_w)H_g^* &= \sum_{j=0}^{n} \sum_{|m|=j} \lambda_{j,m} H_f T_{\varphi_w^m} T_{\overline{\varphi_w}^m} H_g^* \\ &= \sum_{j=0}^{n} \sum_{|m|=j} \lambda_{j,m} \mathcal{S}_{\varphi_w^m} \left(H_f H_g^*\right) \mathcal{S}_{\overline{\varphi_w}^m} = \mathscr{S}_w \left(H_f H_g^*\right). \end{aligned} \tag{5.4}$$

By Proposition 2.2, we have $\left\|\mathcal{S}_{\varphi_w^m}\right\| = \left\|\mathcal{S}_{\overline{\varphi_w}^m}\right\| = \|\varphi_w^m\|_\infty \le 1$. Thus, we infer that

$$\begin{aligned} \left\|H_f(k_w \otimes k_w)H_g^*\right\| &\le \sum_{j=0}^{n} \sum_{|m|=j} |\lambda_{j,m}| \left\|\mathcal{S}_{\varphi_w^m}\right\| \left\|H_f H_g^*\right\| \left\|\mathcal{S}_{\overline{\varphi_w}^m}\right\| \\ &\le \left\|H_f H_g^*\right\| \sum_{j=0}^{n} \sum_{|m|=j} |\lambda_{j,m}| < \infty; \end{aligned} \tag{5.5}$$

whence, the theorem is proved. □

The following result gives a necessary condition for the compactness of a Hankel product $H_f H_g^*$. Notice that the compactness matter in the "dual case" of $H_f^* H_g$ has been considered by J. Xia [12]. In that paper, J. Xia proves that Condition (5.6) fails to be necessary for the compactness of $H_f^* H_g$. Later on, T. Le [7] provided a certain progress in this direction.

Theorem 5.2. *Let f and g be in $L^2(\mathbb{S}_n)$. If the Hankel product $H_f H_g^*$ is compact, then*

$$\lim_{w \to \mathbb{S}_n} \|f \circ \varphi_w - \mathcal{P}(f \circ \varphi_w)\|_2 \, \|g \circ \varphi_w - \mathcal{P}(g \circ \varphi_w)\|_2 = 0. \tag{5.6}$$

Proof. By Equations (5.2), (5.3) and (5.4), we see that

$$\|f \circ \varphi_w - \mathcal{P}(f \circ \varphi_w)\|_2 \, \|g \circ \varphi_w - \mathcal{P}(g \circ \varphi_w)\|_2 = \left\|\mathscr{S}_w \left(H_f H_g^*\right)\right\|. \tag{5.7}$$

Consequently, if $H_f H_g^*$ is compact, we see by Theorem 3.4 that

$$\lim_{w \to \mathbb{S}_n} \left\|\mathscr{S}_w \left(H_f H_g^*\right)\right\| = 0,$$

and the claimed assertion follows. □

Similar characterizations of bounded and compact mixed Haplitz products $T_f H_g^*$ and $H_g T_f$ are given as follows:

Theorem 5.3. *Let f be in $\mathcal{H}^2(\mathbb{S}_n)$ and g be in $L^2(\mathbb{S}_n)$. If one of the mixed Haplitz products $T_f H_g^*$ or $H_g T_{\overline{f}}$ is bounded, then*

$$\sup_{w \in \mathbb{B}_n} \|f \circ \varphi_w\|_2 \, \|g \circ \varphi_w - \mathcal{P}(g \circ \varphi_w)\|_2 < \infty.$$

Proof. Relying on the fact that $\varphi_w \in \mathcal{H}^\infty(\mathbb{S}_n)$ and owing to the analyticity of f, we see by Lemma 2.1 that $T_f T_{\varphi_w} = T_{\varphi_w} T_f$ and $T_{\overline{\varphi_w}} H_g^* = H_g^* \mathcal{S}_{\overline{\varphi_w}}$. Thus, as in the proof of Theorem 5.1, inserting T_f and H_g^* into Formula (3.7), we see that

$$T_f(k_w \otimes k_w)H_g^* = \sum_{j=0}^{n} \sum_{|m|=j} \lambda_{j,m} T_f T_{\varphi_w^m} T_{\overline{\varphi_w}^m} H_g^* = \sum_{j=0}^{n} \sum_{|m|=j} \lambda_{j,m} T_{\varphi_w^m} \left(T_f H_g^*\right) \mathcal{S}_{\overline{\varphi_w}^m}. \tag{5.8}$$

Estimating the norms of Toeplitz and dual Toeplitz operators with automorphic symbols, we get $\|T_{\varphi_w^m}\| \leq 1$ and $\|\mathcal{S}_{\overline{\varphi_w}^m}\| \leq 1$. Thus, if $T_f H_g^*$ is bounded, we infer that

$$\left\|T_f(k_w \otimes k_w)H_g^*\right\| \leq \left\|T_f H_g^*\right\| \sum\nolimits_{j=0}^{n} \sum\nolimits_{|m|=j} |\lambda_{j,m}| < \infty. \tag{5.9}$$

Whence, as in Equations (5.2) and (5.3), we obtain the claimed estimate. Similar argument can be used to handel the remaining case. □

Compact mixed Haplitz products can also be characterized similarly:

Theorem 5.4. *Let $f \in \mathcal{H}^\infty(\mathbb{S}_n)$ and $g \in L^2(\mathbb{S}_n)$. If one of the mixed Haplitz products $T_f H_g^*$ or $H_g T_{\overline{f}}$ is compact, then*

$$\lim_{w \to \mathbb{S}_n} \|f \circ \varphi_w\|_2 \, \|g \circ \varphi_w - \mathcal{P}(g \circ \varphi_w)\|_2 = 0.$$

Proof. As in the proof of Theorem 3.4, for any operator $T : \left(\mathcal{H}^2(\mathbb{S}_n)\right)^\perp \longrightarrow \mathcal{H}^2(\mathbb{S}_n)$, we have

$$\sum_{j=0}^{n} \sum_{|m|=j} \lambda_{j,m} T_{\varphi_w^m} T \mathcal{S}_{\overline{\varphi_w}^m} = \sum_{j=0}^{n-1} \sum_{|m|=j} \lambda_{j,m}^{(n-1)} T_{\varphi_w^m} \left(T - \sum_{i=1}^{n} T_{\varphi_{w,i}} T \mathcal{S}_{\overline{\varphi_{w,i}}}\right) \mathcal{S}_{\overline{\varphi_w}^m}. \tag{5.10}$$

We claim that if such a T is compact, then

$$\lim_{|w| \longrightarrow 1^-} \sum\nolimits_{j=0}^{n} \sum\nolimits_{|m|=j} \lambda_{j,m} T_{\varphi_w^m} T \mathcal{S}_{\overline{\varphi_w}^m} = 0. \tag{5.11}$$

By Identity (5.10), we only need to verify that

$$\left\|T - \sum\nolimits_{i=1}^{n} T_{\varphi_{w,i}} T \mathcal{S}_{\overline{\varphi_{w,i}}}\right\| \longrightarrow 0 \text{ as } |w| \longrightarrow 1^-. \tag{5.12}$$

Using the density of finite rank operators in the set of compact operators, we only need to verify the latter for rank one operators acting from $\left(\mathcal{H}^2(\mathbb{S}_n)\right)^\perp$ into $\mathcal{H}^2(\mathbb{S}_n)$. For let $f \in \mathcal{H}^2(\mathbb{S}_n)$ and $g \in \left(\mathcal{H}^2(\mathbb{S}_n)\right)^\perp$. Then, one has

$$\begin{aligned} &\left\|f \otimes g - \sum\nolimits_{i=1}^{n} T_{\varphi_{w,i}}(f \otimes g)\mathcal{S}_{\overline{\varphi_{w,i}}}\right\| \\ &\leq \sum\nolimits_{i=1}^{n} \left\{\left\|(\zeta_i f - T_{\varphi_{w,i}} f) \otimes (\zeta_i g)\right\| + \left\|(T_{\varphi_{w,i}} f) \otimes (\zeta_i g - \mathcal{S}_{\varphi_{w,i}} g)\right\|\right\}. \end{aligned} \tag{5.13}$$

Now, for $z \in \mathbb{S}_n$ and $w \in \mathbb{B}_n$, observe that $w - \varphi_w(z) \longrightarrow 0$ *a.e.* as $|w| \to 1^-$; and thus componentwise, for $i = 1, 2, 3, \ldots, n$, we have $w_i - \varphi_{w,i}(z) \longrightarrow 0$ *a.e.* as $|w| \to 1^-$. Making use of the dominated convergence theorem, we infer that for

$f \in \mathcal{H}^2(\mathbb{S}_n)$ and $g \in \left(\mathcal{H}^2(\mathbb{S}_n)\right)^{\perp}$ one has

$$\|w_i f - \varphi_{w,i} f\|_2^2 = \int_{\mathbb{S}_n} |w_i f(z) - \varphi_{w,i}(z) f(z)|^2 \, d\sigma(z) \longrightarrow 0 \text{ as } |w| \longrightarrow 1^-,$$

and

$$\|w_i g - \varphi_{w,i} g\|_2^2 = \int_{\mathbb{S}_n} |w_i g(z) - \varphi_{w,i}(z) g(z)|^2 \, d\sigma(z) \longrightarrow 0 \text{ as } |w| \longrightarrow 1^-.$$

Hence, for $i = 1, 2, 3, \ldots, n$, we see that

$$\|\zeta_i f - \varphi_{w,i} f\|_2 \longrightarrow 0 \quad \text{and} \quad \|\zeta_i g - \varphi_{w,i} g\|_2 \longrightarrow 0$$

as $\mathbb{B}_n \ni w \longrightarrow \zeta \in \mathbb{S}_n$. Because of the identities $\mathcal{P}(\zeta_i f(z)) = \zeta_i f(z)$ and $(I - \mathcal{P})(\zeta_i g(z)) = \zeta_i g(z)$, we see that

$$\left\|\zeta_i f - T_{\varphi_{w,i}} f\right\|_2 = \|\mathcal{P}\left(\zeta_i f - \varphi_{w,i} f\right)\|_2 \longrightarrow 0 \text{ as } \mathbb{B}_n \ni w \longrightarrow \zeta \in \mathbb{S}_n,$$

and

$$\left\|\zeta_i g - \mathcal{S}_{\varphi_{w,i}} g\right\|_2 = \|(I - \mathcal{P})\left(\zeta_i g - \varphi_{w,i} g\right)\|_2 \longrightarrow 0 \text{ as } \mathbb{B}_n \ni w \longrightarrow \zeta \in \mathbb{S}_n.$$

Combining the latter two limits together with Inequality (5.13), we infer that

$$\left\| f \otimes g - \sum_{i=1}^{n} T_{\varphi_{w,i}} (f \otimes g) \mathcal{S}_{\overline{\varphi_{w,i}}} \right\| \longrightarrow 0 \text{ as } \mathbb{B}_n \ni w \longrightarrow \zeta \in \mathbb{S}_n;$$

which proves (5.11).

Next, suppose for instance that $T_f H_g^*$ is compact, (the other case related to $H_g T_{\overline{f}}$, can be handled similarly), then by (5.8) and (5.11), we see that

$$\left\|T_f (k_w \otimes k_w) H_g^*\right\| \longrightarrow 0 \text{ as } \mathbb{B}_n \ni w \longrightarrow \zeta \in \mathbb{S}_n.$$

Thus, as in Equations (5.2) and (5.3), we obtain the claimed condition. □

Owing to the alternative representation (2.2) of the commutator of two dual Toeplitz operators, we can characterize its compactness:

Theorem 5.5. *Let f and g be bounded measurable on $\mathbb{S}_n$. If the commutator $[\,\mathcal{S}_f,\ \mathcal{S}_g]$ is compact, then*

$$\left\|(H_g k_w) \otimes \left(H_{\overline{f}} k_w\right) - (H_f k_w) \otimes (H_{\overline{g}} k_w)\right\| \longrightarrow 0 \ \ as \ \ |w| \to 1^-.$$

Proof. Making use of Formulas (2.2) and (5.4), we obtain:

$$\mathscr{S}_w([\,S_f,\ S_g]) = (H_g k_w) \otimes \left(H_{\overline{f}} k_w\right) - (H_f k_w) \otimes \left(H_{\overline{g}} k_w\right).$$

So, if the commutator is compact, then the result follows from Theorem 3.4. □

Acknowledgment. The author would like to thank both of the ICTP (International Center For Theoretical Physics, Trieste, Italy) and the College of Science's Research Center of King Saud University for their invaluable support.

References

[1] A.B. Aleksandrov, *Function theory in the ball*, in: G.M. Khenkin, A.G. Vitushkin (eds.), Encyclopaedia Math. Sci. 8 (Several Complex Variables. II), Springer-Verlag, Berlin, 1994, pp. 107–178.

[2] Y. Chen and T. Yu, *Essentially commuting dual Toeplitz operators on the unit ball.* Adv. Math. (China), **38**, (2009), (4), 453–464.

[3] B.R. Choe; H.W. Koo and Y.J. Lee, *Toeplitz products with pluriharmonic symbols on the Hardy space over the ball.* J. Math. Anal. Appl., **381** (2011), (1), 365–382.

[4] A.M. Davie and N.P. Jewell, *Toeplitz operators in several complex variables.* J. Funct. Anal., **26** (1977), 356–368.

[5] H. Guediri, *Dual Toeplitz operators on the sphere.* Acta. Math. Sinica, Eng. Ser., 2013. DOI:10.1007/s10114-013-1717-z

[6] M. Hamada, *Remark on application of distribution function inequality for Toeplitz and Hankel operators.* Hokkaido Math. J.n, **32** (2003), 193–208.

[7] T.L. Le, *Toeplitz operators on the Hardy and Bergman spaces.* Ph.D. Thesis, State Univ. of New York at Buffalo, 2007.

[8] Y.F. Lu and S.X. Shang, *Bounded Hankel products on the Bergman space of the polydisk.* Cand. J. Math., **61** (2009), (1), 190–204.

[9] F.S. Nie, *Products of Toeplitz operators and products of Hankel operators on the unit sphere.* Ph.D. Thesis, State Univ. of New York at Buffalo, 1997.

[10] W. Rudin, *Function Theory in the Unit Ball of* $\mathbb{C}^n$. Springer, Berlin, 1980.

[11] K. Stroethoff and D. Zheng, *Algebraic and spectral properties of dual Toeplitz operators.* Trans. Amer. Math. Soc., **354** (2002), (6), 2495–2520.

[12] J. Xia, *On the compactness of the product of Hankel operators on the sphere.* Proc. Amer. Math. Soc., **136** (2008), (2), 1375–1384.

[13] D. Zheng, *Toeplitz operators and Hankel operators on the Hardy space of the unit sphere.* J. Funct. Anal., **149** (1997), 1–24.

[14] D. Zheng, *The distribution function inequality and products of Toeplitz operators and Hankel operators.* J. Funct. Anal., **138** (1996), 477–501.

[15] K. Zhu, *Spaces of holomorphic functions in the unit ball.* Grad. Texts in Math., Vol. 226, Springer-Verlag, New York, 2005.

Hocine Guediri
Department of Mathematics, College of Science
King Saud University, P.O. Box 2455
Riyadh 11451, Saudi Arabia
e-mail: `hguediri@ksu.edu.sa`, `hguediri@ictp.it`

Operator Theory:
Advances and Applications, Vol. 236, 257–273

Three-dimensional Direct and Inverse Scattering for the Schrödinger Equation with a General Nonlinearity

Markus Harju and Valery Serov

Abstract. We discuss the direct and inverse scattering theory for the nonlinear Schrödinger equation

$$-\Delta u(x) + h(x, |u(x)|)u(x) = k^2 u(x), \quad x \in \mathbb{R}^3,$$

where h is a very general and possibly singular combination of potentials. We prove first that the direct scattering problem has a unique bounded solution. We establish also the asymptotic behaviour of scattering solutions. A uniqueness result and a representation formula is proved for the inverse scattering problem with general scattering data. The method of Born approximation is applied for the recovery of jumps in the unknown function from general scattering data and fixed angle data.

Mathematics Subject Classification (2010). Primary 35P25; Secondary 35R30.

Keywords. Schrödinger operator, inverse problem, nonlinearity, Born approximation.

1. Introduction

We consider the nonlinear Schrödinger equation

$$-\Delta u(x) + h(x, |u(x)|)u(x) = k^2 u(x), \quad x \in \mathbb{R}^3, \tag{1.1}$$

where $\Delta = \sum_{j=1}^{3} \partial^2/\partial x_j^2$ is the Laplacian and h is a function that satisfies certain assumptions that will be mentioned later. The free space equation ($h \equiv 0$) is solved, e.g., by the incident plane wave $u_0(x, k, \theta) = \mathrm{e}^{\mathrm{i}k(x,\theta)}$ with direction $\theta \in \mathbb{S}^2 := \{x \in \mathbb{R}^3 : |x| = 1\}$ and wavenumber $k > 0$. The difference

$$u_{\mathrm{sc}} := u - u_0 \tag{1.2}$$

is called the (outgoing) scattered wave after it is required to satisfy the Sommerfeld radiation condition

$$\lim_{r\to\infty} r\left(\frac{\partial u_{sc}}{\partial r} - iku_{sc}\right) = 0, \quad r = |x| \tag{1.3}$$

uniformly in all directions. The direct scattering theory studies the effects of u_0 to the model (1.1). The inverse scattering theory treats the coefficients of (1.1) as unknowns and attempts to recover them from the knowledge of u_{sc} far away from the origin. See [1] for an overview of direct and inverse scattering.

The standard setting for scattering investigations has been the linear case ($h(x,|u|) = q(x)$) which is usually called the classical quantum mechanical scattering. It has been studied by several authors in various dimensions and under many different types of scattering data, some of them being limited in the sense that the inverse problem is formally well determined. See, e.g., [3, 6, 7, 8, 9, 10, 11] and the references therein to gain an understanding of the current status of the linear problem.

More recently, some nonlinear equations have received similar attention. In one space dimension studies exist on cubic nonlinearity [17], power nonlinearity [18], saturation model [18] and a very general nonlinearity [15]. Such particular cases with locally bounded coefficients can be met, e.g., in nonlinear optics. These problems have been investigated also in two dimensions [13, 14, 19, 20, 21]. In three dimensions we are only aware of [2, 16]. The present work extends [2] to more general nonlinearities h that may appear in (1.1). Our main goal is to show that the high frequency approach works perfectly also for general nonlinear equations.

This work is organized as follows. In Section 2 we study the direct scattering problem by proving that it has a unique bounded solution, whose scattered wave obeys an asymptotic representation giving us data for the inverse problem. Section 3 is devoted to two inverse problems. Firstly, under the general scattering data we prove Saito's formula which implies a uniqueness result and a representation formula for the unknown function $h_0(x) = h(x,1)$. Secondly, the method of Born approximation is considered for the general scattering data as well as fixed incident angle data. Under both data we establish the recovery of certain jumps of h_0.

2. Direct scattering problem

The scattering solutions (1.1)–(1.3) are the unique solutions of the Lippmann–Schwinger equation

$$u(x,k,\theta) = \mathrm{e}^{\mathrm{i}k(x,\theta)} - \int_{\mathbb{R}^3} G_k^+(|x-y|)h(y,|u(y)|)u(y)\mathrm{d}y, \tag{2.1}$$

where G_k^+ is the outgoing fundamental solution of the corresponding Helmholtz equation and is defined as

$$G_k^+(|x|) = \frac{\mathrm{e}^{\mathrm{i}|k||x|}}{4\pi|x|}.$$

So the function G_k^+ is the kernel of the integral operator $\widehat{G_k^+} := (-\Delta - k^2 - \mathrm{i}0)^{-1}$.

Let us assume the following facts about the nonlinearity h:

(A1) for any $v \in L^\infty(\mathbb{R}^3)$ such that $\|v\|_{L^\infty(\mathbb{R}^3)} \le \rho$ we have $|h(x,|v|)| \le c_\rho \alpha(x)$, where α is such that

$$c_\alpha := \sup_{x\in\mathbb{R}^3} \frac{1}{4\pi}\int_{\mathbb{R}^3} \frac{1}{|x-y|}\alpha(y)\mathrm{d}y < \infty$$

(A2) for any $v_1, v_2 \in L^\infty(\mathbb{R}^3)$ we have $|h(x,|v_1|) - h(x,|v_2|)| \le \beta(x)\big||v_1| - |v_2|\big|$, where β is such that $c_\beta < \infty$.

It is easy to check that $c_\alpha < \infty$ if, for example, $\alpha \in L^p(\mathbb{R}^3) \cap L^1(\mathbb{R}^3)$ for some $3/2 < p \le \infty$. Alternatively, one may use the more narrow space $L^p_{2\delta}(\mathbb{R}^3) \subset L^1(\mathbb{R}^3)$ for $2\delta > 3 - 3/p$ and same p to formulate the assumption in terms of decay at infinity.

Existence of forward solutions requires us to pose an additional size constraint for α, β in the sense of next theorem.

Theorem 2.1. *Let h be as above. If $c_\alpha < 1/c_\rho$ and $c_\beta < (1 - c_\rho c_\alpha)^2$ then the Lippmann–Schwinger equation* (2.1) *has a unique solution $u \in L^\infty(\mathbb{R}^3)$.*

Proof. We consider the ball $B_\rho(0) := \{u \in L^\infty(\mathbb{R}^3) : \|u\|_{L^\infty(\mathbb{R}^3)} \le \rho\}$ and the operator

$$(Tu)(x) := u_0 - \int_{\mathbb{R}^3} G_k^+(|x-y|)h(y,|u|)u(y)\mathrm{d}y.$$

As we are looking for solutions of $Tu = u$ our aim is to apply Banach's fixed point theorem [22] for T. We assume that

$$\frac{1}{1 - c_\rho c_\alpha} \le \rho < \frac{1 - c_\rho c_\alpha}{c_\beta}$$

and show that $T : B_\rho(0) \to B_\rho(0)$ and also that T is a contraction. Let $u \in B_\rho(0)$. Then

$$\begin{aligned}
\|Tu\|_{L^\infty(\mathbb{R}^3)} &= \sup_{x\in\mathbb{R}^3} |u_0 - \int_{\mathbb{R}^3} G_k^+(|x-y|)h(y,|u|)u\mathrm{d}y| \\
&\le 1 + \|u\|_{L^\infty(\mathbb{R}^3)} \sup_{x\in\mathbb{R}^3} \int_{\mathbb{R}^3} |G_k^+(|x-y|)||h(y,|u|)|\mathrm{d}y \\
&\le 1 + \rho c_\rho \sup_{x\in\mathbb{R}^3} \frac{1}{4\pi}\int_{\mathbb{R}^3} \frac{1}{|x-y|}\alpha(y)\mathrm{d}y = 1 + \rho c_\rho c_\alpha \le \rho.
\end{aligned}$$

It remains to show that T is a contraction. Let $u_1, u_2 \in B_\rho(0)$. Then

$$\begin{aligned}
\|Tu_1 - Tu_2\|_{L^\infty(\mathbb{R}^3)} &= \left\| \int_{\mathbb{R}^3} G_k^+(|x-y|)\left(h(y,|u_1|)u_1 - h(y,|u_2|)u_2\right)\mathrm{d}y \right\|_{L^\infty(\mathbb{R}^3)} \\
&= \sup_{x\in\mathbb{R}^3} \Big| \int_{\mathbb{R}^3} G_k^+(|x-y|) \\
&\qquad \times \big((h(y,|u_1|) - h(y,|u_2|))u_1 + h(y,|u_2|)(u_1 - u_2)\big)\mathrm{d}y \Big|
\end{aligned}$$

$$
\begin{aligned}
&\leq \sup_{x\in\mathbb{R}^3}\int_{\mathbb{R}^3}|G_k^+(|x-y|)||h(y,|u_1|)-h(y,|u_2|)||u_1|dy\\
&\quad+\sup_{x\in\mathbb{R}^3}\int_{\mathbb{R}^3}|G_k^+(|x-y|)||h(y,|u_2|)||u_1-u_2|dy\\
&\leq \rho\,\|u_1-u_2\|_{L^\infty(\mathbb{R}^3)}\,c_\beta+\|u_1-u_2\|_{L^\infty(\mathbb{R}^3)}\,c_\rho c_\alpha\\
&=(\rho c_\beta+c_\rho c_\alpha)\,\|u_1-u_2\|_{L^\infty(\mathbb{R}^3)}<\|u_1-u_2\|_{L^\infty(\mathbb{R}^3)}\,.
\end{aligned}
$$

Thus we may conclude that T is a contraction from $B_\rho(0)$ into itself. Therefore it has a unique fixed point in $L^\infty(\mathbb{R}^3)$. □

Remark 2.2. In two dimensions existence and uniqueness holds for large k without size constraints, see [20]. In three dimensions that approach is not available since G_k^+ loses its dependance on k after taking the modulus. For the same reason, we are not able to include higher dimensions in this work since in general G_k^+ retains dependance on k after modulus.

Theorem 2.3. *Let h be as in Theorem 2.1. Assume that α,β belong to $L^p_{2\delta}(\mathbb{R}^3)$ with $2\delta>3-3/p$ and $3/2<p\leq\infty$. Then for fixed $k>0$ the solution $u(x,k,\theta)$ admits the representation*

$$
u(x,k,\theta)=\mathrm{e}^{\mathrm{i}k(x,\theta)}-\frac{\mathrm{e}^{\mathrm{i}k|x|}}{4\pi|x|}A(k,\theta',\theta)+o\left(\frac{1}{|x|}\right),\quad |x|\to+\infty \tag{2.2}
$$

uniformly with respect to $\theta\in\mathbb{S}^2$. The function $A(k,\theta',\theta)$ is called the scattering amplitude and is defined as

$$
A(k,\theta',\theta):=\int_{\mathbb{R}^3}\mathrm{e}^{-\mathrm{i}k(\theta',y)}h(y,|u|)u(y,k,\theta)\mathrm{d}y, \tag{2.3}
$$

where $\theta':=x/|x|\in\mathbb{S}^2$ is the direction of observation (measurement).

Proof. The large $|x|$ behaviour of $|x-y|^{-1}\mathrm{e}^{\mathrm{i}k|x-y|}$ can be studied by dividing it into two cases: $|y|\leq|x|^a$ and $|y|>|x|^a$, where $a\geq0$ is a parameter that we can adjust to our liking. In the first case we have

$$
|x-y|=|x|-(\theta',y)+\mathcal{O}\left(|x|^{2a-1}\right),\quad |y|\leq|x|^a
$$

for $a<1$. In a similar fashion,

$$
|x-y|^{-1}=|x|^{-1}\left(1+\mathcal{O}\left(|x|^{a-1}\right)\right),\quad |y|\leq|x|^a
$$

for $a<1$. Here we have used the fact that $(1+w)^s=1+sw+\mathcal{O}(w^2)$ as $w\to0$ by the Maclaurin series of $(1+w)^s$, $s\in\mathbb{R}$. With $a<1/2$ it follows that

$$
|x-y|^{-1}\mathrm{e}^{\mathrm{i}k|x-y|}=|x|^{-1}\mathrm{e}^{\mathrm{i}k|x|}\mathrm{e}^{-\mathrm{i}k(\theta',y)}+\mathcal{O}(|x|^{2a-2}),\quad |y|\leq|x|^a
$$

as $|x| \to \infty$. Now we split

$$u_{\mathrm{sc}}(x,k,\theta) = -\int_{|y|\le|x|^a} G_k^+(|x-y|)h(y,|u|)u(y,k,\theta)\mathrm{d}y$$
$$-\int_{|y|>|x|^a} G_k^+(|x-y|)h(y,|u|)u(y,k,\theta)\mathrm{d}y =: I_1 + I_2.$$

Consider I_1 first. By the above discussion we may conclude that, for $0 < a < 1/2$, we have

$$\begin{aligned} I_1 &= -\frac{\mathrm{e}^{\mathrm{i}k|x|}}{4\pi|x|}\int_{|y|\le|x|^a} \mathrm{e}^{-\mathrm{i}k(\theta',y)}h(y,|u|)u(y,k,\theta)\mathrm{d}y + \mathcal{O}(|x|^{2a-2}) \\ &= -\frac{\mathrm{e}^{\mathrm{i}k|x|}}{4\pi|x|}\int_{\mathbb{R}^3} \mathrm{e}^{-\mathrm{i}k(\theta',y)}h(y,|u|)u(y,k,\theta)\mathrm{d}y + \mathcal{O}(|x|^{2a-2}) \\ &\quad + \frac{\mathrm{e}^{\mathrm{i}k|x|}}{4\pi|x|}\int_{|y|>|x|^a} \mathrm{e}^{-\mathrm{i}k(\theta',y)}h(y,|u|)u(y,k,\theta)\mathrm{d}y, \quad |x|\to\infty, \end{aligned}$$

where the last term is $o(|x|^{-1})$ since u is bounded and $\alpha \in L^1(\mathbb{R}^3)$. The same is true about $\mathcal{O}(|x|^{2a-2})$ for $a < 1/2$. It means that I_1 is of the form

$$I_1 = -\frac{\mathrm{e}^{\mathrm{i}k|x|}}{4\pi|x|}\int_{\mathbb{R}^3} \mathrm{e}^{-\mathrm{i}k(\theta',y)}h(y,|u|)u(y,k,\theta)\mathrm{d}y + o(|x|^{-1}), \quad |x|\to\infty.$$

Turning now to I_2 we have by Hölder's inequality

$$\begin{aligned} |I_2| &\le C\int_{|y|>|x|^a} \frac{1}{|x-y|}|h(y,|u|)|\mathrm{d}y \\ &\le C\left(\int_{|y|>|x|^a} \frac{(1+|y|)^{-2\delta p'}\mathrm{d}y}{|x-y|^{p'}}\right)^{1/p'} \|\alpha\|_{L^p_{2\delta}(\mathbb{R}^3)}. \end{aligned}$$

Let us prove that the first term on the right-hand side is $o(|x|^{-1})$, or equivalently, that

$$\int_{|y|>|x|^a} \frac{(1+|y|)^{-2\delta p'}\mathrm{d}y}{|x-y|^{p'}}$$

is $o(|x|^{-p'})$. The region $|y| > |x|^a$ is divided into two subregions: $|x|^a < |y| < |x|/2$ and $|y| > |x|/2$. If $|x|^a < |y| < |x|/2$ then $|x-y| \ge |x| - |y| > |x|/2$ and thus

$$\int_{|x|^a<|y|<|x|/2} \frac{(1+|y|)^{-2\delta p'}\mathrm{d}y}{|x-y|^{p'}} \le C|x|^{-p'}\int_{|x|^a<|y|<|x|/2} \frac{\mathrm{d}y}{|y|^{2\delta p'}}.$$

In the region $|y| > |x|/2$ we have, after denoting $z := y/|x|$, that

$$\begin{aligned} \int_{|y|>|x|/2} \frac{(1+|y|)^{-2\delta p'}\mathrm{d}y}{|x-y|^{p'}} &\le C\int_{|y|>|x|/2} \frac{|y|^{-2\delta p'}\mathrm{d}y}{|x-y|^{p'}} = C\int_{|z|>1/2} \frac{|xz|^{-2\delta p'}|x|^3\mathrm{d}z}{|\theta'-z|^{p'}|x|^{p'}} \\ &= C|x|^{3-2\delta p'-p'}\int_{|z|>1/2} \frac{|z|^{-2\delta p'}\mathrm{d}z}{|\theta'-z|^{p'}}. \end{aligned}$$

The last integral converges for $2\delta > 2 - 3/p$ since

$$\int_{2>|z|>1/2} \frac{|z|^{-2\delta p'}\mathrm{d}z}{|\theta' - z|^{p'}} \le C \int_{2>|z|>1/2} \frac{\mathrm{d}z}{|\theta' - z|^{p'}} < \infty$$

and

$$\int_{|z|>2} \frac{|z|^{-2\delta p'}\mathrm{d}z}{|\theta' - z|^{p'}} \le C \int_{|z|>2} \frac{\mathrm{d}z}{|z|^{p'+2\delta p'}} < \infty.$$

Here we have used the basic estimate $|\theta' - z| \ge |z| - 1 > |z|/2$ for $|z| > 2$. These considerations allow us to conclude that for $2\delta > 2 - 3/p$ we have

$$\int_{|y|>|x|^a} \frac{(1+|y|)^{-2\delta p'}\mathrm{d}y}{|x-y|^{p'}} \le C|x|^{-p'} \int_{|x|^a<|y|} \frac{\mathrm{d}y}{|y|^{2\delta p'}} + C|x|^{3-2\delta p'-p'}.$$

But this is $o(|x|^{-p'})$ for $2\delta > 3/p' = 3 - 3/p$. This finishes the proof. □

Remark 2.4. Sometimes (2.3) is taken as a definition of the scattering amplitude without a rigorous proof of its presence in (2.2). We also prefer to avoid assuming that the potentials have compact support or have some strong, pointwise decay at infinity.

We will finish off this section with some simple results that are employed in Section 3.

Lemma 2.5. *Under the same assumptions as in Theorem 2.3 we have*

$$\|u_{\mathrm{sc}}\|_{L^\infty} \to 0, \quad k \to \infty.$$

Proof. Write

$$u_{\mathrm{sc}}(x,k,\theta) = C \int_{\mathbb{R}^3} \mathrm{e}^{\mathrm{i}k|x-y|} \frac{h(y,|u(y)|)u(y)}{|x-y|}\mathrm{d}y,$$

and apply Riemann–Lebesgue lemma. □

The following result from [5] allows us to control u_{sc} in weighted Lebesgue spaces in terms of k. This will be fruitful in the sequel.

Proposition 2.6. *Let $3/2 < p \le \infty$. Then for all $|k| \ge 1$ the limit*

$$(-\Delta - k^2 - \mathrm{i}0)^{-1} := \lim_{\varepsilon\to+0} (-\Delta - k^2 - \mathrm{i}\varepsilon)^{-1}$$

exists in the uniform operator topology from $L_\delta^{\frac{2p}{p+1}}(\mathbb{R}^3)$ to $L_{-\delta}^{\frac{2p}{p-1}}(\mathbb{R}^3)$ with the norm estimate

$$\|(-\Delta - k^2 - \mathrm{i}0)^{-1} f\|_{L_{-\delta}^{\frac{2p}{p-1}}(\mathbb{R}^3)} \le \frac{C}{|k|^\gamma} \|f\|_{L_\delta^{\frac{2p}{p+1}}(\mathbb{R}^3)},$$

where $\gamma = 2 - 3/p$ and $\delta = 0$ if $3/2 < p \le 2$ and $\gamma = 1 - 1/p$ and $\delta > 1/2 - 1/p$ if $2 < p \le \infty$.

By Hölder's inequality $L_{2\delta}^p(\mathbb{R}^3) \cap L^1(\mathbb{R}^3) \subset L_\delta^{\frac{2p}{p+1}}(\mathbb{R}^3)$ for $1 \le p \le \infty$ and $\delta \in \mathbb{R}$. It means that we may apply Proposition 2.6 in the frame of Theorem 2.3.

Corollary 2.7. *Let α, β be as in Theorem 2.3. Then*

$$\|u_{\mathrm{sc}}\|_{L^{\frac{2p}{p-1}}_{-\delta}(\mathbb{R}^3)} \le \frac{C}{|k|^\gamma},$$

where γ is as in Proposition 2.6.

Proof. The claim follows from

$$\begin{aligned}\|u_{\mathrm{sc}}\|_{L^{\frac{2p}{p-1}}_{-\delta}(\mathbb{R}^3)} &= \left\|(-\Delta - k^2 - \mathrm{i}0)^{-1}(hu)\right\|_{L^{\frac{2p}{p-1}}_{-\delta}(\mathbb{R}^3)} \\ &\le c_\rho \|u\|_{L^\infty(\mathbb{R}^3)} \frac{C}{|k|^\gamma} \|\alpha\|_{L^{\frac{2p}{p+1}}_{\delta}(\mathbb{R}^3)}.\end{aligned}$$

□

Corollary 2.8. *Let α, β be as in Theorem 2.3. If $f \in L^p_{2\delta}(\mathbb{R}^3)$ then*

$$\left\||f|^{1/2} u_{\mathrm{sc}}\right\|_{L^2(\mathbb{R}^3)} \le \frac{C}{|k|^\gamma},$$

where γ is as in Proposition 2.6.

Proof. Hölder's inequality and Corollary 2.7. □

3. Inverse scattering problems

The inverse problems that are considered in this section is to recover features of the unknown function

$$h_0(x) := h(x, 1)$$

from the scattering amplitude $A(k, \theta', \theta)$. We consider both the general scattering data

$$D := \{A(k, \theta', \theta) : k > 0, \theta, \theta' \in \mathbb{S}^2\}$$

and the fixed angle scattering data

$$D_{\mathrm{A}} := \{A(k, \theta', \theta) : k > 0, \theta = \theta_0 \text{ fixed}, \theta' \in \mathbb{S}^2\}.$$

In this section we assume that the nonlinearity h possesses the Taylor expansion

$$h(x, 1 + s) = h(x, 1) + h'_s(x, 1)s + h''_s(x, s^*)s^2/2,$$

where $1 < s^* < 1 + s$ and

$$|h'_s(x, 1)| \le \eta_1(x), \quad |h''_s(x, s^*)| \le \eta_2(x)$$

uniformly in $s \in (0, s_0)$ with some $s_0 > 0$ and with $\eta_1, \eta_2 \in L^p_{2\delta}(\mathbb{R})$, where p, δ are as in Theorem 2.3. Then, using

$$|u| = (u\overline{u})^{1/2} = (1 + u_0\overline{u_{\mathrm{sc}}} + u_{\mathrm{sc}}\overline{u_0} + |u_{\mathrm{sc}}|^2)^{1/2} = 1 + \frac{1}{2}u_0\overline{u_{\mathrm{sc}}} + \frac{1}{2}u_{\mathrm{sc}}\overline{u_0} + \mathcal{O}(|u_{\mathrm{sc}}|^2)$$

we may write

$$h(x, |u|) = h_0 + h'_s(x, 1)\left(\frac{1}{2}u_0\overline{u_{\mathrm{sc}}} + \frac{1}{2}u_{\mathrm{sc}}\overline{u_0} + \mathcal{O}(|u_{\mathrm{sc}}|^2)\right) + h''_s(x, s^*)\mathcal{O}(|u_{\mathrm{sc}}|^2).$$

So we obtain

$$h(x,|u|)u = h_0(x)u_0 + g_1(x)u_{sc} + g_2(x)u_0^2\overline{u_{sc}} + \eta(x)\mathcal{O}(|u_{sc}|^2), \qquad (3.1)$$

where $g_2(x) = h_s'(x,1)/2, g_1(x) = h_0(x) + g_2(x)$ and $\eta \in L^p_{2\delta}(\mathbb{R}^3)$ with p, δ as in Theorem 2.3.

3.1. A uniqueness result and a representation formula

For the over-determined data D we obtain uniqueness of the inverse scattering problem as a consequence of the following result.

Theorem 3.1 (Saito's formula). *Under the same assumptions as in Theorem 2.3 the limit*

$$\lim_{k\to\infty} k^2 \int_{\mathbb{S}^2\times\mathbb{S}^2} \mathrm{e}^{-\mathrm{i}k(\theta-\theta',x)} A(k,\theta',\theta)\mathrm{d}\theta\mathrm{d}\theta' = 8\pi^2 \int_{\mathbb{R}^3} \frac{h_0(y)}{|x-y|^2}\mathrm{d}y, \quad x \in \mathbb{R}^3 \qquad (3.2)$$

holds in the sense of distributions for $2 < p < \infty$ and uniformly for $p = \infty$.

Proof. Split the integral of interest as

$$\begin{aligned} I &:= k^2 \int_{\mathbb{S}^2\times\mathbb{S}^2} \mathrm{e}^{-\mathrm{i}k(\theta-\theta',x)} A(k,\theta',\theta)\mathrm{d}\theta\mathrm{d}\theta' \\ &= k^2 \int_{\mathbb{S}^2\times\mathbb{S}^2} \mathrm{e}^{-\mathrm{i}k(\theta-\theta',x)} \int_{\mathbb{R}^3} \mathrm{e}^{-\mathrm{i}k(\theta',y)} h(y,|u|)u(y)\mathrm{d}y\mathrm{d}\theta\mathrm{d}\theta' \\ &= k^2 \int_{\mathbb{S}^2\times\mathbb{S}^2} \mathrm{e}^{-\mathrm{i}k(\theta-\theta',x)} \int_{\mathbb{R}^3} \mathrm{e}^{-\mathrm{i}k(\theta'-\theta,y)} h_0(y)\mathrm{d}y\mathrm{d}\theta\mathrm{d}\theta' \\ &\quad + k^2 \int_{\mathbb{S}^2\times\mathbb{S}^2} \mathrm{e}^{-\mathrm{i}k(\theta-\theta',x)} \int_{\mathbb{R}^3} \mathrm{e}^{-\mathrm{i}k(\theta',y)} [h(y,|u|)u(y) - h_0u_0]\mathrm{d}y\mathrm{d}\theta\mathrm{d}\theta' := I_1 + I_2. \end{aligned}$$

We have [2]

$$I_1 = k^2 \int_{\mathbb{R}^3} h_0(y)\mathrm{d}y \int_{\mathbb{S}^2\times\mathbb{S}^2} \mathrm{e}^{\mathrm{i}k(\theta-\theta',y-x)}\mathrm{d}\theta\mathrm{d}\theta' = 8\pi^3k^2 \int_{\mathbb{R}^3} h_0(y)\mathrm{d}y \frac{J^2_{1/2}(k|x-y|)}{k|x-y|},$$

where $J_{1/2}$ is the Bessel function of the first kind and order 1/2. The presence of $J_{1/2}$ suggests us to divide this integral into two parts according to small and large argument of $J_{1/2}$. The part I_1' over $k|x-y| < 1$ can be estimated as

$$\begin{aligned} |I_1'| &\le Ck^2 \int_{k|x-y|<1} |h_0(y)|\mathrm{d}y \\ &\le Ck^2 \left(\int_{k|x-y|<1} |h_0(y)|^p\mathrm{d}y\right)^{1/p} \left(\int_{k|x-y|<1} 1\mathrm{d}y\right)^{1/p'} = Ck^{2-3/p'} \to 0 \end{aligned}$$

for $3 < p \leq \infty$ by the Hölder inequality. The part over $k|x-y| > 1$ becomes

$$I_1'' = 8\pi^3 k \int_{|x-y|>k^{-1}} \frac{h_0(y)}{|x-y|} \times \left[\sqrt{\frac{2}{\pi k|x-y|}} \cos\left(k|x-y| - \frac{\pi}{2}\right) + \mathcal{O}\left(\frac{1}{(k|x-y|)^{3/2}}\right)\right]^2 \mathrm{d}y$$

$$= 8\pi^3 k \int_{|x-y|>k^{-1}} \frac{h_0(y)\mathrm{d}y}{|x-y|} \left[\frac{2\cos^2(k|x-y| - \frac{\pi}{2})}{\pi k|x-y|} + \mathcal{O}\left(\frac{1}{(k|x-y|)^2}\right)\right]$$

$$= 8\pi^2 \int_{|x-y|>k^{-1}} \frac{h_0(y)\mathrm{d}y}{|x-y|^2} + 8\pi^2 \int_{|x-y|>k^{-1}} \frac{h_0(y)}{|x-y|^2} \cos(2k|x-y| - \pi)\,\mathrm{d}y + \frac{1}{k}\int_{|x-y|>k^{-1}} \frac{|h_0(y)|\mathcal{O}(1)}{|x-y|^3}\mathrm{d}y =: I_1^{(1)} + I_1^{(2)} + I_1^{(3)}.$$

The part $I_1^{(1)}$ tends to right-hand side of (3.2) as $k \to \infty$ for $3 < p \leq \infty$ because

$$\int_{\mathbb{R}^3} \frac{h_0(y)\mathrm{d}y}{|x-y|^2} \leq C \int_{\mathbb{R}^3} \frac{\alpha(y)\mathrm{d}y}{|x-y|^2} < \infty$$

in that case. For the same reason the part $I_1^{(2)}$ tends to zero by Riemann–Lebesgue lemma as $k \to \infty$ for $3 < p \leq \infty$. The part $I_1^{(3)}$ satisfies the estimate

$$|I_1^{(3)}| \leq \frac{C}{k} \int_{|x-y|>k^{-1}} \frac{\alpha(y)}{|x-y|^3}\mathrm{d}y \leq \frac{C}{k^{1-\varepsilon}} \int_{|x-y|>k^{-1}} \frac{\alpha(y)}{|x-y|^{3-\varepsilon}}\mathrm{d}y, \quad \varepsilon > 0.$$

By Hölder's inequality the latter integral converges if we choose $3/p < \varepsilon < 3$. Hence $I_1^{(3)} \to 0$ as $k \to \infty$ upon choosing $3/p < \varepsilon < 1$. Thus

$$\lim_{k\to\infty} I_1 = 8\pi^2 \int_{\mathbb{R}^3} \frac{h_0(y)}{|x-y|^2}\mathrm{d}y$$

for $3 < p \leq \infty$ uniformly in x.

It remains to consider

$$I_2 = k^2 \int_{\mathbb{S}^2\times\mathbb{S}^2} \mathrm{e}^{-\mathrm{i}k(\theta-\theta',x)} \int_{\mathbb{R}^3} \mathrm{e}^{-\mathrm{i}k(\theta',y)}[h(y,|u|)u(y) - h_0u_0]\mathrm{d}y\mathrm{d}\theta\mathrm{d}\theta'$$

$$= Ck^2 \int_{\mathbb{S}^2}\int_{\mathbb{R}^3} \mathrm{e}^{-\mathrm{i}k(\theta,x)} \frac{J_{1/2}(k|x-y|)}{\sqrt{k|x-y|}}[h(y,|u|)u(y) - h_0u_0]\mathrm{d}y\mathrm{d}\theta.$$

Making use of (3.1) leads to

$$I_2 = Ck^2 \int_{\mathbb{S}^2}\int_{\mathbb{R}^3} \mathrm{e}^{-\mathrm{i}k(\theta,x)} \frac{J_{1/2}(k|x-y|)}{\sqrt{k|x-y|}} \times [g_1(y)u_{\mathrm{sc}}(y) + g_2(y)u_0(y)^2\overline{u_{\mathrm{sc}}(y)} + \eta(y)\mathcal{O}(|u_{\mathrm{sc}}|^2)]\mathrm{d}y\mathrm{d}\theta := I_2' + I_2'' + I_2'''.$$

Here the first term is

$$\begin{aligned}I_2' &= Ck^2 \int_{\mathbb{S}^2}\int_{\mathbb{R}^3} \mathrm{e}^{-\mathrm{i}k(\theta,x)} \frac{J_{1/2}(k|x-y|)}{\sqrt{k|x-y|}} g_1(y) u_{\mathrm{sc}}(y)\mathrm{d}y\mathrm{d}\theta \\ &= -Ck^2 \int_{\mathbb{S}^2}\int_{\mathbb{R}^3} \mathrm{e}^{-\mathrm{i}k(\theta,x)} \frac{J_{1/2}(k|x-y|)}{\sqrt{k|x-y|}} g_1(y) \\ &\quad \times \int_{\mathbb{R}^3} G_k^+(|y-z|) h(z,|u(z)|)u(z)\mathrm{d}z\mathrm{d}y\mathrm{d}\theta.\end{aligned}$$

Another application of (3.1) yields

$$\begin{aligned}I_2' &= -Ck^2 \int_{\mathbb{S}^2}\int_{\mathbb{R}^3} \mathrm{e}^{-\mathrm{i}k(\theta,x)} \frac{J_{1/2}(k|x-y|)}{\sqrt{k|x-y|}} g_1(y) \int_{\mathbb{R}^3} G_k^+(|y-z|) h_0(z) u_0(z)\mathrm{d}z\mathrm{d}y\mathrm{d}\theta \\ &- Ck^2 \int_{\mathbb{S}^2}\int_{\mathbb{R}^3} \mathrm{e}^{-\mathrm{i}k(\theta,x)} \frac{J_{1/2}(k|x-y|)}{\sqrt{k|x-y|}} g_1(y) \int_{\mathbb{R}^3} G_k^+(|y-z|) g_1(z) u_{\mathrm{sc}}(z)\mathrm{d}z\mathrm{d}y\mathrm{d}\theta \\ &- Ck^2 \int_{\mathbb{S}^2}\int_{\mathbb{R}^3} \mathrm{e}^{-\mathrm{i}k(\theta,x)} \frac{J_{1/2}(k|x-y|)}{\sqrt{k|x-y|}} g_1(y) \int_{\mathbb{R}^3} G_k^+(|y-z|) g_2(z) u_0^2(z)\overline{u_{\mathrm{sc}}(z)}\mathrm{d}z\mathrm{d}y\mathrm{d}\theta \\ &- Ck^2 \int_{\mathbb{S}^2}\int_{\mathbb{R}^3} \mathrm{e}^{-\mathrm{i}k(\theta,x)} \frac{J_{1/2}(k|x-y|)}{\sqrt{k|x-y|}} g_1(y) \int_{\mathbb{R}^3} G_k^+(|y-z|) \eta(z)\mathcal{O}(|u_{\mathrm{sc}}(z)|^2)\mathrm{d}z\mathrm{d}y\mathrm{d}\theta \\ &=: I_2^{(1)} + I_2^{(2)} + I_2^{(3)} + I_2^{(4)}.\end{aligned}$$

Now

$$\begin{aligned}I_2^{(1)} &= -Ck^2 \int_{\mathbb{S}^2}\int_{\mathbb{R}^3} \mathrm{e}^{-\mathrm{i}k(\theta,x-z)} \frac{J_{1/2}(k|x-y|)}{\sqrt{k|x-y|}} g_1(y) \int_{\mathbb{R}^3} G_k^+(|y-z|) h_0(z)\mathrm{d}z\mathrm{d}y\mathrm{d}\theta \\ &= -Ck^2 \int_{\mathbb{R}^3} \frac{J_{1/2}(k|x-y|)}{\sqrt{k|x-y|}} g_1(y) \int_{\mathbb{R}^3} \frac{J_{1/2}(k|x-z|)}{\sqrt{k|x-z|}} G_k^+(|y-z|) h_0(z)\mathrm{d}z\mathrm{d}y \\ &= -Ck^2 \int_{\mathbb{R}^3} \frac{J_{1/2}(k|x-y|)}{\sqrt{k|x-y|}} g_1(y) \widehat{G_k^+}\left[h_0 \frac{J_{1/2}(k|x-\cdot|)}{\sqrt{k|x-\cdot|}}\right](y)\mathrm{d}y.\end{aligned}$$

Writing $f_{1/2} = |f|^{1/2}\operatorname{sgn} f$ and $f = f_{1/2}|f|^{1/2}$ we may conclude that

$$\begin{aligned}I_2^{(1)} &= -Ck^2 \int_{\mathbb{R}^3} \frac{J_{1/2}(k|x-y|)}{\sqrt{k|x-y|}} g_{1,1/2}(y)|g_1(y)|^{1/2} \\ &\quad \times \widehat{G_k^+}\left[h_{0,1/2}|h_0|^{1/2} \frac{J_{1/2}(k|x-\cdot|)}{\sqrt{k|x-\cdot|}}\right](y)\mathrm{d}y \\ &= -Ck^2 \int_{\mathbb{R}^3} \frac{J_{1/2}(k|x-y|)}{\sqrt{k|x-y|}} g_{1,1/2}(y)\widehat{K}\left[|h_0|^{1/2} \frac{J_{1/2}(k|x-\cdot|)}{\sqrt{k|x-\cdot|}}\right](y)\mathrm{d}y,\end{aligned}$$

where $\widehat{K}$ is the integral operator with the kernel

$$K(x,y) = |g_1(x)|^{1/2} G_k^+(|x-y|) h_{0,1/2}(y).$$

It follows easily from Proposition 2.6 that $\widehat{K} : L^2(\mathbb{R}^3) \to L^2(\mathbb{R}^3)$ has the same norm estimate $Ck^{-\gamma}$ as $\widehat{G_k^+}$. Hence, by Hölder's inequality,

$$
\begin{aligned}
|I_2^{(1)}| &\le Ck^2 \int_{\mathbb{R}^3} \frac{|J_{1/2}(k|x-y|)|}{\sqrt{k|x-y|}} |g_1|^{1/2}(y) \left| \widehat{K}\left[|h_0|^{1/2} \frac{J_{1/2}(k|x-\cdot|)}{\sqrt{k|x-\cdot|}} \right](y) \right| \mathrm{d}y \\
&\le Ck^{2-\gamma} \left(\int_{\mathbb{R}^3} \frac{J_{1/2}^2(k|x-y|)}{k|x-y|} |g_1(y)| \mathrm{d}y \right)^{\frac{1}{2}} \left(\int_{\mathbb{R}^3} |h_0(z)| \frac{J_{1/2}^2(k|x-z|)}{k|x-z|} \mathrm{d}z \right)^{\frac{1}{2}}.
\end{aligned}
$$

Each of the latter integrals can be estimated as

$$
\begin{aligned}
\int_{\mathbb{R}^3} \frac{J_{1/2}^2(k|x-y|)}{k|x-y|} |g_1(y)| \mathrm{d}y &= C \int_{\mathbb{R}^3} \frac{\sin^2(k|x-y|)}{(k|x-y|)^2} |g_1(y)| \mathrm{d}y \\
&\le C \int_{\mathbb{R}^3} \frac{|g_1(y)| \mathrm{d}y}{(k|x-y|)^2} = \frac{C}{k^2}
\end{aligned}
$$

for $3 < p \le \infty$ as we have seen above. So

$$
|I_2^{(1)}| \le Ck^{-\gamma} \to 0, \quad k \to \infty.
$$

In a somewhat similar fashion,

$$
\begin{aligned}
I_2^{(2)} &= -Ck^2 \int_{\mathbb{S}^2} \int_{\mathbb{R}^3} \mathrm{e}^{-\mathrm{i}k(\theta,x)} \frac{J_{1/2}(k|x-y|)}{\sqrt{k|x-y|}} g_1(y) \widehat{G_k^+}[g_1 u_{\mathrm{sc}}](y) \mathrm{d}y \mathrm{d}\theta \\
&= -Ck^2 \int_{\mathbb{S}^2} \int_{\mathbb{R}^3} \mathrm{e}^{-\mathrm{i}k(\theta,x)} \frac{J_{1/2}(k|x-y|)}{\sqrt{k|x-y|}} g_{1,1/2}(y) \widehat{T}[|g_1|^{1/2} u_{\mathrm{sc}}](y) \mathrm{d}y \mathrm{d}\theta,
\end{aligned}
$$

where $\widehat{T}$ is the integral operator with the kernel

$$
T(x,y) = |g_1(x)|^{1/2} G_k^+(|x-y|) g_{1,1/2}(y)
$$

and with the same mapping property as $\widehat{K}$. Hence

$$
\begin{aligned}
|I_2^{(2)}| &\le Ck^2 \int_{\mathbb{S}^2} \int_{\mathbb{R}^3} \frac{|J_{1/2}(k|x-y|)|}{\sqrt{k|x-y|}} |g_1|^{1/2}(y) \left| \widehat{T}[|g_1|^{1/2} u_{\mathrm{sc}}](y) \right| \mathrm{d}y \mathrm{d}\theta \\
&\le Ck^{2-\gamma} \int_{\mathbb{S}^2} \left(\int_{\mathbb{R}^3} \frac{J_{1/2}^2(k|x-y|)}{k|x-y|} |g_1(y)| \mathrm{d}y \right)^{1/2} \left(\int_{\mathbb{R}^3} |g_1| |u_{\mathrm{sc}}(y)|^2 \mathrm{d}y \right)^{1/2} \mathrm{d}\theta \\
&\le Ck^{1-2\gamma} \to 0, \quad k \to \infty
\end{aligned}
$$

for $3 < p \le \infty$. The term $I_2^{(3)}$ can be estimated in similar manner for $3 < p \le \infty$. For the term $I_2^{(4)}$ the first step is

$$
I_2^{(4)} = -Ck^2 \int_{\mathbb{S}^2} \int_{\mathbb{R}^3} \mathrm{e}^{-\mathrm{i}k(\theta,x)} \frac{J_{1/2}(k|x-y|)}{\sqrt{k|x-y|}} g_{1,1/2}(y) \widehat{U}[|\eta|^{1/2} \mathcal{O}(|u_{\mathrm{sc}}(\cdot)|^2)](y) \mathrm{d}y \mathrm{d}\theta,
$$

where $\widehat{U}$ is the integral operator with the kernel

$$
U(x,y) = |g_1(x)|^{1/2} G_k^+(|x-y|) \eta_{1/2}(y)
$$

and with the same mapping property as $\widehat{K}$ and $\widehat{T}$. By Hölder's inequality and Corollary 2.8,

$$|I_2^{(4)}| \le Ck^{2-\gamma} \int_{\mathbb{S}^2} \left(\int_{\mathbb{R}^3} \frac{J_{1/2}^2(k|x-y|)}{k|x-y|} |g_1(y)| \mathrm{d}y \right)^{\frac{1}{2}} \left(\int_{\mathbb{R}^3} |\eta(y)| \mathcal{O}(|u_{\mathrm{sc}}(y)|^4) \mathrm{d}y \right)^{\frac{1}{2}} \mathrm{d}\theta$$
$$\le Ck^{2-2\gamma} \left(\int_{\mathbb{R}^3} \frac{J_{1/2}^2(k|x-y|)}{k|x-y|} |g_1(y)| \mathrm{d}y \right)^{\frac{1}{2}}.$$

As above $|I_2^{(4)}| \to 0$ as $k \to \infty$ for $3 < p \le \infty$.

The analysis of I_2'' is analogous to I_2'. For the term I_2''' we obtain

$$|I_2'''| \le Ck^2 \left\| u_{\mathrm{sc}} \right\|_{L^\infty} \int_{\mathbb{S}^2} \int_{\mathbb{R}^3} \frac{|J_{1/2}(k|x-y|)|}{\sqrt{k|x-y|}} |\eta(y)| \mathcal{O}(|u_{\mathrm{sc}}|) \mathrm{d}y \mathrm{d}\theta$$
$$\le Ck^2 \left\| u_{\mathrm{sc}} \right\|_{L^\infty} \int_{\mathbb{S}^2} \left(\int_{\mathbb{R}^3} \frac{J_{1/2}^2(k|x-y|)}{k|x-y|} |\eta(y)| \mathrm{d}y \right)^{1/2}$$
$$\times \left(\int_{\mathbb{R}^3} |\eta(y)| \mathcal{O}(|u_{\mathrm{sc}}|^2) \mathrm{d}y \right)^{1/2} \mathrm{d}\theta$$
$$f \le Ck^{1-\gamma} \left\| u_{\mathrm{sc}} \right\|_{L^\infty} \to 0, \quad k \to \infty$$

by Lemma 2.5 for $p = \infty$. It is this term that forces us to restrict to the case $p = \infty$ in the proof of uniform limit.

In the case $2 < p < \infty$ one may use similar techniques as above to prove convergence in the sense of distributions. See [2] for details. □

Saito's formula (3.2) implies immediately the following uniqueness result and the representation formula for the unknown function h_0.

Corollary 3.2 (Uniqueness). *Consider the scattering problems for two sets of potentials h and $\widetilde{h}$. Under the same assumptions as in Theorem 2.3, if the scattering amplitudes coincide for some sequence $k_j \to \infty$ and for all $\theta, \theta' \in \mathbb{S}^2$ then*

$$h_0(x) = \widetilde{h_0}(x)$$

holds in the sense of distributions for $2 < p \le \infty$.

Proof. The claim follows in standard manner from (3.2), see [12, Thm. 5.4]. □

Corollary 3.3 (Representation formula). *Under the same assumptions as in Theorem 2.3, the representation*

$$h_0(x) = \lim_{k \to \infty} \frac{k^3}{16\pi^4} \int_{\mathbb{S}^2 \times \mathbb{S}^2} \mathrm{e}^{-\mathrm{i}k(\theta - \theta', x)} A(k, \theta', \theta) |\theta - \theta'| \mathrm{d}\theta \mathrm{d}\theta'$$

holds in the sense of distributions for $2 < p \le \infty$.

Proof. See [4, Corollary 3.1]. □

3.2. Born approximation

Recall that

$$A(k,\theta',\theta) = \int_{\mathbb{R}^3} e^{-ik(\theta',y)} h(y,|u|)u(y)dy.$$

Substituting u_0 for u we arrive at the function

$$A_0(k,\theta',\theta) := \int_{\mathbb{R}^3} e^{-ik(\theta'-\theta,y)} h_0(y)dy = F(h_0)(k(\theta-\theta')),$$

where F is the Fourier transform

$$F(f)(\xi) = \int_{\mathbb{R}^3} e^{i(\xi,y)} f(y)dy.$$

We know from previous considerations that, for k large,

$$A(k,\theta',\theta) \approx F(h_0)(k(\theta-\theta')).$$

This relation can be inverted by applying the inverse Fourier transform

$$F^{-1}(f)(x) = \frac{1}{(2\pi)^3}\int_{\mathbb{R}^3} e^{-i(\xi,x)} f(\xi)d\xi$$

in some suitable coordinates. To this end we fix $\theta = \theta_0$ and switch to polar coordinates via $\xi = k(\theta-\theta')$. Then we can get k and θ' back by

$$k_\theta(\xi) := \frac{|\xi|}{2(\theta,\widehat{\xi})}, \quad \theta'_\theta(\xi) := \theta - 2(\theta,\widehat{\xi})\widehat{\xi}, \quad \widehat{\xi} = \xi/|\xi|.$$

Since the Jacobian of this transformation is $J = J(k,\theta,\theta') = k^2|\theta-\theta'|^2/2$ it follows that

$$F^{-1}(f)(x) = \frac{1}{2(2\pi)^3}\int_0^\infty k^2 \int_{\mathbb{S}^2} e^{-ik(\theta-\theta',x)} f(k(\theta-\theta'))|\theta-\theta'|^2 d\theta' dk.$$

These considerations give rise to the following definition.

Definition 3.4. The inverse Born approximation $q_B(x)$ and the inverse fixed angle Born approximation $q_B^0(x)$ of the function $h_0(x)$ are defined by

$$q_B(x) := \frac{1}{64\pi^4}\int_0^\infty k^2 \int_{\mathbb{S}^2\times\mathbb{S}^2} e^{-ik(\theta-\theta',x)} A(k,\theta',\theta)|\theta-\theta'|^2 d\theta d\theta' dk$$

and

$$q_B^0(x) := \frac{1}{16\pi^3}\int_0^\infty k^2 \int_{\mathbb{S}^2} e^{-ik(\theta_0-\theta',x)} A(k,\theta',\theta_0)|\theta_0-\theta'|^2 d\theta' dk,$$

respectively.

Now we substitute (3.1) into q_B and get, with $C_B^{-1} = 64\pi^4$,

$$\begin{aligned}
q_B(x) &= C_B \int_0^\infty \int_{\mathbb{S}^2\times\mathbb{S}^2} e^{-ik(\theta-\theta',x)} k^2|\theta-\theta'|^2 dk d\theta d\theta' \int_{\mathbb{R}^3} e^{-ik(\theta',y)} h(y,|u|)u(y)dy \\
&= C_B \int_0^\infty \int_{\mathbb{S}^2\times\mathbb{S}^2} e^{-ik(\theta-\theta',x)} k^2|\theta-\theta'|^2 dk d\theta d\theta' \int_{\mathbb{R}^3} e^{-ik(\theta',y)} h_0(y)u_0(y)dy \\
&\quad + C_B \int_1^\infty \int_{\mathbb{S}^2\times\mathbb{S}^2} e^{-ik(\theta-\theta',x)} k^2|\theta-\theta'|^2 dk d\theta d\theta' \\
&\quad\quad \times \int_{\mathbb{R}^3} e^{-ik(\theta',y)} (g_1(y)u_{sc}(y) + g_2(y)u_0^2(y)\overline{u_{sc}})dy \\
&\quad + C_B \int_0^1 \int_{\mathbb{S}^2\times\mathbb{S}^2} e^{-ik(\theta-\theta',x)} k^2|\theta-\theta'|^2 dk d\theta d\theta' \\
&\quad\quad \times \int_{\mathbb{R}^3} e^{-ik(\theta',y)} (g_1(y)u_{sc}(y) + g_2(y)u_0^2(y)\overline{u_{sc}} + \eta(y)\mathcal{O}(|u_{sc}|^2))dy \\
&\quad + C_B \int_1^\infty \int_{\mathbb{S}^2\times\mathbb{S}^2} e^{-ik(\theta-\theta',x)} k^2|\theta-\theta'|^2 dk d\theta d\theta' \int_{\mathbb{R}^3} e^{-ik(\theta',y)} \eta(y)\mathcal{O}(|u_{sc}|^2)dy \\
&=: q_0(x) + q_1(x) + q_\infty(x) + q_r(x).
\end{aligned}$$

Here the first term is actually the unknown function since

$$\begin{aligned}
q_0(x) &= \frac{1}{64\pi^4} \int_0^\infty \int_{\mathbb{S}^2\times\mathbb{S}^2} e^{-ik(\theta-\theta',x)} k^2|\theta-\theta'|^2 dk d\theta d\theta' \int_{\mathbb{R}^3} e^{-ik(\theta'-\theta,y)} h_0(y)dy \\
&= \frac{1}{64\pi^4} \int_{\mathbb{R}^3} h_0(y) \int_0^\infty \int_{\mathbb{S}^2\times\mathbb{S}^2} e^{-ik(\theta-\theta',x)} k^2|\theta-\theta'|^2 dk d\theta d\theta' e^{-ik(\theta'-\theta,y)} dy \\
&= \frac{1}{32\pi^4} \int_{\mathbb{R}^3} h_0(y) \int_{\mathbb{R}^3\times\mathbb{S}^2} e^{-i(\xi,x)} d\xi d\theta e^{i(\xi,y)} dy \\
&= \frac{1}{8\pi^3} \int_{\mathbb{R}^3} h_0(y) \int_{\mathbb{R}^3} e^{-i(\xi,x)} d\xi e^{i(\xi,y)} dy = F^{-1}(Fh_0)(x) = h_0(x).
\end{aligned}$$

It means that our problem is reduced to the study of right-hand side of

$$q_B(x) - h_0(x) = q_1(x) + q_\infty(x) + q_r(x).$$

Proceeding similarly for q_B^0 allows us to conclude that

$$q_B^0(x) - h_0(x) = q_1^0(x) + q_\infty^0(x) + q_r^0(x),$$

where $q_1^0(x), q_\infty^0(x)$ and $q_r^0(x)$ are similar to $q_1(x), q_\infty(x)$ and $q_r(x)$ but without integration over θ.

Noting that $q_\infty, q_\infty^0 \in C^\infty(\mathbb{R}^3)$ as inverse Fourier transforms of compactly supported tempered distributions we study q_1, q_r, q_1^0 and q_r^0.

The proof of the following fact makes use of the integral operator

$$A_0(k)f(\theta') = \int_{\mathbb{R}^3} e^{-ik(\theta',y)} f(y)dy$$

which has a norm estimate

$$\|A_0(k)f\|^2_{L^2(\mathbb{S}^2)} \le \frac{C}{|k|^{1+\gamma}} \|f\|^2_{L_\delta^{\frac{2p}{p+1}}(\mathbb{R}^3)},$$

where γ is as in Proposition 2.6.

Lemma 3.5. *Under the same assumptions as in Theorem 2.3 the terms q_1, q_r, q_1^0 and q_r^0 belong to the Sobolev space $H^t(\mathbb{R}^3)$ for $t < 1/2 - 3/(2p)$ if $3 < p \le \infty$.*

Proof. We start by writing q_r as an inverse Fourier transform. Indeed,

$$\begin{aligned} q_r(x) &= C\int_0^\infty \int_{\mathbb{S}^2\times\mathbb{S}^2} (1-\chi(k))\mathrm{e}^{-ik(\theta-\theta',x)}k^2|\theta-\theta'|^2 \mathrm{d}k\mathrm{d}\theta\mathrm{d}\theta' \\ &\quad\times \int_{\mathbb{R}^3} \mathrm{e}^{-ik(\theta',y)}\eta(y)\mathcal{O}(|u_{\mathrm{sc}}|^2)\mathrm{d}y \\ &= C\int_{\mathbb{S}^2}\int_{\mathbb{R}^3} (1-\chi(k_\theta(\xi)))\mathrm{e}^{-\mathrm{i}(\xi,x)}\mathrm{d}\xi\mathrm{d}\theta \int_{\mathbb{R}^3} \mathrm{e}^{-ik_\theta(\xi)(\theta'_\theta(\xi),y)}\eta(y)\mathcal{O}(|u_{\mathrm{sc}}|^2)\mathrm{d}y \\ &= CF^{-1}\int_{\mathbb{S}^2}(1-\chi(k_\theta(\xi)))\mathrm{d}\theta \int_{\mathbb{R}^3} \mathrm{e}^{-ik_\theta(\xi)(\theta'_\theta(\xi),y)}\eta(y)\mathcal{O}(|u_{\mathrm{sc}}|^2)\mathrm{d}y, \end{aligned}$$

where $\chi(k)$ is the characteristic function of the interval $[0,1]$. Then

$$\begin{aligned} \|q_r\|^2_{H^t(\mathbb{R}^3)} &= \left\|(1+|\cdot|^2)^{t/2}Fq_r\right\|^2_{L^2(\mathbb{R}^3)} = \int_{\mathbb{R}^3}(1+|\xi|^2)^t|Fq_r(\xi)|^2\mathrm{d}\xi \\ &= C\int_{\mathbb{R}^3}(1+|\xi|^2)^t\mathrm{d}\xi \\ &\quad\times\left|\int_{\mathbb{S}^2}(1-\chi(k_\theta(\xi)))\mathrm{d}\theta\int_{\mathbb{R}^3}\mathrm{e}^{-ik_\theta(\xi)(\theta'_\theta(\xi),y)}\eta(y)\mathcal{O}(|u_{\mathrm{sc}}|^2)\mathrm{d}y\right|^2 \\ &\le C\int_{\mathbb{R}^3}(1+|\xi|^2)^t\mathrm{d}\xi\int_{\mathbb{S}^2}(1-\chi(k_\theta(\xi)))^2\mathrm{d}\theta \\ &\quad\times\left|\int_{\mathbb{R}^3}\mathrm{e}^{-ik_\theta(\xi)(\theta'_\theta(\xi),y)}\eta(y)\mathcal{O}(|u_{\mathrm{sc}}|^2)\mathrm{d}y\right|^2 \end{aligned}$$

by Hölder's inequality. Going back to angular variables k, θ' yields, for $t \ge 0$,

$$\begin{aligned} \|q_r\|^2_{H^t(\mathbb{R}^3)} &\le C\int_1^\infty\int_{\mathbb{S}^2} k^2(1+|k|^2)^t\mathrm{d}k\mathrm{d}\theta'\int_{\mathbb{S}^2}\mathrm{d}\theta\left|\int_{\mathbb{R}^3}\mathrm{e}^{-ik(\theta',y)}\eta(y)\mathcal{O}(|u_{\mathrm{sc}}|^2)\mathrm{d}y\right|^2 \\ &\le C\int_1^\infty\int_{\mathbb{S}^2} k^2(1+|k|^2)^t\mathrm{d}k\mathrm{d}\theta'\int_{\mathbb{S}^2}\mathrm{d}\theta\,|A_0(k)\eta(y)\mathcal{O}(|u_{\mathrm{sc}}|^2)|^2 \\ &\le C\int_1^\infty\int_{\mathbb{S}^2} k^{1-\gamma}(1+|k|^2)^t\mathrm{d}k\mathrm{d}\theta\,\|\eta(\cdot)\mathcal{O}(|u_{\mathrm{sc}}|^2)\|^2_{L_\delta^{\frac{2p}{p+1}}(\mathbb{R}^3)} \\ &\le C\int_1^\infty\int_{\mathbb{S}^2} k^{1-\gamma}(1+|k|^2)^t\mathrm{d}k\mathrm{d}\theta\,\|\eta\|^2_{L^p_{2\delta}(\mathbb{R}^3)}\left\||u_{\mathrm{sc}}|^2\right\|^2_{L_{-\delta}^{\frac{2p}{p-1}}(\mathbb{R}^3)} \\ &\le C\int_1^\infty k^{2t+1-3\gamma}\mathrm{d}k < \infty \end{aligned}$$

where γ is as in Proposition 2.6. The last integral converges if and only if

$$3\gamma - 1 - 2t > 1$$

or $t < 3\gamma/2 - 1 = 1/2 - 3/(2p)$ for $3 < p \leq \infty$. The analogous proofs for q_1, q_1^0 and q_r^0 are left to the reader. □

Now our main result concerning the method of Born approximation follows immediately from Lemma 3.5.

Theorem 3.6. *Under the same assumptions as in Theorem* 2.3,

$$q_{\mathrm{B}}^0 - h_0, q_{\mathrm{B}} - h_0 \in H^t_{\mathrm{loc}}(\mathbb{R}^3),$$

where t is as in Lemma 3.5.

For finite p we cannot recover any singularities of h_0 from q_{B}. In the case $p = \infty$ we can, however, obtain the following application of q_{B}. If $s_0 < 1/2$ is fixed and $h_0 \in H^s(\mathbb{R}^3) \cap L^\infty_{2\delta}(\mathbb{R}^3)$, $s \leq s_0$ and $h_0 \notin H^{s+\varepsilon}(\mathbb{R}^3)$ for any $\varepsilon > 0$ then we can recover jumps of h_0, if any, from q_{B}. Similarly for q_{B}^0.

Acknowledgment

This work was supported by the Academy of Finland (application number 213476, Finnish Programme for Centres of Excellence in Research 2006-2011).

References

[1] D. Colton and R. Kress, *Inverse Acoustic and Electromagnetic Scattering Theory*, Springer-Verlag, Berlin, Heidelberg, 1998.

[2] M. Harju, *On the direct and inverse scattering problems for a nonlinear three-dimensional Schrödinger equation*, Doctoral Dissertation, University of Oulu, 2010.

[3] P. Ola, L. Päivärinta and V. Serov, *Recovering singularities from backscattering in two dimensions.* Commun. Partial Differ. Equ. **26** (2001), 697–715.

[4] L. Päivärinta and V. Serov, *Recovery of singularities of a multidimensional scattering potential.* SIAM J. Math. Anal. **29** (1998), 697–711.

[5] L. Päivärinta and V. Serov, *New mapping properties for the resolvent of the Laplacian and recovery of singularities of a multi-dimensional scattering potential.* Inverse Problems **17** (2001), 1321–1326.

[6] L. Päivärinta and V. Serov, *Recovery of jumps and singularities in the multidimensional Schrödinger operator from limited data.* Inverse Probl. Imaging **1** (2007), 525–535.

[7] L. Päivärinta, V. Serov and E. Somersalo, *Reconstruction of singularities of a scattering potential in two dimensions.* Adv. Appl. Math. **15** (1994), 97–113.

[8] L. Päivärinta and E. Somersalo, *Inversion of discontinuities for the Schrödinger equation in three dimensions.* SIAM J. Math. Anal. **22** (1991), 480–499.

[9] J.M. Reyes, *Inverse backscattering for the Schrödinger equation in 2D.* Inverse Problems **23** (2007), 625–643.

[10] A. Ruiz, *Recovery of the singularities of a potential from fixed angle scattering data.* Commun. Partial Differ. Equ. **26** (2001), 1721–1738.

[11] A. Ruiz and A. Vargas, *Partial recovery of a potential from backscattering data.* Commun. Partial Differ. Equ. **30** (2005), 67–96.

[12] Y. Saito, *Some properties of the scattering amplitude and the inverse scattering problem.* Osaka J. Math. **19** (1982), 527–547.

[13] V. Serov, *Inverse Born approximation for the nonlinear two-dimensional Schrödinger operator.* Inverse Problems **23** (2007), 1259–1270.

[14] V. Serov, *Inverse Born approximation for the generalized nonlinear Schrödinger operator in two dimensions.* Modern Phys. Lett. B **22** (2008), 2257–2275.

[15] V. Serov, *An inverse Born approximation for the general nonlinear Schrödinger operator on the line.* J. Phys. A: Math. Theor. **42** (2009), 332002.

[16] V. Serov, *Inverse fixed energy scattering problem for the generalized nonlinear Schrödinger operator.* Inverse Problems **28** (2012), 025002.

[17] V. Serov and M. Harju, *Reconstruction of discontinuities in the nonlinear one-dimensional Schrödinger equation from limited data.* Inverse Problems **23** (2007), 493–506.

[18] V. Serov and M. Harju, *Partial recovery of potentials in generalized nonlinear Schrödinger equations on the line.* J. Math. Phys. **48** (2007), 083512.

[19] V. Serov and M. Harju, *A uniqueness theorem and reconstruction of singularities for a two-dimensional nonlinear Schrödinger equation.* Nonlinearity **21** (2008), 1323–1337.

[20] V. Serov, M. Harju and G. Fotopoulos, *Direct and inverse scattering for nonlinear Schrödinger equation in 2D*, J. Math. Phys. **53** (2012), 123522.

[21] V. Serov and J. Sandhu, *Inverse backscattering problem for the generalized nonlinear Schrödinger operator in two dimensions.* J. Phys. A: Math. Theor. **43** (2010), 325206.

[22] E. Zeidler, *Applied Functional Analysis: Part I. Applications to Mathematical Physics*, Springer, New York, 1995.

Markus Harju and Valery Serov
Department of Mathematical Sciences
University of Oulu
PO Box 3000
FIN-90014 Oulu, Finland
e-mail: markus.harju@oulu.fi
vserov@cc.oulu.fi

[16] A. Ruiz, *Recovery of the singularities of a potential from fixed angle scattering data*, Comm. in Partial Differ. Equ. 26 (2001), 1721–1738.

[17] A. Ruiz and A. Vargas, *Partial recovery of a potential from backscattering data*, Comm. Partial Differ. Equ. 30 (2005), 67–96.

[illegible]

[illegible]

[illegible]

[illegible]

[illegible]

Operator Theory:
Advances and Applications, Vol. 236, 275–291

The Cauchy Singular Integral Operator on Weighted Variable Lebesgue Spaces

Alexei Yu. Karlovich and Ilya M. Spitkovsky

Abstract. Let $p : \mathbb{R} \to (1, \infty)$ be a globally log-Hölder continuous variable exponent and $w : \mathbb{R} \to [0, \infty]$ be a weight. We prove that the Cauchy singular integral operator S is bounded on the weighted variable Lebesgue space $L^{p(\cdot)}(\mathbb{R}, w) = \{f : fw \in L^{p(\cdot)}(\mathbb{R})\}$ if and only if the weight w satisfies

$$\sup_{-\infty<a<b<\infty} \frac{1}{b-a} \|w\chi_{(a,b)}\|_{p(\cdot)} \|w^{-1}\chi_{(a,b)}\|_{p'(\cdot)} < \infty \quad (1/p(x) + 1/p'(x) = 1).$$

Mathematics Subject Classification (2010). Primary 42A50; Secondary 42B25, 46E30.

Keywords. Weighted variable Lebesgue space, log-Hölder continuous variable exponent, Cauchy singular integral operator, Hardy–Littlewood maximal operator.

1. Introduction

Let $p : \mathbb{R} \to [1, \infty]$ be a measurable a.e. finite function. By $L^{p(\cdot)}(\mathbb{R})$ we denote the set of all complex-valued functions f on $\mathbb{R}$ such that

$$I_{p(\cdot)}(f/\lambda) := \int_{\mathbb{R}} |f(x)/\lambda|^{p(x)} dx < \infty$$

for some $\lambda > 0$. This set becomes a Banach space when equipped with the norm

$$\|f\|_{p(\cdot)} := \inf \{\lambda > 0 : I_{p(\cdot)}(f/\lambda) \le 1\}.$$

It is easy to see that if p is constant, then $L^{p(\cdot)}(\mathbb{R})$ is nothing but the standard Lebesgue space $L^p(\mathbb{R})$. The space $L^{p(\cdot)}(\mathbb{R})$ is referred to as a *variable Lebesgue space*.

A measurable function $w : \mathbb{R} \to [0, \infty]$ is referred to as a *weight* whenever $0 < w(x) < \infty$ a.e. on $\mathbb{R}$. Given a variable exponent $p : \mathbb{R} \to [1, \infty]$ and a weight

The first author is partially supported by FCT project PEstOE/MAT/UI4032/2011 (Portugal).

$w : \mathbb{R} \to [0, \infty]$, we define the weighted variable exponent space $L^{p(\cdot)}(\mathbb{R}, w)$ as the space of all measurable complex-valued functions f such that $fw \in L^{p(\cdot)}(\mathbb{R})$. The norm on this space is naturally defined by

$$\|f\|_{p(\cdot),w} := \|fw\|_{p(\cdot)}.$$

Given $f \in L^1_{\mathrm{loc}}(\mathbb{R})$, the Hardy–Littlewood maximal operator is defined by

$$Mf(x) := \sup_{Q \ni x} \frac{1}{|Q|} \int_Q |f(y)|dy$$

where the supremum is taken over all intervals $Q \subset \mathbb{R}$ containing x. The Cauchy singular integral operator S is defined for $f \in L^1_{\mathrm{loc}}(\mathbb{R})$ by

$$(Sf)(x) := \frac{1}{\pi i} \int_{\mathbb{R}} \frac{f(\tau)}{\tau - x} d\tau \quad (x \in \mathbb{R}),$$

where the integral is understood in the principal value sense.

Following [4, Section 2] or [6, Section 4.1], one says that $\alpha : \mathbb{R} \to \mathbb{R}$ is locally log-Hölder continuous if there exists $c_1 > 0$ such that

$$|\alpha(x) - \alpha(y)| \le \frac{c_1}{\log(e + 1/|x - y|)}$$

for all $x, y \in \mathbb{R}$. Further, α is said to satisfy the log-Hölder decay condition if there exist $\alpha_\infty \in \mathbb{R}$ and a constant $c_2 > 0$ such that

$$|\alpha(x) - \alpha_\infty| \le \frac{c_2}{\log(e + |x|)}$$

for all $x \in \mathbb{R}$. One says that α is globally log-Hölder continuous on $\mathbb{R}$ if it is locally log-Hölder continuous and satisfies the log-Hölder decay condition. Put

$$p_- := \operatorname*{ess\,inf}_{x \in \mathbb{R}} p(x), \quad \operatorname*{ess\,sup}_{x \in \mathbb{R}} p(x) =: p_+.$$

As usual, we use the convention $1/\infty := 0$ and denote by $\mathcal{P}^{\log}(\mathbb{R})$ the set of all variable exponents such that $1/p$ is globally log-Hölder continuous. If $p \in \mathcal{P}^{\log}(\mathbb{R})$, then the limit

$$\frac{1}{p(\infty)} := \lim_{|x| \to \infty} \frac{1}{p(x)}$$

exists. If $p_+ < \infty$, then $p \in \mathcal{P}^{\log}(\mathbb{R})$ if and only if p is globally log-Hölder continuous.

By [6, Theorem 4.3.8], if $p \in \mathcal{P}^{\log}(\mathbb{R})$ with $p_- > 1$, then the Hardy–Littlewood maximal operator M is bounded on $L^{p(\cdot)}(\mathbb{R})$. Notice, however, that the condition $p \in \mathcal{P}^{\log}(\mathbb{R})$ is not necessary, there are even discontinuous exponents p such that M is bounded on $L^{p(\cdot)}(\mathbb{R})$. Corresponding examples were first constructed by Lerner and they are contained in [6, Section 5.1].

In this paper we will mainly suppose that

$$1 < p_-, \quad p_+ < \infty. \tag{1.1}$$

Under these conditions, the space $L^{p(\cdot)}(\mathbb{R})$ is separable and reflexive, and its Banach space dual $[L^{p(\cdot)}(\mathbb{R})]^*$ is isomorphic to $L^{p'(\cdot)}(\mathbb{R})$, where

$$1/p(x) + 1/p'(x) = 1 \quad (x \in \mathbb{R})$$

(see [6, Chap. 3]). If, in addition, $w\chi_E \in L^{p(\cdot)}(\mathbb{R})$ and $\chi_E/w \in L^{p'(\cdot)}(\mathbb{R})$ for any measurable set $E \subset \mathbb{R}$ of finite measure, then $L^{p(\cdot)}(\mathbb{R}, w)$ is a Banach function space and $[L^{p(\cdot)}(\mathbb{R}, w)]^* = L^{p(\cdot)}(\mathbb{R}, w^{-1})$. Here and in what follows, χ_E denotes the characteristic function of the set E.

Probably, one of the simplest weights is the following power weight

$$w(x) := |x - i|^{\lambda_\infty} \prod_{j=1}^{m} |x - x_j|^{\lambda_j}, \tag{1.2}$$

where $-\infty < x_1 < \cdots < x_m < +\infty$ and $\lambda_1, \ldots, \lambda_m, \lambda_\infty \in \mathbb{R}$. Kokilashvili, Paatashvili, and Samko studied the boundedness of the operators M and S on $L^{p(\cdot)}(\mathbb{R}, w)$ with power weights (1.2). From [12, Theorem A] and [15, Theorem B] one can extract the following result.

Theorem 1.1. *Let $p \in \mathcal{P}^{\log}(\mathbb{R})$ satisfy* (1.1) *and w be a power weight* (1.2).

(a) **(Kokilashvili, Samko).** *Suppose, in addition, that p is constant outside an interval containing $x_1, \ldots, x_m$. Then the Hardy–Littlewood maximal operator M is bounded on $L^{p(\cdot)}(\mathbb{R}, w)$ if and only if*

$$0 < \frac{1}{p(x_j)} + \lambda_j < 1 \text{ for } j \in \{1, \ldots, m\}, \quad 0 < \frac{1}{p(\infty)} + \lambda_\infty + \sum_{j=1}^{m} \lambda_j < 1. \tag{1.3}$$

(b) **(Kokilashvili, Paatashvili, Samko).** *The Cauchy singular integral operator S is bounded on $L^{p(\cdot)}(\mathbb{R}, w)$ if and only if* (1.3) *is fulfilled.*

Further, the sufficiency portion of this result was extended in [13, 14] to radial oscillating weights of the form $\prod_{j=1}^{m} \omega_j(|x - x_j|)$, where $\omega_j(t)$ are continuous functions for $t > 0$ that may oscillate near zero and whose Matuszewska–Orlicz indices can be different. Notice that the Matuszewska–Orlicz indices of $\omega_j(t) = t^{\lambda_j}$ are both equal to λ_j.

Very recently, Cruz-Uribe, Diening, and Hästö [6, Theorem 1.3] generalized part (a) of Theorem 1.1 to the case of general weights. To formulate their result, we will introduce the following generalization of the classical Muckenhoupt condition (written in the symmetric form). We say that a weight $w : \mathbb{R} \to [0, \infty]$ belongs to the class $\mathcal{A}_{p(\cdot)}(\mathbb{R})$ if

$$\sup_{-\infty < a < b < \infty} \frac{1}{b - a} \|w\chi_{(a,b)}\|_{p(\cdot)} \|w^{-1}\chi_{(a,b)}\|_{p'(\cdot)} < \infty.$$

This condition goes back to Berezhnoi [3] (in the more general setting of Banach function spaces), it was studied by the first author [9] (in the case of Banach function spaces defined on Carleson curves) and Kopaliani [16].

Theorem 1.2 (Cruz-Uribe, Diening, Hästö). *Let* $p \in \mathcal{P}^{\log}(\mathbb{R})$ *satisfy* (1.1) *and* $w : \mathbb{R} \to [0, \infty]$ *be a weight. The Hardy–Littlewood maximal operator* M *is bounded on the weighted variable Lebesgue space* $L^{p(\cdot)}(\mathbb{R}, w)$ *if and only if* $w \in \mathcal{A}_{p(\cdot)}(\mathbb{R})$.

Another proof of Theorem 1.2 was given by Cruz-Uribe, Fiorenza and Neugebauer [5, Theorem 1.5].

The aim of this paper is to generalize part (b) of Theorem 1.1 to the case of general weights. We will prove the following.

Theorem 1.3 (Main result). *Let* $p \in \mathcal{P}^{\log}(\mathbb{R})$ *satisfy* (1.1) *and* $w : \mathbb{R} \to [0, \infty]$ *be a weight. The Cauchy singular integral operator* S *is bounded on the weighted variable Lebesgue space* $L^{p(\cdot)}(\mathbb{R}, w)$ *if and only if* $w \in \mathcal{A}_{p(\cdot)}(\mathbb{R})$.

From this theorem, by using standard techniques, we derive also the following.

Theorem 1.4. *Let* $p \in \mathcal{P}^{\log}(\mathbb{R})$ *satisfy* (1.1) *and* $w \in \mathcal{A}_{p(\cdot)}(\mathbb{R})$. *Then* $S^2 = I$ *on the space* $L^{p(\cdot)}(\mathbb{R}, w)$ *and* $S^* = S$ *on the space* $L^{p'(\cdot)}(\mathbb{R}, w^{-1})$.

The paper is organized as follows. In Section 2 we collect necessary facts on Banach function spaces $X(\mathbb{R})$ in the sense of Luxemburg and discuss weighted Banach functions spaces $X(\mathbb{R}, w) = \{f : fw \in X(\mathbb{R})\}$. A special attention is paid to conditions implying that $X(\mathbb{R}, w)$ is a Banach function space itself, to separability and reflexivity of $X(\mathbb{R}, w)$, and to density of smooth compactly supported functions in $X(\mathbb{R}, w)$ and in its dual space $X'(\mathbb{R}, w^{-1})$. In Section 3.2 we prepare the proof of a sufficient condition for the boundedness of the operator S and formulate two key estimates by Lerner [17] and Álvarez and Pérez [1]. On the basis of these results, in Section 3.3 we prove that if $X(\mathbb{R})$ is a separable Banach function space and the Hardy–Littlewood maximal function is bounded on the weighted Banach function spaces $X(\mathbb{R}, w)$ and $X'(\mathbb{R}, w^{-1})$, then S is bounded on $X(\mathbb{R}, w)$ and $S^2 = I$. Moreover, if $X(\mathbb{R})$ is reflexive, then S^* coincides with S on $X'(\mathbb{R}, w^{-1})$. In Section 3.4 we prove that if S is bounded on the weighted Banach function spaces $X(\mathbb{R}, w)$, then

$$\sup_{-\infty<a<b<\infty} \frac{1}{b-a} \|w\chi_{(a,b)}\|_{X(\mathbb{R})} \|w^{-1}\chi_{(a,b)}\|_{X'(\mathbb{R})} < \infty$$

where $X'(\mathbb{R})$ is the associate space for $X(\mathbb{R})$. Finally, in Section 3.5 we explain that Theorems 1.3 and 1.4 follow from results of Sections 3.3–3.4 and Theorem 1.2 because $L^{p(\cdot)}(\mathbb{R})$ is a Banach function space, which is separable and reflexive whenever p satisfies (1.1).

2. Weighted Banach function spaces

2.1. Banach function spaces

The set of all Lebesgue measurable complex-valued functions on $\mathbb{R}$ is denoted by $\mathcal{M}$. Let $\mathcal{M}^+$ be the subset of functions in $\mathcal{M}$ whose values lie in $[0, \infty]$. The characteristic function of a measurable set $E \subset \mathbb{R}$ is denoted by χ_E and the Lebesgue measure of E is denoted by $|E|$.

Definition 2.1 ([2, Chap. 1, Definition 1.1]). A mapping $\rho : \mathcal{M}^+ \to [0, \infty]$ is called a *Banach function norm* if, for all functions f, g, f_n $(n \in \mathbb{N})$ in $\mathcal{M}^+$, for all constants $a \ge 0$, and for all measurable subsets E of $\mathbb{R}$, the following properties hold:

$$
\begin{aligned}
&\text{(A1)} \quad && \rho(f) = 0 \Leftrightarrow f = 0 \text{ a.e.}, \quad \rho(af) = a\rho(f), \quad \rho(f+g) \le \rho(f) + \rho(g),\\
&\text{(A2)} \quad && 0 \le g \le f \text{ a.e.} \;\Rightarrow\; \rho(g) \le \rho(f) \quad \text{(the lattice property)},\\
&\text{(A3)} \quad && 0 \le f_n \uparrow f \text{ a.e.} \;\Rightarrow\; \rho(f_n) \uparrow \rho(f) \quad \text{(the Fatou property)},\\
&\text{(A4)} \quad && |E| < \infty \Rightarrow \rho(\chi_E) < \infty,\\
&\text{(A5)} \quad && |E| < \infty \Rightarrow \int_E f(x)\,dx \le C_E \rho(f)
\end{aligned}
$$

with $C_E \in (0, \infty)$ which may depend on E and ρ but is independent of f.

When functions differing only on a set of measure zero are identified, the set $X(\mathbb{R})$ of all functions $f \in \mathcal{M}$ for which $\rho(|f|) < \infty$ is called a *Banach function space*. For each $f \in X(\mathbb{R})$, the norm of f is defined by

$$\|f\|_{X(\mathbb{R})} := \rho(|f|).$$

The set $X(\mathbb{R})$ under the natural linear space operations and under this norm becomes a Banach space (see [2, Chap. 1, Theorems 1.4 and 1.6]).

If ρ is a Banach function norm, its associate norm ρ' is defined on $\mathcal{M}^+$ by

$$\rho'(g) := \sup\left\{ \int_{\mathbb{R}} f(x)g(x)\,dx \;:\; f \in \mathcal{M}^+,\ \rho(f) \le 1 \right\}, \quad g \in \mathcal{M}^+.$$

It is a Banach function norm itself [2, Chap. 1, Theorem 2.2]. The Banach function space $X'(\mathbb{R})$ determined by the Banach function norm ρ' is called the *associate space* (*Köthe dual*) of $X(\mathbb{R})$. The associate space $X'(\mathbb{R})$ is a subspace of the dual space $[X(\mathbb{R})]^*$. The construction of the associate space implies the following Hölder inequality for Banach function spaces.

Lemma 2.2 ([2, Chap. 1, Theorem 2.4]). *Let $X(\mathbb{R})$ be a Banach function space and $X'(\mathbb{R})$ be its associate space. If $f \in X(\mathbb{R})$ and $g \in X'(\mathbb{R})$, then fg is integrable and*

$$\|fg\|_{L^1(\mathbb{R})} \le \|f\|_{X(\mathbb{R})} \|g\|_{X'(\mathbb{R})}.$$

The next result provides a useful converse to the integrability assertion of Lemma 2.2.

Lemma 2.3 ([2, Chap. 1, Lemma 2.6]). *Let $X(\mathbb{R})$ be a Banach function space. In order that a measurable function g belong to the associate space $X'(\mathbb{R})$, it is necessary and sufficient that fg be integrable for every f in $X(\mathbb{R})$.*

2.2. Weighted Banach function spaces

Let $X(\mathbb{R})$ be a Banach function space generated by a Banach function norm ρ. We say that $f \in X_{\rm loc}(\mathbb{R})$ if $f\chi_E \in X(\mathbb{R})$ for any measurable set $E \subset \mathbb{R}$ of finite measure. A measurable function $w : \mathbb{R} \to [0, \infty]$ is referred to as a *weight* if

$0 < w(x) < \infty$ a.e. on $\mathbb{R}$. Define the mapping $\rho_w : \mathcal{M}^+ \to [0,\infty]$ and the set $X(\mathbb{R},w)$ by

$$\rho_w(f) := \rho(fw) \quad (f \in \mathcal{M}^+), \qquad X(\mathbb{R},w) := \{f \in \mathcal{M}^+ : \; fw \in X(\mathbb{R})\}.$$

Lemma 2.4. *Let $X(\mathbb{R})$ be a Banach function space generated by a Banach function norm ρ, let $X'(\mathbb{R})$ be its associate space, and let $w : \mathbb{R} \to [0,\infty]$ be a weight.*

(a) *The mapping ρ_w satisfies Axioms* (A1)–(A3) *in Definition* 2.1 *and $X(\mathbb{R},w)$ is a linear normed space with respect to the norm*

$$\|f\|_{X(\mathbb{R},w)} := \rho_w(|f|) = \rho(|fw|) = \|fw\|_{X(\mathbb{R})}.$$

(b) *If $w \in X_{\rm loc}(\mathbb{R})$ and $1/w \in X'_{\rm loc}(\mathbb{R})$, then ρ_w is a Banach function norm and $X(\mathbb{R},w)$ is a Banach function space generated by ρ_w.*

(c) *If $w \in X_{\rm loc}(\mathbb{R})$ and $1/w \in X'_{\rm loc}(\mathbb{R})$, then $X'(\mathbb{R},w^{-1})$ is the associate space for the Banach function space $X(\mathbb{R},w)$.*

Proof. The proof is analogous to that one of [9, Lemma 2.5].

Part (a) follows from Axioms (A1)–(A3) for the Banach function norm ρ and the fact that $0 < w(x) < \infty$ almost everywhere on $\mathbb{R}$.

(b) If $w \in X_{\rm loc}(\mathbb{R})$, then $w\chi_E \in X(\mathbb{R})$ for every measurable set $E \subset \mathbb{R}$ of finite measure. Therefore $\rho_w(\chi_E) = \rho(w\chi_E) < \infty$. Then ρ_w satisfies Axiom (A4).

Since $1/w \in X'_{\rm loc}(\mathbb{R})$, we have $C_E := \rho'(\chi_E/w) < \infty$ for every measurable set $E \subset \mathbb{R}$ of finite measure. On the other hand, by Axiom (A2), for $f \in \mathcal{M}^+$ we have $\rho(fw\chi_E) \le \rho(fw) = \rho_w(f)$. By Hölder's inequality for ρ (Lemma 2.2), we obtain

$$\int_E f(x)\,dx = \int_{\mathbb{R}} f(x)w(x)\chi_E(x)\cdot\frac{\chi_E(x)}{w(x)}\,dx \le \rho(fw\chi_E)\rho'(\chi_E/w) \le C_E\rho_w(f).$$

Thus ρ_w satisfies Axiom (A5), that is, $X(\mathbb{R},w)$ is a Banach function space. Part (b) is proved.

(c) For $g \in \mathcal{M}^+$, we have

$$\begin{aligned}(\rho_w)'(g) &= \sup\left\{\int_{\mathbb{R}} f(x)g(x)\,dx : \; f \in \mathcal{M}^+, \; \rho_w(f) \le 1\right\}\\ &= \sup\left\{\int_{\mathbb{R}} (f(x)w(x))\left(\frac{g(x)}{w(x)}\right)dx : \; f \in \mathcal{M}^+, \; \rho(fw) \le 1\right\}\\ &= \sup\left\{\int_{\mathbb{R}} h(x)\left(\frac{g(x)}{w(x)}\right)dx : \; h \in \mathcal{M}^+, \; \rho(h) \le 1\right\}\\ &= \rho'(g/w).\end{aligned}$$

Hence $(X(\mathbb{R},w))' = X'(\mathbb{R},w^{-1})$. □

From Lemma 2.4 and the Lorentz–Luxemburg theorem (see, e.g., [2, Chap. 1, Theorem 2.7]) we obtain the following.

Lemma 2.5. *Let $X(\mathbb{R})$ be a Banach function space and $w : \mathbb{R} \to [0, \infty]$ be a weight such that $w \in X_{\rm loc}(\mathbb{R})$ and $1/w \in X'_{\rm loc}(\mathbb{R})$. Then*

$$\|f\|_{X(\mathbb{R},w)} = \sup\left\{ \int_{\mathbb{R}} |f(x)g(x)|\,dx \ :\ g \in X'(\mathbb{R}, w^{-1}),\ \|g\|_{X'(\mathbb{R},w^{-1})} \le 1 \right\} \quad (2.1)$$

for all $f \in X(\mathbb{R}, w)$ and

$$\|g\|_{X'(\mathbb{R},w^{-1})} = \sup\left\{ \int_{\mathbb{R}} |f(x)g(x)|\,dx \ :\ f \in X(\mathbb{R}, w),\ \|f\|_{X(\mathbb{R},w)} \le 1 \right\} \quad (2.2)$$

for all $g \in X'(\mathbb{R}, w^{-1})$.

2.3. Reflexivity of weighted Banach function spaces

A function f in a Banach function space $X(\mathbb{R})$ is said to have *absolutely continuous norm* in $X(\mathbb{R})$ if $\|f\chi_{E_n}\|_{X(\mathbb{R})} \to 0$ for every sequence $\{E_n\}_{n=1}^{\infty}$ of measurable sets on $\mathbb{R}$ satisfying $\chi_{E_n} \to 0$ a.e. on $\mathbb{R}$ as $n \to \infty$. If all functions $f \in X(\mathbb{R})$ have this property, then the space $X(\mathbb{R})$ itself is said to have *absolutely continuous norm* (see [2, Chap. 1, Section 3]).

Lemma 2.6 ([2, Chap. 1, Lemma 3.4]). *Let $X(\mathbb{R})$ be a Banach function space. If $f \in X(\mathbb{R})$ has absolutely continuous norm, then to each $\varepsilon > 0$ there corresponds $\delta > 0$ such that $|E| < \delta$ implies $\|f\chi_E\|_{X(\mathbb{R})} < \varepsilon$.*

Lemma 2.7. *Let $X(\mathbb{R})$ be a Banach function space and $w : \mathbb{R} \to [0, \infty]$ be a weight such that $w \in X_{\rm loc}(\mathbb{R})$ and $1/w \in X'_{\rm loc}(\mathbb{R})$. If $X(\mathbb{R})$ has absolutely continuous norm, then $X(\mathbb{R}, w)$ has absolutely continuous norm too.*

Proof. The proof is a literal repetition of that one of [9, Proposition 2.6]. By Lemma 2.4(b), $X(\mathbb{R}, w)$ is a Banach function space. If $f \in X(\mathbb{R}, w)$, then $fw \in X(\mathbb{R})$ has absolutely continuous norm in $X(\mathbb{R})$. Therefore,

$$\|f\chi_{E_n}\|_{X(\mathbb{R},w)} = \|fw\chi_{E_n}\|_{X(\mathbb{R})} \to 0$$

for every sequence $\{E_n\}_{n=1}^{\infty}$ of measurable sets on $\mathbb{R}$ satisfying $\chi_{E_n} \to 0$ a.e. on $\mathbb{R}$ as $n \to \infty$. Thus, $f \in X(\mathbb{R}, w)$ has absolutely continuous norm in $X(\mathbb{R}, w)$. □

From Lemma 2.4 and [2, Chap. 1, Corollaries 4.3, 4.4] we obtain the following.

Lemma 2.8. *Let $X(\mathbb{R})$ be a Banach function space and $w : \mathbb{R} \to [0, \infty]$ be a weight such that $w \in X_{\rm loc}(\mathbb{R})$ and $1/w \in X'_{\rm loc}(\mathbb{R})$.*

(a) *The Banach space dual $[X(\mathbb{R}, w)]^*$ of the weighted Banach function space $X(\mathbb{R}, w)$ is isometrically isomorphic to the associate space $X'(\mathbb{R}, w^{-1})$ if and only if $X(\mathbb{R}, w)$ has absolutely continuous norm. If this is the case, then the general form of a linear functional on $X(\mathbb{R}, w)$ is given by*

$$G(f) := \int_{\mathbb{R}} f(x)\overline{g(x)}\,dx \quad \text{for} \quad g \in X'(\mathbb{R}, w^{-1})$$

and $\|G\|_{[X(\mathbb{R},w)]^*} = \|g\|_{X'(\mathbb{R},w^{-1})}$.

(b) *The weighted Banach function space $X(\mathbb{R}, w)$ is reflexive if and only if both $X(\mathbb{R}, w)$ and $X'(\mathbb{R}, w^{-1})$ have absolutely continuous norm.*

Corollary 2.9. *Let $X(\mathbb{R})$ be a Banach function space and $w : \mathbb{R} \to [0, \infty]$ be a weight such that $w \in X_{\mathrm{loc}}(\mathbb{R})$ and $1/w \in X'_{\mathrm{loc}}(\mathbb{R})$. If $X(\mathbb{R})$ is reflexive, then $X(\mathbb{R}, w)$ is reflexive.*

Proof. The proof is a literal repetition of that one of [9, Corollary 2.8]. If $X(\mathbb{R})$ is reflexive, then, by [2, Chap. 1, Corollary 4.4], both $X(\mathbb{R})$ and $X'(\mathbb{R})$ have absolutely continuous norm. In that case, due to Lemma 2.7, both $X(\mathbb{R}, w)$ and $X'(\mathbb{R}, w^{-1})$ have absolutely continuous norm. By Lemma 2.8(b), $X(\mathbb{R}, w)$ is reflexive. □

2.4. Density of smooth compactly supported functions

For a subset Y of $L^\infty(\mathbb{R})$, let Y_0 denote the set of all compactly supported functions in Y.

Lemma 2.10. *Let $X(\mathbb{R})$ be a Banach function space.*

(a) $L_0^\infty(\mathbb{R}) \subset X(\mathbb{R})$.

(b) *If $X(\mathbb{R})$ has absolutely continuous norm, then $L_0^\infty(\mathbb{R})$, $C_0(\mathbb{R})$, and $C_0^\infty(\mathbb{R})$ are dense in $X(\mathbb{R})$.*

Proof. Part (a) follows from the definition of a Banach function space.

(b) From [2, Chap. 1, Proposition 3.10 and Theorem 3.11] it follows that $L_0^\infty(\mathbb{R})$ is dense in $X(\mathbb{R})$.

Let us show that each function $u \in L_0^\infty(\mathbb{R})$ can be approximated by a function from $C_0(\mathbb{R})$ in the norm of $X(\mathbb{R})$. We have $\operatorname{supp} u \subset Q$ and $|u(x)| \le a$ for almost all $x \in \mathbb{R}$, where Q is some finite closed segment and $a > 0$. By Axiom (A4), $\chi_Q \in X(\mathbb{R})$ and χ_Q has absolutely continuous norm by the hypothesis. From Lemma 2.6 it follows that for every $\varepsilon > 0$ there is a $\delta > 0$ such that $|E| < \delta$ implies that $\|\chi_Q \chi_E\|_{X(\mathbb{R})} < \varepsilon$. By Luzin's theorem, for such a $\delta > 0$ there is a continuous function v supported in Q such that $|v(x)| \le a$ and the measure of the set $\widetilde{Q} := \{x \in Q : u(x) \ne v(x)\}$ is less than δ. Then

$$|u(x) - v(x)| \le 2a\chi_{\widetilde{Q}}(x) \quad (x \in \mathbb{R}).$$

Therefore, by Axiom (A2),

$$\|u - v\|_{X(\mathbb{R})} \le 2a\|\chi_Q \chi_{\widetilde{Q}}\|_{X(\mathbb{R})} < 2a\varepsilon.$$

Hence, each function $u \in L_0^\infty(\mathbb{R})$ can be approximated by a function from $C_0(\mathbb{R})$ in the norm of $X(\mathbb{R})$. Thus, $C_0(\mathbb{R})$ is dense in $X(\mathbb{R})$.

Now let us prove that each function $v \in C_0(\mathbb{R})$ can be approximated by a function from $C_0^\infty(\mathbb{R})$ in the norm of $X(\mathbb{R})$. Let $a \in C_0^\infty(\mathbb{R})$ and $\int_{\mathbb{R}} a(x)dx = 1$. Consider

$$v_t(x) = \frac{1}{t}\int_{\mathbb{R}} a\left(\frac{y}{t}\right) v(x - y)\, dy \quad (t > 0).$$

It is easy to see that $v_t \in C_0^\infty(\mathbb{R})$. Fix an interval Q containing the supports of v and v_t. Then for every $\varepsilon > 0$ there is a $t > 0$ such that $|v_t(x) - v(x)| < \varepsilon$ for all $x \in Q$. Hence,

$$\|v_t - v\|_{X(\mathbb{R})} = \|(v_t - v)\chi_Q\|_{X(\mathbb{R})} < \varepsilon\|\chi_Q\|_{X(\mathbb{R})},$$

that is, $v \in C_0(\mathbb{R})$ can be approximated by a function from $C_0^\infty(\mathbb{R})$ in the norm of $X(\mathbb{R})$. Thus, $C_0^\infty(\mathbb{R})$ is dense in $X(\mathbb{R})$. □

From [2, Chap. 1, Corollary 5.6] one can extract the following.

Lemma 2.11. *A Banach function space $X(\mathbb{R})$ is separable if and only if it has absolutely continuous norm.*

Gathering the results mentioned above, we arrive at the next result.

Lemma 2.12. *Let $X(\mathbb{R})$ be a Banach function space and $w : \mathbb{R} \to [0, \infty]$ be a weight such that $w \in X_{\rm loc}(\mathbb{R})$ and $1/w \in X'_{\rm loc}(\mathbb{R})$.*

(a) *If $X(\mathbb{R})$ is separable, then $L_0^\infty(\mathbb{R})$, $C_0(\mathbb{R})$, and $C_0^\infty(\mathbb{R})$ are dense in the weighted Banach function space $X(\mathbb{R}, w)$.*
(b) *If $X(\mathbb{R})$ is reflexive, then $L_0^\infty(\mathbb{R})$, $C_0(\mathbb{R})$, and $C_0^\infty(\mathbb{R})$ are dense in the weighted Banach function spaces $X(\mathbb{R}, w)$ and $X'(\mathbb{R}, w^{-1})$.*

Proof. (a) If $X(\mathbb{R})$ is separable, then by Lemma 2.11, $X(\mathbb{R})$ has absolutely continuous norm. Therefore $X(\mathbb{R}, w)$ has absolutely continuous norm too, in view of Lemma 2.7. Hence, from Lemma 2.10(b) we derive that $L_0^\infty(\mathbb{R})$, $C_0(\mathbb{R})$, and $C_0^\infty(\mathbb{R})$ are dense in $X(\mathbb{R}, w)$. Part (a) is proved.

(b) If $X(\mathbb{R})$ is reflexive, then by [2, Chap. 1, Corollary 4.4], both $X(\mathbb{R})$ and $X'(\mathbb{R})$ have absolutely continuous norm. Hence both $X(\mathbb{R}, w)$ and $X'(\mathbb{R}, w^{-1})$ have absolutely continuous norm in view of Lemma 2.7. Thus, from Lemma 2.10(b) we get that $L_0^\infty(\mathbb{R})$, $C_0(\mathbb{R})$, and $C_0^\infty(\mathbb{R})$ are dense in $X(\mathbb{R}, w)$ and in $X'(\mathbb{R}, w^{-1})$. □

3. Boundedness of the Cauchy singular integral operator on weighted Banach function spaces

3.1. Well-known properties of the Cauchy singular integral operator

The following results are proved in many standard texts on Harmonic Analysis, see, e.g., [2, Chap. 3, Theorem 4.9(b)] or [7, pp. 51–52].

Theorem 3.1 (M. Riesz). *The Cauchy singular integral operator S is bounded on $L^p(\mathbb{R})$ for every $p \in (1, \infty)$.*

Theorem 3.2. *If $f, g \in L^2(\mathbb{R})$, then*

$$(S^2 f)(x) = f(x) \quad (x \in \mathbb{R}), \tag{3.1}$$

$$\int_{\mathbb{R}} (Sf)(x)\overline{g(x)}\, dx = \int_{\mathbb{R}} f(x)\overline{(Sg)(x)}\, dx. \tag{3.2}$$

3.2. Pointwise estimates for sharp maximal functions

For $\delta > 0$ and $f \in L^\delta_{\rm loc}(\mathbb{R})$, set

$$f^\#_\delta(x) := \sup_{Q \ni x} \inf_{c \in \mathbb{R}} \left(\frac{1}{|Q|} \int_Q |f(y) - c|^\delta dy \right)^{1/\delta}.$$

The non-increasing rearrangement (see, e.g., [2, Chap. 2, Section 1]) of a measurable function f on $\mathbb{R}$ is defined by

$$f^*(t) := \inf \{\lambda > 0 : |\{x \in \mathbb{R} : |f(x)| > \lambda\}| \le t\} \quad (0 < t < \infty).$$

For a fixed $\lambda \in (0,1)$ and a given measurable function f on $\mathbb{R}$, consider the local sharp maximal function $M^\#_\lambda f$ defined by

$$M^\#_\lambda f(x) := \sup_{Q \ni x} \inf_{c \in \mathbb{R}} \left((f - c)\chi_Q\right)^* (\lambda|Q|).$$

In all above definitions the suprema are taken over all intervals $Q \subset \mathbb{R}$ containing x.

The following result was proved by Lerner [17, Theorem 1] for the case of $\mathbb{R}^n$.

Theorem 3.3 (Lerner). *For a function $g \in L^1_{\rm loc}(\mathbb{R})$ and a measurable function φ satisfying*

$$|\{x \in \mathbb{R} : |\varphi(x)| > \alpha\}| < \infty \quad \textit{for all} \quad \alpha > 0, \tag{3.3}$$

one has

$$\int_{\mathbb{R}} |\varphi(x) g(x)|\, dx \le C_L \int_{\mathbb{R}} M^\#_\lambda \varphi(x) M g(x)\, dx,$$

where $C_L > 0$ and $\lambda \in (0,1)$ are some absolute constants.

The sharp maximal functions can be related as follows.

Lemma 3.4 ([10, Proposition 2.3]). *If $\delta > 0$, $\lambda \in (0,1)$, and $f \in L^\delta_{\rm loc}(\mathbb{R})$, then*

$$M^\#_\lambda f(x) \le (1/\lambda)^{1/\delta} f^\#_\delta(x) \quad (x \in \mathbb{R}).$$

The following estimate was proved in [1, Theorem 2.1] for the case of Calderón–Zygmund singular integral operators with standard kernels in the sense of Coifman and Meyer on $\mathbb{R}^n$. It is well known that the Cauchy kernel is an archetypical example of a standard kernel (see, e.g., [7, p. 99]).

Theorem 3.5 (Álvarez–Pérez). *If $0 < \delta < 1$, then for every $f \in C_0^\infty(\mathbb{R})$,*

$$(Sf)^\#_\delta(x) \le C_\delta M f(x) \quad (x \in \mathbb{R})$$

where $C_\delta > 0$ is some constant depending only on δ.

3.3. Sufficient condition

The set of all bounded sublinear operators on a Banach function space $Y(\mathbb{R})$ will be denoted by $\widetilde{\mathcal{B}}(Y(\mathbb{R}))$ and its subset of all bounded linear operators will be denoted by $\mathcal{B}(Y(\mathbb{R}))$.

Theorem 3.6. *Let $X(\mathbb{R})$ be a separable Banach function space and $w : \mathbb{R} \to [0,\infty]$ be a weight such that $w \in X_{\mathrm{loc}}(\mathbb{R})$ and $1/w \in X'_{\mathrm{loc}}(\mathbb{R})$. Suppose the Hardy–Littlewood maximal operator M is bounded on $X(\mathbb{R},w)$ and on $X'(\mathbb{R},w^{-1})$. Assume that $0<\delta<1$ and T is an operator such that*

(a) *T is bounded on some $L^p(\mathbb{R})$ with $p \in (1,\infty)$;*
(b) *for each $f \in C_0^\infty(\mathbb{R})$,*

$$(Tf)^{\#}_{\delta}(x) \le C_\delta Mf(x) \quad (x \in \mathbb{R})$$

where C_δ is a positive constant depending only on δ.

Then $T \in \mathcal{B}(X(\mathbb{R},w))$ and

$$\|T\|_{\mathcal{B}(X(\mathbb{R},w))} \le (1/\lambda)^{\delta} C_L \|M\|_{\widetilde{\mathcal{B}}(X(\mathbb{R},w))} \|M\|_{\widetilde{\mathcal{B}}(X'(\mathbb{R},w^{-1}))} C_\delta, \tag{3.4}$$

where $\lambda \in (0,1)$ and $C_L>0$ are the constants from Theorem 3.3.

Proof. The idea of the proof is borrowed from [10, Theorem 2.7]. By Lemma 2.4, $X(\mathbb{R},w)$ is a Banach function space whose associate space is $X'(\mathbb{R},w^{-1})$. Let $f \in C_0^\infty(\mathbb{R})$ and $g \in X'(\mathbb{R},w^{-1}) \subset L^1_{\mathrm{loc}}(\mathbb{R})$. From the boundedness of T on $L^p(\mathbb{R})$ and the Chebyshev inequality it follows that

$$|\{x \in \mathbb{R} : |(Tf)(x)| > \alpha\}| \le \frac{1}{\alpha^p} \int_{\mathbb{R}} |(Tf)|^p \, dx \le \left(\frac{\|T\|_{\mathcal{B}(L^p(\mathbb{R}))}}{\alpha} \|f\|_{L^p(\mathbb{R})} \right)^p$$

for all $\alpha>0$. Hence Tf satisfies (3.3). From Theorem 3.3 we get that there exist constants $\lambda \in (0,1)$ and $C_L>0$ independent of f and g such that

$$\int_{\mathbb{R}} |(Tf)(x)g(x)| \, dx \le C_L \int_{\mathbb{R}} M^{\#}_{\lambda}(Tf)(x) Mg(x) \, dx. \tag{3.5}$$

Since T is bounded on some standard Lebesgue space $L^p(\mathbb{R})$ for $1<p<\infty$ and $L^s(J) \subset L^r(J)$ whenever $0<r<s<\infty$ and J is a finite interval, we see that $Tf \in L^{\delta}_{\mathrm{loc}}(\mathbb{R})$ for each $\delta \in (0,p]$. From Lemma 3.4 and hypothesis (b) it follows that

$$M^{\#}_{\lambda}(Tf)(x) \le (1/\lambda)^{1/\delta} (Tf)^{\#}_{\delta}(x) \le (1/\lambda)^{1/\delta} C_\delta Mf(x) \quad (x \in \mathbb{R}) \tag{3.6}$$

for some $\delta \in (0,1)$. Combining (3.5) and (3.6) with Hölder's inequality (see Lemma 2.2), we obtain

$$\begin{aligned} \int_{\mathbb{R}} |(Tf)(x)g(x)| \, dx &\le C_1 \int_{\mathbb{R}} Mf(x) Mg(x) \, dx \\ &\le C_1 \|Mf\|_{X(\mathbb{R},w)} \|Mg\|_{X'(\mathbb{R},w^{-1})}, \end{aligned} \tag{3.7}$$

where $C_1 := (1/\lambda)^{1/\delta} C_\delta C_L > 0$ is independent of $f \in C_0^\infty(\mathbb{R})$ and $g \in X'(\mathbb{R}, w^{-1})$. Taking into account that M is bounded on $X(\mathbb{R}, w)$ and on $X'(\mathbb{R}, w^{-1})$, from (3.7) we get

$$\int_{\mathbb{R}} |(Tf)(x)g(x)|\, dx \le C_2 \|f\|_{X(\mathbb{R},w)} \|g\|_{X'(\mathbb{R},w^{-1})},$$

where $C_2 := C_1 \|M\|_{\widetilde{\mathcal{B}}(X(\mathbb{R},w))} \|M\|_{\widetilde{\mathcal{B}}(X'(\mathbb{R},w^{-1}))}$. From this inequality and (2.1) we obtain

$$\begin{aligned}\|Tf\|_{X(\mathbb{R},w)} &= \sup\left\{ \int_{\mathbb{R}} |(Tf)(x)g(x)|\, dx \ :\ g \in X'(\mathbb{R}, w^{-1}),\ \|g\|_{X'(\mathbb{R},w^{-1})} \le 1 \right\} \\ &\le C_2 \|f\|_{X(\mathbb{R},w)}\end{aligned}$$

for all $f \in C_0^\infty(\mathbb{R})$. Taking into account that $C_0^\infty(\mathbb{R})$ is dense in $X(\mathbb{R}, w)$ in view of Lemma 2.12(a), from the latter inequality it follows that T is bounded on $X(\mathbb{R}, w)$ and (3.4) holds. □

Remark 3.7. The proof of this result without changes extends to the case of $\mathbb{R}^n$.

Theorem 3.8. *Let $X(\mathbb{R})$ be a Banach function space and $w : \mathbb{R} \to [0, \infty]$ be a weight such that $w \in X_{\mathrm{loc}}(\mathbb{R})$ and $1/w \in X'_{\mathrm{loc}}(\mathbb{R})$. Suppose the Hardy–Littlewood maximal operator M is bounded on $X(\mathbb{R}, w)$ and on $X'(\mathbb{R}, w^{-1})$.*

(a) *If the space $X(\mathbb{R})$ is separable, then the Cauchy singular integral operator S is bounded on the space $X(\mathbb{R}, w)$ and $S^2 = I$.*
(b) *If the space $X(\mathbb{R})$ is reflexive, then the Cauchy singular integral operator S is bounded on the spaces $X(\mathbb{R}, w)$ and $X'(\mathbb{R}, w^{-1})$ and its adjoint S^* coincides with S on the space $X'(\mathbb{R}, w^{-1})$.*

Proof. From Theorems 3.1 and 3.5 it follows that all hypotheses of Theorem 3.6 are fulfilled. Hence, the operator S is bounded on $X(\mathbb{R}, w)$.

Let now $\varphi \in X(\mathbb{R}, w)$. Then there exists a sequence $f_n \in L_0^\infty(\mathbb{R})$ such that $f_n \to \varphi$ in $X(\mathbb{R}, w)$ as $n \to \infty$. From (3.1) we get $S^2 f_n = f_n$ because $L_0^\infty(\mathbb{R}) \subset L^2(\mathbb{R})$. Hence

$$\begin{aligned}\|S^2\varphi - \varphi\|_{X(\mathbb{R},w)} &\le \|S^2\varphi - f_n\|_{X(\mathbb{R},w)} + \|f_n - \varphi\|_{X(\mathbb{R},w)} \\ &= \|S^2(\varphi - f_n)\|_{X(\mathbb{R},w)} + \|\varphi - f_n\|_{X(\mathbb{R},w)} \\ &\le (\|S^2\|_{\mathcal{B}(X(\mathbb{R},w))} + 1)\|\varphi - f_n\|_{X(\mathbb{R},w)} \to 0\end{aligned}$$

as $n \to \infty$. Thus $S^2\varphi = \varphi$. Part (a) is proved.

(b) From (3.2) it follows that

$$\int_{\mathbb{R}} (Sf)(x)\overline{g(x)}\, dx = \int_{\mathbb{R}} f(x)\overline{(Sg)(x)}\, dx.$$

for all $f, g \in L_0^\infty(\mathbb{R})$. From this equality and Lemmas 2.8 and 2.12(b) it follows that S is a self-adjoint and densely defined operator on $X(\mathbb{R}, w)$ and $X'(\mathbb{R}, w^{-1})$. By the standard argument (see [11, Chap. III, Section 5.5]), one can show that $S = S^* \in \mathcal{B}(X'(\mathbb{R}, w^{-1}))$ because $S \in \mathcal{B}(X(\mathbb{R}, w))$ by part (a). □

3.4. Necessary condition

Let $X(\mathbb{R})$ be a Banach function space and $X'(\mathbb{R})$ be its associate space. We say that a weight $w\colon \mathbb{R} \to [0,\infty]$ belongs to the class $A_X(\mathbb{R})$ if

$$\sup_{-\infty<a<b<\infty} \frac{1}{b-a} \|w\chi_{(a,b)}\|_{X(\mathbb{R})} \|w^{-1}\chi_{(a,b)}\|_{X'(\mathbb{R})} < \infty.$$

Theorem 3.9. *Let $X(\mathbb{R})$ be a Banach function space and $w\colon \mathbb{R} \to [0,\infty]$ be a weight. If the operator S is bounded on the space $X(\mathbb{R},w)$, then*

(a) *$w \in X_{\mathrm{loc}}(\mathbb{R})$ and $1/w \in X'_{\mathrm{loc}}(\mathbb{R})$;*
(b) *$X(\mathbb{R},w)$ is a Banach function space;*
(c) *$w \in A_X(\mathbb{R})$.*

Proof. (a) The idea of the proof is borrowed from [8, Lemma 3.3]. Let $E \subset \mathbb{R}$ be a measurable set of finite measure. Then there exist $a,b \in \mathbb{R}$ such that $E \subset (a,b) =: J$. It is clear that

$$w(x)\chi_E(x) \le w(x)\chi_J(x), \quad \chi_E(x)/w(x) \le \chi_J(x)/w(x)$$

for almost all $x \in \mathbb{R}$. Then by Axiom (A2),

$$\|w\chi_E\|_{X(\mathbb{R})} \le \|w\chi_J\|_{X(\mathbb{R})}, \quad \|\chi_E/w\|_{X'(\mathbb{R})} \le \|\chi_J/w\|_{X'(\mathbb{R})}.$$

Thus, it is sufficient to prove that $w\chi_J \in X(\mathbb{R})$ and $\chi_J/w \in X'(\mathbb{R})$.

Obviously, the operator $(Vf)(x) = \chi_J(x)xf(x)$ is bounded on $X(\mathbb{R},w)$ and

$$((SV - VS)f)(x) = \frac{1}{\pi i}\int_J f(y)\,dy$$

for almost all $x \in \mathbb{R}$. Since the operator $SV - VS$ is bounded on $X(\mathbb{R},w)$, there exists a constant $C_1 > 0$ such that

$$\left\| \frac{1}{\pi i}\int_J f(y)dy \right\|_{X(\mathbb{R},w)} \le C_1\|f\|_{X(\mathbb{R},w)} \quad \text{for all} \quad f \in X(\mathbb{R},w). \tag{3.8}$$

On the other hand,

$$\left\| \frac{1}{\pi i}\int_J f(y)dy \right\|_{X(\mathbb{R},w)} = \frac{1}{\pi}\left|\int_J f(y)dy\right| \|w\chi_J\|_{X(\mathbb{R})}. \tag{3.9}$$

Since $w(x) > 0$ a.e. on $\mathbb{R}$, we have $\|w\chi_J\|_{X(\mathbb{R})} > 0$. Hence, from (3.8)–(3.9) it follows that

$$\left|\int_J f(y)dy\right| \le \frac{C_1\pi}{\|w\chi_J\|_{X(\mathbb{R})}}\|f\|_{X(\mathbb{R},w)}.$$

Therefore,

$$\left|\int_{\mathbb{R}} f(y)w(y)\cdot\frac{\chi_J(y)}{w(y)}dy\right| \le \frac{C_1\pi}{\|w\chi_J\|_{X(\mathbb{R})}}\|fw\|_{X(\mathbb{R})}$$

for all measurable functions f such that $fw \in X(\mathbb{R})$. By Lemma 2.3, we have $\chi_J/w \in X'(\mathbb{R})$.

Let us show that there exists a function $g_0 \in X(\mathbb{R})$ such that

$$C_2 := \frac{1}{\pi}\left|\int_J \frac{g_0(y)}{w(y)}dy\right| > 0. \tag{3.10}$$

Assume the contrary. Then, taking into account Lemma 2.10(a), we obtain

$$\int_J \frac{g(y)}{w(y)}dy = 0 \tag{3.11}$$

for all g continuous on $\overline{J}$. By Axiom (A5), $(1/w)|_J \in L^1(J)$. Without loss of generality, assume that $|J| = 2\pi$. Let $\eta : [0, 2\pi] \to \overline{J}$ be a homeomorphism such that $|\eta'(x)| = 1$ for almost all $x \in [0, 2\pi]$. From (3.11) we get

$$\int_0^{2\pi} \frac{\varphi(x)}{w(\eta(x))}dx = 0 \quad \text{for all} \quad \varphi \in C[0, 2\pi]. \tag{3.12}$$

Taking $\varphi(x) = e^{inx}$ with $n \in \mathbb{Z}$, we see from (3.12) that all Fourier coefficients of $1/(w \circ \eta)$ vanish. This implies that $1/w(\eta(x)) = 0$ for almost all $x \in [0, 2\pi]$. Consequently, $w(y) = \infty$ almost everywhere on J. This contradicts the assumption that w is a weight. Thus, $C_2 > 0$.

Clearly, $f_0 = g_0/w \in X(\Gamma, w)$. Then from (3.8)–(3.10) it follows that

$$\|w\chi_J\|_{X(\mathbb{R})} \le \frac{C_1\pi}{C_2}\|f_0\|_{X(\mathbb{R},w)},$$

that is, $w\chi_J \in X(\mathbb{R})$. Part (a) is proved.

Part (b) follows from part (a) and Lemma 2.4(b).

(c) The idea of the proof is borrowed from [8, Theorem 3.2]. By part(b), $X(\mathbb{R}, w)$ is a Banach function space.

Let Q be an arbitrary interval and Q_1, Q_2 be its two halves. Take a function $f \ge 0$ supported in Q_1. Then for $\tau \in Q_1$ and $x \in Q_2$ we have $|\tau - x| \le |Q|$. Therefore,

$$\begin{aligned}|(Sf)(x)| &= \frac{1}{\pi}\left|\int_{Q_1} \frac{f(\tau)}{\tau - x}d\tau\right| = \frac{1}{\pi}\int_{Q_1} \frac{f(\tau)}{|\tau - x|}d\tau \\ &\ge \frac{1}{\pi|Q|}\int_{Q_1} f(\tau)\,d\tau = \frac{1}{2\pi|Q_1|}\int_{Q_1} f(\tau)\,d\tau.\end{aligned}$$

Thus,

$$|(Sf)(x)|\chi_{Q_2}(x) \ge \frac{1}{2\pi|Q_1|}\left(\int_{Q_1} f(\tau)\,d\tau\right)\chi_{Q_2}(x) \quad (x \in \mathbb{R}).$$

Then, by Axioms (A1) and (A2),

$$\|Sf\|_{X(\mathbb{R},w)} \ge \|(Sf)\chi_{Q_2}\|_{X(\mathbb{R},w)} \ge \frac{1}{2\pi|Q_1|}\left(\int_{Q_1} f(\tau)\,d\tau\right)\|\chi_{Q_2}\|_{X(\mathbb{R},w)}. \tag{3.13}$$

On the other hand, since S is bounded on $X(\mathbb{R}, w)$, we get

$$\|Sf\|_{X(\mathbb{R},w)} \le \|S\|_{\mathcal{B}(X(\mathbb{R},w))}\|f\|_{X(\mathbb{R},w)} = \|S\|_{\mathcal{B}(X(\mathbb{R},w))}\|f\chi_{Q_1}\|_{X(\mathbb{R},w)}. \tag{3.14}$$

Combining (3.13) and (3.14), we arrive at

$$\frac{1}{|Q_1|}\left(\int_{Q_1} f(\tau)\,d\tau\right)\|w\chi_{Q_2}\|_{X(\mathbb{R})} \le 2\pi\|S\|_{\mathcal{B}(X(\mathbb{R},w))}\|f\chi_{Q_1}\|_{X(\mathbb{R},w)}. \tag{3.15}$$

Taking $f=\chi_{Q_1}$, from (3.15) we get

$$\|w\chi_{Q_2}\|_{X(\mathbb{R})} \le 2\pi\|S\|_{\mathcal{B}(X(\mathbb{R},w))}\|w\chi_{Q_1}\|_{X(\mathbb{R})}.$$

Analogously one can obtain

$$\|w\chi_{Q_1}\|_{X(\mathbb{R})} \le 2\pi\|S\|_{\mathcal{B}(X(\mathbb{R},w))}\|w\chi_{Q_2}\|_{X(\mathbb{R})}. \tag{3.16}$$

From (3.15) and (3.16) it follows that

$$\frac{1}{|Q_1|}\left(\int_{Q_1} f(\tau)\,d\tau\right)\|w\chi_{Q_1}\|_{X(\mathbb{R})} \le C\|f\chi_{Q_1}\|_{X(\mathbb{R},w)}, \tag{3.17}$$

where $C:=\left(2\pi\|S\|_{\mathcal{B}(X(\mathbb{R},w))}\right)^2$. Let

$$Y:=\left\{g\in X(\mathbb{R},w):\|g\|_{X(\mathbb{R},w)}\le 1\right\}.$$

If $g\in Y$, then $|g|\chi_{Q_1}\ge 0$ is supported in Q_1. Then from (3.17) we obtain

$$\|w\chi_{Q_1}\|_{X(\mathbb{R})}\int_{\mathbb{R}}|g(\tau)|\chi_{Q_1}(\tau)\,d\tau \le C|Q_1| \tag{3.18}$$

for all $g\in Y$. From (2.2) we get

$$\|w^{-1}\chi_{Q_1}\|_{X'(\mathbb{R})} = \|\chi_{Q_1}\|_{X'(\mathbb{R},w^{-1})} = \sup_{g\in Y}\int_{\mathbb{R}}|g(\tau)|\chi_{Q_1}(\tau)\,d\tau. \tag{3.19}$$

From (3.18) and (3.19) it follows that

$$\|w\chi_{Q_1}\|_{X(\mathbb{R})}\|w^{-1}\chi_{Q_1}\|_{X'(\mathbb{R})} \le C|Q_1|.$$

Since $Q_1\subset\mathbb{R}$ is an arbitrary interval, we conclude that $w\in A_X(\mathbb{R})$. □

3.5. The case of weighted variable Lebesgue spaces

We start this subsection with the following well-known fact.

Theorem 3.10 ([6, Theorems 3.2.13 and 3.4.7]). *Let $p:\mathbb{R}\to[1,\infty]$ be a measurable a.e. finite function satisfying* (1.1). *Then $L^{p(\cdot)}(\mathbb{R})$ is a separable and reflexive Banach function space whose associate space is isomorphic to $L^{p'(\cdot)}(\mathbb{R})$.*

Now we are in a position to give a proof of Theorem 1.3.

Proof of Theorem 1.3. *Necessity.* Theorem 3.10 immediately implies that if p satisfies (1.1), then $L^{p(\cdot)}(\mathbb{R})$ is a Banach function space and

$$\mathcal{A}_{p(\cdot)}(\mathbb{R}) = A_{L^{p(\cdot)}}(\mathbb{R}).$$

From Theorem 3.9 it follows that if S is bounded on the space $L^{p(\cdot)}(\mathbb{R},w)$, then $w\in\mathcal{A}_{p(\cdot)}(\mathbb{R})$. The necessity portion is proved.

Sufficiency. From Theorem 3.10 we know that $L^{p(\cdot)}(\mathbb{R})$ is a separable and reflexive Banach function space. If $w\in\mathcal{A}_{p(\cdot)}(\mathbb{R})$, then $w\in L^{p(\cdot)}_{\mathrm{loc}}(\mathbb{R})$, $1/w\in L^{p'(\cdot)}_{\mathrm{loc}}(\mathbb{R})$, and $1/w \in \mathcal{A}_{p'(\cdot)}(\mathbb{R})$. Further, it is easy to see that p is globally log-Hölder continuous if and only if so is p'. Hence, by Theorem 1.2, the Hardy–Littlewood maximal operator is bounded on $L^{p(\cdot)}(\mathbb{R}, w)$ and on $L^{p'(\cdot)}(\mathbb{R}, w^{-1})$. Applying Theorem 3.8(a), we see that the operator S is bounded on $L^{p(\cdot)}(\mathbb{R}, w)$. This finishes the proof of Theorem 1.3. □

Theorem 1.4 follows immediately from Theorems 1.2, 3.8, and 3.10.

References

[1] J. Álvarez and C. Pérez, *Estimates with A_∞ weights for various singular integral operators.* Boll. Un. Mat. Ital. A (7) **8** (1994), 123–133.

[2] C. Bennett and R. Sharpley, *Interpolation of Operators.* Academic Press, New York, 1988.

[3] E.I. Berezhnoi, *Two-weighted estimations for the Hardy–Littlewood maximal function in ideal Banach spaces.* Proc. Amer. Math. Soc. **127** (1999), 79–87.

[4] D. Cruz-Uribe, L. Diening, and P. Hästö, *The maximal operator on weighted variable Lebesgue spaces.* Frac. Calc. Appl. Anal. **14** (2011), 361–374.

[5] D. Cruz-Uribe, A. Fiorenza, and C.J. Neugebauer, *Weighted norm inequalities for the maximal operator on variable Lebesgue spaces.* J. Math. Anal. Appl. **394** (2012), 744–760.

[6] L. Diening, P. Harjulehto, P. Hästö, and M. Růžička, *Lebesgue and Sobolev Spaces with Variable Exponents.* Lecture Notes in Mathematics **2017**, Springer, Berlin, 2011.

[7] J. Duoandikoetxea, *Fourier Analysis.* Graduate Studies in Mathematics **29**. American Mathematical Society, Providence, RI, 2001.

[8] A.Yu. Karlovich, *Singular integral operators with piecewise continuous coefficients in reflexive rearrangement-invariant spaces.* Integral Equations and Operator Theory **32** (1998), 436–481.

[9] A.Yu. Karlovich, *Fredholmness of singular integral operators with piecewise continuous coefficients on weighted Banach function spaces.* J. Integral Equations Appl. **15** (2003), no. 3, 263–320.

[10] A.Yu. Karlovich and A.K. Lerner, *Commutators of singular integrals on generalized Lp spaces with variable exponent.* Publ. Mat. **49** (2005), 111–125.

[11] T. Kato, *Perturbation Theory for Linear Operators.* Reprint of the 1980 edition. Springer-Verlag, Berlin, 1995.

[12] V. Kokilashvili, V. Paatashvili, and S. Samko, *Boundedness in Lebesgue spaces with variable exponent of the Cauchy singular operator on Carleson curves.* In: "Modern operator theory and applications". Operator Theory: Advances and Applications **170**. Birkhäuser Verlag, Basel, 2006, pp. 167–186.

[13] V. Kokilashvili, N. Samko, and S. Samko, *The maximal operator in weighted variable spaces $L^{p(\cdot)}$.* J. Funct. Spaces Appl. **5** (2007), 299–317.

[14] V. Kokilashvili, N. Samko, and S. Samko, *Singular operators in variable spaces $L^{p(\cdot)}(\Omega, \rho)$ with oscillating weights.* Math. Nachr. **280** (2007), 1145–1156.

[15] V. Kokilashvili and S. Samko, *Boundedness of maximal operators and potential operators on Carleson curves in Lebesgue spaces with variable exponent.* Acta Math. Sin. (Engl. Ser.) **24** (2008), 1775–1800.

[16] T. Kopaliani, *Infimal convolution and Muckenhoupt $A_{p(\cdot)}$ condition in variable L^p spaces.* Arch. Math. (Basel) **89** (2007), 185–192.

[17] A.K. Lerner, *Weighted norm inequalities for the local sharp maximal function.* J. Fourier Anal. Appl. **10** (2004), 465–474.

Alexei Yu. Karlovich
Departamento de Matemática
Faculdade de Ciências e Tecnologia
Universidade Nova de Lisboa
Quinta da Torre
P-2829–516 Caparica, Portugal
e-mail: oyk@fct.unl.pt

Ilya M. Spitkovsky
Department of Mathematics
College of William & Mary
Williamsburg, VA 23187-8795, USA
e-mail: ilya@math.wm.edu

Operator Theory:
Advances and Applications, Vol. 236, 293–298

Extension of Certain Distributions on Weighted Hölder Space and the Riemann Boundary Value Problem for Non-rectifiable Curves

Boris A. Kats

Abstract. Let Γ be a non-rectifiable curve on the complex plane $\mathbb{C}$. We extend distributional derivative $\overline{\partial}F$ of a function F, which is holomorphic in domain $\overline{\mathbb{C}} \setminus \Gamma$, up to continuous functional on the weighted Hölder space and apply this extension for solution of the Riemann boundary value problem on the curve Γ.

Mathematics Subject Classification (2010). Primary 30E25; secondary 30E20.

Keywords. Hölder space, non-rectifiable arc, Cauchy transform, Riemann boundary value problem.

Introduction

In the papers [1], [2], [3] the author considered the following construction.

Let function $F(z)$ be holomorphic in domain $\overline{\mathbb{C}}\setminus\Gamma$ where Γ is a non-rectifiable curve on the complex plane $\mathbb{C}$. If $F(z)$ is locally integrable in the complex plane, we identify it with distribution

$$F : C_0^\infty(\mathbb{C}) \ni \phi \mapsto \iint_D F(\zeta)\phi(\zeta)d\zeta d\overline{\zeta},$$

where, as usually, $C_0^\infty(D)$ stands for the space of all infinitely smooth functions with compact support in D. Its distributional derivative $\overline{\partial}F$ can be considered as operation of weighted integration over Γ. Then we extend this distribution on the Hölder spaces. This extension generates certain new distributions of integration type. Finally, the Cauchy transform of these distributions is a generalization of the Cauchy type integral for non-rectifiable curves, and it allows us to solve the

This work was partially supported by Russian Fund for Basic Researches (grants 12-01-00636-a and 12-01-97015-r-povolzhie-a).

Riemann boundary value problem with boundary data from the Hölder spaces on non-rectifiable curves.

In the present paper we extend $\overline{\partial}F$ to the weighted Hölder space and apply this extension for solution of the Riemann boundary value problem on non-rectifiable curves in the case of coefficients with singularities.

In Sections 1 and 2 we describe the desired extension and boundary properties of its Cauchy transform. Section 3 contains application of these results for solution of the Riemann boundary value problem.

1. Extension of $\overline{\partial}F$

Let A be a compact set on the complex plane. The Hölder space $H_\nu(A)$ consists of all defined on A functions f with finite Hölder coefficient

$$h_\nu(f,A) := \sup\{\frac{|f(t')-f(t'')|}{|t'-t''|^\nu} : t',t'' \in A, t' \neq t''\},$$

where $\nu \in (0,1]$. It is the Banach space with norm $\|f\|_{H_\nu(A)} := \|f\|_{C(A)} + h_\nu(f,A)$, where $\|f\|_{C(A)} = \sup\{|f(\zeta)| : \zeta \in A\}$. Let us denote $H^*_\nu(A) := \bigcup_{\mu>\nu} H_\mu(A)$. If we fix a sequence of exponents $\{\nu_j\}$ such that $1 > \nu_1 > \nu_2 > \cdots > \nu_j > \nu_{j+1} > \cdots$ and $\lim_{j\to\infty} \nu_j = \nu$, then the semi-norms $\{h_{\nu_j}(\cdot, A)\}$, $j = 1,2,\ldots$ and $\|f\|_{C(A)}$ turn $H^*_\nu(A)$ into countably normed space. If a function $w(t)$ (weight) is defined on A, then we put

$$H_\nu(A,w) := \{f : wf \in H_\nu(A)\}, \quad H^*_\nu(A,w) := \{f : wf \in H^*_\nu(A)\}.$$

We equip these spaces by intrinsic norms.

We consider a non-rectifiable Jordan curve $\Gamma \subset \mathbb{C}$ of null plane measure, and a holomorphic in $\overline{\mathbb{C}} \setminus \Gamma$ and integrable in a neighborhood of Γ function F such that $F(\infty) = 0$. Let $Q \subset \mathbb{C}$ be sufficiently large dyadic square such that $\Gamma \subset Q$.

In the present section we estimate the distribution

$$\langle \overline{\partial}F, \phi\rangle = -\langle F, \overline{\partial}\phi\rangle$$

in the norm of space $H_\nu(\overline{Q}, w)$, where

$$w(t) := \prod_{j=1}^{m} |t-t_j|^{p_j}, \quad 0 < p_j < 1, \quad j = 1,\ldots m. \tag{1}$$

We consider the Whitney decomposition (see, for instance, [4]) of the set $\mathbb{C} \setminus \Gamma$. It consists of non-overlapping dyadic squares such that $\operatorname{dist}\{Q_j, \Gamma\} \asymp \operatorname{diam} Q_j$. Let $Q_1, Q_2, \ldots$ be squares of this family lying inside Q. For $\phi \in C_0^\infty(Q)$ we have

$$\langle \overline{\partial}F, \phi\rangle = -\iint F(\zeta)\frac{\partial\phi}{\partial\overline{\zeta}}d\zeta d\overline{\zeta} = -\sum_{j=1}^{\infty}\iint_{Q_j} F(\zeta)\frac{\partial\phi}{\partial\overline{\zeta}}d\zeta d\overline{\zeta} = \sum_{j=1}^{\infty}\int_{\partial Q_j} F(\zeta)\phi(\zeta)d\zeta.$$

Let $\Gamma^* := \left(\overline{\bigcup_{j=1}^{\infty} \partial Q_j}\right) \cup \partial Q$, and $\psi := \mathcal{E}_0(w\phi|_{\Gamma^*})$, where $\mathcal{E}_0$ is the Whitney extension operator (see, for instance, [4], Ch. VI, s. 2.2). Below we need the following property of this operator: if A is compact set on the complex plane, $f \in H_\nu(A)$, and $\mathcal{E}_0$ is the Whitney extension operator from the set A, then the extension $\mathcal{E}_0 f$ is differentiable in $\mathbb{C} \setminus A$ and

$$|\nabla \mathcal{E}_0 f(z)| \leq C \operatorname{dist}^{\nu-1}(z, A), \tag{2}$$

where C is a constant. The function $\phi_0 := w^{-1}\psi$ is continuation of restriction $\phi|_{\Gamma^*}$ on the whole complex plane, and we obtain

$$\langle \overline{\partial} F, \phi \rangle = -\sum_{j=1}^{\infty} \iint_{Q_j} F(\zeta)\psi(\zeta)\frac{\partial w^{-1}}{\partial \overline{\zeta}} d\zeta d\overline{\zeta} - \sum_{j=1}^{\infty} \iint_{Q_j} F(\zeta) w^{-1}(\zeta)\frac{\partial \psi}{\partial \overline{\zeta}} d\zeta d\overline{\zeta}. \tag{3}$$

Obviously,

$$\left| \sum_{j=1}^{\infty} \iint_{Q_j} F(\zeta)\psi(\zeta)\frac{\partial w^{-1}}{\partial \overline{\zeta}} d\zeta d\overline{\zeta} \right| \leq \|w\phi\|_{C(Q)} \iint_{Q} \left| F(\zeta)\frac{\partial w^{-1}}{\partial \overline{\zeta}} d\zeta d\overline{\zeta} \right|,$$

and the last integral converges.

Now we assume that F is integrable near Γ with arbitrarily large degree, and introduce characteristics

$$d := \limsup_{\varepsilon \to 0} \frac{\log N(\varepsilon)}{-\log \varepsilon},$$

where $N(\varepsilon)$ is the least number of disks of radius ε necessary for covering of Γ. This characteristics is known as upper metric dimension of the set Γ (see [5]) or its box dimension (see [6]). If $d < 2$, then Γ has null plane measure.

Obviously, $w\phi \in H_\nu(\overline{Q})$ for any ν. Then, as shown in [1], [2] (see also [7]) the estimate (2) implies that $\frac{\partial \psi}{\partial \overline{\zeta}} \in L^p_{loc}(\mathbb{C})$ for any

$$p < \frac{2-d}{1-\nu},$$

and for $p \in [1, \frac{2-d}{1-\nu}]$ we have

$$\iint_{Q} \left| \frac{\partial \psi}{\partial \overline{\zeta}} \right|^p |d\zeta d\overline{\zeta}| \leq C h^p_\nu(w\phi, Q),$$

where constant C depends only on Γ and ν. Consequently,

$$\left| \sum_{j=1}^{\infty} \iint_{Q_j} F(\zeta) w^{-1}(\zeta)\frac{\partial \psi}{\partial \overline{\zeta}} d\zeta d\overline{\zeta} \right| \leq C^{1/p} h_\nu(w\phi, Q) \left(\iint_{Q} |F(\zeta) w^{-1}(\zeta)|^q |d\zeta d\overline{\zeta}| \right)^{1/q},$$

where $\frac{1}{p} + \frac{1}{q} = 1$. The last integral converges for

$$p_j < \frac{2(1+\nu-d)}{2-d}, j = 1, 2, \ldots, m. \tag{4}$$

As a result, we obtain from (3)

Theorem 1.1. *Let Γ have upper metric dimension $d < 2$, $F(z)$ be holomorphic in $\overline{\mathbb{C}} \setminus \Gamma$ and integrable with any power near Γ, $\nu > d - 1$ and the weight (1) satisfy restriction (4). Then any function $\phi \in C_0^\infty(Q)$ satisfies the bound*

$$|\langle \overline{\partial} F, \phi \rangle| \leq C \|\phi\|_{H_\nu(Q,w)}.$$

Thus, under assumptions of the theorem distribution $\overline{\partial} F$ is extendable up to continuous functional on the space $H_{d-1}^*(Q, w)$. We keep notation $\overline{\partial} F$ for this functional and define for any $f \in H_{d-1}^*(Q, w)$ distribution $f \overline{\partial} F$:

$$\langle f \overline{\partial} F, \phi \rangle := \langle \overline{\partial} F, f\phi \rangle.$$

2. The Cauchy transform of products $f\overline{\partial}F$

The Cauchy transform of distribution φ with compact support S is defined by equality

$$\operatorname{Cau} \varphi := \frac{1}{2\pi i} \left\langle \varphi, \frac{1}{\zeta - z} \right\rangle,$$

where $z \notin S$, and φ is applied to the Cauchy kernel $\frac{1}{2\pi i(\zeta - z)}$ as to function of variable ζ. The function

$$\Phi(z) := \operatorname{Cau} f \overline{\partial} F(z) \tag{5}$$

is holomorphic in $\overline{\mathbb{C}} \setminus \Gamma$.

Theorem 2.1. *Let the following assumptions be valid:*

- *Γ is directed curve of upper metric dimension $d < 2$;*
- *the function F is holomorphic in $\overline{\mathbb{C}} \setminus \Gamma$, integrable with arbitrarily large power in a neighborhood of Γ, and has boundary values from the left $F^+(t)$ and from the right $F^-(t)$ at any point $t \in \Gamma \setminus E$, where the set E consists of the points $t_1, t_2, \ldots, t_m$ and end points of Γ if this curve is not closed;*
- *$f \in H_\nu(Q, w)$, where w is the weight (1) and $\nu > \frac{d}{2}$.*

Then the function (5) has continuous boundary values $\Phi^+(t)$ and $\Phi^-(t)$ from the left and from the right correspondingly at any point $t \in \Gamma \setminus E$,

$$\Phi^+(t) - \Phi^-(t) = (F^+(t) - F^-(t)) f(t), t \in \Gamma \setminus E, \tag{6}$$

and near the points $t_j, j = 1, 2, \ldots, m$ it satisfies the bounds

$$|\Phi(z)| \leq C|z - t_j|^{-r_j}, j = 1, 2, \ldots, m, \tag{7}$$

where exponents r_j are any numbers satisfying inequalities

$$r_j > p_j + \frac{1 - \nu}{2 - d}, j = 1, 2, \ldots, m. \tag{8}$$

The proof of relation (6) is analogous to the proof of Theorem 2.1 in [3] (note that for $\nu > \frac{d}{2}$ the right side of (4) is lesser than 1), and the bound (7) follows from well-known integral inequalities.

3. The Riemann boundary value problem on non-rectifiable curves

Let Γ be closed non-rectifiable Jordan curve directed counter-clockwise. We consider the Riemann boundary value problem, i.e., the problem on evaluation of holomorphic in $\overline{\mathbb{C}} \setminus \Gamma$ function $\Phi(z)$ satisfying equality

$$\Phi^+(t) = G(t)\Phi^-(t) + g(t), \quad t \in \Gamma \setminus \{t_1, t_2, \dots, t_m\}, \tag{9}$$

and bounds

$$|\Phi(z)| \le C|z - t_j|^{-\gamma}, \quad \gamma = \gamma(\Phi) < 1, \quad j = 1, 2, \dots, m. \tag{10}$$

If Γ is piecewise smooth, then solution of this problem in terms of the Cauchy type integral is well known (see [8, 9]). In the present paper we show that the Cauchy transforms can be used here instead of the Cauchy type integrals if the curve is not rectifiable.

First we consider the case $G \equiv 1$, $g \in H_\nu(\Gamma, w)$, i.e., the so-called jump problem

$$\Phi^+(t) - \Phi^-(t) = g(t), \quad t \in \Gamma \setminus \{t_1, t_2, \dots, t_m\}. \tag{11}$$

Let function $\chi(z)$ be equal to unit inside of Γ and to zero outside of it. Then $\chi^+(t) - \chi^-(t) = 1$ for $t \in \Gamma$, and by virtue of Theorem 2.1 the Cauchy transform Cau $\left(w^{-1}\mathcal{E}_0(wg)\right)\overline{\partial}\chi$ satisfies the boundary value condition (11). Here $\mathcal{E}_0(wg)$ stands for the Whitney extension of wg from Γ into the whole complex plane. It remains to satisfy the condition (10). If $\nu > \frac{d}{2}$, then $\frac{1-\nu}{2-d} < \frac{1}{2}$, and the left side of (8) is lesser than unit for

$$p_j < 1/2, \quad j = 1, 2, \dots, m. \tag{12}$$

Thus, under assumptions of Theorem 2.1 (with g instead of f) and additional condition (12) the jump problem (11) has a solution satisfying condition (10).

If Γ is non-rectifiable Jordan arc beginning at point a_1 and ending at point a_2, then we replace $\chi(z)$ by function

$$k_\Gamma(z) = \frac{1}{2\pi i} \log \frac{z - a_2}{z - a_1},$$

where the branch of logarithm is determined by means of the cut along Γ and condition $k_\Gamma(\infty) = 0$, and obtain analogous result.

Now let us return to the Riemann boundary value problem (9) on closed curve Γ in the class of functions satisfying condition (10). We assume that $G \in H_\nu(\Gamma)$, $g \in H_\nu(\Gamma, w)$ and $G(t) \neq 0$ for $t \in \Gamma$. Then $G(t)$ is representable in the form $G(t) = (z - z_0)^\kappa \exp f(t)$, $f \in H_\nu(\Gamma)$, where z_0 is a fixed point inside Γ and κ is a integer number (see [8], [9]). We put $\Theta(z) = \mathrm{Cau}\,(\mathcal{E}_0 f)\overline{\partial}\chi$ and introduce function $X(z)$ equaling to $\exp\Theta(z)$ inside of Γ and $(z - z_0)^{-\kappa}\exp\Theta(z)$ outside of it. The customary factorization technique reduces the Riemann boundary value problem to the jump problem

$$\frac{\Phi^+(t)}{X^+(t)} - \frac{\Phi^-(t)}{X^-(t)} = \frac{g(t)}{X^+(t)}, \quad t \in \Gamma \setminus \{t_1, t_2, \dots, t_m\}. \tag{13}$$

Its solution is given by the Cauchy transform

$$\Phi_0(z) = \mathrm{Cau}\left(w^{-1}\mathcal{E}_0(wg)\right)\overline{\partial}(\chi/X)(z).$$

As a result, we obtain

Theorem 3.1. *Let the following assumptions be valid:*

- Γ *is directed closed Jordan curve of upper metric dimension* $d < 2$;
- $G \in H_\nu(\Gamma)$ *and* $G(t) \neq 0$ *for* $t \in \Gamma$, *where* $\nu > d/2$;
- $g \in H_\nu(Q, w)$, *where* w *is the weight* (1) *satisfying condition* (12).

Then:

- *if* $\kappa \geq 0$, *then the problem* (9) *has in the class* (10) *family of solutions* $\Phi(z) = X(z)(\Phi_0(z) + P_\kappa(z))$, *where* $P_\kappa(z)$ *is arbitrary polynomial of degree no larger than* κ;
- *if* $\kappa \leq 0$, *then the function* $\Phi(z) = X(z)\Phi_0(z)$ *is solution of the problem* (9) *in the class* (10) *under conditions*

$$\left\langle \left(w^{-1}\mathcal{E}_0(wg)\right)\overline{\partial}\frac{\chi}{X}, t^{k-1}\right\rangle = 0, \quad k = 1, 2, \ldots, -\kappa - 1.$$

The uniqueness of these solutions can be described in terms of the Hausdorff dimension of Γ (see, for instance, [7]).

References

[1] B.A. Kats, *The Cauchy transform of certain distributions with application*, Complex Analysis and Operator Theory, 2010, 1–18 (electronic publication).

[2] B.A. Kats, *The Riemann boundary value problem on a closed nonrectifiable curve and the Cauchy transform*, Lobachevskii Journal of Mathematics, 2011, v. 32, no. 4, 438–445.

[3] B.A. Kats, *The Riemann Boundary Value Problem on Non-rectifiable Arcs and the Cauchy Transform*, Operator Theory: Advances and Applications, 2012, Vol. 221, 451–459

[4] E.M. Stein, *Singular integrals and differential properties of functions*, Mir Publishers, Moscow, 1973.

[5] A.N. Kolmogorov, V.M. Tikhomirov, *ε-entropy and capacity of set in functional spaces*, Uspekhi Math. Nauk, **14** (1959), 3–86.

[6] I. Feder, *Fractals*, Mir Publishers, Moscow, 1991.

[7] B.A. Kats, *The Cauchy integral over non-rectifiable paths*, Contemporary Mathematics, 455 (2008), 183–196.

[8] N.I. Muskhelishvili, *Singular integral equations*, Nauka publishers, Moscow, 1962.

[9] F.D. Gakhov, *Boundary value problems*, Nauka publishers, Moscow, 1977.

Boris A. Kats
Kazan Federal University
Kremlevskaya str. 18
Kazan, 420008, Russia
e-mail: katsboris877@gmail.com

Operator Theory:
Advances and Applications, Vol. 236, 299–316

An Analogue of the Spectral Mapping Theorem for Condition Spectrum

G. Krishna Kumar and S.H. Kulkarni

Abstract. For $0 < \epsilon < 1$, the ϵ-condition spectrum of an element a in a complex unital Banach algebra A is defined as,

$$\sigma_\epsilon(a) = \left\{\lambda \in \mathbb{C} : \lambda - a \text{ is not invertible or } \|\lambda - a\|\|(\lambda - a)^{-1}\| \geq \frac{1}{\epsilon}\right\}.$$

This is a generalization of the idea of spectrum introduced in [5]. This is expected to be useful in dealing with operator equations. In this paper we prove a mapping theorem for condition spectrum, extending an earlier result in [5]. Let f be an analytic function in an open set Ω containing $\sigma_\epsilon(a)$. We study the relations between the sets $\sigma_\epsilon(\tilde{f}(a))$ and $f(\sigma_\epsilon(a))$. In general these two sets are different. We define functions $\phi(\epsilon), \psi(\epsilon)$ (that take small values for small values of ϵ) and prove that $f(\sigma_\epsilon(a)) \subseteq \sigma_{\phi(\epsilon)}(\tilde{f}(a))$ and $\sigma_\epsilon(\tilde{f}(a)) \subseteq f(\sigma_{\psi(\epsilon)}(a))$. The classical Spectral Mapping Theorem is shown as a special case of this result. We give estimates for these functions in some special cases and finally illustrate the results by numerical computations.

Mathematics Subject Classification (2010). Primary 47A60, 15A60, 46H05.

Keywords. Spectrum, analytic function, condition spectrum, Spectral Mapping Theorem.

1. Introduction

The Spectral Mapping Theorem is a fundamental result in functional analysis of great importance. Let A be a complex algebra with unit 1. We shall identify $\lambda.1$ with λ. We recall that the spectrum of an element $a \in A$ is defined as

$$\sigma(a) = \{\lambda \in \mathbb{C} : \lambda - a \notin A^{-1}\},$$

where A^{-1} is the set of all invertible elements of A [9]. The Spectral Mapping Theorem says that if f is an analytic function on an open set containing $\sigma(a)$, then

$$f(\sigma(a)) = \sigma(\tilde{f}(a)).$$

There are several generalizations of the concept of the spectrum in literature such as Ransford spectrum [8], pseudospectrum [11], n-pseudospectrum [2, 3], condition spectrum [5] etc. It is natural to ask whether there are any results similar to the Spectral Mapping Theorem for these sets. It is known that similar results hold if f is an affine function, that is, $f(z) = \alpha + \beta z$ for some $\alpha, \beta \in \mathbb{C}$. (see Theorem 2.7 [5], Theorem 2.2 [6]). However it is not true, if f is an arbitrary analytic function (see Example 1). In [6], the author gives an analogue of the Spectral Mapping Theorem for pseudospectrum in the matrix algebra. The author carries forward this work in his recent paper [7]. The aim of this paper is to obtain an analogue of the Spectral Mapping Theorem for condition spectra of elements in a Banach algebra. We begin with the definition of condition spectrum.

Definition 1.1. (ϵ-condition spectrum) Let A be a complex unital Banach algebra with unit 1 and $0 < \epsilon < 1$. The *ϵ-condition spectrum* of an element $a \in A$, denoted by $\sigma_\epsilon(a)$, is defined as,

$$\sigma_\epsilon(a) = \left\{ \lambda \in \mathbb{C} : \ \|\lambda - a\| \|(\lambda - a)^{-1}\| \geq \frac{1}{\epsilon} \right\}$$

with the convention that $\|\lambda - a\| \|(\lambda - a)^{-1}\| = \infty$, if $\lambda - a$ is not invertible. Note that because of this convention $\sigma(a) \subseteq \sigma_\epsilon(a)$.

Suppose X is a Banach space and $T : X \to X$ is a bounded linear map. Then $\lambda \notin \sigma_\epsilon(T)$ means that the operator equation $Tx - \lambda x = y$ has a stable solution for every $y \in X$. This fact makes the ϵ-condition spectrum a potentially useful tool in the numerical solutions of operator equations. See [5] for examples and elementary properties of the condition spectrum.

Let f be an analytic function on some open set Ω containing $\sigma_\epsilon(a)$. Since $\sigma(a) \subseteq \sigma_\epsilon(a) \subseteq \Omega$, $\tilde{f}(a)$ can be defined by functional calculus as,

$$\tilde{f}(a) = \frac{1}{2\pi i} \int_\Gamma f(z)(z - a)^{-1} dz,$$

where Γ is any contour that surrounds $\sigma(a)$ in Ω [9]. If f is a polynomial, then $\tilde{f}(a) = f(a)$([9], Theorem 10.25). In view of this, some authors use the notation $f(a)$ in place of $\tilde{f}(a)$. We use the notation $\tilde{f}$ as in [9]. Our aim is to study the relations between the sets $f(\sigma_\epsilon(a))$ and $\sigma_\epsilon(\tilde{f}(a))$. Note that, in general we can not expect $f(\sigma_\epsilon(a)) = \sigma_\epsilon(\tilde{f}(a))$ (see Example 1 below). In other words, the verbatim analogue of the Spectral Mapping Theorem is not true. Hence we define functions ϕ, ψ such that $\lim_{\epsilon \to 0} \phi(\epsilon) = 0 = \lim_{\epsilon \to 0} \psi(\epsilon)$ and prove that $f(\sigma_\epsilon(a)) \subseteq \sigma_{\phi(\epsilon)}(\tilde{f}(a))$ and $\sigma_\epsilon(\tilde{f}(a)) \subseteq f(\sigma_{\psi(\epsilon)}(a))$. These functions ϕ and ψ depend on f and a. If for some f, $\phi(\epsilon) = \epsilon = \psi(\epsilon)$, then we would get $f(\sigma_\epsilon(a)) = \sigma_\epsilon(\tilde{f}(a))$ for that f. This happens when f is an affine function.

The following is an outline of the paper. In Section 2, the general theorem in the form of two set inclusions is stated and proved (Theorem 2.1). It is shown that the set inclusions reduce to an equality if the mapping is an affine function

(Remark 2.4). It is also shown that the usual Spectral Mapping Theorem as well as the pseudospectral mapping theorem of Lui [6] are special cases of our result (Remark 2.7). In Section 3, a weak version of the theorem is proved in a Banach algebra with some additional property (Theorem 3.4). In Section 4, we present some numerical experiments which illustrate the theory developed in the earlier sections.

2. Main theorem

First we give an example to show that $f(\sigma_\epsilon(a)) \neq \sigma_\epsilon(\tilde{f}(a))$ in general. Next we give an analogue of the Spectral Mapping Theorem for condition spectrum for complex analytic functions. The theorem is an easy consequence of the definition of the functions defined in the statement of the theorem.

Example 1. Let $A = \mathbb{C}^{2\times 2}$, the algebra of all 2×2 matrices with the operator norm $\|\cdot\|_2$. Let

$$P = \begin{bmatrix} -1 & 0 \\ 0 & 1 \end{bmatrix}$$

and $f(z) = z^2$, then

$$\tilde{f}(P) = P^2 = \begin{bmatrix} 1 & 0 \\ 0 & 1 \end{bmatrix} = I$$

Hence $\sigma_\epsilon(\tilde{f}(P)) = \{1\}$. On the other hand $\sigma_\epsilon(P)$ contains complex numbers different from -1 and 1 (see Corollary 3.4, [5]). Hence $f(\sigma_\epsilon(P))$ contains complex numbers different from 1.

Theorem 2.1. *Let A be a complex Banach algebra with unit 1. For $a \in A$, $0 < \epsilon < 1$ sufficiently small, Ω a bounded open subset of $\mathbb{C}$ containing $\sigma_\epsilon(a)$ and f an analytic function on Ω, define*

$$\phi(\epsilon) = \sup_{\lambda \in \sigma_\epsilon(a)} \left\{ \frac{1}{\|f(\lambda) - \tilde{f}(a)\| \|[f(\lambda) - \tilde{f}(a)]^{-1}\|} \right\}.$$

If $\tilde{f}(a)$ is not a scalar multiple of unit, then $\phi(\epsilon)$ is well defined, $0 \leq \phi(\epsilon) \leq 1$, $\lim_{\epsilon\to 0} \phi(\epsilon) = 0$ and for ϵ satisfying $\phi(\epsilon) < 1$, we have

$$f(\sigma_\epsilon(a)) \subseteq \sigma_{\phi(\epsilon)}(\tilde{f}(a)).$$

Further suppose f is injective on Ω and there exists ϵ_0 with $0 < \epsilon_0 < 1$ such that $\sigma_{\epsilon_0}(\tilde{f}(a)) \subseteq f(\Omega)$. For $0 < \epsilon \leq \epsilon_0$ define

$$\psi(\epsilon) = \sup_{\mu \in f^{-1}(\sigma_\epsilon(\tilde{f}(a)))} \left\{ \frac{1}{\|\mu - a\| \|(\mu - a)^{-1}\|} \right\}.$$

Then $\psi(\epsilon)$ is well defined, $0 \leq \psi(\epsilon) \leq 1$, $\lim_{\epsilon\to 0} \psi(\epsilon) = 0$ and for ϵ satisfying $\psi(\epsilon) < 1$, we have

$$\sigma_\epsilon(\tilde{f}(a)) \subseteq f(\sigma_{\psi(\epsilon)}(a)).$$

Proof. First, we show that for each $a \in A$, $\phi(\epsilon)$ is well defined. Define $g:\mathbb{C}\to\mathbb{R}$ by,

$$g(\lambda) = \frac{1}{\|f(\lambda) - \tilde{f}(a)\|\|[f(\lambda) - \tilde{f}(a)]^{-1}\|}.$$

We claim that g is continuous. Clearly g is continuous on $\mathbb{C} \setminus \sigma(a)$. Let $\lambda \in \sigma(a)$. Then by the Spectral Mapping Theorem,

$$f(\lambda) \in f(\sigma(a)) = \sigma(\tilde{f}(a)).$$

Thus by our convention $g(\lambda) = 0$. To complete the proof of the claim, we need to show the following. If $\lambda_n \in \mathbb{C} \setminus \sigma(a)$, $\lambda_n \to \lambda \in \sigma(a)$, then $g(\lambda_n) \to 0$. Let $\{\lambda_n\}$ be such a sequence. Then $f(\lambda_n) - \tilde{f}(a) \to f(\lambda) - \tilde{f}(a)$. Hence $\{f(\lambda_n) - \tilde{f}(a)\}$ is a bounded sequence. On the other hand, since $f(\lambda) \in \sigma(\tilde{f}(a))$, $\|(f(\lambda_n)-\tilde{f}(a))^{-1}\| \to \infty$ (Lemma 10.17 of [9]). Hence $g(\lambda_n) \to 0$. This proves the claim. Next for $0 < \epsilon < 1$, $\sigma_\epsilon(a)$ is a compact set [5] and $\phi(\epsilon) = \sup\{g(\lambda) : \lambda \in \sigma_\epsilon(a)\}$. Hence $\phi(\epsilon)$ is well defined, that is, finite.

Next we prove $\lim_{\epsilon\to 0} \phi(\epsilon) = 0$. Let $\epsilon_n > 0$ be a sequence converging to 0. By compactness of $\sigma_{\epsilon_n}(a)$ there exist $\lambda_n \in \sigma_{\epsilon_n}(a)$ such that $g(\lambda_n) = \phi(\epsilon_n)$. Now λ_n is a bounded sequence and hence has a convergent subsequence $\{\lambda_{n_k}\}$ converging to λ. Hence $\{\lambda_{n_k} - a\}$ is a bounded sequence. On the other hand, $\|\lambda_{n_k} - a\|\|(\lambda_{n_k} - a)^{-1}\| \geq \frac{1}{\epsilon_{n_k}}$ for all n_k. Thus $\|(\lambda_{n_k} - a)^{-1}\| \to \infty$ as $n_k \to \infty$. This imply that $\lambda - a$ is not invertible. Thus $\lambda \in \sigma(a)$ and $f(\lambda) \in \sigma(\tilde{f}(a))$. Now $\{f(\lambda_{n_k})-\tilde{f}(a)\}$ converges to $f(\lambda) - \tilde{f}(a)$. Hence $\{f(\lambda_{n_k}) - \tilde{f}(a)\}$ is bounded and $\|(f(\lambda_{n_k}) - \tilde{f}(a))^{-1}\| \to \infty$. This gives $\phi(\epsilon_{n_k}) = g(\lambda_{n_k}) \to 0$. Since $\phi(\epsilon_n)$ is monotonically increasing $\phi(\epsilon_n) \to 0$. Now let ϵ be sufficiently small so that $0 \leq \phi(\epsilon) < 1$ and let $\lambda \in \sigma_\epsilon(a)$. Then $g(\lambda) \leq \phi(\epsilon)$. Hence

$$\|f(\lambda) - \tilde{f}(a)\|\|[f(\lambda) - \tilde{f}(a)]^{-1}\| = \frac{1}{g(\lambda)} \geq \frac{1}{\phi(\epsilon)}.$$

This means that $f(\lambda) \in \sigma_{\phi(\epsilon)}(\tilde{f}(a))$. Thus

$$f(\sigma_\epsilon(a)) \subseteq \sigma_{\phi(\epsilon)}(\tilde{f}(a)).$$

Next we assume that f is injective on Ω and there exists ϵ_0 with $0 < \epsilon_0 < 1$ such that $\sigma_{\epsilon_0}(\tilde{f}(a)) \subseteq f(\Omega)$ and we show that for each $a \in A$ and $0 < \epsilon \leq \epsilon_0$, $\psi(\epsilon)$ is well defined. Define $h : \mathbb{C} \to \mathbb{R}$ by,

$$h(\mu) = \frac{1}{\|\mu - a\|\|(\mu - a)^{-1}\|}$$

We claim that h is continuous. Clearly h is continuous on $\mathbb{C} \setminus \sigma(a)$. Let $\mu \in \sigma(a)$, by our convention $h(\mu) = 0$. To complete the proof of the claim we need to show the following. If $\mu_n \in \mathbb{C} \setminus \sigma(a)$, $\mu_n \to \mu \in \sigma(a)$, then $h(\mu_n) \to 0$. Let $\{\mu_n\}$ be such a sequence. Then $\mu_n - a \to \mu - a$. Hence $\{\mu_n - a\}$ is a bounded sequence. On the other hand, since $\mu \in \sigma(a)$, $\|(\mu_n - a)^{-1}\| \to \infty$ (Lemma 10.17 of [9]). Hence $h(\mu_n) \to 0$. This proves the claim. Since $h(\mu) \leq 1$ for all $\mu \in \mathbb{C}$, $\psi(\epsilon)$ is well defined and $0 \leq \psi(\epsilon) \leq 1$.

Next we prove $\lim_{\epsilon\to 0}\psi(\epsilon)=0$. Let $\epsilon_n>0$ be a sequence converging to 0. Since

$$\psi(\epsilon_n)=\sup_{\mu\in f^{-1}(\sigma_{\epsilon_n}(\tilde{f}(a)))} h(\mu),$$

and $f^{-1}(\sigma_{\epsilon_n}(\tilde{f}(a)))$ is closed and bounded, hence compact, there exists $\mu_n\in f^{-1}(\sigma_{\epsilon_n}(\tilde{f}(a)))$ such that $\psi(\epsilon_n)=h(\mu_n)$. Since each $\mu_n\in\Omega$, which is bounded, it has a convergent subsequence $\{\mu_{n_k}\}$ converging to μ. On the other hand, since $f(\mu_{n_k})\in\sigma_{\epsilon_{n_k}}(\tilde{f}(a))$, we have

$$\|f(\mu_{n_k})-\tilde{f}(a)\|\|[f(\mu_{n_k})-\tilde{f}(a)]^{-1}\|\geq\frac{1}{\epsilon_{n_k}}$$

for all n_k. Thus $\|[f(\mu_{n_k})-\tilde{f}(a)]^{-1}\|\to\infty$ as $n_k\to\infty$. This implies that $f(\mu)-\tilde{f}(a)$ is not invertible. Thus $f(\mu)\in\sigma(\tilde{f}(a))$. Since f is injective $\mu\in\sigma(a)$ and $h(\mu)=0$. Since h is continuous $\psi(\epsilon_{n_k})=h(\mu_{n_k})\to h(\mu)=0$. Finally since ψ is monotonically increasing $\psi(\epsilon_n)\to 0$.

Now let ϵ be sufficiently small so that $0\leq\psi(\epsilon)<1$. Let $\lambda\in\sigma_\epsilon(\tilde{f}(a))\subseteq\sigma_{\epsilon_0}(\tilde{f}(a))\subseteq f(\Omega)$. Consider $\mu\in\Omega$ such that $\lambda=f(\mu)$. Then $\mu\in f^{-1}(\sigma_\epsilon(\tilde{f}(a)))$, hence $h(\mu)\leq\psi(\epsilon)$, that is,

$$\|\mu-a\|\|(\mu-a)^{-1}\|\geq\frac{1}{\psi(\epsilon)}.$$

Thus $\mu\in\sigma_{\psi(\epsilon)}(a)$. Hence $\lambda=f(\mu)\in f(\sigma_{\psi(\epsilon)}(a))$. This proves

$$\sigma_\epsilon(\tilde{f}(a))\subseteq f(\sigma_{\psi(\epsilon)}(a)).$$

□

Remark 2.2. Combining the two inclusions, we get

$$f(\sigma_\epsilon(a))\subseteq\sigma_{\phi(\epsilon)}(\tilde{f}(a))\subseteq f(\sigma_{\psi(\phi(\epsilon))}(a)).$$

and

$$\sigma_\epsilon(\tilde{f}(a))\subseteq f(\sigma_{\psi(\epsilon)}(a))\subseteq\sigma_{\phi(\psi(\epsilon))}(\tilde{f}(a)).$$

Remark 2.3. Since for every $a\in A$, $\lim_{\epsilon\to 0}\phi(\epsilon)=0=\lim_{\epsilon\to 0}\psi(\epsilon)$, $\sigma(a)=\bigcap_{0<\epsilon<1}\sigma_\epsilon(a)$ and ϕ,ψ are monotonically increasing functions, the usual Spectral Mapping Theorem can be deduced from Theorem 2.1. However, it may be noted that the proof of Theorem 2.1 uses the Spectral Mapping Theorem.

Remark 2.4. Let $a\in A$ and $f(z)=\alpha+\beta z$ where α,β are complex numbers with $\beta\neq 0$. Then

$$\begin{aligned}\phi(\epsilon)&=\sup_{\lambda\in\sigma_\epsilon(a)}\frac{1}{\|\beta\lambda-\beta a\|\|(\beta\lambda-\beta a)^{-1}\|}\\&=\sup_{\lambda\in\sigma_\epsilon(a)}\frac{1}{\|\lambda-a\|\|(\lambda-a)^{-1}\|}\\&=\epsilon\end{aligned}$$

In a similar way we have $\psi(\epsilon) = \epsilon$. Thus $\sigma_\epsilon(\alpha + \beta a) = \alpha + \beta\sigma_\epsilon(a)$ (see (7) of Theorem 2.7 in [5]), that is $\sigma_\epsilon(\tilde{f}(a)) = f(\sigma_\epsilon(a))$. This leads to the following question.

Question 2.5. Let f be a non-constant analytic function defined on a nonempty open set Ω in the complex plane. Suppose

$$f(\sigma_\epsilon(a)) = \sigma_{\phi(\epsilon)}(\tilde{f}(a))$$

for all $a \in A$ with $\sigma(a) \subset \Omega$. Then does it follow that $\phi(\epsilon) = \epsilon$ and $f(z) = \alpha + \beta z$ for some $\alpha, \beta \in \mathbb{C}$?

Remark 2.6. The hypothesis that $\tilde{f}(a)$ is not a scalar multiple of unity cannot be dropped from Theorem 2.1. Let f and P be as in Example 1. Since $\tilde{f}(P) = I$, we have $\sigma_{\phi(\epsilon)}(\tilde{f}(P)) = \{1\}$, On the other hand we have noted in Example 1 that $\sigma_\epsilon(P)$ contains complex numbers different from -1 and 1. Hence $f(\sigma_\epsilon(P))$ contains complex numbers different from 1. Thus

$$f(\sigma_\epsilon(P)) \nsubseteq \sigma_{\phi(\epsilon)}(\tilde{f}(P)).$$

Remark 2.7. Let $\Lambda_\epsilon(a) := \{\lambda \in \mathbb{C} : \|(\lambda - a)^{-1}\| \geq 1/\epsilon\}$ denote the pseudospectrum of a. (See [11] for examples and applications of pseudospectrum.) It was shown in [4] that if a is not a scalar multiple of 1, then there exist positive numbers α, β depending on a, such that $\sigma_\epsilon(a) \subseteq \Lambda_{\alpha\epsilon}(a)$ and $\Lambda_\epsilon(a) \subseteq \sigma_{\beta\epsilon}(a)$. (See [4] for exact values of α, β.) Now from Theorem 2.1

$$f(\Lambda_\epsilon(a)) \subseteq f(\sigma_{\beta\epsilon}(a)) \subseteq \sigma_{\phi(\beta\epsilon)}(\tilde{f}(a)) \subseteq \Lambda_{\alpha\phi(\beta\epsilon)}(\tilde{f}(a)).$$

$$\Lambda_\epsilon(\tilde{f}(a)) \subseteq \sigma_{\beta\epsilon}(\tilde{f}(a)) \subseteq f(\sigma_{\psi(\beta\epsilon)}(a)) \subseteq f(\Lambda_{\alpha\psi(\beta\epsilon)}(a)).$$

This is a more general form of the pseudospectral mapping theorem given in [6].

3. Weak versions

The functions ϕ and ψ defined in the last section are continuous and monotonically increasing but it appears to be difficult to find the values of these functions explicitly in a Banach algebra. In this section, we replace these functions ϕ, ψ with the functions $\gamma_\epsilon, \delta_\epsilon$ respectively that are relatively easier to estimate. The results using these functions are weaker in the following sense

1. We need to assume some additional property for Banach algebras.
2. We need to take a bigger neighborhood Ω.

The next lemma describes this additional property.

Lemma 3.1. *Let A be a complex unital Banach algebra with the following property:*

$$\forall\ a \in A^{-1},\ \exists\ b \in A \setminus A^{-1} \ \textit{ such that } \ \|a - b\| = \frac{1}{\|a^{-1}\|} \tag{3.1}$$

Then for every $a \in A$ such that a is not a scalar multiple of unity and $\lambda \in \sigma_\epsilon(a)$, there exists an element $b \in A$ such that

$$\lambda \in \sigma(a+b) \quad \textit{with} \quad \|b\| \leq \epsilon \|\lambda - a\|.$$

Proof. We refer to [5] for a proof of this result. □

The article [5] contains examples of Banach algebras satisfying Property 3.1. In particular the uniform algebras and matrix algebras satisfy this property (see Examples 2.18, 2.20 in [5]).

Lemma 3.2. *Let A be a complex Banach algebra with unit 1. Let $0 < \epsilon < 1$ and $a \in A$ be such that a is not a scalar multiple of unit. Let $m = \inf\{\|z.1 - a\| : z \in \mathbb{C}\}$. Then*

$$\bigcup_{\|b\| \leq m\epsilon} \sigma(a+b) \subseteq \sigma_\epsilon(a).$$

Further if A has Property 3.1 *stated in* Lemma 3.1 *then*

$$\sigma_\epsilon(a) \subseteq \bigcup_{\|b\| \leq \frac{2\epsilon}{1-\epsilon}\|a\|} \sigma(a+b).$$

Thus for such algebras

$$\bigcup_{\|b\| \leq m\epsilon} \sigma(a+b) \subseteq \sigma_\epsilon(a) \subseteq \bigcup_{\|b\| \leq \frac{2\epsilon}{1-\epsilon}\|a\|} \sigma(a+b).$$

Proof. Let $\lambda \in \sigma(a+b)$ with $b \in A$ and $\|b\| \leq \epsilon m$. Since

$$m = \inf\{\|z.1 - a\| : z \in \mathbb{C}\} \leq \|\lambda - a\|,$$

we have $\|b\| \leq \epsilon \|\lambda - a\|$. Hence by Theorem 2.16 of [5], we obtain

$$\sigma(a+b) \subseteq \sigma_\epsilon(a).$$

Next suppose A has Property 3.1 mentioned in Lemma 3.1. Let $\lambda \in \sigma_\epsilon(a)$. Then by Theorem 2.9 of [5], $|\lambda| \leq \dfrac{1+\epsilon}{1-\epsilon}\|a\|$.
Also by Lemma 3.1, $\lambda \in \sigma(a+b)$ for some $b \in A$ with $\|b\| \leq \epsilon\|\lambda - a\|$. Now

$$\|b\| \leq \epsilon\|\lambda - a\| \leq \epsilon(|\lambda| + \|a\|) \leq \frac{2\epsilon}{1-\epsilon}\|a\|.$$

This proves the second relation. □

Theorem 3.3. *Let A be a complex Banach algebra with unit* 1 *satisfying Property* 3.1 *stated in* Lemma 3.1. *Let $a \in A$ and Ω be an open set containing $\sigma(a)$. Then there exist $0 < \epsilon < 1$ such that $\sigma_\epsilon(a) \subseteq \Omega$.*

Proof. Recall that the map $a \longmapsto \sigma(a)$ is upper semicontinuous [1]. Hence there exist $\delta > 0$ such that $\sigma(a+b) \subseteq \Omega$ for all $b \in A$ with $\|b\| \leq \delta$ (see Theorem 10.20 of [9]). Now take $\epsilon = \dfrac{\delta}{\delta + 2\|a\|}$. Lemma 3.2 gives $\sigma_\epsilon(a) \subseteq \Omega$. □

The following theorem is the weak version of Theorem 2.1

Theorem 3.4. *Let A be a complex Banach algebra with unit 1 satisfying Property 3.1 mentioned in* Lemma 3.1. *Let $a \in A$, $0 < \epsilon < 1$ sufficiently small, Ω be an open subset of $\mathbb{C}$ containing* $\bigcup_{\|b\| \leq \frac{2\epsilon}{1-\epsilon}\|a\|} \sigma(a+b)$. *Let f be an injective analytic function defined on Ω. Assume that $a, \tilde{f}(a)$ are not scalar multiples of unity. Define*

$$\gamma_\epsilon := \sup\Big\{\|\tilde{f}(a+p) - \tilde{f}(a)\| : \|p\| \leq \frac{2\epsilon}{1-\epsilon}\|a\|\Big\}.$$

$$m := \inf\{\|z.1 - a\| : z \in \mathbb{C}\} > 0.$$

$$\delta_\epsilon := \sup\Big\{\|q\| : \|\tilde{f}(a+q) - \tilde{f}(a)\| \leq \frac{2\epsilon}{1-\epsilon}\|\tilde{f}(a)\|\Big\}.$$

$$m' := \inf\big\{\|z.1 - \tilde{f}(a)\| : z \in \mathbb{C}\big\} > 0.$$

Then $\lim_{\epsilon\to 0} \gamma_\epsilon = 0 = \lim_{\epsilon \to 0} \delta_\epsilon$.

1. Let $\epsilon > 0$ be such that $\frac{\gamma_\epsilon}{m'} < 1$. Then $f(\sigma_\epsilon(a)) \subseteq \sigma_{\frac{\gamma_\epsilon}{m'}}(\tilde{f}(a))$.
2. Let $\epsilon > 0$ be such that $\frac{\delta_\epsilon}{m} < 1$. Then $\sigma_\epsilon(\tilde{f}(a)) \subseteq f(\sigma_{\frac{\delta_\epsilon}{m}}(a))$.

Proof. Since the map $x \longmapsto \tilde{f}(x)$ is continuous, we obtain $\lim_{\epsilon\to 0} \gamma_\epsilon = 0$. Let $g : f(\Omega) \to \Omega$ be the inverse of f. Using the continuity of the map $y \longmapsto \tilde{g}(y)$, we obtain $\lim_{\epsilon \to 0} \gamma_\epsilon = 0$. Next let $\epsilon > 0$ be such that $\frac{\gamma_\epsilon}{m'} < 1$ and let $\lambda \in \sigma_\epsilon(a)$. By Lemma 3.2 there exist $b \in A$ with $\|b\| \leq \dfrac{2\epsilon}{1-\epsilon}\|a\|$ such that $\lambda \in \sigma(a+b)$. Then by the Spectral Mapping Theorem, $f(\lambda) \in \sigma(\tilde{f}(a+b))$. Let $c = \tilde{f}(a+b) - \tilde{f}(a)$, then $\|c\| \leq \gamma_\epsilon$ and by the above lemma,

$$f(\lambda) \in \sigma(\tilde{f}(a) + c) \subseteq \sigma_{\frac{\gamma_\epsilon}{m'}}(\tilde{f}(a)).$$

This proves 1.

Let $\lambda \in \sigma_\epsilon(\tilde{f}(a))$. Then by Lemma 3.2, $\lambda \in \sigma(\tilde{f}(a) + d)$ for some $d \in A$ with $\|d\| \leq \dfrac{2\epsilon}{1-\epsilon}\|\tilde{f}(a)\|$. By the inverse mapping theorem, [9], there exist $p \in A$ and $\epsilon_1 > 0$ such that $\|p\| \leq \epsilon_1$ and $\tilde{f}(a+p) = \tilde{f}(a) + d$. Thus by the Spectral Mapping Theorem there exist $\mu \in \sigma(a+p)$ such that,

$$f(\mu) = \lambda \in \sigma(\tilde{f}(a+p)) = \sigma(\tilde{f}(a) + d).$$

Claim: $\mu \in \sigma_{\frac{\delta_\epsilon}{m}}(a)$.

$$\|d\| = \|\tilde{f}(a+p) - \tilde{f}(a)\| \leq \frac{2\epsilon}{1-\epsilon}\|\tilde{f}(a)\|.$$

Hence

$$\|p\| \leq \delta_\epsilon := \sup\Big\{\|q\| : \|\tilde{f}(a+q) - \tilde{f}(a)\| \leq \frac{2\epsilon}{1-\epsilon}\|\tilde{f}(a)\|\Big\}.$$

Now by Lemma 3.2, $\mu \in \sigma_{\frac{\delta_\epsilon}{m}}(a)$. This proves the claim. Hence $\lambda = f(\mu) \in f(\sigma_{\frac{\delta_\epsilon}{m}}(a))$. This proves 2. □

Remark 3.5. If $\tilde{f}$ has a bounded Fréchet derivative in a neighborhood Ω containing $\sigma_\epsilon(a)$, then γ_ϵ can be estimated as follows. Let A be a complex unital Banach algebra, $a \in A$ and $0 < \epsilon < 1$. Let $(D\tilde{f})_x$ denote the Fréchet derivative of $\tilde{f}$ at $x \in A$. Let

$$L_\epsilon := \sup\left\{\|(D\tilde{f})_x\| : \; x \in A, \; \|x - a\| \le \frac{2\epsilon}{1-\epsilon}\|a\|\right\}.$$

Then, for $b \in A$ with $\|b\| \le \frac{2\epsilon}{1-\epsilon}\|a\|$, we have by the Mean Value Theorem [10],

$$\|\tilde{f}(a+b) - \tilde{f}(a)\| \le L_\epsilon \|b\| \le \frac{2\epsilon}{1-\epsilon} L_\epsilon \|a\|.$$

Thus

$$\gamma_\epsilon \le \frac{2\epsilon}{1-\epsilon} L_\epsilon \|a\|.$$

Remark 3.6. Let A be a complex unital Banach algebra, $a \in A$ and $0 < \epsilon < 1$. Let f be an injective analytic function defined on an open set Ω containing $\sigma_\epsilon(a)$. Let $g : f(\Omega) \to \Omega$ be the inverse of f. If $\tilde{g}$ has a bounded Fréchet derivative in a neighborhood of $\sigma_\epsilon(\tilde{f}(a))$, then δ_ϵ can be estimated as follows. Let

$$L'_\epsilon := \sup\left\{\|(D\tilde{g})_x\| : \; x \in A, \; \|x - \tilde{f}(a)\| \le \frac{2\epsilon}{1-\epsilon}\|\tilde{f}(a)\|\right\}.$$

Then, for $d' \in A$ with $\|d' - \tilde{f}(a)\| \le \frac{2\epsilon}{1-\epsilon}\|a\|$, we have by the Mean Value Theorem [10],

$$\|\tilde{g}(d') - \tilde{g}(\tilde{f}(a))\| \le L'_\epsilon \|d' - \tilde{f}(a)\| \le \frac{2\epsilon}{1-\epsilon} L'_\epsilon \|\tilde{f}(a)\|.$$

Thus

$$\delta_\epsilon \le \frac{2\epsilon}{1-\epsilon} L'_\epsilon \|\tilde{f}(a)\|.$$

In the following examples we give estimates for $\gamma_\epsilon, \delta_\epsilon$ for the functions $f(z) = z^2$, $f(z) = z^3$ and $f(z) = e^z$.

Example 2. Let $A = (C[1,2], \|\cdot\|_\infty)$, $0 < \epsilon < 1$ sufficiently small and $a \in A$ is defined by $a(x) = x$ for all $x \in [1,2]$. Then $\|a\|_\infty = 2$, $\sigma(a) = [1,2]$.

$$\begin{aligned} m &:= \inf\{\|z - a\|_\infty : z \in \mathbb{C}\}. \\ &= \inf\{\sup\{|z - x| : x \in [1,2]\} : z \in \mathbb{C}\}. \\ &= \inf\{\max\{|z-1|, |z-2|\} : z \in \mathbb{C}\}. \\ &= \frac{1}{2}. \end{aligned}$$

$$\begin{aligned} m' &:= \inf\{\|z - a^2\|_\infty : z \in \mathbb{C}\}. \\ &= \inf\{\sup\{|z - x^2| : x \in [1,2]\} : z \in \mathbb{C}\}. \\ &= \inf\{\max\{|z-1|, |z-4|\} : z \in \mathbb{C}\}. \\ &= \frac{3}{2}. \end{aligned}$$

$$\begin{aligned} \gamma_\epsilon &:= \sup\left\{\|(a+p)^2 - a^2\|_\infty : \|p\|_\infty \leq \frac{4\epsilon}{1-\epsilon}\right\}. \\ &= \sup\left\{\|2ap + p^2\|_\infty : \|p\|_\infty \leq \frac{4\epsilon}{1-\epsilon}\right\}. \\ &\leq \left\{4\|p\|_\infty + \|p\|_\infty^2 : \|p\|_\infty \leq \frac{4\epsilon}{1-\epsilon}\right\} \leq \frac{16\epsilon}{(1-\epsilon)^2}. \end{aligned}$$

Hence $\sigma_\epsilon(a)^2 \subseteq \sigma_{\epsilon_1}(a^2)$, (by Theorem 3.4). Where $\epsilon_1 = \frac{32\epsilon}{3(1-\epsilon)^2}$

$$\begin{aligned} \delta_\epsilon &:= \sup\left\{\|q\|_\infty : \|(a+q)^2 - a^2\|_\infty \leq \frac{8\epsilon}{1-\epsilon}\right\}. \\ &= \sup\left\{\|q\|_\infty : \|2qa + q^2\|_\infty \leq \frac{8\epsilon}{1-\epsilon}\right\}. \\ &= \sup\left\{\|q\|_\infty : \|2q + q^2\|_\infty \leq \frac{8\epsilon}{1-\epsilon}\right\}. \\ &\leq \sup\left\{\|q\|_\infty : 2\|q\|_\infty - \|q\|_\infty^2 \leq \frac{8\epsilon}{1-\epsilon}\right\}. \\ &\leq 1 - \sqrt{1-9\epsilon}. \end{aligned}$$

This gives $\sigma_\epsilon(a^2) \subseteq \sigma_{\epsilon_1}(a)^2$, where $\epsilon_1 = 2(1-\sqrt{1-9\epsilon})$.

Next for $f(z) = z^3$.

$$\begin{aligned} m' &:= \inf\{\|z - a^3\|_\infty : z \in \mathbb{C}\}. \\ &= \inf\{\sup\{|z - x^3| : x \in [1,2]\} : z \in \mathbb{C}\}. \\ &= \inf\{\max\{|z-1|, |z-8|\} : z \in \mathbb{C}\}. \\ &= \frac{7}{2}. \end{aligned}$$

$$\begin{aligned} \gamma_\epsilon &:= \sup\left\{\|(a+p)^3 - a^3\|_\infty : \|p\|_\infty \leq \frac{4\epsilon}{1-\epsilon}\right\}. \\ &= \sup\left\{\|3a^2p + 3ap^2 + p^3\|_\infty : \|p\|_\infty \leq \frac{4\epsilon}{1-\epsilon}\right\}. \\ &\leq \left\{3\|a\|_\infty^2\|p\|_\infty + 3\|a\|\|p\|_\infty^2 + \|p\|_\infty^3 : \|p\|_\infty \leq \frac{4\epsilon}{1-\epsilon}\right\}. \\ &\leq \frac{16\epsilon(3 - 3\epsilon - 2\epsilon^2)}{(1-\epsilon)^3}. \end{aligned}$$

Hence $\sigma_\epsilon(a)^3 \subseteq \sigma_{\epsilon_1}(a^3)$ (by Theorem 3.4). Where $\epsilon_1 = \frac{32\epsilon(3-3\epsilon-2\epsilon^2)}{7(1-\epsilon)^3}$

$$\begin{aligned}
\delta_\epsilon &:= \sup\left\{\|q\|_\infty : \|(a+q)^3 - a^3\|_\infty \le \frac{8\epsilon}{1-\epsilon}\right\}.\\
&= \sup\left\{\|q\|_\infty : \|3qa^2 + 3q^2a + q^3\|_\infty \le \frac{8\epsilon}{1-\epsilon}\right\}.\\
&= \sup\left\{\|q\|_\infty : \|3q + 3q^2 + q^3\|_\infty \le \frac{8\epsilon}{1-\epsilon}\right\}.\\
&\le \sup\left\{\|q\|_\infty : 3\|q\|_\infty - 3\|q\|_\infty^2 - \|q\|_\infty^3 \le \frac{8\epsilon}{1-\epsilon}\right\}.\\
&\le 8\epsilon.
\end{aligned}$$

This gives $\sigma_\epsilon(a^3) \subseteq \sigma_{16\epsilon}(a)^3$.

For $f(z) = e^z$

$$\begin{aligned}
m' &:= \inf\{\|z - \exp(a)\|_\infty : z \in \mathbb{C}\}.\\
&= \inf\{\sup\{|z - \exp(x)| : x \in [1,2]\} : z \in \mathbb{C}\}.\\
&= \inf\{\max\{|z - e|, |z - e^2|\} : z \in \mathbb{C}\}.\\
&= \frac{e(e-1)}{2}.\\
\gamma_\epsilon &:= \sup\left\{\|\exp(a+p) - \exp(a)\|_\infty : \|p\|_\infty \le \frac{4\epsilon}{1-\epsilon}\right\}.\\
&= \sup\left\{\|\exp(a)(\exp(p) - 1)\|_\infty : \|p\|_\infty \le \frac{4\epsilon}{1-\epsilon}\right\}.\\
&\le \left\{e^{\|a\|_\infty}\|\exp(p) - 1\|_\infty : \|p\|_\infty \le \frac{4\epsilon}{1-\epsilon}\right\}.\\
&\le \left\{e^2\|\sum_{k=1}^{\infty}\frac{p^k}{k!}\|_\infty : \|p\|_\infty \le \frac{4\epsilon}{1-\epsilon}\right\}.\\
&\le e^2(e^{\frac{4\epsilon}{1-\epsilon}} - 1).
\end{aligned}$$

Hence $\exp(\sigma_\epsilon(a)) \subseteq \sigma_{\epsilon_1}(\exp(a))$ (by Theorem 3.4). Where $\epsilon_1 = \dfrac{2e(e^{4\epsilon/1-\epsilon} - 1)}{e-1}$.

$$\delta_\epsilon := \sup\left\{\|q\|_\infty : \|\exp(a+q) - \exp(a)\|_\infty \le \frac{2e^2\epsilon}{1-\epsilon}\right\}.$$

Let $\exp(a) = c, \exp(a+q) = d = c + b$. Then

$$\begin{aligned}
\delta_\epsilon &= \sup\left\{\|\log(c+b) - \log(c)\|_\infty : \|b\|_\infty \le \frac{2e^2\epsilon}{1-\epsilon}\right\}.\\
&= \sup\left\{\|\log(1 + c^{-1}b)\|_\infty : \|b\|_\infty \le \frac{2e^2\epsilon}{1-\epsilon}\right\}.
\end{aligned}$$

$$\leq \sup\left\{\log(1+\|c^{-1}\|\|b\|_\infty) : \|b\|_\infty \leq \frac{2e^2\epsilon}{1-\epsilon}\right\}.$$
$$\leq \log\left(1+\|c^{-1}\|\frac{2e^2\epsilon}{1-\epsilon}\right)$$
$$\leq \log\left(1+\frac{2e^3\epsilon}{1-\epsilon}\right).$$

This gives $\sigma_\epsilon(\exp(a)) \subseteq \exp(\sigma_{\epsilon_1}(a))$. Where $\epsilon_1 = 2\log\left(1+\frac{2e^3\epsilon}{1-\epsilon}\right)$.

Example 3. Let $A = BL(l^\infty, \|\cdot\|)$, $0 < \epsilon < 1$ sufficiently small and $T \in A$ is defined by $T(x)(i) = x(i+1)$ for all $x \in l^\infty$, the left shift operator.

Consider $f(z) = z^3$. From Example 2.14 of [5] we have,

$$\sigma_\epsilon(T) = \left\{\lambda \in \mathbb{C} : |\lambda| \leq \frac{1+\epsilon}{1-\epsilon}\right\}.$$

From Theorem 2.1 we have,

$$\phi(\epsilon) = \sup_{\lambda\in\sigma_\epsilon(T)} g(\lambda), \quad \text{where } g(\lambda) = \frac{1}{\|\lambda^3 - T^3\|\|(\lambda^3-T^3)^{-1}\|}.$$

Also it is well known that $\sigma(T) = \{\lambda : |\lambda| \leq 1\}$ [9]. Hence $g(\lambda) = 0$ for $|\lambda| \leq 1$. Next let $1 < |\lambda| \leq \frac{1+\epsilon}{1-\epsilon}$. Then

$$\|\lambda^3 - T^3\| = 1 + |\lambda|^3, \|(\lambda^3 - T^3)^{-1}\| = \frac{1}{|\lambda|^3 - 1}.$$

Hence,

$$g(\lambda) = \frac{1}{\|\lambda^3 - T^3\|\|(\lambda^3-T^3)^{-1}\|} = \frac{|\lambda|^3-1}{|\lambda|^3+1}$$
$$\leq \frac{(\frac{1+\epsilon}{1-\epsilon})^3 - 1}{2} = \frac{6\epsilon + 2\epsilon^3}{2(1-\epsilon)^3} = \frac{\epsilon(3+\epsilon^2)}{(1-\epsilon)^3}$$

Thus $\phi(\epsilon) \leq \frac{\epsilon(3+\epsilon^2)}{(1-\epsilon)^3}$. Note that,

$$\sigma_\epsilon(T)^3 = \left\{\lambda \in \mathbb{C} : |\lambda| \leq \frac{(1+\epsilon)^3}{(1-\epsilon)^3}\right\} \subseteq \sigma_{\phi(\epsilon)}(T^3)$$
$$\subseteq \left\{\lambda \in \mathbb{C} : |\lambda| \leq \frac{1+\phi(\epsilon)}{1-\phi(\epsilon)}\right\}$$
$$\subseteq \left\{\lambda \in \mathbb{C} : |\lambda| \leq \frac{(1-\epsilon)^3 + \epsilon(3+\epsilon^2)}{(1-\epsilon)^3 - \epsilon(3+\epsilon^2)}\right\}$$

Next $\sigma_\epsilon(T^3) = \{\lambda \in \mathbb{C} : |\lambda| \leq \frac{1+\epsilon}{1-\epsilon}\}$. From Theorem 2.1, we have $\psi(\epsilon) = \sup\{h(\mu) : \mu^3 \in \sigma_\epsilon(T^3)\}$, where $h(\mu) = \frac{1}{\|\mu - T\|\|(\mu-T)^{-1}\|}$. Since $\sigma(T^3) = \{\lambda \in \mathbb{C} : |\lambda| \leq 1\}$, hence $h(\mu) = 0$ for $|\mu| \leq 1$. Consider $1 < |\mu^3| \leq \dfrac{1+\epsilon}{1-\epsilon}$.

$$\begin{aligned} h(\mu) &= \frac{1}{\|\mu - T\|\|(\mu-T)^{-1}\|} \leq \frac{|\mu|-1}{\mu+1} \\ &\leq \frac{(\frac{1+\epsilon}{1-\epsilon})^{1/3}-1}{2} = \frac{1}{2}\left[\left(1+\frac{2\epsilon}{1-\epsilon}\right)^{1/3} - 1\right] \leq \frac{\epsilon}{3(1-\epsilon)}. \end{aligned}$$

Thus $\psi(\epsilon) \leq \dfrac{\epsilon}{3(1-\epsilon)}$. Hence,

$$\begin{aligned} \sigma_\epsilon(T^3) &= \left\{\lambda \in \mathbb{C} : |\lambda| \leq \frac{1+\epsilon}{1-\epsilon}\right\} \\ &\subseteq \sigma_{\psi(\epsilon)}(T)^3 \\ &\subseteq \left\{\lambda \in \mathbb{C} : |\lambda| \leq \left(\frac{1+\psi(\epsilon)}{1-\psi(\epsilon)}\right)^3\right\} \\ &\leq \left\{\lambda \in \mathbb{C} : |\lambda| \leq \left(\frac{3-2\epsilon}{3-4\epsilon}\right)^3\right\}. \end{aligned}$$

4. Numerical results

In this section, we report the results of some numerical experiments done using matlab.

Let $A = (C[1,2], \|\cdot\|_\infty)$, $\epsilon = 0.1$ and $a \in A$ be defined by $a(x) = x$ for all $x \in [1,2]$ as in Example 2. If $f(z) = z^2$. Then $\tilde{f}(a) = a^2$ defined by $a^2(x) = a(x)a(x) = x^2$. The ϵ-condition spectrum of a can be calculated as follows, Let $z = \alpha + \beta i$. Then there are four cases.

- $\alpha < 1$. In this case $\|z-a\|_\infty = \sqrt{(\alpha-2)^2+\beta^2}$ and $\|(z-a)^{-1}\|_\infty = 1/\sqrt{(\alpha-1)^2+\beta^2}$
- $1 \leq \alpha < 1.5$. In this case $\|z-a\|_\infty = \sqrt{(\alpha-2)^2+\beta^2}$ and $\|(z-a)^{-1}\|_\infty = 1/|\beta|$
- $1.5 \leq \alpha < 2$. In this case $\|z-a\|_\infty = \sqrt{(\alpha-1)^2+\beta^2}$ and $\|(z-a)^{-1}\|_\infty = 1/|\beta|$
- $\alpha \geq 2$. In this case $\|z-a\|_\infty = \sqrt{(\alpha-1)^2+\beta^2}$ and $\|(z-a)^{-1}\|_\infty = 1/\sqrt{(\alpha-2)^2+\beta^2}$

Thus ϵ-condition spectrum can be calculated explicitly using the definition. In a similar way ϵ-condition spectrum of $\tilde{f}(a)$ also can be calculated. To calculate approximate value of $\phi(\epsilon)$ we choose a certain number of uniformly distributed points in $\sigma_\epsilon(a)$, compute $\|z^2-a^2\|_\infty\|(z^2-a^2)^{-1}\|_\infty$ at each of these points and take the maximum of these values as an approximation of $\phi(\epsilon)$. Similarly $\psi(\epsilon)$ is computed. For $\epsilon = 0.1$, these computed values turn out to be $\phi(\epsilon) = 0.1332$ and

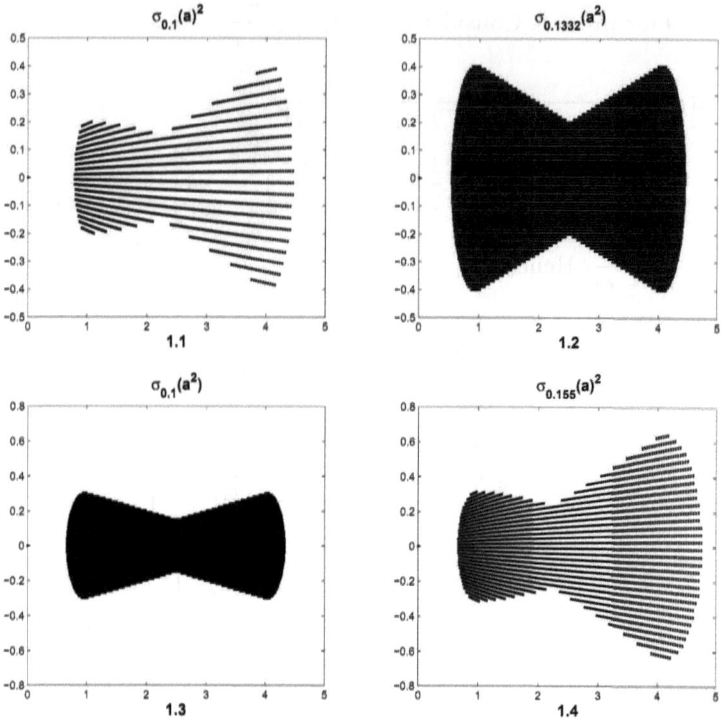

FIGURE 1

$\psi(\epsilon) = 0.155$. From Theorem 2.1 we have the following inclusions

$$\sigma_{0.1}(a)^2 \subseteq \sigma_{0.1332}(a^2).$$

$$\sigma_{0.1}(a^2) \subseteq \sigma_{0.155}(a)^2.$$

The following figures are obtained using matlab. Figure 1.1 shows $\sigma_{0.1}(a)^2$, Figure 1.2 shows $\sigma_{0.1332}(a^2)$, Figure 1.3 shows $\sigma_{0.1}(a^2)$, and Figure 1.4 shows $\sigma_{0.155}(a)^2$.

The condition spectrum of an $n \times n$ matrix T can be computed as follows. It is proved in [5] (Theorem 2.9) that,

$$|\lambda| \leq \frac{1+\epsilon}{1-\epsilon}\|T\| \quad \text{for all} \quad \lambda \in \sigma_\epsilon(T).$$

We can consider certain number of uniformly distributed points in the disc

$$\left\{ z \in \mathbb{C} : |z| \leq \frac{1+\epsilon}{1-\epsilon}\|T\| \right\},$$

evaluate $\|z - T\|\|(z-T)^{-1}\|$ at each of these points and include and save those z for which

$$\|z - T\|\|(z-T)^{-1}\| \geq \frac{1}{\epsilon}.$$

This gives $\sigma_\epsilon(T)$. We plot the points to the complex plane using matlab and obtain the figure for $\sigma_\epsilon(T)$. For each such chosen points z in $\sigma_\epsilon(T)$, we compute

$$\frac{1}{\|f(z) - \tilde{f}(T)\| \|(f(z) - \tilde{f}(T))^{-1}\|}$$

and take the maximum value as an approximation of $\phi(\epsilon)$ defined in Theorem 2.1. Similarly we calculate $\psi(\epsilon)$. As in the case of pseudospectrum [11], condition spectrum of a matrix also can be computed using different algorithms. Since our aim is only to illustrate our results, we have used a very basic algorithm. We do not make any claim about the efficiency of this algorithm.

We have considered $(\mathbb{C}^{10\times 10}, \|\cdot\|_2)$ and the following 10×10 Toeplitz matrix.

$$T = \begin{bmatrix} 1 & 1 & 0 & 0 & \dots & 0 \\ 0 & 1 & 1 & 0 & \dots & 0 \\ . & . & . & . & \dots & . \\ . & . & . & . & \dots & . \\ 0 & \dots & . & 0 & 1 & 1 \\ 0 & \dots & \dots & . & 0 & 1 \end{bmatrix}_{10\times 10}$$

(1) Let $f(z) = z^2$ and $\epsilon = 0.1$. Then $\tilde{f}(T) = T^2$ is also a Toeplitz matrix

$$T^2 = \begin{bmatrix} 1 & 2 & 1 & 0 & \dots & 0 \\ 0 & 1 & 2 & 1 & \dots & 0 \\ . & . & . & . & \dots & . \\ . & . & . & . & \dots & . \\ 0 & \dots & . & 0 & 1 & 2 \\ 0 & \dots & \dots & . & 0 & 1 \end{bmatrix}_{10\times 10}$$

Using the algorithm explained above we obtain $\phi(\epsilon) = 0.1662$ and $\psi(\epsilon) = 0.1602$. From Theorem 2.1 we have the following inclusions

$$\sigma_{0.1}(T)^2 \subseteq \sigma_{0.1662}(T^2).$$
$$\sigma_{0.1}(T^2) \subseteq \sigma_{0.1602}(T)^2.$$

The figures obtained using matlab computations are given in Figure 2. Figure 2.1 shows $\sigma_{0.1}(T)^2$, Figure 2.2 shows $\sigma_{0.1662}(T^2)$, Figure 2.3 shows $\sigma_{0.1}(T^2)$, and Figure 2.4 shows $\sigma_{0.1602}(T)^2$.

(2) Let $f(z) = e^z$ and $\epsilon = 0.01$. Then $\tilde{f}(T) = \exp(T)$ is also a Toeplitz matrix.

$$\exp(T) = \begin{bmatrix} e & e & 1.3591 & 0.4530 & \dots & 0.000 \\ 0 & e & e & 1.3591 & \dots & 0.001 \\ . & . & . & . & \dots & . \\ . & . & . & . & \dots & . \\ 0 & \dots & . & 0 & e & e \\ 0 & \dots & \dots & . & 0 & e \end{bmatrix}_{10\times 10}$$

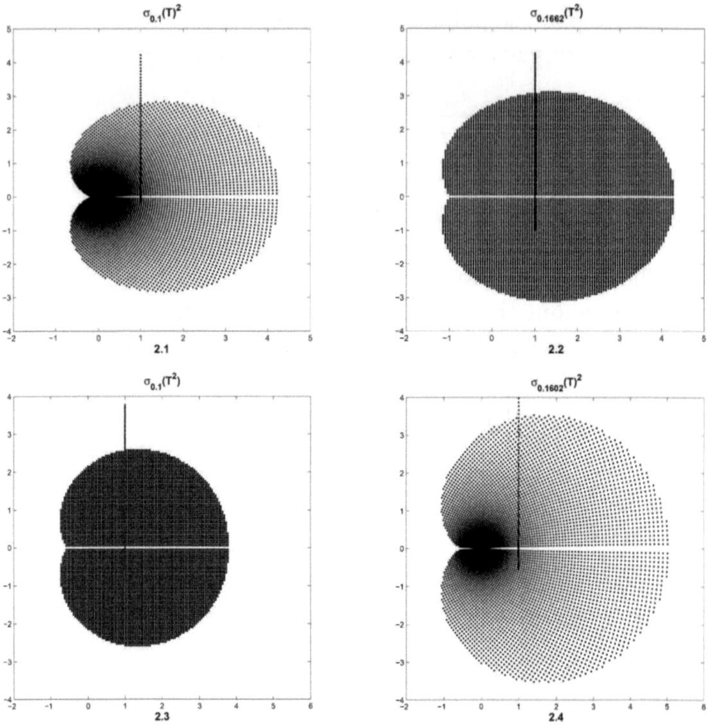

FIGURE 2

Using the same algorithm we obtain $\phi(\epsilon) = 0.0195$ and $\psi(\epsilon) = 0.0258$. Thus by Theorem 2.1, we have the following two inclusions

$$e^{\sigma_{0.01}(T)} \subseteq \sigma_{0.0195}(\exp(T)).$$

$$\sigma_{0.01}(\exp(T)) \subseteq e^{\sigma_{0.0258}(T)}.$$

The figures obtained using matlab computations are given in Figure 3. Figure 3.1 shows $e^{\sigma_{0.01}(T)}$, Figure 3.2 shows $\sigma_{0.0195}(\exp(T))$, Figure 3.3 shows $\sigma_{0.01}(\exp(T))$, and Figure 3.4 shows $e^{\sigma_{0.0258}(T)}$.

(3) In the next example we consider a random matrix J of order 3×3.

$$J = \begin{bmatrix} 0.5 & 1 & -1 \\ 1.5 & -0.5 & 0.25 \\ 0.75 & 1.5 & 1.25 \end{bmatrix}_{3\times 3}$$

Let $f(z) = z^3$ and $\epsilon = 0.01$, we have $\tilde{f}(J) = J^3$ is given by

$$J^3 = \begin{bmatrix} 0.125 & -3.5 & -0.25 \\ 2.6719 & 3.1562 & 1.703 \\ -2.3906 & -0.0938 & 2.8906 \end{bmatrix}_{3\times 3}$$

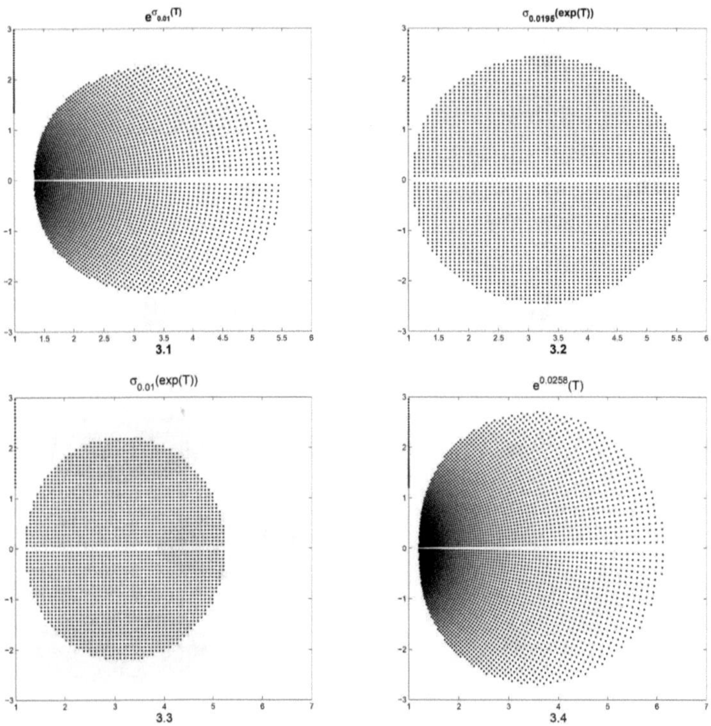

FIGURE 3

As above we obtain $\phi(\epsilon) = 0.0381$ and $\psi(\epsilon) = 0.1945$ Thus by Theorem 2.1, we have the following two inclusions

$$\sigma_{0.01}(J)^3 \subseteq \sigma_{0.0381}(J^3), \qquad \sigma_{0.01}(J^3) \subseteq \sigma_{0.1411}(J)^3.$$

The figures obtained using matlab computations are given in Figure 4. Figure 4.1 shows $\sigma_{0.01}(J)^3$, Figure 4.2 shows $\sigma_{0.0381}(J^3)$, Figure 4.3 shows $\sigma_{0.01}(J^3)$, and Figure 4.4 shows $\sigma_{0.1411}(J)^3$.

References

[1] R. Hagen, S. Roch and B. Silbermann, *C^*-algebras and numerical analysis*, Monographs and Textbooks in Pure and Applied Mathematics, 236, Dekker, New York, 2001.

[2] A.C. Hansen, *On the solvability complexity index, the n-pseudospectrum and approximations of spectra of operators*, J. Amer. Math. Soc. **24** (2011), no. 1, 81–124.

[3] A.C. Hansen, *Infinite-dimensional numerical linear algebra: theory and applications*, Proc. R. Soc. Lond. Ser. A Math. Phys. Eng. Sci. **466** (2010), no. 2124, 3539–3559.

[4] G. Krishna Kumar and S.H. Kulkarni, *Linear maps preserving pseudospectrum and condition spectrum*, Banach J. Math. Anal. **6** (2012), no. 1, (45–60).

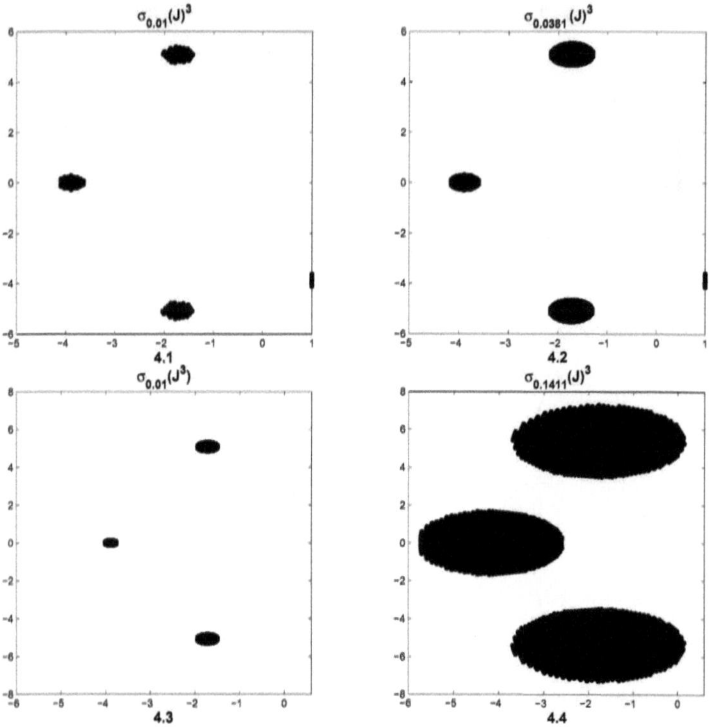

FIGURE 4

[5] S.H. Kulkarni and D. Sukumar, *The condition spectrum*, Acta Sci. Math. (Szeged) **74** (2008), no. 3-4, 625–641.

[6] S.H. Lui, *A pseudospectral mapping theorem*, Math. Comp. **72** (2003), no. 244, 1841–1854 (electronic).

[7] S.H. Lui, *Pseudospectral mapping theorem II*, Electron. Trans. Numer. Anal. **38** (2011), 168–183.

[8] T.J. Ransford, *Generalized spectra and analytic multivalued functions*, J. London Math. Soc. (2) **29** (1984), no. 2, 306–322.

[9] W. Rudin, *Functional analysis*, McGraw-Hill, New York, 1973.

[10] E.S. Şuhubi, *Functional analysis*, Kluwer Acad. Publ., Dordrecht, 2003.

[11] L.N. Trefethen and M. Embree, *Spectra and pseudospectra*, Princeton Univ. Press, Princeton, NJ, 2005.

G. Krishna Kumar and S.H. Kulkarni
Department of Mathematics
Indian Institute of Technology Madras
Chennai-600 036, India
e-mail: `krishna.math@gmail.com`
`shk@iitm.ac.in`

Operator Theory:
Advances and Applications, Vol. 236, 317–336

Commutative Algebras of Toeplitz Operators on the Super Upper Half-plane: Quasi-hyperbolic and Quasi-parabolic Cases

M. Loaiza and A. Sánchez-Nungaray

Abstract. In this paper we study Toeplitz operators acting on the super Bergman space on the upper half-plane. We consider the super subgroups of isometries $\mathbb{R} \times S^1$ and $\mathbb{R}_+ \times S^1$ and we prove that the C^*-algebras generated by Toeplitz operators whose symbols are invariant under the action of these groups are commutative.

Mathematics Subject Classification (2010). Primary 47B35; Secondary 47L80, 32A36, 58A50.

Keywords. Toeplitz operators, commutative C^*-algebras, Bergman spaces, supermanifolds and graded manifolds.

1. Introduction

Commutative C^*-algebras generated by Toeplitz operators acting on the (weighted) Bergman space on the unit disk have been recently an important object of study.

In [19, 20] Vasilevski discovered a family of commutative C^*-algebras of Toeplitz operators on the unit disk. These algebras can be classified as follows: Each pencil of hyperbolic geodesics determines a set of symbols consisting of functions which are constant on the corresponding cycles, the orthogonal trajectories to geodesics forming the pencil. The C^*-algebra generated by all Toeplitz operators with such kind of symbols turns out to be commutative.

In [7] Grudski, Quiroga and Vasilevski proved that the *C^*-algebra generated by Toeplitz operators is commutative on each weighted Bergman space if and only if there is a pencil of hyperbolic geodesics such that the symbols of the Toeplitz operators are constant on the cycles of this pencil.* In fact, the cycles are the orbits of a one-parameter subgroup of isometries for the hyperbolic geometry on the unit disk. This provides us with the following result: *the C^*-algebra generated by Toeplitz operators is commutative on each weighted Bergman space on the unit disk* *if and only if there is a maximal commutative subgroup of Möbius transformations such that the symbols of the Toeplitz operators are invariant under the action of this subgroup.*

In [2, 3] Borthwick, Klimek, Lesniewski, and Rinaldi introduced a general theory of non-perturbative quantization of a class of hermitian symmetric super-manifolds. The quantization idea in these two papers is based on the notion of super Toeplitz operator defined on a suitable $\mathbb{Z}_2$-graded Hilbert space of super-holomorphic functions. The quantized super-manifold arises as the C^*-algebra generated by these operators.

Super Lie groups act on super-manifolds as linear transformations, in the sense we will define later on the paper. In order to extend results in [7] we find commutative super groups of isometries of the super unit disk and consider as symbols class all functions which are invariant under the action of this group. Once we get this group of isometries we analyze the behavior of the C^*-algebra generated by all super Toeplitz operators whose symbols are invariant under the action of this group.

One step forward in this direction was made in [11]. There we proved that the algebras of Toeplitz operators whose symbols are invariant under the action of the super-unit circle or under the action of the torus are commutative. In order to do so, we found an orthonormal basis for the weighted super Bergman space of the unit disk in which each Toeplitz operator with radial symbol is represented as a diagonal matrix.

Using a unitary operator, given in terms of a Möbius transformation, we transform the Bergman space on the unit disk onto the Bergman space on the upper half-plane. Thus, finding a unitary transformation from the super Bergman space on the unit disk onto the super Bergman space on the upper half-plane would allow us to continue the classification of the commutative algebras of super Toeplitz operators. This step was made in [17]. There, the matrix representation of a super Toeplitz operator, in terms of regular Toeplitz or Toeplitz-like operators, is given. Using this representation it was proved that the C^*-algebra generated by Toeplitz operators, whose symbols are invariant under the action of the super real numbers (we define this action in this paper) is commutative by showing that these operators are unitary equivalent to multiplication operators.

This paper is a continuation of the study made in [17]. Here we find two more maximal abelian super groups, $\mathbb{R} \times S^1$ and $\mathbb{R}_+ \times S^1$, such that the Toeplitz algebra generated by Toeplitz operators whose symbols are invariant under the action of each one of these groups is commutative.

In summary, if we consider classical Toeplitz operators, there are only three different (up to a conjugation) classes of symbols that generate commutative Toeplitz algebras on the unit disk. These Toeplitz classes are: functions which are invariant under elliptic Möbius transformations, functions which are invariant under hyperbolic Möbius transformations and, functions which are invariant under parabolic transformations. In terms of maximal abelian groups these classes correspond to invariant classes under the action of the unit circle, the real numbers and the positive real numbers respectively. The situation behaves quite different for super Toeplitz operators, acting on the super Bergman space on the super unit disk or on the Bergman space on the super upper half-plane. Until now we have found five different cases of commutative algebras of super Toeplitz operators, which correspond to the action of the super-circle, the 2-torus, the super-real numbers, the product of the real numbers and the unit circle, the product of the positive real numbers and the unit circle. The natural question arising now is the following: are there more classes of conjugation of commutative algebras generated by super Toeplitz operators? The first step in order to give an answer to this question would be finding the classification of commutative Lie supersubgroups of $SU(1,1|1)$.

This paper is organized as follows: in Section 2 we give a review of some results obtained by the authors on Toeplitz operators on the super Bergman space on the unit disk. In sections 3 and 4 we analyze the relation between the super Bergman space on the unit disk and the super Bergman space on the upper half-plane. In Section 5 we include the main results from [17]. In Section 6 we show that the groups $\mathbb{R} \times S^1$ and $\mathbb{R}_+ \times S^1$ are groups of isometries of the super upper plane, and find the form of the functions defined in this set which are invariant under the action of these groups. In this section we also prove that the Toeplitz algebra whose symbols are invariant under the action of $\mathbb{R} \times S^1$ or $\mathbb{R}_+ \times S^1$ is commutative and that these operators are unitarily equivalent to multiplication operators. Moreover, we give conditions for a Toeplitz operator with this kind of symbol in order to be compact or bounded.

2. Super Toeplitz operators on the unit super disk

In [10] super Toeplitz operators on the unit disk were studied. We include here the main results. Let $\mathcal{O}(\mathbb{D})$ denote the algebra of all holomorphic functions $\psi(z)$ in the open unit disk $\mathbb{D}$

$$\mathbb{D} := \{z \in \mathbb{C} : \ |z| < 1\}.$$

Let Λ_1 denote the complex Grassmann algebra with generator ζ, satisfying the relation $\zeta^2 = 0$. Thus

$$\Lambda_1 = \mathbb{C}\langle 1, \zeta\rangle.$$

The tensor product algebra

$$\mathcal{O}(\mathbb{D}^{1|1}) := \mathcal{O}(\mathbb{D}) \otimes \Lambda_1 = \mathcal{O}(\mathbb{D})\langle 1, \zeta\rangle$$

consists of all "super-holomorphic" functions

$$\Psi = \psi_0 + \zeta\psi_1$$

with $\psi_0, \psi_1 \in \mathcal{O}(\mathbb{D})$. We sometimes write

$$\Psi(z, \zeta) = \psi_0(z) + \zeta\psi_1(z)$$

for all $z \in \mathbb{D}$.

For $\nu > 1$, the *weighted Bergman space*

$$H^2_\nu(\mathbb{D}) := \mathcal{O}(\mathbb{D}) \cap L^2(\mathbb{D}, d\mu_\nu)$$

consists of all holomorphic functions on $\mathbb{D}$ which are square-integrable with respect to the probability measure

$$d\mu_\nu(z) = \frac{\nu - 1}{\pi}(1 - |z|^2)^{\nu-2} dz, \tag{2.1}$$

where dz denotes the Lebesgue measure on $\mathbb{C}$. It is well known that $H^2_\nu(\mathbb{D})$ has the reproducing kernel

$$K_\nu(z, w) = (1 - z\overline{w})^{-\nu},$$

where $z, w \in \mathbb{D}$.

Let $\Lambda_1^{\mathbb{C}}$ denote the complex Grassmann algebra with 2 generators $\zeta, \overline{\zeta}$ satisfying the conditions

$$\zeta^2 = \overline{\zeta}^2 = 0, \quad \zeta\overline{\zeta} = -\overline{\zeta}\zeta.$$

Thus

$$\Lambda_1^{\mathbb{C}} = \mathbb{C}\langle 1, \zeta, \overline{\zeta}, \overline{\zeta}\zeta\rangle = \Lambda_1\langle 1, \overline{\zeta}\rangle.$$

By $\mathcal{C}(\overline{\mathbb{D}})$ we denote the algebra of continuous functions on $\overline{\mathbb{D}}$. The tensor product

$$\mathcal{C}(\overline{\mathbb{D}}^{1|1}) := \mathcal{C}(\overline{\mathbb{D}}) \otimes \Lambda_1^{\mathbb{C}} = \mathcal{C}(\overline{\mathbb{D}})\langle 1, \zeta, \overline{\zeta}, \overline{\zeta}\zeta\rangle$$

consists of all "continuous super-functions"

$$F = f_{00} + \overline{\zeta} f_{10} + \zeta f_{01} + \overline{\zeta}\zeta f_{11}, \tag{2.2}$$

where $f_{00}, f_{10}, f_{01}, f_{11} \in \mathcal{C}(\overline{\mathbb{D}})$. The involution on $\mathcal{C}(\overline{\mathbb{D}}^{1|1})$ is given by

$$\overline{F} = \overline{f}_{00} + \zeta \overline{f}_{10} + \overline{\zeta} \overline{f}_{01} + \overline{\zeta}\zeta \overline{f}_{11},$$

where $\overline{f}(z) := \overline{f(z)}$ (pointwise conjugation).

The "fermionic integration" is determined by the rules

$$\int_{\mathbb{C}^{0|1}} d\zeta \cdot \zeta = \int_{\mathbb{C}^{0|1}} d\zeta \cdot \overline{\zeta} = \int_{\mathbb{C}^{0|1}} d\zeta \cdot 1 = 0, \quad \int_{\mathbb{C}^{0|1}} d\zeta \cdot \overline{\zeta}\zeta = 1.$$

Definition 2.1. For any parameter $\nu > 1$ the (weighted) *super-Bergman space*

$$H^2_\nu(\mathbb{D}^{1|1}) \subset \mathcal{O}(\mathbb{D}^{1|1})$$

consists of all super-holomorphic functions $\Psi(z, \zeta)$ which satisfy the square-integrability condition

$$(\Psi|\Psi)_\nu := \frac{1}{\pi} \int_{\mathbb{D}^{1|1}} dz\, d\zeta \, (1 - z\overline{z} - \zeta\overline{\zeta})^{\nu-1} \overline{\Psi(z, \zeta)} \, \Psi(z, \zeta) < +\infty.$$

Proposition 2.2. *For $\Psi = \psi_0 + \zeta\,\psi_1 \in \mathcal{O}(\mathbb{D}^{1|1})$ we have*

$$\frac{1}{\pi}\int\limits_{\mathbb{D}^{1|1}} dz\,d\zeta\;(1 - z\overline{z} - \zeta\overline{\zeta})^{\nu-1}\overline{\Psi}(z,\zeta)\,\Psi(z,\zeta) = (\psi_0|\psi_0)_\nu + \frac{1}{\nu}\,(\psi_1|\psi_1)_{\nu+1}\,,$$

i.e., there is an orthogonal decomposition

$$H^2_\nu(\mathbb{D}^{1|1}) = H^2_\nu(\mathbb{D}) \oplus [H^2_{\nu+1}(\mathbb{D}) \otimes \Lambda^1(\mathbb{C}^1)]$$

onto a sum of weighted Bergman spaces, where $\Lambda^1(\mathbb{C}^1)$ is the one-dimensional vector space with basis vector ζ.

For $\Psi = \psi_0 + \zeta\,\psi_1 \in H^2_\nu(\mathbb{D}^{1|1})$ we have the reproducing kernel property

$$\Psi(z,\zeta) = \frac{1}{\pi}\int\limits_{\mathbb{D}^{1|1}} dw\,d\omega\;(1 - w\overline{w} - \omega\overline{\omega})^{\nu-1}(1 - z\overline{w} - \zeta\overline{\omega})^{-\nu}\Psi(w,\omega),$$

i.e., $H^2_\nu(\mathbb{D}^{1|1})$ has the reproducing kernel

$$K_\nu(z,\zeta,w,\omega) = (1 - z\overline{w} - \zeta\overline{\omega})^{-\nu}.$$

For $F \in \mathcal{C}(\overline{\mathbb{D}}^{1|1})$, the *super-Toeplitz operator* $T^{(\nu)}_F$ on $H^2_\nu(\mathbb{D}^{1|1})$ is defined as

$$T^{(\nu)}_F\Psi = P^{(\nu)}(F\Psi),$$

where $P^{(\nu)}$ denotes the orthogonal projection onto $H^2_\nu(\mathbb{D}^{1|1})$.

Theorem 2.3. *With respect to the decomposition $\Psi = \psi_0 + \zeta\,\psi_1$, the super-Toeplitz operator $T^{(\nu)}_F$ on $H^2_\nu(\mathbb{D}^{1|1})$ is given by the block matrix*

$$T^{(\nu)}_F \;=\; \begin{pmatrix} T^\nu_\nu\left(f_{00} + \frac{1-w\overline{w}}{\nu-1}\,f_{11}\right) & T^{\nu+1}_\nu\left(\frac{1-w\overline{w}}{\nu-1}\,f_{10}\right) \\ T^\nu_{\nu+1}\,(f_{01}) & T^{\nu+1}_{\nu+1}\,(f_{00}) \end{pmatrix}. \tag{2.3}$$

Here $T^{\nu+j}_{\nu+i}(f)$, for $0 \le i,j \le 1$, denotes the Toeplitz type operator from $H^2_{\nu+j}(\mathbb{D})$ to $H^2_{\nu+i}(\mathbb{D})$ defined by

$$T^{\nu+j}_{\nu+i}(f)\,\psi := P_{\nu+i}(f\psi)$$

for $\psi \in H^2_{\nu+j}(\mathbb{D})$ and $P_{\nu+i}$ is the orthogonal projection from $L^2_{\nu+i}(\mathbb{D})$ onto $H^2_{\nu+i}(\mathbb{D})$.

2.1. Commutative algebras of super Toeplitz operators on the unit super disk

In [11] we studied commutative algebras of super Toeplitz operators on the unit super disk. We found two different groups of isometries such that Toeplitz operators whose symbols are invariant under the action of these groups (one algebra for each group) generate commutative algebras. We recall here the main results from [11].

The Berezinian is defined by the following formula

$$\mathrm{Ber}\begin{pmatrix} A & B \\ C & D \end{pmatrix} = \det(A - BD^{-1}C)\det D^{-1},$$

where A, D are even, D is invertible and B, C are odd. The super Lie Group $SU(1,1|1)$ is the supermanifold $SU(1,1)$ and its structure sheaf is generated by the 3×3 matrices (s_{ij}) and $(\overline{s}_{ij})$, where the element s_{ij} (resp. $\overline{s}_{ij}$) is even if

$1 \le i, j \le 2$ or $i = j = 3$ and odd otherwise. In addition we must have $s^* J s = J$, where

$$J = \begin{pmatrix} 1 & 0 & 0 \\ 0 & -1 & 0 \\ 0 & 0 & -1 \end{pmatrix}$$

and that $\operatorname{Ber} s = 1$.

The super Lie Group $SU(1,1|1)$ acts on the unit super disk as follows. Given a matrix $\begin{pmatrix} a & b & \alpha \\ c & d & \beta \\ \gamma & \delta & e \end{pmatrix} \in SU(1,1|1)$ the corresponding transformation is given by the formulas

$$\begin{aligned} z &\mapsto \frac{az + b + \alpha\zeta}{cz + d + \beta\zeta}, \\ \zeta &\mapsto \frac{\gamma z + \delta + e\zeta}{cz + d + \beta\zeta}. \end{aligned} \tag{2.4}$$

Consider the form

$$\begin{aligned} g = {} & \frac{\partial^2 log(1 - z\bar{z} - \zeta\bar{\zeta})}{\partial z \partial \bar{z}} dz d\bar{z} + \frac{\partial^2 log(1 + z\bar{z} - \zeta\bar{\zeta})}{\partial z \partial \bar{\zeta}} dz d\bar{\zeta} \\ & + \frac{\partial^2 log(1 + z\bar{z} - \zeta\bar{\zeta})}{\partial \zeta \partial \bar{z}} d\zeta d\bar{z} + \frac{\partial^2 log(1 + z\bar{z} - \zeta\bar{\zeta})}{\partial \zeta \partial \bar{\zeta}} d\zeta d\bar{\zeta}, \end{aligned}$$

where

$$\begin{aligned} \frac{\partial^2 log(1 - z\bar{z} - \zeta\bar{\zeta})}{\partial z \partial \bar{z}} &= \frac{1}{(1 - z\bar{z})^2} + \frac{(1 + z\bar{z})\zeta\bar{\zeta}}{(1 - z\bar{z})^3}, \\ \frac{\partial^2 log(1 + z\bar{z} - \zeta\bar{\zeta})}{\partial z \partial \bar{\zeta}} &= \frac{\bar{z}\zeta}{(1 - z\bar{z})^2}, \\ \frac{\partial^2 log(1 + z\bar{z} - \zeta\bar{\zeta})}{\partial \zeta \partial \bar{z}} &= \frac{z\bar{\zeta}}{(1 - z\bar{z})^2}, \\ \frac{\partial^2 log(1 + z\bar{z} - \zeta\bar{\zeta})}{\partial \zeta \partial \bar{\zeta}} &= \frac{1}{(1 - z\bar{z})}. \end{aligned}$$

In [2] it was proved that g is an invariant form under the action of the super Lie group $SU(1,1|1)$. This form produces a metric (Bergman super metric), for the unit super disk, which is invariant under the action of $SU(1,1|1)$.

The unit super circle is defined as the super manifold

$$S^{1|1} = \{(a = a_1 + ia_2, \tau = \eta_1 + i\eta_2) |\ a_1^2 + a_2^2 = 1, \ \ \eta_1 a_1 + \eta_2 a_2 = 0\},$$

where η_1, η_2 are real Grassmann variables. This definition is equivalent to the definition of the unit super circle as the set of matrices

$$A = \begin{pmatrix} a & \tau \\ \tau & a \end{pmatrix},$$

such that $A^*A = I$, and $\operatorname{Ber} A = 1$. The supermanifold $S^{1|1}$ is a subgroup of $SU(1,1|1)$ and the corresponding action of $S^{1|1}$ on the unit super disk is given by the formulas

$$z \mapsto w = az - \tau\zeta, \tag{2.5}$$

$$\zeta \mapsto \eta = \tau z + a\zeta. \tag{2.6}$$

A smooth function f, defined in the unit super disk, is invariant under the action of $S^{1|1}$ if and only if f has the form

$$f(z,\zeta) = f_0(r) + f_1(r)z\bar{\zeta} + f_1(r)\bar{z}\zeta - \frac{f_0'(r)}{2r}\bar{\zeta}\zeta,$$

where $r = |z|$ and f_0 and f_1 are radial functions. We define a *radial super function* as an invariant function under the action of $S^{1|1}$.

Theorem 2.4. *Let f be a radial super function. Then the Toeplitz operator T_f is diagonal.*

Proof. The detailed proof is given in [11]. Here we sketch only the main ideas of the proof. If f is a radial super function, the space generated by z^0, i.e., all constant functions, and the space generated by $\{z^n, z^{n-1}\zeta\}, n \geq 0$, are invariant subspaces for the operator T_f. For $n \geq 1$ the restriction of T_f to the bidimensional space generated by $\{z^n, z^{n-1}\zeta\}$ can be written in matrix form as follows

$$\frac{2\Gamma(n+\nu)}{(n-1)!\Gamma(\nu)} \begin{pmatrix} \int_0^1 (1-r^2)^{\nu-1} r^{2n-1} f_0(r)dr & \int_0^1 (1-r^2)^{\nu-1} r^{2n+1} f_1(r)dr/n \\ \int_0^1 (1-r^2)^{\nu-1} r^{2n+1} f_1(r)dr & \int_0^1 (1-r^2)^{\nu-1} r^{2n-1} f_0(r)dr \end{pmatrix}.$$

This 2×2 matrix has the following eigenvalues

$$\frac{2\Gamma(n+\nu)}{(n-1)!\Gamma(\nu)} \left(\int_0^1 (1-r^2)^{\nu-1} r^{2n-1} f_0(r)dr) \pm \int_0^1 (1-r^2)^{\nu-1} r^{2n+1} f_1(r)dr/\sqrt{n} \right).$$

The corresponding eigenvectors are

$$\omega_{n_1} = \frac{z^n}{\sqrt{n}} + z^{n-1}\zeta,$$

$$\omega_{n_2} = -\frac{z^n}{\sqrt{n}} + z^{n-1}\zeta,$$

whose norms are

$$\|\omega_{n_1}\| = \|\omega_{n_2}\| = \sqrt{\frac{(n-1)!\Gamma(\nu)}{\Gamma(n+\nu)} + \frac{n!\Gamma(\nu)}{(n+\nu)\Gamma(n+\nu)}}.$$

Let $\{1, \omega_{n_1}, \omega_{n_2} | n \geq 0\}$ be the orthonormal basis obtained by normalizing all of the eigenvectors. Then the super Toeplitz operator with radial symbol f is the diagonal operator of the form

$$T_f(w_{n_i}) = \frac{2\Gamma(n+\nu)}{(n-1)!\Gamma(\nu)} \left(\int_0^1 (1-r^2)^{\nu-1} r^{2n-1} f_0(r) dr \right. \\ \left. + (-1)^{i+1} \int_0^1 (1-r^2)^{\nu-1} r^{2n+1} f_1(r) dr/\sqrt{n} \right) w_{n_i}, \tag{2.7}$$

for $i = 1, 2, n = 1, \ldots$ □

An immediate consequence of the last theorem is the fact that the C^*-algebra generated by all Toeplitz operators with radial symbols is commutative. Furthermore, since the second factor in formula (2.7) has the term $1/\sqrt{n}$, for a bounded function f_1, the super Toeplitz operator T_f is bounded if and only if

$$\left(\frac{2\Gamma(n+\nu)}{(n-1)!\Gamma(\nu)} \int_0^1 (1-r^2)^{\nu-1} r^{2n-1} f_0(r) dr \right)$$

is a bounded sequence and T_f is compact if and only if the sequence

$$\left(\frac{2\Gamma(n+\nu)}{(n-1)!\Gamma(\nu)} \int_0^1 (1-r^2)^{\nu-1} r^{2n-1} f_0(r) dr \right),$$

tends to zero when $n \to \infty$.

In the definition of a radial function given above we excluded all complex-valued radial functions. In order to see the classical case of Toeplitz operators with radial symbols as elements of a commutative algebra of super Toeplitz operators we introduce the following group of elements of $SU(1|1|1)$. For two complex numbers $z = e^{it}, w = e^{is}$ in the unit circle, consider the element of $SU(1,1|1)$ given by the following matrix

$$T_{s,t} = \begin{pmatrix} e^{is} & 0 & 0 \\ 0 & e^{it} & 0 \\ 0 & 0 & e^{i(t+s)} \end{pmatrix}.$$

It is easy to prove that $T^2 = \{T_{s,t} | s, t \in [0, 2\pi)\}$ is a subgroup of $SU(1,1|1)$.

Let f be a bounded function defined in the unit super disk. If f is invariant under the action of T^2 then f has the form

$$f(z,\zeta) = f_0(r) + f_1(r)\bar{\zeta}\zeta, \tag{2.8}$$

where f_0 and f_1 are radial functions.

By Theorem 2.3 the super Toeplitz operator with symbol $f(z,\zeta) = f_0(r) + f_1(r)\bar{\zeta}\zeta$ has the form

$$\begin{pmatrix} T_\nu^\nu \left(f_0 + \frac{1-w\overline{w}}{\nu-1} f_1 \right) & 0 \\ 0 & T_{\nu+1}^{\nu+1}(f_0) \end{pmatrix}. \tag{2.9}$$

This implies that the Toeplitz algebra generated by all super Toeplitz operators whose symbols are invariant under the action of T^2 is commutative. Here we used the fact that a Toeplitz operator with radial symbol, acting on the Bergman space, is diagonal.

3. The super upper half-plane and its Bergman space

Denote by $\Lambda(\mathbb{C})$ the exterior algebra over $\mathbb{C} = \mathbb{R}^2$. The super upper half-plane $\mathbb{H}^{1|1}$ is the supermanifold $(\mathbb{H}, \mathcal{O})$, where $\mathbb{H} = \{z \in \mathbb{C} : \operatorname{Im} z > 0\}$ and $\mathcal{O}$ is the sheaf of superalgebras on $\mathbb{H}$ whose space of global sections is $C^\infty(\mathbb{H}^{1|1}) = C^\infty(\mathbb{H}) \otimes \Lambda(\mathbb{C})$. Denote by η and $\bar{\eta}$ the standard generators of $\Lambda(\mathbb{C})$. An element $f \in C^\infty(\mathbb{H}^{1|1})$ has the form

$$f(z, \zeta, \bar{\zeta}) = f_{00}(z) + f_{01}(z)\eta + f_{10}(z)\bar{\eta} + f_{11}(z)\bar{\eta}\eta,$$

where $f_{ij} \in C^\infty(\mathbb{H})$.

The Lie super-group $SL_{(2|2)}(\mathbb{R})$ is defined as follows. Its base manifold is $SL_2(\mathbb{R})$ and its structure sheaf is generated by γ_{ij} and $\bar{\gamma}_{ij}$ for $1 \leq i, j \leq 3$, with the following parity assignments:

$$|\gamma_{ij}| = |\bar{\gamma}_{ij}| = \begin{cases} 0 & \text{if } 1 \leq i, j \leq 2 \text{ and } i = j = 3, \\ 1 & \text{otherwise.} \end{cases} \tag{3.1}$$

This means that if $|\gamma_{ij}| = 0$ then γ_{ij} is an even super-number, and γ_{ij} is odd otherwise.

Let $\gamma = (\gamma_{ij})$ denote the super-matrix with entries γ_{ij} and, let γ^* be its hermitian adjoint, where $\gamma^*_{ij} = \bar{\gamma}_{ji}$.

For $\gamma \in SL_{(2|2)}(\mathbb{R})$ we assume that

$$\gamma^* I \gamma = I, \tag{3.2}$$

where

$$I = \begin{pmatrix} 0 & -i & 0 \\ i & 0 & 0 \\ 0 & 0 & -1 \end{pmatrix}, \tag{3.3}$$

and that

$$\operatorname{Ber}\gamma = 1, \tag{3.4}$$

where Ber denotes the Berezinian.

The above conditions are the relations that define the structure sheaf of $SL_{(2|2)}(\mathbb{R})$. We define an action of $SL_{(2|2)}(\mathbb{R})$ on $\mathbb{H}^{1|1}$ as follows:

$$z \mapsto z' := \frac{\gamma_{11}z + \gamma_{12} + \gamma_{13}\theta}{\gamma_{21}z + \gamma_{22} + \gamma_{23}\theta}, \tag{3.5}$$

$$\theta \mapsto \theta' := \frac{\gamma_{31}z + \gamma_{32} + \gamma_{33}\theta}{\gamma_{21}z + \gamma_{22} + \gamma_{23}\theta}. \tag{3.6}$$

The expression $(\gamma_{21}z + \gamma_{22} + \gamma_{23}\theta)^{-1}$ is defined in terms of the Taylor series for super-functions (see [1]) by

$$(\gamma_{21}z + \gamma_{22} + \gamma_{23}\theta)^{-1} = \frac{1}{\gamma_{21}z + \gamma_{22}} - \frac{\gamma_{23}}{(\gamma_{21}z + \gamma_{22})^2}\theta.$$

With a slight abuse of notation, we write equations (3.5), (3.6) jointly by $Z' = (z', \theta') = \gamma(Z)$.

If γ is a morphism between super domains, we define

$$\gamma'(Z) = \mathrm{Ber}\begin{pmatrix} \frac{\partial z'}{\partial z} & \frac{\partial \theta'}{\partial z} \\ \frac{\partial z'}{\partial \theta} & \frac{\partial \theta'}{\partial \theta} \end{pmatrix} = \mathrm{Ber}\frac{\partial Z'}{\partial Z}, \tag{3.7}$$

for more details see [1].

Definition 3.1. For $\nu > 1$, the *weighted Bergman space*

$$H^2_\nu(\mathbb{H}) := \mathcal{O}(\mathbb{H}) \cap L^2(\mathbb{H}, d\omega_\nu)$$

consists of all holomorphic functions on $\mathbb{H}$ which are square-integrable with respect to the probability measure

$$d\omega_\nu(z) = \frac{\nu - 1}{\pi}(z - \bar{z})^{\nu-2} dz. \tag{3.8}$$

Here dz denotes Lebesgue measure on $\mathbb{C}$.

It is well known that $H^2_\nu(\mathbb{H})$ has the reproducing kernel

$$K_\nu(z, w) = (z - \bar{w})^{-\nu},$$

where $z, w \in \mathbb{H}$.

Let $\mathcal{C}(\overline{\mathbb{H}})$ denote the algebra of continuous functions on $\overline{\mathbb{H}}$. The tensor product

$$\mathcal{C}(\overline{\mathbb{H}}^{1|1}) := \mathcal{C}(\overline{\mathbb{H}}) \otimes \Lambda(\mathbb{C})$$

consists of all "continuous super-functions" of the form

$$F = f_{00} + \bar{\zeta} f_{10} + \zeta f_{01} + \bar{\zeta}\zeta f_{11}, \tag{3.9}$$

where $f_{00}, f_{10}, f_{01}, f_{11} \in \mathcal{C}(\overline{\mathbb{H}})$.

Definition 3.2. For any parameter $\nu > 1$ the (weighted) *super-Bergman space*

$$H^2_\nu(\mathbb{H}^{1|1}) \subset \mathcal{O}(\mathbb{H}^{1|1})$$

consists of all super-holomorphic functions $\Psi(z, \zeta)$ which satisfy the square-integrability condition

$$(\Psi|\Psi)_\nu := \frac{1}{\pi}\int_{\mathbb{H}^{1|1}} dz\, d\zeta\ (2\,\mathrm{Im}(z) - \zeta\bar{\zeta})^{\nu-1}\, \overline{\Psi(z,\zeta)}\, \Psi(z, \zeta) < +\infty.$$

Proposition 3.3. *For $\Psi = \psi_0 + \zeta\,\psi_1 \in \mathcal{O}(\mathbb{H}^{1|1})$ we have*

$$\frac{1}{\pi}\int_{\mathbb{H}^{1|1}} dz\, d\zeta\ (2\,\mathrm{Im}(z) - \zeta\bar{\zeta})^{\nu-1}\overline{\Psi}(z,\zeta)\, \Psi(z, \zeta) = (\psi_0|\psi_0)_\nu + \frac{1}{\nu}(\psi_1|\psi_1)_{\nu+1},$$

i.e., there is an orthogonal decomposition

$$H^2_\nu(\mathbb{H}^{1|1}) = H^2_\nu(\mathbb{H}) \oplus [H^2_{\nu+1}(\mathbb{H}) \otimes \Lambda^1(\mathbb{C}^1)]$$

onto a sum of weighted Bergman spaces, where $\Lambda^1(\mathbb{C}^1)$ is the one-dimensional vector space with basis vector ζ.

Proposition 3.4. *For* $\Psi = \psi_0 + \eta\psi_1 \in H^2_\nu(\mathbb{H}^{1|1})$ *we have the reproducing kernel property*

$$\Psi(w,\eta) = \frac{1}{\pi}\int_{\mathbb{H}^{1|1}} dx\, d\xi\ (2\,\mathrm{Im}(x) - \xi\bar{\xi})^{\nu-1}\left(\frac{w-\bar{x}}{i} - \eta\bar{\xi}\right)^{-\nu}\Psi(x,\xi),$$

i.e., $H^2_\nu(\mathbb{H}^{1|1})$ *has the reproducing kernel*

$$K_\nu(w,\eta,x,\xi) = (\frac{w-\bar{x}}{i} - \eta\bar{\xi})^{-\nu}.$$

4. Relation between the Bergman space on the super unit disk and the Bergman space on the super upper half-plane

In order to study Toeplitz operators on the Bergman space on the unit disk we can pass to the upper half-plane by using a Möbius transformation and then constructing a unitary operator that transforms the Bergman space on the unit disk onto the Bergman space on the upper half-plane. In this section we include some results from [17] that show how to transform the Bergman space on the unit super disk onto the Bergman space on the upper half-plane.

Consider the super matrix

$$\psi = \begin{pmatrix} \frac{1}{\sqrt{2}} & \frac{i}{\sqrt{2}} & 0 \\ \frac{i}{\sqrt{2}} & \frac{1}{\sqrt{2}} & 0 \\ 0 & 0 & 1 \end{pmatrix},$$

with $\mathrm{Ber}(\psi) = 1$. This matrix induces a linear transformation from the unit super disk $\mathbb{D}^{1|1}$ to the super upper half-plane $\mathbb{H}^{1|1}$ defined by $\psi(z,\zeta) = (w,\eta)$, where

$$w = \frac{z+i}{iz+1} \quad \text{and} \quad \eta = \frac{\zeta\sqrt{2}}{iz+1}, \tag{4.1}$$

with the inverse $\psi^{-1}(w,\eta) = (z,\theta)$ given by

$$z = \frac{w-i}{-iw+1} \quad \text{and} \quad \zeta = \frac{\eta\sqrt{2}}{-iw+1}. \tag{4.2}$$

The super group $SL_{(2|2)}(\mathbb{R})$ is isomorphic to $SU(1,1|1)$ (for details see [17]). The isomorphism is given by

$$\gamma \mapsto \psi\gamma\psi^{-1}. \tag{4.3}$$

Lemma 4.1. *Let* $Z_1 = (z_1,\zeta_1)$, $Z_2 = (z_2,\zeta_2)$, $\psi(Z_1) = (w_1,\eta_1) = W_1$ *and* $\psi(Z_2) = (w_2,\eta_2) = W_2$ *where* ψ *is given by* (4.1) *then,*

$$\frac{w_1 - \bar{w_2}}{i} - \eta_1\bar{\eta_2} = (1 + z_1\bar{z_2} - \zeta_1\bar{\zeta_2})\psi'(Z_1)\overline{\psi'(Z_2)}, \tag{4.4}$$

$$(1 + z_1\bar{z_2} - \zeta_1\bar{\zeta_2}) = \left(\frac{w_1 - \bar{w_2}}{i} - \eta_1\bar{\eta_2}\right)(\psi^{-1})'(W_1)\overline{(\psi^{-1})'(W_2)}. \tag{4.5}$$

In [2] it was proved that the invariant measure under the action of $SU(1,1|1)$ has the form

$$\frac{1}{\pi}(1 - z\bar{z} - \theta\bar{\theta})^{-1} dz d\bar{z} d\theta d\bar{\theta}. \tag{4.6}$$

By Lemma 4.1 and Equation (4.3), we have that the invariant measure of the unit super disk corresponds to the measure of the super upper half-plane as follows

$$\frac{1}{\pi}(1 - z\bar{z} - \theta\bar{\theta})^{-1} dz d\bar{z} d\theta d\bar{\theta} = \frac{1}{\pi}(2\operatorname{Im} w - \eta\bar{\eta})^{-1} dw d\bar{w} d\eta d\bar{\eta}, \tag{4.7}$$

where $\psi(z,\theta) = (w,\eta)$. Moreover, the measure of the upper half-plane is invariant under the action of the group $SL_{(2|2)}(\mathbb{R})$.

For $F \in C(\overline{\mathbb{H}}^{1|1})$, the *super Toeplitz operator* $T_F^{(\nu)}$ on $H_\nu^2(\mathbb{H}^{1|1})$ is defined as

$$T_F^{(\nu)}\Psi = P^{(\nu)}(F\Psi), \tag{4.8}$$

where $P^{(\nu)}$ denotes the orthogonal projection onto $H_\nu^2(\mathbb{H}^{1|1})$.

Theorem 4.2. *With respect to the decomposition $\Psi = \psi_0 + \eta\,\psi_1$, the super-Toeplitz operator $T_F^{(\nu)}$ on $H_\nu^2(\mathbb{H}^{1|1})$ is given by the block matrix*

$$T_F^{(\nu)} = \begin{pmatrix} T_\nu^\nu\left(f_{00} + \frac{2\operatorname{Im}(w)}{\nu-1} f_{11}\right) & T_\nu^{\nu+1}\left(\frac{2\operatorname{Im}(w)}{\nu-1} f_{10}\right) \\ T_{\nu+1}^\nu(f_{01}) & T_{\nu+1}^{\nu+1}(f_{00}) \end{pmatrix}. \tag{4.9}$$

Here for $0 \le i,j \le 1$, $T_{\nu+i}^{\nu+j}(f)$ denotes the Toeplitz type operator from $H_{\nu+j}^2(\mathbb{H})$ to $H_{\nu+i}^2(\mathbb{H})$ defined by

$$T_{\nu+i}^{\nu+j}(f)\,\psi := P_{\nu+i}(f\psi),$$

for $\psi \in H_{\nu+j}^2(\mathbb{H})$ and $P_{\nu+i}$ is the orthogonal projection from $L_{\nu+i}^2(\mathbb{H})$ onto $H_{\nu+i}^2(\mathbb{H})$.

5. Toeplitz operators on the super upper half-plane: Super parabolic case

The super group $\mathbb{R}^{1|1}$ can be seen as the subgroup of $SL_{(2|2)}(\mathbb{R})$ consisting of all matrices

$$M(h,\tau) = \begin{pmatrix} 1 & h & \tau \\ 0 & 1 & 0 \\ 0 & -i\tau & 1 \end{pmatrix},$$

where $\tau^* = \tau$ and $h \in \mathbb{R}$.

The action of $\mathbb{R}^{1|1}$ on $\mathbb{H}^{1|1}$ is given by $M(h,\tau)(z,\zeta) = (w,\eta)$ where

$$w = z + h + \tau\zeta \text{ and } \eta = -i\tau + \zeta.$$

Let f be a smooth function on the super upper half-plane. If f is invariant under the action of $\mathbb{R}^{1|1}$, then f has the form

$$f(z,\zeta) = f_0(y) + f_1(y)\zeta + f_1(y)\bar{\zeta} + \frac{f_0'(y)}{2}\bar{\zeta}\zeta. \tag{5.1}$$

In [4] it was proved that $\psi \in H^2_\nu(\mathbb{H})$ has a representation in the form of the Fourier integral

$$\psi(z) = \frac{1}{\sqrt{\Gamma(\nu)}} \int_{\mathbb{R}_+} t^{\frac{\nu-1}{2}} \phi(t) e^{itz} dt, \tag{5.2}$$

where $\phi \in L_2(\mathbb{R}_+)$.

Assume that F is a bounded super function which is invariant under the action of $\mathbb{R}^{1|1}$ of the form (5.1), then we have that the Toeplitz operator with symbol F evaluated on each element Ψ in $H^2_\nu(\mathbb{H}^{1|1})$,

$$\Psi(w,\eta) = \int_{\mathbb{R}_+} \left(\frac{t^{\frac{\nu-1}{2}}}{\sqrt{\Gamma(\nu)}} \right) \phi_0(t) e^{itw} dt + \int_{\mathbb{R}_+} \left(\frac{t^{\frac{\nu}{2}}\eta}{\sqrt{\Gamma(\nu+1)}} \right) \phi_1(t) e^{itw} dt,$$

where ϕ_0 and ϕ_1 are elements of $L_2(\mathbb{R}_+)$, has the following form

$$\begin{aligned} T^\nu_F(\Psi)(w,\eta) &= \frac{1}{\sqrt{\Gamma(\nu)}} \int_{\mathbb{R}_+} t^{\frac{\nu-1}{2}} \left(\phi_0(t)\gamma_{[f_0,\nu]}(t) + \frac{\phi_1(t)\gamma_{[f_1,\nu]}(t)}{t^{\frac{1}{2}}\nu^{\frac{1}{2}}} \right) e^{itw} dt \\ &+ \frac{\eta}{\sqrt{\Gamma(\nu+1)}} \int_{\mathbb{R}_+} t^{\frac{\nu}{2}} \left(\frac{\nu^{\frac{1}{2}}\phi_0(t)\gamma_{[f_1,\nu]}(t)}{t^{\frac{1}{2}}} + \phi_1(t)\gamma_{[f_0,\nu)]}(t) \right) e^{itw} dt \end{aligned} \tag{5.3}$$

where

$$\gamma_{[f,\nu]}(t) = \frac{t^\nu}{\Gamma(\nu)} \int_{\mathbb{R}_+} f(\frac{y}{2}) e^{-ty} y^{\nu-1} dy, \text{ with } t > 0.$$

It is clear that if f is bounded then $\gamma_{[f,\nu]}(t)$ is also bounded over $\mathbb{R}_+$. Therefore, $\gamma_{[f_i,\nu]}\phi_j \in L_2(\mathbb{R}_+)$ for $i,j = 0,1$.

On the other hand, if we consider the elements of $H^2_\nu(\mathbb{H}^{1|1})$ of the form

$$\Psi_+(w,\eta) = \int_{\mathbb{R}_+} \left(\frac{t^{\frac{\nu-1}{2}}}{\sqrt{\Gamma(\nu)}} + \nu^{\frac{1}{2}} \frac{t^{\frac{\nu}{2}}\eta}{\sqrt{\Gamma(\nu+1)}} \right) \phi(t) e^{itw} dt,$$

$$\Psi_-(w,\eta) = \int_{\mathbb{R}_+} \left(\frac{t^{\frac{\nu-1}{2}}}{\sqrt{\Gamma(\nu)}} - \nu^{\frac{1}{2}} \frac{t^{\frac{\nu}{2}}\eta}{\sqrt{\Gamma(\nu+1)}} \right) \phi(t) e^{itw} dt,$$

where $\phi \in L_2(\mathbb{R})$, it is clear that $\{\Psi_+, \Psi_-\}$ generate $H^2_\nu(\mathbb{H}^{1|1})$ and the Toeplitz operator with symbol F over Ψ_+ and Ψ_- is a multiplication operator

$$T^\nu_F(\Psi_+)(w,\eta) = \int_{\mathbb{R}_+} \left(\frac{t^{\frac{\nu-1}{2}}}{\sqrt{\Gamma(\nu)}} + \nu^{\frac{1}{2}} \frac{t^{\frac{\nu}{2}}\eta}{\sqrt{\Gamma(\nu+1)}} \right) [\gamma_{[f_0,\nu]}(t) + \gamma_{[f_1,\nu]}(t)]\phi(t) e^{itw} dt,$$

$$T^\nu_F(\Psi_-)(w,\eta) = \int_{\mathbb{R}_+} \left(\frac{t^{\frac{\nu-1}{2}}}{\sqrt{\Gamma(\nu)}} - \nu^{\frac{1}{2}} \frac{t^{\frac{\nu}{2}}\eta}{\sqrt{\Gamma(\nu+1)}} \right) [\gamma_{[f_0,\nu]}(t) - \gamma_{[f_1,\nu]}(t)]\phi(t) e^{itw} dt,$$

where $t > 0$.

Theorem 5.1. *The Toeplitz algebra generated by all super Toeplitz operators whose symbols are invariant under the action of* $\mathbb{R}^{1|1}$ *is commutative.*

6. Toeplitz operators on the super upper half-plane: Quasi-hyperbolic and Quasi-parabolic cases

6.1. The super groups $\mathbb{R} \times S^1$ and $\mathbb{R}_+ \times S^1$

The super group $\mathbb{R} \times S^1$ can be seen as the group of matrices

$$A(h,t) = \begin{pmatrix} e^{it} & e^{it}h & 0 \\ 0 & e^{it} & 0 \\ 0 & 0 & e^{2it} \end{pmatrix},$$

where $t, h \in \mathbb{R}$.

Since $A(h,t) \in \mathbb{R} \times S^1$, $A(h,t)^* I A(h,t) = I$, where I is given in (3.3). In consequence we have that $A(h,t) \in SL_{(2|2)}(\mathbb{R})$.

Theorem 6.1. *Let f be a smooth function on the super Lobachevsky plane (super upper half-plane). If f is invariant under the action of $\mathbb{R} \times S^1$, then f has the form*

$$f(z,\zeta) = f_{00}(y) + f_{11}(y)\bar{\zeta}\zeta,$$

where $y = \mathrm{Im}(z)$.

Proof. A smooth function f on the super upper half-plane has the form

$$f(z,\zeta) = f_{00}(z) + f_{10}(z)\zeta + f_{01}(z)\bar{\zeta} + f_{11}(z)\bar{\zeta}\zeta,$$

where f_{ij} are smooth functions.

The function $f(z,\zeta)$ is invariant under the action of $\mathbb{R} \times S^1$ if and only if

$$f(z,\zeta) = f(z+h, e^{it}\zeta).$$

In particular $f(z,\zeta)$ must be invariant under the action of elements in $\mathbb{R} \times S^1$ of the form $A(h,0)$, i.e., it must satisfy the following equation

$$\begin{aligned} &f_{00}(z) + f_{10}(z)\zeta + f_{01}(z)\bar{\zeta} + f_{11}(z)\bar{\zeta}\zeta \\ &\quad = f_{00}(z+h) + f_{10}(z+h)\zeta + f_{01}(z+h)\bar{\zeta} + f_{11}(z+h)\bar{\zeta}\zeta. \end{aligned}$$

Last equation implies that every function f_{ij} depends only on $y = \mathrm{Im}(z)$. Thus $f(z,\zeta)$ has the form

$$f(z,\zeta) = f_{00}(y) + f_{10}(y)\zeta + f_{01}(y)\bar{\zeta} + f_{11}(y)\bar{\zeta}\zeta. \tag{6.1}$$

Additionally $f(z,\zeta)$ must be invariant under the action of all elements in $\mathbb{R} \times S^1$ of the form $A(0,t)$, i.e.,

$$f(z,\zeta) = f(z, e^{it}\zeta).$$

Using last equation and equation (6.1) we get

$$f_{00}(y) + f_{10}(y)\zeta + f_{01}(y)\bar{\zeta} + f_{11}(y)\bar{\zeta}\zeta = f_{00}(y) + f_{10}(y)\zeta e^{it} + f_{01}(y)\bar{\zeta}e^{-it} + f_{11}(y)\bar{\zeta}\zeta.$$

Then $f_{10} = f_{01} = 0$. □

The super group $\mathbb{R}_+ \times S^1$ can be seen as the group of matrices

$$B(a,t) = \begin{pmatrix} ae^{it} & 0 & 0 \\ 0 & a^{-1}e^{it} & 0 \\ 0 & 0 & e^{2it} \end{pmatrix},$$

where $t \in \mathbb{R}$ and $a \in \mathbb{R}_+$.

Since $B(a,t) \in \mathbb{R}_+ \times S^1$ then $B(a,t)^* IB(a,t) = I$, where I is given in (3.3). Therefore $B(a,t) \in SL_{(2|2)}(\mathbb{R})$.

Theorem 6.2. *Let f be a smooth function on the super upper half-plane. If f is invariant under the action of $\mathbb{R}_+ \times S^1$, then f has the form*

$$f(z,\zeta) = f_{00}(\vartheta) + \frac{f_{11}(\vartheta)\bar{\zeta}\zeta}{r},$$

where $\vartheta = \arg(z)$ and $r = |z|$.

Proof. A smooth function f on the super upper half-plane has the form

$$f(z,\zeta) = f_{00}(z) + f_{10}(z)\zeta + f_{01}(z)\bar{\zeta} + f_{11}(z)\bar{\zeta}\zeta,$$

where f_{ij} are smooth functions.

A function $f(z,\zeta)$ is invariant under the action of $\mathbb{R} \times S^1$ on the super plane if and only if

$$f(z,\zeta) = f(a^2 z, ae^{it}\zeta).$$

The function $f(z,\zeta)$ must be invariant under the action of the elements of $\mathbb{R} \times S^1$ of the form $B(a,0)$. Then f satisfies the following equation

$$\begin{aligned} &f_{00}(z) + f_{10}(z)\zeta + f_{01}(z)\bar{\zeta} + f_{11}(z)\bar{\zeta}\zeta \\ &\quad = f_{00}(a^2 z) + f_{10}(a^2 z)a\zeta + f_{01}(a^2 z)a\bar{\zeta} + f_{11}(a^2 z)a^2\bar{\zeta}\zeta. \end{aligned}$$

By the last equation $f(z,\zeta)$ has the form

$$f(z,\zeta) = f_{00}(\vartheta) + \frac{f_{10}(\vartheta)}{r^{1/2}}\zeta + \frac{f_{01}(\vartheta)}{r^{1/2}}\bar{\zeta} + \frac{f_{11}(\vartheta)}{r}\bar{\zeta}\zeta,$$

where ϑ is the argument of z.

The function $f(z,\zeta)$ must be also invariant under the action of the elements in $\mathbb{R} \times S^1$ of the form $B(1,t)$. Then

$$f(z,\zeta) = f(z, e^{it}\zeta) = f_{00}(\vartheta) + \frac{f_{10}(\vartheta)}{r^{1/2}}e^{it}\zeta + \frac{f_{01}(\vartheta)}{r^{1/2}}e^{-it}\bar{\zeta} + \frac{f_{11}(\vartheta)}{r}\bar{\zeta}\zeta.$$

Thus $f_{10} = f_{01} = 0$. □

6.2. Toeplitz operators whose symbols are invariant under the action of $\mathbb{R} \times S^1$
Consider a Toeplitz operator with $\mathbb{R} \times S^1$-invariant symbol $F(z,\zeta) = f_{00}(y) + f_{11}(y)\bar{\zeta}\zeta$, where $y = \mathrm{Im}(z)$. Thus this operator has the form

$$\begin{aligned} T_F^{(\nu)}(\Psi)(w,\eta) &= \frac{\nu-1}{\pi}\int_{\mathbb{H}} \left(f_{00}(y) + \frac{2yf_{11}(y)}{\nu-1}\right)\psi_0(z)(2\,\mathrm{Im}(z))^{\nu-2}\left(\frac{w-\bar{z}}{i}\right)^{-\nu} dz \\ &\quad + \left(\frac{\nu}{\pi}\int_{\mathbb{H}} f_{00}(y)\psi_1(z)(2\,\mathrm{Im}(z))^{\nu-1}\left(\frac{w-\bar{z}}{i}\right)^{-(\nu+1)} dz\right)\eta \\ &= T_\nu^\nu\left(f_{00}(y) + \frac{2yf_{11}(y)}{\nu-1}\right)[\psi_0](w) + T_{\nu+1}^{\nu+1}(f_{00}(y))[\psi_1](w)\eta, \quad (6.2) \end{aligned}$$

where $\Psi = \psi_0 + \psi_1\zeta \in H_\nu^2(\mathbb{H}^{1|1})$.

Theorem 6.3. *Let F be a $\mathbb{R} \times S^1$-invariant function, then the Toeplitz operator $T_F^{(\nu)}$ is a multiplication operator.*

Proof. Let $\Psi = \psi_0 + \psi_1\zeta \in H_\nu^2(\mathbb{H}^{1|1})$, since $\psi_0 \in H_\nu^2(\mathbb{H})$ and $\psi_1 \in H_{\nu+1}^2(\mathbb{H})$ and using (5.2) they can be written in the form

$$\psi_0(z) = \frac{1}{\sqrt{\Gamma(\nu)}}\int_{\mathbb{R}_+} t^{\frac{\nu-1}{2}}\phi_0(t)e^{itz}dt, \quad (6.3)$$

$$\psi_1(z)\zeta = \frac{1}{\sqrt{\Gamma(\nu+1)}}\int_{\mathbb{R}_+} t^{\frac{\nu}{2}}\phi_1(t)e^{itz}dt\zeta, \quad (6.4)$$

where $\phi_0, \phi_1 \in L_2(\mathbb{R}_+)$.
In [4] it was proved that the operators $T_\nu^\nu\left(f_{00}(y) + \frac{2yf_{11}(y)}{\nu-1}\right)$ and $T_{\nu+1}^{\nu+1}(f_{00}(y))$ are multiplication operators. In fact,

$$\begin{aligned} T_\nu^\nu&\left(f_{00}(y) + \frac{2yf_{11}(y)}{\nu-1}\right)[\psi_0](z) \\ &= \frac{1}{\sqrt{\Gamma(\nu)}}\int_{\mathbb{R}_+} t^{\frac{\nu-1}{2}}\gamma_{[f_{00}(y)+\frac{2yf_{11}(y)}{\nu-1},\nu-1]}(t)\phi_0(t)e^{itz}dt, \\ T_{\nu+1}^{\nu+1}(f_{00}(y))[\psi_1](z) &= \frac{1}{\sqrt{\Gamma(\nu+1)}}\int_{\mathbb{R}_+} t^{\frac{\nu}{2}}\gamma_{[f_{00},\nu]}(t)\phi_1(t)e^{itz}dt, \end{aligned}$$

where

$$\gamma_{[f,\nu]}(t) = \frac{t^\nu}{\Gamma(\nu)}\int_{\mathbb{R}_+} f(\frac{y}{2})e^{-ty}y^{\nu-1}dy, \text{ with } t > 0. \quad (6.5)$$

□

Theorem 6.4. *The algebra generated by Toeplitz operators with invariant symbols under the action of $\mathbb{R} \times S^1$ is commutative.*

Proof. Let $F(z,\zeta) = f_{00}(y) + f_{11}(y)\bar{\zeta}\zeta$ and $G(z,\zeta) = g_{00}(y) + g_{11}(y)\bar{\zeta}\zeta$ be $\mathbb{R} \times S^1$-invariant functions, if we evaluate $T_G^{(\nu)} \circ T_F^{(\nu)}$ on the elements of the form (6.3) and (6.4) then we have

$$T_G^{(\nu)}(T_F^{(\nu)})(\psi_0(w))(z,\zeta)$$
$$= \frac{1}{\sqrt{\Gamma(\nu)}} \int_{\mathbb{R}_+} t^{\frac{\nu-1}{2}} \gamma_{[f_{00}(y)+\frac{2yf_{11}(y)}{\nu-1},\nu-1]} \gamma_{[g_{00}(y)+\frac{2yg_{11}(y)}{\nu-1},\nu-1]} \phi_0(t) e^{itz} dt,$$
$$T_G^{(\nu)}(T_F^{(\nu)})(\psi_1(w)\eta)(z,\zeta) = \frac{1}{\sqrt{\Gamma(\nu+1)}} \int_{\mathbb{R}_+} t^{\frac{\nu}{2}} \gamma_{[f_{00},\nu]} \gamma_{[g_{00},\nu]} \phi_1(t) e^{itz} dt \zeta.$$

Therefore, the C^*-algebra generated by all Toeplitz operators with $\mathbb{R} \times S^1$-invariant symbols is commutative. □

Corollary 6.5. *Let F be a $\mathbb{R} \times S^1$-invariant function, then the spectrum of the Toeplitz operator $T_a^{(\nu)}$ is given by*

$$spT_F^{(\nu)} = \left\{ \gamma_{[f_{00},\nu-1]}(t) + \frac{1}{t}\gamma_{[f_{11},\nu]}(t) \text{ and } \gamma_{[f_{00},\nu]}(t) : \ t \in \mathbb{R}_+ \right\},$$

where $\gamma_{[f,\nu]}(t)$ is given by (6.5).

6.3. Toeplitz operator with invariant symbols under the action of $\mathbb{R}_+ \times S^1$

Consider a Toeplitz operator with $\mathbb{R}_+ \times S^1$-invariant symbol $F(z,\zeta) = f_{00}(\vartheta) + r^{-1} f_{11}(\vartheta)\bar{\zeta}\zeta$, then the operator $T_F^{(\nu)}$ has the form

$$T_F^{(\nu)}(\Psi)(w,\eta) = \frac{\nu-1}{\pi} \int\limits_{\mathbb{H}} \left(f_{00}(\vartheta) + \frac{2yf_{11}(\vartheta)}{r(\nu-1)} \right) \psi_0(z)(2\,\mathrm{Im}(z))^{\nu-2} \left(\frac{w-\bar{z}}{i}\right)^{-\nu} dz$$
$$+ \left(\frac{\nu}{\pi} \int\limits_{\mathbb{H}} f_{00}(\vartheta)\psi_1(z)(2\,\mathrm{Im}(z))^{\nu-1} \left(\frac{w-\bar{z}}{i}\right)^{-(\nu+1)} dz \right) \eta$$
$$= T_\nu^\nu \left(f_{00}(\vartheta) + \frac{2\sin(\vartheta) f_{11}(\vartheta)}{\nu-1} \right) [\psi_0](w) + T_{\nu+1}^{\nu+1}(f_{00}(\vartheta))[\psi_1](w)\eta,$$

where $\Psi = \psi_0 + \psi_1\zeta \in H_\nu^2(\mathbb{H}^{1|1})$ and $T_\nu^\nu\left(f_{00}(\vartheta) + \frac{2\sin(\vartheta)f_{11}(\vartheta)}{\nu-1}\right)$, $T_{\nu+1}^{\nu+1}(f_{00}(\vartheta))$ are the Toeplitz operators over $H_\nu^2(\mathbb{H})$ and $H_{\nu+1}^2(\mathbb{H})$, respectively.

Theorem 6.6. *Let F be a $\mathbb{R}_+ \times S^1$-invariant function, then the Toeplitz operator $T_F^{(\nu)}$ is a multiplication operator.*

Proof. In [5] it was proved that if $\psi \in H_\nu^2(\mathbb{H})$ then it has the form

$$\psi(z) = \frac{1}{\sqrt{2}} \int_{\mathbb{R}} z^{it-(\nu/2)} \frac{|\Gamma(\frac{\nu}{2}+it)|}{\sqrt{\pi\Gamma(\nu)}} e^{\pi t/2} \phi(t) dt,$$

where $\phi \in L_2(\mathbb{R})$.

Let $\Psi = \psi_0 + \psi_1\zeta \in H^2_\nu(\mathbb{H}^{1|1})$, since $\psi_0 \in H^2_\nu(\mathbb{H})$ and $\psi_1 \in H^2_{\nu+1}(\mathbb{H})$ they have the representation

$$\psi_0(z) = \frac{1}{\sqrt{2}}\int_{\mathbb{R}} z^{it-(\nu/2)}\frac{|\Gamma(\frac{\nu}{2}+it)|}{\sqrt{\pi\Gamma(\nu)}}e^{\pi t/2}\phi_0(t)dt, \tag{6.6}$$

$$\psi_1(z)\zeta = \frac{1}{\sqrt{2}}\int_{\mathbb{R}} z^{it-(\nu+1)/2}\frac{|\Gamma(\frac{\nu+1}{2}+it)|}{\sqrt{\pi\Gamma(\nu+1)}}e^{\pi t/2}\phi_1(t)dt\zeta. \tag{6.7}$$

In [5] it was proved that the operators $T^\nu_\nu\left(f_{00}(y) + \frac{2yf_{11}(y)}{\nu-1}\right)$ and $T^{\nu+1}_{\nu+1}(f_{00}(y))$ are multiplication operators. In fact,

$$\begin{aligned} &T^\nu_\nu\left(f_{00}(\vartheta) + \frac{2\sin(\vartheta)f_{11}(\vartheta)}{\nu-1}\right)[\psi_0](z) \\ &\quad= \frac{1}{\sqrt{2}}\int_{\mathbb{R}} z^{it-(\nu/2)}\frac{|\Gamma(\frac{\nu}{2}+it)|}{\sqrt{\pi\Gamma(\nu)}}e^{\pi t/2}\beta_{[f_{00}(\vartheta)+\frac{2\sin(\vartheta)f_{11}(\vartheta)}{\nu-1},\nu-1]}(t)\phi_0(t)dt,\end{aligned}$$

$$T^{\nu+1}_{\nu+1}(f_{00}(\vartheta))[\psi_1](z) = \frac{1}{\sqrt{2}}\int_{\mathbb{R}} z^{it-(\nu+1)/2}\frac{|\Gamma(\frac{\nu+1}{2}+it)|}{\sqrt{\pi\Gamma(\nu+1)}}e^{\pi t/2}\beta_{[f_{00}(\vartheta),\nu]}(t)\phi_1(t)dt,$$

where

$$\beta_{[f(\vartheta),\nu]}(t) = 2^{\nu-1}\nu\left(\frac{|\Gamma(\frac{\nu+1}{2}+it)|}{\sqrt{\pi\Gamma(\nu+1)}}\right)^2 e^{\pi t}\int_0^\pi f(\vartheta)e^{-2t\vartheta}\sin^{\nu-1}(\vartheta)d\vartheta \text{ and } t\in\mathbb{R}. \tag{6.8}$$

□

Theorem 6.7. *The algebra generated by Toeplitz operators with invariant symbols under the action of $\mathbb{R}_+\times S^1$ is commutative.*

Proof. If $F(z,\zeta) = f_{00}(\vartheta) + r^{-1}f_{11}(\vartheta)\bar{\zeta}\zeta$ and $G(z,\zeta) = g_{00}(\vartheta) + r^{-1}g_{11}(\vartheta)\bar{\zeta}\zeta$ are $\mathbb{R}\times S^1$-invariant functions then we have

$$\begin{aligned} T^{(\nu)}_G(T^{(\nu)}_F)(\psi_0(w))(z,\zeta) &= \frac{1}{\sqrt{2}}\int_{\mathbb{R}} z^{it-(\nu/2)}\frac{|\Gamma(\frac{\nu}{2}+it)|}{\sqrt{\pi\Gamma(\nu)}}e^{\pi t/2} \\ &\quad\times \beta_{[f_{00}(\vartheta)+\frac{2\sin(\vartheta)f_{11}(\vartheta)}{\nu-1},\nu-1]}(t) \\ &\quad\times \beta_{[g_{00}(\vartheta)+\frac{2\sin(\vartheta)g_{11}(\vartheta)}{\nu-1},\nu-1]}(t)\phi_0(t)dt,\end{aligned}$$

$$\begin{aligned} T^{(\nu)}_G(T^{(\nu)}_F)(\psi_1(w)\eta)(z,\zeta) &= \frac{1}{\sqrt{2}}\int_{\mathbb{R}} z^{it-(\nu+1)/2}\frac{|\Gamma(\frac{\nu+1}{2}+it)|}{\sqrt{\pi\Gamma(\nu+1)}}e^{\pi t/2}\beta_{[f_{00}(\vartheta),\nu]}(t) \\ &\quad\times \beta_{[g_{00}(\vartheta),\nu]}(t)\phi_1(t)dt\ \zeta.\end{aligned}$$

□

Corollary 6.8. *Let $F(z,\zeta) = f_{00}(\vartheta)+r^{-1}f_{11}(\vartheta)\bar{\zeta}\zeta$ be a $\mathbb{R}_+\times S^1$-invariant function, then the spectrum of the Toeplitz operator $T^{(\nu)}_F$ is given by*

$$spT^{(\nu)}_F = \left\{\beta_{[f_{00},\nu-1]}(t) + \left|\frac{\Gamma(\frac{\nu}{2}+it)}{\Gamma(\frac{\nu+1}{2}+it)}\right|^2\beta_{[f_{11},\nu]}(t) \text{ and } \beta_{[f_{00},\nu]}(t) : t\in\mathbb{R}\right\},$$

where $\beta_{[f,\nu]}(t)$ is given by (6.8).

Until now we have found five commutative algebras of super Toeplitz operators whose symbols are invariant under the action of super-subgroups of isometries. At the same time in the case of classic Toeplitz operators there are only three different cases. The following question remains open: Are there more cases of commutative algebras of super Toeplitz operators? This question is closely related to the classification of all maximal super subgroups of isometries for the unit super disk or the super upper half-plane.

References

[1] F.A. Berezin, *Introduction to Superanalysis*, Reidel, Dordrecht, 1987.

[2] D. Borthwick, S. Klimek, A. Lesniewski, M. Rinaldi, *Super Toeplitz operators and non-perturbative deformation quantization of supermanifolds*, Commun. Math. Phys. **153** (1993), 49–76.

[3] Borthwick D., Klimek S., Lesniewski A., and Rinaldi M., *Matrix Cartan Superdomains, super upper Operators, and Quantization*, ArXiv:help-th/9406050 v1, 9 Jun., 1994.

[4] S. Grudsky, A. Karapetyants, N. Vasilevski, *Dynamics of properties of Toeplitz Operators on the upper half-plane: Parabolic case*, J. Operator Theory 52(1) (2004), 185–204.

[5] S. Grudsky, A. Karapetyants, N. Vasilevski,*Dynamics of properties of Toeplitz Operators on the upper half-plane: Hyperbolic case*, Bol. Soc. Mat. Mexicana 10 (2004), 119–138.

[6] S. Grudsky, A. Karapetyants, N. Vasilevski, *Toeplitz operators on the unit ball in* $\mathbb{C}^n$ *with radial symbols*, Journal of Operator Theory 49 (2003), 325–346.

[7] S. Grudsky, R. Quiroga, N. Vasilevski,*Commutative* C^**-algebras of Toeplitz operators and quantization on the unit disk*, Journal of Functional Analysis 234 (2006), 1–44.

[8] I.S. Gradshteyn, I.M. Ryzhik *Table of Integrals, Series and Products*, Academic Press, New York, 1980.

[9] H. Hedenmalm, B. Korenblum, K. Zhu, *Theory of Bergman Spaces*, Springer-Verlag, New York, 2000.

[10] M. Loaiza, H. Upmeier, *Toeplitz* C^**-algebras on super Cartan domains*, Revista Matemática Complutense, vol. 21 (2), 489–518.

[11] M. Loaiza, A. Sánchez-Nungaray, *On* C^**-Algebras of super Toeplitz operators with radial symbols*, Operator Theory: Advances and Applications, 2010, Vol. 210, 175–188.

[12] E. Prieto-Zanabria, *Phd Desertation* 2007, CINVESTAV, Mexico.

[13] R. Quiroga-Barranco, N. Vasilevski, *Commutative algebras of Toeplitz operators on Reinhardt domains*, Integral Equations and Operator Theory 59, 67–98, 2007.

[14] R. Quiroga-Barranco R., N. Vasilevski, *Commutative* C^**-algebra of Toeplitz operators on the unit ball, I. Bargmann type transform and spectral representations of Toeplitz operators*, Integral Equations and Operator Theory, 59(3):379–419, 2007.

[15] R. Quiroga-Barranco, N. Vasilevski, *Commutative C*-algebra of Toeplitz operators on the unit ball, II. Geometry of the level sets of symbols*, Integral Equations and Operator Theory, 59(1):89–132, 2008.

[16] A. Sánchez-Nungaray, *Commutative Algebras of Toeplitz Operators on the Supersphere of dimension* (2|2), Phd Desertation 2009, CINVESTAV, Mexico.

[17] A. Sánchez-Nungaray, *Commutative algebras of Toeplitz operators on the super upper half-plane: Super parabolic case*, Operator Theory: Advances and Applications. 2012. Vol. 220, 263–280.

[18] K. Zhu, *Spaces of Holomorphic Functions in the Unit Ball*, Springer-Verlag, New York, 2005.

[19] N. Vasilevski,*Toeplitz operators on the Bergman spaces: Inside-the-domain effects* , Contemp. Math. 289:(2001),79–146,2001.

[20] Vasilevski N., *Bergman space structure, commutative algebras of Toeplitz operators and hyperbolic geometry*, Integral Equations and Operator Theory, 46, 235–251,2003.

M. Loaiza
Departamento de Matemáticas
CINVESTAV
Apartado Postal 14-740 07000
México, MEXICO
e-mail: `mloaiza@math.cinvestav.mx`

A. Sánchez-Nungaray
Facultad de Matemáticas
Universidad Veracruzana
Lomas del estadio s/n, Zona Universitaria
Xalapa, Veracruz, MEXICO
e-mail: `sancheznungaray@gmail.com`

Operator Theory:
Advances and Applications, Vol. 236, 337–356

Computing the Hilbert Transform in Wavelet Bases on Adaptive Grids

Frank Martin and Elias Wegert

Abstract. We propose an algorithm for the efficient numerical computation of the periodic Hilbert transform. The function to be transformed is represented in a basis of spline wavelets in Sobolev spaces. The underlying grids have a hierarchical structure which is locally refined during computation according to the behavior of the involved functions. Under appropriate assumptions, we prove that the algorithm can deliver a result with prescribed accuracy. Several test examples illustrate how the method works in practice.

Mathematics Subject Classification (2010). 65T60, 65R10, 44A15.

Keywords. Hilbert transform, wavelets, adaptive methods.

1. Introduction

The Hilbert transform is one of the fundamental operators in (complex) analysis and plays an important role in various applications, in particular in signal processing (see [14], [16]) and optics (see [16]).

Numerical methods for computing the Hilbert operator are based on different principles. A straightforward approach evaluates the singular integral defining the operator by quadrature formulas. This leads to discrete versions of the Hilbert transform, like the so-called Wittich operator (see [11], [13], [15]).

More efficient methods are based on the fact that the Hilbert transform is a diagonal operator in Fourier space, which allows for applying fast Fourier techniques. As long as the function to be transformed is given (or can be evaluated) on an appropriate uniform grid this is certainly the fastest and most reliable computational method.

On the other hand, there are applications where one is mainly interested in the behavior of the functions on a small subset of the domain, which makes uniform grids inefficient. Similarly, when the functions are mostly 'tame' except

Both authors are supported by the Deutsche Forschungsgemeinschaft, grant We 1704/8-2.

in a small region with 'wild' behavior, high precision approximation would require unreasonably large uniform grids.

These observations motivate the search for appropriate methods which allow one to compute the Hilbert transform of functions on non-uniform grids. There is quite a number of different approaches which have been proposed in the literature to solve related problems, like multipole methods (Dutt and Rokhlin [9], [10]), Fourier methods involving non-equispaced fast Fourier transform NFFT (Kunis and Potts [17]), panel clustering (Börm and Hackbusch [2]), and wavelet methods (Dahmen, Prößdorf and Schneider [6], [7], Rathsfeld [22]).

Our preference to wavelet methods has a simple reason: wavelets also allow for *detecting* the regions where the behavior of the solution requires refinement of the underlying grid. This feature is of special importance in applications of the Hilbert transform to conformal mapping ([25], [26]) and iterative methods for solving nonlinear Riemann–Hilbert problems ([23], [24]). In the iterative process the solution may develop specific 'singularities' with a priori unknown location, so that the grid has to be adapted appropriately. The purpose of this paper is to describe an algorithm which *automatically* refines the grid during the computation of the Hilbert transform.

This paper is a continuation of [20], where we studied the Hilbert transform for biorthogonal spline wavelets on a uniform mesh, following Dahmen, Prößdorf and Schneider [6], [7]. It is organised as follows: In Section 2 we introduce notation and summarise relevant facts from [20]. In the next section we describe the structure of the hierarchical grids which will be constructed in the adaptation process. Section 4 is devoted to the computation of the Hilbert transform in the corresponding wavelet spaces. In Section 5 we propose algorithms which simultaneously compute the Hilbert transform and adapt the grid according to the behavior of the functions involved. In the last section we present the results of some test calculations.

2. Preliminaries

We denote by $L_2(\mathbb{R})$ and $L_2(\mathbb{T})$ the Lebesgue spaces of square integrable functions on the real line $\mathbb{R}$ and the unit circle $\mathbb{T} := \{z \in \mathbb{C} : |z| = 1\}$, respectively. These spaces are equipped with the usual norms and scalar products. The Sobolev spaces of order s with $s \in \mathbb{N}$ are denoted by $W_2^s(\mathbb{R})$ and $W_2^s(\mathbb{T})$, their norms are given by

$$\|u\|_{W_2^s(\mathbb{R})} = \|u\|_{L_2(\mathbb{R})} + \|D_x^s u\|_{L_2(\mathbb{R})}$$
$$\|u\|_{W_2^s(\mathbb{T})} = \|u\|_{L_2(\mathbb{T})} + \|D_t^s u\|_{L_2(\mathbb{T})},$$

respectively, with the differential operators

$$D_x u(x) := \frac{\mathrm{d}}{\mathrm{d}x} u(x), \qquad D_t u\left(\mathrm{e}^{\mathrm{i}\tau}\right) := -\mathrm{i}\frac{\mathrm{d}}{\mathrm{d}\tau} u\left(\mathrm{e}^{\mathrm{i}\tau}\right).$$

Setting $\widetilde{u}(\tau) = u\left(\mathrm{e}^{2\pi\mathrm{i}\tau}\right)$ we identify functions on $\mathbb{T}$ with 1-periodic functions on $\mathbb{R}$. The corresponding periodic Sobolev spaces are denoted by $W_2^s\left(-\frac{1}{2}, \frac{1}{2}\right)$.

The Hilbert transform of a function $u \in L_2(\mathbb{T})$ is given by

$$Hu\left(e^{i\tau}\right) := \frac{1}{2\pi} \int_{-\pi}^{\pi} \cot \frac{\sigma - \tau}{2} u\left(e^{i\sigma}\right) d\sigma,$$

where the integral is interpreted in the Cauchy principal value sense. The Hilbert transform H is a bounded linear operator in $W_2^s(\mathbb{T})$ for every $s \in \mathbb{N}$.

The algorithm which we propose for computing the Hilbert transform is based on an approach which has been analysed in a more general context by Dahmen, Prößdorf and Schneider in [6] and [7]. The starting point is a representation of the function to be transformed in a biorthogonal wavelet basis, generated by the scaling function φ^1 and the wavelet ψ^1,

$$u(x) = \sum_{k=0}^{2^{j_0}-1} a_{j_0,k} \varphi_{j_0,k}^1(x) + \sum_{j=j_0}^{\infty} \sum_{k=0}^{2^j-1} b_{j,k} \psi_{j,k}^1(x). \tag{2.1}$$

Here the scaled translates $\varphi_{j,k}^1$ and $\psi_{j,k}^1$ of φ^1 and ψ^1 are defined by

$$\varphi_{j,k}^1(x) := 2^{\frac{j}{2}} \varphi^1(2^j x - k), \quad \psi_{j,k}^1(x) := 2^{\frac{j}{2}} \psi^1(2^j x - k),$$

respectively. The coefficients $a_{j_0,k}$ and $b_{j,k}$ are given by $a_{j_0,k} = \langle u, \widehat{\varphi}_{j_0,k}^1 \rangle$ and $b_{j,k} = \langle u, \widehat{\psi}_{j,k}^1 \rangle$, respectively, where $\widehat{\varphi}^1$ and $\widehat{\psi}^1$ are the dual functions to φ^1 and ψ^1, and $\langle .,. \rangle$ denotes the scalar product in L_2. Applying the Hilbert transform to this representation and expanding the result in the same or another biorthogonal wavelet basis generated by φ^2 and ψ^2, we get

$$Hu(x) = \sum_{k=0}^{2^{j_0}-1} \overline{a}_{j_0,k} \varphi_{j_0,k}^2(x) + \sum_{j=j_0}^{\infty} \sum_{k=0}^{2^j-1} \overline{b}_{j,k} \psi_{j,k}^2(x)$$

with $\overline{a}_{j_0,k} = \langle Hu, \widehat{\varphi}_{j_0,k}^2 \rangle$ and $\overline{b}_{j,k} = \langle Hu, \widehat{\psi}_{j,k}^2 \rangle$. In order to compute the coefficients $\overline{a}_{j_0,k}$ and $\overline{b}_{j,k}$ of Hu from given coefficients $a_{j_0,k}$ and $b_{j,k}$ of u, one has to evaluate the scalar products

$$\langle H\varphi_{j_0,k}^1, \widehat{\varphi}_{j_0,l}^2 \rangle, \qquad \langle H\psi_{j,k}^1, \widehat{\varphi}_{j_0,l}^2 \rangle, \qquad \langle H\varphi_{j_0,k}^1, \widehat{\psi}_{\nu,l}^2 \rangle, \qquad \langle H\psi_{j,k}^1, \widehat{\psi}_{\nu,l}^2 \rangle. \tag{2.2}$$

Since the wavelets have compact support and a number of vanishing moments, most of the scalar products (2.2) are small. Replacing those entries by zero which have absolute values below some threshold, we get a sparse matrix representation approximating H in the wavelet basis, which allows for efficient computation of the Hilbert transform.

With hindsight to the applications we have in mind, we somewhat modify the above scheme. So we work in Sobolev spaces W_2^s with $s \geq 1$ rather than in L_2. Functions in these spaces are continuous, and we project them to the corresponding (finite-dimensional) wavelet spaces by spline interpolation, instead of using projection with respect to the L_2 scalar product. Because we are mainly interested in pointwise evaluation of the transformed function at the grid points, we choose Dirac functionals in the range space of H.

In this setting, algorithms for the computation of the Hilbert transform of spline wavelets are described and investigated in [19] and [20]. We have chosen biorthogonal B-Spline wavelets to represent the function u in the form (2.1) (see [4] and [8], for example). The scaling function is a centered B-Spline N_m of order m with odd m (typically with $m = 1$ or $m = 3$). These splines are defined recursively by

$$N_0(x) := \begin{cases} 1, & \text{if } -\frac{1}{2} \le x < \frac{1}{2} \\ 0, & \text{elsewhere} \end{cases}, \qquad N_m(x) := \int_{-\frac{1}{2}}^{\frac{1}{2}} N_{m-1}(x-t)\,\mathrm{d}t.$$

We denote the scaling function and the corresponding wavelet by $\widetilde{\varphi}^m$ and $\widetilde{\psi}^m$, respectively, and set

$$\widetilde{\varphi}^m_{j,k}(x) := 2^{\frac{j}{2}}\widetilde{\varphi}^m\left(2^j x - k\right), \quad \widetilde{\psi}^m_{j,k}(x) := 2^{\frac{j}{2}}\widetilde{\psi}^m\left(2^j x - k\right),$$

for $j, k \in \mathbb{Z}$. In order to periodise these functions we define

$$\varphi^m_{j,k}(x) := \sum_{n=-\infty}^{\infty} \widetilde{\varphi}^m_{j,k}(x+n), \quad \psi^m_{j,k}(x) := \sum_{n=-\infty}^{\infty} \widetilde{\psi}^m_{j,k}(x+n).$$

Note that these sums are finite because $\widetilde{\varphi}^m_{j,k}$ and $\widetilde{\psi}^m_{j,k}$ have compact support. We point out that the set

$$\{\varphi^m_{j_0,k} : k = 0, \dots, 2^{j_0} - 1\} \cup \{\psi^m_{j,k} : k = 0, 1, \dots, 2^j - 1; j = j_0, j_0 + 1, \dots\}$$

is a Riesz basis of $W_2^s(\mathbb{T})$, which is reflected by the norm estimates

$$\begin{aligned} \|u\|^2_{W_2^s} &\le C_s \left(\sum_{k=0}^{2^{j_0}-1} |a_{j_0,k}|^2 + \sum_{j=j_0}^{\infty} 2^{2js} \sum_{k=0}^{2^j-1} |b_{j,k}|^2 \right), \\ \|u\|^2_{W_2^s} &\ge c_s \left(\sum_{k=0}^{2^{j_0}-1} |a_{j_0,k}|^2 + \sum_{j=j_0}^{\infty} 2^{2js} \sum_{k=0}^{2^j-1} |b_{j,k}|^2 \right), \end{aligned} \tag{2.3}$$

with Riesz constants C_s and c_s ([3], [5], [18]).

When we are now going to consider non-uniform and adaptive grids, we first of all need an algorithm to calculate the wavelet coefficients. The efficiency of the computations can be improved when these grids have a special structure which is described in the next section.

3. Grid structure

All grids considered in this paper are subsets of an underlying *fine grid* $\mathcal{F}_J$ (with $J \in \mathbb{N}$), which is a uniform subdivision of the interval $\left[-\frac{1}{2}, \frac{1}{2}\right)$ with mesh size

$h := 2^{-J}$. In the following we only admit a subset of grids with special hierarchical structure.

Definition 3.1. The grid $\mathcal{G} \subseteq \mathcal{F}_J$ is said to be *admissible* if the following two conditions are satisfied:

- there exists $j_0 \le J$ such that the *coarse grid*
$$\mathcal{G}_{j_0} := \left\{ l2^{-j_0} : l = -2^{j_0-1}, \ldots, 2^{j_0-1} - 1 \right\}$$
belongs to $\mathcal{G}$, i.e., $\mathcal{G}_{j_0} \subseteq \mathcal{G}$,
- if $x = a\,2^{-j} \in \mathcal{G}$ and a is odd, then the points $x_+ := (a-1)\,2^{-j}$ and $x_- := (a+1)\,2^{-j}$ are in $\mathcal{G}$.

The second condition in Definition 3.1 implies that every point in $\mathcal{G}$ is the midpoint of two points which also belong to $\mathcal{G}$.

We say that a grid point $x \in \mathcal{F}_J$ belongs to the level j if x can be written as $x = a2^{-j}$ with some integer a. Note that all points of level $j < J$ also belong to level $j+1$. The *canonical representation* of a grid point is $x = a2^{-j}$ with a odd. We denote by $\mathcal{G}_j$ the set of all points in $\mathcal{G}$ which belong to the level j but not to the level $j-1$. So $\mathcal{G}_j$ consists of all points in $\mathcal{G}$ with canonical representation $x = a2^{-j}$, and $\mathcal{G}$ is a disjoint union

$$\mathcal{G} = \mathcal{G}_{j_0} \cup \mathcal{G}_{j_0+1} \cup \cdots \cup \mathcal{G}_{J-1} \cup \mathcal{G}_J.$$

Figure 3.1 shows one admissible grid $\mathcal{G}$. The sets $\mathcal{G}_j$ are distinguished by the lengths of the tick marks.

FIGURE 3.1. Example of an admissible mesh

4. Hilbert transform

Let u be a function in the periodic Sobolev space $W_2^s\left(-\frac{1}{2}, \frac{1}{2}\right)$. Assuming that the values of this function are known on an admissible grid $\mathcal{G}$, we would like to compute its Hilbert transform on the same or on another grid $\widetilde{\mathcal{G}}$.

If the values of u were given on the fine grid $\mathcal{F}_J$, we could interpolate them by a linear combination of the scaling functions $\varphi_{J,k}$, perform a wavelet transformation to get the wavelet coefficients, and apply the matrix representation $\mathcal{H}$ of the Hilbert transform H in the wavelet basis. This approach has complexity $\mathcal{O}\left(J2^J\right)$, which can be reduced to $\mathcal{O}\left(2^J\right)$ by using the algorithm from [1] for interpolating the function on the fine grid $\mathcal{F}_J$.

The following algorithm avoids computations on the fine grid, starting the interpolation procedure on the coarse grid and working 'top-down' so that at the finer levels only relevant grid points are involved.

Input:	j_0 ... coarse level
	J ... fine level
	m ... order of the B-Spline
	$\mathcal{G}$ ... admissible grid
	u ... values of function on $\mathcal{G}$
Step 0:	Set $j = j_0$.
Step 1:	Set u to zero on $\{l2^j : l = -2^{j-1}, \dots, 2^{j-1}-1\} \setminus \mathcal{G}_j$ and interpolate the function u on $\mathcal{G}_j$ with the scaling functions $\varphi_{j,k}$, $k = 0, \dots, 2^j - 1$. Denote the interpolating function by $L_j u$.
Step 2:	Apply a wavelet decomposition to $L_j u$.
Step 3:	Set $u := u - L_j u$.
Step 4:	Increase j by one and go to Step 1 as long as $j \leq J$.
Step 5:	Add up the functions $L_j u$ for $j = j_0, \dots, J$.
Output:	wavelet coefficient $a_{j_0,k}$ $(k = 0, \dots, 2^{j_0} - 1)$ and $b_{j,k}$ $(j = j_0, \dots, J-1$, $k = 0, \dots, 2^j - 1)$.

Algorithm 1: Computation of the wavelet coefficient on an admissible grid

As was shown in [19], this algorithm has complexity $\mathcal{O}(JN)$ with $N = |\mathcal{G}|$.

Once the wavelet decomposition of u is determined, we can apply the (truncated) matrix representation $\mathcal{H}$ of the Hilbert transform H in the corresponding bases to obtain the approximate values $\widetilde{H}u$ of Hu on the grid $\widetilde{\mathcal{G}}$.

Input:	$\mathcal{G}$ ... admissible grid where the function u is given
	$\widetilde{\mathcal{G}}$ ... admissible grid where the Hilbert transform is to be computed
	u ... values of function on $\mathcal{G}$
Step 1:	Compute the wavelet coefficient of u on $\mathcal{G}$ by Algorithm 1.
Step 2:	Apply the matrix representation $\mathcal{H}$ to the wavelet coefficients.
Output:	values of $\widetilde{H}u$ on the grid $\widetilde{\mathcal{G}}$

Algorithm 2: Computation of the Hilbert transform on an admissible grid

The application of $\mathcal{H}$ can be done with complexity $\mathcal{O}(J\widetilde{N})$, where $\widetilde{N}$ is the number of grid points of $\widetilde{\mathcal{G}}$, so that the complete algorithm has complexity $\mathcal{O}(J(N + \widetilde{N}))$.

5. Adaptive grid generation

So far we supposed that the grids $\mathcal{G}$ and $\widetilde{\mathcal{G}}$ are given. In this section we describe an algorithm for *adaptive construction* of admissible grids on which the function and its Hilbert transform can be represented with some prescribed accuracy.

Basically the absolute values of the wavelet coefficients tell us where the function is less regular, so that a finer grid is needed in those regions. In the first approach we construct an admissible grid starting at the fine level. We compute the wavelet coefficients of the function to be transformed and decide which grid points can be omitted without increasing the error too much.

5.1. From fine to coarse

We consider a function $u \in W_2^s\left(-\frac{1}{2}, \frac{1}{2}\right)$ with wavelet representation

$$u(x) = \sum_{k=0}^{2^{j_0}-1} a_{j_0,k}\varphi_{j_0,k}^m(x) + \sum_{j=j_0}^{\infty}\sum_{k=0}^{2^j-1} b_{j,k}\psi_{j,k}^m(x).$$

In order to construct an appropriate grid on which u will be represented, we assign to each wavelet $\psi_{j,k}^m$ a set of grid points. Since we are dealing with 1-periodic functions, while the grid points lie in the interval $\left[-\frac{1}{2}, \frac{1}{2}\right)$, we write $y := x \mod 1$ for the point $y \in \left[-\frac{1}{2}, \frac{1}{2}\right)$ with $x - y \in \mathbb{Z}$. Since the support of $\widetilde{\psi}^m$ is contained in $[-m, m+1]$, we now associate the set

$$I_{j,k} := \left\{-\frac{m}{2^j} + \frac{k}{2^j} + \frac{l}{2^{j+1}} \mod 1 : \ l = 0, \ldots, 4m+2\right\}$$

with the wavelet $\psi_{j,k}^m$. According to the size of the wavelet coefficients of u we build an index set

$$I \subseteq \left\{(j,k) : \ k = 0, \ldots, 2^j - 1; j = j_0, \ldots, J-1\right\}$$

and form the grid

$$\mathcal{G} = \mathcal{G}_{j_0} \cup \left(\bigcup_{(j,k)\in I} I_{j,k}\right). \tag{5.1}$$

The decision which elements (j,k) are included in I will depend on a threshold for the associated wavelet coefficients. The next theorem describes how this threshold can be chosen in order to guarantee that the truncation error is sufficiently small.

Theorem 5.1. *Let $m \in 2\mathbb{N}_0 + 1$ and $u \in W_2^s\left(-\frac{1}{2}, \frac{1}{2}\right)$ with*

$$u(x) = \sum_{k=0}^{2^{j_0}-1} a_{j_0,k}\varphi_{j_0,k}^m(x) + \sum_{j=j_0}^{J-1}\sum_{k=0}^{2^j-1} b_{j,k}\psi_{j,k}^m(x).$$

For $\varepsilon > 0$ we define

$$\widetilde{b}_{j,k} := \begin{cases} b_{j,k}, & |b_{j,k}| \geq \delta_s(j) := \dfrac{\varepsilon}{\sqrt{C_s}2^{(2s+1)\frac{j}{2}}\sqrt{J-j_0}} \\ 0, & \text{elsewhere} \end{cases} \tag{5.2}$$

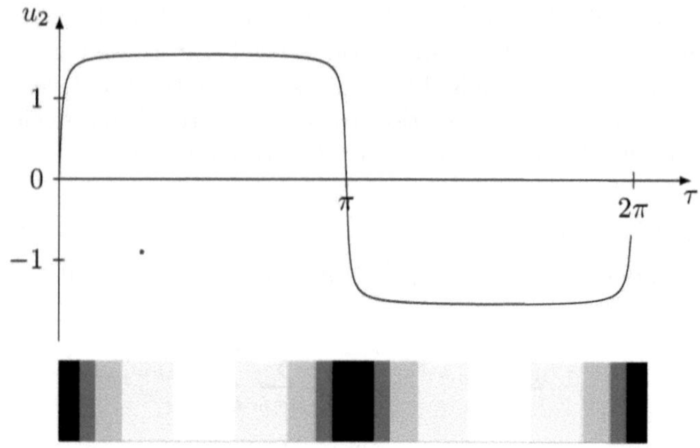

FIGURE 5.1. graph of u_2 and distribution of grid points

with the Riesz constant C_s from (2.3) *and*

$$\widetilde{u}(x) := \sum_{k=0}^{2^{j_0}-1} a_{j_0,k}\varphi_{j_0,k}^m(x) + \sum_{j=j_0}^{J-1}\sum_{k=0}^{2^j-1} \widetilde{b}_{j,k}\psi_{j,k}^m(x).$$

Then we have for $0 \le s < m + \frac{1}{2}$

$$\|u - \widetilde{u}\|_{W_2^s} \le \varepsilon.$$

Proof. Using the Riesz inequality (2.3) we can estimate

$$\begin{aligned}\|u - \widetilde{u}\|_{W_2^s}^2 &\le C_s \sum_{j=j_0}^{J-1} 2^{2js} \sum_{k=0}^{2^j-1} \left|b_{j,k} - \widetilde{b}_{j,k}\right|^2 \le C_s \sum_{j=j_0}^{J-1} 2^{2js} \sum_{k=0}^{2^j-1} \delta_s(j)^2 \\ &= C_s \sum_{j=j_0}^{J-1} 2^{2js} 2^j \delta_s(j)^2 = C_s \sum_{j=j_0}^{J-1} 2^{2js} 2^j \frac{\varepsilon^2}{C_s 2^{(2s+1)j}(J-j_0)} = \varepsilon^2. \quad \square\end{aligned}$$

In order to build the index set I we consider the wavelet coefficient $b_{j,k}$ of a function $u \in W_2^s\left(-\frac{1}{2}, \frac{1}{2}\right)$. The index (j,k) is added to the set I if the absolute value $|b_{j,k}|$ is greater than the threshold $\delta_s(j)$ in (5.2),

$$I := \{(j,k) : |b_{j,k}| > \delta_s(j)\}. \tag{5.3}$$

If necessary, we enlarge the set I in order to get an admissible mesh.

Figure 5.1 visualises the grid generated for the function

$$u_2(t) = \operatorname{Re}\left(\mathrm{i}\log\left(\frac{1-pt}{1+pt}\right)\right), \quad t \in \mathbb{T},$$

with $p = 0.97$, $m = 3$, $j_0 = 8$, $J = 12$ and $\delta_0(j) = 10^{-7}$. Beneath the graph of the function the distribution of the grid points is depicted. Darker colors correspond to a higher density of grid points.

5.2. From coarse to fine

The approach described above has a serious drawback: the starting point is the wavelet representation of the function u on the fine mesh $\mathcal{F}_J$. So we first have to compute *all* coefficients in order to decide afterwards which can be omitted. Instead of using this 'bottom-up' strategy, it would be more efficient to work 'top-down', beginning on the coarse level and adding the required points level by level. This is achieved by the following algorithm, which simultaneously refines the grid and computes approximations to the Hilbert transform.

Input: j_0 ... coarse level
J ... fine level
m ... order of the B-Splines
ε ... accuracy
u ... function in $W_2^s\left(-\frac{1}{2}, \frac{1}{2}\right)$

Step 0: Set $\mathcal{G} := \mathcal{G}_{j_0}$.

Step 1: Add the point $x \in \mathcal{F}_J$ to the grid $\mathcal{G}$ if $x_- \in \mathcal{G}$ and $x_+ \in \mathcal{G}$.

Step 2: Evaluate the function u on $\mathcal{G}$ and compute the Hilbert transform Hu by Algorithm 2.

Step 3: Update the grid $\mathcal{G}$ according to (5.1) and (5.3) with respect to the wavelet coefficients of u and Hu.

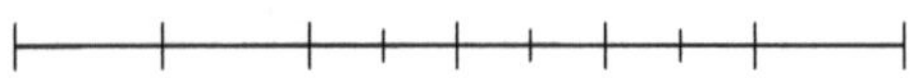

Step 4: Go to Step 1, but break after $J - j_0 + 1$ steps.

Output: admissible grid $\mathcal{G}$ and approximate Hilbert transform $\widetilde{H}u$ on $\mathcal{G}$

Algorithm 3: Adaptive computation of the Hilbert transform

Of course one cannot expect that this algorithm works well for all functions. If, for example, the function u vanishes on the coarse and the next level, the algorithm cannot detect whether u vanishes everywhere or just lives on the finer levels.

On the other hand, there are classes of functions which are more or less uniformly distributed over all levels. What is meant by this vague formulation is made precise in the following definition. Here the parameter M depends on the wavelets considered and has to be chosen such that the support of $\psi_{j,0}$ has a non-empty intersection with the supports of $\psi_{j-1,l}$ for $l = -M, \ldots, M$. For $m = 1$ and $m = 3$ we have $M = 3$ and $M = 6$, respectively.

Definition 5.2. A function $u \in W_2^s\left(-\frac{1}{2}, \frac{1}{2}\right)$ is called *q-balanced* if its wavelet coefficients $b_{j,k}$ fulfill the following condition for $j = j_0 + 1, \ldots,\; k = 0, \ldots, 2^j - 1$,

$$|b_{j,k}|^2 \le \frac{q}{2}\frac{1}{2M+1} \sum_{l=\lfloor \frac{k}{2} \rfloor - M}^{l=\lfloor \frac{k}{2} \rfloor + M} |b_{j-1,l}|^2 . \tag{5.4}$$

The condition (5.4) expresses the fact that large wavelet coefficients on level j already manifest themselves in reasonably large wavelet coefficients on the coarser level $j-1$. So one could hope that all relevant coefficients will be maintained by the algorithm.

Unfortunately this is not true. By adding a new level to the grid all previously calculated wavelet coefficient are changed, and it is hard to see how the estimated coefficients are related to the original coefficients. In fact one can construct non-zero functions which vanish on the coarse and the next finer level and satisfy the condition from Definition 5.2.

For this reason we modify the algorithm and introduce another type of balanced functions. Analysing Algorithm 3 in the piecewise linear case $m = 1$, we recognize that the wavelet coefficients do indeed change, but the coefficients of the hat-functions, which result from interpolation at the different levels in Step 1 of Algorithm 1, do not change. To explore this further, we introduce the set of hat-functions visualised in Figure 5.2,

$$\left\{\varphi_{j_0,k}^1 : k = 0, \ldots, 2^{j_0} - 1\right\} \cup \left\{\varphi_{j,2k+1}^1 : j = j_0 + 1, \ldots; k = 0, \ldots, 2^{j-1} - 1\right\} \tag{5.5}$$

and investigate the representation of a function u by

$$u(x) = \sum_{k=0}^{2^{j_0}-1} a_{j_0,k}\varphi_{j_0,k}^1(x) + \sum_{j=j_0+1}^{\infty} \sum_{k=0}^{2^{j-1}-1} a_{j,k}\varphi_{j,2k+1}^1(x) . \tag{5.6}$$

When the infinite sum on the right-hand side is replaced by the finite sum over j from j_0+1 to J, then (5.6) represents the piecewise linear spline which interpolates u on the fine grid.

The modified algorithm which we shall propose afterwards is based on the following result.

Theorem 5.3. *The functions* (5.5) *form a Riesz basis for* $W_2^1\left(-\frac{1}{2}, \frac{1}{2}\right)$, *i.e., if a function* $u \in W_2^1\left(-\frac{1}{2}, \frac{1}{2}\right)$ *has the representation* (5.6), *then*

$$\|u\|_{W_2^1}^2 \le C\left(\sum_{k=0}^{2^{j_0}-1} |a_{j_0,k}|^2 + \sum_{j=j_0+1}^{\infty} 2^{2j} \sum_{k=0}^{2^{j-1}-1} |a_{j,k}|^2\right),$$

$$\|u\|_{W_2^1}^2 \ge c\left(\sum_{k=0}^{2^{j_0}-1} |a_{j_0,k}|^2 + \sum_{j=j_0+1}^{\infty} 2^{2j} \sum_{k=0}^{2^{j-1}-1} |a_{j,k}|^2\right).$$

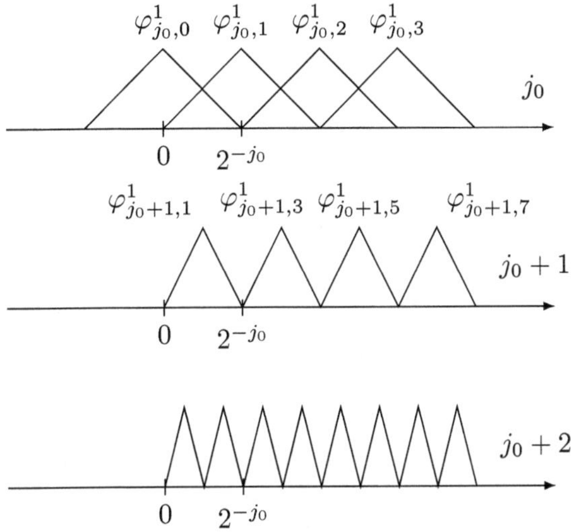

FIGURE 5.2. The set of functions (5.5)

Proof. We analyse the eigenvalues of the Gram matrix of the normalised system $\varphi^1_{j,k}$ in the Sobolev space W^1_2. In the first step we omit the functions $\varphi^1_{j_0,k}$ on the coarse level and consider the finite system

$$\{\varphi^1_{j,2k+1} : j = j_0+1,\dots,J;\ k = 0,\dots,2^{j-1}-1\}. \tag{5.7}$$

We arrange these functions in the following way

$$\begin{array}{llllll}
\varphi^1_{j_0+1,1}, & \varphi^1_{j_0+2,1}, & \varphi^1_{j_0+2,3}, & \varphi^1_{j_0+3,1}, & \cdots & \varphi^1_{J,2^{J-j_0}-1}, \\
\varphi^1_{j_0+1,3}, & \varphi^1_{j_0+2,5}, & \varphi^1_{j_0+2,7}, & \varphi^1_{j_0+3,9}, & \cdots & \varphi^1_{J,2^{J-j_0+1}-1}, \\
\cdots & \cdots & \cdots & \cdots & \cdots & \cdots \\
\varphi^1_{j_0+1,2^{j_0+1}-1}, & \varphi^1_{j_0+2,2^{j_0+2}-3}, & \varphi^1_{j_0+2,2^{j_0+2}-1}, & \varphi^1_{j_0+3,2^{j_0+3}-7}, & \cdots & \varphi^1_{J,2^J-1}.
\end{array}$$

Every row of this table starts with a function $\varphi^1_{j_0+1,2k+1}$ from level j_0+1. Then we select all functions $\varphi^1_{j,k}$ with $j \geq j_0+2$ whose support is contained in the support of $\varphi^1_{j_0+1,2k+1}$ and arrange them lexicographically. It is clear that every element of (5.7) appears exactly once.

In the next step we normalise the functions $\varphi^1_{j,k}$ with respect to the W^1_2-norm. A straightforward calculation yields that

$$\Delta_j^2 := \|\varphi^1_{j,k}\|^2_{W^1_2} = 2\left(\frac{1}{3} + 2^{2j}\right), \quad j = j_0+1,\dots,J, \quad k = 0,\dots,2^{j-1}-1.$$

Since the supports of two functions in different rows of the table have no common interior points, the Gram matrix G of the normalized system $\varphi^1_{j,k}/\Delta_j$ has the block structure

$$G = \begin{pmatrix} G_1 & 0 & \cdots & 0 \\ 0 & G_2 & \cdots & 0 \\ \vdots & \vdots & \ddots & \vdots \\ 0 & 0 & \cdots & G_{2^{j_0}} \end{pmatrix},$$

where G_k is the Gram matrix of the normalized system in the kth row of the table. Since the functions in the kth row differ from the functions in the first row just by a translation, we further have $G_1 = G_2 = \cdots = G_{2^{j_0}}$, so that we only need to consider the eigenvalues of G_1.

Preparing the application of Gershgorin's theorem, we estimate the sum of the absolute values of the non-diagonal entries of the rows of G_1. It is not hard to see that the largest sum is attained in the first row.

In order to compute the scalar products of the functions $\varphi^1_{j,k}/\Delta_j$ in the first row, which are the entries of the matrix G_1, we remark that the derivatives of these functions form an orthogonal system in L_2, so that outside the main diagonal of G_1 only the L_2-part of the scalar product is relevant. By a simple calculation we get, for $j = j_0+2,\ldots,J$ and $k = 0,\ldots,2^{j-j_0-2}-1$,

$$\begin{aligned} \langle \varphi^1_{j_0+1,1}, \varphi^1_{j,2k+1} \rangle &= \sqrt{2}^{3(j_0-j+1)}\,(2k+1)\,, \\ \langle \varphi^1_{j_0+1,1}, \varphi^1_{j,2^{j-j_0-1}-2k-1} \rangle &= \sqrt{2}^{3(j_0-j+1)}\,(2k+1)\,. \end{aligned} \tag{5.8}$$

Taking into account normalisation and using that $\Delta_j > \sqrt{2}\,2^j$, we estimate the sum of (the absolute values of) the non-diagonal entries

$$\begin{aligned} &\sum_{j=j_0+2}^{J} \sum_{k=0}^{2^{j-j_0-1}-1} \left\langle \frac{1}{\Delta_{j_0+1}} \varphi^1_{j_0+1,1}, \frac{1}{\Delta_j} \varphi^1_{j,2k+1} \right\rangle \\ &\qquad = \sum_{j=j_0+2}^{J} 2 \sum_{k=0}^{2^{j-j_0-2}-1} \frac{\sqrt{2}^{3(j_0-j+1)}}{\Delta_{j_0+1}\Delta_j}\,(2k+1) \\ &\qquad = \sum_{j=j_0+2}^{J} \frac{2}{\Delta_{j_0+1}\Delta_j} \sqrt{2}^{3(j_0-j+1)}\, 2^{2(j-j_0-2)} \\ &\qquad < \sum_{j=j_0+2}^{J} \frac{\sqrt{2}^{j-j_0-3}}{\sqrt{2}^{2j+1}\,\sqrt{2}^{2j_0+3}} \\ &\qquad < \sum_{j=j_0+2}^{\infty} \sqrt{2}^{-j-3j_0-7} = (\sqrt{2}+1)\,2^{-2j_0-4} < 1. \end{aligned}$$

Hence all eigenvalues of the Gram matrix G are located in the interval

$$(A,B) := \left(1-(\sqrt{2}+1)\,2^{-2j_0-4}, 1+(\sqrt{2}+1)\,2^{-2j_0-4}\right).$$

Note that A and B do not depend on the choice of J. Taking the limit $J \to \infty$, we obtain that any function u in the closure W_{j_0} of the subspace of W_2^1 which is spanned by the system

$$\left\{\varphi_{j,2k+1}^1 : j = j_0+1, j_0+2, \ldots; \; k = 0, \ldots, 2^{j-1}-1\right\}$$

satisfies a Riesz estimate: if u is represented by

$$u = \sum_{j=j_0+1}^{\infty} \sum_{k=0}^{2^{j-1}-1} a_{j,k}\varphi_{j,2k+1}^1 = \sum_{j=j_0+1}^{\infty} \sum_{k=0}^{2^{j-1}-1} (\Delta_j a_{j,k}) \frac{\varphi_{j,2k+1}^1}{\Delta_j}$$

then we have

$$A \sum_{j=j_0+1}^{\infty} \sum_{k=0}^{2^{j-1}-1} \Delta_j^2 \left|a_{j,k}\right|^2 \le \|u\|_{W_2^1}^2 \le B \sum_{j=j_0+1}^{\infty} \sum_{k=0}^{2^{j-1}-1} \Delta_j^2 \left|a_{j,k}\right|^2.$$

On account of $2 \cdot 2^{2j} < \Delta_j^2 < 3 \cdot 2^{2j}$ we obtain that

$$2A \sum_{j=j_0+1}^{\infty} 2^{2j} \sum_{k=0}^{2^{j-1}-1} \left|a_{j,k}\right|^2 \le \|u\|_{W_2^1}^2 \le 3B \sum_{j=j_0+1}^{\infty} 2^{2j} \sum_{k=0}^{2^{j-1}-1} \left|a_{j,k}\right|^2. \tag{5.9}$$

In order also to include the coarse part we study the angle between the spaces W_{j_0} and $V_{j_0} := \operatorname{span}\left\{\varphi_{j_0,k}^1 : k = 0, \ldots, 2^{j_0}-1\right\}$. So let f in V_{j_0} and g in W_{j_0} with

$$f(x) = \sum_{k=0}^{2^{j_0}-1} a_{j_0,k}\varphi_{j_0,k}^1(x), \quad g(x) = \sum_{j=j_0+1}^{\infty} \sum_{k=0}^{2^{j-1}-1} a_{j,k}\varphi_{j,k}^1(x).$$

Because the derivatives of $\varphi_{j_0,k}^1$ and $\varphi_{j,k}^1$ with $j > j_0$ form an orthogonal system in L_2, we have $\langle D_x f, D_x g\rangle = 0$, so that

$$\langle f, g\rangle + \langle D_x f, D_x g\rangle = \sum_{k=0}^{2^{j_0}-1} \sum_{j=j_0+1}^{\infty} \sum_{\kappa=0}^{2^{j-1}-1} a_{j_0,k} a_{j,\kappa} \langle \varphi_{j_0,k}^1, \varphi_{j,2\kappa+1}^1\rangle.$$

Applying the Cauchy–Schwarz inequality we arrive at

$$\begin{aligned} \left|\langle f, g\rangle + \langle D_x f, D_x g\rangle\right|^2 \le \left(\sum_{k=0}^{2^{j_0}-1} \sum_{j=j_0+1}^{\infty} \sum_{\kappa=0}^{2^{j-1}-1} 2^{-2j} \left|\langle \varphi_{j_0,k}^1, \varphi_{j,2\kappa+1}^1\rangle\right|^2\right) \cdot \\ \cdot \left(\sum_{k=0}^{2^{j_0}-1} \left|a_{j_0,k}\right|^2\right) \left(\sum_{j=j_0+1}^{\infty} \sum_{\kappa=0}^{2^{j-1}-1} 2^{2j} \left|a_{j,\kappa}\right|^2\right). \end{aligned}$$

Using (2.3), (5.8), (5.9), and $A = 1 - 2^{-2j_0-4}(\sqrt{2}+1) > 3/4$, we get

$$|\langle f, g\rangle + \langle D_x f, D_x g\rangle|^2 \le \frac{3\cdot 2^{j_0}}{A}\,\|f\|_{L_2}^2\,\|g\|_{W_2^1}^2 \sum_{j=j_0+1}^{\infty}\sum_{k=0}^{2^{j-j_0-1}-1} 2^{-2j}\frac{2^{3j_0}}{2^{3j}}\,(2k+1)^2$$
$$\le \frac{2^{1-j_0}}{15\,A}\,\|f\|_{W_2^1}^2\,\|g\|_{W_2^1}^2 \le \frac{1}{5}\,\|f\|_{W_2^1}^2\,\|g\|_{W_2^1}^2\,.$$

Consequently the angle α between V_{j_0} and W_{j_0} is bounded from below by a positive constant. The statement now follows from (5.9) and

$$(1-\cos\alpha)\left(\|f\|_{W_2^1}^2 + \|g\|_{W_2^1}^2\right) \le \|f+g\|_{W_2^1}^2 \le 2\left(\|f\|_{W_2^1}^2 + \|g\|_{W_2^1}^2\right) \qquad \square$$

To investigate the algorithm with piecewise linear splines, we now modify Definition 5.2 as follows.

Definition 5.4. Let $m = 1$ and $M = 3$. A function $u \in W_2^1\left(-\frac{1}{2},\frac{1}{2}\right)$ is called *linearly balanced*, if the coefficients $a_{j,k}$ in the representation

$$u\,(x) = \sum_{k=0}^{2^{j_0}-1} a_{j_0,k}\varphi_{j_0,k}^1\,(x) + \sum_{j=j_0+1}^{\infty}\sum_{k=0}^{2^{j-1}-1} a_{j,k}\varphi_{j,2k+1}^1\,(x)\,.$$

satisfy the condition

$$|a_{j,k}|^2 \le \frac{1}{8}\frac{1}{2M+1}\sum_{l=\lfloor\frac{k}{2}\rfloor-M}^{l=\lfloor\frac{k}{2}\rfloor+M} |a_{j-1,l}|^2\,, \tag{5.10}$$

for all $j \ge j_0+2$ and $k = 0,\dots,2^{j-1}-1$.

Remark 5.5. The factor 1/8 in (5.10) is chosen such that $|a_{j-1,l}| \le \delta_1\,(j-2)$ for $l = \lfloor k/2\rfloor - M,\dots,\lfloor k/2\rfloor + M$ implies $|a_{j,k}| \le \delta_1\,(j-1)$, where $\delta_1\,(j)$ is the threshold from (5.2) for $s = 1$.

Now we change Algorithm 3 and adapt the grid not according to the wavelet coefficients $b_{j,k}$ but with respect to the coefficients $a_{j,k}$ in the representation (5.6). The index set I in (5.1) is now constructed according to the criterion

$$I := \{(j,k) : |a_{j+1,k}| \ge \delta_1\,(j)\}.$$

The modified algorithm for $m = 1$ indeed works in the following sense: For any linearly balanced function the algorithm produces an admissible mesh and an approximation of the Hilbert transform with prescribed accuracy. A more precise statement is given in the next theorem.

Theorem 5.6. *There exists a constant D (given below) such that the following holds: If $u \in W_2^s\left(-\frac{1}{2},\frac{1}{2}\right)$ with $s > 1$ is linearly balanced, and*

$$\|u - L_J u\|_{W_2^1} < \varepsilon_1 := \varepsilon/D$$

the modified Algorithm 3 *with piecewise linear splines and threshold* ε_1 *provides an admissible mesh* $\mathcal{G}$, *a function* $\widehat{u}$ *and an approximate Hilbert transform* $\widetilde{H}\widehat{u}$ *such that*

$$\left\|u - \widehat{u}\right\|_{W_2^1} < \varepsilon, \qquad \|Hu - \widetilde{H}\widehat{u}\|_{W_2^1} < 2\varepsilon.$$

Note that the assumption $\left\|u - L_J u\right\| < \varepsilon/D$ can always be achieved by choosing J large enough, see for instance [12] or [21, page 44].

We only sketch the proof and refer to [19] for the details. The result of the modified Algorithm 3 is the function $\widehat{u}$ with the representation

$$\widehat{u}\,(x) = \sum_{k=0}^{2^{j_0}-1} a_{j_0,k}\varphi^1_{j_0,k}\,(x) + \sum_{j=j_0+1}^{J} \sum_{k\notin\mathcal{I}_j^0} \widehat{a}_{j,k}\varphi^1_{j,2k+1}\,(x)$$

where we define $\mathcal{I}_j^0 := \left\{k \in \mathbb{N}_0 :\ 0 \le k < 2^{j-1} \text{ and } \widehat{a}_{j,k} = 0\right\}$. For $k \notin \mathcal{I}_j^0$ we have $a_{j,k} = \widehat{a}_{j,k}$, so that the error between u and $\widehat{u}$ can be estimated with Theorem 5.3 by

$$\|u - \widehat{u}\|^2_{W_2^1} \le C\left(\sum_{j=j_0+1}^{J} 2^{2j} \sum_{k\in\mathcal{I}_j^0} |a_{j,k}|^2 + \sum_{j=J+1}^{\infty} 2^{2j} \sum_{k=0}^{2^{j-1}-1} |a_{j,k}|^2\right).$$

The last sum is the interpolation error on the fine mesh $\mathcal{F}_J$ and is less than ε_1 by assumption.

In order to estimate the first sum we investigate the absolute value of the coefficients $a_{j,k}$ and verify that these are less than the corresponding threshold $\delta_1\,(j-1)$. Arguing by induction, we assume that this is true for all $a_{j,k}$ with $k \in \mathcal{I}_j^0$ at some level j (which is certainly satisfied for $j = j_0$) and consider the coefficients $a_{j+1,k}$ with $k \in \mathcal{I}_{j+1}^0$.

There are two possible reasons why k belongs to $\mathcal{I}_{j+1}^0$: In the first case the coefficient $a_{j+1,k}$ was calculated but replaced by zero because it was less than the threshold – then there is nothing to prove. In the second case, the coefficient $a_{j+1,k}$ has not been computed because the corresponding grid point did not belong to the actual mesh. This can only happen when at the previous level j all related pairs (j,l) with $\lfloor k/2\rfloor - M \le l \le \lfloor k/2\rfloor + M$ belong to the index set $\mathcal{I}_j^0$. Then, by assumption, all coefficients $a_{j,l}$ satisfy $|a_{j,l}| < \delta_1(j-1)$. Since u is linearly balanced, the condition (5.10) implies that $|a_{j+1,k}| < \delta_1(j)$ (see the remark to Definition 5.2).

Consequently we get for the first sum

$$\begin{aligned}\sum_{j=j_0+1}^{J} 2^{2j} \sum_{k\in\mathcal{I}_j^0} |a_{j,k}|^2 &\le \sum_{j=j_0+1}^{J} 2^{2j} \sum_{k\in\mathcal{I}_j^0} \delta_1\,(j-1)^2 \\ &\le \sum_{j=j_0+1}^{J} 2^{2j} \sum_{k=0}^{2^{j-1}-1} \frac{\varepsilon^2}{C_1 2^{3(j-1)}\,(J-j_0)} \le \frac{4\varepsilon^2}{C_1},\end{aligned}$$

where C_1 is the Riesz constant from (2.3). With $D := \sqrt{C\,(1+4/C_1)}$ we arrive at

$$\|u-\widehat{u}\|_{W_2^1} \le \sqrt{C\,(1+4/C_1)}\,\varepsilon_1 < \varepsilon.$$

The construction of $\widetilde{H}$ can be made in such a way that $\|H\widehat{u}-\widetilde{H}\widehat{u}\|_{W_2^1}$ is less than ε. Taking into account that H is bounded in $W_2^1\left(-\frac{1}{2},\frac{1}{2}\right)$ with norm 1 we get for the Hilbert transform

$$\|Hu-\widetilde{H}\widehat{u}\|_{W_2^1} \le \|H\|\,\|u-\widehat{u}\|_{W_2^1} + \|H\widehat{u}-\widetilde{H}\widehat{u}\|_{W_2^1} \le 2\varepsilon.$$

6. Numerical examples

The Hilbert transform connects real and imaginary part of the boundary values of a holomorphic function $f = u + \mathrm{i}v$ in the unit disk,

$$u = Hv + u(0), \quad v = -Hu + v(0),$$

which provides many examples for testing the algorithm. We have chosen five functions $f_1, \ldots, f_5$ defined for $t \in \mathbb{T}$ by

$$f_0\,(t) = \frac{1-pt}{1+pt}, \qquad f_1\,(t) = \frac{t+p}{1+pt} - p, \qquad f_2\,(t) = \mathrm{i}\log f_0\,(t)$$

$$f_3\,(t) = \exp\left(-f_0\,(t)\right) - 1/\mathrm{e},\ f_4\,(t) = \exp\left(\mathrm{i}\log f_0\,(t)\right) - 1,\ f_5\,(t) = \sqrt{1+pt} - 1.$$

All functions can be holomorphically extended into the unit disk with vanishing real part at the origin, such that $u = Hv$. The real parameter p with $0 \le p < 1$ controls the behavior of the functions. As $p \to 1$ specific singularities arise which are typical for solutions of certain Riemann–Hilbert problems. For $p = 0.99$ real (solid) and imaginary (dashed) part of these functions are shown in the left part of Figures 6.3–6.7.

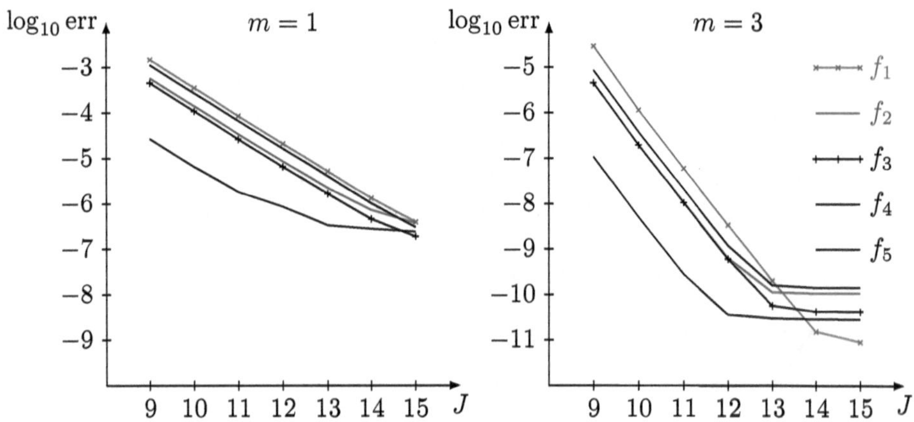

FIGURE 6.1. Logarithm of the error $\|\widetilde{H}v - Hv\|_{L_2(\mathbb{T})}$ for Algorithm 3

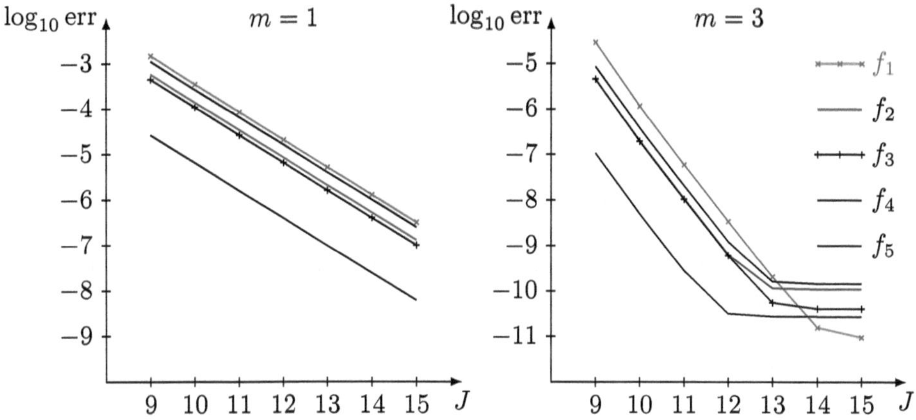

FIGURE 6.2. Logarithm of the error $\|\widetilde{H}v - Hv\|_{L_2(\mathbb{T})}$ using the full grid

The error $\|\widetilde{H}v - Hv\|_{L_2(\mathbb{T})}$ of the approximate Hilbert transform $\widetilde{H}v$ computed with Algorithm 3 is shown in Figure 6.1 for $p = 0.95$, $j_0 = 8$ and various J. The error decreases almost linearly in a logarithmic scale. The slopes of the error curve indicate decay of order h^{-2} for $m = 1$ and of order h^{-4} for $m = 3$, respectively, where h is the step size of the fine mesh. This corresponds to the behavior of the interpolation error for splines on a *uniform* mesh. In the rightmost part there is no further improvement, since then the truncation error takes over (the threshold was chosen so that $\varepsilon \approx 10^{-6}$ (linear case) and $\varepsilon \approx 10^{-10}$ (cubic case)). It is remarkable that there is practically no difference between these results and the experiments in [20], where we computed the Hilbert transform on the complete fine mesh $\mathcal{F}_J$ (for convenience reproduced in Figure 6.2). So Algorithm 3 indeed produces an admissible mesh which reflects the behavior of the test functions very well.

The percentage of 'active' grid points in relation to the fine mesh $\mathcal{F}_{15}$ is listed in Table 6.1. The adapted grids displayed in Figures 6.3–6.7 show clearly that the

	f_1	f_2	f_3	f_4	f_5
$m = 1$	7.15%	16.93%	34.05%	44.88%	11.57%
$m = 3$	6.74%	14.25%	18.6%	16.88%	5.29%

TABLE 6.1. Active part of the grid points of the adapted meshes

points are mainly concentrated in those regions which contain the 'almost singular' parts of the functions.

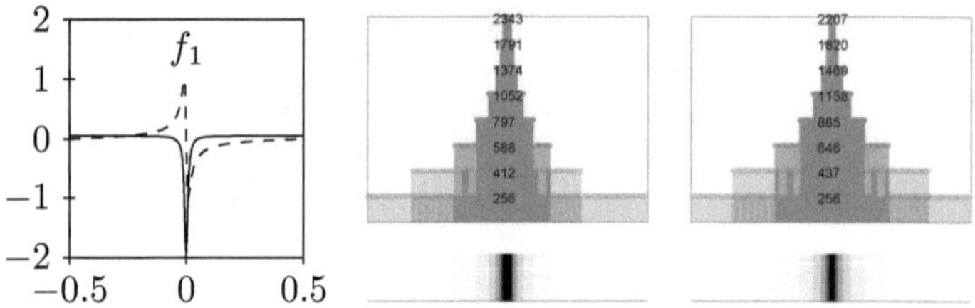

FIGURE 6.3. Real and imaginary part of f_1 and associated admissible meshes of Algorithm 3 (left $m = 1$ and right $m = 3$)

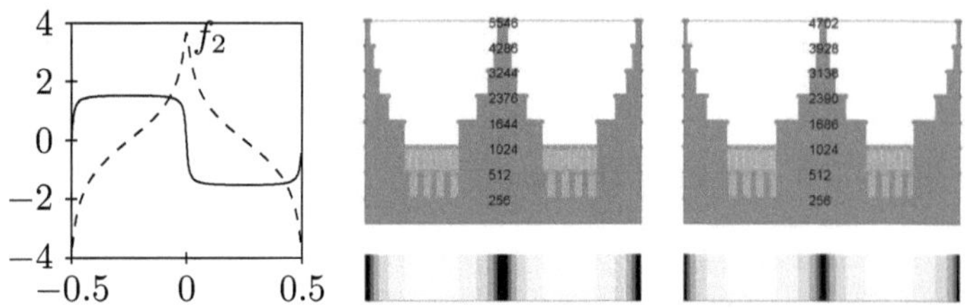

FIGURE 6.4. Real and imaginary part of f_2 and associated admissible meshes of Algorithm 3 (left $m = 1$ and right $m = 3$)

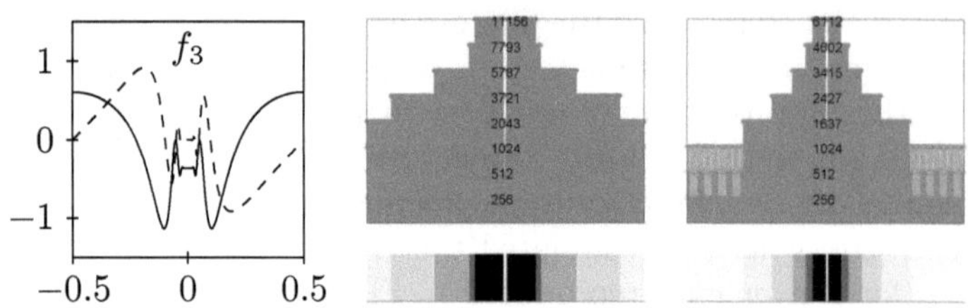

FIGURE 6.5. Real and imaginary part of f_3 and associated admissible meshes of Algorithm 3 (left $m = 1$ and right $m = 3$)

Acknowledgement

The authors are grateful to Daniel Potts and Jürgen Prestin for valuable discussions.

References

[1] K. Bittner. Fast algorithms for periodic spline wavelets on sparse grids. SIAM J. Sci. Comput., 20(4):1192–1213, 1999

[2] S. Börm and W. Hackbusch. Hierarchical quadrature for singular integrals. Computing, 74(2):75–100, 2005

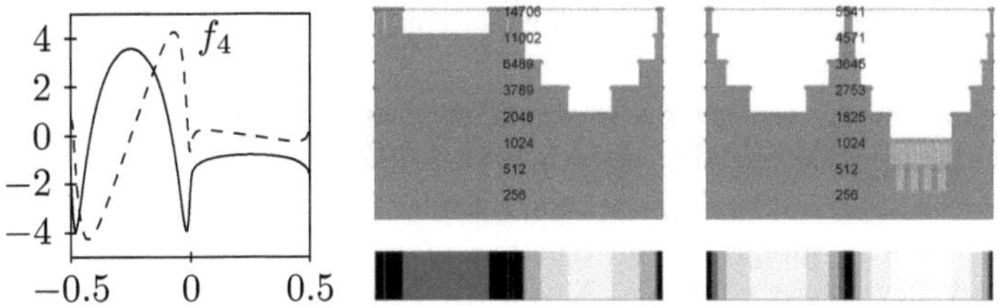

FIGURE 6.6. Real and imaginary part of f_4 and associated admissible meshes of Algorithm 3 (left $m = 1$ and right $m = 3$)

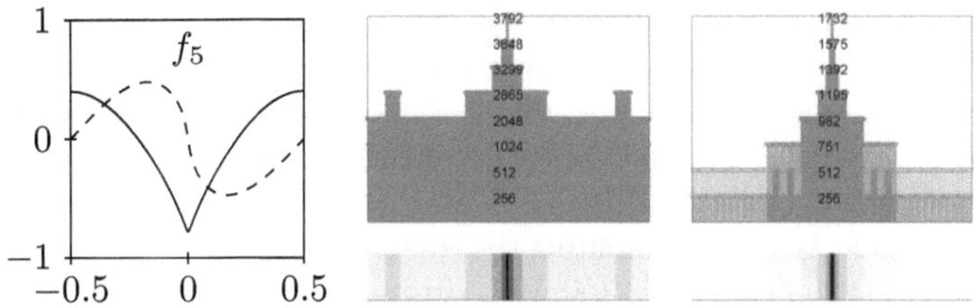

FIGURE 6.7. Real and imaginary part of f_5 and associated admissible meshes of Algorithm 3 (left $m = 1$ and right $m = 3$)

[3] C.K. Chui. An introduction to wavelets. Wavelet Analysis and Its Applications. 1. Boston, MA: Academic Press, 1992

[4] A. Cohen, I. Daubechies, and J.-C. Feauveau. Biorthogonal bases of compactly supported wavelets. Commun. Pure Appl. Math., 45(5):485–560, 1992

[5] W. Dahmen and A. Kunoth. Multilevel preconditioning. Numer. Math., 63(3):315–344, 1992

[6] W. Dahmen, S. Prössdorf, and R. Schneider. Wavelet approximation methods for pseudodifferential equations. II: Matrix compression and fast solution. Adv. Comput. Math., 1(3-4):259–335, 1993

[7] W. Dahmen, S. Prössdorf, and R. Schneider. Wavelet approximation methods for pseudodifferential equations. I: Stability and convergence. Math. Z., 215(4):583–620, 1994

[8] I. Daubechies. Ten lectures on wavelets. SIAM, Society for Industrial and Applied Mathematics, Philadelphia, PA, 1992

[9] A. Dutt and V. Rokhlin. Fast Fourier transforms for nonequispaced data. SIAM J. Sci. Comput., 14(6):1368–1393, 1993

[10] A. Dutt and V. Rokhlin. Fast Fourier transforms for nonequispaced data. II. Appl. Comput. Harmon. Anal., 2(1):85–100, 1995

[11] D. Gaier. Konstruktive Methoden der konformen Abbildung. Berlin – Göttingen – Heidelberg: Springer-Verlag, 1964

[12] M. Golomb. Approximation by periodic spline interpolants on uniform meshes. J. Approximation Theory, 1:26–65, 1968

[13] M.H. Gutknecht. Fast algorithms for the conjugate periodic function. Computing, 22:79–91, 1979

[14] S.L. Hahn. Hilbert transforms in signal processing. The Artech House Signal Processing Library. Boston, MA: Artech House, 1996

[15] P. Henrici. Applied and computational complex analysis. Volume III: Discrete Fourier analysis, Cauchy integrals, construction of conformal maps, univalent functions. Reprint. Wiley Classics Library. New York, NY: Wiley, 1993

[16] F.W. King. Hilbert transforms. Volume 1. Encyclopedia of Mathematics and its Applications 124. Cambridge: Cambridge University Press, 2009

[17] S. Kunis and D. Potts. Time and memory requirements of the nonequispaced FFT. Sampl. Theory Signal Image Process., 7(1):77–100, 2008

[18] A.K. Louis, P. Maaß and A. Rieder. Wavelets. Theorie und Anwendungen. Teubner Studienbücher: Mathematik. Stuttgart: Teubner, 1994

[19] F. Martin. Analytische und numerische Verfahren zur Berechnung der Hilbert Transformation und zur Lösung funktionentheoretischer Randwertaufgaben. Ph.D. thesis, TU Bergakademie Freiberg, 2010

[20] F. Martin and E. Wegert. Computing the Hilbert transform using biorthogonal spline wavelets, J. Math. Sci., 189(1):150–163, 2013.

[21] S. Prössdorf and B. Silbermann. Numerical analysis for integral and related operator equations. Berlin: Akademie-Verlag, 1991

[22] A. Rathsfeld. On the stability of piecewise linear wavelet collocation and the solution of the double layer equation over polygonal curves. In: Golberg, Michael (ed.), Boundary integral methods: numerical and mathematical aspects. Southampton: WIT Press/ Computational Mechanics Publications. Comput Eng. 1, 177–215, 1999

[23] E. Wegert. An iterative method for solving nonlinear Riemann–Hilbert problems. J. Comput. Appl. Math., 29(3):311–327, 1990

[24] E. Wegert. Nonlinear boundary value problems for holomorphic functions and singular integral equations. Mathematical Research. 65. Berlin: Akademie Verlag, 1992

[25] R. Wegmann. Ein Iterationsverfahren zur konformen Abbildung. Numer. Math., 30:453–466, 1978

[26] R. Wegmann. Methods for numerical conformal mapping. In: Kühnau, R. (ed.), Handbook of complex analysis: geometric function theory. Volume 2. Amsterdam: Elsevier/North–Holland. 351–477, 2005

Frank Martin and Elias Wegert
e-mail: `martin3@math.tu-freiberg.de`
`wegert@math.tu-freiberg.de`

Operator Theory:
Advances and Applications, Vol. 236, 357–370

B_σ-Campanato Estimates for Commutators of Calderón–Zygmund Operators

Katsuo Matsuoka

Dedicated to Professor Akihiko Miyachi in celebration of his 60th birthday

Abstract. In the earlier papers [16, 19], the B_σ-function spaces were introduced for the purpose of unifying central Morrey spaces, λ-central mean oscillation spaces and usual Morrey–Campanato spaces.

The purpose of this paper is to establish the B_σ-Campanato estimates for commutators of Calderón–Zygmund operators on B_σ-Morrey spaces.

Mathematics Subject Classification (2010). Primary 42B35; Secondary 46E35, 46E30, 26A33.

Keywords. Morrey space, Campanato space, central Morrey space, λ-central mean oscillation space, B_σ-Morrey–Campanato space, Calderón–Zygmund operator, commutator.

1. Introduction

In 1989, Y. Chen and K. Lau [4] and J. García-Cuerva [10] introduced the central mean oscillation space $\mathrm{CMO}^p(\mathbb{R}^n)$ with its predual, which contains $B^p(\mathbb{R}^n)$ and the bounded mean oscillation space $\mathrm{BMO}(\mathbb{R}^n)$. Here the space $B^p(\mathbb{R}^n)$ was introduced by A. Beurling [3], together with its predual $A^p(\mathbb{R}^n)$, so-called the Beurling algebra. For the homogeneous versions of $\mathrm{CMO}^p(\mathbb{R}^n)$, S.Z. Lu and D.C. Yang [17, 18] used the notation of the central bounded mean oscillation space $\mathrm{CBMO}^p(\mathbb{R}^n)$.

In recent years, J. Alvarez, M. Guzmán-Partida and J. Lakey [2] pointed out that $B^p(\mathbb{R}^n)$ and $\mathrm{BMO}(\mathbb{R}^n)$ are roughly the bad part and the good part of $\mathrm{CMO}^p(\mathbb{R}^n)$, respectively. Moreover, as an extension of $B^p(\mathbb{R}^n)$ and $\mathrm{CMO}^p(\mathbb{R}^n)$, they introduced the non-homogeneous central Morrey space $B^{p,\lambda}(\mathbb{R}^n)$ (cf. J. García-Cuerva and M.J.L. Herrero [11]) and the λ-central mean oscillation space

The author was partially supported by Grant for 2011 Overseas Researcher of Nihon University, Japan.

$\mathrm{CMO}^{p,\lambda}$ $(\mathbb{R}^n)$, respectively. For the homogeneous versions of these spaces, we use the following notation: the central Morrey space $\dot{B}^{p,\lambda}(\mathbb{R}^n)$ and the λ-central bounded mean oscillation space $\mathrm{CBMO}^{p,\lambda}(\mathbb{R}^n)$, respectively.

In [16, 19], in order to unify the spaces $B^{p,\lambda}(\mathbb{R}^n)$, $\dot{B}^{p,\lambda}(\mathbb{R}^n)$, $\mathrm{CMO}^{p,\lambda}(\mathbb{R}^n)$, $\mathrm{CBMO}^{p,\lambda}(\mathbb{R}^n)$, $L_{p,\lambda}(\mathbb{R}^n)$ and $\mathcal{L}_{p,\lambda}(\mathbb{R}^n)$, we introduced new function spaces named B_σ-Morrey–Campanato spaces.

On the other hand, several authors have studied the behavior of various operators such as the Calderón–Zygmund operators, the fractional integral operators, the commutators, and so on, on above spaces. Furthermore, in [16, 19], we established the boundedness of various integral operators on our B_σ-Morrey–Campanato spaces.

Among several operators, in particular, we take up the commutators. The commutator $[b,T]$ generated by $b \in L^1_{\mathrm{loc}}(\mathbb{R}^n)$ and the Calderón–Zygmund operator T is defined by

$$[b,T](f)(x) = b(x)Tf(x) - T(bf)(x), \quad x \in \mathbb{R}^n,$$

for $f \in L^\infty_{\mathrm{comp}}(\mathbb{R}^n)$. In 1976, R.R. Coifman, R. Rochberg and G. Weiss [6] established that if $b \in \mathrm{BMO}(\mathbb{R}^n)$, then $[b,T]$ is bounded on $L^p(\mathbb{R}^n)$ $(p \in (1,\infty))$ (see Theorem 4.2 below). In the above, $L^\infty_{\mathrm{comp}}(\mathbb{R}^n)$ denotes the set of all L^∞ functions with compact support. Later, in 1998, L. Grafakos, X. Li and D. Yang [12] proved the CBMO estimates for $[b,T]$ on the Herz spaces. Further, J. Alvarez, M. Guzmán-Partida and J. Lakey [2] and Y. Komori [14] obtained the λ-CMO estimates for $[b,T]$ on $B^{p,\lambda}(\mathbb{R}^n)$ (cf. Z. Fu, Y. Lin and S. Lu [9]).

Motivated by these results, we will consider the B_σ-Campanato estimates for the commutators of Calderón–Zygmund operators on B_σ-Morrey spaces.

2. B_σ-Morrey–Campanato spaces

For details of the function spaces in this section, we refer to [16, 19].

Let $Q_r \subset \mathbb{R}^n$ denote either the open cube having center 0 and sidelength $2r$, or the open ball having center 0 and radius r, i.e.,

$$Q_r = \left\{ y = (y_1, y_2, \ldots, y_n) \in \mathbb{R}^n : \max_{1\le i\le n} |y_i| < r \right\} \text{ or } Q_r = \{ y \in \mathbb{R}^n : |y| < r \}.$$

And let for $x \in \mathbb{R}^n$,

$$Q(x,r) = x + Q_r = \{ x + y : y \in Q_r \}.$$

For a measurable set $G \subset \mathbb{R}^n$, we denote the Lebesgue measure of G by $|G|$ and the characteristic function of G by χ_G. Further, for a function $f \in L^1_{\mathrm{loc}}(\mathbb{R}^n)$ and a measurable set $G \subset \mathbb{R}^n$ with $|G| > 0$, let

$$f_G = \fint_G f(y)\,dy = \frac{1}{|G|}\int_G f(y)\,dy.$$

First, we state the definitions of Morrey and Campanato spaces.

Definition 2.1. Let $U = \mathbb{R}^n$ or $U = Q_r$ with $r > 0$. For $p \in [1, \infty)$ and $\lambda \in \mathbb{R}$, $L_{p,\lambda}(U)$ and $\mathcal{L}_{p,\lambda}(U)$ are defined to be the sets of all functions f on U such that the following functionals are finite, respectively:

$$\|f\|_{L_{p,\lambda}(U)} = \sup_{Q(\xi,s)\subset U} \frac{1}{s^\lambda} \left(\fint_{Q(\xi,s)} |f(y)|^p \, dy \right)^{1/p}$$

and

$$\|f\|_{\mathcal{L}_{p,\lambda}(U)} = \sup_{Q(\xi,s)\subset U} \frac{1}{s^\lambda} \left(\fint_{Q(\xi,s)} |f(y) - f_{Q(\xi,s)}|^p \, dy \right)^{1/p}.$$

Remark 2.2. In this paper, when $U = \mathbb{R}^n$, we abbreviate $\|f\|_{E(U)}$ to $\|f\|_E$ for the symbol of the norm, while we don't abbreviate $E(\mathbb{R}^n)$ for the symbol of the function space.

Note that we regard $L_{p,\lambda}(U)$ and $\mathcal{L}_{p,\lambda}(U)$ as spaces of functions modulo null-functions. Then $L_{p,\lambda}(U)$ is a Banach space. Let $\mathcal{C}$ be the space of all constant functions and $\cdot/\mathcal{C}$ mean the quotient space by $\mathcal{C}$. Then $\mathcal{L}_{p,\lambda}(\mathbb{R}^n)/\mathcal{C}$ is a Banach space equipped with the norm $\|f\|_{\mathcal{L}_{p,\lambda}}$. For the unit ball Q_1, $\|f\|_{\mathcal{L}_{p,\lambda}} + |f_{Q_1}|$ is a norm and thereby $\mathcal{L}_{p,\lambda}(U)$ is a Banach space.

By the definition, if $\lambda = -n/p$, then $L_{p,\lambda}(U) = L^p(U)$. If $p = 1$ and $\lambda = 0$, then $\mathcal{L}_{1,0}(U)$ is the usual $\mathrm{BMO}(U)$.

Definition 2.3. Let $\sigma \in [0, \infty)$, $p \in [1, \infty)$ and $\lambda \in \mathbb{R}$. For

$$E = L^p, \ L_{p,\lambda} \text{ or } \mathcal{L}_{p,\lambda},$$

let $B_\sigma(E)(\mathbb{R}^n)$ and $\dot{B}_\sigma(E)(\mathbb{R}^n)$ be the sets of all functions f on $\mathbb{R}^n$ such that $\|f\|_{B_\sigma(E)} < \infty$ and $\|f\|_{\dot{B}_\sigma(E)} < \infty$, respectively, where

$$\|f\|_{B_\sigma(E)} = \sup_{r\geq 1} \frac{1}{r^\sigma} \|f\|_{E(Q_r)} \quad \text{and} \quad \|f\|_{\dot{B}_\sigma(E)} = \sup_{r>0} \frac{1}{r^\sigma} \|f\|_{E(Q_r)}.$$

Then $B_\sigma(L^p)(\mathbb{R}^n)$ and $B_\sigma(L_{p,\lambda})(\mathbb{R}^n)$ are Banach spaces, and $B_\sigma(\mathcal{L}_{p,\lambda})(\mathbb{R}^n)/\mathcal{C}$ is a Banach space equipped with the norm $\|f\|_{B_\sigma(\mathcal{L}_{p,\lambda})}$. Moreover, for the unit ball Q_1, $\|f\|_{B_\sigma(\mathcal{L}_{p,\lambda})} + |f_{Q_1}|$ is a norm and thereby $B_\sigma(\mathcal{L}_{p,\lambda})(\mathbb{R}^n)$ is a Banach space.

Example. In the paper by J. García-Cuerva and M.J.L. Herrero [11] and in the paper by J. Alvarez, M. Guzmán-Partida and J. Lakey [2] the authors introduced for $p \in [1, \infty)$ and $\lambda \in \mathbb{R}$, the non-homogeneous central Morrey space $B^{p,\lambda}(\mathbb{R}^n)$, the central Morrey space $\dot{B}^{p,\lambda}(\mathbb{R}^n)$, the λ-central mean oscillation space $\mathrm{CMO}^{p,\lambda}(\mathbb{R}^n)$ and the λ-central bounded mean oscillation space $\mathrm{CBMO}^{p,\lambda}(\mathbb{R}^n)$ with the following norms, respectively:

$$\|f\|_{B^{p,\lambda}} = \sup_{r\geq 1} \frac{1}{r^\lambda} \left(\fint_{Q_r} |f(y)|^p \, dy \right)^{1/p},$$

$$\|f\|_{\dot{B}^{p,\lambda}} = \sup_{r>0} \frac{1}{r^\lambda} \left(\fint_{Q_r} |f(y)|^p \, dy \right)^{1/p},$$

$$\|f\|_{\mathrm{CMO}^{p,\lambda}} = \sup_{r\geq 1} \frac{1}{r^\lambda} \left(\fint_{Q_r} |f(y) - f_{B_r}|^p \, dy \right)^{1/p}$$

and

$$\|f\|_{\mathrm{CBMO}^{p,\lambda}} = \sup_{r>0} \frac{1}{r^\lambda} \left(\fint_{Q_r} |f(y) - f_{B_r}|^p \, dy \right)^{1/p}.$$

Then these spaces are realized as $B_\sigma(E)(\mathbb{R}^n)$ and $\dot{B}_\sigma(E)(\mathbb{R}^n)$ with $E = L^p$ (or $E = L^p/\mathcal{C}$) and $\sigma = n/p + \lambda$.

Remark 2.4. The spaces $B^{p,\lambda}(\mathbb{R}^n)$, $\dot{B}^{p,\lambda}(\mathbb{R}^n)$, $\mathrm{CMO}^{p,\lambda}(\mathbb{R}^n)$ and $\mathrm{CBMO}^{p,\lambda}(\mathbb{R}^n)$ were introduced as an extension of $B^p(\mathbb{R}^n)$, $\dot{B}^p(\mathbb{R}^n)$, $\mathrm{CMO}^p(\mathbb{R}^n)$ and $\mathrm{CBMO}^p(\mathbb{R}^n)$, respectively: When $\lambda = n(1/q - 1)$,

$$B^{p,\lambda}(\mathbb{R}^n) = B^p_q(\mathbb{R}^n) \quad \text{and} \quad \mathrm{CMO}^{p,\lambda}(\mathbb{R}^n) = \mathrm{CMO}^p_q(\mathbb{R}^n).$$

Moreover, when $q = 1$, i.e., $\lambda = 0$,

$$B^{p,\lambda}(\mathbb{R}^n) = B^p(\mathbb{R}^n) \quad \text{and} \quad \mathrm{CMO}^{p,\lambda}(\mathbb{R}^n) = \mathrm{CMO}^p(\mathbb{R}^n).$$

Also, for $\dot{B}^{p,\lambda}(\mathbb{R}^n)$ and $\mathrm{CBMO}^{p,\lambda}(\mathbb{R}^n)$, analogous properties hold. For these spaces, we refer to Y. Chen and K. Lau [4], J. García-Cuerva [10], J. García-Cuerva and M.J.L. Herrero [11] and S.Z. Lu and D.C. Yang [17, 18] (cf. [16, Examples 1 and 2]).

Remark 2.5. We note that $B_\sigma(L_{p,\lambda})(\mathbb{R}^n)$ unifies $L_{p,\lambda}(\mathbb{R}^n)$ and $B^{p,\lambda}(\mathbb{R}^n)$ and that $B_\sigma(\mathcal{L}_{p,\lambda})(\mathbb{R}^n)$ unifies $\mathcal{L}_{p,\lambda}(\mathbb{R}^n)$ and $\mathrm{CMO}^{p,\lambda}(\mathbb{R}^n)$. Actually, we have the following relations:

$$B_0(L_{p,\lambda})(\mathbb{R}^n) = L_{p,\lambda}(\mathbb{R}^n), \;\; B_0(\mathcal{L}_{p,\lambda})(\mathbb{R}^n) = \mathcal{L}_{p,\lambda}(\mathbb{R}^n), \tag{2.1}$$

$$B_{\lambda+n/p}(L_{p,-n/p})(\mathbb{R}^n) = B^{p,\lambda}(\mathbb{R}^n), \;\; B_{\lambda+n/p}(\mathcal{L}_{p,-n/p})(\mathbb{R}^n) = \mathrm{CMO}^{p,\lambda}(\mathbb{R}^n). \tag{2.2}$$

In the above relations, the first three follow immediately from their definitions, and the last one follows from Theorem 2.6 below. We also have the same properties for the function spaces $\dot{B}_\sigma(L_{p,\lambda})(\mathbb{R}^n)$ and $\dot{B}_\sigma(\mathcal{L}_{p,\lambda})(\mathbb{R}^n)$.

Theorem 2.6. *If $p \in [1,\infty)$ and $\lambda \in [-n/p, 0)$, then, for each $r > 0$, $\mathcal{L}_{p,\lambda}(Q_r)/\mathcal{C} \cong L_{p,\lambda}(Q_r)$. More precisely, the map $f \mapsto f - f_{Q_r}$ is bijective and bicontinuous from $\mathcal{L}_{p,\lambda}(Q_r)/\mathcal{C}$ to $L_{p,\lambda}(Q_r)$, i.e., there exists a constant $C > 0$, dependent only on n and λ, such that*

$$C^{-1}\|f\|_{\mathcal{L}_{p,\lambda}(Q_r)} \leq \|f - f_{Q_r}\|_{L_{p,\lambda}(Q_r)} \leq C\|f\|_{\mathcal{L}_{p,\lambda}(Q_r)}.$$

The same conclusion holds on $\mathbb{R}^n$ by using $\lim_{r\to\infty} f_{Q_r}$ instead of f_{Q_r}.

For the proof of this theorem, see [20, 22].

At the end of this section, we recall the following relations among classical function spaces (see [16, Proposition 1]).

Proposition 2.7. *Let $p \in [1,\infty)$.*

(i) (Supercritical case) *If $\lambda \in (-\infty, -n/p)$, then*

$$\begin{cases} B^{p,\lambda}(\mathbb{R}^n) = \dot{B}^{p,\lambda}(\mathbb{R}^n) = L_{p,\lambda}(\mathbb{R}^n) = \{0\}, \\ \mathrm{CMO}^{p,\lambda}(\mathbb{R}^n) = \mathrm{CBMO}^{p,\lambda}(\mathbb{R}^n) = \mathcal{L}_{p,\lambda}(\mathbb{R}^n) = \mathcal{C}. \end{cases}$$

(ii) (Critical case) *If $\lambda = -n/p$, then*

$$\begin{cases} B^{p,\lambda}(\mathbb{R}^n) = \dot{B}^{p,\lambda}(\mathbb{R}^n) = L_{p,\lambda}(\mathbb{R}^n) = L^p(\mathbb{R}^n), \\ \mathrm{CMO}^{p,\lambda}(\mathbb{R}^n)/\mathcal{C} = \mathrm{CBMO}^{p,\lambda}(\mathbb{R}^n)/\mathcal{C} = \mathcal{L}_{p,\lambda}(\mathbb{R}^n)/\mathcal{C} \cong L^p(\mathbb{R}^n). \end{cases}$$

(iii) (Subcritical case) *If $\lambda \in (-n/p, 0)$, then*

$$\begin{cases} \mathrm{CMO}^{p,\lambda}(\mathbb{R}^n)/\mathcal{C} \cong B^{p,\lambda}(\mathbb{R}^n), \\ \mathrm{CBMO}^{p,\lambda}(\mathbb{R}^n)/\mathcal{C} \cong \dot{B}^{p,\lambda}(\mathbb{R}^n), \\ \mathcal{L}_{p,\lambda}(\mathbb{R}^n)/\mathcal{C} \cong L_{p,\lambda}(\mathbb{R}^n). \end{cases}$$

(iv) (Critical case) *If $\lambda = 0$, then*

$$\begin{cases} L_{p,\lambda}(\mathbb{R}^n) = L^\infty(\mathbb{R}^n), \\ \mathcal{L}_{p,\lambda}(\mathbb{R}^n) = \mathrm{BMO}(\mathbb{R}^n). \end{cases}$$

(v) (Subcritical case) *If $\lambda \in (0,1]$, then*

$$\begin{cases} L_{p,\lambda}(\mathbb{R}^n) = \{0\}, \\ \mathcal{L}_{p,\lambda}(\mathbb{R}^n) = \mathrm{Lip}_\lambda(\mathbb{R}^n). \end{cases}$$

(vi) (Supercritical case) *If $\lambda \in (1,\infty)$, then*

$$\begin{cases} L_{p,\lambda}(\mathbb{R}^n) = \{0\}, \\ \mathcal{L}_{p,\lambda}(\mathbb{R}^n) = \mathcal{C}. \end{cases}$$

In the above, for a couple of Banach spaces (A,B), $A \cong B$ means that there exists a bijective and bicontinuous map from A to B, while $A = B$ means that A and B are identical as the set and that their norms are equivalent.

Here, note that $\mathrm{Lip}_\lambda(\mathbb{R}^n)$ ($\lambda \in (0,1]$) stands for the Lipschitz space of order λ on $\mathbb{R}^n$ with the norm

$$\|f\|_{\mathrm{Lip}_\lambda} = \sup_{x,y\in\mathbb{R}^n,\, x\neq y} \frac{|f(x)-f(y)|}{|x-y|^\lambda}.$$

which is regarded as the space of functions defined everywhere on $\mathbb{R}^n$. For the relation between Campanato and Lipschitz spaces, see [16, 19].

Remark 2.8. By this proposition, we need to restrict the range of the parameter λ to $[-n/p, 0]$ for $L_{p,\lambda}(\mathbb{R}^n)$ and to $[-n/p, 1]$ for $\mathcal{L}_{p,\lambda}(\mathbb{R}^n)$ so that the space contains non-constant functions.

3. Main results

We first recall the definitions of Calderón–Zygmund operator and commutator.

Definition 3.1. Let $K \in L^1_{\mathrm{loc}}(\mathbb{R}^n \times \mathbb{R}^n \setminus \{(x,x) : x \in \mathbb{R}^n\})$ be a standard kernel, i.e., it satisfies the following conditions:

(i) (Size condition)

$$|K(x,y)| \le \frac{C}{|x-y|^n} \tag{3.1}$$

for $x \neq y$;

(ii) (Regularity condition)
There exists $\delta > 0$ such that

$$|K(x,y) - K(z,y)| + |K(y,x) - K(y,z)| \le C \frac{|x-z|^\delta}{|x-y|^{n+\delta}}$$

for $|x-y| \ge 2|x-z|$.

And let T be a linear operator from $\mathcal{S}(\mathbb{R}^n)$ to $\mathcal{S}'(\mathbb{R}^n)$. Then we say that T is a Calderón–Zygmund operator (associated with a standard kernel K), if

$$\|Tf\|_{L^2} \le C\|f\|_{L^2}, \quad f \in \mathcal{S}(\mathbb{R}^n),$$

and there is a standard kernel K such that

$$Tf(x) = \int_{\mathbb{R}^n} K(x,y)f(y)dy$$

for all $f \in C^\infty_{\mathrm{comp}}(\mathbb{R}^n)$ and for all $x \notin \operatorname{supp} f$, where $C^\infty_{\mathrm{comp}}(\mathbb{R}^n)$ is the set of all infinitely differentiable functions on $\mathbb{R}^n$ with compact support.

Here, it is well known that the Calderón–Zygmund operator T is bounded on $L^p(\mathbb{R}^n)$, where $p \in (1,\infty)$, and is weak $(1,1)$.

Definition 3.2. Let $b \in L^1_{\mathrm{loc}}(\mathbb{R}^n)$ and T be a Calderón–Zygmund operator T. Then the commutator $[b,T]$ generated by b and T is defined by

$$[b,T]f(x) = b(x)Tf(x) - T(bf)(x), \quad x \in \mathbb{R}^n,$$

for $f \in L^\infty_{\mathrm{comp}}(\mathbb{R}^n)$.

For the boundedness of commutator $[b,T]$ on $B_\sigma(L_{p,\lambda})(\mathbb{R}^n)$, we can show the following:

Theorem 3.3. *Let T be a Calderón–Zygmund operator and $\sigma,\tau,\rho \in [0,\infty)$, $p,q \in (1,\infty)$, $t \in [1,\infty)$, $\lambda \in [-n/p,0)$, $\mu \in [-n/q,0)$, $\nu \in [-n/t,1]$. Assume that*

$$\tau = \sigma + \rho, \quad \mu = \lambda + \nu, \quad \sigma + \lambda < 0, \quad \sigma + \rho + \lambda + \nu < 0$$

and

$$q \le \begin{cases} (\lambda/\mu)p & \text{if } \nu \in [0,1], \\ pt/(p+t) & \text{if } \nu \in [-n/t,0). \end{cases}$$

If $b \in B_\rho(\mathcal{L}_{t,\nu})(\mathbb{R}^n)$, then the commutator $[b,T]$ is bounded from $B_\sigma(L_{p,\lambda})(\mathbb{R}^n)$ to $B_\tau(L_{q,\mu})(\mathbb{R}^n)$, i.e., there exists a constant $C>0$ such that

$$\|[b,T]f\|_{B_\tau(L_{q,\mu})} \le C\|b\|_{B_\rho(\mathcal{L}_{t,\nu})}\|f\|_{B_\sigma(L_{p,\lambda})},$$

where C is independent of $b \in B_\rho(\mathcal{L}_{t,\nu})(\mathbb{R}^n)$ and $f \in B_\sigma(L_{p,\lambda})(\mathbb{R}^n)$.

The same conclusion holds for $\dot{B}_\sigma(L_{p,\lambda})(\mathbb{R}^n)$, if $b \in \dot{B}_\rho(\mathcal{L}_{t,\nu})(\mathbb{R}^n)$.

In the above theorem, if $\sigma=\rho=0$, then the following corollary is obtained by (2.1).

Corollary 3.4. *Let T be a Calderón–Zygmund operator and $p,q \in (1,\infty)$, $t \in [1,\infty)$, $\lambda \in [-n/p,0)$, $\mu \in [-n/q,0)$, $\nu \in [-n/t,1]$. Assume that $\mu=\lambda+\nu$ and*

$$q \le \begin{cases} (\lambda/\mu)p & \text{if } \nu \in [0,1], \\ pt/(p+t) & \text{if } \nu \in [-n/t,0). \end{cases}$$

If $b \in B_\rho(\mathcal{L}_{t,\nu})(\mathbb{R}^n)$, then the commutator $[b,T]$ is bounded from $L_{p,\lambda}(\mathbb{R}^n)$ to $L_{q,\mu}(\mathbb{R}^n)$, i.e., there exists a constant $C>0$ such that

$$\|[b,T]f\|_{L_{q,\mu}} \le C\|b\|_{\mathcal{L}_{t,\nu}}\|f\|_{L_{p,\lambda}},$$

where C is independent of $b \in \mathcal{L}_{t,\nu}(\mathbb{R}^n)$ and $f \in L_{p,\lambda}(\mathbb{R}^n)$.

Remark 3.5. In Corollary 3.4, if $\nu=0$, then $q \le p$ and $b \in \mathrm{BMO}(\mathbb{R}^n)$, so that this corollary contains the result of G. Di Fazio and M.A. Ragusa [7], i.e., Theorem 4.3 below. Hence, furthermore, if $\lambda=-n/p$, then this corollary also contains the result of R.R. Coifman, R. Rochberg and G. Weiss [6], i.e., Theorem 4.2 below.

Also, if $\nu=-n/t$, $\rho=\nu+n/t$ and $\lambda=-n/p$, $\sigma=\lambda+n/p$ in Theorem 3.3, then by (2.2), we obtain the following corollary, which is due to Y. Komori [14] (cf. J. Alvarez, M. Guzmán-Partida and J. Lakey [2]).

Corollary 3.6. *Let T be a Calderón–Zygmund operator and $p,q,t \in (1,\infty)$, $\lambda \in [-n/p,0)$, $\mu \in [-n/q,0)$, $\nu \in [-n/t,\infty)$. Assume that $\mu=\lambda+\nu$ and $1/q = 1/p+1/t$. If $b \in \mathrm{CMO}^{t,\nu}(\mathbb{R}^n)$, then the commutator $[b,T]$ is bounded from $B^{p,\lambda}(\mathbb{R}^n)$ to $B^{q,\mu}(\mathbb{R}^n)$, i.e., there exists a constant $C>0$ such that*

$$\|[b,T]f\|_{B^{q,\mu}} \le C\|b\|_{\mathrm{CMO}^{t,\nu}}\|f\|_{B^{p,\lambda}},$$

where C is independent of $b \in \mathrm{CMO}^{t,\nu}(\mathbb{R}^n)$ and $f \in B^{p,\lambda}(\mathbb{R}^n)$.

The same conclusion holds for $\dot{B}^{p,\lambda}(\mathbb{R}^n)$, if $b \in \mathrm{CBMO}^{t,\nu}(\mathbb{R}^n)$.

4. Preliminary results and lemmas

In what follows, we use the symbol $A \lesssim B$ to denote that there exists a constant $C>0$ such that $A \le CB$. If $A \lesssim B$ and $B \lesssim A$, we then write $A \sim B$.

First, we list the results on boundedness of the Calderón–Zygmund operator T on Morrey spaces and the commutator $[b,T]$ generated by $b \in \mathrm{BMO}(\mathbb{R}^n)$ and T on Lebesgue spaces or Morrey spaces.

Theorem 4.1 (**[5, 21, 23]**). *If $p \in (1,\infty)$ and $\lambda \in [-n/p, 0)$, then the Calderón–Zygmund operator T is bounded on $L_{p,\lambda}(\mathbb{R}^n)$, i.e., there exists a constant $C > 0$ such that*

$$\|Tf\|_{L_{p,\lambda}} \le C\|f\|_{L_{p,\lambda}}, \quad f \in L_{p,\lambda}(\mathbb{R}^n).$$

Theorem 4.2 (**[6]**). *If $p \in (1,\infty)$, T is a Calderón–Zygmund operator and $b \in \mathrm{BMO}(\mathbb{R}^n)$, then the commutator $[b,T]$ is bounded on $L^p(\mathbb{R}^n)$.*

Theorem 4.3 (**[7]**). *If $p \in (1,\infty)$, $\lambda \in (-n/p, 0)$, T is a Calderón–Zygmund operator and $b \in \mathrm{BMO}(\mathbb{R}^n)$, then the commutator $[b,T]$ is bounded on $L_{p,\lambda}(\mathbb{R}^n)$, i.e., there exists a constant $C > 0$ such that*

$$\|[b,T]f\|_{L_{p,\lambda}} \le C\|b\|_{\mathrm{BMO}}\|f\|_{L_{p,\lambda}}, \quad f \in L_{p,\lambda}(\mathbb{R}^n).$$

Next, we invoke several lemmas for B_σ-Morrey–Campanato spaces.

Lemma 4.4 ([19, Lemma 3.5]). *Let $p \in [1,\infty)$ and $r > 0$. If $\lambda \in [-n/p, 0)$, then*

$$\|f\chi_{Q_r}\|_{L_{p,\lambda}} \le \|f\|_{L_{p,\lambda}(Q_{3r})}$$

for all $f \in L^p_{\mathrm{loc}}(\mathbb{R}^n)$ with $\|f\|_{L_{p,\lambda}(Q_{3r})} < \infty$.

Lemma 4.5 ([19, Lemma 4.1]). *Let $p \in [1,\infty)$, $\beta, \lambda \in \mathbb{R}$ and $\sigma \in [0,\infty)$. If $\beta + \sigma + \lambda < 0$, then there exists a constant $C > 0$ such that*

$$\int_{\mathbb{R}^n\setminus Q_r} \frac{|f(y)|}{|y|^{n-\beta}}\,dy \le Cr^{\beta+\sigma+\lambda}\|f\|_{B_\sigma(L_{p,\lambda})} \quad \text{for all } f \in B_\sigma(L_{p,\lambda})(\mathbb{R}^n) \text{ and } r \ge 1,$$

and

$$\int_{\mathbb{R}^n\setminus Q_r} \frac{|f(y)|}{|y|^{n-\beta}}\,dy \le Cr^{\beta+\sigma+\lambda}\|f\|_{\dot{B}_\sigma(L_{p,\lambda})} \quad \text{for all } f \in \dot{B}_\sigma(L_{p,\lambda})(\mathbb{R}^n) \text{ and } r > 0.$$

Lemma 4.6. *Let $t \in [1,\infty)$, $\nu \in \mathbb{R}$, $\rho \in [0,\infty)$ and $j \in \mathbb{N} \cap [2,\infty)$. If $\rho + \nu \ge 0$, then there exists a constant $C > 0$ such that*

$$\left(\fint_{Q_{2^j r}} |b(y) - b_{Q_r}|^t\,dy\right)^{1/t} \le Cj(2^j r)^{\rho+\nu}\|b\|_{B_\rho(\mathcal{L}_{t,\nu})}$$

$$\text{for all } b \in B_\rho(\mathcal{L}_{t,\nu})(\mathbb{R}^n) \text{ and } r \ge 1,$$

and

$$\left(\fint_{Q_{2^j r}} |b(y) - b_{Q_r}|^t\,dy\right)^{1/t} \le Cj(2^j r)^{\rho+\nu}\|b\|_{\dot{B}_\rho(\mathcal{L}_{t,\nu})}$$

$$\text{for all } b \in \dot{B}_\rho(\mathcal{L}_{t,\nu})(\mathbb{R}^n) \text{ and } r > 0.$$

Note that the proof shows that when $\rho + \nu > 0$, we do not need the factor j.

Proof. We prove only the case $f \in B_\rho(\mathcal{L}_{t,\nu})(\mathbb{R}^n)$ and $r \geq 1$. Since $\rho+\nu \geq 0$, it follows that

$$\left(\fint_{Q_{2^j r}} |b(y)-b_{Q_r}|^t\,dy\right)^{1/t} \leq \left(\fint_{Q_{2^j r}} |b(y)-b_{Q_{2^j r}}|^t\,dy\right)^{1/t} + \sum_{k=1}^{j} |b_{Q_{2^k r}} - b_{Q_{2^{k-1} r}}|$$
$$\lesssim (2^j r)^\nu \|b\|_{\mathcal{L}_{t,\nu}(Q_{2^j r})} + \sum_{k=1}^{j} (2^k r)^\nu \|b\|_{\mathcal{L}_{t,\nu}(Q_{2^k r})}$$
$$\lesssim j\,(2^j r)^{\rho+\nu} \|b\|_{B_\rho(\mathcal{L}_{t,\nu})}.$$

The proof for $b \in \dot{B}_\rho(\mathcal{L}_{t,\nu})(\mathbb{R}^n)$ and $r > 0$ is the same as above. □

Lemma 4.7. *Let $p,t \in [1,\infty)$, $\lambda,\nu \in \mathbb{R}$ and $\sigma,\rho \in [0,\infty)$. If $1/p+1/t \leq 1$, $\rho+\nu \geq 0$ and $\beta+\sigma+\rho+\lambda+\nu < 0$, then there exists a constant $C > 0$ such that*

$$\int_{\mathbb{R}^n \setminus Q_r} \frac{|b(y)-b_{Q_r}||f(y)|}{|y|^{n-\beta}}\,dy \leq C r^{\beta+\sigma+\rho+\lambda+\nu} \|b\|_{B_\rho(\mathcal{L}_{t,\nu})} \|f\|_{B_\sigma(L_{p,\lambda})}$$
$$\text{for all } b \in B_\rho(\mathcal{L}_{t,\nu})(\mathbb{R}^n),\ f \in B_\sigma(L_{p,\lambda})(\mathbb{R}^n) \text{ and } r \geq 1,$$

and

$$\int_{\mathbb{R}^n \setminus Q_r} \frac{|b(y)-b_{Q_r}||f(y)|}{|y|^{n-\beta}}\,dy \leq C r^{\beta+\sigma+\rho+\lambda+\nu} \|b\|_{\dot{B}_\rho(\mathcal{L}_{t,\nu})} \|f\|_{\dot{B}_\sigma(L_{p,\lambda})}$$
$$\text{for all } b \in \dot{B}_\rho(\mathcal{L}_{t,\nu})(\mathbb{R}^n),\ f \in \dot{B}_\sigma(L_{p,\lambda})(\mathbb{R}^n) \text{ and } r > 0.$$

Proof. We prove only the case $b \in B_\rho(\mathcal{L}_{t,\nu})(\mathbb{R}^n)$, $f \in B_\sigma(L_{p,\lambda})(\mathbb{R}^n)$ and $r \geq 1$, since the proof for $b \in \dot{B}_\rho(\mathcal{L}_{t,\nu})(\mathbb{R}^n)$, $f \in \dot{B}_\sigma(L_{p,\lambda})(\mathbb{R}^n)$ and $r > 0$ is similar.

It is again significant to prove the assertion when $1/p+1/t = 1$, which we assume. By decomposing $\mathbb{R}^n \setminus Q_r$ dyadically, we obtain

$$\int_{\mathbb{R}^n \setminus Q_r} \frac{|b(y)-b_{Q_r}||f(y)|}{|y|^{n-\beta}}\,dy \lesssim \sum_{j=1}^{\infty} \frac{1}{(2^j r)^{n-\beta}} \int_{Q_{2^j r} \setminus Q_{2^{j-1} r}} |b(y)-b_{Q_r}||f(y)|\,dy$$
$$\lesssim r^\beta \sum_{j=1}^{\infty} 2^{j\beta} \fint_{Q_{2^j r}} |b(y)-b_{Q_r}||f(y)|\,dy.$$

Then, it follows from Hölder's inequality and Lemma 4.6 that

$$\int_{\mathbb{R}^n \setminus Q_r} \frac{|b(y)-b_{Q_r}||f(y)|}{|y|^{n-\beta}}\,dy$$
$$\leq r^\beta \sum_{j=1}^{\infty} 2^{j\beta} \left(\fint_{Q_{2^j r}} |b(y)-b_{Q_r}|^t\,dy\right)^{1/t} \left(\fint_{Q_{2^j r}} |f(y)|^p\,dy\right)^{1/p}$$
$$\lesssim r^{\beta+\sigma+\rho+\lambda+\nu} \|b\|_{B_\rho(\mathcal{L}_{t,\nu})} \|f\|_{B_\sigma(L_{p,\lambda})} \sum_{j=1}^{\infty} j 2^{j(\beta+\sigma+\rho+\lambda+\nu)}.$$

Since $\beta+\sigma+\rho+\lambda+\nu<0$, the series in the most right-hand side converges to a finite value, we conclude

$$\int_{\mathbb{R}^n\setminus Q_r} \frac{|b(y)-b_{Q_r}||f(y)|}{|y|^{n-\beta}}\,dy \lesssim r^{\beta+\sigma+\rho+\lambda+\nu}\|b\|_{B_\rho(\mathcal{L}_{t,\nu})}\|f\|_{B_\sigma(L_{p,\lambda})}.$$

Thus, the proof is complete. □

5. Proof of Theorem 3.3

Let $b\in B_\rho(\mathcal{L}_{t,\nu})(\mathbb{R}^n)$, $f\in B_\sigma(L_{p,\lambda})(\mathbb{R}^n)$ and $r\geq 1$. Then, we prove that for any ball Q_r,

$$\|[b,T](f)\|_{L_{q,\mu}(Q_r)} \lesssim r^{\tau}\|b\|_{B_\rho(\mathcal{L}_{t,\nu})}\|f\|_{B_\sigma(L_{p,\lambda})}.$$

Here $\tau=\sigma+\rho$.

To prove this, we use the decomposition of f

$$f=f\chi_{Q_{2r}}+f(1-\chi_{Q_{2r}})=f^0+f^\infty,\quad \text{say},$$

and then for a Calderón–Zygmund operator T with kernel K, we have

$$Tf=Tf^0+Tf^\infty.$$

Now, for $x\in Q_r$, it follows that

$$\begin{aligned}[b,T](f)(x) &=[b,T](f^0)(x)+(b(x)-b_{Q_{2r}})Tf^\infty(x)-T((b-b_{Q_{2r}})f^\infty)(x)\\ &=I^0(x)+I_1^\infty(x)+I_2^\infty(x),\quad \text{say}.\end{aligned}$$

(i) First, when $\nu\in[0,1]$ and $q\leq(\lambda/\mu)p$, to estimate I^0, let

$$h(x)=\begin{cases}1, & (|x|\leq 1)\\ 0, & (|x|\geq 2)\end{cases}\quad \text{such that}\quad \|h\|_{\mathrm{Lip}_1(\mathbb{R}^n)}\leq 1,$$

and

$$h_r(x)=h(x/r),$$

and then

$$\|I^0\|_{L_{q,\mu}(Q_r)}\leq \|[bh_{4r},T](f^0)\|_{L_{q,\mu}}.$$

If $\nu=0$, then it follows from Theorem 4.3 that

$$\begin{aligned}\|[bh_{4r},T](f^0)\|_{L_{q,\mu}} &\lesssim \|bh_{4r}\|_{\mathrm{BMO}}\|f^0\|_{L_{q,\mu}}\\ &\leq \|bh_{4r}\|_{\mathcal{L}_{t,\nu}}\|f^0\|_{L_{p,\lambda}}.\end{aligned}$$

If $\nu>0$, then it follows from the boundedness of I_ν from $L_{p,\lambda}(\mathbb{R}^n)$ to $L_{q,\mu}(\mathbb{R}^n)$ (see [1]) that

$$\begin{aligned}\|[bh_{4r},T](f^0)\|_{L_{q,\mu}} &\lesssim \|bh_{4r}\|_{\mathrm{Lip}_\nu}\|I_\nu(|f^0|)\|_{L_{q,\mu}}\\ &\lesssim \|bh_{4r}\|_{\mathcal{L}_{t,\nu}}\|f^0\|_{L_{p,\lambda}},\end{aligned}$$

where I_ν is the fractional integral operator defined by

$$I_\nu f(x)=\int_{\mathbb{R}^n}\frac{f(y)}{|x-y|^{n-\nu}}\,dy.$$

Therefore, we obtain

$$\|I^0\|_{L_{q,\mu}(Q_r)} \lesssim r^\tau \|b\|_{B_\rho(\mathcal{L}_{t,\nu})}\|f\|_{B_\sigma(L_{p,\lambda})}.$$

Next, when $\nu \in [-n/t, 0)$ and $q \le pt/(p+t)$, to estimate I^0, we decompose $I^0(x)$ further into the following pieces:

$$\begin{aligned} I^0(x) &= (b(x) - b_{Q_{2r}})Tf^0(x) - T((b - b_{Q_{2r}})f^0)(x) \\ &= I_1^0(x) + I_2^0(x), \quad \text{say.} \end{aligned}$$

To estimate I_1^0, by applying the Hölder inequality for the indices p, q, t satisfying $1/q \ge 1/p + 1/t$, we deduce

$$\begin{aligned} &\|I_1^0\|_{L_{q,\mu}(Q_r)} \\ &\quad \le \sup_{Q(\xi,s)\subset Q_r} \frac{1}{s^\mu}\left(⨍_{Q(\xi,s)} |b(y) - b_{Q_{2r}}|^t\,dy\right)^{1/t}\left(⨍_{Q(\xi,s)} |Tf^0(y)|^p\,dy\right)^{1/p}. \end{aligned}$$

From the assumption $\mu = \lambda + \nu$ and the definitions of the norms, we have

$$\begin{aligned} \|I_1^0\|_{L_{q,\mu}(Q_r)} &\lesssim \|b\|_{\mathcal{L}_{t,\nu}(Q_r)}\|Tf^0\|_{L_{p,\lambda}(Q_r)} \\ &\lesssim r^\rho \|b\|_{B_\rho(\mathcal{L}_{t,\nu})}\|Tf^0\|_{L_{p,\lambda}(Q_r)}. \end{aligned}$$

Now, by the boundedness of T on $L_{p,\lambda}(\mathbb{R}^n)$ (Theorem 4.1) and Lemma 4.4, we have

$$\|Tf^0\|_{L_{p,\lambda}(Q_r)} \lesssim \|f^0\|_{L_{p,\lambda}} \le \|f\|_{L_{p,\lambda}(Q_{6r})} \lesssim r^\sigma \|f\|_{B_\sigma(L_{p,\lambda})}.$$

Hence, we obtain

$$\|I_1^0\|_{L_{q,\nu}(Q_r)} \lesssim r^\tau \|b\|_{B_\rho(\mathcal{L}_{t,\nu})}\|f\|_{B_\sigma(L_{p,\lambda})}.$$

Next, by using the boundedness of T on $L_{q,\mu}(\mathbb{R}^n)$,

$$\begin{aligned} \|I_2^0\|_{L_{q,\nu}(Q_r)} &\le \|T((b - b_{Q_{2r}})f^0)\|_{L_{q,\mu}} \lesssim \|(b - b_{Q_{2r}})f^0\|_{L_{q,\mu}} \\ &= \sup_{\substack{\xi\in\mathbb{R}^n \\ s>0}} \frac{1}{s^\mu}\left(⨍_{Q(\xi,s)} |(b(y) - b_{Q_{2r}})f^0(y)|^q\,dy\right)^{1/q}. \end{aligned}$$

Here we may assume that $Q(\xi, s) \cap Q_{2r} \ne \emptyset$ since f^0 is supported on Q_{2r} and we are now considering the integral over $Q(\xi, s)$. Now, if $s < r/2$, then $Q(\xi, s) \subset Q_{3r}$. Consequently, from the Hölder inequality, Theorem 2.6 and Lemma 4.4, it follows that

$$\begin{aligned} &\sup_{\substack{\xi\in\mathbb{R}^n \\ 0<s<r/2}} \frac{1}{s^\mu}\left(⨍_{Q(\xi,s)} |(b(y) - b_{Q_{2r}})f^0(y)|^q\,dy\right)^{1/q} \\ &\le \sup_{Q(\xi,s)\subset Q_{3r}} \frac{1}{s^\mu}\left(⨍_{Q(\xi,s)} |(b(y) - b_{Q_{2r}})f^0(y)|^q\,dy\right)^{1/q} \lesssim \|b\|_{\mathcal{L}_{t,\nu}(Q_{3r})}\|f^0\|_{L_{p,\lambda}} \\ &\le r^\rho \|b\|_{B_\rho(\mathcal{L}_{t,\nu})}\|f\|_{L_{p,\lambda}(B^{6r})} \lesssim r^\tau \|b\|_{B_\rho(\mathcal{L}_{t,\nu})}\|f\|_{B_\sigma(L_{p,\lambda})}. \end{aligned}$$

Assume $s \ge r/2$ instead. Then we have by Hölder's inequality for the indices p, q, t satisfying $1/q \ge 1/p + 1/t$ and by virtue of Lemma 4.4 again,

$$\sup_{\substack{\xi\in\mathbb{R}^n\\ s\ge r/2}} \frac{1}{s^\mu}\left(\fint_{Q(\xi,s)} |(b(y)-b_{Q_{2r}})f^0(y)|^q\,dy\right)^{1/q} \\ \le \sup_{\substack{\xi\in\mathbb{R}^n\\ s\ge r/2}} \frac{1}{s^{\mu+n/t}}\left(\int_{Q(\xi,s)\cap Q_{2r}} |(b(y)-b_{Q_{2r}})|^t\,dy\right)^{1/t}\left(\fint_{Q(\xi,s)} |f^0(y)|^p\,dy\right)^{1/p}.$$

Consequently, from the definition of the norm $\|\cdot\|_{L_{p,\lambda}}$, it follows that

$$\sup_{\substack{\xi\in\mathbb{R}^n\\ s\ge r/2}} \frac{1}{s^\mu}\left(\fint_{Q(\xi,s)} |(b(y)-b_{Q_{2r}})f^0(y)|^q\,dy\right)^{1/q} \\ \le \sup_{\substack{\xi\in\mathbb{R}^n\\ s\ge r/2}} \frac{(2r)^\nu}{s^\nu}\frac{1}{(2r)^\nu}\left(\frac{|Q_{2r}|}{|Q(\xi,s)|}\fint_{Q_{2r}} |b(y)-b_{Q_{2r}}|^t\,dy\right)^{1/t}\|f^0\|_{L_{p,\lambda}} \\ \lesssim \|b\|_{\mathcal{L}_{t,\nu}(Q_{2r})}\|f\|_{L_{p,\lambda}(Q_{6r})} \\ \lesssim r^\tau\|b\|_{B_\rho(\mathcal{L}_{t,\nu})}\|f\|_{B_\sigma(L_{p,\lambda})}.$$

Hence, we have

$$\|I_2^0\|_{L_{q,\nu}(Q_r)} \lesssim r^\tau\|b\|_{B_\rho(\mathcal{L}_{t,\nu})}\|f\|_{B_\sigma(L_{p,\lambda})}.$$

Therefore, we obtain

$$\|I^0\|_{L_{q,\nu}(Q_r)} \lesssim r^\tau\|b\|_{B_\rho(\mathcal{L}_{t,\nu})}\|f\|_{B_\sigma(L_{p,\lambda})}.$$

(ii) To estimate I_1^∞, by using the same estimate as that of I_1^0, it follows that

$$\|I_1^\infty\|_{L_{q,\mu}(Q_r)} \le \sup_{Q(\xi,s)\subset Q_r} \frac{1}{s^\mu}\left(\fint_{Q(\xi,s)} |b(y)-b_{Q_{2r}}|^t\,dy\right)^{1/t}\left(\fint_{Q(\xi,s)} |Tf^\infty(y)|^p\,dy\right)^{1/p} \\ \lesssim r^\rho\|b\|_{B_\rho(\mathcal{L}_{t,\nu})}\|Tf^\infty\|_{L_{p,\lambda}(Q_r)}.$$

Also, by Lemma 4.5, we have

$$\|Tf^\infty\|_{L_{p,\lambda}(Q_r)} \le r^{-\lambda}\|Tf^\infty\|_{L^\infty(Q_r)} \lesssim r^{-\lambda}\int_{\mathbb{R}^n\setminus Q_{2r}} \frac{|f(y)|}{|y|^n}\,dy \lesssim r^\sigma\|f\|_{B_\sigma(L_{p,\lambda})},$$

since $\sigma + \lambda < 0$.

Therefore, we obtain

$$\|I_1^\infty\|_{L_{q,\nu}(Q_r)} \lesssim r^\tau\|b\|_{B_\rho(\mathcal{L}_{t,\nu})}\|f\|_{B_\sigma(L_{p,\lambda})}.$$

(iii) To estimate I_2^∞, by (3.1) and Lemma 4.7, we get

$$|T((b-b_{Q_{2r}})f^\infty)(x)| \lesssim \int_{\mathbb{R}^n\setminus Q_{2r}} \frac{|b(y)-b_{Q_{2r}}||f(y)|}{|y|^n}\,dy \\ \lesssim r^{\sigma+\rho+\lambda+\nu}\|b\|_{B_\rho(\mathcal{L}_{t,\nu})}\|f\|_{B_\sigma(L_{p,\lambda})}.$$

Therefore, we obtain

$$\|I_2^\infty\|_{L_{q,\mu}(Q_r)} \lesssim \sup_{Q(\xi,s)\subset Q_r} \frac{1}{s^{\lambda+\nu}} r^{\sigma+\rho+\lambda+\nu} \|b\|_{B_\rho(\mathcal{L}_{t,\nu})} \|f\|_{B_\sigma(L_{p,\lambda})}$$
$$\leq r^\tau \|b\|_{B_\rho(\mathcal{L}_{t,\nu})} \|f\|_{B_\sigma(L_{p,\lambda})},$$

since $\lambda + \nu < 0$.

Thus, combining the above estimates of I^0, I_1^∞ and I_2^∞, we have

$$\|[b,T](f)\|_{L_{q,\mu}(Q_r)} \lesssim r^\tau \|b\|_{B_\rho(\mathcal{L}_{t,\nu})} \|f\|_{B_\sigma(L_{p,\lambda})}.$$

Similarly, applying the same argument as above, the proof of the boundedness of $[b,T]$ from $\dot{B}_\sigma(L_{p,\lambda})(\mathbb{R}^n)$ to $\dot{B}_\tau(L_{q,\mu})(\mathbb{R}^n)$, if $b \in \dot{B}_\rho(\mathcal{L}_{t,\nu})(\mathbb{R}^n)$, is obtained.

Acknowledgment

The author spent the period of time for Overseas Researcher of Nihon University in the fall term of 2011 at Claremont McKenna College (CMC), Claremont, California, USA, and the greater part of this work was developed at that time. He would like to express his deep gratitude to Professor Asuman Güven Aksoy of CMC for her support. He also likes to thank the Department of Mathematics of CMC for their hospitality. And he is thankful to the referee for his/her very valuable comments and suggestions about the presentation.

References

[1] D.R. Adams, *A note on Riesz potentials.* Duke Math. J. **42** (1975), no. 4, 765–778.

[2] J. Alvarez, M. Guzmán-Partida and J. Lakey, *Spaces of bounded λ-central mean oscillation, Morrey spaces, and λ-central Carleson measures.* Collect. Math. **51** (2000), 1–47.

[3] A. Beurling, *Construction and analysis of some convolution algebra.* Ann. Inst. Fourier **14** (1964), 1–32.

[4] Y. Chen and K. Lau, *Some new classes of Hardy spaces.* J. Func. Anal. **84** (1989), 255–278.

[5] F. Chiarenza and M. Frasca, *Morrey spaces and Hardy–Littlewood maximal function.* Rend. Mat. Appl. **7** (1987), no. 3-4, 273–279.

[6] R.R. Coifman, R. Rochberg and G. Weiss, *Factorization theorems for Hardy spaces in several variables.* Ann. of Math. **103** (1976), 611–635.

[7] G. Di Fazio and M.A. Ragusa, *Commutators and Morrey spaces.* Bollettino U. M. I. **5-A** (1991), no. 7, 255–278.

[8] H. Feichtinger, *An elementary approach to Wiener's third Tauberian theorem on Euclidean n-space.* Proceedings of Conference at Cortona 1984, Symposia Mathematica **29**, Academic Press, New York, 1987, 267–301.

[9] Z. Fu, Y. Lin and S. Lu, *λ-central BMO estimates for commutators of singular integral operators with rough kernels.* Acta Math. Sin. (Engl. Ser.) **24** (2008), no. 3, 373–386.

[10] J. García-Cuerva, *Hardy spaces and Beurling algebras.* J. London Math. Soc. (2) **39** (1989), 499–513.

[11] J. García-Cuerva and M.J.L. Herrero, *A theory of Hardy spaces associated to the Herz spaces.* Proc. London Math. Soc. **69** (1994), 605–628.

[12] L. Grafakos, X. Li and D. Yang, *Bilinear operators on Herz-type Hardy spaces.* Trans. Amer. Math. Soc. **350** (1998), 1249–1275.

[13] C. Herz, *Lipschitz spaces and Bernstein's theorem on absolutely convergent Fourier transforms.* J. Math. Mech. **18** (1968), 283–324.

[14] Y. Komori, *Notes on singular integrals on some inhomogeneous Herz spaces.* Taiwanese J. Math. **8** (2004), 547–556.

[15] Y. Komori-Furuya and K. Matsuoka, *Strong and weak estimates for fractional integral operators on some Herz-type function spaces.* Proceedings of the Maratea Conference FAAT 2009, Rendiconti del Circolo Mathematico di Palermo, Serie II, Suppl., **82** (2010), 375–385.

[16] Y. Komori-Furuya, K. Matsuoka, E. Nakai and Y. Sawano, *Integral operators on B_σ-Morrey–Campanato spaces.* Rev. Mat. Complut. **26** (2013) 1–32, DOI:10.1007/s13163-011-0091-6.

[17] S.Z. Lu and D.C. Yang, *The Littlewood–Paley function and ϕ-transform characterizations of a new Hardy space HK_2 associated with the Herz space.* Studia Math. **101** (1992), no. 3, 285–298.

[18] S.Z. Lu and D.C. Yang, *The central BMO spaces and Littlewood–Paley operators.* Approx. Theory Appl. **11** (1995), 72–94.

[19] K. Matsuoka and E. Nakai, *Fractional integral operators on $B^{p,\lambda}$ with Morrey–Campanato norms.* Function Spaces IX, Banach Center Publ. **92**, 249–264, Inst. Math., Polish Acad. Sci., Warsawa, 2011.

[20] T. Mizuhara, *Relations between Morrey and Campanato spaces with some growth functions, II.* Proceedings of Harmonic Analysis Seminar 11, pp. 67–74 (1995) (in Japanese).

[21] E. Nakai, *Hardy–Littlewood maximal operator, singular integral operators and the Riesz potentials on generalized Morrey spaces.* Math. Nachr. **166** (1994), 95–103.

[22] E. Nakai, *The Campanato, Morrey and Hölder spaces on spaces of homogeneous type.* Studia Math. **176** (2006), no. 1, 1–19.

[23] J. Peetre, *On convolution operators leaving $L_{p,\lambda}$ spaces invariant.* Ann. Math. Pure Appl. **72** (1966), 295–304.

Katsuo Matsuoka
College of Economics, Nihon University
1-3-2 Misaki-cho, Chiyoda-ku
Tokyo 101-8360, Japan
e-mail: katsu.m@nihon-u.ac.jp

Operator Theory:
Advances and Applications, Vol. 236, 371–376

Composition Operators on Large Fractional Cauchy Transform Spaces

Yusuf Abu Muhanna and El-Bachir Yallaoui

Abstract. For $\alpha > 0$ and z in the unit disk $\mathbf{D}$ the spaces of fractional Cauchy transforms F_α are known as the family of all functions $f(z)$ such that $f(z) = \int_{\mathbf{T}} [K(\overline{x}z)]^\alpha d\mu(x)$ where $K(z) = (1-z)^{-1}$ is the Cauchy kernel, $\mathbf{T}$ is the unit circle and $\mu \in \mathfrak{M}$ the set of complex Borel measure on $\mathbf{T}$. The Banach space F_α may be written as $F_\alpha = (F_\alpha)_a \oplus (F_\alpha)_s$, where $(F_\alpha)_a$ is isomorphic to a closed subspace of $\mathfrak{M}_a$ the subset of absolutely continuous measures of $\mathfrak{M}$, and $(F_\alpha)_s$ is isomorphic to $\mathfrak{M}_s$ the subspace of $\mathfrak{M}$ of singular measures. In this article we show that for $\alpha \geq 1$, the composition operator C_φ is compact on F_α if and only if $C_\varphi [K^\alpha (\overline{x}z)] \subset (F_\alpha)_a$ and in doing so, extend a result due to [1] who showed that C_φ is compact on F_1 if and only if $C_\varphi (F_1) \subset (F_1)_a$.

Mathematics Subject Classification (2010). Primary: 30E20; Secondary: 30D99.

Keywords. Cauchy transforms, composition operators, absolutely continuous measures.

1. Introduction

Let $\mathbf{D} = \{z \in \mathbf{C} : |z| < 1\}$ be the open unit disc in the complex plane $\mathbf{C}$, and $\mathbf{T} = \partial\mathbf{D}$ the boundary of $\mathbf{D}$. We let $H(\mathbf{D})$ denote the class of holomorphic functions on $\mathbf{D}$. $H(\mathbf{D})$ is a locally convex linear topological space with respect to the topology given by uniform convergence on compact subsets of $\mathbf{D}$. We denote by $\mathfrak{M}$ be the set of all complex-valued Borel measures on $\mathbf{T}$.

For $z \in \mathbf{D}$, consider the classical Cauchy kernel $K(z) = 1/(1-z)$. For $\alpha > 0$ we define the space of Fractional Cauchy Transforms F_α as the space of analytic

This research has been supported by a grant from Sultan Qaboos University IG/SCI/DOMS/11/12.

We would like to thank the referee for his valuable comments.

functions $f \in H(\mathbf{D})$ defined in (1.1) for which there exists a measure $\mu \in \mathfrak{M}$ such that

$$f(z) = \int_{\mathbf{T}} K^{\alpha}(\overline{x}z) d\mu(x) \tag{1.1}$$

The norm on F_α defined by

$$\|f\|_{F_\alpha} = \inf_{\mu \in \mathfrak{M}} \left\{ \|\mu\| : f(z) = \int_{\mathbf{T}} K^{\alpha}(\overline{x}z)\, d\mu(x) \right\} \tag{1.2}$$

makes F_α into a Banach space. When $\alpha = 1$, F_1 is known as the classical Cauchy space. The spaces F_α have been studied extensively (see [3] for an extensive list of references).

Let $\mathfrak{M}_a := \{\mu_a \in \mathfrak{M} : \mu_a \ll m\}$ where m is the normalized Lebesgue measure on the unit circle, and $\mathfrak{M}_s := \{\mu_s \in \mathfrak{M} : \mu_s \perp m\}$. According to the Lebesgue decomposition theorem $\mathfrak{M} = \mathfrak{M}_a \oplus \mathfrak{M}_s$ thus any $\mu \in \mathfrak{M}$ can be written as $\mu = \mu_a + \mu_s$ where $\mu_a \in \mathfrak{M}_a$, $\mu_s \in \mathfrak{M}_s$ and $\|\mu\| = \|\mu_a\| + \|\mu_s\|$. The space F_α may be identified with $\mathfrak{M}/\overline{H_0^1}$ the quotient of the Banach space $\mathfrak{M}$ of Borel measures by $\overline{H_0^1}$ the subspace of L^1 consisting of functions with mean value zero whose conjugate belongs the Hardy space H^1. According to the Radon–Nikodým theorem since μ_a is absolutely continuous with respect to the Lebesgue measure there exists a non-negative L^1-valued function of the independent variable x, $x \to g_x$ which is continuous with respect to the L^1 norm such that we can decompose μ as $d\mu(x) = d\mu(e^{it}) = g(e^{it})dt + d\mu_s(e^{it})$ where $g(e^{it}) \in \overline{H_0^1} \subset L^1$ (see [4]).

Consequently the Banach space F_α may be written as $F_\alpha = (F_\alpha)_a \oplus (F_\alpha)_s$, where $(F_\alpha)_a$ is isomorphic to $L^1/\overline{H_0^1}$ a closed subspace of $\mathfrak{M}$ of absolutely continuous measures, and $(F_\alpha)_s$ is isomorphic to $\mathfrak{M}_s$ the closed subspace of $\mathfrak{M}$ of singular measures. If $f \in (F_\alpha)_a$, then the singular part is null and the measure μ for which the integral holds reduces to $d\mu(t) = d\mu(e^{it}) = g(e^{it})dt$ where $g(e^{it}) \in L^1$ and dt is the Lebesgue measure on $\mathbf{T}$. In which case the functions in $(F_\alpha)_a$ may be then written as, $f(z) = \int_{-\pi}^{\pi} K^{\alpha}(e^{-it}z)\, g(e^{it})dt$ where if $g(e^{it})$ is non-negative then $\|f\|_{F_\alpha} = \|g(e^{it})\|_{L^1}$.

If φ is an analytic self map of the unit disc $\mathbf{D}$, we say that φ induces a bounded composition operator C_φ on the Banach space X if there exists a positive constant A such that for all $f \in X$, $\|C_\varphi(f)\|_X = \|(f \circ \varphi)\|_X \le A\|f\|_X$. A bounded operator C_φ will be a compact operator if the image of every bounded set of X is relatively compact (i.e., has compact closure) in X. Equivalently C_φ is a compact operator on X if and only if for every bounded sequence $\{f_n\}$ of X, $\{C_\varphi(f_n)\}$ has a convergent subsequence in X.

The fractional Cauchy spaces F_α have been studied by several authors (see [3] for a complete list of references). It is known from [3] among other things that:

- If $\alpha > 0$ and φ is a conformal automorphism of $\mathbf{D}$ then $C_\varphi(f) \in F_\alpha$ for every $f \in F_\alpha$.
- If $\alpha > 0$ then C_φ is bounded on F_α if and only if $C_\varphi[K^\alpha(\overline{x}z)] \in F_\alpha$ for every $x \in \mathbf{T}$.

- If $\beta > \alpha > 0$ and if C_φ is bounded on F_α then C_φ is also bounded on F_β.
- If $\alpha \geq 1$ and $\varphi \in H(\mathbf{D})$ then $C_\varphi(f) \in F_\alpha$ for every $f \in F_\alpha$.
- If $0 < p \leq 1$ then $H^p \subset F_{1/p}$
- If $0 < \alpha < \beta$ then $F_\alpha \subset (F_\beta)_a$ (see [2]).
- C_φ is compact on F_1 if and only if $C_\varphi(F_1) \subset (F_1)_a$ (see [1]).

In this article we show that for all $\alpha \geq 1$ the composition operator C_φ is compact on F_α if and only if $C_\varphi[K^\alpha(\overline{x}z)] \in (F_\alpha)_a$ for every $x \in \mathbf{T}$ and in doing so generalize the previous result by [1]. Since F_α are Möbius invariant there is no loss of generality in assuming that $\varphi(0) = 0$ whenever we consider the composition operator C_φ acting on F_α and we will do so through out the article.

2. Compactness of the Composition Operator when $\alpha \geq 1$

Let φ be an analytic self map of the unit disc with $\varphi(0) = 0$. As was mentioned in the previous section C_φ is bounded on F_α for $\alpha \geq 1$. In this section we will show that compactness of the composition operator C_φ on F_α is strongly tied with the absolute continuity of the measure that supports it. First observe that $H^\infty \subset F_{1a}$ and thus $H^\infty \cap F_\alpha \subset F_{1a} \cap F_\alpha \subseteq (F_\alpha)_a$ for all $\alpha \geq 1$. Now we give our first main result.

Theorem 2.1. *If C_φ is compact on F_α for $\alpha \geq 1$ then $C_\varphi[K^\alpha(\overline{x}z)] \in (F_\alpha)_a$ for every $x \in \mathbf{T}$ and*

$$C_\varphi[K^\alpha(\overline{x}z)] = \int_{-\pi}^{\pi} g_x(e^{it})\, K^\alpha(e^{-it}z)dt \tag{2.1}$$

where g_x is a non-negative L^1 continuous function of x and $\|g_x\|_{L^1} \leq \|C_\varphi\| < \infty$.

Proof. Assume that C_φ is compact and let $\{f_j\}_{j=1}^\infty$ be a sequence of functions such that

$$f_j(z) = K^\alpha(\rho_j\overline{x}z) = \frac{1}{(1-\rho_j\overline{x}z)^\alpha}$$

where $0 < \rho_j < 1$ and $\lim_{j\to\infty}\rho_j = 1$. Then it is known from [3] that $f_j(z) \in F_\alpha$ for every j,and all $x = e^{it} \in \mathbf{T}$ and $\|f_j(z)\|_{F_\alpha} = 1$. Furthermore there exists a probability measure $\mu_j \in \mathfrak{M}$, such that $\|\mu_j\| = 1$, $d\mu_j > 0$ and can be written as

$$f_j(z) = K^\alpha(\rho_j\overline{x}z) = \frac{1}{(1-\rho_j\overline{x}z)^\alpha} = \int_{\mathbf{T}} \frac{1}{(1-\overline{x}z)^\alpha} d\mu_j(x).$$

Since C_φ is compact on F_α then $C_\varphi(f_j) \in F_\alpha$ and $\|C_\varphi(f_j)\| \leq \|C_\varphi\|\,\|f_j\|_{F_\alpha} = \|C_\varphi\|$ for all j. Furthermore $C_\varphi(f_j) \in H^\infty$, and thus $C_\varphi(f_j) \in H^\infty \cap F_\alpha \subset (F_\alpha)_a$ for every j. Henceforth there exists a non-negative L^1 function g_x^j such that $d\mu_j(x) = g_x^j(e^{it})\,dt$, $\|g_x^j\|_{L^1} \leq \|C_\varphi\|$ and

$$C_\varphi(f_j)(z) = (f_j \circ \varphi)(z) = K^\alpha[\overline{x}\rho_j\varphi(z)] = \int_{-\pi}^{\pi} g_x^j(e^{it})\, K^\alpha(e^{-it}\rho_j z)\,dt.$$

Now since $(F_\alpha)_a$ is closed and C_φ is compact, the sequence $\{C_\varphi(f_j)\}_{j=1}^{\infty}$ has a convergent subsequence $\{C_\varphi(f_{j_k})\}_{j_k}$ that converges to $C_\varphi(f)(z) = K^\alpha[\overline{x}\varphi(z)] \in (F_\alpha)_a$. Therefore, there exits an L^1 non-negative continuous function of the variable $x = e^{it}$, g_x with $\|g_x\|_{L^1} \leq \|C_\varphi\|$ and such that

$$\lim_{k\to\infty}(f_{j_k}\circ\varphi)(z) = \lim_{k\to\infty}(K^\alpha\circ\varphi)(\overline{x}\rho_{j_k}z) = \lim_{k\to\infty}\int_{-\pi}^{\pi} g_x^{j_k}(e^{it})\,K^\alpha(e^{-it}\rho_{j_k}z)\,dt$$

$$= \int_{-\pi}^{\pi} g_x(e^{it})\,K^\alpha(e^{-it}z)\,dt = (K^\alpha\circ\varphi)(\overline{x}z) = C_\varphi[K^\alpha(\overline{x}z)] \in (F_\alpha)_a$$

For the continuity of g_x in L^1 with respect to x where $|x| = 1$, we take a sequence $\{x_k\}$, such that $|x_k| = 1$ and $x_k \to x$. Now since C_φ is compact then

$$\lim_{k\to\infty}(K_\alpha\circ\varphi)(\overline{x}_k z) = (K_\alpha\circ\varphi)(\overline{x}z)$$

which concludes the proof. □

The following lemmas are needed to prove the converse of Theorem 1.

Lemma 2.2. *Suppose $g_x(e^{it})$ is a non-negative L^1continuous function of x and let $\{\mu_n\}$ be a sequence of non-negative Borel measures that are weak* convergent to μ. Define $w_n(t) = \int_{\mathbf{T}} g_x(e^{it})\,d\mu_n(x)$ and $w(t) = \int_{\mathbf{T}} g_x(e^{it})\,d\mu(x)$, then $\|w_n - w\|_{L^1} \longrightarrow 0$.*

The proof of the lemma is easy and left to the reader.

Lemma 2.3. *Let $g_x(e^{it})$ be a non-negative L^1 continuous function of x such that $\|g_x\|_{L^1} \leq a < \infty$ and $g_x(e^{it})$ defines a bounded operator on $\overline{H_0^1}$.*
If $f(z) = \int \frac{1}{(1-\overline{x}z)^\alpha}d\mu(x)$, let L be the operator given by

$$L[f(z)] = \int_{\mathbf{T}}\int_{-\pi}^{\pi} \frac{g_x(e^{it})}{(1-e^{-it}z)^\alpha}\,dt\,d\mu(x)$$

then L is compact operator on $F_\alpha, \alpha \geq 1$.

Proof. First note that the condition that $g_x(e^{it})$ defines a bounded operator on $\overline{H_0^1}$ implies that the L operator is a well-defined function on $\mathbf{F}_\alpha$. Let $\{f_n(z)\}$ be a bounded sequence in F_α and let $\{\mu_n\}$ be the corresponding norm bounded sequence of measures in $\mathfrak{M}$. Since every norm bounded sequence of measures in $\mathfrak{M}$ has a weak star convergent subsequence, let $\{\mu_n\}$ be such subsequence that is convergent to $\mu \in \mathfrak{M}$. We want to show that $\{L(f_n)\}$ has a convergent subsequence in F_α. First, let us assume that $d\mu_n(x) \gg 0$ for all n, and let $w_n(t) = \int g_x(e^{it})\,d\mu_n(x)$ and $w(t) = \int g_x(e^{it})\,d\mu(x)$, then we know from the previous lemma that $w_n(t), w(t) \in L^1$ for all n, and $w_n(t) \to w(t)$ in L^1. Now since $g_x(e^{it})$ is a non-negative continuous function in x and $\{\mu_n\}$ is weak star convergent to μ, then

$$L(f_n(z)) = \int_{\mathbf{T}} \int_{-\pi}^{\pi} \frac{g_x(e^{it})\, d(t)}{(1-e^{-it}z)^{\alpha}} d\mu_n(x) = \int_{-\pi}^{\pi} \frac{w_n(t)}{(1-e^{-it}z)^{\alpha}} dt.$$

$$L(f(z)) = \int_{\mathbf{T}} \int_{-\pi}^{\pi} \frac{g_x(e^{it})\, d(t)}{(1-e^{-it}z)^{\alpha}} d\mu(x) = \int_{-\pi}^{\pi} \frac{w(t)}{(1-e^{-it}z)^{\alpha}} dt.$$

Furthermore because $w_n(t)$ is non-negative then

$$\|L(f_n)\|_{F_\alpha} = \|w_n\|_{L^1}, \qquad \|L(f)\|_{F_\alpha} = \|w\|_{L^1}$$

Now since $\|w_n - w\|_{L^1} \to 0$ then $\|L(f_n) - L(f)\|_{F_\alpha} \to 0$ which shows that $\{L(f_n)\}$ has convergent subsequence in F_α and thus L is a compact operator for the case where μ is a positive measure.

In the case where μ is a complex measure we write

$$d\mu_n(x) = (d\mu_n^1(x) - d\mu_n^2(x)) + i(d\mu_n^3(x) - d\mu_n^4(x)),$$

where each $d\mu_n^j(x) > 0$ and define $w_n^j(t) = \int g_x(e^{it})\, d\mu_n^j(x)$ then

$$w_n(t) = \int g_x(e^{it})\, d\mu_n(x) = \left(w_n^1(t) - w_n^2(t)\right) + i\left(w_n^3(t) - w_n^4(t)\right).$$

Using an argument similar to the one above we get that $w_n^j(t)$, $w^j(t) \in L^1$, and $\left\|w_n^j - w^j\right\|_{L^1} \longrightarrow 0$. Consequently, $\|w_n - w\|_{L^1} \longrightarrow 0$, where

$$w(t) = \left(w^1(t) - w^2(t)\right) + i\left(w^3(t) - w^4(t)\right) = \int g_x(e^{it})\, d\mu(x).$$

Hence, $\|L(f_n) - L(f)\|_{F_\alpha} \le \|w_n - w\|_{L^1} \longrightarrow 0$. Finally, we conclude that the operator is compact. □

The following is the converse of Theorem 2.1.

Theorem 2.4. *For an analytic self-map φ of the unit disc $\mathbf{D}$, if*

$$C_\varphi[K^\alpha(\overline{x}z)] = \frac{1}{(1-\overline{x}\varphi(z))^{\alpha}} = \int_{-\pi}^{\pi} \frac{g_x(e^{it})}{(1-e^{-it}z)^{\alpha}} dt$$

where $g_x \in L^1$, non-negative, $\|g_x\|_{L^1} \le a < \infty$ for all $x \in \mathbf{T}$ and g_x is an L^1 continuous function of x, then C_φ is compact on F_α.

Proof. We want to show that C_φ is compact on F_α. Let $f(z) \in F_\alpha$ then there exists a measure μ in $\mathfrak{M}$ such that for every z in $\mathbf{D}$

$$f(z) = \int_{\mathbf{T}} \frac{1}{(1-\overline{x}z)^{\alpha}} d\mu(x).$$

Using the assumption of the theorem we get that

$$C_\varphi[f(z)] = (f \circ \varphi)(z) = \int_{\mathbf{T}} \frac{1}{(1-\overline{x}\varphi(z))^{\alpha}} d\mu(x) = \int_{\mathbf{T}} \int_{-\pi}^{\pi} \frac{g_x(e^{it})}{(1-e^{-it}z)^{\alpha}} dt\, d\mu(x)$$

which by the previous lemma was shown to be compact on F_α. □

Now we give some examples:

Corollary 2.5. *Let* $\varphi \in H(\mathbf{D})$, *with* $\|\varphi\|_\infty < 1$. *Then* C_φ *is compact on* F_α *for all* $\alpha \geq 1$.

Proof. $C_\varphi(K^\alpha(\overline{x}z)) = 1/(1-\overline{x}\varphi(z))^\alpha \in H^\infty \cap F_\alpha \subset (F_\alpha)_a$ and is subordinate to $1/(1-z)^\alpha$, hence

$$C_\varphi(K_x^\alpha)(z) = \int_{-\pi}^{\pi} K^\alpha(\overline{x}z) g_x\left(e^{it}\right) dt$$

with $g_x\left(e^{it}\right) \geq 0$ and since $1 = C_\varphi(K_x^\alpha)(0) = \int_{-\pi}^{\pi} g_x\left(e^{it}\right) dt$ we get that $\left\|g_x\left(e^{it}\right)\right\|_1 = 1$. Hence $C_\varphi\left(F_\alpha\right) \subset \left(F_\alpha\right)_a$ □

Corollary 2.6. *If* $\alpha \geq 1$, C_φ *is compact on* F_α *and* $\lim_{r\to 1}\left|\varphi(re^{i\theta})\right| = 1$ *then*

$$\left|1/\varphi'(e^{i\theta})\right| = 0.$$

Proof. If C_φ is compact then

$$C_\varphi\left[K^\alpha(\overline{x}z)\right] = \int_{-\pi}^{\pi} K^\alpha(\overline{x}z) g_x\left(e^{it}\right) dt$$

Now if we let $z = e^{i\theta}$ and $\varphi(e^{i\theta}) = x$ then

$$\lim_{r\to 1} \frac{(e^{i\theta} - re^{i\theta})^\alpha}{(1-\overline{x}\varphi(re^{i\theta}))^\alpha} = 0.$$ □

Corollary 2.7. *For all* $\alpha \geq 1$ *if* C_φ *is compact on* F_α *then* C_φ *is a contraction.*

References

[1] J.A. Cima and A. Matheson, Cauchy transforms and composition operators, Illinois J. Math. Vol. 42, No. 1, (1998), 58–69.

[2] R.A. Hibschweiler and E. Nordgren, Cauchy transforms of measures and weighted shift operators on the disc algebra, Rock. Mt. J. Vol. (26), 2, (1996), 627–654.

[3] R.A. Hibschweiler and T.H. Macgregor, Fractional Cauchy Transforms, Chapman and Hall (2005).

[4] W. Rudin, Real and Complex Analysis, McGraw Hill 1986.

Yusuf Abu Muhanna
Department of Mathematics
American University of Sharjah P.O. Box 26666
Sharjah, United Arab Emirates
e-mail: ymuhanna@aus.ac.ae

El-Bachir Yallaoui
Sultan Qaboos University
P.O. Box 36 Al-Khod 123
Muscat, Oman
e-mail: yallaoui@squ.edu.om

Operator Theory:
Advances and Applications, Vol. 236, 377–390

Hankel Operators on Fock Spaces

A. Perälä, A. Schuster and J.A. Virtanen

Abstract. We study Hankel operators on the weighted Fock spaces F^p_γ. The boundedness and compactness of these operators are characterized in terms of BMO and VMO, respectively. Along the way, we also study Berezin transform and harmonic conjugates on the plane. Our results are analogous to Zhu's characterization of bounded and compact Hankel operators on Bergman spaces of the unit disk.

Mathematics Subject Classification (2010). Primary 47B35; Secondary 30H20, 30H35.

Keywords. Hankel operators, Fock spaces, boundedness, compactness.

1. Introduction

Hankel operators have been studied for several decades in the setting of various analytic function spaces. Starting with Hankel matrices, which can be viewed as Hankel operators on Hardy spaces (see [9]), the field has expanded to Hankel operators on Bergman spaces, Dirichlet type spaces, Bergman and Hardy spaces of the unit ball in $\mathbb{C}^n$, of symmetric domains, and Fock spaces. In addition to being a beautiful and rapidly developing part of analysis, Hankel operators have a vast number of applications, which in the case of Hardy spaces are well known and recognized (see, e.g., [9]), while Hankel operators on Bergman and Fock spaces have found applications mainly in quantum mechanics.

We are interested in the basic properties of Hankel operators on Fock spaces, and in particular characterize their boundedness and compactness in terms of the (mean) oscillation of the generating symbols. In the Bergman space setting one is led to the space of bounded mean oscillation BMO^p_∂ and the space of vanishing mean oscillation VMO^p_∂ with respect to the Bergman metric. Due to K. Zhu [12], a characterization of bounded and compact Hankel operators has been known

The first author acknowledges support by the Academy of Finland project no. 75166001. The third author was supported by a Marie Curie International Outgoing Fellowship within the 7th European Community Framework Programme.

for two decades. It is natural to ask whether an analogous result carries over to Fock spaces. Indeed, the question was recently settled in [2] for the Hilbert–Fock space F^2.

For a symbol f (satisfying suitable conditions), we define the Hankel operator H_f by

$$H_f = (I - P)M_f,$$

where P is a projection defined below in (1) and M_f is the operator of multiplication associated with f. In this paper we study Hankel operators on standard Fock spaces F^p_γ with $1 \le p < \infty$ and $\gamma > 0$. We will introduce spaces BMO^p and VMO^p (in the Euclidean metric) and obtain useful characterizations for these spaces. We prove decomposition theorems similar to those in [11, 12]; in particular, we show that these spaces can be characterized in terms of certain Gaussian integrals, where $\gamma > 0$ can be arbitrary.

Note that the John–Nirenberg theorem implies that the classical BMO and VMO spaces are independent of the parameter p. However, as in the case of the Bergman metric, the spaces BMO^p and VMO^p presented here depend on p.

2. The weighted Fock spaces

We will use the definitions from [7]. Let $\gamma > 0$ and $1 \le p < \infty$. The weighted Fock space F^p_γ consists of entire functions f such that

$$\|f\|^p_{p,\gamma} = \int_{\mathbb{C}} |f(z)|^p e^{-(\gamma p/2)|z|^2} dA(z) < \infty.$$

Here $dA(z) = dxdy$ is the standard Lebesgue area measure. Similarly, the space F^∞_γ consists of those entire f, for which

$$\|f\|_{\infty,\gamma} = \sup_{z \in \mathbb{C}} |f(z)| e^{-(\gamma/2)|z|^2}$$

is finite. The respective Lebesgue L^p_γ spaces and their norms are defined in an obvious way.

It is known that F^2_γ is a Hilbert space with inner product

$$\langle f, g \rangle = \int_{\mathbb{C}} f(z)\overline{g(z)} e^{-\gamma|z|^2} dA(z).$$

Remark. The point-evaluation functionals $f \mapsto f(z)$ are bounded $F^p_\gamma \to \mathbb{C}$ are bounded and hence F^2_γ is known to possess reproducing kernels $K_z := K^\gamma_z$; $f(z) = \langle f, K_z \rangle$.

One immediate corollary is that norm convergence implies locally uniform convergence. In other words, if f_n and f are in F^p_γ and $\|f_n - f\|_{p,\gamma} \to 0$ as $n \to \infty$, then $f_n(z) \to f(z)$ uniformly on each compact subset of $\mathbb{C}$. Another corollary is that the space F^p_γ is *complete*; if $\{f_n\}$ is a Cauchy sequence in norm, then $f_n \to f$ in norm for some $f \in F^p_\gamma$.

The reproducing kernels K_z can be explicitly computed; $K_z(w) = e^{\gamma \bar{z} w}$. The Bergman projection $P := P_\gamma$ is given by

$$Pf(z) = \int_{\mathbb{C}} f(w) e^{\gamma z \bar{w}} e^{-\gamma |w|^2} dA(w). \tag{1}$$

It is known that $P : L^p_\gamma \to F^p_\gamma$ is bounded for every $\gamma > 0$ and $p \in [1, \infty]$. Proofs can be found in [5]. We will just write K_z and P, instead of K^γ_z and P_γ; the parameter γ will be clear from context.

A measurable function f is said to belong to $\tau^p = \tau^p_\gamma$ if and only if $fK_z \in L^p_\gamma$ for every $z \in \mathbb{C}$. This requirement is natural, since linear combinations of the kernel functions form a dense subset of F^p_γ. Henceforth, we will usually assume $f \in \tau^p$.

3. BMO and related spaces

For $0 < r < \infty$, let $D(z, r)$ be the Euclidean disk of radius r and center z. For $f \in L^1_{loc}, 0 < r < \infty, z \in \mathbb{C}$, let

$$\widehat{f}_r(z) = \frac{1}{\pi r^2} \int_{D(z,r)} f(w) dA(w).$$

Fix $0 < r < \infty$ and $p \geq 1$. Define BMO^p_r to be the set of L^p_{loc} integrable functions f such that

$$\|f\|_{BMO^p_r} = \sup_{z \in \mathbb{C}} \left\{ \frac{1}{\pi r^2} \int_{D(z,r)} |f(w) - \widehat{f}_r(z)|^p dA(w) \right\}^{\frac{1}{p}} < \infty.$$

Let BO_r be the set of continuous functions in $\mathbb{C}$ such that

$$\|f\|_{BO_r} = \sup_{z \in \mathbb{C}} \omega_r(f)(z) < \infty,$$

where

$$\omega_r(f)(z) = \sup_{w \in D(z,r)} |f(z) - f(w)|.$$

Lemma 3.1. *Let $f \in L^p_{loc}$. Then $f \in BMO^p_r$ if and only if there is a constant $C > 0$ such that for every $z \in \mathbb{C}$ there is a constant λ_z such that*

$$\frac{1}{\pi r^2} \int_{D(z,r)} |f(w) - \lambda_z|^p dA(w) \leq C.$$

Proof. For the proof of the forward direction, let $\lambda_z = \widehat{f}_r(z)$. For the other direction, note that

$$\left\{\frac{1}{\pi r^2}\int_{D(z,r)}|f(w)-\widehat{f}_r(z)|^p dA(w)\right\}^{\frac{1}{p}}$$
$$\leq \left\{\frac{1}{\pi r^2}\int_{D(z,r)}|f(w)-\lambda_z|^p dA(w)\right\}^{\frac{1}{p}} + \left\{\frac{1}{\pi r^2}\int_{D(z,r)}|\lambda_z-\widehat{f}_r(z)|^p dA(w)\right\}^{\frac{1}{p}}$$
$$= \left\{\frac{1}{\pi r^2}\int_{D(z,r)}|f(w)-\lambda_z|^p dA(w)\right\}^{\frac{1}{p}} + |\lambda_z-\widehat{f}_r(z)|.$$

But

$$|\lambda_z-\widehat{f}_r(z)| = \left|\frac{1}{\pi r^2}\int_{D(z,r)}(f(w)-\lambda_z)dA(w)\right| \leq \left\{\frac{1}{\pi r^2}\int_{D(z,r)}|\lambda_z-f(w)|^p dA(w)\right\}^{\frac{1}{p}}$$

Therefore

$$\left\{\frac{1}{\pi r^2}\int_{D(z,r)}|f(w)-\widehat{f}_r(z)|^p dA(w)\right\}^{\frac{1}{p}} \leq 2\left\{\frac{1}{\pi r^2}\int_{D(z,r)}|\lambda_z-f(w)|^p dA(w)\right\}^{\frac{1}{p}}. \qquad \square$$

Lemma 3.2. *Let $s > r > 0$. Then $BMO^p_s \subset BMO^p_r$.*

Proof. Suppose $f \in BMO^p_s$ so that for every $z \in \mathbb{C}$ we have $\lambda_z \in \mathbb{C}$ such that

$$\sup_{z\in\mathbb{C}}\frac{1}{\pi s^2}\int_{D(z,s)}|f(w)-\lambda_z|^p dA(w) = C < \infty.$$

Now

$$\frac{1}{\pi r^2}\int_{D(z,r)}|f(w)-\lambda_z|^p dA(w) \leq \frac{s^2}{r^2}\frac{1}{\pi s^2}\int_{D(z,s)}|f(w)-\lambda_z|^p dA(w) \leq C\frac{s^2}{r^2}$$

for every $z \in \mathbb{C}$. $\square$

Lemma 3.3. *BO_r is independent of r.*

Proof. Let $r < s$. Then $\|f\|_{BO_r} \leq \|f\|_{BO_s}$.

Choose $N \in \mathbb{N}$ such that for any $w \in D(0,s)$, there exists a path $\{0 = z_1, z_2, \dots, z_N = w\}$ in $D(0,s)$ such that $|z_{i-1}-z_i| < r$. Let now $z \in \mathbb{C}$. Then for any $w \in D(z,s)$, we have a path $\{z = \zeta_1, \zeta_2, \dots, \zeta_N = w\}$, where $\zeta_i = z_i + z$, and $|\zeta_{i-1}-\zeta_i| < r$. Therefore

$$|f(z)-f(w)| \leq \sum_{i=1}^{N}|f(\zeta_{i-1})-f(\zeta_i)| \leq N\sup\{w_r(f)(\zeta_i) : i\} \leq N\|f\|_{BO_r}.$$

We now take the supremum over all $w \in D(z,r)$ and then over all $z \in \mathbb{C}$ to obtain the desired result. $\square$

By the above lemma, we shall now refer to $BO = BO_1$.

Lemma 3.4. *Let f be a continuous function on $\mathbb{C}$. Then $f \in BO$ if and only if there is a constant $C > 0$ such that*

$$|f(z) - f(w)| \leq C(|z - w| + 1)$$

for all $z, w \in \mathbb{C}$.

Proof. The backward direction is obviously true. For the forward direction, let $w, z \in \mathbb{C}$. If $f \in BO$, then

$$C \geq \sup_{\alpha \in \mathbb{C}} \omega_1(f)(\alpha) = \sup_{\alpha \in \mathbb{C}} \sup_{\beta \in D(\alpha,1)} |f(\alpha) - f(\beta)|.$$

In other words, if $|z - w| \leq 1$, then $|f(z) - f(w)| \leq C \leq C(|z - w| + 1)$. Suppose now that $|z - w| > 1$. Let $N = [|z - w|] + 1$, where $[x]$ is the greatest integer less than or equal to x. Let $z_0 = z$, z_1 be the point a distance of $|z - w|/N$ along the line from z to w. Let z_2 be the point a distance of $|z - w|/N$ along the line from z_1 to w, and so on, until $z_N = w$. Then

$$|f(z) - f(w)| \leq \sum_{i=1}^{N} |f(z_{i-1}) - f(z_i)| \leq N\|f\|_{BO} \leq \|f\|_{BO}(|z - w| + 1). \quad \square$$

Let BA_r^p denote the space of all functions f on $\mathbb{C}$ such that

$$\|f\|_{BA_r^p} = \sup_{z \in \mathbb{C}} \left\{ \widehat{|f|^p}_r(z) \right\}^{1/p} < \infty.$$

In other words, $f \in BA_r^p$ if

$$\frac{1}{\pi r^2} \int_{D(z,r)} |f(w)|^p dA(w)$$

is bounded independently of $z \in \mathbb{C}$. The notion of BA_r^p is closely related to Carleson measures on Fock spaces, see [7], or [10] for more generality.

Lemma 3.5. *Let $r > 0$. Then $f \in BA_r^p$ if and only if $M_f : F_\gamma^p \to L_\gamma^p$ is bounded for some (and thus all) $\gamma > 0$.*

Proof. Let $d\mu(w) = |f(w)|^p dA(w)$. Then

$$\left\{ \widehat{|f|^p}_r(z) \right\} = \frac{1}{\pi r^2} \int_{D(z,r)} |f(w)|^p dA(w).$$

Then

$$\mu(D(z,r)) = \int_{D(z,r)} d\mu(w) = \int_{D(z,r)} |f(w)|^p dA(w).$$

Of course this implies that μ is a Carleson measure if and only if $f \in BA_r^p$. But this means

$$\begin{aligned} \int_{\mathbb{C}} |g(w)|^p e^{-(\gamma p/2)|w|^2} |f(w)|^p dA(w) &= \int_{\mathbb{C}} |g(w)|^p e^{-(\gamma p/2)|w|^2} d\mu(w) \\ &\leq C \int_{\mathbb{C}} |g(w)|^p e^{-(\gamma p/2)|w|^2} dA(w) \end{aligned}$$

for all $g \in F_\gamma^p$. In other words, $M_f : F_\gamma^p \to L_\gamma^p$ is bounded. $\square$

Lemma 3.6. *If $f \in BMO^p_{2r}$, then $\widehat{f}_r \in BO_r$.*

Proof. Let $f \in BMO^p_{2r}$, and suppose $|w - z| \le r$. Then

$$\begin{aligned}
|\widehat{f}_r(z) - \widehat{f}_r(w)| &\le |\widehat{f}_r(z) - \widehat{f}_{2r}(z)| + |\widehat{f}_{2r}(z) - \widehat{f}_r(w)| \\
&= \left| \frac{1}{\pi r^2} \int_{D(z,r)} f(u) dA(u) - \frac{1}{\pi r^2} \int_{D(z,r)} \widehat{f}_{2r}(z) dA(u) \right| \\
&\quad + \left| \frac{1}{\pi r^2} \int_{D(w,r)} f(u) dA(u) - \frac{1}{\pi r^2} \int_{D(w,r)} \widehat{f}_{2r}(z) dA(u) \right| \\
&\le \frac{1}{\pi r^2} \int_{D(z,r)} |f(u) - \widehat{f}_{2r}(z)| dA(u) \\
&\quad + \frac{1}{\pi r^2} \int_{D(w,r)} |f(u) - \widehat{f}_{2r}(z)| dA(u) \\
&\le 4 \frac{1}{4\pi r^2} \int_{D(z,2r)} |f(u) - \widehat{f}_{2r}(z)| dA(u) \\
&\quad + 4 \frac{1}{4\pi r^2} \int_{D(z,2r)} |f(u) - \widehat{f}_{2r}(z)| dA(u) \\
&\le 4 \left\{ \frac{1}{\pi 4 r^2} \int_{D(z,2r)} |f(u) - \widehat{f}_{2r}(z)|^p dA(u) \right\}^{\frac{1}{p}} \\
&\quad + 4 \left\{ \frac{1}{\pi 4 r^2} \int_{D(z,2r)} |f(u) - \widehat{f}_{2r}(z)|^p dA(u) \right\}^{\frac{1}{p}} \\
&\le 8 \|f\|_{BMO^p_{2r}}.
\end{aligned}$$

The fourth line follows from the fact that $D(z, r) \subset D(z, 2r)$ and $D(w, r) \subset D(z, r)$ and the fifth follows from Hölder's inequality. □

Let $k_z = K_z / \|K_z\|_{2,\gamma}$, so that $k_z(w) = e^{\gamma \bar{z} w - (\gamma/2)|z|^2}$ denote the normalized reproducing kernel of F^2_γ. An easy calculation reveals that $k_z = k_z^\gamma$ is a unit vector on F^p_γ for every $p \in [1, \infty)$.

The Berezin transform (or the heat-transform) of a function f is given by

$$B_\gamma f(z) = \int_{\mathbb{C}} f(w) |k_z^\gamma(w)|^2 e^{-\gamma |w|^2} dA(w).$$

We will omit the γ, when it is clear form the context. In this case we just write Bf.

Lemma 3.7. *Let $f \in \tau^p$. Then the following are equivalent.*

(1) $f \in BA^p_r$;
(2) $\sup_{z \in \mathbb{C}} \int_{\mathbb{C}} |f(z - w)|^p e^{-\gamma |w|^2} dA(w) \le C$ *for some* $\gamma > 0$;
(3) $\sup_{z \in \mathbb{C}} \int_{\mathbb{C}} |f(z - w)|^p e^{-\gamma |w|^2} dA(w) \le C$ *for all* $\gamma > 0$.

Proof. By the definition of BA_r^p, $f \in BA_r^p$ if and only if $\int_{D(z,r)} |f(w)|^p dA(w) \leq C$ if and only if $|f|^p dA$ is a Carleson measure for F_γ^2 for some (and thus for every) $\gamma > 0$ if and only if the Berezin transform $B_\gamma |f|^p$ is bounded. But

$$\begin{aligned} B_\gamma |f|^p(z) &= \int_{\mathbb{C}} |k_z(w)|^2 e^{-\gamma|w|^2} |f(w)|^p dA(w) \\ &= \int_{\mathbb{C}} e^{-\gamma|z|^2 + 2\gamma \Re \bar{z} w - \gamma |w|^2} |f(w)|^p dA(w) \\ &= \int_{\mathbb{C}} e^{-\gamma|z-w|^2} |f(w)|^p dA(w) = \int_{\mathbb{C}} e^{-\gamma|w|^2} |f(z-w)|^p dA(w). \qquad \square \end{aligned}$$

Note that by Lemmas 3.3 and 3.5, both BO_r and BA_r^p are independent of r. In fact, if we combine Lemmas 3.5 and 3.7, we obtain the following lemma.

Lemma 3.8. *Let $f \in \tau^p$. The following conditions are equivalent:*

(1) $f \in BA_r^p$;
(2) $\sup_{z\in\mathbb{C}} \int_{\mathbb{C}} |f(z-w)|^p e^{-\gamma|w|^2} dA(w) < \infty$ *for some (and thus all)* $\gamma > 0$;
(3) $M_f : F_\gamma^p \to L_\gamma^p$ *is bounded for some (and thus all)* $\gamma > 0$.

Lemma 3.9. *If $f \in BMO_{2r}^p$, then $f - \widehat{f}_r \in BA^p$.*

Proof. By assumption and Lemma 3.2, $f \in BMO_r^p$. Let $g = f - \widehat{f}_r$. Then

$$\begin{aligned} \left\{ \widehat{|g|^p}_r(z) \right\}^{\frac{1}{p}} &= \left\{ \frac{1}{\pi r^2} \int_{D(z,r)} |f(u) - \widehat{f}_r(u)|^p dA(u) \right\}^{\frac{1}{p}} \\ &\leq \left\{ \frac{1}{\pi r^2} \int_{D(z,r)} |f(u) - \widehat{f}_r(z)|^p dA(u) \right\}^{\frac{1}{p}} + \left\{ \frac{1}{\pi r^2} \int_{D(z,r)} |\widehat{f}_r(z) - \widehat{f}_r(u)|^p dA(u) \right\}^{\frac{1}{p}} \\ &\leq \|f\|_{BMO_r^p} + \omega_r(\widehat{f}_r)(z). \qquad \square \end{aligned}$$

Lemma 3.10. *Let $r > 0$. Then*

$$BMO_r^p \subset BO_r + BA_r^p.$$

Proof. Let $r = 2s$ and $f \in BMO_r^p = BMO_{2s}^p$. Then Lemmas 3.6 and 3.9 imply that $\widehat{f}_s \in BO_s$ and $f - \widehat{f}_s \in BA_s^p$. Therefore, $f = \widehat{f}_s + f - \widehat{f}_s \in BO_s + BA_s^p = BO_r + BA_r^p$. $\square$

Lemma 3.11. *If $f \in BMO_r^p$, then*

$$\int_{\mathbb{C}} |f(z-w) - B_\gamma f(z)|^p e^{-\gamma|w|^2} dA(w) \leq C$$

for all $z \in \mathbb{C}$ and $\gamma > 0$.

Proof. By Lemma 3.10, it is enough to show the inequality holds for $f \in BA^p$ and $f \in BO$. Suppose first that $f \in BA^p$. By Hölder,

$$|B_\gamma f(z)| \le C\left\{\int_{\mathbb{C}} |f(z-w)|^p e^{-\gamma|w|^2} dA(w)\right\}^{\frac{1}{p}}.$$

Therefore

$$\begin{aligned}
&\left\{\int_{\mathbb{C}} |f(z-w) - B_\gamma f(z)|^p e^{-\gamma|w|^2} dA(w)\right\}^{\frac{1}{p}} \\
&\quad \le \left\{\int_{\mathbb{C}} |f(z-w)|^p e^{-\gamma|w|^2} dA(w)\right\}^{\frac{1}{p}} + |B_\gamma f(z)| \\
&\quad \le (1+C')\left\{\int_{\mathbb{C}} |f(z-w)|^p e^{-\gamma|w|^2} dA(w)\right\}^{\frac{1}{p}} \le C,
\end{aligned}$$

where the last inequality follows from Lemma 3.7.

Suppose next that $f \in BO$. Then

$$\begin{aligned}
&\int_{\mathbb{C}} |f(z-w) - B_\gamma f(z)|^p e^{-\gamma|w|^2} dA(w) \\
&\quad = \int_{\mathbb{C}} \left|f(z-w) - \int_{\mathbb{C}} f(z-u) e^{-\gamma|u|^2} dA(u)\right|^p e^{-\gamma|w|^2} dA(w) \\
&\quad \le C \int_{\mathbb{C}}\int_{\mathbb{C}} |f(z-w) - f(z-u)|^p e^{-\gamma|w|^2} dA(w) e^{-\gamma|u|^2} dA(u).
\end{aligned}$$

Since $f \in BO$, Lemma 3.4 tells us that $|f(z-w) - f(z-u)|^p \le C(|w-u|+1)^p$. Therefore, the last quantity in the last displayed equation is bounded above by

$$C^2 \int_{\mathbb{C}}\int_{\mathbb{C}} (|u-w|+1)^p e^{-\gamma|w|^2} dA(w) e^{-\gamma|u|^2} dA(u),$$

which is a constant. □

Lemma 3.12. *Suppose there exists $\gamma > 0$ such that*

$$\int_{\mathbb{C}} |f(z-w) - B_\gamma f(z)|^p e^{-\gamma|w|^2} dA(w) \le C$$

for all $z \in \mathbb{C}$. Then $f \in BMO^p_r$.

Proof. Let $z \in \mathbb{C}$ and fix $\gamma > 0$. Note that $e^{-\gamma|z-w|^2} \ge c > 0$ for $w \in D(z,r)$. Therefore

$$\begin{aligned}
c\int_{D(z,r)} |f(w) - B_\gamma f(z)|^p dA(w) &\le \int_{D(z,r)} |f(w) - B_\gamma f(z)|^p e^{-\gamma|z-w|^2} dA(w) \\
&\le \int_{\mathbb{C}} |f(w) - B_\gamma f(z)|^p e^{-\gamma|z-w|^2} dA(w) \\
&= \int_{\mathbb{C}} |f(z-w) - B_\gamma f(z)|^p e^{-\gamma|w|^2} dA(w) \le C.
\end{aligned}$$

The result then follows from an application of Lemma 3.1. □

We now have proven that BMO^p_r is independent of r; in what follows, we will write $BMO^p = BMO^p_1$.

Theorem 1. *Let $p \geq 1$. Then the following are equivalent:*

(1) $f \in BMO^p$;

(2) $f \in BO + BA^p$;

(3) $\sup_{z\in\mathbb{C}} \int_{\mathbb{C}} |f(z-w) - B_\gamma f(z)|^p e^{-\gamma|w|^2} dA(w) < \infty$, *for some* $\gamma > 0$;

(3′) $\sup_{z\in\mathbb{C}} \int_{\mathbb{C}} |f(z-w) - B_\gamma f(z)|^p e^{-\gamma|w|^2} dA(w) < \infty$, *for all* $\gamma > 0$;

(4) *There is a constant C and $\gamma > 0$ such that for every $z \in \mathbb{C}$, there is a constant λ_z such that*

$$\int_{\mathbb{C}} |f(z-w) - \lambda_z|^p e^{-\gamma|w|^2} dA(w) \leq C;$$

(4′) *For every $\gamma > 0$ there is a constant C such that for every $z \in \mathbb{C}$, there is a constant λ_z such that*

$$\int_{\mathbb{C}} |f(z-w) - \lambda_z|^p e^{-\gamma|w|^2} dA(w) \leq C.$$

Proof. (1) ⇒ (2) follows from Lemma 3.10. (2) ⇒ (3′) follows from the proof of Lemma 3.11. Obviously (3′) ⇒ (3) and (4′) ⇒ (4). The proofs of (3) ⇔ (4) and (3′) ⇔ (4′) are similar to the proof of Lemma 3.1. (3) ⇒ (1) follows from Lemma 3.12. □

Lemma 3.13. *If $f \in BMO^p$, then $B_\gamma f \in BO$, and $f - B_\gamma f \in BA^p$ for every $\gamma > 0$.*

Proof. Fix $\gamma > 0$. We have

$$\begin{aligned}
|B_\gamma f(z) - \widehat{f}_r(z)| &= |B_\gamma f(z) - \frac{1}{\pi r^2}\int_{D(z,r)} f(w)dA(w)| \\
&= \left| \frac{1}{\pi r^2}\int_{D(z,r)} B_\gamma f(z) dA(w) - \frac{1}{\pi r^2}\int_{D(z,r)} f(w)dA(w)\right| \\
&\leq \frac{1}{\pi r^2}\int_{D(z,r)} |f(w) - B_\gamma f(z)| dA(w) \\
&\leq C\int_{\mathbb{C}} |f(z-w) - B_\gamma f(z)| e^{-\gamma|w|^2} dA(w).
\end{aligned}$$

Here the last inequality follows from the proof of Lemma 3.12. Since $f \in BMO^p$, the last integral is finite. Thus $Bf - \widehat{f}_r$ is a bounded continuous function and so lies in $BO \cap BA^p$. By Lemma 3.6, $\widehat{f}_r \in BO$, so $B_\gamma f = B_\gamma f - \widehat{f}_r + \widehat{f}_r \in BO + BO = BO$. By Lemma 3.9, $f - \widehat{f}_r \in BA^p$, so $f - B_\gamma f = f - \widehat{f}_r + \widehat{f}_r - B_\gamma f \in BA^p + BA^p = BA^p$. □

4. Bounded Hankel operators

We begin with a short discussion of harmonic conjugates. If $f = u + iv$ is entire, then both u and v are harmonic. Conversely, given a harmonic $u : \mathbb{C} \to \mathbb{R}$, there exists a unique harmonic $v : \mathbb{C} \to \mathbb{R}$ such that $f = u + iv$ is entire and $v(0) = 0$.

Lemma 4.1. *Let $u : \mathbb{C} \to \mathbb{R}$ be harmonic. If $u \in L^p_\gamma$ for $p \in (1, \infty)$ and $\gamma > 0$, then $v \in L^p_\gamma$.*

Proof. Looking at the proof of Theorem 4.1 of [6], one obtains for $r < 1$ a $C > 0$ such that

$$\int_0^{2\pi} |v(re^{i\theta})|^p d\theta \leq C \int_0^{2\pi} |u(re^{i\theta})|^p d\theta.$$

But if $r > 1$, consider the dilations $u_R(z) = u(Rz)$ and $v_R(z) = v(Rz)$ for large enough R. Of course, both u and u_R always belong to the Hardy space h^p of the unit circle. Now,

$$\int_0^{2\pi} |v(re^{i\theta})|^p d\theta = \int_0^{2\pi} |v_R(se^{i\theta})|^p d\theta \leq C \int_0^{2\pi} |u_R(se^{i\theta})|^p d\theta = C \int_0^{2\pi} |u(re^{i\theta})|^p d\theta,$$

where R is chosen so that $s := r/R < 1$. Now, inevitably

$$\int_0^{2\pi} |v(re^{i\theta})|^p re^{-(p/2)r^2} d\theta \leq C \int_0^{2\pi} |u(re^{i\theta})|^p re^{-(p/2)r^2} d\theta.$$

The rest follows from evaluating the norms in polar coordinates. □

Corollary 4.2. *Let $p \in (1, \infty)$ and $\gamma > 0$. Suppose $f = u + iv$ is entire and that $u \in L^p_\gamma$. Then $f \in F^p_\gamma$. Moreover, there exists $C > 0$ such that $\|f - f(0)\|_{p,\gamma} \leq \|u\|_{p,\gamma}$.*

In what follows, if the possible values of p are not indicated, we assume that $p \in (1, \infty)$.

Recall that the Bergman projection P is given by

$$Pg(z) = \int_{\mathbb{C}} g(w) e^{\gamma z \bar{w}} e^{-|w|^2} dA(w).$$

If $f \in \tau^p$, then the Hankel operator with symbol f is given for $g \in F^p_\gamma$ by

$$H_f g(z) = (I - P)(fg)(z).$$

Note that we can also write

$$H_f g(z) = \int_{\mathbb{C}} (f(z) - f(w)) g(w) e^{\gamma z \bar{w}} e^{-\gamma |w|^2} dA(w).$$

Lemma 4.3. *If $f \in BA^p$, then H_f is bounded on F^p_γ.*

Proof. By Lemma 3.8, M_f is bounded $F^p_\gamma \to L^p_\gamma$. Since P is bounded, we obtain the desired result. □

Lemma 4.4. *If $f \in BO$, then H_f is bounded on F^p_γ for every $p \in [1, \infty]$.*

Proof. By Lemma 3.4

$$|H_f(g)(z)| \leq C \int_{\mathbb{C}} (|z-w|+1)|e^{\gamma z\bar{w}} g(w)| e^{-\gamma|w|^2} dA(w).$$

There are $C > 0$ and $\epsilon > 0$ such that

$$|e^{\gamma z\bar{w}}| \leq Ce^{(\gamma/2)|z|^2+(\gamma/2)|w|^2-\epsilon|z-w|}.$$

Therefore, we arrive at

$$\begin{aligned}
&|H_f(g)(z)|^p e^{-(p\gamma/2)|z|^2}\\
&\quad \leq Ce^{-((p-1)\gamma/2)|z|^2} \left| \int_{\mathbb{C}} (|z-w|+1)|g(w)|e^{-(\gamma/2)|w|^2} e^{-\epsilon|z-w|} dA(w) \right|^p\\
&\quad \leq Ce^{-((p-1)\gamma/2)|z|^2} \left\{ \int_{\mathbb{C}} |g(w)|^p e^{-(\gamma p/2)|w|^2} dA(w) \right\}\\
&\qquad\qquad \times \left\{ \int_{\mathbb{C}} |z-w|^q e^{-q\epsilon|z-w|} dA(w) \right\}^{p/q}\\
&\quad \leq Ce^{-((p-1)\gamma/2)|z|^2} \|g\|_{p,\gamma}^p.
\end{aligned}$$

If $1 < p < \infty$ and $1/p + 1/q = 1$, we get the desired result by integrating with respect to z.

If $p = 1$, we use the above reasoning together with Fubini and proceed as follows.

$$\begin{aligned}
&\int_{\mathbb{C}} |H_f(g)(z)|^p e^{-(\gamma/2)|z|^2} dA(z)\\
&\quad \leq C \int_{\mathbb{C}} |g(w)|e^{-(\gamma/2)|w|^2} dA(w) \int_{\mathbb{C}} (|z-w|+1)e^{-\epsilon|z-w|} dA(z)\\
&\quad \leq C\|g\|_{1,\gamma}.
\end{aligned}$$

By similar arguments, one can also show that

$$|H_f(g)(z)|e^{-(\gamma/2)|z|^2} \leq \int_{\mathbb{C}} (|z-w|+1)e^{-\epsilon|z-w|} dA(z) \|g\|_{\infty,\gamma} \leq C\|g\|_{\infty,\gamma}.$$

The result is now proven for all $p \in [1,\infty]$. □

Theorem 2. *Let $f \in \tau^p$. Then $f \in BMO^p$ if and only if the operators H_f and $H_{\bar{f}}$ are both bounded $F_\gamma^p \to L_\gamma^p$.*

Proof. If $f \in BMO^p$, then so is $\bar{f}$ and it follows from the previous two lemmas that H_f and $H_{\bar{f}}$ are bounded.

Suppose now that H_f and $H_{\bar{f}}$ are both bounded. Without loss of generality, we may then assume that f is real valued. Recall that $k_z(w) = e^{\gamma\bar{z}w-(\gamma/2)|z|^2}$ are unit vectors in F_γ^p and so we have $C > 0$ such that

$$\|fk_z - P(fk_z)\|_{p,\gamma} = \|H_f(k_z)\|_{p,\gamma} \leq C.$$

Note that $k_z(z-w) = 1/k_z(w)$ and

$$e^{-\gamma(p/2)|z-w|^2} = e^{-\gamma(p/2)|z|^2 - \gamma(p/2)|w|^2 + \gamma(p/2)z\bar{w} + \gamma(p/2)\bar{z}w}.$$

Thus, by a change of variables $w \mapsto z - w$, one obtains

$$\begin{aligned} C^p &\geq \|fk_z - P(fk_z)\|_{p,\gamma}^p \\ &= \int_{\mathbb{C}} |f(z-w) - e^{-(\gamma/2)|z|^2} P(fk_z)(z-w)|^p e^{-(p\gamma/2)|w|^2} dA(w) \end{aligned}$$

Setting $g_z(w) = e^{-(\gamma/2)|z|^2} P(fk_z)(z-w)$, one obtains

$$\sup_{z\in\mathbb{C}} \int_{\mathbb{C}} |f(z-w) - g_z(w)|^p e^{-(p\gamma/2)|w|^2} dA(w) \leq C^p.$$

Since f is real-valued, then the imaginary part of g_z must belong L^p_γ and so

$$\|g_z - g_z(0)\|_{p,\gamma} \leq M$$

for every $z \in \mathbb{C}$ and some $M > 0$. Applying triangle inequality and the main theorem of the previous section with $\lambda_z = g_z(0)$, one sees that $f \in BMO^p$. □

5. VMO and compact Hankel operators

In this section we study VMO and compactness of Hankel operators. The results and their proofs are completely analogous to the results of the previous two sections. A great deal of details is therefore omitted and left for the reader to verify.

Define VMO^p_r to be the set of L^p_{loc} integrable functions f such that

$$\lim_{z\to\infty} \frac{1}{\pi r^2} \int_{D(z,r)} |f(w) - \widehat{f}_r(z)|^p dA(w) = 0.$$

Let $VO_r \subset BO_r$ be the set of continuous functions in $\mathbb{C}$ such that

$$\lim_{z\to\infty} \omega_r(f) = 0.$$

Let VA^p_r be the set of functions f on $\mathbb{C}$ such that $\lim_{z\to\infty} \widehat{f}_r(z) = 0$. The space VA^p_r is related to the space of vanishing Carleson measures on Fock spaces, see [7], [10].

Similarly to Section 3, it can be shown that VO_r and VA^p_r are independent of r and we will write VO and VA^p, respectively. The following results are also analogous to the BMO-setting.

Lemma 5.1. *Let $f \in VMO^p$. Then*

(1) *$B_\gamma f \in BO$ for every $\gamma > 0$;*
(2) *$\widehat{f}_r \in BO$ for every $r > 0$;*
(3) *$f - B_\gamma f \in BA^p$ for every $\gamma > 0$;*
(4) *$f - \widehat{f}_r \in BA^p$ for every $r > 0$.*

Theorem 3. *Let $p \geq 1$. Then the following are equivalent:*

(1) $f \in VMO^p$;

(2) $f \in VO + VA^p$;

(3) $\lim_{z\to\infty} \int_{\mathbb{C}} |f(z-w) - B_\gamma f(z)|^p e^{-\gamma|w|^2} dA(w) = 0$, *for some* $\gamma > 0$;

(3′) $\lim_{z\to\infty} \int_{\mathbb{C}} |f(z-w) - B_\gamma f(z)|^p e^{-\gamma|w|^2} dA(w) = 0$, *for all* $\gamma > 0$;

(4) *There is a* $\gamma > 0$ *such that for every* $z \in \mathbb{C}$, *there is a constant* λ_z *such that*

$$\lim_{z\to\infty} \int_{\mathbb{C}} |f(z-w) - \lambda_z|^p e^{-\gamma|w|^2} dA(w) = 0;$$

(4′) *For every* $\gamma > 0$ *and every* $z \in \mathbb{C}$, *there is a constant* λ_z *such that*

$$\lim_{z\to\infty} \int_{\mathbb{C}} |f(z-w) - \lambda_z|^p e^{-\gamma|w|^2} dA(w) = 0.$$

Theorem 4. *Let $f \in \tau^p$. Then the operators H_f and $H_{\bar f}$ are compact if and only if $f \in VMO^p$.*

Proof. Suppose first that $f \in VA^p$. But then $|f|^p dA$ is vanishing Carleson, so the multiplication operators M_f and $M_{\bar f}$ are compact $F^p_\gamma \to L^p_\gamma$. From the boundedness of the projection P, it follows that H_f and $H_{\bar f}$ are both compact.

If $f \in VO$, we refer to Lemma 5.1 of [2]. It follows that both H_f and $H_{\bar f}$ can be approximated in norm by Hankel operators with symbols having a compact support. Therefore, both operators are compact. In conclusion, we have shown that if $f \in VMO^p$, then H_f and $H_{\bar f}$ are compact.

As for the other direction. Note that $k_z \to 0$ weakly, as $z \to \infty$. But now

$$\|H_f k_z\|_{p,\gamma} \to 0 \text{ and } \|H_{\bar f} k_z\|_{p,\gamma} \to 0,$$

as $z \to \infty$. By reasoning similar to that in Theorem 2, it follows that

$$\int_{\mathbb{C}} |f(z-w) - g_z(0)|^p e^{-(\gamma p/2)|w|^2} dA(w) \to 0,$$

as $z \to \infty$, so $f \in VMO^p$. □

References

[1] Bauer, W., *Hilbert–Schmidt Hankel operators on the Segal–Bargmann space.* Proc. Amer. Math. Soc. 132 (2004), no. 10, 2989–2996 (electronic).

[2] Bauer, W., *Mean Oscillation and Hankel Operators on the Segal–Bargmann Space.* Integr. equ. oper. theory 52 (2005), 1–15.

[3] Berndtsson, B.; Ortega Cerdà, J., *On interpolation and sampling in Hilbert spaces of analytic functions.* J. Reine Angew. Math. 464 (1995), 109–128.

[4] Bommier-Hato, H.; Youssfi, E., *Hankel operators on weighted Fock spaces.* Integral Equations Operator Theory 59 (2007), no. 1, 1–17.

[5] Dostanic, M.; Zhu, K., *Integral operators induced by the Fock kernel.* Integral Equations and Operator Theory 60 (2008), 217–236.

[6] Duren, P., *Theory of H^p spaces.* Academic Press, New York, 1970.
[7] Isralowitz, J.; Zhu, K.; *Toeplitz operators on the Fock space.* Integral Equations and Operator Theory 66 (2010), no. 4, 593–611.
[8] Janson, S.; Peetre, J.; Rochberg, R., *Hankel forms and the Fock space.* Rev. Mat. Iberoamericana 3 (1987), no. 1, 61–138.
[9] Peller, V., *Hankel operators and their applications.* Springer-Verlag, New York, 2003.
[10] Schuster, A.; Varolin, D., *Toeplitz Operators and Carleson Measures on Generalized Bargmann–Fock Spaces.* Integral Equations and Operator Theory, Volume 72, Number 3 (2012), 363–392.
[11] Taskinen, J.; Virtanen, J.A., *Weighted BMO and Toeplitz operators on the Bergman space A^1.* (to appear in Journal of Operator Theory).
[12] Zhu, K., *BMO and Hankel operators on Bergman spaces.* Pacific J. Math. 155 (1992), 377–395.

A. Perälä
Department of Mathematics
University of Helsinki
Helsinki 00014, Finland
e-mail: antti.i.perala@helsinki.fi

A. Schuster
Department of Mathematics
San Francisco State University
San Francisco, CA 94132, USA

and

Department of Mathematics and Actuarial Science
The American University in Cairo
AUC Avenue PO Box 74
New Cairo Egypt 11835
e-mail: schuster@sfsu.edu

J.A. Virtanen
Department of Mathematics
University of Bristol
Bristol, England BS8 1TW, UK
e-mail: virtanen@cims.nyu.edu

Operator Theory:
Advances and Applications, Vol. 236, 391–404

Evolutionary Problems Involving Sturm–Liouville Operators

Rainer Picard and Bruce A. Watson

Abstract. The purpose of this paper is to further exemplify an approach to evolutionary problems originally developed in [3], [4] for a special case and extended to more general evolutionary problems, see [7], compare the survey article [5]. The ideas there are utilized for $(1+1)$-dimensional evolutionary problems, which in a particular case results in a hyperbolic partial differential equation with a Sturm–Liouville type spatial operator constrained by an impedance type boundary condition.

Mathematics Subject Classification (2010). Primary 34B24, 35F10, Secondary 35A22, 47G20, 34L40, 35K90, 35L90.

Keywords. Evolution equations, Sturm–Liouville operator, partial differential equations, causality, impedance type boundary condition, memory, delay.

0. Introduction

A canonical form of many linear evolutionary equations of mathematical physics is given by a dynamic system of equations

$$\partial_0 V + AU = F$$

completed by a so-called material law

$$V = \mathcal{M}U.$$

Here we have used the term *evolutionary* equations, since the term *evolution equations* is usually reserved for a special case of the class of evolutionary equations considered here. The symbol ∂_0 denotes differentiation with respect to the time variable, A is a usually unbounded operator containing spatial derivatives and $\mathcal{M}$ is a continuous linear operator. Here we inspect more closely the very specific situation where the space dimension is 1, i.e., A is an ordinary differential operator in the space variable for fixed time. Although this is a rather particular case it

B.A. Watson is funded in part by NRF grant IFR2011032400120.

has the advantage that an impedance type boundary condition, which we wish to consider, can be considered in a more "tangible" way without incurring regularity assumptions on coefficients and data. In the higher-dimensional case for the acoustic equations, which can also be discussed with no further regularity requirements, the constraints on the impedance type boundary condition are much less explicit, compare [7], [5]. Moreover, we hope to gain access to a class of problems, which are closely linked to Sturm–Liouville operators, yielding a generalization of such operators. That we are discussing the direct time-dependent problem rather than an associated spectral problem will actually be advantageous, since it provides a simpler discussion of well-posedness.

More specifically we want to consider

$$A = \begin{pmatrix} 0 & 0 & \partial \\ 0 & 0 & 0 \\ \partial & 0 & 0 \end{pmatrix}$$

as a differential operator on the unit interval $]-1/2, 1/2[$ with an impedance type boundary condition of the form

$$\partial_0 \alpha\,(\pm 1/2 \mp 0)\, s\,(\,\cdot\,, \pm 1/2 \mp 0) - v\,(\,\cdot\,, \pm 1/2 \mp 0) = 0$$

holding on the real time-line $\mathbb{R}$ as a constraint characterizing $\begin{pmatrix} s \\ w \\ v \end{pmatrix}$ in the domain $D\,(A)$. Here ∂ denotes the spatial derivative and α is a coefficient operator specified more precisely later. We shall focus here on the time-translation invariant, i.e., autonomous, case. This means that time-translation and consequently time-differentiation commutes with α, $\mathcal{M}$ and A.

Our discussion is embedded into an abstract setting, which we will develop in Section 1 first. In Section 2 we will then discuss our problem of interest as an application of the abstract solution theory.

1. The abstract solution framework

1.1. Sobolev chains associated with the time-derivative

A particular instance of the construction of Sobolev chains is the one based on the time-derivative ∂_0. We recall, e.g., from [4, 5], that differentiation considered in the complex Hilbert space $H_{\nu,0}(\mathbb{R}) := \{f \in L^2_{\mathrm{loc}}(\mathbb{R}) | (x \mapsto \exp(-\nu x) f(x)) \in L^2(\mathbb{R})\}$, $\nu \in \mathbb{R} \setminus \{0\}$, with inner product

$$(f, g) \mapsto \langle f, g \rangle_{\nu,0} := \int_{\mathbb{R}} f(x)^* g(x) \, \exp(-2\nu x) \, dx$$

can indeed be established as a normal operator, which we denote by $\partial_{0,\nu}$, with

$$\mathfrak{Re}\, \partial_{0,\nu} = \nu.$$

For $\mathfrak{Im}\,\partial_{0,\nu}$ we have as a spectral representation the Fourier–Laplace transform $\mathcal{L}_\nu : H_{\nu,0}(\mathbb{R}) \to L^2(\mathbb{R})$ given by the unitary extension of

$$\mathring{C}_\infty(\mathbb{R}) \subseteq H_{\nu,0}(\mathbb{R}) \to L^2(\mathbb{R})$$
$$\phi \mapsto \left(x \mapsto \frac{1}{\sqrt{2\pi}} \int_{\mathbb{R}} \exp(-\mathrm{i}xy)\, \exp(-\nu y)\, \phi(y)\, dy \right).$$

In other words, we have the unitary equivalence

$$\mathfrak{Im}\,\partial_{0,\nu} = \mathcal{L}_\nu^{-1} m\, \mathcal{L}_\nu,$$

where m denotes the selfadjoint multiplication-by-argument operator in $L^2(\mathbb{R})$. Since 0 is in the resolvent set of $\partial_{0,\nu}$ we have that $\partial_{0,\nu}^{-1}$ is an element of the Banach space $L(H_{\nu,0}(\mathbb{R}), H_{\nu,0}(\mathbb{R}))$ of continuous (left-total) linear mappings in $H_{\nu,0}(\mathbb{R})$. Denoting generally the operator norm of the Banach space $L(X,Y)$ by $\|\cdot\|_{L(X,Y)}$, we get

$$\left\|\partial_{0,\nu}^{-1}\right\|_{L(H_{\nu,0}(\mathbb{R}),H_{\nu,0}(\mathbb{R}))} = \frac{1}{|\nu|}.$$

Not too surprisingly, we find for $\nu > 0$

$$\left(\partial_{0,\nu}^{-1}\varphi\right)(x) = \int_{-\infty}^{x} \varphi(t)\, dt$$

and for $\nu < 0$

$$\left(\partial_{0,\nu}^{-1}\varphi\right)(x) = -\int_{x}^{\infty} \varphi(t)\, dt$$

for all $\varphi \in H_{\nu,0}(\mathbb{R})$ and $x \in \mathbb{R}$. Since we are interested in the forward causal situation, we assume $\nu > 0$ throughout. Moreover, in the following we shall mostly write ∂_0 for $\partial_{0,\nu}$ if the choice of ν is clear from the context.

Thus, we obtain a chain $(H_{\nu,k}(\mathbb{R}))_{k\in\mathbb{Z}}$ of Hilbert spaces, where $H_{\nu,k}(\mathbb{R})$ is the completion of the inner product space $D\left(\partial_0^k\right)$ with norm $|\cdot|_{\nu,k}$ given by

$$\phi \mapsto \left|\partial_0^k \phi\right|_{\nu,0}.$$

Similarly, for $\mathrm{i}m + \nu$ as a normal operator in $L^2(\mathbb{R})$ we construct the chain of polynomially weighted $L^2(\mathbb{R})$-spaces

$$\left(L_k^2(\mathbb{R})\right)_{k\in\mathbb{Z}}$$

with

$$L_k^2(\mathbb{R}) := \left\{ f \in L^2_{\mathrm{loc}}(\mathbb{R}) \,\middle|\, (\mathrm{i}m+\nu)^k f \in L^2(\mathbb{R}) \right\}$$

for $k \in \mathbb{Z}$.

Since the unitarily equivalent operators $\partial_{0,\nu}$ and $\mathrm{i}m + \nu$ (via the Fourier–Laplace transform) can canonically be lifted to the X-valued case, X an arbitrary complex Hilbert space, we are led to corresponding chains $(H_{\nu,k}(\mathbb{R},X))_{k\in\mathbb{Z}}$ and

$\left(L_k^2(\mathbb{R}, X)\right)_{k\in\mathbb{Z}}$ of X-valued generalized functions. The Fourier–Laplace transform can also be lifted to the X-valued case yielding

$$\begin{aligned} H_{\nu,k}(\mathbb{R}, X) &\to L_k^2(\mathbb{R}, X) \\ f &\mapsto \mathcal{L}_\nu f \end{aligned}$$

as a unitary mapping for $k \in \mathbb{N}$ and by continuous extension, keeping the notation $\mathcal{L}_\nu$ for the extension, also for $k \in \mathbb{Z}$. Since $\mathcal{L}_\nu$ has been constructed from a spectral representation of $\mathfrak{Im}\,\partial_{0,\nu}$, we can utilize the corresponding operator function calculus for functions of $\mathfrak{Im}\,\partial_{0,\nu}$. Noting that $\partial_0 = \mathrm{i}\,\mathfrak{Im}\,\partial_{0,\nu} + \nu$ is a function of $\mathfrak{Im}\,\partial_{0,\nu}$ we can define operator-valued functions of ∂_0.

Definition 1.1. Let $r > \frac{1}{2\nu} > 0$ and $M : B_\mathbb{C}(r, r) \to L(H, H)$ be bounded and analytic, H a Hilbert space, where $B_\mathbb{C}(a, b)$ denotes the open ball in $\mathbb{C}$ of radius b centred at a. Then define

$$M\left(\partial_0^{-1}\right) := \mathcal{L}_\nu^* \, M\left(\frac{1}{im + \nu}\right) \mathcal{L}_\nu,$$

where

$$M\left(\frac{1}{im + \nu}\right)\phi(t) := M\left(\frac{1}{it + \nu}\right)\phi(t) \quad (t \in \mathbb{R})$$

for $\phi \in \mathring{C}_\infty(\mathbb{R}, H)$.

Remark 1.2. The definition of $M(\partial_0^{-1})$ is largely independent of the choice of ν in the sense that the operators for two different parameters ν_1, ν_2 coincide on the intersection of the respective domains.

Simple examples are polynomials in ∂_0^{-1} with operator coefficients. A more exotic example of an analytic and bounded function of ∂_0^{-1} is the delay operator, which itself is a special case of the time translation:

Example. Let $r > 0$, $\nu > \frac{1}{2r}$, $h \in \mathbb{R}$ and $u \in H_{\nu,0}(\mathbb{R}, X)$. We define

$$\tau_h u := u(\,\cdot\, + h).$$

The operator $\tau_h \in L(H_{\nu,0}(\mathbb{R}, X), H_{\nu,0}(\mathbb{R}, X))$ is called a *time-translation operator.* If $h < 0$ the operator τ_h is also called a *delay operator.* In the latter case the function

$$B_\mathbb{C}(r, r) \ni z \mapsto M(z) := \exp(z^{-1}h)$$

is analytic and uniformly bounded for every $r \in \mathbb{R}_{>0}$ (considered as an $L(X, X)$-valued function). An easy computation shows for $u \in H_{\nu,0}(\mathbb{R}, X)$ that

$$u(\,\cdot\, + h) = \mathcal{L}_\nu^* \exp((im + \nu)h)\mathcal{L}_\nu u = M(\partial_0^{-1})u = \exp(\left(\partial_0^{-1}\right)^{-1} h)\, u.$$

Thus

$$\tau_h = \exp\left(\left(\partial_0^{-1}\right)^{-1} h\right) = \exp\left(h\partial_0\right).$$

Another class of interesting bounded analytic functions of ∂_0^{-1} are mappings produced by a temporal convolution with a suitable operator-valued integral kernel.

1.2. Abstract solution theory

We shall discuss equations of the form

$$\left(\partial_0 M\left(\partial_0^{-1}\right) + A\right) U = \mathcal{J}. \tag{1.1}$$

Here we shall assume that A and A^* are commuting with ∂_0 and non-negative in the Hilbert space $H_{\nu,0}(\mathbb{R}, H)$, H a given Hilbert space, in the sense that

$$\mathfrak{Re}\,\langle U|AU\rangle_{\nu,0} \geq 0,\ \mathfrak{Re}\,\langle V|A^*V\rangle_{\nu,0} \geq 0$$

for all $U \in D(A)$, $V \in D(A^*)$, and M is a material law in the sense of [3, 6]. More specifically we assume that M is of the form

$$M(z) = M_0 + zM_1 + z^2 M_2(z)$$

where $M_2(z)$ is an analytic and bounded $L(H,H)$-valued function in the ball $B_{\mathbb{C}}(r,r)$ for some $r \in \mathbb{R}_{>0}$ and M_0 is a continuous, selfadjoint and non-negative operator in H. The operator $M_1 \in L(H,H)$ is such that

$$\nu M_0 + \mathfrak{Re}\, M_1 \geq c_0 > 0 \tag{1.2}$$

for all sufficiently large $\nu \in \mathbb{R}_{>0}$. The operator $M\left(\partial_0^{-1}\right)$ is then to be understood in the sense of the operator-valued function calculus associated with the selfadjoint operator $\mathfrak{Im}\,(\partial_0) = \frac{1}{2\mathrm{i}}\left(\partial_0 - \partial_0^*\right)$.

The appropriate setting turns out to be the Sobolev chain

$$\left(H_{\nu,k}(\mathbb{R},\ H)\right)_{\nu,k\in\mathbb{Z}}.$$

From [3, 6] we paraphrase the following solution result.

Theorem 1.3. *For $\mathcal{J} \in H_{\nu,k}(\mathbb{R},\ H)$ the problem (1.1) has a unique solution $U \in H_{\nu,k}(\mathbb{R},\ H)$. Moreover,*

$$F \mapsto \left(\partial_0 M\left(\partial_0^{-1}\right) + A\right)^{-1} F$$

is a linear mapping in $L\left(H_{\nu,k}(\mathbb{R},\ H), H_{\nu,k}(\mathbb{R},\ H)\right)$, $k \in \mathbb{Z}$. These mappings are causal in the sense that if $F \in H_{\nu,k}(\mathbb{R},\ H)$ vanishes on the time interval $]-\infty, a]$, then so does $\left(\partial_0 M\left(\partial_0^{-1}\right) + A\right)^{-1} F$, $a \in \mathbb{R}$, $k \in \mathbb{Z}$.

Remark 1.4. If $U \in H_{\nu,k}(\mathbb{R},\ H)$ and $\mathcal{J} \in H_{\nu,k}(\mathbb{R},\ H)$ equation (1.1) actually makes sense in $H_{\nu,k-1}(\mathbb{R},\ H)$. Initially the solution theory is for the closure of $\left(\partial_0 M\left(\partial_0^{-1}\right) + A\right)$ as a closed operator in $H_{\nu,k-1}(\mathbb{R},\ H)$, but

$$\partial_0 M\left(\partial_0^{-1}\right) U + AU = \overline{\left(\partial_0 M\left(\partial_0^{-1}\right) + A\right)}\, U$$

in $H_{\nu,k-1}(\mathbb{R},\ H)$ (although the right-hand side is in $H_{\nu,k}(\mathbb{R}, H)$). Indeed, for $\phi \in H_{\nu,k}(\mathbb{R}, H) \cap D(A^*)$ we have

$$\begin{aligned}\left\langle \phi \middle| \overline{\left(\partial_0 M\left(\partial_0^{-1}\right) + A\right)}\, U \right\rangle_{\nu,k-1,0} &= \left\langle \left(\partial_0^* M\left(\partial_0^{-1}\right)^* + A^*\right)\phi \middle| U \right\rangle_{\nu,k-1,0} \\ &= \left\langle \phi \middle| M\left(\partial_0^{-1}\right)\partial_0 U\right\rangle_{\nu,k-1,0} + \left\langle A^*\phi | U\right\rangle_{\nu,k-1,0}\end{aligned}$$

and we read off that $U \in D(A)$ if A is considered in $H_{\nu,k-1}(\mathbb{R}, H)$ (rather than $H_{\nu,k}(\mathbb{R}, H)$) giving

$$\begin{aligned} AU &= \overline{\left(\partial_0 M\left(\partial_0^{-1}\right) + A\right)} U - M\left(\partial_0^{-1}\right) \partial_0 U, \\ &= \overline{\left(\partial_0 M\left(\partial_0^{-1}\right) + A\right)} U - \partial_0 M\left(\partial_0^{-1}\right) U. \end{aligned}$$

The rigorous argument is somewhat more involved, see [5, 4]. This observation, however, motivates dropping the closure bar throughout.

2. An application: An evolutionary problem involving a Sturm–Liouville type operator with an impedance type boundary condition

A Sturm–Liouville boundary value problem with impedance type boundary conditions will be posed as a $(1+1)$-dimensional example, i.e., having one time and one space parameter. This exemplifies the theory outlined above. Note that a 3×3-system representation has been chosen, rather than an alternative 2×2-system formulation, since conservativity is more apparent in this setting, see Remark 2.4.

Consider

$$A = \begin{pmatrix} 0 & 0 & \partial \\ 0 & 0 & 0 \\ \partial & 0 & 0 \end{pmatrix}$$

with an impedance type boundary condition implemented in the domain of A given by

$$D(A) = \left\{ \begin{pmatrix} s \\ w \\ v \end{pmatrix} \in H_{\nu,0}\left(\mathbb{R}, H(\partial, I) \oplus L^2(I) \oplus H(\partial, I)\right) \,\middle|\, a\left(\partial_0^{-1}\right) s - \partial_0^{-1} v \in H_{\nu,0}\left(\mathbb{R}, H\left(\mathring{\partial}, I\right)\right) \right\},$$

where $\mathring{\partial}$ denotes the closure of ∂ restricted to smooth function with compact support in $I =]-1/2, 1/2[$ and $H\left(\mathring{\partial}, I\right)$ denotes its domain, which is a Hilbert space with respect to the graph norm of the derivative operator ∂. The space $H(\partial, I)$ is the domain of the adjoint, which we denote again by ∂, of $\mathring{\partial}$ also equipped with the corresponding graph norm. Here we focus on the finite interval, I, case, however, the same reasoning would apply for the half infinite interval case, say $I = \mathbb{R}_{>0}$. Indeed this case would, in a sense, be simpler since only one boundary point would need to be considered.

We now give the assumptions which we shall require of $a(x, z)$ in the specification of $D(A)$.

Assumption I. For some $r > 0$, the map $a : I \times B_{\mathbb{C}}(0, 2r) \to \mathbb{C}$ with $(x, z) \mapsto a(x, z)$ is bounded on $I \times B_{\mathbb{C}}(0, 2r)$, analytic in z for each $x \in I$ and uniformly continuous in x for $z \in B_{\mathbb{C}}(0, 2r)$.

Remark 2.1. Under Assumption I, $(z \mapsto a(x,z))_{x\in I}$ is also a bounded family of analytic functions in $B_{\mathbb{C}}(r,r)$ and for $\nu > \frac{1}{2r}$ we have a continuous linear and causal mapping:

$$a\left(\partial_0^{-1}\right) : H_{\nu,0}\left(\mathbb{R}, L^2(I)\right) \to H_{\nu,0}\left(\mathbb{R}, L^2(I)\right)$$
$$\varphi \mapsto \left(t \mapsto \left(x \mapsto a\left(x, \partial_0^{-1}\right)\varphi(t,x)\right)\right).$$

Assumption II. The x-distributional derivative $a'(x,z)$ of $a(x,z)$ is a bounded function defined on $(I\backslash N) \times B_{\mathbb{C}}(0,2r)$ where N is of Lebesgue measure zero, and for each $x \in I\backslash N$, $a'(x,z)$ is analytic on $B_{\mathbb{C}}(0,2r)$.

Remark 2.2. Again we have that

$$a'\left(\partial_0^{-1}\right) : H_{\nu,0}\left(\mathbb{R}, L^2(I)\right) \to H_{\nu,0}\left(\mathbb{R}, L^2(I)\right)$$

is a bounded linear and causal mapping and the product rule

$$\partial\left(a\left(\partial_0^{-1}\right) s\right) = a'\left(\partial_0^{-1}\right) s + a\left(\partial_0^{-1}\right)\partial s$$

holds for $s \in D(\partial)$.

Assumption III. The map a is real in the sense that

$$a(x,z)^* = a(x,z^*)$$

for $x \in I$ and $z \in B_{\mathbb{C}}(0,2r)$.

Assumption IV. At each $x \in I$ denote the three term Taylor expansion of $a(x,z)$ with respect to z by

$$a(x,z) = a_0(x) + a_1(x)z + a_2(x)z^2 + a_3(x,z)z^3$$

where $a_3(x,z), z \in B_{\mathbb{C}}(0,2r)$, is bounded for each $x \in I$. We assume

$$\pm a_0(\pm 1/2) \geq 0, \tag{2.1}$$

and that

$$\pm\nu a_0(\pm 1/2) \pm a_1(\pm 1/2) \geq c_0 > 0 \tag{2.2}$$

for $\nu \in \mathbb{R}_{>0}$ sufficiently large.

Remark 2.3. If $a\left(\partial_0^{-1}\right) = a_0 + a_1\partial_0^{-1}$ it is sufficient to require

$$\pm\nu a_0(\pm 1/2) \pm a_1(\pm 1/2) \geq 0$$

for all sufficiently large $\nu \in \mathbb{R}_{>0}$.

Such an operator A combined with a suitable material law yields an evolutionary problem of the form

$$\left(\partial_0 M\left(\partial_0^{-1}\right) + A\right)U = \mathcal{J}. \tag{2.3}$$

Here we consider material law operators of the form

$$M\left(\partial_0^{-1}\right) = \begin{pmatrix} \kappa_0 & 0 & 0 \\ 0 & \kappa_1 & -\mu_0^*\partial_0^{-1} \\ 0 & \mu_0\partial_0^{-1} & \varepsilon + \eta\,\partial_0^{-1} + \mu_1\partial_0^{-2} \end{pmatrix},$$

where $\varepsilon : L^2(I) \to L^2(I)$, $\kappa_0 : L^2(I) \to L^2(I)$, $\kappa_1 : L^2(I) \to L^2(I)$ are suitable continuous, selfadjoint, non-negative mappings and $\mu_0 : L^2(I) \to L^2(I)$, $\mu_1 : L^2(I) \to L^2(I)$, $\eta : L^2(I) \to L^2(I)$ are continuous and linear.

Assumption V: We assume that the coefficient operators are such that (1.2) is satisfied, i.e.,

$$\nu\varepsilon + \mathfrak{Re}\,\eta \geq c_0 > 0$$

for some $c_0 \in \mathbb{R}$ and all sufficiently large $\nu \in \mathbb{R}_{>0}$.

Remark 2.4. This implies a polynomial type material law operator of the form

$$M\left(\partial_0^{-1}\right) = \begin{pmatrix} \kappa_0 & 0 & 0 \\ 0 & \kappa_1 & 0 \\ 0 & 0 & \varepsilon \end{pmatrix} + \partial_0^{-1} \begin{pmatrix} 0 & 0 & 0 \\ 0 & 0 & -\mu_0^* \\ 0 & \mu_0 & \eta \end{pmatrix} + \partial_0^{-2} \begin{pmatrix} 0 & 0 & 0 \\ 0 & 0 & 0 \\ 0 & 0 & \mu_1 \end{pmatrix}.$$

We see that for κ_0, κ_1, ε strictly positive continuous linear operators, $\mu_1 = 0$ and $\eta = 0$ and A skew-selfadjoint, e.g., if $a\left(\partial_0^{-1}\right) = 0$, we have a conservative system since then $M^{(2)}\left(\partial_0^{-1}\right) = 0$ and

$$M_1 = \begin{pmatrix} 0 & 0 & 0 \\ 0 & 0 & -\mu_0^* \\ 0 & \mu_0 & \eta \end{pmatrix}$$

is skew-selfadjoint making $A + M_1$ and so also $\sqrt{M_0^{-1}}\,(A + M_1)\sqrt{M_0^{-1}}$ skew-selfadjoint. Consequently, $\sqrt{M_0^{-1}}\,(A + M_1)\sqrt{M_0^{-1}}$ generates a unitary 1-parameter group and "energy" conservation holds in the sense that for the solution U of a pure initial value problem we have for $t \in \mathbb{R}_{>0}$

$$\left|\sqrt{M_0}U(t)\right|_H = \left|\sqrt{M_0}U(0+)\right|_H$$

or if one prefers to underscore the "energy" metaphor

$$E(t) := \frac{1}{2}\left|\sqrt{M_0}U(t)\right|_H^2 = \frac{1}{2}\left|\sqrt{M_0}U(0+)\right|_H^2 = E(0+).$$

Such material laws are suggested by models of linear acoustics, see, e.g., [2], or by the so-called Maxwell–Cattaneo–Vernotte law [1, 3] describing heat propagation. In the one-dimensional case, focused on here, this special material law operator can be reduced to a wave or heat equation type partial differential operator with a Sturm–Liouville type operator as spatial part. Indeed, assuming additionally that κ_0 and κ_1 are strictly positive, two elementary row operations applied to

$$\begin{pmatrix} \kappa_0\partial_0 & 0 & \partial \\ 0 & \kappa_1\partial_0 & -\mu_0^* \\ \partial & \mu_0 & \varepsilon\partial_0 + \eta + \mu_1\partial_0^{-1} \end{pmatrix}$$

yield formally

$$\begin{pmatrix} \kappa_0\partial_0 & 0 & \partial \\ 0 & \kappa_1\partial_0 & -\mu_0^* \\ 0 & 0 & \partial_0^{-1}\left(\varepsilon\partial_0^2+\eta\partial_0+\left(\mu_0\kappa_1^{-1}\mu_0+\mu_1\right)-\partial\kappa_0^{-1}\partial\right) \end{pmatrix}.$$

Remark 2.5. A more common point of view for this operation would be to think of new unknowns being introduced. Indeed, if the system unknowns are $\begin{pmatrix} cs \\ w \\ v \end{pmatrix}$ then letting

$$y := \partial_0^{-1} v$$

we would get from a line by line inspection of the system

$$\begin{pmatrix} \kappa_0\partial_0 & 0 & \partial \\ 0 & \kappa_1\partial_0 & -\mu_0^* \\ \partial & \mu_0 & \varepsilon\partial_0+\eta+\mu_1\partial_0^{-1} \end{pmatrix}\begin{pmatrix} s \\ w \\ v \end{pmatrix} = \begin{pmatrix} 0 \\ 0 \\ f \end{pmatrix}$$

that

$$\begin{aligned} s &= -\kappa_0^{-1}\partial y \\ w &= \kappa_1^{-1}\mu_0^* y \end{aligned}$$

and so that

$$\varepsilon\partial_0^2 y+\eta\partial_0 y+\mu_1 y+\mu_0\kappa_1^{-1}\mu_0^* y-\partial\kappa_0^{-1}\partial y = f.$$

Clearly, applying the temporal Fourier–Laplace transform to the second-order expression $\left(\varepsilon\partial_0^2+\eta\partial_0+\left(\mu_0\kappa_1^{-1}\mu_0^*+\mu_1\right)-\partial\kappa_0^{-1}\partial\right)$ we obtain point-wise, writing $\sqrt{\lambda}$ instead of $(\mathrm{i}m+\nu)$,

$$\left(\varepsilon\lambda+\eta\sqrt{\lambda}+\left(\mu_0\kappa_1^{-1}\mu_0+\mu_1\right)-\partial\kappa_0^{-1}\partial\right),$$

which for vanishing "damping" η is indeed a Sturm–Liouville operator

$$r\lambda+q-\partial p\partial$$

with

$$\begin{aligned} r &:= \varepsilon, \\ q &:= \mu_0\kappa_1^{-1}\mu_0+\mu_1, \\ p &:= \kappa_0^{-1}. \end{aligned}$$

For our purposes we may allow for general material laws in the problem (2.3).

Remark 2.6. If $\varepsilon = 0$ and η has a strictly positive definite symmetric part, i.e., the selfadjoint $\mathfrak{Re}\,\eta$ is strictly positive, we arrive at the parabolic type operator

$$\eta\partial_0+q-\partial p\partial$$

and writing λ instead of $(\mathrm{i}m+\nu)$ we get again a Sturm–Liouville type operator

$$\eta\lambda+q-\partial p\partial,$$

where now η plays the role of r.

Theorem 2.7. *Under assumptions I to V (above), for every $\mathcal{J} \in H_{\nu,k}(\mathbb{R}, H)$ the problem* (2.3) *has a unique solution $U \in H_{\nu,k}(\mathbb{R}, H)$. The solution operator $\left(\partial_0 M\left(\partial_0^{-1}\right) + A\right)^{-1} : H_{\nu,k}(\mathbb{R}, H) \to H_{\nu,k}(\mathbb{R}, H)$ is continuous and causal for every $k \in \mathbb{Z}$ and any sufficiently large $\nu \in \mathbb{R}_{>0}$.*

Proof. Denoting the inner product and norm of $H_{\nu,0}\left(\mathbb{R}, L^2(I) \oplus L^2(I) \oplus L^2(I)\right)$ by $\langle \cdot | \cdot \rangle_{\nu,0,0}$ and $|\cdot|_{\nu,0,0}$, respectively, we calculate

$$
\begin{aligned}
&\mathfrak{Re}\left\langle \chi_{]-\infty,0]}(m_0)\begin{pmatrix} s \\ w \\ v \end{pmatrix} \middle| A\begin{pmatrix} s \\ w \\ v \end{pmatrix}\right\rangle_{\nu,0,0} \\
&= \mathfrak{Re}\left(\left\langle \chi_{]-\infty,0]}(m_0)\, s|\partial v\right\rangle_{\nu,0,0} + \left\langle \partial s|\chi_{]-\infty,0]}(m_0)\, v\right\rangle_{\nu,0,0}\right) \\
&= \mathfrak{Re}\left\langle \chi_{]-\infty,0]}(m_0)\, s|\mathring{\partial}\left(v - \partial_0 a\left(\partial_0^{-1}\right) s\right)\right\rangle_{\nu,0,0} \\
&\quad + \mathfrak{Re}\left\langle \chi_{]-\infty,0]}(m_0)\, s|\partial\partial_0 a\left(\partial_0^{-1}\right) s\right\rangle_{\nu,0,0} + \mathfrak{Re}\left\langle \partial s|\chi_{]-\infty,0]}(m_0)\, v\right\rangle_{\nu,0,0} \\
&= -\mathfrak{Re}\left\langle \partial s|\chi_{]-\infty,0]}(m_0)\left(v - \partial_0 a\left(\partial_0^{-1}\right) s\right)\right\rangle_{\nu,0,0} \\
&\quad + \mathfrak{Re}\left\langle \chi_{]-\infty,0]}(m_0)\, s|\partial\partial_0 a\left(\partial_0^{-1}\right) s\right\rangle_{\nu,0,0} + \mathfrak{Re}\left\langle \partial s|\chi_{]-\infty,0]}(m_0)\, v\right\rangle_{\nu,0,0} \\
&= \mathfrak{Re}\left\langle \partial \chi_{]-\infty,0]}(m_0)\, s|\partial_0 a\left(\partial_0^{-1}\right) s\right\rangle_{\nu,0,0} + \mathfrak{Re}\left\langle \chi_{]-\infty,0]}(m_0)\, s|\partial\partial_0 a\left(\partial_0^{-1}\right) s\right\rangle_{\nu,0,0} \\
&= \mathfrak{Re}\left\langle \chi_{]-\infty,0]}(m_0)\, s(\cdot, +1/2)\,|\partial_0 a\left(+1/2, \partial_0^{-1}\right) s(\cdot, +1/2)\right\rangle_{\nu,0} \\
&\quad - \mathfrak{Re}\left\langle \chi_{]-\infty,0]}(m_0)\, s(\cdot, -1/2)\,|\partial_0 a\left(-1/2, \partial_0^{-1}\right) s(\cdot, -1/2)\right\rangle_{\nu,0}. \qquad (2.4)
\end{aligned}
$$

As noted earlier

$$a\left(\partial_0^{-1}\right) = a_0 + a_1\partial_0^{-1} + a_2\partial_0^{-2} + \partial_0^{-3}a^{(3)}\left(\partial_0^{-1}\right),$$

where $a^{(3)}\left(\partial_0^{-1}\right)$ is bounded. From this we obtain that

$$
\begin{aligned}
&\mathfrak{Re}\left\langle \chi_{]-\infty,0]}(m_0)\,\varphi| \pm \partial_0 a_0(\pm 1/2)\,\varphi\right\rangle_{\nu,0} \\
&\quad = \pm\int_{-\infty}^{0} \varphi(t)^* \left(\partial_0 a_0(\pm 1/2)\,\varphi\right)(t)\, \exp(-2\nu t)\; dt \\
&\quad = \pm\nu\int_{-\infty}^{0} a_0(\pm 1/2)\,|\varphi(t)|^2 \exp(-2\nu t)\; dt \pm \frac{1}{2}a_0(\pm 1/2)\,|\varphi(0)|^2
\end{aligned}
$$

which is non-negative by (2.1). Similarly

$$
\begin{aligned}
&\mathfrak{Re}\left\langle \chi_{]-\infty,0]}(m_0)\,\varphi| \pm a_1(\pm 1/2)\,\varphi\right\rangle_{\nu,0} \\
&\quad = \pm\int_{-\infty}^{0} \varphi(t)^*\, a_1(\pm 1/2)\,\varphi(t)\, \exp(-2\nu t)\; dt.
\end{aligned}
$$

Now from (2.2) we obtain

$$\begin{aligned}&\mathfrak{Re}\left\langle \chi_{]-\infty,0]}\left(m_0\right)\varphi \middle| \pm \partial_0 a\left(\pm 1/2, \partial_0^{-1}\right)\varphi\right\rangle_{\nu,0}\\ &\quad\geq \left(\pm\nu a_0\left(\pm 1/2\right) \pm a_1\left(\pm 1/2\right)\right)\left|\chi_{]-\infty,0]}\left(m_0\right)\varphi\right|_{\nu,0}^2\\ &\qquad+\mathfrak{Re}\left\langle \chi_{]-\infty,0]}\left(m_0\right)\varphi \middle| \pm \partial_0^{-1} a^{(2)}\left(\pm 1/2, \partial_0^{-1}\right)\varphi\right\rangle_{\nu,0}.\end{aligned}$$

Due to causality we have

$$\begin{aligned}&\left|\mathfrak{Re}\left\langle \chi_{]-\infty,0]}\left(m_0\right)\varphi \middle| \pm \partial_0^{-1} a^{(2)}\left(\pm 1/2, \partial_0^{-1}\right)\varphi\right\rangle_{\nu,0}\right|\\ &\quad= \left|\mathfrak{Re}\left\langle \chi_{]-\infty,0]}\left(m_0\right)\varphi \middle| \pm a^{(2)}\left(\pm 1/2, \partial_0^{-1}\right)\chi_{]-\infty,0]}\left(m_0\right)\partial_0^{-1}\varphi\right\rangle_{\nu,0}\right|\\ &\quad\leq C_1\left|\chi_{]-\infty,0]}\left(m_0\right)\varphi\right|_{\nu,0}\left|\chi_{]-\infty,0]}\left(m_0\right)\partial_0^{-1}\varphi\right|_{\nu,0}\\ &\quad\leq C_1\left|\chi_{]-\infty,0]}\left(m_0\right)\varphi\right|_{\nu,0}\left|\chi_{]-\infty,0]}\left(m_0\right)\partial_0^{-1}\chi_{]-\infty,0]}\left(m_0\right)\varphi\right|_{\nu,0}\\ &\quad\leq C_1\left|\chi_{]-\infty,0]}\left(m_0\right)\varphi\right|_{\nu,0}\left|\partial_0^{-1}\chi_{]-\infty,0]}\left(m_0\right)\varphi\right|_{\nu,0}\\ &\quad\leq C_1\nu^{-1}\left|\chi_{]-\infty,0]}\left(m_0\right)\varphi\right|_{\nu,0}^2.\end{aligned}$$

Under Assumptions I–III we therefore have found that

$$\mathfrak{Re}\left\langle \chi_{]-\infty,0]}\left(m_0\right)U \middle| AU\right\rangle_{\nu,0,0} \geq 0 \tag{2.5}$$

for all $U \in D(A)$ if $\nu \in \mathbb{R}_{>0}$ is sufficiently large. Note that by the time-translation invariance this is the same as saying

$$\mathfrak{Re}\left\langle \chi_{]-\infty,\tau]}\left(m_0\right)U \middle| AU\right\rangle_{\nu,0,0} \geq 0 \tag{2.6}$$

for all $U \in D(A)$ and all $\tau \in \mathbb{R}$. Letting $\tau \to \infty$ we obtain from this

$$\mathfrak{Re}\left\langle U \middle| AU\right\rangle_{\nu,0,0} \geq 0 \tag{2.7}$$

for all $U \in D(A)$.

We need to find the adjoint of A. It must satisfy

$$-\begin{pmatrix} 0 & 0 & \mathring{\partial} \\ 0 & 0 & 0 \\ \mathring{\partial} & 0 & 0\end{pmatrix} \subseteq A^* \subseteq -\begin{pmatrix} 0 & 0 & \partial \\ 0 & 0 & 0 \\ \partial & 0 & 0\end{pmatrix}$$

in the sense of extensions. We now show that $D(A^*)$ is given by

$$\left\{\begin{pmatrix} s \\ w \\ v\end{pmatrix} \in H_{\nu,0}(\mathbb{R}, H(\partial, I) \oplus L^2(I) \oplus H(\partial, I)) \,\middle|\, a\left(\left(\partial_0^{-1}\right)^*\right)s + \left(\partial_0^{-1}\right)^* v \in H_{\nu,0}(\mathbb{R}, H\left(\mathring{\partial}, I\right)\right\}.$$

Indeed, for

$$\begin{pmatrix} s \\ w \\ v\end{pmatrix} \in D(A)$$

we have

$$\begin{pmatrix} 1 & 0 \\ -a\left(\partial_0^{-1}\right) & \partial_0^{-1} \end{pmatrix} \begin{pmatrix} s \\ v \end{pmatrix} \in H_{\nu,0}\left(\mathbb{R}, H(\partial, I) \oplus H\left(\mathring{\partial}, I\right)\right).$$

Direct computation gives

$$\begin{aligned} \begin{pmatrix} 0 & \mathring{\partial} \\ \partial & 0 \end{pmatrix} \begin{pmatrix} 1 & 0 \\ -a\left(\partial_0^{-1}\right) & \partial_0^{-1} \end{pmatrix} &= \begin{pmatrix} 0 & \partial \\ \partial & 0 \end{pmatrix} \begin{pmatrix} 1 & 0 \\ 0 & \partial_0^{-1} \end{pmatrix} + \begin{pmatrix} 0 & \partial \\ 0 & 0 \end{pmatrix} \begin{pmatrix} 0 & 0 \\ -a\left(\partial_0^{-1}\right) & 0 \end{pmatrix} \\ &= \begin{pmatrix} \partial_0^{-1} & 0 \\ 0 & 1 \end{pmatrix} \begin{pmatrix} 0 & \partial \\ \partial & 0 \end{pmatrix} - \begin{pmatrix} a'\left(\partial_0^{-1}\right) & 0 \\ 0 & 0 \end{pmatrix} + \\ &\quad - \begin{pmatrix} a\left(\partial_0^{-1}\right)\partial & 0 \\ 0 & 0 \end{pmatrix} \\ &= \begin{pmatrix} \partial_0^{-1} & -a\left(\partial_0^{-1}\right) \\ 0 & 1 \end{pmatrix} \begin{pmatrix} 0 & \partial \\ \partial & 0 \end{pmatrix} - \begin{pmatrix} a'\left(\partial_0^{-1}\right) & 0 \\ 0 & 0 \end{pmatrix}. \end{aligned}$$

Thus, we have

$$\begin{aligned} \begin{pmatrix} 0 & \partial \\ \partial & 0 \end{pmatrix} \partial_0^{-1} \begin{pmatrix} s \\ v \end{pmatrix} &= \begin{pmatrix} 1 & a\left(\partial_0^{-1}\right) \\ 0 & \partial_0^{-1} \end{pmatrix} \begin{pmatrix} 0 & \mathring{\partial} \\ \partial & 0 \end{pmatrix} \begin{pmatrix} 1 & 0 \\ -a\left(\partial_0^{-1}\right) & \partial_0^{-1} \end{pmatrix} \begin{pmatrix} s \\ v \end{pmatrix} + \\ &\quad + \begin{pmatrix} 1 & a\left(\partial_0^{-1}\right) \\ 0 & \partial_0^{-1} \end{pmatrix} \begin{pmatrix} a'\left(\partial_0^{-1}\right) & 0 \\ 0 & 0 \end{pmatrix} \begin{pmatrix} s \\ v \end{pmatrix} \\ &= \begin{pmatrix} 1 & a\left(\partial_0^{-1}\right) \\ 0 & \partial_0^{-1} \end{pmatrix} \begin{pmatrix} 0 & \mathring{\partial} \\ \partial & 0 \end{pmatrix} \begin{pmatrix} 1 & 0 \\ -a\left(\partial_0^{-1}\right) & \partial_0^{-1} \end{pmatrix} \begin{pmatrix} s \\ v \end{pmatrix} + \\ &\quad + \begin{pmatrix} a'\left(\partial_0^{-1}\right) & 0 \\ 0 & 0 \end{pmatrix} \begin{pmatrix} s \\ v \end{pmatrix}. \end{aligned}$$

Letting $\begin{pmatrix} 1 & 0 \\ -a\left(\partial_0^{-1}\right) & \partial_0^{-1} \end{pmatrix} \begin{pmatrix} s \\ v \end{pmatrix} = W$ we have for $\begin{pmatrix} v_0 \\ v_1 \\ v_2 \end{pmatrix} \in D(A^*)$ and for every $W \in H_{\nu,0}\left(\mathbb{R}, H(\partial, I) \oplus H\left(\mathring{\partial}, I\right)\right)$,

$$\begin{aligned} 0 &= \left\langle \begin{pmatrix} 0 & \partial \\ \partial & 0 \end{pmatrix} \begin{pmatrix} \partial_0^{-1} & 0 \\ a\left(\partial_0^{-1}\right) & 1 \end{pmatrix} W \middle| \begin{pmatrix} v_0 \\ v_2 \end{pmatrix} \right\rangle_{\nu,0,0} \\ &\quad + \left\langle \begin{pmatrix} \partial_0^{-1} & 0 \\ a\left(\partial_0^{-1}\right) & 1 \end{pmatrix} W \middle| \begin{pmatrix} 0 & \partial \\ \partial & 0 \end{pmatrix} \begin{pmatrix} v_0 \\ v_2 \end{pmatrix} \right\rangle_{\nu,0,0} \\ &= \left\langle \begin{pmatrix} 1 & a\left(\partial_0^{-1}\right) \\ 0 & \partial_0^{-1} \end{pmatrix} \begin{pmatrix} 0 & \mathring{\partial} \\ \partial & 0 \end{pmatrix} W \middle| \begin{pmatrix} v_0 \\ v_2 \end{pmatrix} \right\rangle_{\nu,0,0} \\ &\quad + \left\langle \begin{pmatrix} a'\left(\partial_0^{-1}\right) & 0 \\ 0 & 0 \end{pmatrix} W \middle| \begin{pmatrix} v_0 \\ v_2 \end{pmatrix} \right\rangle_{\nu,0,0} \\ &\quad + \left\langle \begin{pmatrix} \partial_0^{-1} & 0 \\ a\left(\partial_0^{-1}\right) & 1 \end{pmatrix} W \middle| \begin{pmatrix} 0 & \partial \\ \partial & 0 \end{pmatrix} \begin{pmatrix} v_0 \\ v_2 \end{pmatrix} \right\rangle_{\nu,0,0} \end{aligned}$$

$$= \left\langle \begin{pmatrix} 0 & \mathring{\partial} \\ \partial & 0 \end{pmatrix} W \middle| \begin{pmatrix} 1 & 0 \\ a\left(\left(\partial_0^{-1}\right)^*\right) & \left(\partial_0^{-1}\right)^* \end{pmatrix} \begin{pmatrix} v_0 \\ v_2 \end{pmatrix} \right\rangle_{\nu,0,0}$$
$$+ \left\langle W \middle| \begin{pmatrix} a'\left(\left(\partial_0^{-1}\right)^*\right) & 0 \\ 0 & 0 \end{pmatrix} \begin{pmatrix} v_0 \\ v_2 \end{pmatrix} \right\rangle_{\nu,0,0}$$
$$+ \left\langle W \middle| \begin{pmatrix} \left(\partial_0^{-1}\right)^* & a\left(\left(\partial_0^{-1}\right)^*\right) \\ 0 & 1 \end{pmatrix} \begin{pmatrix} 0 & \partial \\ \partial & 0 \end{pmatrix} \begin{pmatrix} v_0 \\ v_2 \end{pmatrix} \right\rangle_{\nu,0,0}.$$

This implies that

$$\begin{pmatrix} 1 & 0 \\ a\left(\left(\partial_0^{-1}\right)^*\right) & \left(\partial_0^{-1}\right)^* \end{pmatrix} \begin{pmatrix} v_0 \\ v_2 \end{pmatrix} \in H_{\nu,0}\left(\mathbb{R}, H(\partial, I) \oplus H\left(\mathring{\partial}, I\right)\right),$$

which is the above characterization of $D(A^*)$. Moreover,

$$\begin{pmatrix} 0 & \mathring{\partial} \\ \partial & 0 \end{pmatrix} \begin{pmatrix} 1 & 0 \\ a\left(\left(\partial_0^{-1}\right)^*\right) & \left(\partial_0^{-1}\right)^* \end{pmatrix} \begin{pmatrix} v_0 \\ v_2 \end{pmatrix}$$
$$= \begin{pmatrix} a'\left(\left(\partial_0^{-1}\right)^*\right) & 0 \\ 0 & 0 \end{pmatrix} \begin{pmatrix} v_0 \\ v_2 \end{pmatrix} + \begin{pmatrix} \left(\partial_0^{-1}\right)^* & a\left(\left(\partial_0^{-1}\right)^*\right) \\ 0 & 1 \end{pmatrix} \begin{pmatrix} 0 & \partial \\ \partial & 0 \end{pmatrix} \begin{pmatrix} v_0 \\ v_2 \end{pmatrix}.$$

As a consequence of the similarity between A and A^* we find by analogous reasoning that we have not only (2.7) but also, indeed more straight-forwardly,

$$\mathfrak{Re}\,\langle V|A^*V\rangle_{\nu,0,0} \geq 0 \tag{2.8}$$

for all $V \in D(A^*)$. The calculation is similar to (2.4) but without the cut-off with $\chi_{]-\infty,0}(m_0)$. Thus we have indeed that

$$\partial_0 M\left(\partial_0^{-1}\right) + A$$

is continuously invertible with causal inverse $\left(\partial_0 M\left(\partial_0^{-1}\right) + A\right)^{-1} : H_{\nu,k}(\mathbb{R}, H) \to H_{\nu,k}(\mathbb{R}, H)$ for every $k \in \mathbb{Z}$ and any sufficiently large $\nu \in \mathbb{R}_{>0}$. □

References

[1] C. Cattaneo. Sur une forme de l'équation de la chaleur éliminant le paradoxe d'une propagation instantané. *C.R. Acad. Sci. Paris I*, 247:431, 1958.

[2] R. Leis. *Initial boundary value problems in mathematical physics.* John Wiley & Sons Ltd. and B.G. Teubner; Stuttgart, 1986.

[3] R. Picard. A Structural Observation for Linear Material Laws in Classical Mathematical Physics. *Math. Methods Appl. Sci.*, 32(14):1768–1803, 2009.

[4] R. Picard. On a Class of Linear Material Laws in Classical Mathematical Physics. *Int. J. Pure Appl. Math.*, 50(2):283–288, 2009.

[5] R. Picard. An Elementary Hilbert Space Approach to Evolutionary Partial Differential Equations. *Rend. Istit. Mat. Univ. Trieste*, 42 suppl.:185–204, 2010.

[6] R. Picard and D.F. McGhee. *Partial Differential Equations: A unified Hilbert Space Approach*, volume 55 of *De Gruyter Expositions in Mathematics*. De Gruyter. Berlin, New York. 518 pp., 2011.

[7] R. Picard. A class of evolutionary problems with an application to acoustic waves with impedance type boundary conditions. *Proc. Appl. Math. Mech.*, 11(1):687–688, 2011.

Rainer Picard
Institut für Analysis, Fachrichtung Mathematik
Technische Universität Dresden
Dresden, Germany
e-mail: `rainer.picard@tu-dresden.de`

Bruce A. Watson
School of Mathematics
University of the Witwatersrand, Johannesburg
South Africa
e-mail: `bruce.watson@wits.ac.za`

Operator Theory:
Advances and Applications, Vol. 236, 405–420

Crystal Frameworks, Matrix-valued Functions and Rigidity Operators

S.C. Power

Abstract. An introduction and survey is given of some recent work on the infinitesimal dynamics of *crystal frameworks*, that is, of translationally periodic discrete bond-node structures in $\mathbb{R}^d$, for $d = 2, 3, \ldots$. We discuss the rigidity matrix, a fundamental object from finite bar-joint framework theory, rigidity operators, matrix-function representations and low energy phonons. These phonons in material crystals, such as quartz and zeolites, are known as rigid unit modes, or RUMs, and are associated with the relative motions of rigid units, such as SiO_4 tetrahedra in the tetrahedral polyhedral bond-node model for quartz. We also introduce semi-infinite crystal frameworks, bi-crystal frameworks and associated multi-variable Toeplitz operators.

Mathematics Subject Classification (2010). Primary 52C75; Secondary 46T20.

Keywords. Crystal framework, rigidity operator, matrix function, rigid unit mode.

1. Introduction

A survey is given of some recent work on the infinitesimal dynamics of *crystal frameworks*, by which we mean translationally periodic discrete bar-joint frameworks in $\mathbb{R}^d$. This includes a discussion of rigidity operators, matrix symbol function representations and the connections with models for low energy phonon modes in various material crystals. These modes are also known as rigid unit modes, or RUMs, reflecting their origin in the relative motion of rigid units in the crystalline structure. I also introduce briefly the contexts of semi-infinite crystal frameworks and bicrystal frameworks and indicate how their rigidity operators involve multivariable Toeplitz operators whose symbol functions are matrices over multivariable trigonometric polynomials on the d-torus.

The topic of infinite bar-joint frameworks, whether periodic or not, can be pursued as a purely mathematical endeavour and many aspects of deformability and rigidity remain to be understood. The main perspectives below and related issues are developed in Owen and Power [21], [22], [24] and Power [25], [26].

Translationally periodic bond/node bar-joint frameworks or networks are ubiquitous in mathematics (periodic tilings for example), solid state physics (crystal lattices, graphene), solid state chemistry (zeolites) and material science (microporous metal organic frameworks). So there is no lack of interesting examples. I shall illustrate a number of concepts with three examples derived from tilings seen in Seville at the Alcazaar and the Cathedral.

2. Models for material crystals and low energy phonons

We begin by outlining one particular motivation from material science. A crystal framework $\mathcal{C}$ in $\mathbb{R}^3$ can serve as a mathematical model for the essential geometry of the disposition of atoms and bonds in a material crystal $\mathcal{M}$. In the model of interest to us the vertices correspond to certain atoms while the edges correspond in some way to strong bonds. Also the identification of strongly bonded "units" in $\mathcal{M}$ imply a polyhedral net structure and it is this that gives the relevant abstract framework $\mathcal{C}$. A fundamental example of this kind is quartz, SiO_2, in which each silicon atom lies at the centre of a strongly bonded SiO_4 unit, which in turn may be modeled as a tetrahedron with an oxygen atom at each vertex. In this way the material crystal quartz provides a mathematical crystal framework of pairwise connected tetrahedra with a particular connectedness and geometry.

Material scientists are interested in the manifestation and explanation of various forms of low energy motion and oscillation in materials. Of particular interest are the rigid unit modes in aluminosilicate crystals and zeolites, where quite complicated tetrahedral net models are relevant. These low energy (long wavelength) phonon modes are observed in neutron scattering experiments and have been shown to correlate closely with the modes observed in computational simulations. There is now a considerable body of literature tabulating the (reduced) wave vectors of RUMs of various crystals as subsets of the unit cube (Brillouin zone) and it has become evident that the primary determinant is the geometric structure of the abstract frameworks $\mathcal{C}$. See, for example, Dove et al. [8], Hammond et al. [12], [13], Giddy et al. [10] and Swainson and Dove [31]. Particularly intriguing is the simulation study in Dove et al. [8] which gives a range of pictures of the RUM spectrum and multiplicities for various idealized crystal types.

In the experiments and in the simulations the background mathematical model is classical lattice dynamics and rigid unit modes are observed where the phonon dispersion curves indicate vanishing energy. However, one can also identify such limiting cases through a direct linear approach as we outline below and from this it follows that these sets (at least in simulations) may be viewed as real algebraic varieties. See Theorem 5.4 below, [24], [25] and Wegner [33]. It is convenient to define the RUM dimension to be the dimension of this algebraic variety. (See Section 5.) In 3D it takes the values $0, 1, 2, 3$.

3. An illustrative example

The following simple example will serve well to illustrate the notation scheme for general crystal frameworks in d dimensions that we adopt. The example is also of interest in its own right, as we see later.

Figure 1 indicates a translationally periodic bar-joint framework $\mathcal{C} = (G, p)$ determined by a sequence $p = (p_k)$ in $\mathbb{R}^2$. The framework edges $[p_i, p_j]$, associated with the edges of the underlying graph G, are viewed as inextensible bars connected at the framework vertices p_k but otherwise unconstrained.

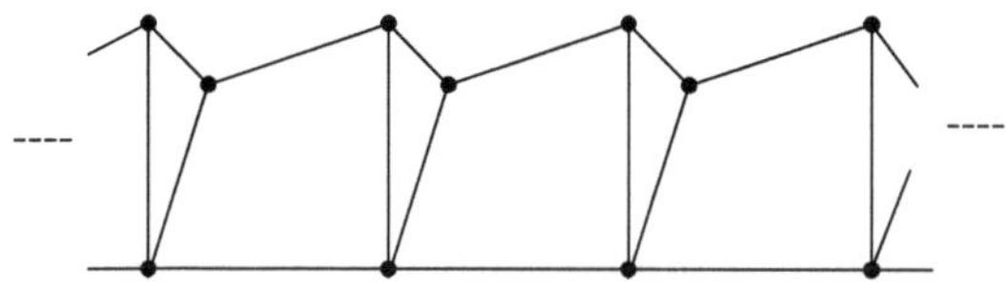

FIGURE 1. An infinite bar-joint framework.

Let the scaling be such that

$$p_1 = p_{1,0} = (0,0),\ p_2 = p_{2,0} = (0,4),\ p_3 = p_{3,0} = (1,3)$$

are three framework vertices of a triangular subframework. Write their translates as

$$p_{\kappa,k} = T_k p_{\kappa,0}, \qquad \text{for } \kappa \in \{1,2,3\}, k \in \mathbb{Z},$$

where T_k is the isometry $T_k : (x, y) \to (x + 4k, y)$ od $\mathbb{R}^2$. The translation group $\mathcal{T} = \{T_k : k \in \mathbb{Z}\}$ is also used to define a natural periodic labelling of the framework edges:

$$e_1 = e_{1,0} = [p_1, p_2],\ e_2 = e_{2,0} = [p_2, p_3],\ e_3 = e_{3,0} = [p_1, p_3,],$$

$$e_4 = e_{4,0} = [p_{1,0}, p_{1,1}],\ e_5 = e_{5,0} = [p_{3,0}, p_{2,1}]$$

and

$$e_{j,k} = T_k e_j, \qquad \text{for } j \in \{1,2,3,4,5\}, k \in \mathbb{Z}.$$

Thus, the pair of finite sets

$$F_v = \{p_1, p_2, p_3\}, \quad F_e = \{e_1, \ldots, e_5\}$$

have disjoint translates under $\mathcal{T}$ and the set of all such translates determines $\mathcal{C}$.

In an exactly similar way a translationally periodic bar-joint framework $\mathcal{C}$ in $\mathbb{R}^d$ is determined by a triple $(F_v, F_e, \mathcal{T})$ where we refer to the finite set pair $\mathcal{M} = (F_v, F_e)$ as a *motif* for $\mathcal{C}$. Of particular interest for applications are the cases $d = 2, 3$ in which $\mathcal{T} = \{T_k : k \in \mathbb{Z}^d\}$ and $\mathcal{T}$ has full rank. For $d = 3$ "full rank" means that the so-called *period vectors*

$$a_1 = T_{(1,0,0)}(0), \quad a_2 = T_{(0,1,0)}(0), \quad a_3 = T_{(0,0,1)}(0)$$

are linearly independent, in which case the framework vertices, if they are distinct, form a discrete set in $\mathbb{R}^3$. We call such discrete bar-joint frameworks *crystal frameworks*.

We now introduce a key dynamical ingredient, namely the notion of an *infinitesimal flex*. This definition is the same as that for finite bar-joint frameworks being a specification of velocity vectors at the nodes which, to first order, do not change edge lengths.

Definition 3.1. Let $\mathcal{C}$ be crystal framework, with framework vertices $p_{\kappa,k}$ as above. An infinitesimal flex of $\mathcal{C}$ is a set of vectors $u_{\kappa,k}$ (velocity vectors) such that for each edge $e = [p_{\kappa,k}, p_{\tau,l}]$

$$\langle p_{\kappa,k} - p_{\tau,l}, u_{\kappa,k} - u_{\tau,l} \rangle = 0.$$

The linear equations required for an infinitesimal flex $u = (u_{\kappa,k})$ translate to a single equation $R(\mathcal{C})u = 0$ where $R(\mathcal{C})$ is the so-called *rigidity matrix* for the framework and where u is a vector in the direct product vector space $\mathcal{H}_v = \prod_{\kappa,k} \mathbb{R}^d$, regarded as a composite vector of instantaneous velocities. The rigidity matrix is sparse with rows labelled by edges and columns labelled by the Euclidean coordinate labels (κ, k, σ) of the framework vertices, with $\sigma \in \{1, \dots, d\}$; the row for framework edge $e = [p_i, p_j]$ has the entry $(p_{\kappa,k} - p_{\tau,l})_\sigma$ for column (κ, k, σ), and has the negative of this entry for column (τ, l, σ). Thus for $d = 3$ row e appears as

$$[0 \dots 0 \; v_e \; 0 \dots 0 \; -v_e \; 0 \dots 0]$$

where the vector $v_e = p_{\kappa,k} - p_{\tau,l}$ (resp $-v_e$) is distributed in the columns for (κ, k, σ) (resp. (τ, l, σ)) with $\sigma \in \{x, y, z\}$.

From various viewpoints, such as phase-periodic velocity vectors on the one hand, or square-summable velocity vectors on the other hand, with the introduction of complex scalars and functional representations of vector spaces, the rigidity matrix $R(\mathcal{C})$ leads to a matrix-valued function $\Phi(z)$ with $|F_e|$ rows and $d|F_e|$ columns. The entries are scalar-valued functions on the d-torus of points $z = (z_1, \dots, z_d)$ in $\mathbb{C}^d$ with $|z_i| = 1$.

We define this matrix function below in Definition 5.3 and one can check that the strip framework of Figure 1 has associated matrix function

$$\Phi(z) = \begin{bmatrix} 0 & -4 & 0 & 4 & 0 & 0 \\ 0 & 0 & -1 & 1 & 1 & -1 \\ -1 & -3 & 0 & 0 & 1 & 3 \\ -4(1-\overline{z}) & 0 & 0 & 0 & 0 & 0 \\ 0 & 0 & 3\overline{z} & \overline{z} & -3 & -1 \end{bmatrix}.$$

3.1. Examples from Seville

The next three frameworks are based on some simple two-dimensional tessellations that are suggested by tilings found in Seville cathedral (Figures 2 and 3) and in the Alcazaar in Seville (Figure 4). All three are in *Maxwell counting equilibrium* in the sense that the average number of edge constraints per vertex is 2, matching the degrees of freedom of each vertex, while each finite subframework is not "overconstrained", in the sense that the number of edges does not exceed twice the number of vertices.

A motif for the framework $\mathcal{C}_{\mathrm{sev}_1}$ is shown in Figure 2 together with the period vectors (dotted). The motif edges consist of a square of edges together with two vertical edges whose (equal) lengths fix the geometry up to a global scaling. (The "rigid units" of this framework are the single vertex subframeworks.)

Note that there is an evident (edge-length preserving) continuous flex, or deformation, $p(t) = (p_{\kappa,k}(t))$ of $\mathcal{C}_{\mathrm{sev}_1}$ which is associated with an expansion in the x direction and a matching contraction in the y direction. We remark that in the case of the geometry with all edge lengths equal this deformation passes through the framework composed of congruent rhombs which is reciprocal (in the lattice sense [6]) to the well-known kagome framework, indicated in Section 5.1. From the first instant of the deformation, so to speak, one obtains an infinitesimal flex $u = p'(0)$ which (unlike infinitesimal translation flexes) is unbounded. Less evident are various nontrivial bounded infinitesimal flexes, but we see below that there are plenty of these and that the RUM dimension is 1.

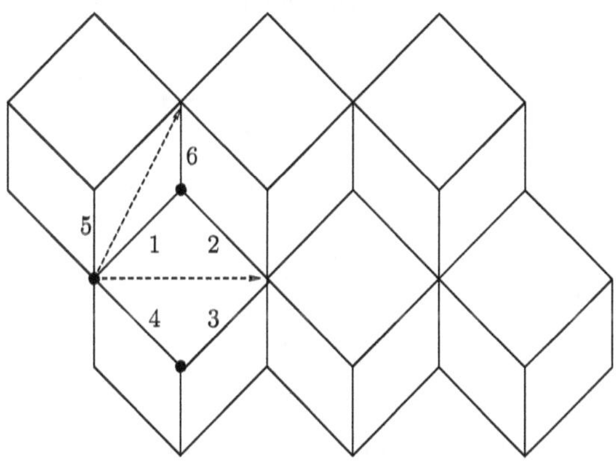

FIGURE 2. The crystal framework $\mathcal{C}_{\mathrm{sev}_1}$.

The framework $\mathcal{C}_{\mathrm{sev}_2}$ in Figure 3 has triangular rigid units and an infinitesimal flex is indicated which is finitely nonzero, with four nonzero velocity vector components. Such a flex is also called a local or internal infinitesimal flex. It is a general principle, as we note further below, that such a local phenomenon makes the framework maximally flexible from a RUM point of view.

The planar graph, or "topology" of this framework is rather interesting, being a network of 4-rings of triangles connected "square-wise", so that the "holes" are 8-cycles and 4-cycles. One can make similar constructions with equilateral triangles with this topology although now the local infinitesimal flex is lost. We remark that periodic networks of pairwise corner-joined congruent equilateral triangles provide the 2D variants of the tetrahedral nets associated with zeolites. (That the hole cycles of 2D zeolites can be arbitrarily large follows from the substitution move

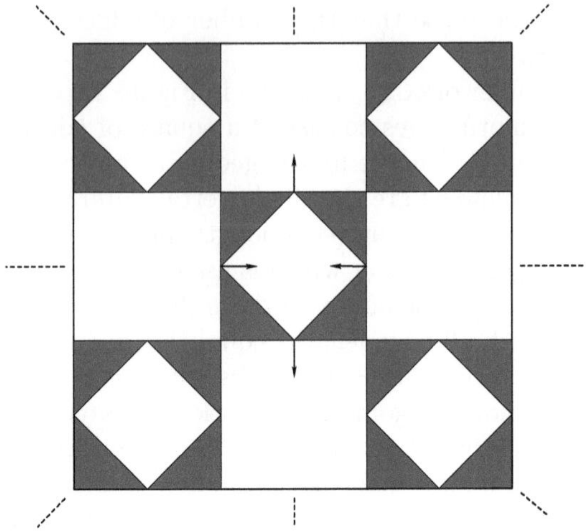

FIGURE 3. $\mathcal{C}_{\text{sev}_2}$, with a local infinitesimal flex.

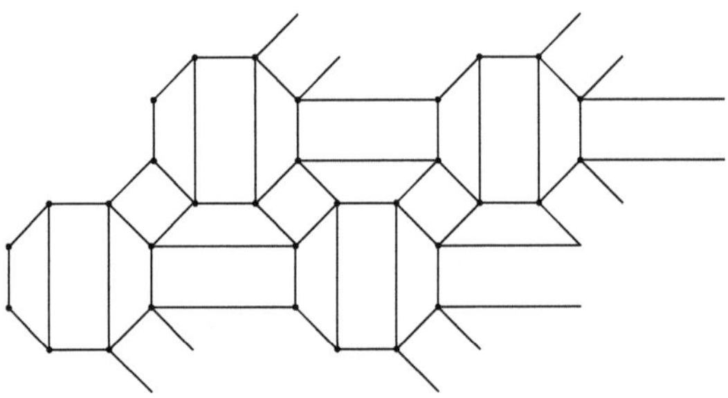

FIGURE 4. Part of $\mathcal{C}_{\text{sev}_3}$, a homeomorph of $\mathcal{C}_{\mathbb{Z}^2}$.

which replaces each rigid unit triangle by a 3-ring of smaller triangles with edge length halved.)

The framework $\mathcal{C}_{\text{sev}_3}$ is derived from a tiling in the Alcazaar in Seville. A moment's thought reveals it to have the same topology (underlying graph) as that for the basic square grid framework $\mathcal{C}_{\mathbb{Z}^2}$. The infinitesimal flexibility is less evident than is the case for $\mathcal{C}_{\mathbb{Z}^2}$ but from the RUM viewpoint they turn out to be equally flexible with RUM dimension 1.

4. Bar-joint frameworks – a very brief overview

4.1. Watt, Peaucellier, Cauchy, Euler, Kempe, Maxwell, Laman

Informally, a "linkage" is a bar-joint framework with one degree of essential flexibility. In 1784 James Watt designed a bar-joint linkage which transformed circular motion into approximate linear motion. This was rather important for steam engine transmission. The mechanism was approximate and was superseded by Peaucellier's exact linear motion linkage eighty years later. In 1876 Kempe [18] solved the general inverse problem by showing that any finite algebraic curve can be simulated by a linkage. The rigidity of geometric frameworks was also of interest to Euler and to Cauchy who in particular were concerned with the rigidity of polyhedra with hinged faces; a beautiful classical result is the infinitesimal rigidity of all convex triangle-faced polyhedra.

James Clerk Maxwell initiated combinatorial aspects with the observation that a graph $G = (V, E)$ with a minimally rigid generic framework realisation in the plane must satisfy the simple counting rule $2|V| - |E| = 3$ together with the inequalities $2|V'| - |E'| \geq 3$ for all subgraphs. The number 3 represents the number of independent global infinitesimal motions for the plane. In 1970 Laman [19] obtained the fundamental result that Maxwell's conditions are sufficient for generic rigidity and this result also anticipated the advent of matroid theory in rigidity theory. While corresponding counting rules are necessary in three dimensions they fail to be sufficient and no necessary and sufficient combinatorial conditions for generic rigidity are known! For further information see [1], [2], [11].

4.2. Some recent work

Laman's theorem concerns *generic* frameworks with a particular graph. One can expect that special frameworks with global symmetries may have more flexibility and this is a topic of current interest. See for example, Connelley et al. [5], Owen and Power [22] and Shulze [28], [29]. Understanding constraint systems of geometric objects with symmetry present is also of significance for algorithms for CAD software [22].

Laman's theorem is also concerned with *finite* frameworks. A natural generalisation, also of significance for applications, are periodic frameworks in the sense of the crystal frameworks above. See also Whiteley [32], Borcea and Streinu [4], Malestein and Theran [20], and Ross, Schulze and Whiteley [27].

The theory of general infinite bar-joint frameworks, from the point of view of rigidity, is a novel topic and perhaps a rather curious one. We point out in Owen and Power [23], [21] that it is possible to generalise Kempe's theorem to the effect that any *continuous* curve (i.e., continuous image of $[0, 1]$ in $\mathbb{R}^2$) may be simulated by an infinite linkage. (In [23] this is achieved with three vertices of infinite degree but in fact infinite degree vertices are not necessary.)

In material science the microporous flexing materials known as zeolites are on the one hand important for industrial applications, as filters, and on the other hand present diverse tetrahedral rigid unit frameworks. The degree of continuous flexibility of such idealised zeolites is investigated in Kapko et al. [17].

5. The RUM spectrum and RUM dimension of $\mathcal{C}$

Let $\mathcal{C} = (F_v, F_e, \mathcal{T})$ be a crystal framework in $\mathbb{R}^d$ and let $\mathcal{K}_v$ be the vector space $\prod_{\kappa,k} \mathbb{C}^{d|F_v|}$ consisting of infinitesimal velocity vectors. Let $\mathbb{T}^d$ be the d-torus of points $\omega = (\omega_1, \dots, \omega_d)$ and for $k \in \mathbb{Z}^d$ write ω^k for the unimodular complex number $\omega_1^{k_1} \dots \omega_d^{k_d}$.

Definition 5.1 ([24], [25]).

(a) A velocity vector $\tilde{u}$ in $\mathcal{K}_v$ is periodic-modulo-phase for the (multi-)phase factor $\omega \in \mathbb{T}^d$ if there exists a vector $u = (u_\kappa)$ in $\mathbb{C}^{|F_v|}$ such that

$$\tilde{u}_{\kappa,k} = \omega^k u_\kappa, \quad \kappa \in F_v, k \in \mathbb{Z}^d.$$

Also $\mathcal{K}_v^\omega$ denotes the associated vector subspace of such vectors.

(b) A periodic-modulo-phase infinitesimal flex (or, for brevity, a wave flex) is a vector $\tilde{u}$ in $\mathcal{K}_v^\omega$ which is an infinitesimal flex for $\mathcal{C}$.

(c) The rigid unit mode spectrum, or RUM spectrum, of $\mathcal{C}$ (with specified translation group $\mathcal{T}$) is the set $\Omega(\mathcal{C})$ of phases ω for which there exists a nonzero wave flex.

To each multiphase ω there exists a unique wave vector $\mathbf{k} = (\mathbf{k}_1, \dots, \mathbf{k}_d)$ in $[0,1)^d$ such that $\omega_j = e^{2\pi i\ \mathbf{k}_j}, 1 \leq j \leq d$. Ignoring bond constraints for the moment, recall that the framework points of $\mathcal{C}$ undergo harmonic motion with wave vector $\mathbf{k}$ when the vertex positions at time t satisfy equations of the form

$$p_{\kappa,k}(t) = p_{\kappa,k} + \exp(2\pi i\ \mathbf{k} \cdot k)\exp(i\alpha t)v_\kappa,$$

where $\alpha/2\pi$ is the frequency of the oscillation and where $\mathbf{k} \cdot k$ is the inner product $\mathbf{k}_1 \cdot k_1 + \dots + \mathbf{k}_d \cdot k_d$. Such pure motions appear as basic solutions in lattice dynamics, under harmonic approximation, with general solutions obtained by linear superposition. (See Dove [7] for example.)

The following theorem from [25] provides an explanation for the connection between low energy oscillation modes alluded to in Section 2 and infinitesimal wave flexes.

Theorem 5.2. *Let $\mathcal{C}$ be a crystal framework, with specified periodicity, and let $\mathbf{k}$ be a wave vector with point $\omega \in \mathbb{T}^d$. Then the following assertions are equivalent.*

(i) $(\omega^k u_\kappa)_{\kappa,k}$ *is a nonzero periodic-modulo-phase infinitesimal (complex) flex for $\mathcal{C}$.*

(ii) *For the vertex wave motion*

$$p_{\kappa,k}(t) = p_{\kappa,k} + u_\kappa \exp(2\pi i\ \mathbf{k} \cdot k)\exp(i\alpha t),$$

and a given time interval, $t \in [0, T]$, the bond length changes

$$|p_{\kappa,k}(t) - p_{\tau,l}(t)| - |p_{\kappa,k}(0) - p_{\tau,l}(0)|,$$

for the edges e tend to zero uniformly, in both t and e, as the wavelength $2\pi/\alpha$ tends to infinity.

Next we define the matrix function $\Phi_{\mathcal{C}}(z)$ of a crystal framework $\mathcal{C}$ with given period vectors. For the multi-index $k = (k_1, \dots, k_d)$ write z^k for the usual monomial function on $\mathbb{T}^d$.

Definition 5.3. Let $\mathcal{C}$ be a crystal framework in $\mathbb{R}^d$, with motif sets

$$F_v = \{p_{\kappa,0} : 1 \le \kappa \le |F_v|\}, \quad F_e = \{e_i : 1 \le i \le |F_e|\},$$

and for each edge $e = [p_{\kappa,k}, p_{\tau,l}]$ in F_e let v_e be the edge vector $p_{\kappa,k} - p_{\tau,l}$. The matrix-valued function $\Phi_{\mathcal{C}}(z)$ has rows labelled by the edges $e \in F_e$ and has $d|F_v|$ columns labelled by κ and the coordinate index $\sigma \in \{1, \dots, d\}$. If $\kappa \neq \tau$ for edge e in F_e then

$$(\Phi_{\mathcal{C}}(z))_{e,(\kappa,\sigma)} = (v_e)_\sigma \bar{z}^k,$$
$$(\Phi_{\mathcal{C}}(z))_{e,(\tau,\sigma)} = -(v_e)_\sigma \bar{z}^l,$$

while for each reflexive edge $[p_{\kappa,k}, p_{\tau,l}]$

$$(\Phi_{\mathcal{C}}(z))_{e,(\kappa,\sigma)} = (v_e)_\sigma (\bar{z}^k - \bar{z}^l),$$

with the remaining entries in each row equal to zero.

The next theorem gives one connection between $\Phi_{\mathcal{C}}(z)$ and the infinitesimal flex properties of $\mathcal{C}$. Here we view the rigidity matrix $R(\mathcal{C})$ as a linear transformation from the product vector space $\mathcal{K}_v = \prod_{\kappa,k} \mathbb{C}^d$ to the edge vector space $\mathcal{K}_e = \prod_{\text{edges}} \mathbb{C} = \prod_{e,k} \mathbb{C}$.

Theorem 5.4. *The restriction of the rigidity matrix $R(\mathcal{C})$ to the finite-dimensional vector space $\mathcal{K}_v^\omega$ has representing matrix $\Phi_{\mathcal{C}}(\overline{\omega})$ with respect to natural vector space bases.*

Proof. Let $\tilde{u}$ be a velocity vector in $\mathcal{K}_v^\omega$ determined by $u \in \mathbb{C}^{d|F_v|}$ as above. Let e in F_e be an edge of the form $[p_{\kappa,k}, p_{\tau,l}]$. Let $\langle \cdot, \cdot \rangle$ denote the bilinear form on $\mathbb{C}^d$. The $(e,k')^{th}$ entry of $R(\mathcal{C})\tilde{u}$ can be written as

$$\begin{aligned}(R(\mathcal{C})\tilde{u})_{e,k'} &= \langle v_e, \tilde{u}_{\kappa,k'+k}\rangle - \langle v_e, \tilde{u}_{\tau,k'+l}\rangle = \langle v_e, \omega^{k'+k} u_\kappa\rangle - \langle v_e, \omega^{k'+l} u_\tau\rangle \\ &= \omega^{k'}(\langle \omega^k v_e, u_\kappa\rangle + \langle -\omega^l v_e, u_\tau\rangle).\end{aligned}$$

This agrees with $\omega^{k'}(\Phi_{\mathcal{C}}(\overline{\omega})u)_e$, both in the case $\kappa \neq \tau$ and in the reflexive case $\kappa = \tau$, as required. □

It now follows that the *RUM spectrum* of $\mathcal{C}$ is identifiable as the algebraic variety in $\mathbb{T}^d$ given by

$$\Omega(\mathcal{C}) = \{z : \operatorname{rank} \Phi_{\mathcal{C}}(z) < |F_e|\}.$$

This set does depend on the choice of translation group. One could define the *primitive RUM spectrum* to correspond to the translation group for a primitive unit cell, and this is then well defined, up to coordinate permutations. If one doubles the period vectors, and hence the unit cell, then (see [25]) the new spectrum is obtained as the range of the old spectrum under the doubling map $(w_1, w_2, w_3) \to (w_1^2, w_2^2, w_3^2)$.

While we have given the multiphase form of the RUM spectrum the convention in material science is to indicate such a spectrum (in three dimensions) as the set of (reduced) wave vectors $\mathbf{k}$ in the unit cube $[0,1)^3$. For calculations of RUM spectrum by different methods see Dove et al. [8] (simulation calculations), Wegner [33] (computer algebra calculations), Owen and Power [24] and Power [25] (direct calculations).

The algebraic variety perspective of RUMs appears to be new and opens the way for new methods and terminology for understanding the curious curved surfaces in [8]. For example, it is natural to define the *RUM dimension* of $\mathcal{C}$ to be the topological dimension of $\Omega(\mathcal{C})$ as a real algebraic variety. By the comments above on unit cell doubling it follows that this quantity is independent of the translation group. (See [25] for details.)

Tetrahedral net frameworks in 3 dimensions, with pairwise vertex connection, satisfy Maxwell counting equilibrium and in the periodic case, with no penetrating tetrahedra, these are sometimes referred to as hypothetical zeolites [9]. (In material crystalline zeolites the rigidly bonded SiO_4 units make up such a bond-node framework.) Even in this case all possibilities occur for the RUM dimension, namely $0, 1, 2, 3$, and this depends, roughly speaking, on the degree of symmetry of the framework. In particular as we note below the framework for the cubic form of sodalite indicated below has full RUM spectrum, corresponding to dimension 3. (This so-called order N property of sodalite was first observed experimentally. See [12].)

For crystal frameworks in Maxwell counting equilibrium the matrix function is square and the RUM spectrum is revealed, in theory at least, as the intersection of the zero set of the multi-variable polynomial $\det \Phi_{\mathcal{C}}(z)$ with the d-torus $\mathbb{T}^d$. In fact, after fixing a monomial order on the d indeterminates $z_1, \dots, z_d$ one may formally define the *crystal polynomial* $p_{\mathcal{C}}(z)$, associated with $\mathcal{C}$. (See also [25].) This is given as the product $\alpha z^\gamma \det(\Phi_{\mathcal{C}}(z))$ where the monomial exponent γ and the scalar α are chosen so that

(i) $p_{\mathcal{C}}(z)$ is a linear combination of non-negative power monomials,

$$p_{\mathcal{C}}(z) = \sum_{\alpha \in \mathbb{Z}_+^d} a_\alpha z^\alpha,$$

(ii) $p_{\mathcal{C}}(z)$ has minimum total degree, and
(iii) $p_{\mathcal{C}}(z)$ has leading monomial with coefficient 1.

The RUM spectrum certainly has symmetry reflecting the crystallographic group symmetries of the crystal framework. Even so the point group may be trivial and the following abstract inverse problem (a Kempe theorem for RUMS?) may well have an affirmative answer.

Problem. Let $q(z, w)$ be a polynomial with real coefficients with $q(1,1) = 0$. Is there a crystal framework with crystal polynomial $p(z, w)$ whose zero set on the 2-torus is the same as that for $q(z, w)$?

5.1. Examples

The tiling-derived framework of Figure 3 has vanishing crystal polynomial. Indeed, this can be predicted from the existence of a local infinitesimal flex. Such a flex allows the construction of infinitesimal phase-periodic flexes for all phases and so the zero set of the polynomial includes the entire 2-torus.

The tiling-derived framework of Figure 2 is also in Maxwell counting equilibrium, its symbol function is 6×6 and one can show by direct calculation that if we write the indeterminates in this case as z, w, then the crystal polynomial is

$$(z-1)(w-1)(z-w)$$

The *kagome framework* is the framework $\mathcal{C}_{\text{kag}}$ formed by pairwise corner connected equilateral triangles in regular hexagonal arrangement. Its symbol function is a 6×6 matrix and the crystal polynomial is also

$$(z-1)(w-1)(z-w).$$

There is a natural $3D$ variant of the kagome lattice known as the *kagome net*. The corresponding crystal framework $\mathcal{C}_{\text{knet}}$ has period vectors formed by three edges of a parallelapiped at pairwise angles of $\pi/3$, and each parallelapiped contains two tetrahedral rigid units such that the three planar slices of $\mathcal{C}_{\text{knet}}$ for each pair of period vectors, is a copy of $\mathcal{C}_{\text{kag}}$. The crystal polynomial takes the form

$$p(z,w,u) = (z-1)(w-1)(u-1)(z-w)(w-u)(z-u).$$

The factorisations in these examples makes evident the nature of the RUM spectrum as a union of lines and a union of surfaces, respectively. In fact the individual factors can be predicted in terms of the identification of infinitesimal flexes that are supported within a linear band (for the $2D$ case) or a linear tube (in the $3D$ case). See [25]. For considerably more complicated polynomials with nonlinear "exotic" spectrum see Wegner [33] and Power [25].

Figure 4 shows a 4-ring of tetrahedra, three copies of which, placed on three adjacent sides of an imaginary cube, provide the edges and vertices for a motif (F_v, F_e) for the framework $\mathcal{C}_{\text{SOD}}$ for the cubic form of sodalite. (From a mathematical perspective, this structure is arguably the most elegant of the naturally occurring zeolite framework types [3].)

A full set of eight 4-rings forms a so-called sodalite cage. With the 24 outer vertices of this cage fixed there is nevertheless an infinitesimal flex of the structure and so a local infinitesimal flex of $\mathcal{C}_{\text{SOD}}$ exists. This is in analogy with the framework $\mathcal{C}_{\text{sev}_2}$. It follows that the determinant of the symbol function (a 72×72 sparse function matrix) vanishes identically and that the sodalite framework $\mathcal{C}_{\text{SOD}}$ has RUM dimension 3.

6. Flexes with decay and Toeplitz operators

Let $\mathcal{C}$ be a crystal framework with an implicit choice of translational periodicity. Write $\mathcal{K}_v^2$ and $\mathcal{K}_e^2$ for the Hilbert spaces of square-summable sequences in $\mathcal{K}_v$ and

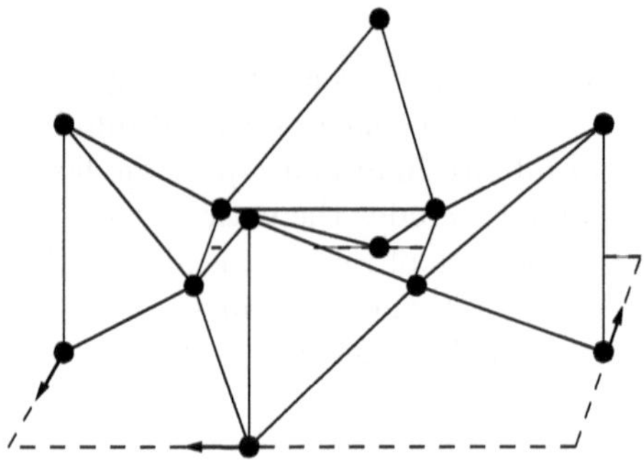

FIGURE 5. A 4-ring building unit from $\mathcal{C}_{\mathrm{SOD}}$.

$\mathcal{K}_e$. Then $R(\mathcal{C})$ determines a bounded Hilbert space operator from $\mathcal{K}_v^2$ to $\mathcal{K}_e^2$. The natural Hilbert space basis for $\mathcal{K}_v^2$, associated with the given periodicity, may be denoted $\{\xi_{\kappa,\sigma,k}\}$, where σ ranges from 1 to d. Similarly, the basis for $\mathcal{K}_e^2$ is $\{\eta_{e,k}\}$, with $e \in F_e, k \in \mathbb{Z}^d$.

Regarding such square-summable sequences as the Fourier series of square integrable vector-valued functions one obtains unitary operators

$$U_v : \mathcal{K}_v^2 \to L^2(\mathbb{T}^d) \otimes \mathbb{C}^{d|F_v|}$$

and $U_e : \mathcal{K}_e^2 \to L^2(\mathbb{T}^d) \otimes \mathbb{C}^{|F_e|}$. In the next theorem, from [24], the unitary equivalence referred to is two-sided and the equivalence in question is the operator identity $U_e^* R(\mathcal{C}) U_v = M_{\Phi_{\mathcal{C}}}$. That $U_e^* R(\mathcal{C}) U_v$ has the form of a multiplication operator M_Ψ follows from standard operator theory, since this operator intertwines the canonical shift operators. Borrowing operator terminology we refer to $\Phi_{\mathcal{C}}$ as the *symbol function* of $\mathcal{C}$.

Theorem 6.1. *The infinite rigidity matrix $R(\mathcal{C})$ of the crystal framework $\mathcal{C}$ in $\mathbb{R}^d$ determines a Hilbert space operator which is unitarily equivalent to the multiplication operator*

$$M_{\Phi_{\mathcal{C}}} : L^2(\mathbb{T}^d) \otimes \mathbb{C}^{d|F_v|} \to L^2(\mathbb{T}^d) \otimes \mathbb{C}^{|F_e|},$$

where $\Phi_{\mathcal{C}}$ is the matrix function for $\mathcal{C}$.

The following corollary follows from elementary operator theory.

Corollary 6.2. *A crystal framework has a square-summable infinitesimal flex if and only if its symbol function has reduced column rank on a set of positive measure.*

It is natural to ask whether crystal frameworks possess infinitesimal flexes which decay to zero at infinity. Note that if an infinite linear subframework has an infinitesimal flex with such decay then the flex velocities must be orthogonal to the direction of this subframework. (For otherwise there must be identical nozero velocity components in that direction on all the subframework points.) For this reason it follows that the kagome framework and similar "linear" frameworks have no asymptotically vanishing flexes and in particular, no square-summable flexes. In fact one can exploit the matrix function formalism to obtain the following much more general fact.

Theorem 6.3. [24] *The following are equivalent for a crystal framework $\mathcal{C}$ with Maxwell counting equilibrium.*

(i) *$\mathcal{C}$ has a nonzero local infinitesimal flex.*
(ii) *$\mathcal{C}$ has a nonzero summable infinitesimal flex.*
(iii) *$\mathcal{C}$ has a nonzero square-summable infinitesimal flex.*

6.1. Semi-infinite and bi-crystal frameworks

We may define a *semi-infinite crystal framework* $\mathcal{D}$ as a subframework of a crystal framework $\mathcal{C}$ with an exposed face. More formally $\mathcal{D}$ is supported by the framework vertices that lie in a half-space which is invariant under a subsemigroup of an underlying translation group for $\mathcal{C}$. In the case of planar frameworks this may be specified in the form of a triple $(F_v, F_e, \mathcal{T}_+)$ where (F_v, F_e) is an appropriate motif and $\mathcal{T}_+$ is a subsemigroup of $\mathcal{T}$ isomorphic to one of $\mathbb{Z}_+ \times \mathbb{Z}$, $\mathbb{Z}_- \times \mathbb{Z}$, $\mathbb{Z} \times \mathbb{Z}_+$, $\mathbb{Z} \times \mathbb{Z}_-$. It is not hard to verify that the rigidity operators of semi-infinite crystal frameworks can be identified with various Toeplitz operators derived from $M_{\Phi_\mathcal{C}}$ by compression to Hardy space Hilbert spaces, such as $H^2(\mathbb{T}) \otimes L^2(\mathbb{T})$ in the case of $\mathbb{Z}_+ \times \mathbb{Z}$.

We remark that semi-infinite frameworks have rigidity matrices that feature as block submatrices of the rigidity matrices of *bi-crystal frameworks*. By this we mean (for example) a framework obtained in three dimensions by identifying two semi-infinite frameworks at their common surface of vertices, when this is possible. It seems that Toeplitz operators could provide a useful formalism for their analysis.[1]

For semi-infinite crystal frameworks the equivalence of the previous theorem no longer holds and we illustrate this with a simple variant of the strip framework in Figure 1 whose matricial symbol function $\Phi(z)$ is as given in Section 2. Consider the submatrix function $\Phi_0(z)$ obtained on removing the first two columns, corresponding to the "supporting" framework vertices, and removing the row corresponding to the "base" edges. The degeneracies of this matrix function correspond to the phases of periodic-modulo-phase infinitesimal flexes which do not deflect the "supporting" vertices. We have

$$\Phi_0(z) = \begin{bmatrix} 0 & 4 & 0 & 0 \\ -1 & 1 & 1 & -1 \\ 0 & 0 & 1 & 3 \\ 3\overline{z} & \overline{z} & -3 & -1 \end{bmatrix}.$$

The determinant, $16(2-3\overline{z})$, does not vanish on $\mathbb{T}$ and so the "base-rooted" framework is infinitesimally rigid from the point of view of phase-periodic infinitesimal

[1] Added October 2011. It appears, from [30] for example, that there is extensive interest in material bicrystals.

flexes. Since the determinant certainly does not vanish on a set of positive measure on $\mathbb{T}$, there are no square-summable infinitesimal flexes which fix the baseline vertices of the framework. On the other hand there is an unbounded infinitesimal flex, corresponding to a two-way infinite geometric series for $z = 2/3$ and this unbounded flex reflects the concatenated lever structure of the framework.

This analysis also applies to the strip framework in Figure 6. One can readily check that up to scalar multiplication there is a unique proper (unbounded) infinitesimal flex. (Note incidentally, that this flex does not extend to a continuous flex. Put another way, each finite strip subframework here is continuously flexible but the complete two-way infinite framework is not. This kind of phenomenon is referred to as *vanishing flexibility* in [24] and can occur in more subtle ways.)

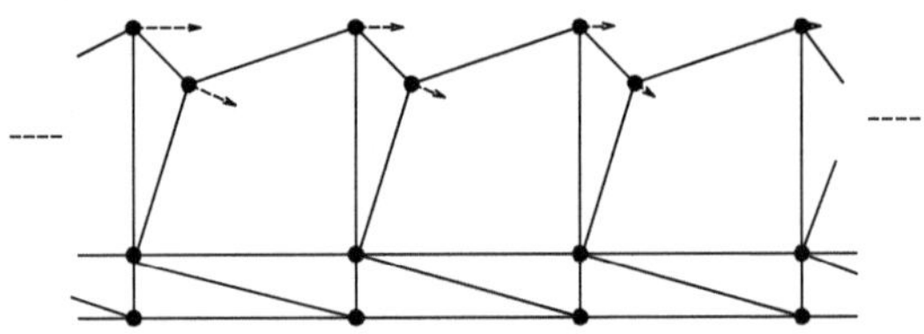

FIGURE 6. An unbounded infinitesimal flex.

There are two natural semi-infinite frameworks $\mathcal{C}_+, \mathcal{C}_-$ associated with the strip framework of Figure 6, namely the right strip and the left strip. Each has a infinite triangulated rigid base framework which supports linked triangles. The former has a square-summable flex while the latter does not. This fact is evident by elementary direct analysis and in fact can be viewed as a reflection of the contrasting nature of analytic and coanalytic Toeplitz operators. We expect such operator theory to play a useful role in the analysis of more complex examples with larger unit cells, and in the analysis of surface phonons and surface phenomena in semi-infinite structures.

References

[1] L. Asimow and B. Roth, The rigidity of graphs, Trans. Amer. Math. Soc., 245 (1978), 279–289.

[2] L. Asimow and B. Roth, Rigidity of graphs II, J. Math. Anal. Appl. 68 (1979) 171–190.

[3] Ch. Baerlocher and L.B. McCusker, Database of Zeolite Structures: http://www.iza-structure.org/databases/

[4] C.S. Borcea and I. Streinu, Periodic frameworks and flexibility, Proc. R. Soc. A 2010 466, 2633–2649.

[5] R.Connelly, P.W. Fowler, S.D. Guest, B. Schulze, W.J. Whiteley, "When is a pin-jointed framework isostatic?" International J. of Solids and Structures, 46 (2009), 762–773.

[6] H.S.M. Coxeter, Regular polytopes, Dover, New York, 1973.

[7] M.T. Dove, Introduction to lattice dynamics, Cambridge topics in Mineral Physics and Chemistry, C.U.P., 1993.

[8] M.T. Dove, A.K.A. Pryde, V. Heine and K.D. Hammonds. Exotic distributions of rigid unit modes in the reciprocal spaces of framework aluminosilicates, J. Phys., Condens. Matter 19 (2007) doi:10.1088/0953-8984/19/27/275209.

[9] M.D. Foster and M.M.J. Treacy, A database of hypothetical zeolite structures: http://www.hypotheticalzeolites.net

[10] A.P. Giddy, M.T. Dove, G.S. Pawley, V. Heine, The determination of rigid unit modes as potential soft modes for displacive phase transitions in framework crystal structures. Acta Crystallogr., A49 (1993), 697–703.

[11] J. Graver, B. Servatius and H. Servatius, Combinatorial rigidity, Graduate Texts in Mathematics, vol 2, Amer. Math. Soc., 1993.

[12] K.D. Hammonds, H. Deng, V. Heine, and M.T. Dove, How floppy modes give rise to adsorption sites in zeolites, PRL 78 (1997), 3701–3704.

[13] K.D. Hammonds, V. Heine, and M.T. Dove, Rigid-Unit Modes and the quantitative determination of the flexibility possessed by zeolite frameworks, J. Phys. Chem. 102 (1998), 1759–1767.

[14] P.W. Fowler and S.D. Guest, A symmetry extension of Maxwell's rule for rigidity of frames, International Journal of Solids and Structures 37 (2000), 1793–1804.

[15] S.D. Guest and J.W. Hutchinson, On the determinacy of repetitive structures, Journal of the Mechanics and Physics of Solids, 51 (2003), 383–391.

[16] R.G. Hutchinson and N.A. Fleck, The structural performance of the periodic truss, Journal of the Mechanics and Physics of Solids, 54 (2006) 756–782.

[17] V. Kapko, C. Dawson, M.M.J. Treacy and M.F. Thorpe, Flexibility of ideal zeolite frameworks, Physical Chemistry Chemical Physics, DOI: 10.1039/c003977b.

[18] A.B. Kempe, On a general method of describing plane curves of the nth degree by linkwork, Proc. London Math. Soc. 7 (1876), 213–216.

[19] G. Laman, On graphs and the rigidity of plane skeletal structures, J. Engineering Mathematics, 4 (1970), 331–340.

[20] J. Malestein and L. Theran, Generic combinatorial rigidity of periodic frameworks, Advances in Mathematics, 233(1), 291–331, 2013.

[21] J.C. Owen and S.C. Power, Infinite bar-joint frameworks, Proceedings of the Symposium in Applied Computing, (SAC 2009) March 8–12, 2009, Honolulu, Hawaii, USA.

[22] J.C. Owen and S.C. Power, Frameworks, symmetry and rigidity, Inter. J. Computational Geometry and Applications, 20, (2010), 723–750.

[23] J.C. Owen and S.C. Power, Continuous curves from infinite Kempe linkages, Bull. London Math. Soc., 41 (2009) 1105–1111.

[24] J.C. Owen and S.C. Power, Infinite bar-joint frameworks, crystals and operator theory, New York J. Math., 17 (2011) 445–490.

[25] S.C. Power, Polynomials for crystal frameworks and the rigid unit mode spectrum, Phil. Trans. of Roy. Soc. A, 2013, to appear.

[26] S.C. Power, Crystal frameworks, symmetry and affinely periodic flexes, arXiv: 1103.1914, 2011.

[27] E. Ross, B. Shulze and W. Whiteley, Finite motions from periodic frameworks with added symmetry, Intern. J. of Solids and Structures, 42 (2011), 1711–1728.

[28] B. Schulze, Symmetric versions of Laman's Theorem, Discrete & Computational Geometry, 44 (2010), 946–972.

[29] B. Schulze, Block-diagonalised rigidity matrices of symmetric frameworks and applications, Contributions to Algebra and Geometry 51 (2010), 427–466.

[30] A.P. Sutton and R.W. Balluffi, Interfaces in crystalline materials, Monographs on the physics and chemistry of materials 51, Oxford University Press, 1995.

[31] I.P. Swainson and M.T. Dove, Low-frequency floppy modes in β-cristobalite, Phys. Rev. Letters, 71 (1993), 193–196.

[32] W. Whiteley, The union of matroids and the rigidity of frameworks, Siam J. Discrete Math. Vol. 1 (1988), 237–255.

[33] F. Wegner, Rigid-unit modes in tetrahedral crystals, Journal of Physics: Condensed Matter, Volume 19, Issue 40 (2007).

S.C. Power
Department of Mathematics and Statistics
Lancaster University
Lancaster, LA1 4YF, UK
e-mail: s.power@lancaster.ac.uk

Operator Theory:
Advances and Applications, Vol. 236, 421–454

Refined Size Estimates for Furstenberg Sets via Hausdorff Measures: A Survey of Some Recent Results

Ezequiel Rela

Abstract. In this survey we collect and discuss some recent results on the so-called "Furstenberg set problem", which in its classical form concerns the estimates of the Hausdorff dimension ($\dim_H$) of the sets in the F_α-class: for a given $\alpha \in (0,1]$, a set $E \subseteq \mathbb{R}^2$ is in the F_α-class if for each $e \in \mathbb{S}$ there exists a unit line segment ℓ_e in the direction of e such that $\dim_H(\ell \cap E) \geq \alpha$. For $\alpha = 1$, this problem is essentially equivalent to the "Kakeya needle problem". Define $\gamma(\alpha) = \inf\{\dim_H(E) : E \in F_\alpha\}$. The best-known results on $\gamma(\alpha)$ are the following inequalities:

$$\max\{1/2 + \alpha; 2\alpha\} \leq \gamma(\alpha) \leq (1 + 3\alpha)/2.$$

In this work we approach this problem from a more general point of view, in terms of a generalized Hausdorff measure $\mathcal{H}^h$ associated with the dimension function h. We define the class F_h of Furstenberg sets associated to a given dimension function h. The natural requirement for a set E to belong to F_h, is that $\mathcal{H}^h(\ell_e \cap E) > 0$ for each direction. We generalize the known results in terms of "logarithmic gaps" and obtain analogues to the estimates given above. Moreover, these analogues allow us to extend our results to the endpoint $\alpha = 0$. For the upper bounds we exhibit an explicit construction of F_h-sets which are small enough. To that end we adapt and prove some results on Diophantine Approximation about the dimension of a set of "well-approximable numbers".

We also obtain results about the dimension of Furstenberg sets in the class $F_{\alpha\beta}$, defined analogously to the class F_α but only for a fractal set $L \subset \mathbb{S}$ of directions such that $\dim_H(L) \geq \beta$. We prove analogous inequalities reflecting the interplay between α and β. This problem is also studied in the general scenario of Hausdorff measures.

Mathematics Subject Classification (2010). Primary 28A78, 28A80.

Keywords. Furstenberg sets, Hausdorff dimension, dimension function, Kakeya sets, Jarník's theorems.

This work was completed with the support of the Departmento de Análisis Matemático at the Facultad de Matemáticas, Universidad de Sevilla, Spain.

1. Introduction

In many situations in geometric measure theory, one wants to determine the size of a given set or a given class of sets identified by some geometric properties. Throughout this survey, size will mean Hausdorff dimension, denoted by $\dim_H$. The main purpose of the present expository work is to exhibit some recent results on the study of dimension estimates for *Furstenberg sets*, most of them contained in [MR10], [MR12] and [MR13]. Some related topics and the history of this problem are also presented. As far as we know, there is no other work in the literature collecting the known results about this problem. We begin with the definition of the *Furstenberg classes*.

Definition 1.1. For α in $(0,1]$, a subset E of $\mathbb{R}^2$ is called a *Furstenberg set* or a F_α-set if for each direction e in the unit circle there is a line segment ℓ_e in the direction of e such that the Hausdorff dimension of the set $E \cap \ell_e$ is equal to or greater than α.

We will also say that such set E belongs to the class F_α. It is known that for any F_α-set $E \subseteq \mathbb{R}^2$ the Hausdorff dimension must satisfy the inequality $\dim_H(E) \geq \max\{2\alpha, \alpha + \frac{1}{2}\}$. On the other hand, there are examples of F_α-sets E with $\dim_H(E) \leq \frac{1}{2} + \frac{3}{2}\alpha$. If we denote the example by

$$\gamma(\alpha) = \inf\{\dim_H(E) : E \in F_\alpha\},$$

then the Furstenberg problem is to determine $\gamma(\alpha)$. The best-known bounds on $\gamma(\alpha)$ so far are

$$\max\left\{2\alpha; \frac{1}{2} + \alpha\right\} \leq \gamma(\alpha) \leq \frac{1}{2} + \frac{3}{2}\alpha, \qquad \alpha \in (0,1]. \tag{1.1}$$

1.1. History and related problems

The Furstenberg problem appears for the first time in the work of Harry Furstenberg in [Fur70], regarding the problem of estimating the size of the intersection of fractal sets. Main references on this matter are [Wol99b], [Wol99a] and [Wol02]. See also [KT01] for a discretized version of this problem. In this last article, the authors study some connections between the Furstenberg problem and two other very famous problems: the Falconer distance problem and the Erdös ring problem.

Originally, in [Fur70] Furstenberg dealt with the problem of transversality of sets. Briefly, two closed subsets $A, B \subset \mathbb{R}$ are called *transverse* if

$$\dim_H(A \cap B) \leq \max\{\dim_H A + \dim_H B - 1, 0\}.$$

In addition, they will be called *strongly transverse* if every translate $A + t$ of A is transverse to B. More generally, the problem of the transversality between the dilations uA of A and B was considered. In this case the relevant quantity is $\dim_H(uA + t \cap B)$. This is where the connection pops in, since the dimension of this intersection can be seen as the dimension of the set $(A \times B) \cap \ell_{ut}$, where the line ℓ_{ut} in $\mathbb{R}^2$ is defined by the equation $y = ux + t$. In addition, Furstenberg proves, with some *invariance* hypothesis on A and B, the following: if the product

$A \times B$ intersects one (and it suffices with only one) line in some direction on a set of dimension at least α, then for almost all directions the set $A \times B$ intersects a line in that direction, also in a set of dimension at least α. Therefore, in that case the product is an F_α-set. Hence, any non-trivial lower bound on the class F_α implies a lower bound for the dimension of the product $A \times B$ in this particular case.

We now make the connection between the Furstenberg problem and the Falconer and Erdös problems more precise. We begin with the formulation of the Falconer distance problem. For a compact set $K \subseteq \mathbb{R}^2$, define the *distance set* $\operatorname{dist}(K)$ by

$$\operatorname{dist}(K) := \{|x-y| : x, y \in K\}.$$

The conjecture here is that $\dim_H(\operatorname{dist}(K)) = 1$ whenever $\dim_H(K) \geq 1$. In the direction of proving this conjecture, it was shown by Bourgain in [Bou94] that the conclusion holds for any K of $\dim_H(K) \geq \frac{13}{9}$, improved later by Wolff in [Wol99a] to $\dim_H(K) \geq \frac{4}{3}$. On the other hand, Mattila shows in [Mat87] that if we assume that $\dim_H(K) \geq 1$, then $\dim_H(\operatorname{dist}(K)) \geq \frac{1}{2}$. One may ask if there is an absolute constant $c > 0$ such that $\dim_H(\operatorname{dist}(K)) \geq \frac{1}{2} + c_0$ whenever K is compact and satisfies $\dim_H(K) \geq 1$. The Erdös ring problem, roughly speaking, asks about the existence of a Borel subring R of $\mathbb{R}$ such that $0 < \dim_H(R) < 1$.

The connection has been established only for some discretized version of the above three problems (see [KT01]). Consider the special case of Furstenberg sets belonging to the $F_{\frac{1}{2}}$ class. Note that for this family the two lower bounds for the Hausdorff dimension of Furstenberg sets coincide to become $\gamma(\frac{1}{2}) \geq 1$. Essentially, the existence of the constant c_0 in the Falconer distance problem mentioned above is equivalent to the existence of another constant c_1 such that any $F_{\frac{1}{2}}$-set E must have $\dim_H(E) \geq 1 + c_1$. In addition, any of these two conditions would imply the non-existence of a Borel subring of R of Hausdorff dimension exactly $\frac{1}{2}$.

In addition, Wolff states without proof in [Wol99a] that there is a relation between the Furstenberg problem and the rate of decay of circular means of Fourier transforms of measures. Later, in [Wol02], appears the proof of the following fact. Let μ be a measure and define the s-dimensional energy $I_s(\mu)$ as

$$I_s(\mu) := \int \frac{d\mu(x)d\mu(y)}{|x-y|^s}.$$

Define also $\sigma_1(s)$ to be the supremum of all the numbers σ such that there exists a constant C with

$$\int_{-\pi}^{\pi} \left|\hat{\mu}(Re^{i\theta})d\theta\right| \leq CR^{-\sigma}\sqrt{I_s(\mu)},$$

for all positive measures with finite s-energy supported in the unit disc and all $R \geq 1$. Then the following relation holds:

$$1 - \alpha - \gamma(\alpha) - 4\sigma_1(\gamma(\alpha)) \geq 1.$$

For the particular case of $\alpha = 1$, when we require the set to contain a whole line segment in each direction, we actually are in the presence of the much more famous Kakeya problem. A Kakeya set (or Besicovitch set) is a compact set $E \subseteq \mathbb{R}^n$ that contains a unit line segment in every possible direction. The question here is about the minimal size for the class of Kakeya sets. Besicovitch [Bes19] proved that for all $n \geq 2$, there exist Besicovitch sets of Lebesgue measure zero in $\mathbb{R}^n$.

Originally, Kakeya [FK17] asks what is the possible minimal area that permits us to *continuously* rotate a unit line segment in the plane; in [Bes28] Besicovitch actually shows that the continuous movement can be achieved using an arbitrarily small area by the method known as *shifting triangles* or *Perron's trees.*

The next question, which is relevant for our work, is the *unsolved* "Kakeya conjecture" which asserts that these sets, although they can be small with respect to the Lebesgue measure, must have full Hausdorff dimension. The conjecture was proven by Davies [Dav71] in $\mathbb{R}^2$: all Kakeya sets in $\mathbb{R}^2$ have dimension 2. In higher dimensions the Kakeya problem is still open, and one of the best-known bounds appears in [Wol99b] and states that any Kakeya set $E \subseteq \mathbb{R}^n$ must satisfy the bound $\dim_H(E) \geq \frac{n+2}{2}$.

These kinds of geometric-combinatorial problems have deep implications in many different areas of general mathematics. Some of the connections to other subjects include Bochner–Riesz multipliers, restrictions estimates for the Fourier transform and also partial differential equations. For example, it has been shown that a positive answer to the Restriction Conjecture for the sphere $\mathbb{S}^{n-1}$ would imply that any Kakeya set in $\mathbb{R}^n$ must have full dimension, and therefore solve the Kakeya conjecture (see for example [Wol99b]).

1.2. Our approach

In this work we study the Furstenberg problem using generalized Hausdorff measures. This approach is motivated by the well-known fact that knowing the value of the dimension of a given set does not yet tell us anything about the corresponding measure at this critical dimension. In fact, if $\mathcal{H}^s$ is the Hausdorff s measure of an s-dimensional set E, $\mathcal{H}^s(E)$ can be 0, ∞ or finite. The case of a set E with $0 < \mathcal{H}^s(E) < +\infty$ is of special interest. We refer to it as an s-set, considering it as *truly* s-dimensional. For, if a set E with $\dim_H(E) = s$ has non σ-finite $\mathcal{H}^s$-measure, it is still too big to be correctly measured by $\mathcal{H}^s$. Analogously, the case of null measure reflects that the set is too thin to be measured by $\mathcal{H}^s$. To solve (partially) this problem, the appropriate tools are the "generalized Hausdorff Measures" introduced by Felix Hausdorff in his seminal paper [Hau18] in 1918. For any *dimension function*, i.e., a function belonging to the set

$$\mathbb{H} := \{h : [0, \infty) \to [0 : \infty), \text{non-decreasing, continuous, } h(0) = 0\},$$

he defines

$$\mathcal{H}^h_\delta(E) = \inf \left\{ \sum_i h(\operatorname{diam}(E_i)) : E \subset \bigcup_i^{\infty} E_i, \operatorname{diam}(E_i) < \delta \right\}$$

and

$$\mathcal{H}^h(E) = \sup_{\delta > 0} \mathcal{H}^h_\delta(E).$$

Note that if $h_\alpha(x) := x^\alpha$, we actually recover the previous measure since $\mathcal{H}^{h_\alpha} = \mathcal{H}^\alpha$. We now have a finer criterion to classify sets by a notion of size. If one only looks at the power functions, there is a natural total order given by the exponents. In $\mathbb{H}$ we also have a natural notion of order, but we can only obtain a *partial* order.

Definition 1.2. Let g, h be two dimension functions. We will say that g is dimensionally smaller than h and write $g \prec h$ if and only if

$$\lim_{x\to 0^+} \frac{h(x)}{g(x)} = 0.$$

We note that the speed of convergence to zero can be seen as a notion of distance between g and h. The important subclass of those $h \in \mathbb{H}$ that satisfy a doubling condition will be denoted by $\mathbb{H}_d$:

$$\mathbb{H}_d := \{h \in \mathbb{H} : h(2x) \leq Ch(x) \text{ for some } C > 0\}.$$

We will be interested in the special subclass of dimension functions that allow us to classify zero-dimensional sets.

Definition 1.3. A function $h \in \mathbb{H}$ will be called a "zero-dimensional dimension function" if $h \prec x^\alpha$ for any $\alpha > 0$. We denote by $\mathbb{H}_0$ the subclass of those functions. As a model to keep in mind, consider the family $h_\theta(x) = \frac{1}{\log^\theta(\frac{1}{x})}$.

Now, given an α-dimensional set E that is not an α-set, one could expect to find in the class $\mathbb{H}$ an appropriate function h to detect the precise "size" of it. By that we mean that $0 < \mathcal{H}^h(E) < \infty$, and in this case E is referred to as an h-set. In order to illustrate the main difficulties, we start with a simple observation. The Hausdorff dimension of a set $E \subseteq \mathbb{R}^n$ is the unique real number s characterized by the following properties:

- $\mathcal{H}^r(E) = +\infty$ for all $r < s$.
- $\mathcal{H}^t(E) = 0$ for all $s < t$.

Therefore, to prove that some set has dimension s, it suffices to prove the preceding two properties, and this is independent of the possible values of $\mathcal{H}^s(E)$. It is always true, no matter if $\mathcal{H}^s(E)$ is zero, finite and positive, or infinite.

The above observation could lead to the conjecture that in the wider scenario of dimension functions the same kind of reasoning can be made. In fact, Eggleston claims in [Egg52] that for any $A \subseteq \mathbb{R}^n$, one of the following three possibilities holds.

1. For all $h \in \mathbb{H}$, $\mathcal{H}^h(A) = 0$.
2. There is a function $h_0 \in \mathbb{H}$, such that if $h \succ h_0$ then $\mathcal{H}^h(A) = 0$, whilst if $h \prec h_0$, then $\mathcal{H}^h(A) = +\infty$.
3. For all $h \in \mathbb{H}$, $\mathcal{H}^h(A) = +\infty$.

Note that the most interesting situation is the one on item 2, since it is saying that the correct notion of size for the set A is represented by the function h_0. Clearly, this is the case when we are dealing with an h-set. However, this claim is false, in the sense that there are situations where none of the above three cases applies. The problems arise from two results due to Besicovitch (see [Bes56a] and [Bes56b], also [Rog70] and references therein). The first says that if a set E has null $\mathcal{H}^h$-measure for some $h \in \mathbb{H}$, then there exists a function $g \prec h$ such that $\mathcal{H}^g(E) = 0$. Symmetrically, the second says that if a *compact* set E has non-σ-finite $\mathcal{H}^h$ measure, then there exists a function $g \succ h$ such that E has also non-σ-finite $\mathcal{H}^g$ measure. These results imply that if a *compact* set E satisfies the existence of a function h_0 such that $\mathcal{H}^h(E) > 0$ for any $h \prec h_0$ and $\mathcal{H}^h(E) = 0$ for any $h \succ h_0$, then it must be the case that $0 < \mathcal{H}^{h_0}(E)$ and E has σ-finite $\mathcal{H}^h$-measure.

Consider now the set $\mathbb{L}$ of Liouville numbers. It is known that this set is dimensionless, which means that it is not an h-set for any $h \in \mathbb{H}$. In that direction, further improvements are due to Elekes and Keleti [EK06]. There the authors prove much more than that there is no exact Hausdorff-dimension function for the set $\mathbb{L}$ of Liouville numbers: they prove that for any translation invariant Borel measure $\mathbb{L}$ is either of measure zero or has non-sigma-finite measure. In addition, it is shown in [OR06] that there are two proper nonempty subsets $\mathbb{L}_0, \mathbb{L}_\infty \subseteq \mathbb{H}$ of dimension functions such that $\mathcal{H}^h(\mathbb{L}) = 0$ for all $h \in \mathbb{L}_0$ and $\mathcal{H}^h(\mathbb{L}) = \infty$ for all $h \in \mathbb{L}_\infty$. It follows that the Liouville numbers $\mathbb{L}$ must satisfy condition 2 in the classification of Eggleston. But suppose that h_0 is the claimed dimension function in that case. The discussion in the above paragraph implies that the set $\mathbb{L}$ is an h_0-set, which is a contradiction.

Since in the present work we are interested in estimates for the size of general Furstenberg sets, we have to consider dimension functions that are a true *step down* or *step up* from the critical one. The natural generalization of the class of Furstenberg sets to the wider scenario of dimension functions is the following.

Definition 1.4. Let $\mathfrak{h}$ be a dimension function. A set $E \subseteq \mathbb{R}^2$ is a Furstenberg set of type $\mathfrak{h}$, or an $F_{\mathfrak{h}}$-set, if for each direction $e \in \mathbb{S}$ there is a line segment ℓ_e in the direction of e such that $\mathcal{H}^{\mathfrak{h}}(\ell_e \cap E) > 0$.

Note that this hypothesis is stronger than the one used to define the original Furstenberg F_α-sets. However, the hypothesis $\dim_H(E \cap \ell_e) \geq \alpha$ is equivalent to $\mathcal{H}^\beta(E \cap \ell_e) > 0$ for any β smaller than α. If we use the wider class of dimension functions introduced above, the natural way to define $F_{\mathfrak{h}}$-sets would be to replace the parameters $\beta < \alpha$ with two dimension functions satisfying the relation $h \prec \mathfrak{h}$. But requiring $E \cap \ell_e$ to have positive $\mathcal{H}^h$ measure for any $h \prec \mathfrak{h}$ implies that it has also positive $\mathcal{H}^{\mathfrak{h}}$ measure. It will be useful to introduce also the following subclass of F_α:

Definition 1.5. A set $E \subseteq \mathbb{R}^2$ is an F_α^+-set if for each $e \in \mathbb{S}$ there is a line segment ℓ_e such that $\mathcal{H}^\alpha(\ell_e \cap E) > 0$.

From the preceding discussion, it follows that there is an unavoidable need to study a notion of "gap" between dimension functions. We will show that if E is a set in the class $F_{\mathfrak{h}}$, and h is a dimension function that is *much smaller* than $\mathfrak{h}^2$ or $\sqrt{\cdot}\,\mathfrak{h}$, then $\mathcal{H}^h(E) = \infty$ (Theorem 3.6 and Theorem 3.12 respectively). We further exhibit a very *small* Furstenberg set F in $F_{\mathfrak{h}}$, for some particular choices of $\mathfrak{h}$ and show that for this set, if $\sqrt{\cdot}\,\mathfrak{h}^{3/2}$ is *much smaller* than h, then $\mathcal{H}^h(F) = 0$ (Theorem 5.6). This generalizes the result of the classical setting given in (1.1).

The $\mathfrak{h} \to \mathfrak{h}^2$ bound strongly depends on the known estimates for the Kakeya maximal operator: for an integrable function f on $\mathbb{R}^n$, the Kakeya maximal operator at scale δ applied to f, $\mathcal{K}_\delta(f) : \mathbb{S}^{n-1} \to \mathbb{R}$, is

$$\mathcal{K}_\delta(f)(e) = \sup_{x\in\mathbb{R}^n} \frac{1}{|T_e^\delta(x)|} \int_{T_e^\delta(x)} |f(x)|\, dx \qquad e \in \mathbb{S}^{n-1},$$

where $T_e^\delta(x)$ is a $1 \times \delta$-tube (by this we mean a tube of length 1 and cross section of radius δ) centred at x in the direction of $e \in \mathbb{S}^{n-1} \subset \mathbb{R}^n$. It is well known that in $\mathbb{R}^2$ the Kakeya maximal function satisfies the bound (see [Wol99b])

$$\left\|\mathcal{K}_\delta(f)\right\|_2^2 \le C \log(\frac{1}{\delta}) \|f\|_2^2. \tag{1.2}$$

Our proof of Theorem 3.6 relies on an optimal use of these estimates for the Kakeya maximal function, exploiting the logarithmic factor in the above bound, which is necessary (see [Kei99]), because of the existence of Kakeya sets of zero measure. The other lower bound, which is the relevant bound near the zero-dimensional case, depends on some combinatorial arguments that we extended to this general setting. In addition, our techniques allow us to extend the bounds in (1.1) to "zero-dimensional" classes. At the endpoint $\alpha = 0$ we can show that for $\mathfrak{h} \in \mathbb{H}_0$ defined by $\mathfrak{h}(x) = \frac{1}{\log(\frac{1}{x})}$, any $F_{\mathfrak{h}}$-set E must satisfy $\dim_H(E) \ge \frac{1}{2}$.

For the upper bounds the aim is to explicitly exhibit constructions of reasonably small Furstenberg sets. To achieve these optimal constructions, we needed a suited version of Jarník's theorems on Diophantine Approximation. We exhibit an $F_{\mathfrak{h}}$-set whose dimension function can not be much larger (in terms of logarithmic gaps) than $\sqrt{\cdot}\,\mathfrak{h}^{3/2}$ for the classical case of $\mathfrak{h}(x) = x^\alpha$. We also show in Section 5 a particular set $E \in F_{\mathfrak{h}}$ for $\mathfrak{h}(x) = \frac{1}{\log(\frac{1}{x})}$ satisfying $\dim_H(E) \le \frac{1}{2}$.

We also consider another related problem, both in the classical and generalized setting. We analyze the role of the dimension of the set of directions in the Furstenberg problem. We consider the class of $F_{\alpha\beta}$ sets, defined in the same way as the F_α class but with the directions taken in a subset L of the unit circle such that $\dim_H(L) \ge \beta$. We are able to prove that if E is any $F_{\alpha\beta}$-set, then

$$\dim_H(E) \ge \max\left\{2\alpha + \beta - 1; \frac{\beta}{2} + \alpha\right\}, \qquad \alpha, \beta > 0. \tag{1.3}$$

For the proof of one of the lower bounds we needed estimates for the Kakeya maximal function but for more general measures. The other lower bound uses the δ-*entropy* of the set L of directions, which is the maximal possible cardinality of a δ-separated subset. Our results are proved in the context of the general Hausdorff measures and we obtain (1.3) as a corollary. The only previously known bounds in this setting are for the particular case of $\alpha = 1$, $\beta \in (0, 1]$ (see [Mit02]). The author there obtains that if E is an A-Kakeya set (that is, a planar set with a unit line segment in any direction $e \in A$ for a set $A \subseteq \mathbb{S}$), the $\dim_H(E) \geq 1 + \dim_H(A)$ (this is only one of the lower bounds).

This paper is organized as follows: In Section 2 we provide some extra examples and remarks about Hasudorff measures and dimension functions. In Section 3 we study the problem of finding lower bounds for the size of generalized Furstenberg sets. In Section 4 we study the same problem for a more general class of Furstenberg sets associated to a fractal set of directions. Finally, in Section 5 we study the upper bounds.

As usual, we will use the notation $A \lesssim B$ to indicate that there is a constant $C > 0$ such that $A \leq CB$, where the constant is independent of A and B. By $A \sim B$ we mean that both $A \lesssim B$ and $B \lesssim A$ hold.

2. Preliminaries on Hausdorff measures and dimension functions

In this section we introduce some preliminaries on dimension functions and Hausdorff measures. Moreover, we discuss some additional features of the problem of finding appropriate notions of size for fractal sets. Extra examples are also included.

2.1. Dimension Partition

For a given set $E \subseteq \mathbb{R}^n$, we introduce the notion of dimension partition (see [CHM10]).

Definition 2.1. By the *Dimension Partition* of a set E we mean a partition of $\mathbb{H}$ into (three) sets: $\mathcal{P}(E) = E_0 \cup E_1 \cup E_\infty$ with

- $E_0 = \{h \in \mathbb{H} : \mathcal{H}^h(E) = 0\}$,
- $E_1 = \{h \in \mathbb{H} : 0 < \mathcal{H}^h(E) < \infty\}$,
- $E_\infty = \{h \in \mathbb{H} : \mathcal{H}^h(E)$ has non-σ-finite $\mathcal{H}^h$-measure$\}$.

It is very well known that E_1 could be empty, reflecting the dimensionless nature of E. A classical example of this phenomenon is the set $\mathbb{L}$ of Liouville numbers. On the other hand, E_1 is never empty for an h-set, but it is not easy to determine this partition in the general case. We also remark that it is possible to find non-comparable dimension functions g, h and a set E with the property of being a g-set and an h-set simultaneously. Consider the following example:

Example 2.2. There exists a set E and two dimension functions $g, h \in \mathbb{H}$ which are not comparable and such that E is a g-set and also an h-set.

Proof. We will use the results of [CMMS04]. The set E will be the Cantor set C_a associated to a non-negative decreasing sequence $a = \{a_i\}$ such that $\sum a_i = 1$. We start by removing an interval of length a_1. Then we remove an interval of length a_2 from the left and of length a_3 at the right. Following this scheme, we end up with a perfect set of zero measure. If we define $b_n = \frac{1}{n}\sum_{i\geq n} a_i$, then the main result of the cited work is that

$$\lim_{n\to\infty} nh(b_n) \sim \mathcal{H}^h(C_a), \tag{2.1}$$

for all $h \in \mathbb{H}$. The authors prove that it is possible to construct a spline-type dimension function $h = h_a$ that makes C_a an h-set. Further, the function h satisfies that $h(b_n) = \frac{1}{n}$. Now we want to define g. Consider the sequence $x_n = b_{n!}$ and take g satisfying the following properties:

1. $g(x) \geq h(x)$ for all $x > 0$.
2. $g(x_n) = h(x_n)$ for all $n \in \mathbb{N}$.
3. g is a polygonal spline (same as h), but it is constant in each interval $[b_{n!-1}, b_{(n-1)!}]$ and drops abruptly on $[b_{n!}, b_{n!-1}]$ (we are building up g from the right approaching the origin). More precisely, for each $n \in \mathbb{N}$,

$$g(x) = \begin{cases} \frac{1}{(n-1)!} & \text{if} \quad x \in [b_{n!-1}, b_{(n-1)!}] \\ \frac{1}{n!} & \text{if} \quad x = b_{n!} \end{cases}$$

and it is linear on $[b_{n!}, b_{n!-1}]$.

Conditions 1 and 2 imply that $\overline{\lim}_{x\to 0} \frac{h(x)}{g(x)} = 1 < \infty$. Note that we also have that $\underline{\lim}_{x\to 0} \frac{h(x)}{g(x)} = 0$, since

$$\frac{h(b_{n!-1})}{g(b_{n!-1})} = \frac{(n-1)!}{n!-1} \sim \frac{1}{n} \to 0.$$

It follows that h and g are not comparable. To see that C_a is also a g-set, we use again the characterization (2.1). Since

$$\underline{\lim}_{n\to\infty} ng(b_n) \leq \underline{\lim}_{n\to\infty} n!g(b_{n!}) = \underline{\lim}_{n\to\infty} n!h(b_{n!}) < \infty,$$

we obtain that $\mathcal{H}^g(C_a) < \infty$. In addition, $g(x) \geq h(x)$ for all x, hence $g(b_n) \geq h(b_n)$ for all $n \in \mathbb{N}$ and it follows that

$$\underline{\lim}_{n\to\infty} ng(b_n) \geq \underline{\lim}_{n\to\infty} nh(b_n) > 0$$

and therefore $\mathcal{H}^g(C_a) > 0$. □

We refer the reader to [GMS07] for a detailed study of the problem of equivalence between dimension functions and Cantor sets associated to sequences. The authors also study Packing measures and premeasures of those sets. For the construction of h-sets associated to certain sequences see the work of Cabrelli *et al.* [CMMS04].

It follows from Example 2.2 that even for h-sets the dimension partition, and in particular E_1, is not completely determined. Note that the results of Rogers cited above imply that, for compact sets, E_0 and E_∞ can be thought of as open components of the partition, and E_1 as the "border" of these open components. An interesting problem is then to determine some criteria to classify the functions in $\mathbb{H}$ into those classes. To detect where this "border" is, we will introduce the notion of *chains* in $\mathbb{H}$. This notion allows us to refine the notion of Hausdorff dimension by using an ordered family of dimension functions. More precisely, we have the following definition.

Definition 2.3. A family $\mathcal{C} \subset \mathbb{H}$ of dimension functions will be called a *chain* if it is of the form

$$\mathcal{C} = \{h_t \in \mathbb{H} : t \in \mathbb{R}, h_s \prec h_t \iff s < t\}.$$

That is, a totally ordered one-parameter family of dimension functions.

Suppose that $h \in \mathbb{H}$ belongs to some chain $\mathcal{C}$ and satisfies that, for any $g \in \mathcal{C}$, $\mathcal{H}^g(E) > 0$ if $g \prec h$ and $\mathcal{H}^g(E) = 0$ if $g \succ h$. Then, even if $h \notin E_1$, in this chain, h does measure the size of E. It can be thought of as being "near the frontier" of both E_0 and E_∞. For example, if a set E has Hausdorff dimension α but $\mathcal{H}^\alpha(E) = 0$ or $\mathcal{H}^\alpha(E) = \infty$, take $h(x) = x^\alpha$ and $\mathcal{C}_H = \{x^t : t \geq 0\}$. In this chain, x^α is the function that best measures the size of E.

We look for finer estimates, considering chains of dimension functions that yield "the same Hausdorff dimension". Further, for zero-dimensional sets, this approach allows us to classify them by some notion of dimensionality.

2.2. The exact dimension function for a class of sets

In the previous section we dealt with the problem of detecting an appropriate dimension function for a given set or, more generally, the problem of determining the dimension partition of that set. Now we introduce another related problem, which concerns the analogous problem but for a whole class of sets defined, in general, by geometric properties. We mention one example: As we mentioned before, all Kakeya sets in $\mathbb{R}^2$ have full dimension, but even in that case, there are several distinct types of two-dimensional sets (for instance, with positive or null Lebesgue measure). Hence, one would like to associate a dimension function to the whole class. A dimension function $h \in \mathbb{H}$ will be called the exact Hausdorff dimension function of the class of sets $\mathcal{A}$ if

- For every set E in the class $\mathcal{A}$, $\mathcal{H}^h(E) > 0$.
- There are sets $E \in \mathcal{A}$ with $\mathcal{H}^h(E) < \infty$.

In the direction of finding the exact dimension of the class of Kakeya sets in $\mathbb{R}^2$, Keich has proven in [Kei99] that the exact dimension function h must decrease to zero at the origin faster than $x^2 \log(\frac{1}{x}) \log\log(\frac{1}{x})^{2+\varepsilon}$ for any given $\varepsilon > 0$, but slower than $x^2 \log(\frac{1}{x})$. This notion of speed of convergence tells us precisely that h is between those two dimension functions (see Definition 1.2). More precisely, the author explicitly constructs a small Kakeya set, which is small enough to have finite g measure for $g(x) = x^2 \log(\frac{1}{x})$. Therefore, for h to be an exact dimension function for the class of Kakeya sets, it cannot be dimensionally greater than g. But this last condition is not sufficient to ensure that *any* Kakeya set has positive h-measure. The partial result from [Kei99] is that for any $\varepsilon > 0$ and any Kakeya set E, we have that $\mathcal{H}^{h_\varepsilon}(E) > 0$, where $h_\varepsilon = x^2 \log(\frac{1}{x}) \log\log(\frac{1}{x})^{2+\varepsilon}$.

3. Lower bounds for Furstenberg sets

In this section we deal with the problem of finding sharp lower bounds for the generalized dimension of Furstenberg type sets. Let us begin with some remarks about this problem and the techniques involved.

3.1. Techniques

We start with a *uniformization* procedure. Given an $F_{\mathfrak{h}}$-set E for some $\mathfrak{h} \in \mathbb{H}$, it is always possible to find two constants $m_E, \delta_E > 0$ and a set $\Omega_E \subseteq \mathbb{S}$ of positive σ-measure such that

$$\mathcal{H}^{\mathfrak{h}}_{\delta}(\ell_e \cap E) > m_E > 0 \qquad \forall \delta < \delta_E\,, \quad \forall e \in \Omega_E.$$

For each $e \in \mathbb{S}$, there is a positive constant m_e such that $\mathcal{H}^{\mathfrak{h}}(\ell_e \cap E) > m_e$. Now consider the following pigeonholing argument. Let $\Lambda_n = \{e \in \mathbb{S} : \frac{1}{n+1} \leq m_e < \frac{1}{n}\}$. At least one of the sets must have positive measure, since $\mathbb{S} = \cup_n \Lambda_n$. Let Λ_{n_0} be such set and take $0 < 2m_E < \frac{1}{n_0+1}$. Hence $\mathcal{H}^{\mathfrak{h}}(\ell_e \cap E) > 2m_E > 0$ for all $e \in \Lambda_{n_0}$. Finally, again by pigeonholing, we can find $\Omega_E \subseteq \Lambda_{n_0}$ of positive measure and $\delta_E > 0$ such that

$$\mathcal{H}^{\mathfrak{h}}_{\delta}(\ell_e \cap E) > m_E > 0 \qquad \forall e \in \Omega_E \quad \forall \delta < \delta_E. \tag{3.1}$$

To simplify notation throughout the remainder of the chapter, since inequality (3.1) holds for any Furstenberg set and we will only use the fact that m_E, δ_E and $\sigma(\Omega_E)$ are positive, it will be enough to consider the following definition of $F_{\mathfrak{h}}$-sets:

Definition 3.1. Let $\mathfrak{h}$ be a dimension function. A set $E \subseteq \mathbb{R}^2$ is a Furstenberg set of type $\mathfrak{h}$, or an $F_{\mathfrak{h}}$-set, if for each $e \in \mathbb{S}$ there is a line segment ℓ_e in the direction of e such that $\mathcal{H}^{\mathfrak{h}}_{\delta}(\ell_e \cap E) > 1$ for all $\delta < \delta_E$ for some $\delta_E > 0$.

The following technique is a standard procedure in this area. The lower bounds for the Hausdorff dimension of a given set E, both in the classical and general setting, are achieved by bounding uniformly from below the size of the coverings of E. More precisely, the h-size of a covering $\mathcal{B} = \{B_j\}$ is $\sum_j h(r_j)$. Our aim will be then to prove essentially that $\sum_j h(r_j) \gtrsim 1$, provided that h is a small enough dimension function. We introduce the following notation:

Definition 3.2. Let $\mathfrak{b} = \{b_k\}_{k\in\mathbb{N}}$ be a decreasing sequence with $\lim b_k = 0$. For any family of balls $\mathcal{B} = \{B_j\}$ with $B_j = B(x_j; r_j)$, $r_j \leq 1$, and for any set E, we define

$$J_k^{\mathfrak{b}} := \{j \in \mathbb{N} : b_k < r_j \leq b_{k-1}\}, \tag{3.2}$$

and

$$E_k := E \cap \bigcup_{j \in J_k^{\mathfrak{b}}} B_j.$$

In the particular case of the dyadic scale $\mathfrak{b} = \{2^{-k}\}$, we will omit the superscript and denote

$$J_k := \{j \in \mathbb{N} : 2^{-k} < r_j \leq 2^{-k+1}\}.$$

The idea will be to use the dyadic partition of the covering to obtain that $\sum_{j\geq 0} h(r_j) \gtrsim \sum_{k\geq 0} h(2^{-k})\#J_k$. The lower bounds we need will be obtained if we can prove lower bounds on the quantity J_k in terms of the function h but independent of the covering. The next lemma introduces a technique we borrow from [Wol99b] to decompose the set of all directions.

Lemma 3.3. *Let E be an $F_{\mathfrak{h}}$-set for some $\mathfrak{h} \in \mathbb{H}$ and $\mathfrak{a} = \{a_k\}_{k\in\mathbb{N}} \in \ell^1$ a non-negative sequence. Let $\mathcal{B} = \{B_j\}$ be a δ-covering of E with $\delta < \delta_E$ and let E_k and J_k be as above. Define*

$$\Omega_k := \left\{ e \in \mathbb{S} : \mathcal{H}^{\mathfrak{h}}_{\delta}(\ell_e \cap E_k) \geq \frac{a_k}{2\|\mathfrak{a}\|_1} \right\}.$$

Then $\mathbb{S} = \cup_k \Omega_k$.

Proof. It follows directly from the summability of $\mathfrak{a}$. □

We will need in the next section the main result of [Mit02], which is the following proposition.

Proposition 3.4. *Let μ be a Borel probability measure on $\mathbb{S}$ such that $\mu(B(x,r)) \lesssim \varphi(r)$ for some non-negative function φ for all $r \ll 1$. Define the Kakeya maximal operator $\mathcal{K}_\delta$ as usual:*

$$\mathcal{K}_\delta(f)(e) = \sup_{x\in\mathbb{R}^n} \frac{1}{|T_e^\delta(x)|} \int_{T_e^\delta(x)} |f(x)|\, dx, \qquad e \in \mathbb{S}^{n-1}.$$

Then we have the estimate

$$\|\mathcal{K}_\delta\|^2_{L^2(\mathbb{R}^2)\to L^2(\mathbb{S},d\mu)} \lesssim C(\delta) = \int_\delta^1 \frac{\varphi(u)}{u^2} du. \tag{3.3}$$

Remark 3.5. It should be noted that if we choose $\varphi(x) = x^s$, then we obtain as a corollary that

$$\|\mathcal{K}_\delta\|^2_{L^2(\mathbb{R}^2)\to L^2(\mathbb{S},d\mu)} \lesssim \delta^{s-1}.$$

In the special case of $s = 1$, the bound has the known logarithmic growth:

$$\|\mathcal{K}_\delta\|^2_{L^2(\mathbb{R}^2)\to L^2(\mathbb{S},d\mu)} \sim \log(\frac{1}{\delta}).$$

3.2. The $\mathfrak{h} \to \mathfrak{h}^2$ bound

In this section we generalize the first inequality of (1.1), that is, $\dim_H(E) \geq 2\alpha$ for any F_α-set. For this, given a dimension function $h \prec \mathfrak{h}^2$, we impose some sufficient growth conditions on the gap $\frac{\mathfrak{h}^2}{h}$ to ensure that $\mathcal{H}^h(E) > 0$. We have the following theorem:

Theorem 3.6. *Let $\mathfrak{h} \in \mathbb{H}_d$ be a dimension function and let E be an $F_{\mathfrak{h}}$-set. Let $h \in \mathbb{H}$ such that $h \prec \mathfrak{h}^2$. If $\sum_{k\geq 0} \sqrt{k \frac{\mathfrak{h}^2}{h}(2^{-k})} < \infty$, then $\mathcal{H}^h(E) > 0$.*

Proof. By Definition 3.1, since $E \in F_{\mathfrak{h}}$, we have $\mathcal{H}^{\mathfrak{h}}_{\delta}(\ell_e \cap E) > 1$ for all $e \in \mathbb{S}$ and for any $\delta < \delta_E$. Let $\{B_j\}_{j\in\mathbb{N}}$ be a covering of E by balls with $B_j = B(x_j; r_j)$. We need to bound $\sum_j h(2r_j)$ from below. Since h is non-decreasing, it suffices to obtain the bound $\sum_j h(r_j) \gtrsim 1$ for any $h \in \mathbb{H}$ satisfying the hypothesis of the theorem. Clearly we can restrict ourselves to δ-coverings with $\delta < \frac{\delta_E}{5}$. Define $\mathfrak{a} = \{a_k\}$ with $a_k = \sqrt{k\frac{\mathfrak{h}^2}{h}(2^{-k})}$. By hypothesis, $\mathfrak{a} \in \ell^1$. Also define, as in the previous section, for each $k \in \mathbb{N}$, $J_k = \{j \in \mathbb{N} : 2^{-k} < r_j \leq 2^{-k+1}\}$ and $E_k = E \cap \cup_{j\in J_k} B_j$. Since $\mathfrak{a} \in \ell^1$, we can apply Lemma 3.3 to obtain the decomposition $\mathbb{S} = \bigcup_k \Omega_k$ associated to this choice of $\mathfrak{a}$.

We will apply the maximal function inequality to a weighted union of indicator functions. For each k, let $F_k = \bigcup\limits_{j\in J_k} B_j$ and define the function

$$f := \mathfrak{h}(2^{-k})2^k \chi_{F_k}.$$

We will use the L^2 norm estimates for the maximal function. The L^2 norm of f can be easily estimated as follows:

$$\|f\|_2^2 = \mathfrak{h}^2(2^{-k})2^{2k} \int_{\cup_{J_k} B_j} dx \lesssim \mathfrak{h}^2(2^{-k})2^{2k} \sum_{j\in J_k} r_j^2 \lesssim \mathfrak{h}^2(2^{-k})\#J_k,$$

since $r_j \leq 2^{-k+1}$ for $j \in J_k$. Hence,

$$\|f\|_2^2 \lesssim \#J_k \mathfrak{h}^2(2^{-k}). \tag{3.4}$$

Now fix k and consider the Kakeya maximal function $\mathcal{K}_\delta(f)$ of level $\delta = 2^{-k+1}$ associated to the function f defined for this value of k.

In Ω_k we have the following pointwise lower estimate for the maximal function. Let ℓ_e be the line segment such that $\mathcal{H}^{\mathfrak{h}}_{\delta}(\ell_e \cap E) > 1$, and let T_e be the rectangle of width 2^{-k+2} around this segment. Define, for each $e \in \Omega_k$,

$$J_k(e) := \{j \in J_k : \ell_e \cap E \cap B_j \neq \emptyset\}.$$

With the aid of the Vitali covering lemma, we can select a subset of disjoint balls $\widetilde{J}_k(e) \subseteq J_k(e)$ such that

$$\bigcup_{j\in J_k(e)} B_j \subseteq \bigcup_{j\in \widetilde{J}_k(e)} B(x_j; 5r_j).$$

Note that every ball B_j, $j \in J_k(e)$, intersects ℓ_e and therefore at least half of B_j is contained in the rectangle T_e, yielding $|T_e \cap B_j| \geq \frac{1}{2}\pi r_j^2$. Hence, by definition of the maximal function, using that $r_j \geq 2^{-k+1}$ for $j \in J_k(e)$,

$$\begin{aligned}|\mathcal{K}_{2^{-k+1}}(f)(e)| &\geq \frac{1}{|T_e|}\int_{T_e} f\,dx = \frac{\mathfrak{h}(2^{-k})2^k}{|T_e|}\left|T_e \cap \cup_{J_k(e)} B_j\right| \\ &\gtrsim \mathfrak{h}(2^{-k})2^{2k}\left|T_e \cap \cup_{\widetilde{J}_k(e)} B_j\right| \\ &\gtrsim \mathfrak{h}(2^{-k})2^{2k}\sum_{j\in\widetilde{J}_k(e)} r_j^2 \\ &\gtrsim \mathfrak{h}(2^{-k})\#\widetilde{J}_k(e) \\ &\gtrsim \sum_{\widetilde{J}_k(e)} \mathfrak{h}(r_j).\end{aligned}$$

Now, since

$$\ell_e \cap E_k \subseteq \bigcup_{j\in J_k(e)} B_j \subseteq \bigcup_{j\in\widetilde{J}_k(e)} B(x_j; 5r_j)$$

and for $e \in \Omega_k$ we have $\mathcal{H}_\delta^{\mathfrak{h}}(\ell_e \cap E_k) \gtrsim a_k$, we obtain

$$|\mathcal{K}_{2^{-k+1}}(f)(e)| \gtrsim \sum_{\widetilde{J}_k(e)} \mathfrak{h}(r_j) \gtrsim \sum_{j\in\widetilde{J}_k(e)} \mathfrak{h}(5r_j) \gtrsim a_k.$$

Therefore we have the estimate

$$\|\mathcal{K}_{2^{-k+1}}(f)\|_2^2 \gtrsim \int_{\Omega_k} |f^*_{2^{-k+1}}(e)|^2\,d\sigma \gtrsim a_k^2\,\sigma(\Omega_k) = \sigma(\Omega_k)k\frac{\mathfrak{h}^2}{h}(2^{-k}). \tag{3.5}$$

Combining (3.4), (3.5) and using the maximal inequality (1.2), we obtain

$$\sigma(\Omega_k)k\frac{\mathfrak{h}^2}{h}(2^{-k}) \lesssim \|f^*_{2^{-k+1}}\|_2^2 \lesssim \log(2^k)\|f\|_2^2 \lesssim k\#J_k\mathfrak{h}^2(2^{-k}).$$

Now let h be a dimension function satisfying the hypothesis of Theorem 3.6. We have

$$\sum_{j\geq 0} h(r_j) \geq \sum_{k\geq 0} h(2^{-k})\#J_k \gtrsim \sum_{k\geq 0}\sigma(\Omega_k) \geq \sigma(\mathbb{S}) > 0. \qquad \square$$

Applying this theorem to the class F_α^+, we obtain a sharper lower bound on the generalized Hausdorff dimension:

Corollary 3.7. *Let E an F_α^+-set. If h is any dimension function satisfying the relation $h(x) \geq Cx^{2\alpha}\log^{1+\theta}(\frac{1}{x})$ for $\theta > 2$ then $\mathcal{H}^h(E) > 0$.*

Remark 3.8. At the endpoint $\alpha = 1$, this estimate is worse than the one due to Keich. He obtained, using strongly the full dimension of a ball in $\mathbb{R}^2$, that if E is an F_1^+-set and h is a dimension function satisfying the bound $h(x) \geq Cx^2\log(\frac{1}{x})\left(\log\log(\frac{1}{x})\right)^\theta$ for $\theta > 2$, then $\mathcal{H}^h(E) > 0$.

Remark 3.9. Note that the proof above relies essentially on the L^1 and L^2 size of the ball in $\mathbb{R}^2$, not on the dimension function $\mathfrak{h}$. Moreover, we only use the "gap" between h and $\mathfrak{h}^2$ (measured by the function $\frac{\mathfrak{h}^2}{h}$). This last observation leads to conjecture that this proof can not be used to prove that an $F_{\mathfrak{h}}$-set has positive $\mathfrak{h}^2$ measure, since in the case of $\mathfrak{h}(x) = x$, as we remarked in the introduction, this would contradict the existence of Kakeya sets of zero measure in $\mathbb{R}^2$.

Also note that the absence of conditions on the function $\mathfrak{h}$ allows us to consider the "zero-dimension" Furstenberg problem. However, this bound does not provide any substantial improvement, since the zero-dimensionality property of the function $\mathfrak{h}$ is shared by the function $\mathfrak{h}^2$. This is because the proof above, in the case of the F_α-sets, gives the worse bound ($\dim_H(E) \geq 2\alpha$) when the parameter α is in $(0, \frac{1}{2})$.

3.3. The $\mathfrak{h} \to \mathfrak{h}\sqrt{\cdot}$ bound, positive dimension

Now we will turn our attention to those functions h that satisfy the bound $h(x) \lesssim x^\alpha$ for $\alpha \leq \frac{1}{2}$. For these functions we are able to improve on the previously obtained bounds. We need to impose some growth conditions on the dimension function $\mathfrak{h}$. These conditions can be thought of as imposing a lower bound on the dimensionality of $\mathfrak{h}$ to keep it away from the zero-dimensional case.

The next lemma is from [MR10] and says that we can split the $\mathfrak{h}$-dimensional mass of a set E contained in an interval I into two sets that are positively separated.

Lemma 3.10. *Let $\mathfrak{h} \in \mathbb{H}$, $\delta > 0$, I an interval and $E \subseteq I$. Let $\eta > 0$ be such that $\mathfrak{h}^{-1}(\frac{\eta}{8}) < \delta$ and $\mathcal{H}^{\mathfrak{h}}_\delta(E) \geq \eta > 0$. Then there exist two subintervals I^-, I^+ that are $\mathfrak{h}^{-1}(\frac{\eta}{8})$-separated and with $\mathcal{H}^{\mathfrak{h}}_\delta(I^\pm \cap E) \gtrsim \eta$.*

The key geometric ingredient is contained in the following lemma. The idea is from [Wol99b], but the general version needed here is from [MR10]. This lemma will provide an estimate for the number of lines with a certain separation property that intersect two balls of a given size.

Lemma 3.11. *Let $\mathfrak{b} = \{b_k\}_{k\in\mathbb{N}}$ be a decreasing sequence with $\lim b_k = 0$. Given a family of balls $\mathcal{B} = \{B(x_j; r_j)\}$, we define $J^{\mathfrak{b}}_k$ as in (3.2) and let $\{e_i\}_{i=1}^{M_k}$ be a b_k-separated set of directions. Assume that for each i there are two line segments $I^+_{e_i}$ and $I^-_{e_i}$ lying on a line in the direction e_i that are s_k-separated for some given s_k Define $\Pi_k = J^{\mathfrak{b}}_k \times J^{\mathfrak{b}}_k \times \{1, \dots, M_k\}$ and $\mathcal{L}^{\mathfrak{b}}_k$ by*

$$\mathcal{L}^{\mathfrak{b}}_k := \left\{(j_+, j_-, i) \in \Pi_k : I^-_{e_i} \cap B_{j_-} \neq \emptyset \; I^+_{e_i} \cap B_{j_+} \neq \emptyset\right\}.$$

If $\frac{1}{5}s_k > b_{k-1}$ for all k, then

$$\#\mathcal{L}^{\mathfrak{b}}_k \lesssim \frac{b_{k-1}}{b_k}\frac{1}{s_k}\left(\#J^{\mathfrak{b}}_k\right)^2.$$

Proof. Consider a fixed pair j_-, j_+ and its associated B_{j_-} and B_{j_+} We will use as distance between two balls the distance between the centres, and for simplicity we write $d(j_-, j_+) = d(B_{j_-}, B_{j_+})$. If $d(j_-, j_+) < \frac{3}{5}s_k$ then there is no i such that (j_-, j_+, i) belongs to $\mathcal{L}^{\mathfrak{b}}_k$.

Now, for $d(j_-, j_+) \geq \frac{3}{5}s_k$, we will look at the special configuration given by Figure 1 when we have $r_{j_-} = r_{j_+} = b_{k-1}$ and the balls are tangent to the ends of I^- and I^+. This will give a bound for any possible configuration, since in any other situation the cone of allowable directions is narrower.

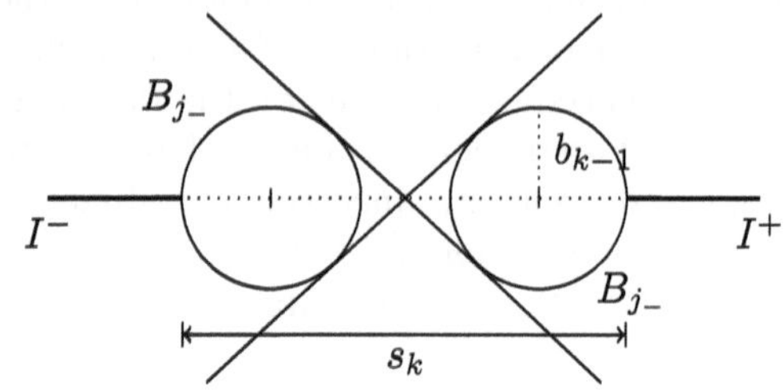

FIGURE 1. Cone of allowable directions I

Let us focus on one half of the cone (Figure 2). Let θ be the width of the cone. In this case, we have to look at $\frac{\theta}{b_k}$ directions that are b_k-separated. Further, we note that $\theta = \frac{2\theta_k}{s_k}$, where θ_k is the bold arc at distance $s_k/2$ from the center of the cone. Let us see that $\theta_k \sim b_{k-1}$.

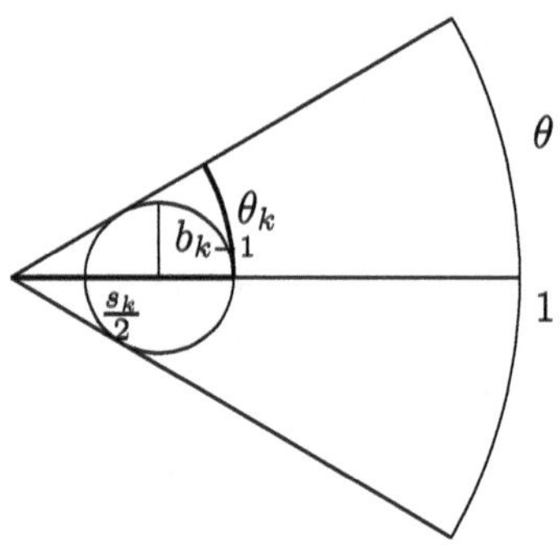

FIGURE 2. Cone of allowable directions II

If we use the notation of Figure 3, we have to prove that $\theta_k \lesssim b_{k-1}$ for $a \in (0, +\infty)$. We have $\theta_k = \theta(a + 2b_{k-1})$. Also $\theta < \tan^{-1}(\frac{b_{k-1}}{a})$, so

$$\theta_k < \tan^{-1}(\frac{b_{k-1}}{a})(a + 2b_{k-1}) \sim b_{k-1}.$$

We conclude that $\theta_k \sim b_{k-1}$, and therefore the number D of lines in b_k-separated directions with non-empty intersection with B_{j_-} and B_{j_+} has to satisfy $D \leq \frac{\theta}{b_k} = \frac{2\theta_k}{s_k b_k} \sim \frac{b_{k-1}}{b_k}\frac{1}{s_k}$. The lemma follows by summing on all pairs (j_-, j_+). $\square$

Now we can present the main result of this section.

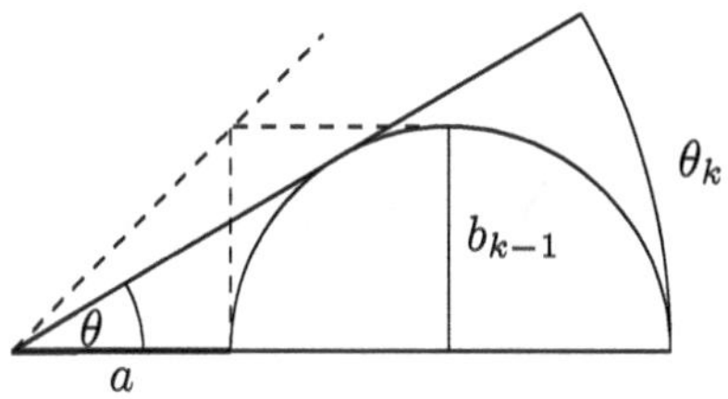

FIGURE 3. The arc θ_k is comparable to b_{k-1}

Theorem 3.12. *Let $\mathfrak{h} \in \mathbb{H}_d$ be a dimension function such that $\mathfrak{h}(x) \lesssim x^\alpha$ for some $0 < \alpha < 1$ and E be an $F_{\mathfrak{h}}$-set. Let $h \in \mathbb{H}$ with $h \prec \mathfrak{h}$. If $\sum_{k\geq 0} \frac{\mathfrak{h}}{h}(2^{-k})^{\frac{2\alpha}{2\alpha+1}} < \infty$, then $\mathcal{H}^{h\sqrt{\cdot}}(E) > 0$.*

Proof. We begin in the same way as in the previous section. Again by Definition 3.1, since $E \in F_{\mathfrak{h}}$, we have $\mathcal{H}^{\mathfrak{h}}_\delta(\ell_e \cap E) > 1$ for all $e \in \mathbb{S}$ for any $\delta < \delta_E$.

Consider the sequence $\mathfrak{a} = \left\{\frac{\mathfrak{h}}{h}(2^{-k})^{\frac{2\alpha}{2\alpha+1}}\right\}_k$. Let k_0 be such that

$$\mathfrak{h}^{-1}\left(\frac{a_k}{16\|\mathfrak{a}\|_1}\right) < \delta_E \qquad \text{for any } k \geq k_0. \tag{3.6}$$

Now take any δ-covering $\mathcal{B} = \{B_j\}$ of E by balls with $\delta < \min\{\delta_E, 2^{-k_0}\}$. Using Lemma 3.3 we obtain $\mathbb{S} = \bigcup_k \Omega_k$ with

$$\Omega_k = \left\{e \in \Omega : \mathcal{H}^{\mathfrak{h}}_\delta(\ell_e \cap E_k) \geq \frac{a_k}{2\|\mathfrak{a}\|_1}\right\}. \tag{3.7}$$

Again we have $E_k = E \cap \bigcup_{j\in J_k} B_j$, but by our choice of δ, the sets E_k are empty for $k < k_0$. Therefore the same holds trivially for Ω_k and we have that $\mathbb{S} = \bigcup_{k\geq k_0} \Omega_k$. Since for each $e \in \Omega_k$ we have the inequality in (3.7), we can apply Lemma 3.10 with $\eta = \frac{a_k}{2\|\mathfrak{a}\|_1}$ to $\ell_e \cap E_k$. Therefore we obtain two intervals I_e^- and I_e^+, contained in ℓ_e with

$$\mathcal{H}^{\mathfrak{h}}_\delta(I_e^\pm \cap E_k) \gtrsim a_k$$

that are $\mathfrak{h}^{-1}(ra_k)$-separated for $r = \frac{1}{16\|\mathfrak{a}\|_1}$. Let $\{e_j^k\}_{j=1}^{M_k}$ be a 2^{-k}-separated subset of Ω_k. Therefore $M_k \gtrsim 2^k\sigma(\Omega_k)$. Define $\Pi_k := J_k \times J_k \times \{1,..,M_k\}$ and

$$\mathcal{T}_k := \left\{(j_-, j_+, i) \in \Pi_k : I_{e_i}^- \cap E_k \cap B_{j_-} \neq \emptyset \; I_{e_i}^+ \cap E_k \cap B_{j_+} \neq \emptyset\right\}.$$

We will count the elements of $\mathcal{T}_k$ in two different ways. First, fix j_- and j_+ and count for how many values of i the triplet (j_-, j_+, i) belongs to $\mathcal{T}_k$. For this, we will apply Lemma 3.11 for the choice $\mathfrak{b} = \{2^{-k}\}$. The estimate we obtain is the number of 2^{-k}-separated directions e_i, that intersect simultaneously the balls B_{j_-} and B_{j_+}, given that these balls are separated. We obtain

$$\#\mathcal{T}_k \lesssim \frac{1}{\mathfrak{h}^{-1}(ra_k)} (\#J_k)^2.$$

Second, fix i. In this case, we have by hypothesis that $\mathcal{H}^{\mathfrak{h}}_{\delta}(I^+_{e_i} \cap E_k) \gtrsim a_k$, so $\sum_{j_+} \mathfrak{h}(r_{j_+}) \gtrsim a_k$. Therefore,

$$a_k \lesssim \sum_{(j_-,j_+,i)\in\mathcal{T}_k} \mathfrak{h}(r_{j_+}) \leq K\mathfrak{h}(2^{-k}),$$

where K is the number of elements of the sum. Therefore $K \gtrsim \frac{a_k}{\mathfrak{h}(2^{-k})}$. The same holds for j_-, so

$$\#\mathcal{T}_k \gtrsim M_k \left(\frac{a_k}{\mathfrak{h}(2^{-k})}\right)^2.$$

Combining the two bounds,

$$\#J_k \gtrsim M_k^{1/2} \frac{a_k}{\mathfrak{h}(2^{-k})}\mathfrak{h}^{-1}(ra_k)^{1/2} \gtrsim 2^{\frac{k}{2}}\sigma(\Omega_k)^{1/2}\frac{a_k}{\mathfrak{h}(2^{-k})}\mathfrak{h}^{-1}(ra_k)^{1/2}.$$

Consider now a dimension function $h \prec \mathfrak{h}$ as in the hypothesis of the theorem. Then again

$$\sum_j h(r_j)r_j^{1/2} \geq \sum_k \frac{\mathfrak{h}(2^{-k})2^{-\frac{k}{2}}\#J_k}{\frac{\mathfrak{h}}{h}(2^{-k})} \gtrsim \sum_{k\geq k_0} \sigma(\Omega_k)^{1/2}\frac{a_k\mathfrak{h}^{-1}(ra_k)^{1/2}}{\frac{\mathfrak{h}}{h}(2^{-k})}. \tag{3.8}$$

To bound this last expression, we use first that there exists $\alpha \in (0,1)$ with $\mathfrak{h}(x) \lesssim x^{\alpha}$ and therefore $\mathfrak{h}^{-1}(x) \gtrsim x^{\frac{1}{\alpha}}$. We then recall the definition of the sequence $\mathfrak{a}$, $a_k = \frac{\mathfrak{h}}{h}(2^{-k})^{\frac{2\alpha}{2\alpha+1}}$ to obtain

$$\sum_j h(r_j)r_j^{1/2} \gtrsim \sum_{k\geq k_0} \sigma(\Omega_k)^{1/2}\frac{a_k^{\frac{2\alpha+1}{2\alpha}}}{\frac{\mathfrak{h}}{h}(2^{-k})} = \sum_{k\geq k_0} \sigma(\Omega_k)^{1/2} \gtrsim 1. \qquad \square$$

The next corollary follows from Theorem 3.12 in the same way as Corollary 3.7 follows from Theorem 3.6.

Corollary 3.13. *Let E be an F_{α}^{+}-set. If h is a dimension function satisfying the relation $h(x) \geq Cx^{\alpha}\sqrt{x}\log^{\theta}(\frac{1}{x})$ for $\theta > \frac{2\alpha+1}{2\alpha}$ then $\mathcal{H}^h(E) > 0$.*

3.4. The $\mathfrak{h} \to \mathfrak{h}\sqrt{\cdot}$ bound, dimension zero

In this section we look at a class of very small Furstenberg sets. We will study, roughly speaking, the extremal case of F_0-sets and ask ourselves if inequality (1.1) can be extended to this class. Our approach to the problem, using dimension functions, allows us to tackle the problem about the dimensionality of these sets in some cases. We study the case of $F_{\mathfrak{h}}$-sets associated to one particular choice of $\mathfrak{h}$. We will look at the function $\mathfrak{h}(x) = \dfrac{1}{\log(\frac{1}{x})}$ as a model of "zero-dimensional" function. Our next theorem will show that in this case inequality (1.1) can indeed be extended. The trick here will be to replace the dyadic scale on the radii in J_k with a faster decreasing sequence $\mathfrak{b} = \{b_k\}_{k\in\mathbb{N}}$.

The main difference will be in the estimate of the quantity of lines in b_k-separated directions that intersect two balls of level J_k with a fixed distance s_k

between them. This estimate is given by Lemma 3.11. Note that the problem in the above bound is the rapid decay of $\mathfrak{h}^{-1}$, which is solved by the positivity assumption. In this case, since we are dealing with a zero-dimensional function $\mathfrak{h}$, the inverse involved decays dramatically to zero. Therefore the strategy cannot be the same as before, where we choose optimally the sequence $\mathfrak{a}$. In this case, we will obtain a result by choosing an appropriate sequence of scales.

Theorem 3.14. *Let* $\mathfrak{h}(x) = \frac{1}{\log(\frac{1}{x})}$ *and let* E *be an* $F_{\mathfrak{h}}$*-set. Then* $\dim_H(E) \geq \frac{1}{2}$.

Proof. Take a non-negative sequence $\mathfrak{b}$ which will be determined later. We will apply the splitting Lemma 3.10 as in the previous section. For this, take k_0 as in (3.6) associated to the sequence $\mathfrak{a} = \{k^{-2}\}_{k\in\mathbb{N}}$. Now, for a given generic δ-covering of E with $\delta < \min\{\delta_E, 2^{-k_0}\}$, we use Lemma 3.3 to obtain a decomposition $\mathbb{S} = \bigcup_{k\geq k_0} \Omega_k$ with

$$\Omega_k = \left\{e \in \mathbb{S} : \mathcal{H}^{\mathfrak{h}}_\delta(\ell_e \cap E_k) \geq ck^{-2}\right\},$$

where $E_k = E \cap \bigcup_{J_k^{\mathfrak{b}}} B_j$, $J_k^{\mathfrak{b}}$ is the partition of the radii associated to $\mathfrak{b}$ and $c > 0$ is a suitable constant. The same calculations as in Theorem 3.12 yield

$$\#J_k^{\mathfrak{b}} \gtrsim \left(\frac{\sigma(\Omega_k)}{b_{k-1}}\right)^{1/2} \frac{\mathfrak{h}^{-1}(ck^{-2})^{1/2}}{k^2\mathfrak{h}(b_{k-1})} \geq \left(\frac{\sigma(\Omega_k)}{b_{k-1}}\right)^{1/2} \frac{e^{-ck^2}}{k^2}.$$

Now we estimate a sum like (3.8). For $\beta < \frac{1}{2}$ we have

$$\sum_{j\geq 0} r_j^\beta \geq \sum_{k\geq 0} \sigma(\Omega_k)^{1/2} \frac{b_k^\beta}{b_{k-1}^{\frac{1}{2}}} \frac{e^{-ck^2}}{k^2} \gtrsim \sqrt{\sum_{k\geq 0} \sigma(\Omega_k) \frac{b_k^{2\beta}}{b_{k-1}} \frac{1}{e^{ck^2}k^4}}.$$

In the last inequality we use that the terms are all non-negative. The goal now is to take some rapidly decreasing sequence such that the factor $\frac{b_k^{2\beta}}{b_{k-1}}$ beats the factor $k^{-4}e^{-ck^2}$. Let us take $0 < \varepsilon < \frac{1-2\beta}{2\beta}$ and consider the hyperdyadic scale $b_k = 2^{-(1+\varepsilon)^k}$. With this choice, we have

$$\frac{b_k^{2\beta}}{b_{k-1}} = 2^{(1+\varepsilon)^{k-1}-(1+\varepsilon)^k 2\beta} = 2^{(1+\varepsilon)^k(\frac{1}{1+\varepsilon}-2\beta)}.$$

We obtain that

$$\left(\sum_{j\geq 0} r_j^\beta\right)^2 \geq \sum_{k\geq 0} \sigma(\Omega_k) \frac{2^{(1+\varepsilon)^k(\frac{1}{1+\varepsilon}-2\beta)}}{e^{ck^2}k^4}.$$

Finally, since by the positivity of $\frac{1}{1+\varepsilon} - 2\beta$ the double exponential in the numerator grows faster than the denominator, we obtain that

$$\left(\sum_{j\geq 0} r_j^\beta\right)^2 \gtrsim \sum_{k\geq 0} \sigma(\Omega_k) \gtrsim 1. \qquad \square$$

Corollary 3.15. *Let* $\theta > 0$. *If* E *is an* $F_{\mathfrak{h}}$*-set with* $\mathfrak{h}(x) = \frac{1}{\log^\theta(\frac{1}{x})}$ *then* $\dim_H(E) \geq \frac{1}{2}$.

This shows that there is a whole class of F_0-sets that must be at least $\frac{1}{2}$-dimensional.

We want to remark that, shortly after [MR10] was published, we were notified indirectly by Tamás Keleti and András Máthé that Theorem 3.14 can actually be improved. The same result holds for the choice of $\mathfrak{h}(x) = \frac{1}{\log\log(\frac{1}{x})}$ if we use a slightly faster hyperdyadic scale, namely $b_k = 2^{(1+\varepsilon)^{k^3}}$. The improved theorem is the following.

Theorem 3.16. *Let* $\mathfrak{h}(x) = \frac{1}{\log\log(\frac{1}{x})}$ *and let* E *be an* $F_{\mathfrak{h}}$*-set. Then* $\dim_H(E) \geq \frac{1}{2}$.

But this is as far as we can go. They have found, for any $h \prec \mathfrak{h}$, an explicit construction of a set $E \in F_h$ such that $\dim_H(E) = 0$.

4. Fractal sets of directions

In this section we will apply our techniques to a more general problem. Consider now the class of Furstenberg sets but defined by a fractal set of directions. Precisely, we have the following definition.

Definition 4.1. For α, β in $(0, 1]$, a subset E of $\mathbb{R}^2$ will be called an $F_{\alpha\beta}$-set if there is a subset L of the unit circle such that $\dim_H(L) \geq \beta$ and, for each direction e in L, there is a line segment ℓ_e in the direction of e such that the Hausdorff dimension of the set $E \cap \ell_e$ is equal to or greater than α.

This generalizes the classical definition of Furstenberg sets, when the whole circle is considered as a set of directions. The purpose here is to study how the parameter β affects the bounds above. From our results we will derive the following proposition.

Proposition 4.2. *For any set* $E \in F_{\alpha\beta}$*, we have that*

$$\dim_H(E) \geq \max\left\{2\alpha + \beta - 1; \frac{\beta}{2} + \alpha\right\}, \qquad \alpha, \beta > 0. \tag{4.1}$$

It is not hard to prove Proposition 4.2 directly, but we will study this problem in a wider scenario and derive it as a corollary. Moreover, by using general Hausdorff measures, we will extend inequalities (4.1) to the zero-dimensional case.

Definition 4.3. Let $\mathfrak{h}$ and $\mathfrak{g}$ be two dimension functions. A set $E \subseteq \mathbb{R}^2$ is a Furstenberg set of type $\mathfrak{h}\mathfrak{g}$, or an $F_{\mathfrak{h}\mathfrak{g}}$-set, if there is a subset L of the unit circle such that $\mathcal{H}^{\mathfrak{g}}(L) > 0$ and, for each direction e in L, there is a line segment ℓ_e in the direction of e such that $\mathcal{H}^{\mathfrak{h}}(\ell_e \cap E) > 0$.

Note that this is the natural generalization of the $F_{\alpha\beta}^{+}$ class:

Definition 4.4. For each pair $\alpha, \beta \in (0, 1]$, a set $E \subseteq \mathbb{R}^2$ will be called an $F_{\alpha\beta}^{+}$-set if there is a subset L of the unit circle such that $\mathcal{H}^{\beta}(L) > 0$ and, for each direction e in L, there is a line segment ℓ_e in the direction of e such that $\mathcal{H}^{\alpha}(\ell_e \cap E) > 0$.

Following the intuition suggested by Proposition 4.2, one could conjecture that if E belongs to the class $F_{\mathfrak{h}\mathfrak{g}}$ then an appropriate dimension function for E should be dimensionally greater than $\frac{\mathfrak{h}^2\mathfrak{g}}{\mathrm{id}}$ and $\mathfrak{h}\sqrt{\mathfrak{g}}$. This will be the case, indeed, and we will provide some estimates on the gap between those conjectured dimension functions and a generic test function $h \in \mathbb{H}$ to ensure that $\mathcal{H}^h(E) > 0$, and also illustrate with some examples. We will consider the two results separately. Namely, for a given pair of dimension functions $\mathfrak{h}, \mathfrak{g} \in \mathbb{H}$, in Section 4.1 we obtain sufficient conditions on a test dimension function $h \in \mathbb{H}$, $h \prec \frac{\mathfrak{h}^2\mathfrak{g}}{\mathrm{id}}$ to ensure that $\mathcal{H}^h(E) > 0$ for any set $E \in F_{\mathfrak{h}\mathfrak{g}}$. In Section 4.2 we consider the analogous problem for $h \prec \mathfrak{h}\sqrt{\mathfrak{g}}$. It turns out that one relevant feature of the set of directions is related to the notion of δ-entropy:

Definition 4.5. Let $E \subset \mathbb{R}^n$ and $\delta \in \mathbb{R}_{>0}$. The δ-entropy of E is the maximal possible cardinality of a δ-separated subset of E. We will denote this quantity with $\mathcal{N}_\delta(E)$.

The main idea is to relate the δ-entropy to some notion of size of the set. Clearly, the entropy is essentially the box dimension or the packing dimension of a set (see [Mat95] or [Fal03] for the definitions) since both concepts are defined in terms of separated δ balls with centres in the set. However, for our proof we will need to relate the entropy of a set to some quantity that has the property of being (in some sense) stable under countable unions. One choice is therefore the notion of Hausdorff content, which enjoys the needed properties: it is an outer measure, it is finite, and it reflects the entropy of a set in the following manner. Recall that the $\mathfrak{g}$-dimensional Hausdorff content of a set E is defined as

$$\mathcal{H}^{\mathfrak{g}}_\infty(E) = \inf\left\{\sum_i \mathfrak{g}(\mathrm{diam}(U_i) : E \subset \bigcup_i U_i\right\}.$$

Note that the $\mathfrak{g}$-dimensional Hausdorff content $\mathcal{H}^{\mathfrak{g}}_\infty$ is clearly not the same as the Hausdorff measure $\mathcal{H}^{\mathfrak{g}}$. In fact, they are the measures obtained by applying Method I and Method II (see [Mat95]) respectively to the premeasure that assigns to a set A the value $\mathfrak{g}(\mathrm{diam}(A))$. For future reference, we state the following estimate for the δ-entropy of a set with positive $\mathfrak{g}$-dimensional Hausdorff content as a lemma.

Lemma 4.6. *Let $\mathfrak{g} \in \mathbb{H}$ and let A be any set. Let $\mathcal{N}_\delta(A)$ be the δ-entropy of A. Then $\mathcal{N}_\delta(A) \geq \frac{\mathcal{H}^{\mathfrak{g}}_\infty(A)}{\mathfrak{g}(\delta)}$.*

Of course, this result is meaningful when $\mathcal{H}^{\mathfrak{g}}_\infty(A) > 0$. We will use it in the case in which $\mathcal{H}^{\mathfrak{g}}(A) > 0$, which is equivalent to $\mathcal{H}^{\mathfrak{g}}_\infty(A) > 0$. Note that the lemma above only requires the finiteness and the subadditivity of the Hausdorff content. The relevant feature that will be needed in our proof is the σ-subadditivity, which is a property that the Box dimension does not share. Following the notation of Definition 3.2 we have the following analogue of Lemma 3.3:

Lemma 4.7. *Let E be an $F_{\mathfrak{h}\mathfrak{g}}$-set for some $\mathfrak{h}, \mathfrak{g} \in \mathbb{H}$ with the directions in $L \subset \mathbb{S}$ and let $\mathfrak{a} = \{a_k\}_{k\in\mathbb{N}} \in \ell^1$ be a non-negative sequence. Let $\mathcal{B} = \{B_j\}$ be a δ-covering of E with $\delta < \delta_E$ and let E_k and J_k be as above. Define*

$$L_k := \left\{ e \in \mathbb{S} : \mathcal{H}^{\mathfrak{h}}_\delta(\ell_e \cap E_k) \geq \frac{a_k}{2\|\mathfrak{a}\|_1} \right\}.$$

Then $L = \cup_k L_k$.

4.1. The Kakeya type bound

Now we will prove a generalized version of the bound $\dim_H(E) \geq 2\alpha + \beta - 1$ for $E \in F_{\alpha\beta}$. We have the following theorem.

Theorem 4.8 ($\mathfrak{h}\mathfrak{g} \to \frac{\mathfrak{h}^2\mathfrak{g}}{\mathrm{id}}$). *Let $\mathfrak{g} \in \mathbb{H}$ and $\mathfrak{h} \in \mathbb{H}_d$ be two dimension functions and let E be an $F_{\mathfrak{h}\mathfrak{g}}$-set. Let $h \in \mathbb{H}$ such that $h \prec \frac{\mathfrak{h}^2\mathfrak{g}}{\mathrm{id}}$. For $\delta > 0$, let $C(\delta)$ be as in (3.3). If $\sum_{k\geq 0} \sqrt{\frac{\mathfrak{h}^2(2^{-k})C(2^{-k+1})}{h(2^{-k})}} < \infty$, then $\mathcal{H}^h(E) > 0$.*

Proof. Let $E \in F_{\mathfrak{h}\mathfrak{g}}$ and let $\{B_j\}_{j\in\mathbb{N}}$ be a covering of E by balls with $B_j = B(x_j; r_j)$. Define $\mathfrak{a} = \{a_k\}$ by $a_k^2 = \frac{\mathfrak{h}^2(2^{-k})C(2^{-k+1})}{h(2^{-k})}$. Therefore, by hypothesis $\mathfrak{a} \in \ell^1$. Also define, as in the previous section, for each $k \in \mathbb{N}$, $J_k = \{j \in \mathbb{N} : 2^{-k} < r_j \leq 2^{-k+1}\}$ and $E_k = E \cap \cup_{j\in J_k} B_j$. Since $\mathfrak{a} \in \ell^1$, we can apply Lemma 3.3 to obtain the decomposition of the set of directions as $L = \bigcup_k L_k$ associated to this choice of $\mathfrak{a}$. We proceed as in the F_α-class and apply the maximal function inequality to a weighted union of indicator functions:

$$f := \mathfrak{h}(2^{-k})2^k \chi_{F_k}.$$

As before,

$$\|f\|_2^2 \lesssim \#J_k \mathfrak{h}^2(2^{-k}). \tag{4.2}$$

The same arguments used in the proof of Theorem 3.6 in Section 3 allows us to obtain a lower bound for the maximal function. Essentially, the maximal function is pointwise bounded from below by the average of f over the tube centred on the line segment ℓ_e for any $e \in L_k$. Therefore, we have the following bound for the (L^2, μ) norm. Here, μ is a measure supported on L that obeys the law $\mu(B(x,r) \leq \mathfrak{g}(r)$ for any ball $B(x,r)$ given by Frostman's lemma.

$$\|\mathcal{K}_{2^{-k+1}}(f)\|^2_{L^2(d\mu)} \gtrsim a_k^2 \mu(L_k) = \frac{\mu(L_k)\mathfrak{h}^2(2^{-k})C(2^{-k})}{h(2^{-k})}. \tag{4.3}$$

Combining (4.3) with the maximal inequality (3.3), we obtain

$$\frac{\mu(L_k)\mathfrak{h}^2(2^{-k})C(2^{-k})}{h(2^{-k})} \lesssim \|\mathcal{K}_{2^{-k+1}}(f)\|_2^2 \lesssim C(2^{-k+1})\|f\|_2^2 \leq C(2^{-k})\|f\|_2^2.$$

We also have the bound (4.2), which implies that $\frac{\mu(L_k)}{h(2^{-k})} \lesssim \#J_k$, which easily yields the desired result. $\square$

Corollary 4.9. *Let E an $F^+_{\alpha\beta}$-set. If h is any dimension function satisfying $h(x) \geq Cx^{2\alpha+\beta-1}\log^\theta(\frac{1}{x})$ for $\theta > 2$, then $\mathcal{H}^h(E) > 0$.*

Remark 4.10. Note that the bound $\dim(E) \geq 2\alpha + \beta - 1$ for $E \in F_{\alpha\beta}$ follows directly from this last corollary.

4.2. The combinatorial bound

In this section we deal with the bound $\mathfrak{h}\mathfrak{g} \to \mathfrak{h}\sqrt{\mathfrak{g}}$, which is the significant bound near the endpoint $\alpha = \beta = 0$ and generalizes the bound $\dim_H(E) \geq \frac{\beta}{2} + \alpha$ for $E \in F_{\alpha\beta}$. Note that the second bound in (4.1) is meaningless for small values of α and β. We will again consider separately the cases of $\mathfrak{h}$ being zero-dimensional or positive-dimensional. In the next theorem, the additional condition on $\mathfrak{h}$ reflects the positivity of the dimension function. We will use again the two relevant lemmas from Section 3. Lemma 3.10 is the "splitting lemma" and Lemma 3.11 is the combinatorial ingredient in the proof of both Theorem 4.11 and Theorem 4.14. We have the following theorem. Recall that $h_\alpha(x) = x^\alpha$.

Theorem 4.11 ($\mathfrak{h}\mathfrak{g} \to \mathfrak{h}\sqrt{\mathfrak{g}}$, $\mathfrak{h} \succ h_\alpha$). *Let $\mathfrak{g} \in \mathbb{H}$, $\mathfrak{h} \in \mathbb{H}_d$ be two dimension functions such that $\mathfrak{h}(x) \lesssim x^\alpha$ for some $0 < \alpha < 1$ and let E be an $F_{\mathfrak{h}\mathfrak{g}}$-set. Let $h \in \mathbb{H}$ with $h \prec \mathfrak{h}\sqrt{\mathfrak{g}}$. If $\sum_{k\geq 0} \left(\frac{\mathfrak{h}(2^{-k})\sqrt{\mathfrak{g}}(2^{-k})}{h(2^{-k})}\right)^{\frac{2\alpha}{2\alpha+1}} < \infty$, then $\mathcal{H}^h(E) > 0$.*

Proof. Let $E \in F_{\mathfrak{h}\mathfrak{g}}$ and let $\{B_j\}_{j\in\mathbb{N}}$ be a covering of E by balls with $B_j = B(x_j; r_j)$. Consider the sequence $\mathfrak{a}$ defined as $\mathfrak{a} = \left\{\left(\frac{\mathfrak{h}\sqrt{\mathfrak{g}}}{h}(2^{-k})\right)^{\frac{2\alpha}{2\alpha+1}}\right\}_{k\geq 1}$. Also define, as in the previous section, for each $k \in \mathbb{N}$, $J_k = \{j \in \mathbb{N} : 2^{-k} < r_j \leq 2^{-k+1}\}$ and $E_k = E \cap \bigcup_{j\in J_k} B_j$. Since by hypothesis $\mathfrak{a} \in \ell^1$, we can apply Lemma 3.3 to obtain the decomposition of the set of directions as $L = \bigcup_k L_k$ associated to this choice of $\mathfrak{a}$, where L_k is defined as

$$L_k := \left\{e \in \mathbb{S} : \mathcal{H}^{\mathfrak{h}}_\delta(\ell_e \cap E_k) \geq \frac{a_k}{2\|\mathfrak{a}\|_1}\right\}.$$

Now, let $\{e^k_j\}_{j=1}^{N_k}$ be a 2^{-k}-separated subset of L_k. Taking into account the estimate for the entropy given in Lemma 4.6. We obtain then that

$$N_k \gtrsim \frac{\mathcal{H}^{\mathfrak{g}}_\infty(L_k)}{\mathfrak{g}(2^{-k})}. \tag{4.4}$$

We can proceed as in the previous section to obtain that

$$\#J_k \gtrsim a_k \mathfrak{h}^{-1}(ra_k)^{1/2}\frac{N_k^{1/2}}{\mathfrak{h}(2^{-k})}.$$

Therefore, for any $h \in \mathbb{H}$ as in the hypothesis of the theorem, we have the estimate

$$\sum_{j\geq 0} h(r_j) \gtrsim \sum_{k\geq 0} h(2^{-k})\#J_k \gtrsim \sum_{k\geq 0} a_k \mathfrak{h}^{-1}(ra_k)^{1/2} N_k^{1/2}\frac{\sqrt{\mathfrak{g}}(2^{-k})}{(\frac{\mathfrak{h}\sqrt{\mathfrak{g}}}{h})(2^{-k})}.$$

Recall now that from (4.4) we have $\sqrt{\mathfrak{g}}(2^{-k})N_k^{\frac{1}{2}} \gtrsim \mathcal{H}_\infty^{\mathfrak{g}}(L_k)^{\frac{1}{2}}$. We obtain

$$\sum_{j\geq 0} h(r_j) \gtrsim \sum_{k\geq 0} \frac{\mathcal{H}_\infty^{\mathfrak{g}}(L_k)^{1/2} a_k^{\frac{2\alpha+1}{2\alpha}}}{(\frac{\mathfrak{h}\sqrt{\mathfrak{g}}}{h})(2^{-k})} = \sum_{k\geq 0} \mathcal{H}_\infty^{\mathfrak{g}}(L_k)^{1/2} \gtrsim 1.$$

We use again that $\mathfrak{h}(x) \lesssim x^\alpha$ implies that $\mathfrak{h}^{-1}(x) \gtrsim x^{\frac{1}{\alpha}}$. In the last inequality, we used the σ-subadditivity of $\mathcal{H}_\infty^{\mathfrak{g}}$. □

Corollary 4.12. *Let E be an $F_{\alpha\beta}^{+}$-set for $\alpha, \beta > 0$. If h is a dimension function satisfying $h(x) \geq Cx^{\frac{\beta}{2}+\alpha}\log^\theta(\frac{1}{x})$ for $\theta > \frac{2\alpha+1}{2\alpha}$, then $\mathcal{H}^h(E) > 0$.*

Remark 4.13. Note that again the bound $\dim(E) \geq \alpha + \frac{\beta}{2}$ for $E \in F_{\alpha\beta}$ follows directly from this last corollary.

In the next theorem we consider the case of a family of very small Furstenberg sets. More precisely, we deal with a family that corresponds to the case $\alpha = 0$, $\beta \in (0,1]$ in the classical setting.

Theorem 4.14 ($\mathfrak{h}\mathfrak{g} \to \mathfrak{h}\sqrt{\mathfrak{g}}$, $\mathfrak{h}$ zero-dimensional, $\mathfrak{g}$ positive). *Let $\beta > 0$ and define $\mathfrak{g}(x) = x^\beta, \mathfrak{h}(x) = \frac{1}{\log\log(\frac{1}{x})}$. If E is an $F_{\mathfrak{h}\mathfrak{g}}$-set, then $\dim(E) \geq \frac{\beta}{2}$.*

The proof follows from the same ideas as in Theorem 3.14 in Section 3, with the natural modifications. We have the following immediate corollary.

Corollary 4.15. *Let $\theta > 0$. If E is an $F_{\mathfrak{h}\mathfrak{g}}$-set with $\mathfrak{h}(x) = \frac{1}{\log\log^\theta(\frac{1}{x})}$ and $\mathfrak{g}(x) = x^\beta$, then $\dim(E) \geq \frac{\beta}{2}$.*

The next question would be: What should be the expected dimension function for an $F_{\mathfrak{h}\mathfrak{g}}$-set if $\mathfrak{h}(x) = \mathfrak{g}(x) = \frac{1}{\log(\frac{1}{x})}$? The preceding results lead us to the following conjecture:

Conjecture 4.16. Let $\mathfrak{h}(x) = \mathfrak{g}(x) = \frac{1}{\log(\frac{1}{x})}$ and let E be an $F_{\mathfrak{h}\mathfrak{g}}$-set. Then $\frac{1}{\log^{\frac{3}{2}}(\frac{1}{x})}$ should be an appropriate dimension function for E, in the sense that a logarithmic gap can be estimated.

We do not know, however, how to prove this.

4.3. A remark on the notion of size for the set of directions

We have emphasized that the relevant ingredient for the combinatorial proof in Section 4.2 is the notion of δ-entropy of a set. In addition, we have discussed the possibility of considering the box dimension as an adequate notion of size to detect this quantity. In this section we present an example that shows that in fact the notion of Packing measure is also inappropriate. We want to remark here that none of them will give any further (useful) information to this problem and therefore there is no chance to obtain similar results in terms of those notions of dimensions. To make it clear, consider the classical problem of proving the bound

$\dim_H(E) \geq \alpha + \frac{\beta}{2}$ for any $E \in F_{\alpha\beta}$ where β is the Box or Packing dimension of the set L of directions.

We illustrate this remark with the extreme case of $\beta = 1$. It is absolutely trivial that nothing meaningful can be said if we only know that the Box dimension ($\dim_B$) of L is 1, since any countable dense subset L of $\mathbb{S}$ satisfies $\dim_B(L) = 1$ but in that case, since L is countable, we can only obtain that $\dim_H(E) \geq \alpha$.

For the Packing dimension ($\dim_P$), it is also easy to see that if we only know that $\dim_P(L) = 1$ we do not have any further information about the Hausdorff dimension of the set E. To see why, consider the following example. Let C_α be a regular Cantor set such that $\dim_H(C_\alpha) = \dim_B(C_\alpha) = \alpha$. Let L be a set of directions with $\dim_H(L) = 0$ and $\dim_P(L) = 1$. Now, we build the Furstenberg set E in polar coordinates as

$$E := \{(r, \theta) : r \in C_\alpha, \theta \in L\}.$$

This can be seen as a "Cantor target", but with a fractal set of directions instead of the whole circle. By the Hausdorff dimension estimates, we know that $\dim_H(E) \geq \alpha$. We show that in this case we also have that $\dim_H(E) \leq \alpha$, which implies that in the general case this is the best that one could expect, even with the additional information about the Packing dimension of L. For the upper bound, consider the function $f : \mathbb{R}^2 \to \mathbb{R}^2$ defined by $f(x, y) = (x \cos y, x \sin y)$. Clearly $E = f(C_\alpha \times L)$. Therefore, by the known product formulae that can be found, for example, in [Fal03], we have that

$$\dim_H(E) = \dim_H(f(C_\alpha \times L)) \leq \dim_H(C_\alpha \times L) = \dim_B(C_\alpha) + \dim_H(L) = \alpha.$$

5. Upper bounds

In this section we look at a refinement of the upper bound for the dimension of Furstenberg sets. Since we are looking for upper bounds on a class of Furstenberg sets, the aim will be to explicitly construct a very small set belonging to the given class.

We first consider the classical case of power functions, x^α, for $\alpha > 0$. Recall that for this case, the known upper bound implies that, for any positive α, there is a set $E \in F_\alpha$ such that $\mathcal{H}^{\frac{1+3\alpha}{2}+\varepsilon}(E) = 0$ for any $\varepsilon > 0$. By looking closer at Wolff's arguments, it can be seen that in fact it is true that $\mathcal{H}^g(E) = 0$ for any dimension function g of the form

$$g(x) = x^{\frac{1+3\alpha}{2}} \log^{-\theta}\left(\frac{1}{x}\right), \qquad \theta > \frac{3(1+3\alpha)}{2} + 1. \tag{5.1}$$

Further, that argument can be modified (Theorem 5.3) to sharpen on the logarithmic gap, and therefore improving (5.1) by proving the same result for any g of the form

$$g(x) = x^{\frac{1+3\alpha}{2}} \log^{-\theta}\left(\frac{1}{x}\right), \qquad \theta > \frac{1+3\alpha}{2}. \tag{5.2}$$

However, this modification will not be sufficient for our main objective, which is to reach the zero-dimensional case. More precisely, we will focus at the endpoint $\alpha = 0$, and give a complete answer about the *exact dimension* of a class of Furstenberg sets. We will prove in Theorem 5.6 that, for any given $\gamma > 0$, there exists a set $E_\gamma \subseteq \mathbb{R}^2$ such that

$$E_\gamma \in F_{\mathfrak{h}_\gamma} \text{ for } \mathfrak{h}_\gamma(x) = \frac{1}{\log^\gamma(\frac{1}{x})} \quad \text{and} \quad \dim_H(E_\gamma) \le \frac{1}{2}. \tag{5.3}$$

This result, together with the results from [MR10] mentioned above, shows that $\frac{1}{2}$ is sharp for the class $F_{\mathfrak{h}_\gamma}$. In fact, for this family both inequalities in (1.1) are in fact the equality $\Phi(F_{\mathfrak{h}_\gamma}) = \frac{1}{2}$.

In order to be able to obtain (5.3), it is not enough to simply "refine" the construction of Wolff. He achieves the desired set by choosing a specific set as the fiber in each direction. This set is known to have the correct dimension. To be able to reach the zero-dimensional case, we need to handle the delicate issue of choosing an analogue zero-dimensional set on each fiber. The main difficulty lies in being able to handle *simultaneously* Wolff's construction and the proof of the fact that the fiber satisfies the stronger condition of having positive measure for the correct dimension function.

We will also focus at the endpoint $\alpha = 0$, and give a complete answer about the size of a class of Furstenberg sets by proving that (Theorem 5.6), for any given $\gamma > 0$, there exists a set $E_\gamma \subseteq \mathbb{R}^2$ such that $E_\gamma \in F_{\mathfrak{h}_\gamma}$ for $\mathfrak{h}_\gamma(x) = \frac{1}{\log^\gamma(\frac{1}{x})}$ and $\dim_H(E_\gamma) \le \frac{1}{2}$.

5.1. Upper bounds for classical Furstenberg-type sets

We begin with a preliminary lemma about a very well-distributed (mod 1) sequence.

Lemma 5.1. *For $n \in \mathbb{N}$ and any real number $x \in [0,1]$, there is a pair $0 \le j,k \le n-1$ such that*

$$\left| x - \left(\sqrt{2}\frac{k}{n} - \frac{j}{n}\right)\right| \le \frac{\log(n)}{n^2}.$$

This lemma is a consequence of Theorem 3.4 of [KN74], p125, in which an estimate is given about the discrepancy of the fractional part of the sequence $\{n\alpha\}_{n\in\mathbb{N}}$ where α is a irrational of a certain type. We also need to introduce the notion of G-sets, a common ingredient in the construction of Kakeya and Furstenberg sets.

Definition 5.2. A G-set is a compact set $E \subseteq \mathbb{R}^2$ which is contained in the strip $\{(x,y) \in \mathbb{R}^2 : 0 \le x \le 1\}$ such that for any $m \in [0,1]$ there is a line segment contained in E connecting $x = 0$ with $x = 1$ of slope m.

Given a line segment $\ell(x) = mx + b$, we define the δ-tube associated to ℓ as

$$S_\ell^\delta := \{(x,y) \in \mathbb{R}^2 : 0 \le x \le 1; |y - (mx+b)| \le \delta\}.$$

Theorem 5.3. *For $\alpha \in (0,1]$ and $\theta > 0$, define $h_\theta(x) = x^{\frac{1+3\alpha}{2}} \log^{-\theta}(\frac{1}{x})$. Then, if $\theta > \frac{1+3\alpha}{2}$, there exists a set $E \in F_\alpha$ with $\mathcal{H}^{h_\theta}(E) = 0$.*

Proof. Fix $n \in \mathbb{N}$ and let n_j be a sequence such that $n_{j+1} > n_j^j$. We consider T to be the set defined as follows:

$$T = \left\{ x \in \left[\frac{1}{4}, \frac{3}{4}\right] : \forall j \; \exists \; p, q \; ; q \leq n_j^\alpha ; \left| x - \frac{p}{q} \right| < \frac{1}{n_j^2} \right\}.$$

It can be seen that $\dim_H(T) = \alpha$. This is a version of Jarník's theorem on Diophantine Approximation (see [Wol99b], p. 10 and [Fal86], p. 134, Theorem 8.16(b)). If $\varphi(t) = \frac{1-t}{t\sqrt{2}}$ and $D = \varphi^{-1}\left(\left[\frac{1}{4}, \frac{3}{4}\right]\right)$, we have that $\varphi : D \to \left[\frac{1}{4}, \frac{3}{4}\right]$ is bi-Lipschitz. Therefore the set

$$T' = \left\{ t \in \mathbb{R} : \frac{1-t}{t\sqrt{2}} \in T \right\} = \varphi^{-1}(T)$$

also has Hausdorff dimension α. The main idea of our proof is to construct a set for which we have, essentially, a copy of T' in each direction and simultaneously keep some optimal covering property. Define, for each $n \in \mathbb{N}$,

$$\Gamma_n := \left\{ \frac{p}{q} \in \left[\frac{1}{4}, \frac{3}{4}\right], q \leq n^\alpha \right\} \quad \text{and} \quad Q_n = \left\{ t : \frac{1-t}{\sqrt{2}t} = \frac{p}{q} \in \Gamma_n \right\} = \varphi^{-1}(\Gamma_n).$$

To count the elements of Γ_n (and Q_n), we take into account that

$$\sum_{j=1}^{\lfloor n^\alpha \rfloor} j \leq \frac{1}{2} \lfloor n^\alpha \rfloor (\lfloor n^\alpha \rfloor + 1) \lesssim \lfloor n^\alpha \rfloor^2 \leq n^{2\alpha}.$$

Therefore, $\#(Q_n) \lesssim n^{2\alpha}$. For $0 \leq j, k \leq n-1$, define the line segments

$$\ell_{jk}(x) := (1-x)\frac{j}{n} + x\sqrt{2}\frac{k}{n} \text{ for } x \in [0,1],$$

and their δ_n-tubes $S_{\ell_{jk}}^{\delta_n}$ with $\delta_n = \frac{\log(n)}{n^2}$. We will use during the proof the notation S_{jk}^n instead of $S_{\ell_{jk}}^{\delta_n}$. Also define

$$G_n := \bigcup_{jk} S_{jk}^n. \tag{5.4}$$

Note that, by Lemma 5.1, all the G_n are G-sets. For each $t \in Q_n$, we look at the points $\ell_{jk}(t)$, and define the set $S(t) := \{\ell_{jk}(t)\}_{j,k=1}^n$. Clearly, $\#(S(t)) \leq n^2$. But if we note that, if $t \in Q_n$, then

$$0 \leq \frac{\ell_{jk}(t)}{t\sqrt{2}} = \frac{1-t}{t\sqrt{2}}\frac{j}{n} + \frac{k}{n} = \frac{p}{q}\frac{j}{n} + \frac{k}{n} = \frac{pj+kq}{nq} < 2,$$

we can bound $\#(S(t))$ by the number of non-negative rationals smaller than 2 of denominator qn. Since $q \leq n^\alpha$, we have $\#(S(t)) \leq n^{1+\alpha}$. Considering *all* the elements of Q_n, we obtain $\#\left(\bigcup_{t\in Q_n} S(t)\right) \lesssim n^{1+3\alpha}$. Let us define

$$\Lambda_n := \left\{(x,y)\in G_n : |x-t| \le \frac{\sqrt{2}}{n^2} \text{ for some } t\in Q_n\right\}. \tag{5.5}$$

Claim 5.4. *For each n, take $\delta_n = \frac{\log(n)}{n^2}$. Then Λ_n can be covered by L_n balls of radio δ_n with $L_n \lesssim n^{1+3\alpha}$.*

To see this, it suffices to set a parallelogram on each point of $S(t)$ for each t in Q_n. The lengths of the sides of the parallelogram are of order n^{-2} and $\frac{\log(n)}{n^2}$, so their diameter is bounded by a constant times $\frac{\log(n)}{n^2}$, which proves the claim.

We can now begin with the recursive construction that leads to the desired set. Let F_0 be a G-set written as

$$F_0 = \bigcup_{i=1}^{M_0} S_{\ell_i^0}^{\delta^0},$$

(the union of M_0 δ^0-thickened line segments $\ell_i^0 = m_i^0 + b_i^0$ with appropriate orientation). Each F_j to be constructed will be a G-set of the form

$$F_j := \bigcup_{i=1}^{M_j} S_{\ell_i^j}^{\delta^j}, \qquad \text{with } \ell_i^j = m_i^j + b_i^j.$$

Having constructed F_j, consider the M_j affine mappings

$$A_i^j : [0,1]\times[-1,1] \to S_{\ell_i^j}^{\delta^j} \qquad 1\le i\le M_j,$$

defined by

$$A_i^j \begin{pmatrix} x \\ y \end{pmatrix} = \begin{pmatrix} 1 & 0 \\ m_i^j & \delta^j \end{pmatrix}\begin{pmatrix} x \\ y \end{pmatrix} + \begin{pmatrix} 0 \\ b_i^j \end{pmatrix}.$$

Here is the key step: by the definition of T, we can choose the sequence n_j to grow as fast as we need (this will not be the case in the next section). For example, we can choose n_{j+1} large enough to satisfy

$$\log\log(n_{j+1}) > M_j \tag{5.6}$$

and apply A_i^j to the sets $G_{n_{j+1}}$ defined in (5.4) to obtain

$$F_{j+1} = \bigcup_{i=1}^{M_j} A_i^j(G_{n_{j+1}}).$$

Since $G_{n_{j+1}}$ is a union of thickened line segments, we have that

$$F_{j+1} = \bigcup_{i=1}^{M_{j+1}} S_{\ell_i^{j+1}}^{\delta^{j+1}},$$

for an appropriate choice of M_{j+1}, δ_{j+1} and M_{j+1} line segments ℓ_i^{j+1}. From the definition of the mappings A_i^j and since the set $G_{n_{j+1}}$ is a G-set, we conclude that F_{j+1} is also a G-set. Define

$$E_j := \{(x,y)\in F_j : x\in T'\}.$$

To cover E_j, we note that if $(x,y) \in E_j$, then $x \in T'$, and therefore there exists a rational $\frac{p}{q} \in \Gamma_{n_j}$ with

$$\frac{1}{n_j^2} > \left|\frac{1-x}{x\sqrt{2}} - \frac{p}{q}\right| = |\varphi(x) - \varphi(r)| \geq \frac{|x-r|}{\sqrt{2}}, \qquad \text{for some } r \in Q_{n_j}.$$

Therefore $(x,y) \in \bigcup_{i=1}^{M_{j-1}} A_i^{j-1}(\Lambda_{n_j})$, so we conclude that E_j can be covered by $M_{j-1}n_j^{1+3\alpha}$ balls of diameter at most $\frac{\log(n_j)}{n_j^2}$. Since we chose n_j such that $\log\log(n_j) > M_{j-1}$, we obtain that E_j admits a covering by $\log\log(n_j)n_j^{1+3\alpha}$ balls of the same diameter. Therefore, if we set $F = \bigcap_j F_j$ and $E := \{(x,y) \in F : x \in T'\}$ we obtain that

$$\begin{aligned}
\mathcal{H}_{\delta_j}^{h_\theta}(E) &\lesssim n_j^{1+3\alpha} \log\log(n_j) h_\theta\left(\frac{\log(n_j)}{n_j^2}\right) \\
&\lesssim n_j^{1+3\alpha} \log\log(n_j) \left(\frac{\log(n_j)}{n_j^2}\right)^{\frac{1+3\alpha}{2}} \log^{-\theta}\left(\frac{n_j^2}{\log(n_j)}\right) \\
&\lesssim \log\log(n_j) \log(n_j)^{\frac{1+3\alpha}{2}-\theta} \lesssim \log^{\frac{1+3\alpha}{2}+\varepsilon-\theta}(n_j)
\end{aligned}$$

for large enough j. Therefore, for any $\theta > \frac{1+3\alpha}{2}$, the last expression goes to zero. In addition, F is a G-set, so it must contain a line segment in each direction $m \in [0,1]$. If ℓ is such a line segment, then

$$\dim_H(\ell \cap E) = \dim_H(T') \geq \alpha.$$

The final set of the proposition is obtained by taking eight copies of E, rotated to achieve *all* the directions in $\mathbb{S}$. □

5.2. Upper bounds for very small Furstenberg-type sets

In this section we will focus on the class F_α at the endpoint $\alpha = 0$. Note that all preceding results involved only the case for which $\alpha > 0$. Introducing the generalized Hausdorff measures, we are able to handle an important class of Furstenberg type sets in F_0.

The idea is to follow the proof of Theorem 5.3. But in order to do that, we need to replace the set T by a generalized version of it. A naïve approach would be to replace the α power in the definition of T by a slower increasing function, like a logarithm. But in this case it is not clear that the set T fulfills the condition of having positive measure for the corresponding dimension function (recall that we want to construct a set in $F_{\mathfrak{h}_\gamma}$). More precisely, we will need the following lemma.

Lemma 5.5. *Let $r > 1$ and consider the sequence $\mathfrak{n} = \{n_j\}$ defined by $n_j = e^{\frac{1}{2}n_{j-1}^{\frac{4}{r}j}}$, the function $\mathfrak{f}(x) = \log(x^2)^{\frac{r}{2}}$ and the set*

$$T = \left\{x \in \left[\frac{1}{4}, \frac{3}{4}\right] \setminus \mathbb{Q} : \forall j\ \exists\ p,q\,; q \leq \mathfrak{f}(n_j); |x - \frac{p}{q}| < \frac{1}{n_j^2}\right\}.$$

Then we have that $\mathcal{H}^{\mathfrak{h}}(T) > 0$ for $\mathfrak{h}(x) = \frac{1}{\log(\frac{1}{x})}$.

This is the essential lemma for our construction. It is trivial that T is a set of Hausdorff dimension zero, but in order to use this set in each fiber of an $F_{\mathfrak{h}}$-set, we have to prove that T has positive $\mathcal{H}^{\mathfrak{h}}$-mass. This is the really difficult part. The proof is a more technical version of a classical result that can be found in [Fal03]. For the proof of our lemma, we refer to [MR12]. Both classical and generalized results are examples of Diophantine Approximation. We emphasize the following fact: in this case, the construction of this new set T, does not allow us, as in (5.6), to freely choose the sequence n_j. On one hand we need the sequence to be quickly increasing to prove that the desired set is small enough, but not arbitrarily fast, since on the other hand, we need to impose some control to be able to prove that the fiber has the appropriate *positive* measure.

With this lemma, we are able to prove the main result of this section. We have the next theorem.

Theorem 5.6. *Let* $\mathfrak{h} = \frac{1}{\log(\frac{1}{x})}$. *There exists a set* $E \in F_{\mathfrak{h}}$ *such that* $\dim_H(E) \leq \frac{1}{2}$.

Proof. We will use essentially a copy of T in each direction in the construction of the desired set to fulfill the conditions required to be an $F_{\mathfrak{h}}$-set. Let T be the set defined in Lemma 5.5. Define $T' = \varphi^{-1}(T)$, where φ is the same bi-Lipschitz function from the proof of Theorem 5.3. Then T' has positive $\mathcal{H}^{\mathfrak{h}}$-measure.

Let us define the corresponding sets of Theorem 5.3 for this generalized case. For $\mathfrak{f}(x) = \mathfrak{f}(x) = \log(x^2)^{\frac{r}{2}}$, define

$$\Gamma_n := \left\{\frac{p}{q} \in \left[\frac{1}{4}, \frac{3}{4}\right], q \leq \mathfrak{f}(n)\right\}, \qquad Q_n = \left\{t : \frac{1-t}{\sqrt{2}t} = \frac{p}{q} \in \Gamma_n\right\} = \varphi^{-1}(\Gamma_n).$$

Now the estimate is $\#(Q_n) \lesssim \mathfrak{f}^2(n) = \log^r(n^2) \sim \log^r(n)$. For each $t \in Q_n$, define $S(t) := \{\ell_{jk}(t)\}_{j,k=1}^{n}$. If $t \in Q_n$, following the previous ideas, we obtain that $\#(S(t)) \lesssim n \log^{\frac{r}{2}}(n)$, and therefore

$$\#\Bigg(\bigcup_{t\in Q_n} S(t)\Bigg) \lesssim n \log(n)^{\frac{3r}{2}}.$$

Now we estimate the size of a covering of the set Λ_n in (5.5). For each n, take $\delta_n = \frac{\log(n)}{n^2}$. As before, the set Λ_n can be covered with L_n balls of radio δ_n with $L_n \lesssim n \log(n)^{\frac{3r}{2}}$.

Once again, define F_j, F, E_j and E as before. Now the sets F_j can be covered by less than $M_{j-1} n_j \log(n_j)^{\frac{3r}{2}}$ balls of diameter at most $\frac{\log(n_j)}{n_j^2}$. Now we can verify that, since each G_n consist of n^2 tubes, we have that $M_j = M_0 n_1^2 \cdots n_j^2$. We can also verify that the sequence $\{n_j\}$ satisfies the relation $\log n_{j+1} \geq M_j = M_0 n_1^2 \cdots n_j^2$, and therefore we have the bound

$$\dim_H(E) \leq \underline{\dim}_B(E) \leq \underline{\lim}_j \frac{\log\left(\log(n_j) n_j \log(n_j)^{\frac{3r}{2}}\right)}{\log\left(n_j^2 \log^{-1}(n_j)\right)} = \frac{1}{2},$$

where $\underline{\dim}_B$ stands for the lower box dimension. Finally, for any $m \in [0,1]$ we have a line segment ℓ with slope m contained in F. It follows that $\mathcal{H}^{\mathfrak{h}}(\ell \cap E) = \mathcal{H}^{\mathfrak{h}}(T') > 0$. □

We remark that the argument in this particular result is essentially the same needed to obtain the family of Furstenberg sets $E_\gamma \in F_{\mathfrak{h}_\gamma}$ for $\mathfrak{h}_\gamma(x) = \frac{1}{\log^\gamma(\frac{1}{x})}$, $\gamma \in \mathbb{R}_+$, such that $\dim_H(E_\gamma) \leq \frac{1}{2}$ announced in the introduction.

5.3. The case $\alpha = 0$, K points

Let us begin with the definition of the class F^K.

Definition 5.7. For $K \in \mathbb{N}$, $K \geq 2$, a set will be an F^K-set or a Furstenberg set of type K if for any direction $e \in \mathbb{S}$, there are at least K points contained in E lined up in the direction of e.

Already in [MR10] we proved that there is an F^2-set with zero Hausdorff dimension (see also [Fal03], Example 7.8). We will generalize this example to obtain even smaller F^2 sets. Namely, for any $h \in \mathbb{H}_0$, there exists G in F^2 such that $\mathcal{H}^h(G) = 0$. It is clear that the set G will depend on the choice of h.

Example 5.8. Given a function $h \in \mathbb{H}$, we will construct two small sets $E, F \subseteq [0,1]$ with $\mathcal{H}^h(E) = \mathcal{H}^h(F) = 0$ and such that $[0,1] \subseteq E + F$. Consider now $G = E \times \{1\} \cup -F \times \{0\}$. Clearly, we have that $\mathcal{H}^h(G) = 0$, and contains two points in every direction $\theta \in [0; \frac{\pi}{4}]$. For, if $\theta \in [0; \frac{\pi}{4}]$, let $c = \tan(\theta)$, so $c \in [0,1]$. By the choice of E and F, we can find $x \in E$ and $y \in F$ with $c = x + y$. The points $(-y, 0)$ and $(x, 1)$ belong to G and determine a segment in the direction θ.

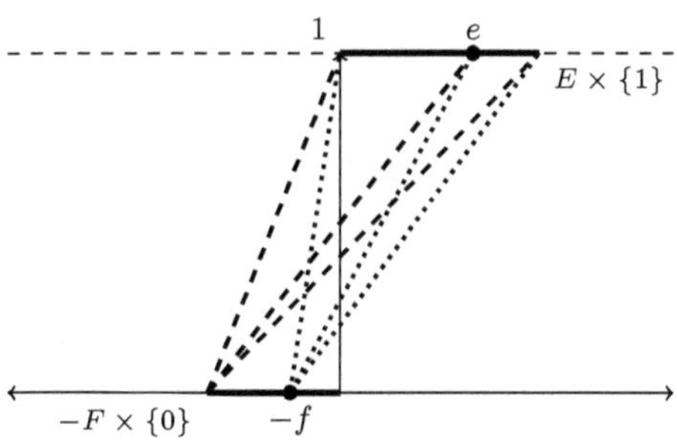

FIGURE 4. An F^2-set of zero $\mathcal{H}^h$-measure

For $x \in [0,1]$, we consider its binary representation $x = \sum_{j \geq 1} r_j 2^{-j}$, $r_j \in \{0,1\}$. We define $E := \{x \in [0,1] : r_j = 0 \text{ if } m_k + 1 \leq j \leq m_{k+1}; k \text{ even}\}$ and $F := \{x \in [0,1] : r_j = 0 \text{ if } m_k + 1 \leq j \leq m_{k+1}; k \text{ odd}\}$. Here $\{m_k; m_0 = 0\}_k$ is an increasing sequence such that $m_k \to +\infty$. Now we estimate the size of the set E. Given $k \in \mathbb{N}$, k even, define $\ell_k = m_k - m_{k-1} + \cdots + m_2 - m_1$. It is clear that E can be covered by 2^{l_k} intervals of length $2^{-m_{k+1}}$. Therefore, if the sequence m_k increases fast enough, then $\dim_H(E) \le \underline{\dim}_B(E) \le \underline{\lim}_k \frac{\log(2^{\ell_k})}{\log(2^{m_{k+1}})} \lesssim \underline{\lim}_k \frac{2^{\ell_k}}{2^{m_{k+1}}} = 0$. Since the same argument shows that $\dim_H(F) = 0$, this estimate proves that the set G has Hausdorff dimension equal to zero. Now, for the finer estimate on the $\mathcal{H}^h$-measure of the set, we must impose a more restrictive condition on the sequence $\{m_k\}$. Recall that the covering property implies that, for a given $h \in \mathbb{H}$, we have $\mathcal{H}^h(E) \le 2^{l_k} h(2^{-m_{k+1}})$. Therefore we need to choose a sequence $\{m_j\}$, depending on h, such that the above quantity goes to zero with k. Since $\ell_k \le m_k$, we can define recursively the sequence $\{m_k\}$ to satisfy the relation $2^{m_k} h(2^{-m_{k+1}}) = \frac{1}{k}$. This last condition is equivalent to $m_{k+1} = \log\left(\frac{1}{h^{-1}(\frac{1}{k2^{m_k}})}\right)$. As a concrete example, take $h(x) = \frac{1}{\log(\frac{1}{x})}$. In this case we obtain that the sequence $\{m_k\}$ can be defined as $m_{k+1} = k2^{m_k}$.

5.4. Remark about the packing dimension for small Furstenberg sets

It is worth noting here that if we were to measure the size of Furstenberg sets with the packing dimension, the situation would be absolutely different. More precisely, for $K \ge 2$, any F^K-set $E \subset \mathbb{R}^2$ must have $\dim_P(E) \ge \frac{1}{2}$. For, if E is an F^2 set, then the map φ defined by $\varphi(a,b) = \frac{a-b}{\|a-b\|}$ is Lipschitz when restricted to $G_\varepsilon := E \times E \setminus \{(x,y) \in E \times E : \|(x,y) - (a,a)\| < \varepsilon; a \in E\}$. Roughly, we are considering the map that recovers the set of directions but restricted "off the diagonal". It is clear that we can assume without loss of generality that all the pairs are the endpoints of unit line segments. Therefore, since E is an F^K-set, $\varphi(G_\varepsilon) = \mathbb{S}$ if ε is small enough. We obtain the inequality

$$1 = \dim_H(\mathbb{S}) \le \dim_H(G_\varepsilon) \le \dim_H(E \times E).$$

The key point is the product formulae for Hausdorff and Packing dimensions. We obtain that

$$1 \le \dim_H(E \times E) \le \dim_H(E) + \dim_P(E) \le 2\dim_P(E)$$

and then $\dim_P(E) \ge \frac{1}{2}$. It also follows that if we achieve small Hausdorff dimension then the Packing dimension is forced to increase. In particular, the F^2-set constructed in [MR10] has Hausdorff dimension 0 and therefore it has Packing dimension 1.

The following construction can be understood as optimal in the sense of obtaining the smallest possible dimensions, both Hausdorff and Packing. There is an F^2 set E such that $\dim_H(E) = \frac{1}{2} = \dim_P(E)$:

Example 5.9. The construction is essentially the same as in Example 5.8, but we use two different sets to obtain all directions. Let A be the set of all the numbers whose expansion in base 4 use only the digits 0 and 1. On the other hand, let B be the set of those numbers which only use the digits 0 and 2. Both sets have Packing and Hausdorff dimension equal to $\frac{1}{2}$ and $[0,1] \subseteq A + B$. The construction follows then the same pattern as in the previous example.

Acknowledgement

This expository article was completed during my stay at the Departamento de Análisis Matemático, Universidad de Sevilla. I am deeply grateful in particular to Professor Carlos Pérez Moreno for his hospitality. I would also like to thank my Ph.D. advisor Ursula Molter from Universidad de Buenos Aires for her guidance and support.

References

[Bes19] A.S. Besicovitch, *Sur deux questions d'intégrabilité des fonctions*, J. Soc. Phys.-Math. (Perm') **2** (1919), 105–123.

[Bes28] ______, *On Kakeya's problem and a similar one*, Math. Z. **27** (1928), no. 1, 312–320.

[Bes56a] ______, *On density of perfect sets*, J. London Math. Soc. **31** (1956), 48–53.

[Bes56b] ______, *On the definition of tangents to sets of infinite linear measure*, Proc. Cambridge Philos. Soc. **52** (1956), 20–29.

[Bou94] Jean Bourgain, *Hausdorff dimension and distance sets*, Israel J. Math. **87** (1994), no. 1-3, 193–201.

[CHM10] Carlos A. Cabrelli, Kathryn E. Hare, and Ursula M. Molter, *Classifying Cantor sets by their fractal dimensions*, Proc. Amer. Math. Soc. **138** (2010), no. 11, 3965–3974.

[CMMS04] Carlos Cabrelli, Franklin Mendivil, Ursula Molter, and Ronald Shonkwiler, *On the Hausdorff h-measure of Cantor sets*, Pacific J. Math. **217** (2004), no. 1, 45–59.

[Dav71] Roy O. Davies, *Some remarks on the Kakeya problem*, Proc. Cambridge Philos. Soc. **69** (1971), 417–421.

[Egg52] H.G. Eggleston, *Sets of fractional dimensions which occur in some problems of number theory*, Proc. London Math. Soc. (2) **54** (1952), 42–93.

[EK06] Márton Elekes and Tamás Keleti, *Borel sets which are null or non-σ-finite for every translation invariant measure*, Adv. Math. **201** (2006), no. 1, 102–115.

[Fal86] K. Falconer, *The geometry of fractal sets*, Cambridge Tracts in Mathematics, vol. 85, Cambridge University Press, Cambridge, 1986.

[Fal03] Kenneth Falconer, *Fractal geometry*, second ed., John Wiley & Sons Inc., Hoboken, NJ, 2003, Mathematical foundations and applications.

[FK17] M. Fujiwara and S. Kakeya, *On some problems of maxima and minima for the curve of constant breadth and the in-revolvable curve of the equilateral triangle*, Tohoku Mathematical Journal **11** (1917), 92–110.

[Fur70] Harry Furstenberg, *Intersections of Cantor sets and transversality of semigroups*, Problems in analysis (Sympos. Salomon Bochner, Princeton Univ., Princeton, N.J., 1969), Princeton Univ. Press, Princeton, N.J., 1970, pp. 41–59.

[GMS07] Ignacio Garcia, Ursula Molter, and Roberto Scotto, *Dimension functions of Cantor sets*, Proc. Amer. Math. Soc. **135** (2007), no. 10, 3151–3161 (electronic).

[Hau18] Felix Hausdorff, *Dimension und äußeres Maß*, Math. Ann. **79** (1918), no. 1-2, 157–179.

[Kei99] U. Keich, *On L^p bounds for Kakeya maximal functions and the Minkowski dimension in* $\mathbf{R}^2$, Bull. London Math. Soc. **31** (1999), no. 2, 213–221.

[KN74] L. Kuipers and H. Niederreiter, *Uniform distribution of sequences*, Wiley-Interscience [John Wiley & Sons], New York, 1974, Pure and Applied Mathematics.

[KT01] Nets Katz and Terence Tao, *Some connections between Falconer's distance set conjecture and sets of Furstenburg type*, New York J. Math. **7** (2001), 149–187 (electronic).

[Mat87] Pertti Mattila, *Spherical averages of Fourier transforms of measures with finite energy; dimension of intersections and distance sets*, Mathematika **34** (1987), no. 2, 207–228.

[Mat95] ———, *Geometry of sets and measures in Euclidean spaces*, Cambridge Studies in Advanced Mathematics, vol. 44, Cambridge University Press, Cambridge, 1995, Fractals and rectifiability.

[Mit02] Themis Mitsis, *Norm estimates for the Kakeya maximal function with respect to general measures*, Real Anal. Exchange **27** (2001/02), no. 2, 563–572.

[MR10] Ursula Molter and Ezequiel Rela, *Improving dimension estimates for Furstenberg-type sets*, Adv. Math. **223** (2010), no. 2, 672–688.

[MR12] ———, *Furstenberg sets for a fractal set of directions*, Proc. Amer. Math. Soc. **140** (2012), no. 8, 2753–2765.

[MR13] ———, *Small Furstenberg sets*, J. Math. Anal. Appl. **400** (2013), no. 2, 475–486.

[OR06] L. Olsen and Dave L. Renfro, *On the exact Hausdorff dimension of the set of Liouville numbers. II*, Manuscripta Math. **119** (2006), no. 2, 217–224.

[Rog70] C.A. Rogers, *Hausdorff measures*, Cambridge University Press, London, 1970.

[Wol99a] Thomas Wolff, *Decay of circular means of Fourier transforms of measures*, Internat. Math. Res. Notices (1999), no. 10, 547–567.

[Wol99b] ———, *Recent work connected with the Kakeya problem*, Prospects in mathematics (Princeton, NJ, 1996), Amer. Math. Soc., Providence, RI, 1999, pp. 129–162.

[Wol02] ———, *Addendum to: "Decay of circular means of Fourier transforms of measures" [Internat. Math. Res. Notices* **1999**, *no.* 10, 547–567], J. Anal. Math. **88** (2002), 35–39, Dedicated to the memory of Tom Wolff.

Ezequiel Rela
Departamento de Análisis Matemático
Facultad de Matemáticas
Universidad de Sevilla
E-41080 Sevilla, Spain
e-mail: `erela@us.es`

Operator Theory:
Advances and Applications, Vol. 236, 455–461

Singular Degenerate Problems Occurring in Atmospheric Dispersion of Pollutants

Aida Sahmurova and Veli B. Shakhmurov

Abstract. The boundary value problems for singular degenerate linear and regular degenerate nonlinear differential-operator equations are studied. We prove the well-posedeness of the linear problem and optimal regularity result for the nonlinear problem which occur in fluid mechanics, environmental engineering and in the atmospheric dispersion of pollutants.

Mathematics Subject Classification (2010). Primary 34G10; Secondary 35J25, 35J70.

Keywords. Differential-operator equations, semigroups of operators, Banach-valued function spaces, separability, degenerate equations, interpolation of Banach spaces, atmospheric dispersion of pollutants.

1. Introduction, notations and background

The maximal regularity properties of boundary value problems (BVPs) for linear differential-operator equations (DOEs) have been studied by many researchers (see, e.g., [1 – 5] and the references therein).

Let $\gamma = \gamma(x)$, $x = (x_1, x_2, \ldots, x_n)$ be a positive measurable function on a domain $\Omega \subset R^n$. Let $L_{p,\gamma}(\Omega; E)$ denote the space of strongly measurable E-valued functions that are defined on Ω with the norm

$$\|f\|_{L_{p,\gamma}} = \|f\|_{L_{p,\gamma}(\Omega;E)} = \left(\int \|f(x)\|_E^p \gamma(x)\,dx\right)^{\frac{1}{p}}, \; 1 \leq p < \infty.$$

For $\gamma(x) \equiv 1$, $L_{p,\gamma}(\Omega; E)$ will be denoted by $L_p = L_p(\Omega; E)$. Let $\mathbf{C}$ be the set of the complex numbers and

$$S_\varphi = \{\lambda; \quad \lambda \in \mathbf{C}, \; |\arg \lambda| \leq \varphi\} \cup \{0\}, \; 0 \leq \varphi < \pi.$$

A linear operator A is said to be φ-positive in a Banach space E with bound $M > 0$ if $D(A)$ is dense on E and $\left\|(A + \lambda I)^{-1}\right\|_{B(E)} \leq M(1 + |\lambda|)^{-1}$ for any

$\lambda \in S_\varphi$, $0 \leq \varphi < \pi$, where I is the identity operator in E and $B(E)$ is the space of bounded linear operators in E.

The φ-positive operator A is said to be R-positive in a Banach space E if the set $L_A = \left\{\xi (A + \xi I)^{-1} : \xi \in S_\varphi\right\}$, $0 \leq \varphi < \pi$ is R-bounded (see, e.g., [2]). Let $E(A)$ denote the space $D(A)$ with norm

$$\|u\|_{E(A)} = \left(\|u\|^p + \|Au\|^p\right)^{\frac{1}{p}}, 1 \leq p < \infty.$$

Let E_0 and E be two Banach spaces and E_0 continuously and densely embedded into E. Let $\gamma = \gamma(x)$ be a weight function on (a, b) and $u^{[i]} = \left(\gamma(x) \frac{d}{dx}\right)^i u(x)$.

We define the following E-valued function spaces

$$W_{p,\gamma}^{[m]}(a, b; E_0, E) = \Big\{u \in L_p(a, b; E_0), \quad u^{[2]} \in L_p(a, b; E),$$
$$\|u\|_{W_{p,\gamma}^{[m]}(a,b;E_0,E)} = \|u\|_{L_p(a,b;E_0)} + \left\|u^{[2]}\right\|_{L_p(a,b;E)}\Big\},$$
$$W_{p,\gamma}^{m}(a, b; E_0, E) = \Big\{u \in L_{p,\gamma}(a, b; E_0), \quad u^{(2)} \in L_{p,\gamma}(a, b; E),$$
$$\|u\|_{W_{p,\gamma}^{m}(a,b;E_0,E)} = \|u\|_{L_{p,\gamma}(a,b;E_0)} + \left\|u^{(2)}\right\|_{L_{p,\gamma}(a,b;E)}\Big\}.$$

2. Linear degenerate DOEs

Consider the BVP for the singular degenerate differential-operator equation

$$-x^{2\alpha}\frac{\partial^2 u}{\partial x^2} - x^{2\beta}\frac{\partial^2 u}{\partial y^2} + Au + \lambda u = f(x, y), \tag{2.1}$$
$$L_1 u = \sum_{i=0}^{1} \delta_{1i} u_x^{[i]}(a, y) = 0,$$
$$L_2 u = \sum_{i=0}^{1} \delta_{2i} u_y^{[i]}(x, b) = 0$$

on the domain $G = (0, a) \times (0, b)$, where

$$u = u(x, y), \quad u_x^{[i]} = \left[x^\alpha \frac{\partial}{\partial x}\right]^i u, \quad u_y^{[i]} = \left[y^\beta \frac{\partial}{\partial y}\right]^i u;$$

δ_{ki} are complex numbers, λ is a complex parameter, A is a linear operator in a Banach space E. Let $W_{p,\alpha,\beta}^{[2]}(G; E(A), E)$ be an E-valued function space defined by

$$W_{p,\alpha,\beta}^{[2]}(G; E(A), E) = \Big\{u \in L_p(G; E(A)), \ u_x^{[2]} \in L_p(G; E), \ u_y^{[2]} \in L_p(G; E),$$
$$\|u\|_{W_{p,\alpha,\beta}^{[2]}(G;E(A),E)} = \|u\|_{L_p(G;E(A))} + \left\|u_x^{[2]}\right\|_{L_p(G;E)} + \left\|u_y^{[2]}\right\|_{L_p(G;E)} < \infty\Big\}.$$

Let $\delta_{k1} \neq 0$. The main result is the following:

Theorem (1). *Let E be an UMD space (see[2]), A be an R-positive operator in E and $1+\frac{1}{p} < \alpha, \beta < \frac{(p-1)}{2}$. Then, problem (1) has a unique solution $u \in W^{[2]}_{p,\alpha,\beta}(G; E(A), E)$ for all $f \in L_p(G; E)$ and $|\arg \lambda| \leq \varphi$ with sufficiently large $|\lambda|$. Moreover, the following coercive uniform estimate holds*

$$\sum_{i=0}^{2} |\lambda|^{1-\frac{i}{2}} \left[\left\| x^{i\alpha} \frac{\partial^i u}{\partial x^i} \right\|_{L_p(G;E)} + \left\| y^{i\beta} \frac{\partial^i u}{\partial y^i} \right\|_{L_p(G;E)} \right] + \|Au\|_{L_p(G;E)} \leq M \|f\|_{L_p(G;E)} . \tag{2.2}$$

For proving the main theorem, consider at first the BVP for the singular degenerate ordinary DOE

$$(L+\lambda) u = -u^{[2]}(x) + (A+\lambda) u(x) = f, \quad L_1 u = \sum_{i=0}^{1} \delta_i u^{[i]}(a) = 0, \tag{2.3}$$

where $u^{[i]} = \left[x^\alpha \frac{d}{dx}\right]^i u(x)$, δ_i are complex numbers and A is a linear operator in E, $\delta_1 \neq 0$.

In a similar way as [4, Theorem 5.1] we obtain

Theorem ($\mathbf{A_1}$). *Suppose E is an UMD space, A is an R positive operator in E, $1+\frac{1}{p} < \alpha < \frac{(p-1)}{2}$, $p \in (1, \infty)$. Then, problem (3) has a unique solution $u \in W^{[2]}_{p,\alpha}(0, a; E(A), E)$ for all $f \in L_p(G; E)$ and $|\arg \lambda| \leq \varphi$ with sufficiently large $|\lambda|$. Moreover, the following uniform coercive estimate holds*

$$\sum_{i=0}^{2} |\lambda|^{1-\frac{i}{2}} \left\|u^{[i]}\right\|_{L_p(0,a;E)} + \|Au\|_{L_p(0,a;E)} \leq C \|f\|_{L_p(0,a;E)} . \tag{2.4}$$

Let B denote the operator generated by problem (3), i.e.,

$$D(B) = \left\{u \in W^{[2]}_{p,\alpha}(0, a; E(A), E) \ L_1 u = 0\right\}, \quad Bu = -u^{[2]} + Au.$$

Theorem A_1 implies that the operator B is positive so, by the Yosida theorem (see, e.g., [3, § 5, Theorem 5.6] $-B^{\frac{1}{2}}$ is a generator of an analytic semigroup in $L_p(0, a; E)$. By reasoning as in [3, Theorem 3.1] we obtain:

Theorem ($\mathbf{A_2}$). *Let all conditions of Theorem A_1 be satisfied. Then, the operator B is R-positive in $L_p(0, a; E)$.*

Consider now the following degenerate DOE with the boundary conditions (3):

$$-x^{2\alpha} u^{(2)}(x) + (A+\lambda) u(x) = f, \ L_1 u = 0. \tag{2.5}$$

Theorem ($\mathbf{A_3}$). *Assume all conditions of Theorem* A_1 *are satisfied. Then, problem* (5) *has a unique solution* $u \in W^{[2]}_{p,\alpha}(0,a;E(A),E)$ *for all* $f \in L_p(0,a;E)$ *and* $|\arg\lambda| \leq \varphi$ *with sufficiently large* $|\lambda|$. *Moreover, the following coercive estimate holds*

$$\sum_{i=0}^{2} |\lambda|^{1-\frac{i}{2}} \left\| x^{i\alpha} u^{(i)} \right\|_{L_p(0,a;E)} + \|Au\|_{L_p(0,a;E)} \leq C \|f\|_{L_p(0,a;E)}. \tag{2.6}$$

Proof. Since $\alpha > 1$, by [4, Theorem 2.3] for every $\varepsilon > 0$ there exists a constant $C(\varepsilon)$ such that

$$\left\| \alpha x^{\alpha-1} u^{[1]} \right\|_{L_p(0,a;E)} \leq \varepsilon \|u\|_{W^{[2]}_{p,\alpha}(0,a;E(A),E)} + C(\varepsilon) \|u\|_{L_p(0,a;E)}. \tag{2.7}$$

Then in view of (7) and Theorem A_1 we get

$$\left\| \alpha x^{\alpha-1} u^{[1]} \right\|_{L_p(0,a;E)} \leq \varepsilon \|(B+\lambda)u\|_{L_p(0,a;E)} + C(\varepsilon) \|u\|_{L_p(0,a;E)}.$$

Moreover, due to positivity of operator B we have

$$\|u\|_{L_p(0,a;E)} \leq \frac{M}{|\lambda|} \|(B+\lambda)u\|_{L_p(0,a;E)}.$$

Then, for a sufficiently large $|\lambda|$ by choosing $\frac{C(\varepsilon)M}{|\lambda|} < \varepsilon$ we obtain from the above two estimates

$$\left\| \alpha x^{\alpha-1} u^{[1]} \right\|_{L_p(0,a;E)} \leq 2\varepsilon \|(B+\lambda)u\|_{L_p(0,a;E)}. \tag{2.8}$$

Since $-x^{2\alpha}u^{(2)} = -u^{[2]} + \alpha x^{\alpha-1}u^{[1]}$, the assertion is obtained from Theorem A_1 and the estimate (8). □

Let Q denote the operator generated by problem (5), i.e.,

$$D(Q) = \left\{ u \in W^{[2]}_{p,\alpha}((0,a;E(A),E)),\ L_1 u = 0 \right\}, Qu = -x^{2\alpha}u^{(2)} + Au.$$

Theorem ($\mathbf{A_4}$). *Assume all conditions of Theorem* A_1 *are satisfied. Then, the operator* $Q + \mu$ *is* R*-positive in* $L_p(0,a;E)$ *for a sufficiently large* $\mu > 0$.

Proof. Indeed, since $(Q+\mu)u = (B+\mu)u + \alpha x^{\alpha-1}u^{[1]}$ from the Theorem A_2, estimate (8) and by perturbation properties of R-positive operators (e.g., see [2, Proposition 4,3] we obtain that the operator $Q+\mu$ is R-positive for the sufficiently large μ. □

Proof of Theorem 1. Since $L_p(0,b;L_p(0,a;E)) = L_p(G;E)$ the problem (1) can be expressed as:

$$-x^{2\beta}\frac{d^2u}{dy^2} + (B+\lambda)u(y) = f(x),\ L_2 u = 0. \tag{2.9}$$

By virtue of [1, Theorem 4.5.2], $F = L_p(0,b;E) \in UMD$ provided $E \in UMD$, $p \in (1,\infty)$. By Theorem A_2, the operator B is R-positive in F. Then by virtue of Theorem A3, for $f \in L_p(0,a;F) = L_p(G;E)$ problem (9) has a unique solution $u \in W^{[2]}_{p,\alpha,\beta} = W^2_{p,\beta}(0,a;D(S),F)$ and the operator Q generated by problem (8) has a bounded inverse from $L_p(G;E)$ to $W^{[2]}_{p,\alpha,\beta}$. So, we obtain the assertion. □

3. Nonlinear degenerate DOE

Consider now the following regular degenerate nonlinear problem

$$-D^{[2]}_x u - D^{[2]}_y u + \Phi\left(x,y,u,u^{[1]}_x,u^{[1]}_y\right) u = F\left(x,y,u,u^{[1]}_x,u^{[1]}_y\right), \tag{3.1}$$

$$L_{1k}u = \sum_{i=0}^{m_{1k}} \alpha_{1ki} u^{[i]}_x(0,y) + \beta_{1ki} u^{[i]}_x(a,y) = 0,$$

$$L_{2k}u = \sum_{i=0}^{m_{2k}} \alpha_{2ki} u^{[i]}_y(x,0) + \beta_{2ki} u^{[i]}_y(x,b) = 0$$

on the domain $G = (0,a)\times(0,b)$, $a \in (0,a_0)$, $b \in (0,b_0)$, where $m_{jk} \in \{0,1\}$ and $k,\ j = 1,2$. Let

$$Y = W^{[2]}_{p,\alpha,\beta}(G;E(A),E), \quad X_1 = L_p(0,a;E),$$

$$Y_1 = W^{[2]}_{p,\alpha}(0,a;E(A),E), \quad X_2 = L_p(0,b;E),$$

$$Y_2 = W^{[2]}_{p,\beta}(0,b;E(A),E), \quad E_k = (Y_{3-k},X_{3-k})_{\theta_k,p},$$

$$E_0 = E_1^2 \times E_2, \quad \theta_1 = \frac{1}{2} + \frac{1}{2p(1-\alpha)}, \quad \theta_2 = \frac{1}{2} + \frac{1}{2p(1-\beta)},$$

$$0 < \alpha,\beta < 1 - \frac{1}{p}, \quad \alpha_{kj} = \alpha_{kjm_k}, \quad \beta_{kj} = \beta_{kjm_k}, \quad k,j = 1,2,$$

where $(Y_i,X_i)_{\theta,p}$, $\theta \in (0,1)$ and $p \in [1,\infty]$ denote the real interpolation spaces [6, §1.3.2].

Remark 3.1. By using J. Lions-I. Petree trace theorem (see, e.g., [6, § 1.8.]) we obtain that there are constants M_k such that

$$\sup_{x\in[0,a]} \|w\|_{E_1} \le M_1 \|w\|_{W^{[2]}_{p,\alpha,\beta}(G;E(A),E)},$$

$$\sup_{x\in[0,a]} \left\|w^{[1]}_x\right\|_{E_1} \le M_3 \|w\|_{W^{[2]}_{p,\alpha,\beta}(G;E(A),E)},$$

$$\sup_{y\in[0,b]} \left\|w^{[1]}_y\right\|_{E_2} \le M_2 \|w\|_{W^{[2]}_{p,\alpha,\beta}(G;E(A),E)}.$$

Condition 1. Assume the following satisfied:

(1) E is an UMD space, for $V = \{\upsilon_0,\upsilon_1,\upsilon_2\}$, $\upsilon_0 \in E_1$, $\upsilon_k \in E_k$, $\Phi(x,y,V) = A(x,y) = A$ is a R-positive operator in E uniformly with respect to V and $x,y \in G_0$, where $D(A(x,y))$ does not depend on V and x,y;

(2) $\Phi : G_0 \times E_0 \to B(E(A), E)$ is continuous and
$$(-1)^{m_{k1}} \alpha_{k1}\beta_{k2} - (-1)^{m_{k2}} \alpha_{k2}\beta_{k1} \neq 0,\ k = 1, 2;$$
(3) the function $F\left(x, y, u(x,y), u_x^{[1]}(x,y), u_y^{[1]}(x,y)\right)$ is a measurable function for $u \in Y$; $F(x, y, .)$ is continuous with respect to $x, y \in G_0$ and $F(x,y,0) \in L_p(G; E)$. Moreover, for each $r > 0$ there exists $\mu(r)$ such that
$$\left\|F(x,U) - F\left(x,\bar{U}\right)\right\|_E \leq \mu(r)\left\|U - \bar{U}\right\|_{E_0}$$
for a.a. $x, y \in G_0$, $U = \{u_0, u_1, u_2\}$, $\bar{U} = \{\bar{u}_0, \bar{u}_1, \bar{u}_2\}$, $v_0, \bar{v}_0 \in E_1$, $u_k, \bar{u}_k \in E_k$, $k = 1, 2$ and $\|U\|_{E_0} \leq r$, $\left\|\bar{U}\right\|_{E_0} \leq r$, $0 < \alpha, \beta < 1 - \frac{1}{p}$, $1 < p < \infty$;
(4) moreover, for each $r > 0$ there is a positive constant $L(r)$ such that
$$\left\|\left[\Phi(x,y,U) - \Phi\left(x,y,\bar{U}\right)\right] v\right\|_E \leq L(r)\left\|U - \bar{U}\right\|_{E_0} \|Av\|_E$$
for $x, y \in G_0$, $\|U\|_{E_0}, \left\|\bar{U}\right\|_{E_0} \leq r$ and $v \in D(A(x,y))$.

We prove here the existence and uniqueness of the nonlinear problem (11).

Theorem (2). *Let Condition* 1 *hold. Then there is* $a \in (0\ a_0]$, $b \in (0\ b_0]$ *such that problem* (11) *has a unique solution that belongs to* $W^{[2]}_{p,\alpha,\beta}((G; E(A), E)$.

Proof. By virtue of [4, Theorem 3] the regular degenerate linear problem $-D_x^{[2]}u - D_y^{[2]}u + Au = f, L_{kj}u = 0$ is maximal regular in L_p, where $L_{kj}u$ are the same boundary conditions as in (11). We solve the problem (11) locally by means of maximal regularity of this linear problem via the contraction mapping theorem. Consider a ball $B_r = \{v \in Y,\ \|v\|_Y \leq r\}$. Define a map U on B_r by $Uv = w$, where w is a solution of the linearized problem. By using the coercivity of the above linear problem and the Remark 1 we show that $U(B_r) \subset B_r$ and U is a contraction operator in Y. This fact implies the assertion. □

Remark 3.2. There are a lot of positive operators in concrete Banach spaces. Therefore, putting concrete Banach spaces instead of E and concrete differential, pseudo differential operators, or finite, infinite matrices, etc. instead of operator A on equations (1) and (11) we can obtain the maximal regularity properties of different class of linear and nonlinear differential equations or system of equations by virtue of Theorem 1 and Theorem 2.

Acknowledgment

The authors would like to express a deep gratitude to the Reviewer for his useful advices and extensive reports.

References

[1] Amann H., *Linear and quasi-linear equations, 1*, Birkhäuser, Basel 1995.

[2] Denk R., Hieber M., Prüss J., *R-boundedness, Fourier multipliers and problems of elliptic and parabolic type*, Mem. Amer. Math. Soc. **166** (2003), n. 788.

[3] Shakhmurov V.B., Shahmurova A., *Nonlinear abstract boundary value problems atmospheric dispersion of pollutants*, Nonlinear Analysis, World Applications, **11** no. 2 (2010), 932–951.

[4] Shakhmurov V.B., *Coercive boundary value problems for regular degenerate differential-operator equations*, Journal of Math. Anal. Appl., **292** no. 2 (2004), 605–620.

[5] Shakhmurov V.B., *Linear and nonlinear abstract equations with parameters*, Nonlinear Analysis, Method and Applications, **73** (2010), 2383–2397.

[6] Triebel H., *Interpolation theory, Function spaces, Differential operators*, North-Holland, Amsterdam, 1978.

Aida Sahmurova
Department of Environmental Sciences
Okan University
Akfirat, Tuzla
34959 Istanbul, Turkey
e-mail: `aida.sahmurova@okan.edu.tr`

Veli B. Shakhmurov
Engineering Faculty, Okan University
Akfirat, Tuzla
34959 Istanbul, Turkey
e-mail: `veli.sahmurov@okan.edu.tr`

[illegible]

[illegible] Nonlinear Analysis, Real World Applications [illegible]

[illegible] Math. Anal. Appl. [illegible]

[illegible]

[illegible]

Operator Theory:
Advances and Applications, Vol. 236, 463–473

Harmonic Spheres Conjecture

Armen Sergeev

Abstract. We discuss the harmonic spheres conjecture, relating the space of harmonic maps of the Riemann sphere into the loop space of a compact Lie group G with the moduli space of Yang–Mills G-fields on four-dimensional Euclidean space.

Mathematics Subject Classification (2010). Primary 58E20, 53C28,32L25.

Keywords. Harmonic spheres, Yang–Mills fields, instantons, loop spaces, Atiyah theorem, Donaldson theorem, Hilbert–Schmidt Grassmannian.

Introduction

Harmonic spheres conjecture relates two kinds of mathematical objects, emerging in theoretical physics – harmonic spheres, arising in sigma-model theory, and Yang–Mills fields, studied in gauge field theory. This conjecture is motivated by the Atiayh theorem [1], establishing a one-to-one correspondence between the moduli space of G-instantons on four-dimensional Euclidean space $\mathbb{R}^4$ and the space of holomorphic spheres in the loop space ΩG of a compact Lie group G, given by holomorphic maps of the Riemann sphere into ΩG. Harmonic spheres conjecture asserts that it should exist a one-to-one correspondence between the moduli space of Yang–Mills G-fields on $\mathbb{R}^4$ and the space of based holomorphic spheres in the loop space ΩG. We formulate this conjecture in Section 7 and give an idea of its proof.

Brief content of the paper. In the first Sections 1 and 2 we introduce the objects related by the conjecture, namely, harmonic spheres and Yang–Mills fields. The main motivation behind the conjecture comes from the twistor theory and the next Sections 3 and 4 are devoted to twistor interpretations of the introduced

---

This paper is an exposition of the talk presented at IWOTA-2011, based on the author's paper, published in "Theor. Math. Physics" **164**(2010), 1140–1150.

While preparing this paper the author was partly supported by the RFBR grants 10-01-00178, 11-01-12033-ofi-m-2011, the program of supporting the Leading Scientific Schools (grant NSh-7675.2010.1), and Scientific Program of Russian Academy of Sciences "Nonlinear dynamics".

objects. In particular, twistor interpretation of harmonic spheres allows to obtain a description of harmonic spheres in complex Grassmann manifolds, presented in Section 5. In Section 6 we formulate Atiyah theorem and give an idea of its proof. This theorem, combined with a theorem of Donaldson, yields an interpretation of G-instantons on $\mathbb{R}^4$ in terms of holomorphic spheres in ΩG. Harmonic spheres conjecture, formulated in Section 7, may be considered as a "realification" of this result. The idea of its proof, proposed in Section 9, is the following. Instead of proving the original conjecture one may try to prove its twistor analogue obtained by "pulling up" the objects, considered in the conjecture, to the associated twistor spaces. Then we are back to holomorphic situation and can use for the proof of twistor conjecture the methods, developed by Atiyah and Donaldson.

1. Harmonic spheres

We consider smooth maps $\varphi : S^2 \to N$ from the 2-sphere into oriented Riemannian manifolds N. The 2-sphere S^2 is identified with the Riemann sphere $\mathbb{P}^1$ provided with the standard complex structure.

Definition 1.1. A smooth map $\varphi : \mathbb{P}^1 \to N$ is called *harmonic* if it is extremal with respect to the energy functional given by the Dirichlet integral

$$E(\varphi) = \frac{1}{2}\int_{\mathbb{C}} |d\varphi|^2_N \frac{|dz \wedge d\bar{z}|}{(1+|z|^2)^2}$$

where the modulus of differential $d\varphi$ is computed with respect to the Riemannian metric of N.

If the manifold N is Kähler, i.e., has a complex structure, compatible with the Riemannian metric, then holomorphic and anti-holomorphic maps $\varphi : \mathbb{P}^1 \to N$ realize local minima of the energy $E(\varphi)$. In the case of $N = \mathbb{P}^1$ such maps exhaust all harmonic maps $\varphi : \mathbb{P}^1 \to \mathbb{P}^1$. More precisely, all harmonic maps $\varphi : \mathbb{P}^1 \to \mathbb{P}^1$ are given by rational holomorphic or anti-holomorphic maps φ. However, for $\dim_{\mathbb{C}} N > 1$ there exist usually also non-minimal harmonic maps.

2. Instantons and Yang–Mills fields

Let G be a compact Lie group and A is a G-connection (*gauge potential*) on $\mathbb{R}^4$ given by a 1-form of type

$$A = \sum_{\mu=1}^{4} A_\mu(x)dx_\mu$$

with smooth coefficients $A_\mu(x)$, taking values in the Lie algebra $\mathfrak{g}$ of G. Denote by F_A the curvature of A (*gauge field*) given by the 2-form

$$F_A = \sum_{\mu,\nu=1}^{4} F_{\mu\nu}(x)dx_\mu \wedge dx_\nu$$

with coefficients, computed by the formula

$$F_{\mu\nu} = \partial_\mu A_\nu - \partial_\nu A_\mu + [A_\mu, A_\nu]$$

where $\partial_\mu := \partial/\partial x_\mu$, $\mu = 1,2,3,4$, and $[\cdot,\cdot]$ denotes the commutator in Lie algebra $\mathfrak{g}$.

Define the *Yang–Mills action* functional by the formula

$$S(A) = \frac{1}{2}\int_{\mathbb{R}^4} \mathrm{tr}(F_A \wedge *F_A)$$

where $*$ is the Hodge operator on $\mathbb{R}^4$, and the trace tr is computed with the help of a fixed invariant inner product on the Lie algebra $\mathfrak{g}$. The functional $S(A)$ is invariant under *gauge transformations* given by

$$A \longmapsto A_g := g^{-1}dg + g^{-1}Ag$$

where $g : \mathbb{R}^4 \to G$ is a smooth map, and G acts on its Lie algebra $\mathfrak{g}$ by adjoint representation. It follows from the invariance of $S(A)$ under gauge transformations that the functional $S(A)$ depends in fact only on the class of the connection A modulo gauge transformations.

Definition 2.1. Gauge fields with finite action $S(A) < \infty$, which are extremal for the functional $S(A)$, are called the *Yang–Mills fields.* Local minima of this functional are called *instantons* or *anti-instantons* depending on the sign of their topological charge given by the formula

$$k(A) = \frac{1}{8\pi^2}\int_{\mathbb{R}^4} \mathrm{tr}(F_A \wedge F_A).$$

Comparing harmonic maps with Yang-Mills fields we notice a formal analogy between:

$$\{\text{(anti)holomorphic maps}\} \longleftrightarrow \{\text{(anti)instantons}\}$$

and

$$\{\text{harmonic maps}\} \longleftrightarrow \{\text{Yang–Mills fields}\}\ .$$

We shall see in Sections 6 and 7 that this formal analogy has, in fact, a deep meaning.

3. Twistor interpretation of instantons

There is a well-known *twistor bundle* over the 4-sphere S^4

$$\pi : \mathbb{P}^3 \xrightarrow{\mathbb{P}^1} S^4 \tag{3.1}$$

where $\mathbb{P}^3$ is the three-dimensional complex projective space. This bundle may be considered as a complex analogue of the Hopf bundle

$$S^7 \xrightarrow{S^3} S^4.$$

To define (3.1) one should identify S^4 with the quaternion projective line $\mathbb{HP}^1$ which consists of pairs $[z_1 + z_2 j, z_1' + z_2' j]$ of quaternions (not equal to zero simultaneously), defined up to multiplication from the right by a nonzero quaternion. Then the twistor bundle

$$\pi : \mathbb{P}^3 \xrightarrow{\mathbb{P}^1} \mathbb{HP}^1$$

will be given by a tautological formula

$$[z_1, z_2, z_3, z_4] \longmapsto [z_1 + z_2 j, z_3 + z_4 j]$$

where the 4-tuple $[z_1, z_2, z_3, z_4]$ of complex numbers is defined up to multiplication by a nonzero complex number while the pair $[z_1 + z_2 j, z_3 + z_4 j]$ of quaternions is defined up to multiplication by a nonzero quaternion.

The restriction of the twistor bundle (3.1) to the Euclidean space $\mathbb{R}^4 = S^4 \setminus \infty$ is the twistor bundle

$$\pi : \mathbb{P}^3 \setminus \mathbb{P}^1_\infty \longrightarrow \mathbb{R}^4 \tag{3.2}$$

over $\mathbb{R}^4$ where the eliminated projective line $\mathbb{P}^1_\infty$ is identified with the fibre of (3.1) at infinity.

According to Atiyah–Hitchin–Singer [2], the fibre of (3.2) at a point $x \in \mathbb{R}^4$ can be identified with the space of complex structures on the tangent space $T_x\mathbb{R}^4 \cong \mathbb{R}^4$, compatible with metric and orientation. Respectively, smooth sections of (3.2) are interpreted as almost complex structures on $\mathbb{R}^4$.

In terms of the twistor bundle $\pi : \mathbb{P}^3 \setminus \mathbb{P}^1 \to \mathbb{R}^4$ the *moduli space of G-instantons*, i.e., the quotient of the space of all G-instantons on $\mathbb{R}^4$ modulo gauge transformations, admits the following interpretation given by *Atiyah–Ward theorem*:

$$\left\{\begin{matrix}\text{moduli space of } G\text{-}\\ \text{instantons on } \mathbb{R}^4\end{matrix}\right\} \longleftrightarrow \left\{\begin{matrix}\text{(based) equivalence classes of holomor-}\\ \text{phic } G^{\mathbb{C}}\text{-bundles over } \mathbb{P}^3\text{, holomorphi-}\\ \text{cally trivial on } \pi\text{-fibers}\end{matrix}\right\}.$$

Here, the term "based" means that the transformations, defining the equivalence of holomorphic $G^{\mathbb{C}}$-bundles over $\mathbb{P}^3$, should be identical on $\mathbb{P}^1_\infty$.

This result has the following two-dimensional reduction to the space $\mathbb{P}^1 \times \mathbb{P}^1$ given by *Donaldson theorem*:

$$\left\{\begin{matrix}\text{moduli space of } G\text{-}\\ \text{instantons on } \mathbb{R}^4\end{matrix}\right\} \longleftrightarrow \left\{\begin{matrix}\text{(based) equivalence classes of holomor-}\\ \text{phic } G^{\mathbb{C}}\text{-bundles over } \mathbb{P}^1 \times \mathbb{P}^1\text{, holomor-}\\ \text{phically trivial on the union } \mathbb{P}^1_\infty \cup \mathbb{P}^1_\infty\end{matrix}\right\}.$$

4. Twistor interpretation of harmonic spheres

Using an interpretation of twistor bundle $\mathbb{P}^3 \to S^4$, given in Sec.3, we can define a *twistor bundle* over any even-dimensional oriented Riemannian manifold N. By definition, it is the bundle of complex structures on the manifold N, compatible with metric and orientation. In other words, $\pi : Z \to N$ is the bundle, associated with the bundle of oriented orthonormal frames on N, with fibre at $x \in N$ given by the space of complex structures J_x on the tangent space T_xN, compatible with metric and orientation. This space can be identified with the homogeneous space $SO(2n)/U(n)$ where $2n$ is the dimension of N. Due to [2], the twistor space Z can be always provided with a natural almost complex structure, denoted by $\mathcal{J}^1$. This almost complex structure is integrable if the manifold N is conformally flat.

However, for the description of harmonic spheres in N we have to use another almost complex structure defined in the following way. The Levi-Civita connection on N generates a connection on the twistor bundle $\pi : Z \to N$. By definition, the new almost complex structure on Z, denoted by $\mathcal{J}^2$, is equal to $\mathcal{J}^1$ in the directions, horizontal with respect to the introduced connection, and to $-\mathcal{J}^1$ in vertical directions. This structure was introduced in [5] and is always non integrable. Harmonic spheres in N have the following interpretation in its terms.

Theorem 4.1 (Eells–Salamon [5]). *Projections $\varphi = \pi \circ \psi$ of the maps $\psi : \mathbb{P}^1 \to Z$, which are holomorphic with respect to the almost complex structure $\mathcal{J}^2$, are harmonic.*

This theorem allows us to construct harmonic spheres in N from almost holomorphic spheres in the twistor space Z. So the original "real" problem of construction of harmonic spheres in the Riemannian manifold N is partially reduced to a "complex" problem of construction of holomorphic spheres in the almost complex manifold Z. It seems from the first glance that the latter problem is in no sense easier than the original one, especially taking into account that the almost complex structure $\mathcal{J}^2$ is never integrable. And it is well known that such a complex structure, for example, may have no non-constant holomorphic functions even locally. However, we are dealing not with holomorphic functions $f : Z \to \mathbb{C}$ but rather with dual objects given by maps $\psi : \mathbb{C} \to Z$, holomorphic with respect to the almost complex structure $\mathcal{J}^2$. Such maps are solutions of the $\bar{\partial}_J$-equation on $\mathbb{C}$ where $J := \psi^*(\mathcal{J}^2)$ is an almost complex structure on $\mathbb{C}$, induced by the map ψ (which is integrable as any almost complex structure on a Riemann surface). In this way, the construction of holomorphic spheres in the space $(Z, \mathcal{J}^2)$ is reduced to the solution of a nonlinear Cauchy–Riemann equation on $\mathbb{C}$ with respect to the complex structure J. In particular, such an equation has many local solutions.

5. Harmonic spheres in complex Grassmann manifolds

Let us take for the target manifold N the complex Grassmann manifold $G_r(\mathbb{C}^d)$ of r-planes in $\mathbb{C}^d$ and apply the twistor approach to the description of harmonic spheres in $G_r(\mathbb{C}^d)$. In this case it is natural to choose for the twistor bundle over $G_r(\mathbb{C}^d)$ the bundle of complex structures on $G_r(\mathbb{C}^d)$ which are invariant under the action of the unitary group $U(d)$. Such bundles coincide with the flag bundles over $G_r(\mathbb{C}^d)$ which will be defined next.

Definition 5.1. The *flag manifold* $F_{\mathbf{r}}(\mathbb{C}^d)$ in $\mathbb{C}^d$ of *type* $\mathbf{r} = (r_1, \ldots, r_n)$ with $d = r_1 + \cdots + r_n$ consists of the flags $\mathcal{W} = (W_1, \ldots, W_n)$, i.e., nested sequences of complex subspaces

$$W_1 \subset \cdots \subset W_n = \mathbb{C}^d$$

such that the dimension of the subspace $V_1 := W_1$ is equal to r_1 and dimensions of the subspaces $V_i := W_i \ominus W_{i-1}$ are equal to r_i for $1 < i \le n$.

The flag manifold $F_{\mathbf{r}}(\mathbb{C}^d)$ admits a homogeneous representation of the following type

$$F_{\mathbf{r}}(\mathbb{C}^d) = \mathrm{U}(d)/\,\mathrm{U}(r_1) \times \cdots \times \mathrm{U}(r_n).$$

It is a compact complex manifold which has an $\mathrm{U}(d)$-invariant complex structure, denoted again by $\mathcal{J}^1$.

In order to construct the twistor flag bundle over $G_r(\mathbb{C}^d)$ we fix an ordered subset $\sigma \subset \{1, \ldots, n\}$ such that $\sum_{i \in \sigma} r_i = r$ and define the *flag bundle*

$$\pi_\sigma : F_{\mathbf{r}}(\mathbb{C}^d) \longrightarrow G_r(\mathbb{C}^d)$$

by

$$\pi_\sigma : \mathcal{W} = (W_1, \ldots, W_n) \longmapsto W := \bigoplus_{i \in \sigma} V_i.$$

As in Sec. 4, we can provide the flag bundle π_σ with an almost complex structure $\mathcal{J}^2_\sigma$ so that the following analogue of Theorem 4.1 will hold.

Theorem 5.2 (Burstall–Salamon [3]). *The flag bundle*

$$\pi_\sigma : F_{\mathbf{r}}(\mathbb{C}^d) \longrightarrow G_r(\mathbb{C}^d),$$

provided with the almost complex structure $\mathcal{J}^2_\sigma$, is a twistor bundle, i.e., the projection $\varphi = \pi_\sigma \circ \psi$ of any almost holomorphic sphere $\psi : \mathbb{P}^1 \to F_{\mathbf{r}}(\mathbb{C}^d)$ is a harmonic sphere $\varphi : \mathbb{P}^1 \to G_r(\mathbb{C}^d)$. Moreover, the converse assertion is also true: any harmonic sphere $\varphi : \mathbb{P}^1 \to G_r(\mathbb{C}^d)$ may be obtained in this way from some flag bundle $\pi_\sigma : F_{\mathbf{r}}(\mathbb{C}^d) \to G_r(\mathbb{C}^d)$.

So in this case the problem of construction of harmonic spheres in $G_r(\mathbb{C}^d)$ is completely reduced to the problem of construction of almost holomorphic spheres in flag bundles. Using this reduction, it was shown in [3] that any harmonic sphere in $G_r(\mathbb{C}^d)$ may be constructed by a Bäcklund-type procedure combining holomorphic and anti-holomorphic spheres.

6. Atiyah theorem

Now we take for the target manifold N an infinite-dimensional Kähler manifold given by the loop space of a compact Lie group.

Definition 6.1. The *loop space* of a compact Lie group G is the quotient

$$\Omega G := LG/G$$

of the group $LG = C^\infty(S^1, G)$ of smooth loops in G by the subgroup G of constant maps $S^1 \to g_0 \in G$.

The space ΩG is a Kähler–Fréchet manifold which can be provided with an LG-invariant complex structure induced from the representation of ΩG as the quotient of the complex loop group $LG^{\mathbb{C}}$:

$$\Omega G = LG^{\mathbb{C}}/L_+G^{\mathbb{C}}$$

where $G^{\mathbb{C}}$ is the complexification of the group G and the subgroup $L_+G^{\mathbb{C}}$ consists of the loops $\gamma \in LG^{\mathbb{C}}$ which can be smoothly extended to holomorphic maps of the unit disc Δ into $G^{\mathbb{C}}$.

Switching to Atiyah theorem, recall the formulation of the Donaldson theorem:

$$\left\{\begin{matrix}\text{moduli space of } G\text{-}\\ \text{instantons on } \mathbb{R}^4\end{matrix}\right\} \longleftrightarrow \left\{\begin{matrix}\text{(based) equivalence classes of holomor-}\\ \text{phic } G^{\mathbb{C}}\text{-bundles over } \mathbb{P}^1\times\mathbb{P}^1\text{, holomor-}\\ \text{phically trivial on the union } \mathbb{P}^1_\infty \cup \mathbb{P}^1_\infty\end{matrix}\right\}.$$

Atiyah theorem asserts that the right-hand side of this correspondence may be identified with the space of holomorphic spheres in ΩG. In more detail, there is a one-to-one correspondence between:

$$\left\{\begin{matrix}\text{(based) equivalence classes of holomor-}\\ \text{phic } G^{\mathbb{C}}\text{-bundles over } \mathbb{P}^1\times\mathbb{P}^1\text{, holomor-}\\ \text{phically trivial on the union } \mathbb{P}^1_\infty \cup \mathbb{P}^1_\infty\end{matrix}\right\} \longleftrightarrow \left\{\begin{matrix}\text{based holomorphic spheres}\\ f:\mathbb{P}^1\to\Omega G\text{, sending } \infty\\ \text{into the origin of } \Omega G\end{matrix}\right\}.$$

The proof of Atiyah theorem is based on the following construction. Consider the restriction of a holomorphic $G^{\mathbb{C}}$-bundle over $\mathbb{P}^1\times\mathbb{P}^1$ to a projective line $\mathbb{P}^1_z$ which is parallel to P^1_∞ and goes through the point $\mathbb{P}^1\times\{z\}$. The restricted bundle is determined by a transition function

$$\tilde{f}_z : S^1 \longrightarrow G^{\mathbb{C}}$$

in the covering $\mathbb{P}^1_z = \overline{\Delta}_+ \cup \overline{\Delta}_-$ of the sphere $\mathbb{P}^1_z$ by lower and upper hemispheres which is holomorphic in a neighborhood of the equator $S^1 = \overline{\Delta}_+ \cap \overline{\Delta}_-$. Hence, $\tilde{f}_z \in LG^{\mathbb{C}}$ and we obtain a map

$$f:\mathbb{P}^1 \ni z \longmapsto \tilde{f}_z \in LG^{\mathbb{C}} \longmapsto f(z) \in \Omega G = LG^{\mathbb{C}}/L_+G^{\mathbb{C}}.$$

This map is holomorphic and based if and only if the original $G^{\mathbb{C}}$-bundle over $\mathbb{P}^1\times\mathbb{P}^1$ is holomorphic and trivial on the union $\mathbb{P}^1_\infty \cup \mathbb{P}^1_\infty$.

7. Harmonic spheres conjecture

The Donaldson and Atiyah theorems imply that there is a one-to-one correspondence between:

$$\left\{\begin{matrix}\text{moduli space of } G\text{-}\\ \text{instantons on } \mathbb{R}^4\end{matrix}\right\} \longleftrightarrow \left\{\begin{matrix}\text{based holomorphic}\\ \text{spheres } f:\mathbb{P}^1\to\Omega G\end{matrix}\right\}.$$

In other words, we have a one-to-one correspondence between local minima of two functionals, introduced before, namely the Yang–Mills action, defined on gauge G-fields on $\mathbb{R}^4$, and energy, defined on smooth spheres in the loop space ΩG. Recall that local minima of Yang–Mills action are given by instantons and anti-instantons on $\mathbb{R}^4$ while local minima of energy are given by holomorphic and anti-holomorphic spheres in ΩG. Replacing local minima by critical points of these functionals, we arrive at the *harmonic spheres conjecture* asserting that it should exist a one-to-one correspondence between:

$$\left\{\begin{matrix}\text{moduli space of Yang–}\\ \text{Mills } G\text{-fields on } \mathbb{R}^4\end{matrix}\right\} \longleftrightarrow \left\{\begin{matrix}\text{based harmonic}\\ \text{spheres } f:\mathbb{P}^1\to\Omega G\end{matrix}\right\}.$$

We can also interpret this replacement of local minima of our functionals by their critical points as a "realification" procedure. Indeed, if we replace smooth spheres in the right-hand side of the diagram by smooth functions $f:\mathbb{C}\to\mathbb{C}$ then the above procedure will reduce to the replacement of holomorphic and anti-holomorphic functions by arbitrary harmonic functions (which can be represented as sums of holomorphic and anti-holomorphic functions). In the case of smooth spheres in ΩG this switching from holomorphic and anti-holomorphic spheres to harmonic ones becomes non-trivial because of the non-linear character of Euler–Lagrange equations for the energy.

Unfortunately, a direct generalization of Atiyah–Donaldson proof to the harmonic case is not possible since the proof of Donaldson theorem, using the monad method, is purely holomorphic and does not extend directly to the harmonic case. However, one can try to reduce the proof of harmonic spheres conjecture to the holomorphic case by "pulling-up" the both sides of the correspondence in this conjecture to the associated twistor spaces, thus reducing the proof to the holomorphic case. The problem is that, while having a good description of the twistor space of harmonic spheres in ΩG (presented in the next Section), we do not know such a description of the moduli space of Yang–Mills fields on $\mathbb{R}^4$. Apart from the proof of the harmonic spheres conjecture, it would be also very interesting to obtain the twistor description of this moduli space.

8. Twistor bundle over the loop space

For the construction of the twistor bundle over the loop space ΩG we first embed ΩG into an infinite-dimensional Grassmannian and then construct the twistor bundle over this Grassmannian by analogy with the finite-dimensional case.

We take for this infinite-dimensional Grassmannian the Hilbert–Schmidt Grassmannian of a complex Hilbert space H, provided with a *polarization*, i.e., a decomposition

$$H = H_+ \oplus H_-$$

into the direct orthogonal sum of closed infinite-dimensional subspaces. In the case when H is identified with the Hilbert space $L_0^2(S^1,\mathbb{C})$ of square integrable functions on the circle with zero average we take a polarization given by the subspaces

$$H_\pm = \left\{\gamma\in H : \gamma = \sum_{\pm k>0}\gamma_k e^{ik\theta}\right\}.$$

Definition 8.1. The *Hilbert–Schmidt Grassmannian* $\mathrm{Gr}_{\mathrm{HS}}(H)$ consists of closed subspaces $W \subset H$ such that the orthogonal projection $\pi_+ : W \to H_+$ is Fredholm and the orthogonal projection $\pi_- : W \to H_-$ is Hilbert–Schmidt.

For a given subspace $W \in \mathrm{Gr}_{\mathrm{HS}}(H)$ the Fredholm index of projection $\pi_+ : W \to H_+$ is called the *virtual dimension* of the subspace W.

The Hilbert–Schmidt Grassmannian $\mathrm{Gr}_{\mathrm{HS}}(H)$ admits a homogeneous representation of the form

$$\mathrm{Gr}_{\mathrm{HS}}(H) = \frac{\mathrm{U}_{\mathrm{HS}}(H)}{\mathrm{U}(H_+) \times \mathrm{U}(H_-)}$$

where the *unitary Hilbert–Schmidt group* $\mathrm{U}_{\mathrm{HS}}(H)$ is

$$\mathrm{U}_{\mathrm{HS}}(H) = \{A \in \mathrm{U}(H) : \pi_- \circ A \circ \pi_+ \text{ is Hilbert–Schmidt}\}.$$

It implies that $\mathrm{Gr}_{\mathrm{HS}}(H)$ is a Hilbert–Kähler manifold, consisting of a countable number of connected components of a fixed virtual dimension:

$$\mathrm{Gr}_{\mathrm{HS}}(H) = \bigcup_d G_d(H)$$

where

$$G_d(H) = \{W \in \mathrm{Gr}_{\mathrm{HS}}(H) : \text{virt.dim}\, W = d\}.$$

The virtual flag manifold $F^d_{\mathbf{r}}(H)$ is defined by analogy with the finite-dimensional case.

Definition 8.2. The *virtual flag manifold* $F^d_{\mathbf{r}}(H)$ in H of *type* $\mathbf{r} = (r_1, \dots, r_n)$ with $d = r_1 + \dots + r_n$ consists of flags $\mathcal{W} = (W_1, \dots, W_n)$, i.e., nested sequences of complex subspaces

$$W_1 \subset \dots \subset W_n \subset H$$

such that the virtual dimension of the subspace $V_1 := W_1$ is equal to r_1, and dimensions of subspaces $V_i := W_i \ominus W_{i-1}$ are equal to r_i for $1 < i \le n$.

For the construction of the twistor flag bundle over the Grassmann manifold $G_r(H)$ we fix again an ordered subset $\sigma \subset \{1, \dots, n\}$ so that $\sum_{i \in \sigma} r_i = r$ and define the *virtual flag bundle*

$$\pi_\sigma : F^d_{\mathbf{r}}(H) \longrightarrow G_r(H)$$

by

$$\pi_\sigma : \mathcal{W} = (W_1, \dots, W_n) \longmapsto W := \bigoplus_{i \in \sigma} V_i.$$

As in the finite-dimensional case, we can provide the virtual flag bundle π_σ with an almost complex structure $\mathcal{J}^2_\sigma$ so that the following analogue of Theorem 5.2 will hold.

Theorem 8.3. *The virtual flag bundle*

$$\pi_\sigma : (F^d_{\mathbf{r}}(H), \mathcal{J}^2_\sigma) \longrightarrow G_r(H),$$

provided with the almost complex structure $\mathcal{J}^2_\sigma$, *is a twistor bundle, i.e., the projection* $\varphi = \pi_\sigma \circ \psi$ *of any almost holomorphic sphere* $\psi : \mathbb{P}^1 \to F^{\mathrm{d}}_{\mathrm{r}}(H)$ *to* $G_r(H)$ *is a harmonic sphere* $\varphi : \mathbb{P}^1 \to G_r(H)$ *in* $G_r(H)$.

We believe that the second part of Theorem 5.2, namely, the conversion of the above theorem is also true in this situation.

We construct now an isometric embedding of the loop space ΩG into the Hilbert–Schmidt Grassmannian. Suppose that the compact Lie group G is realized as a subgroup of the unitary group $\mathrm{U}(N)$ and construct an embedding of ΩG into the Grassmannian $\mathrm{Gr_{HS}}(H)$ where we take for the Hilbert space H the space of vector functions $L^2_0(S^1, \mathbb{C}^N)$.

Let us construct first an embedding of the loop group LG into the unitary Hilbert–Schmidt group $\mathrm{U_{HS}}(H)$. For that we associate with a loop γ, belonging to the space $LG = C^\infty(S^1, G) \subset C^\infty(S^1, \mathrm{U}(N))$, a multiplication operator M_γ in the Hilbert space $H = L^2_0(S^1, \mathbb{C}^N)$, acting by the formula:

$$h \in H = L^2_0(S^1, \mathbb{C}^N) \longmapsto M_\gamma h(z) := \gamma(z)h(z), \ z \in S^1.$$

In other words, $M_\gamma h$ is a vector function from $H = L^2_0(S^1, \mathbb{C}^N)$, obtained by the pointwise application of the matrix function $\gamma \in C^\infty(S^1, \mathrm{U}(N))$ to the vector function $h \in H = L^2_0(S^1, \mathbb{C}^N)$. It is easy to check (cf. [6], Sec. 6.3) that the operator M_γ belongs to the unitary group $\mathrm{U_{HS}}(H)$ if $\gamma \in C^\infty(S^1, \mathrm{U}(N))$.

The embedding $LG \hookrightarrow \mathrm{U_{HS}}(H)$ generates an isometric embedding

$$\Omega G \longrightarrow \mathrm{Gr_{HS}}(H).$$

9. The idea of the proof of harmonic spheres conjecture

Using the isometric embedding $\Omega G \hookrightarrow \mathrm{Gr_{HS}}(H)$ from the last section, we can consider an arbitrary harmonic map $\varphi : \mathbb{P}^1 \to \Omega G$ as taking its values in the Grassmannian $\mathrm{Gr_{HS}}(H)$, hence, in one of the connected components $G_r(H)$ of the manifold $\mathrm{Gr_{HS}}(H)$. To describe harmonic maps $\varphi : \mathbb{P}^1 \to G_r(H)$, one can use the twistor method, as in the finite-dimensional case.

For that we should obtain first a harmonic analogue of the Atiyah theorem. It should assert that there is a one-to-one correspondence between:

$$\left\{ \begin{array}{l} \text{(based) equivalence classes of harmonic} \\ G^{\mathbb{C}}\text{-bundles over } \mathbb{P}^1 \times \mathbb{P}^1 \text{, trivial on the} \\ \text{union } \mathbb{P}^1_\infty \cup \mathbb{P}^1_\infty \end{array} \right\} \longleftrightarrow \left\{ \begin{array}{l} \text{based harmonic spheres} \\ \varphi : \mathbb{P}^1 \to \Omega G \text{, sending} \\ \infty \text{ to the origin} \end{array} \right\}.$$

We proceed as in the holomorphic case. Suppose that a harmonic sphere $\varphi : \mathbb{P}^1 \to \Omega G \subset \mathrm{Gr_{HS}}(H)$ is the projection of a harmonic sphere $\tilde{\varphi} : \mathbb{P}^1 \to LG^{\mathbb{C}}$. Then $\tilde{\varphi}(z) \in LG^{\mathbb{C}}$ may be considered as the transition function for some harmonic bundle over $\mathbb{P}^1_z$. This bundle should be the restriction of a harmonic bundle over $\mathbb{P}^1 \times \mathbb{P}^1$, associated with the composite map

$$\varphi : \mathbb{P}^1 \ni z \longmapsto \tilde{\varphi}(z) \in LG^{\mathbb{C}} \longmapsto \varphi(z) \in \Omega G = LG^{\mathbb{C}} / L_+ G^{\mathbb{C}}.$$

In terms of Grassmannian $\mathrm{Gr_{HS}}(H)$ the image $\varphi(z) \in \Omega G \subset \mathrm{Gr_{HS}}(H)$ is identified with the subspace

$$W_z := M_{\tilde{\varphi}(z)} H_+$$

where M is the multiplication operator, introduced at the end of previous section.

The *twistor interpretation* of this construction has the following form. A harmonic sphere in ΩG may be considered as a harmonic sphere in a submanifold $G_r(H) \subset \mathrm{Gr_{HS}}(H)$, consisting of subspaces $W \subset H$ of some fixed virtual dimension r. In terms of the twistor flag bundle the harmonic sphere $\varphi : \mathbb{P}^1 \to G_r(H)$ is the projection of some $\mathcal{J}^2_\sigma$-holomorphic sphere $\psi : \mathbb{P}^1 \to F^d_{\mathbf{r}}(H)$.

The image $\psi(z) = (\psi_1(z), \ldots, \psi_n(z))$ of a point $z \in \mathbb{P}^1$ under the map $\psi : \mathbb{P}^1 \to F^d_{\mathbf{r}}(H)$ coincides with a virtual flag $\mathcal{W}_z = (W^1_z, \ldots, W^n_z)$. If every map $\psi_i : \mathbb{P}^1 \to G_{r_i}(H)$ is the projection of a map $\tilde{\psi}_i : \mathbb{P}^1 \to LG^{\mathbb{C}}$ so that

$$W^i_z = M_{\tilde{\psi}_i} H_+$$

then each of these maps can be considered as the transition function for some bundle over $\mathbb{P}^1_z$. It follows from the description of the almost complex structure $\mathcal{J}^2_\sigma$ on the twistor bundle π_σ that the maps $\tilde{\psi}_i$ determine either holomorphic, or anti-holomorphic bundles over $\mathbb{P}^1 \times \mathbb{P}^1$. By Donaldson theorem such bundles correspond either to instantons, or anti-instantons on $\mathbb{R}^4$. This may be considered as a twistor construction of the moduli space of Yang–Mills fields on $\mathbb{R}^4$, associating with such a field a finite collection of instantons and anti-instantons on $\mathbb{R}^4$.

References

[1] M.F. Atiayh, *Instantons in two and four dimensions.* Comm. Math. Phys. **93** (1984), 437–451.

[2] M.F. Atiayh, N.J. Hitchin, I.M. Singer, *Self-duality in four-dimensional Riemannian geometry.* Proc. Roy. Soc. London **362** (1978), 425–461.

[3] F.E. Burstall, S. Salamon, *Tournaments, flags and harmonic maps.* Math. Ann. **277** (1987), 249–265.

[4] S.K. Donaldson, *Instantons and geometric invariant theory.* Comm. Math. Phys. **93** (1984), 453–460.

[5] J. Eells, S. Salamon, *Twistorial constructions of harmonic maps of surfaces into four-manifolds.* Ann. Scuola Norm. Super. Pisa **12** (1985), 589–640.

[6] A. Pressley, G. Segal, *Loop Groups.* Clarendon Press, 1986.

Armen Sergeev
Steklov Mathematical Institute, Moscow
e-mail: **sergeev@mi.ras.ru**

[illegible] with the subspace

[illegible]

where M is the multiplication operator [illegible] at the end of previous section [illegible]

[illegible]

Operator Theory:
Advances and Applications, Vol. 236, 475–482

Riesz Bases Multipliers

Diana T. Stoeva and Peter Balazs

Abstract. The paper concerns frame multipliers when one of the involved sequences is a Riesz basis. We determine the cases when the multiplier is well defined and invertible, well defined and not invertible, respectively not well defined.

Mathematics Subject Classification (2010). Primary 42C15; Secondary 47A05.

Keywords. Riesz bases, frame multipliers.

1. Introduction

Multipliers are operators which are defined based on two sequences with elements from a Hilbert space and one scalar sequence. Given sequences $\Phi = (\phi_n)_{n=1}^{\infty}$ and $\Psi = (\psi_n)_{n=1}^{\infty}$ with elements from a Hilbert space $\mathcal{H}$, and given complex scalar sequence $m = (m_n)_{n=1}^{\infty}$ (called the *weight* or the *symbol*), the operator $M_{m,\Phi,\Psi}$ defined by

$$M_{m,\Phi,\Psi} f = \sum_{n=1}^{\infty} m_n \langle f, \psi_n \rangle \phi_n, \ \ f \in \mathcal{H},$$

is called a *multiplier*. Multipliers for Bessel sequences were introduced in [1]. Further, multipliers for general sequences were considered in [13, 14, 15, 16].

The interest to the consideration of multipliers has come from practical reasons. In signal processing many methods employ linear time-variant filters, i.e., convolution operators, which can be described as Fourier multipliers [9]. Gabor multipliers [8] are a particular option to represent time-varying filters. They have many applications, for example in acoustical signal processing [3, 7]. Thus, it is interesting to investigate the possibilities for the inversion of multipliers. This is also interesting from a theoretical point of view as an operator theory question.

This work was supported by the WWTF project MULAC ('Frame Multipliers: Theory and Application in Acoustics'; MA07-025) and the Austrian Science Fund (FWF) START-project FLAME ('Frames and Linear Operators for Acoustical Modeling and Parameter Estimation'; Y 551-N13).

The multiplier $M_{m,\Phi,\Psi} : \mathcal{H} \to \mathcal{H}$ is called *invertible* if it has a bounded inverse defined on $\mathcal{H}$. Some results concerning invertibility of multipliers appeared in [1, 2, 12, 13, 14, 15, 16]. In this paper our attention concerns the case when one of the sequences Φ and Ψ is a Riesz basis. It is known [1] that if Φ and Ψ are Riesz bases and m is semi-normalized, then $M_{m,\Phi,\Psi}$ is invertible and its inverse can be written as $M_{1/m,\widetilde{\Psi},\widetilde{\Phi}}$, where $\widetilde{\Psi}$ and $\widetilde{\Phi}$ are the unique biorthogonal sequences to Ψ and Φ, respectively, and $1/m$ is the sequence $(1/m_n)$. Further, invertibility of $M_{m,\Phi,\Psi}$ when at least one of the sequences Φ and Ψ is a Riesz basis is considered in [13] with the sketch of a proof:

Theorem 1.1. *Let Φ be a Riesz basis for $\mathcal{H}$. Then the following holds.*

(i) *If Ψ is a Riesz basis for $\mathcal{H}$, then $M_{m,\Phi,\Psi}$ (resp. $M_{m,\Psi,\Phi}$) is invertible if and only if m is semi-normalized.*

(ii) *If m is semi-normalized, then $M_{m,\Phi,\Psi}$ (resp. $M_{m,\Psi,\Phi}$) is invertible if and only if Ψ is a Riesz basis for $\mathcal{H}$.*

(iii) *If m is not semi-normalized, then $M_{m,\Phi,\Psi}$ (resp. $M_{m,\Psi,\Phi}$) can be invertible only in the following cases:*

($\mathcal{R}_1$): *Ψ is Bessel in $\mathcal{H}$, which is not a frame for $\mathcal{H}$ and not norm-bounded from below, m is norm-bounded from below and $m \notin \ell^\infty$;*

($\mathcal{R}_2$): *Ψ is non-Bessel in $\mathcal{H}$ which is norm-bounded from below and not norm-bounded, $m \in \ell^\infty$, and m is not norm-bounded from below.*

($\mathcal{R}_3$): *Ψ is non-Bessel in $\mathcal{H}$ which is neither norm-bounded from above nor norm-bounded from below, m is not norm-bounded from below, and $m \notin \ell^\infty$.*

In the cases of invertibility, $M^{-1}_{m,\Phi,\Psi} = M_{(1),\widetilde{\overline{m}\Psi},\widetilde{\Phi}}$ and $M^{-1}_{m,\Psi,\Phi} = M_{(1),\widetilde{\Phi},\widetilde{\overline{m}\Psi}}$. For the cases (i) *and* (ii) *this is equivalent to $M^{-1}_{m,\Phi,\Psi} = M_{1/m,\widetilde{\Psi},\widetilde{\Phi}}$ and $M^{-1}_{m,\Psi,\Phi} = M_{1/m,\widetilde{\Phi},\widetilde{\Psi}}$.*

In this paper we give more detail statements, distinguishing cases when the multiplier is well defined and non-invertible, and when the multiplier is not well defined, with a complete proof.

2. Notation and preliminaries

Throughout the paper, $(\mathcal{H}, \langle\cdot,\cdot\rangle)$ denotes a Hilbert space, $(e_n)_{n=1}^\infty$ denotes an orthonormal basis of $\mathcal{H}$, $\Phi = (\phi_n)_{n=1}^\infty$ and $\Psi = (\psi_n)_{n=1}^\infty$ are sequences with elements from $\mathcal{H}$, m denotes a complex scalar sequence $(m_n)_{n=1}^\infty$ and $\overline{m}$ denotes the sequence of the complex conjugates of m_n; $m\Phi$ denotes the sequence $(m_n\phi_n)_{n=1}^\infty$. If the index set of a sequence is omitted, the set $\mathbb{N}$ should be understood.

The sequence Ψ (resp. m) is called

- *norm-bounded from above*, if there exists a positive constant b such that $\|\psi_n\| \leq b$ (resp. $|m_n| \leq b$), $\forall n$;

- *norm-bounded from below*, if there exists a positive constant a such that $0 < a \le \|\psi_n\|$ (resp. $0 < a \le |m_n|$), $\forall n$;
- *norm-semi-normalized* (resp. *semi-normalized*) if Ψ (resp. m) is both norm-bounded from below and norm-bounded from above.

We abbreviate norm-bounded from above with NBA and norm-bounded from below with NBB.

Recall the definition of the basic type of sequences which we are going to use in the paper.

Definition 2.1. The sequence Ψ is called

- *complete in $\mathcal{H}$* if $\overline{\text{span}}\{\Psi\} = \mathcal{H}$;
- *a Bessel sequence in $\mathcal{H}$* (in short, *Bessel in $\mathcal{H}$*) if there exists a positive constant B so that $\sum_{k=1}^{\infty} |\langle f, \psi_k \rangle|^2 \le B\|f\|^2$ for every $f \in \mathcal{H}$;
- *a frame for $\mathcal{H}$* if it is a Bessel sequence in $\mathcal{H}$ and there exists a positive constant A so that $\sum_{n=1}^{\infty} |\langle f, \psi_n \rangle|^2 \ge A\|f\|^2$ for all $f \in \mathcal{H}$;
- *a Riesz sequence* if there exist positive constants A,B so that

$$A \sum_{n=1}^{\infty} |c_n|^2 \le \left\| \sum_{n=1}^{\infty} c_n \psi_n \right\|^2 \le B \sum_{n=1}^{\infty} |c_n|^2$$

  for all finite scalar sequences (c_k) (and hence, for all $(c_k)_{k=1}^{\infty} \in \ell^2$);
- *a Riesz basis for $\mathcal{H}$* if it is a Riesz sequence which is complete in $\mathcal{H}$;
- *an overcomplete frame for $\mathcal{H}$* if it is a frame for $\mathcal{H}$ and not a Riesz basis for $\mathcal{H}$.

For a given Bessel sequence Ψ, we will use the *analysis operator* U_Ψ defined by

$$U_\Psi h = (\langle h, \psi_n \rangle)_{n=1}^{\infty}, \ h \in \mathcal{H},$$

and the *synthesis operator* T_Ψ defined by

$$T_\Psi h = \sum_{n=1}^{\infty} c_n \psi_n, \ (c_n)_{n=1}^{\infty} \in \ell^2.$$

Definition 2.2. For any Φ, Ψ, m, the multiplier $M_{m,\Phi,\Psi} : \mathcal{H} \to \mathcal{H}$ is called

- *well-defined* if the series $\sum_{n=1}^{\infty} m_n \langle f, \psi_n \rangle \phi_n$ converges for every $f \in \mathcal{H}$;
- *injective* if it is well defined and $M_{m,\Phi,\Psi} f = 0$ implies $f = 0$;
- *surjective* if it is well defined and its range coincides with $\mathcal{H}$;
- *invertible* if it is well defined and it has a bounded inverse defined on $\mathcal{H}$.

Note that if a multiplier $M_{m,\Phi,\Psi}$ is well defined, then by Banach–Steinhaus Theorem, $M_{m,\Phi,\Psi}$ is bounded.

We need the following result concerning Riesz bases.

Proposition 2.3. *The sequence $m\Phi$ can be a Riesz basis for $\mathcal{H}$ only in the following cases:*

- *Φ is a Riesz basis for $\mathcal{H}$ and m is semi-normalized;*

- *Φ is non-NBB Bessel in $\mathcal{H}$ which is not a frame for $\mathcal{H}$ and m is NBB but not in ℓ^∞;*
- *Φ is non-NBA non-Bessel in $\mathcal{H}$ and m is non-NBB with $m_n \neq 0$, $\forall n$.*

Proof. When m is a real-valued sequence, the statement is proved in [12]. Similar techniques can be used to prove the statement for complex-valued sequences m. □

3. Properties of Riesz multipliers

Throughout this section we assume that Φ is a Riesz basis for $\mathcal{H}$ and M denotes any one of the multipliers $M_{m,\Phi,\Psi}$ and $M_{m,\Psi,\Phi}$. First we point on necessary and sufficient conditions for M to be well defined, injective, surjective, invertible.

Proposition 3.1. *The following equivalences hold.*

(a) *M is well defined if and only if $m\Psi$ is a Bessel sequence in $\mathcal{H}$.*
(b1) *$M_{m,\Phi,\Psi}$ is injective if and only if $m\Psi$ is a complete Bessel sequence in $\mathcal{H}$.*
(b2) *$M_{m,\Psi,\Phi}$ is injective if and only if $T_{m\Psi}$ is injective.*
(c1) *$M_{m,\Phi,\Psi}$ is surjective if and only if $\overline{m}\Psi$ is a Riesz sequence.*
(c2) *$M_{m,\Psi,\Phi}$ is surjective if and only if $m\Psi$ is frame for $\mathcal{H}$.*
(d) *M is invertible if and only if $m\Psi$ is a Riesz basis for $\mathcal{H}$.*

Proof. First recall that when Φ is a Riesz basis for $\mathcal{H}$, the analysis operator U_Φ is a bounded bijection from $\mathcal{H}$ onto ℓ^2 and the synthesis operator T_Φ is a bounded bijection from ℓ^2 onto $\mathcal{H}$.

(a) This statement can be found in [14, Prop. 3.4].

(b1) For the well-definedness, use (a). Now assume that $m\Psi$ is a Bessel sequence in $\mathcal{H}$. The multiplier $M_{m,\Phi,\Psi} = T_\Phi U_{\overline{m}\Psi}$ is injective if and only if $U_{\overline{m}\Psi}$ is injective if and only if $U_{m\Psi}$ is injective. Further, the analysis operator $U_{m\Psi}$ is injective if and only if $m\Psi$ is complete in $\mathcal{H}$ [4, Prop. 4.1].

(b2) For the well-definedness, use (a). Now assume that $m\Psi$ is a Bessel sequence in $\mathcal{H}$. The multiplier $M_{m,\Psi,\Phi} = T_{m\Psi} U_\Phi$ is injective if and only if $T_{m\Psi}$ is injective.

(c1) For the well-definedness, use (a). Now assume that $m\Psi$ is a Bessel sequence in $\mathcal{H}$ and hence, it satisfies the upper condition for a Riesz sequence. The multiplier $M_{m,\Phi,\Psi} = T_\Phi U_{\overline{m}\Psi}$ is surjective if and only if $U_{\overline{m}\Psi}$ is surjective. By [17, Ch.4 Sec.2], $U_{\overline{m}\Psi}$ is surjective if and only if $\overline{m}\Psi$ satisfies the lower condition for a Riesz sequence. This completes the proof.

(c2) For the well-definedness, use (a). Now assume that $m\Psi$ is a Bessel sequence in $\mathcal{H}$. The multiplier $M_{m,\Psi,\Phi} = T_{m\Psi} U_\Phi$ is surjective if and only if $T_{m\Psi}$ is surjective. Further, $T_{m\Psi}$ is surjective if and only if $m\Psi$ is a frame for $\mathcal{H}$ [6, Theor. 5.5.1].

(d) follows from (b1), (c1) and (b2), (c2). □

Here we extend Theorem 1.1 considering not only the invertibility possibilities, but also the cases of well-definedness without invertibility and the cases of not well-definedness.

Proposition 3.2. *Let m be semi-normalized. Then the following statements hold.*

(a) *If Ψ is non-Bessel in $\mathcal{H}$, then M is not well defined.*
(b) *If Ψ is a Bessel sequence in $\mathcal{H}$ which is not a Riesz basis for $\mathcal{H}$, then M is well defined but not invertible.*
(c) [1, Prop. 7.7] *If Ψ is a Riesz basis for $\mathcal{H}$, then M is invertible.*

Proof. (a) and (b) follow from Proposition 3.1 using the assumption that m is semi-normalized. □

Proposition 3.3. *Let m be non-NBB and $m \in \ell^\infty$. Then the following statements hold.*

(a) *If Ψ is NBA non-Bessel in $\mathcal{H}$, then M can either be well defined or not, but can never be invertible on $\mathcal{H}$.*
(b) *If Ψ is non-NBA, non-NBB, and non-Bessel in $\mathcal{H}$, then M can either be well defined or not, but can never be invertible on $\mathcal{H}$.*
(c) *If Ψ is non-NBA, NBB, and non-Bessel in $\mathcal{H}$, then for M all the three feasible combinations of invertibility and well-definedness are possible: it can be invertible, it can be well defined and not invertible, and it can be not well defined.*
(d) *If Ψ is Bessel in $\mathcal{H}$, then M is well defined, but not invertible.*

Proof. (a) Assume that M is well defined and thus $m\Psi$ is Bessel in $\mathcal{H}$. By Proposition 3.1, M is invertible if and only if $m\Psi$ is a Riesz basis for $\mathcal{H}$. By Proposition 2.3, the sequence $m\Psi$ can not be a Riesz basis for $\mathcal{H}$ under the assumptions of (a).

As an example of a well-defined multiplier, consider the sequences $\Phi = (e_n)$, $\Psi = (e_1, e_2, e_1, e_3, e_1, e_4, \ldots)$, and $m = (\frac{1}{2}, 1, \frac{1}{2^2}, 1, \frac{1}{2^3}, 1, \ldots)$. Since $m\Psi$ is Bessel in $\mathcal{H}$, M is well defined. Now consider the sequences $\Phi = (e_n)$, $\Psi = (e_1, e_2, e_1, e_3, e_1, e_4, \ldots)$, and $m = (1, \frac{1}{2}, 1, \frac{1}{3}, 1, \frac{1}{4}, \ldots)$. In this case M is not well defined, because $m\Psi$ is not Bessel in $\mathcal{H}$.

(b) Assume that M is invertible. By Proposition 3.1, $m\Psi$ is a Riesz basis for $\mathcal{H}$. Hence, there exists $a > 0$ so that $\|m_n\psi_n\| \geq a$, $\forall n \in \mathbb{N}$. Since $m \in \ell^\infty$, it follows that Ψ is NBB, which contradicts to the assumptions.

As an example of a well-defined multiplier, consider the sequences $\Phi = (e_n)$, $\Psi = (e_1, 2e_2, \frac{1}{3}e_3, 4e_4, \frac{1}{5}e_5, 6e_6, \ldots)$, and $m = (1, \frac{1}{2}, 1, \frac{1}{4}, 1, \frac{1}{6}, \ldots)$. Since $m\Psi$ is Bessel in $\mathcal{H}$, both $M_{m,\Phi,\Psi}$ and $M_{m,\Psi,\Phi}$ are well defined. Now consider the same Φ and Ψ, and the sequence $\nu = (1, 1, \frac{1}{3}, 1, \frac{1}{5}, 1, \ldots)$. Both $M_{\nu,\Phi,\Psi}$ and $M_{\nu,\Psi,\Phi}$ are not well defined, because $\nu\Psi$ is not Bessel in $\mathcal{H}$.

(c) As an example of invertible multipliers, consider $M_{(1/n),(e_n),(ne_n)} = M_{(1/n),(ne_n),(e_n)} = I_{\mathcal{H}}$. As an example of multipliers which are well defined and not invertible, consider $M_{(1/n^2),(e_n),(ne_n)} = M_{(1/n^2),(ne_n),(e_n)}$. For multipliers, which are not well defined, see for example $M_{(\frac{1}{n}),(e_n),(n^2e_n)}$ and $M_{(\frac{1}{n}),(n^2e_n),(e_n)}$.

(d) By Proposition 3.1, M is well defined. The non-invertibility of M can be shown in an analogue way as in (a). □

Proposition 3.4. *Let m be NBB and $m \notin \ell^\infty$. Then the following statements hold.*

(a) *If Ψ is non-Bessel in $\mathcal{H}$ or NBB, then M is not well defined.*

(b) *Let Ψ be non-NBB Bessel in $\mathcal{H}$, which is not a frame for $\mathcal{H}$. Then for M all the three feasible combinations of invertibility and well-definedness are possible.*

(c) *Let Ψ be a non-NBB frame for $\mathcal{H}$. Then M can either be well defined or not, but can never be invertible on $\mathcal{H}$.*

Proof. (a) By Proposition 3.1, well-definedness of M requires $m\Psi$ to be Bessel in $\mathcal{H}$, which requires Ψ to be Bessel in $\mathcal{H}$ (because of the NBB-assumption on m).

If Ψ is NBB, the conclusion follows from [14, Prop. 3.4].

(b) As an example of invertible multipliers, consider $M_{(n),(e_n),(\frac{1}{n}e_n)} = M_{(n),(\frac{1}{n}e_n),(e_n)} = I_{\mathcal{H}}$. As an example of multipliers which are well defined and not invertible, take $M_{(n),(e_n),(\frac{1}{n^2}e_n)} = M_{(n),(\frac{1}{n^2}e_n),(e_n)}$.

The multipliers $M_{(n^2),(e_n),(\frac{1}{n}e_n)}$ and $M_{(n^2),(\frac{1}{n}e_n),(e_n)}$ are not well defined.

(c) If M is well defined, the non-invertibility of M can be shown in an analogue way as in Proposition 3.3(a).

Consider $\Phi = (e_n)$ and the sequence $\Psi = (\frac{1}{2}e_1, e_2, \frac{1}{2^2}e_1, e_3, \frac{1}{2^3}e_1, e_4, \ldots)$, which is non-$NBB$ frame for $\mathcal{H}$. For $m = (\sqrt{2}, 1, \sqrt{2^2}, 1, \sqrt{2^3}, 1, \ldots)$, the sequence $m\Psi$ is Bessel in $\mathcal{H}$, which implies that both $M_{m,\Phi,\Psi}$ and $M_{m,\Psi,\Phi}$ are well defined on $\mathcal{H}$. For $\nu = (2, 1, 2^2, 1, 2^3, 1, \ldots)$, the sequence $\nu\Psi$ is not Bessel in $\mathcal{H}$, which implies that both $M_{\nu,\Phi,\Psi}$ and $M_{\nu,\Psi,\Phi}$ are not well defined. □

Proposition 3.5. *Let m be non-NBB and $m \notin \ell^\infty$. Then the following statements hold.*

(a) *If Ψ is NBB, then M is not well defined.*

(b) *If Ψ is non-NBB, non-NBA, and non-Bessel in $\mathcal{H}$, then for M all the three feasible combinations of invertibility and well-definedness are possible.*

(c) *If Ψ is non-NBB and NBA, then M can either be well defined or not, but can never be invertible on $\mathcal{H}$.*

Proof. (a) As in Proposition 3.4, if Ψ is NBB and M is well defined, then m must be in ℓ^∞.

(b) Consider $\Phi = (e_n)$ and the sequence $\Psi = (e_1, \frac{1}{2^2}e_2, 3e_3, \frac{1}{4^2}e_4, 5e_5, \ldots)$. For $m = (1, 2^2, \frac{1}{3}, 4^2, \frac{1}{5}, \ldots)$, we have the invertible multipliers $M_{m,\Phi,\Psi} = M_{m,\Psi,\Phi} = I_{\mathcal{H}}$. For $\nu = (1, 2, \frac{1}{3^2}, 4, \frac{1}{5^2}, \ldots)$, both multipliers $M_{\nu,\Phi,\Psi}$ and $M_{\nu,\Psi,\Phi}$ coincide with $M_{(\frac{1}{n}),(e_n),(e_n)}$ which is well defined and not invertible. For $\mu = (1, 2^3, \frac{1}{3}, 4^3, \frac{1}{5}, \ldots)$, the sequence $\mu\Psi$ is not Bessel in $\mathcal{H}$, which implies that both $M_{\mu,\Phi,\Psi}$ and $M_{\mu,\Psi,\Phi}$ are not well defined.

(c) If M is well defined, the non-invertibility of M can be shown in a similar way as in Proposition 3.3(a), using the non-NBB property of m and the NBA-property of Ψ.

Examples for the case "Ψ-NBA non-NBB Bessel". Let $\Phi = (e_n)$ and $\Psi = (e_1, \frac{1}{2}e_2, \frac{1}{3}e_3, \ldots)$. For $m = (1, 2^2, \frac{1}{3}, 4^2, \frac{1}{5}, \ldots)$, the sequence $m\Psi$ is not Bessel in $\mathcal{H}$, which implies that both $M_{m,\Phi,\Psi}$ and $M_{m,\Psi,\Phi}$ are not well defined. For $\nu = (1, 2, \frac{1}{3}, 4, \frac{1}{5}, \ldots)$, the sequence $\nu\Psi$ is Bessel in $\mathcal{H}$, which implies that both $M_{\nu,\Phi,\Psi}$ and $M_{\nu,\Psi,\Phi}$ are well defined.

Examples for the case "Ψ-NBA non-NBB non-Bessel". Let $\Phi = (e_n)$ and $\Psi = (e_1, \frac{1}{2}e_2, e_1, \frac{1}{3}e_3, e_1, \frac{1}{4}e_4 \ldots)$. For $m = (\frac{1}{2}, 2, \frac{1}{2^2}, 3, \frac{1}{2^3}, 4, \ldots)$, the sequence $m\Psi$ is Bessel in $\mathcal{H}$ and thus, both $M_{m,\Phi,\Psi}$ and $M_{m,\Psi,\Phi}$ are well defined. For $\nu = (\frac{1}{2}, 2^2, \frac{1}{2^2}, 3^2, \frac{1}{2^3}, 4^2, \ldots)$, the sequence $\nu\Psi$ is not Bessel in $\mathcal{H}$ and thus, both $M_{\nu,\Phi,\Psi}$ and $M_{\nu,\Psi,\Phi}$ are not well defined. □

Acknowledgment

The first author is grateful for the hospitality of the Acoustics Research Institute,the support from the projects MULAC and FLAME, and the possibility to participate in the International Conference IWOTA2011. She is also thankful to the Organizers of IWOTA2011 for the nice atmosphere at the conference.

References

[1] P. Balazs, *Basic definition and properties of Bessel multipliers.* J. Math. Anal. Appl. **325**(1) (2007), 571–585,

[2] P. Balazs, J.-P. Antoine, A. Grybos, *Weighted and controlled frames: Mutual relationship and first numerical properties.* Int. J. Wavelets Multiresolut. Inf. Process. **8**(1) (2010), 109–132.

[3] P. Balazs, B. Laback, G. Eckel, W.A. Deutsch, *Time-frequency sparsity by removing perceptually irrelevant components using a simple model of simultaneous masking.* IEEE Trans. Speech Audio Process. **18**(1) (2010), 34–49.

[4] P. Balazs, D.T. Stoeva, J.-P. Antoine, *Classification of General Sequences Frame-Related Operators.* Sampl. Theory Signal Image Process. **10**(1-2) (2011), 151–170.

[5] J. Benedetto, G. Pfander, *Frame expansions for Gabor multipliers.* Appl. Comput. Harmon. Anal. **20**(1) (2006), 26–40.

[6] O. Christensen, *An Introduction to Frames and Riesz Bases.* Birkhäuser, 2003.

[7] P. Depalle, R. Kronland-Martinet, B. Torrésani, *Time-frequency multipliers for sound synthesis.* In: Proceedings of the Wavelet XII conference, SPIE annual Symposium, San Diego, 2007.

[8] H.G. Feichtinger, K. Nowak, *A first survey of Gabor multipliers.* Birkhäuser, Boston, 2003, Ch. 5, pp. 99–128.

[9] H.G. Feichtinger, G. Narimani, Fourier multipliers of classical modulation spaces, Appl. Comput. Harmon. Anal. **21**(3) (2006) 349–359.

[10] C. Heil, *A Basis Theory Primer.* Birkhäuser, Boston, 2011.

[11] G. Matz and F. Hlawatsch, *Linear Time-Frequency Filters: On-line Algorithms and Applications.* Eds. A. Papandreou-Suppappola, Boca Raton (FL): CRC Press, 2002, Ch. 6 in "Application in Time-Frequency Signal Processing", pp. 205–271.

[12] D.T. Stoeva and P. Balazs. *Weighted frames and frame multipliers.* Annual of the University of Architecture, Civil Engineering and Geodesy, Vol. XLIII-XLIV 2004-2009 (Fasc. II) (2012), pp. 33–42.

[13] D.T. Stoeva and P. Balazs. *Invertibility of Multipliers.* Appl. Comput. Harmon. Anal. **33**(2) (2012), 292–299.

[14] D.T. Stoeva and P. Balazs, *Canonical forms of unconditionally convergent multipliers.* J. Math. Anal. Appl. 399(1) (2013), 252–259.

[15] D.T. Stoeva and P. Balazs, *Detailed characterization of unconditional convergence and invertibility of multipliers.* Sampl. Theory Signal Image Process. 12(2) (2013), to appear.

[16] D.T. Stoeva and P. Balazs, *Representation of the inverse of a multiplier as a multiplier.* arXiv:1108.6286, 2011.

[17] R.M. Young, *An Introduction to Nonharmonic Fourier Series.* Academic Press, New York and London, 1980.

Diana T. Stoeva and Peter Balazs
Acoustics Research Institute
Wohllebengasse 12–14
Vienna A-1040, Austria
e-mail: dstoeva@kfs.oeaw.ac.at
peter.balazs@oeaw.ac.at

Operator Theory:
Advances and Applications, Vol. 236, 483–508

Operator Approximation for Processing of Large Random Data Sets

Anatoli Torokhti

Abstract. Suppose K_Y and K_X are large sets of observed and reference signals, respectively, each containing N signals. Is it possible to construct a filter $\mathcal{F} : K_Y \to K_X$ that requires a priori information only on *few signals*, $p \ll N$, from K_X but *performs better* than the known filters based on a priori information on *every* reference signal from K_X? It is shown that the positive answer is achievable under quite unrestrictive assumptions. The device behind the proposed method is based on a special extension of the piecewise linear interpolation technique to the case of random signal sets. The proposed technique provides a single filter to process any signal from the arbitrarily large signal set. The filter is determined in terms of pseudo-inverse matrices so that it always exists.

Mathematics Subject Classification (2010). Primary 94A12; Secondary 65D05.

Keywords. Piecewise interpolation, filtering.

1. Introduction

In the paper, a new method of operator approximation is developed. The considered technique is motivated by a desire to overcome difficulties associated with a processing of large random data sets. This issue is considered in detail in Section 1.1 that follows.

An idealistic filter transforming signal sets K_X and K_Y is interpreted as an operator $\mathcal{F} : K_Y \to K_X$. A purpose of the proposed methodology is to justify its approximating operator $\mathcal{F}^{(p-1)} : K_Y \to K_X$ with p specified below. The device behind the structure of filter (operator) $\mathcal{F}^{(p-1)}$ is quite simple and is based on a special extension of the piecewise linear interpolation approach to the case of random signal sets. At the same time, such a device is not straightforward and requires the careful substantiation presented in Sections 2.3, 3.4, 4.2 and 4.4 below.

1.1. Motivations

The problem under consideration is motivated by the following particular observations.

1.1.1. FILTERING OF LARGE SETS OF SIGNALS; LESS INITIAL INFORMATION FOR BETTER FILTERING. Suppose we need to transform a set of signals K_Y to another set of signals K_X. The signals are represented by *finite* random vectors[1]. A major difficulty and inconvenience common to many known filtering methodologies (see, for example, [1]–[9], [11], [13], [22, 23, 25]) is that they require a priori information on *each reference signal* to be estimated[2]. In particular, the filters in [22, 23, 25] are based on the use of either the reference signal $\mathbf{x} \in K_X$ itself, as in [22, 23], or its estimate, as in [25]. The Wiener filtering approach (see, for example, [1]–[13], [23], [25]) assumes that a covariance matrix formed from a reference signal, $\mathbf{x} \in K_X$, and an observed signal, $\mathbf{y} \in K_Y$, is known or can be estimated. The latter can be done, for instance, from samples of $\mathbf{x}$ and $\mathbf{y}$. In particular, this means that the reference signal $\mathbf{x}$ can be measured.

In the case of processing large signal sets, such restrictions become much more inconvenient.

The major motivating question for this work is as follows. Let $\mathcal{F} : K_Y \to K_X$ denote a filter that estimates a large set of reference signals, K_X, from a large set of observed signals, K_Y. Each set contains N signals. Is it possible to construct a filter $\mathcal{F}$ that requires a priori information only on *few signals*, $p \ll N$, from K_X but *performs better* than the known filters based on a prior information on *every* reference signal from K_X? We denote such a filter by $\mathcal{F}^{(p-1)}$.

It is shown in Sections 2.3 and 4.4 that the positive answer is achievable under quite unrestrictive assumptions. The required features of filter $\mathcal{F}^{(p-1)}$ are satisfied by its special structure described in Sections 2.3, 3.1 and 3.4. The related conditions are also considered in those sections.

1.1.2. FILTERING BASED ON IDEA OF PIECEWISE FUNCTION INTERPOLATION. The specific structure of the proposed filter follows from the extension of piecewise function interpolation [14]. This is because the technique of piecewise function interpolation [14] has significant advantages over the methods of linear and polynomial approximation used in known filtering techniques (such as, for example, those in [5, 9]).

The structure of the proposed filter is presented in Sections 2.3, 3.1 and 4.2 below.

1.1.3. EXPLOITING PSEUDO-INVERSE MATRICES IN THE FILTER MODEL. Most of the known filtering techniques, for example, those ones in [1]–[3], [6]–[8], [11], [23, 25], are based on exploiting inverse matrices in their mathematical models.

[1]We say a random vector $\mathbf{x}$ is finite if its realization has a finite number of components.

[2]To the best of our knowledge, the exception is the methodology in [10, 12] where the filtering techniques exploit information on reference signals in the form of the vector obtained from averaging over reference signal sets.

In the cases of grossly corrupted signals or erroneous measurements those inverse matrices may not exist and, thus, those filters cannot be applied. The examples in Section 5 illustrate this case.

The filter proposed here avoids this drawback since its model is based on exploiting pseudo-inverse matrices. As a result, the proposed filter always exist. That is, it processes any kind of noisy signals. An extension of the filtering techniques to the case of implementation of the pseudo-inverse matrices is done on the basis of theory presented in [5].

1.1.4. COMPUTATIONAL WORK. Let m and n be the number of components of $\mathbf{x} \in K_X$ and of $\mathbf{y} \in K_Y$, respectively, where K_X and K_Y each contains N signals. The known filtering techniques (e.g., see [1]–[8], [11], [23, 25]), applied to $\mathbf{x}$ and $\mathbf{y}$, require the computation of a product of an $m \times n$ matrix and an $n \times n$ matrix, as well as the computation of an $n \times n$ inverse or pseudo-inverse matrix for *each pair* of signals $\mathbf{x} \in K_X$ and $\mathbf{y} \in K_Y$. This requires $O(2mn^2)$ and $O(26n^3)$ flops, respectively [26]. Thus, for the processing of all signals in K_X and K_Y, the filters in [1]–[8], [11], [23, 25] require $O(2mn^2N) + O(26n^3N)$ operations.

Alternatively, K_X and K_Y can be represented by vectors, χ and γ, each with mN and nN components, respectively. In such a case, the techniques in [1]–[8], [11], [23, 25] can be applied to χ and γ as opposed to each signals in K_X and K_Y. The computational requirement is then $O(2mn^2N^2)$ and $O(26n^3N^3)$ operations, respectively [26].

In both cases, but especially when N is large, the computational work associated with the approaches [1]–[8], [11], [23, 25] becomes unreasonable hard.

For the filter $\mathcal{F}^{(p-1)}$ to be introduced below, the associated computational work is substantially less. This is because $\mathcal{F}^{(p-1)}$ requires the computation of only p pseudo-inverse matrices associated with p selected signals in K_X, where p is much less than the number of signals in K_X. Therefore, for processing of the signal sets, K_X and K_Y, $\mathcal{F}^{(p-1)}$ requires only $O(2mn^2p) + O(26n^3p)$ flops where $p \ll N$. This comparison is illustrated in Section 5.

1.2. Relevant works

Some particular filtering techniques relevant to the method proposed below are as follows.

1.2.1. GENERIC OPTIMAL LINEAR (GOL) FILTER [5]. The generic optimal linear (GOL) filter in [5] is a generalization of the Wiener filter to the case when covariance matrix is not invertible and observable signal is arbitrarily noisy (i.e., when, in particular, noise is not necessarily additive and Gaussian). The GOL filter has been developed for processing an *individual* stochastic signal. Some ideas from [5] are used in the proof of Theorem 4.1 below.

1.2.2. SIMPLICIAL CANONICAL PIECEWISE LINEAR FILTER [23]. A complex Wiener adaptive filter was developed in [23] from the two-dimensional complex-valued simplicial canonical piecewise linear filter [24]. The filter in [23] was developed for the processing of an *individual* stochastic signal and can be exploited when the reference signal is known and a 'covariance-like' matrix is invertible. The latter precludes an application to the signal types considered, for example, in Section 5: the matrices used in [23] are not invertible for the signals as those in Section 5. Similarly, the filters studied in [8, 11] were developed for the processing of a single signal when the covariance matrices are invertible.

For the filter proposed here, these restrictions are removed.

1.2.3. ADAPTIVE PIECEWISE LINEAR FILTER [22]. A piecewise linear filter in [22] was proposed for a fixed image denoising (given by a matrix), corrupted by an *additive* Gaussian noise. That is, the method involved a non stochastic reference signal and required its knowledge. No theoretical justification for the filter was given in [22].

1.2.4. AVERAGING POLYNOMIAL FILTER [10, 12]. The averaging polynomial filter proposed in [10, 12] was developed for the purpose of processing infinite signal sets. The filter was based on an argument involving the'averaging' over sets of signals under consideration. This device allows one to determine a single filter for the processing of infinite signal sets. At the same time, it leads to an increase in the associated error when signals differ considerably from each other. This effect is illustrated in Section 5 below.

1.2.5. OTHER RELEVANT FILTERS. The technique developed in [13] is an extension of the GOL filter to the constraint problem with respect to the filter rank. It concerns data compression.

The methods in [6, 7, 15, 16] have been developed for deterministic signals. Motivated by the results achieved in [15, 16], adaptive filters were elaborated in [17]. A theoretical basis for the device proposed in [15, 16] is provided in [18].

We note that the idea of piecewise linear filtering has been used in the literature in several *very different conceptual frameworks*, despite exploiting some very similar terms (as in [15]–[24]). At the same time, a common feature of those techniques is that they were developed for the processing of a single signal, not of large signal sets as in this paper. In particular, piecewise linear filters in [19] have been obtained by arranging linear filters and thresholds in a tree structure. Piecewise linear filters discussed in [20] were developed using so-called threshold decomposition, which is a segmentation operator exploited to split a signal into a set of multilevel components. Filter design methods for piecewise linear systems proposed in [21] were based on a piecewise Lyapunov function.

1.3. Difficulties associated with the known filtering techniques

Basic difficulties associated with applying the known filtering techniques to the case under consideration (i.e., to processing of large signal sets, K_X and K_Y) are that:

(i) they require an information on *each* reference signal (in the form of a sample, for example),

(ii) matrices used in the known filters can be not invertible (as in the simulations considered below in Section 5) and then the filter does not exist, and

(iii) the associated computation work may require a very long time. For example, in some simulations, MATLAB was out of memory for computing the GOL filter [5] when each of sets K_X and K_Y was represented by a long vector (this option has been discussed in Section 1.1.4 above).

1.4. Differences from the known filtering techniques

The differences from the known filtering techniques discussed above are as follows.

(i) We consider a single filter that processes *arbitrarily large* input-output sets of stochastic signal-vectors. The known filters [1]–[9], [11], [13], [15]–[25] have been developed for the processing of an individual signal-vector only. In the case of their application to arbitrarily large signal sets, they imply difficulties described in Sections 1.1 and 1.3 above.

(ii) As a result, our piecewise linear filter model (Section 3), the statement of the problem (Section 3.3 below) and consequently, the device of its solution (Section 4 below) are different from those considered in [15]–[24]. In this regard, see also Section 1.2.5.

(iii) The above naturally leads to a new structure of the filter presented in Sections 3.4 and 4.2 below.

1.5. Contribution

In general, for the processing of large data sets, the proposed filter allows us to achieve better results in comparison with the known techniques in [1]–[25]. In particular, it allows us to

(i) achieve a desired accuracy in signal estimation[3],

(ii) exploit a priori information only on *few reference signals*, p, from the set K_X that contains $N \gg p$ signals or even infinite number of signals,

(iii) find *a single* filter to process any signal from the *arbitrarily large* signal set,

(vi) determine the filter in terms of pseudo-inverse matrices so that the filter always exists, and

(v) decrease the computational load compared to the related known techniques.

2. Some preliminaries

2.1. Notation

The signal sets we consider are, in fact, special representations of time series.

Let (Ω, Σ, μ) be a probability space[4], and K_X and K_Y be arbitrarily large sets of signals such that

$$K_X = \{\mathbf{x}(t, \cdot) \in L^2(\Omega, \mathbb{R}^m) \mid t \in T\} \quad \text{and} \quad K_Y = \{\mathbf{y}(t, \cdot) \in L^2(\Omega, \mathbb{R}^n) \mid t \in T\}$$

---

[3]This means that any desired accuracy is achieved theoretically, as is shown in Section 4.4 below. In practice, of course, the accuracy is increased to a prescribed reasonable level.

[4]As usual, $\Omega = \{\omega\}$ is the set of outcomes, Σ a σ-field of measurable subsets in Ω and $\mu : \Sigma \to [0, 1]$ an associated probability measure on Σ. In particular, $\mu(\Omega) = 1$.

where $T := [a, b] \subseteq \mathbb{R}$. We interpret $\mathbf{x}(t, \cdot)$ as a reference signal and $\mathbf{y}(t, \cdot)$ as an observable signal, an input to the filter $\mathcal{F}$ studied below[5]. The variable $t \in T \subseteq \mathbb{R}$ represents time.[6] Then, for example, the random signal $\mathbf{x}(t, \cdot)$ can be interpreted as an arbitrary stationary time series.

Let $\{t_k\}_1^p \subset T$ be a sequence of fixed time-points such that

$$a = t_1 < \cdots < t_p = b. \tag{2.1}$$

Because of the partition (2.1), the sets K_Y and K_X are divided in 'smaller' subsets $K_{X,1}, \ldots, K_{X,p-1}$ and $K_{Y,1}, \ldots, K_{Y,p-1}$, respectively, so that, for each $j = 1, \ldots, p$,

$$K_{X,j} = \{\mathbf{x}(t, \cdot) \mid t_j \le t \le t_{j+1}\} \quad \text{and} \quad K_{Y,j} = \{\mathbf{y}(t, \cdot) \mid t_j \le t \le t_{j+1}\}. \tag{2.2}$$

Therefore, K_Y and K_X can now be represented as

$$K_X = \bigcup_{j=1}^{p-1} K_{X,j} \quad \text{and} \quad K_Y = \bigcup_{j=1}^{p-1} K_{Y,j}. \tag{2.3}$$

2.2. Brief description of the problem

Given two *arbitrarily large* sets of random signals, K_Y and K_X, find a single filter $\mathcal{F} : K_Y \to K_X$ that estimates the signal $\mathbf{x} \in K_X$ with a controlled, associated error. Note that in our formulation the set K_Y can be finite or infinite.

2.3. Brief description of the method

The solution of the above problem is based on the representation of the proposed filter in the form of a sum with $p-1$ terms $\mathcal{F}_1, \ldots, \mathcal{F}_{p-1}$ where each term, $\mathcal{F}_j$, is interpreted as a particular sub-filter (see (3.1) and (3.2) below). Such a filter is denoted by $\mathcal{F}^{(p-1)} : K_Y \to K_X$.

The sub-filter $\mathcal{F}_j$ transforms signals that belong to 'piece' $K_{Y,j}$ of set K_Y to signals in 'piece' $K_{X,j}$ of K_X, i.e., $\mathcal{F}_j : K_{Y,j} \to K_{X,j}$. Each sub-filter $\mathcal{F}_j$ depends on two parameters, α_j and $\mathcal{B}_j$.

The prime idea is to determine $\mathcal{F}_j$ (i.e., α_j and $\mathcal{B}_j$) separately, for each $j = 1, \ldots, p-1$. The required α_j and $\mathcal{B}_j$ follow from the solutions of the equation (3.8) and an associated minimization problem (3.8) (see Sections 3.4 and 4.2 below). This procedure adjusts $\mathcal{F}_j$ so that the error associated with the estimation of $\mathbf{x}(t, \cdot) \in K_{X,j}$ is minimal.

A motivation for such a structure of the filter $\mathcal{F}^{(p-1)}$ is as follows. The method of determining α_j and $\mathcal{B}_j$ provides an estimate $\mathcal{F}_j[\mathbf{y}(t, \cdot)]$ that interpolates $\mathbf{x}(t, \cdot) \in K_{X,j}$ at $t = t_j$ and $t = t_{j+1}$. In other words, the filter is flexible to variations in the sets of observed and reference signals K_Y and K_X, respectively. Due to this way of determining $\mathcal{F}_j$, it is natural to expect that the processing of a 'smaller' signal

[5]In an intuitive way $\mathbf{y}$ can be regarded as a noise-corrupted version of $\mathbf{x}$. For example, $\mathbf{y}$ can be interpreted as $\mathbf{y} = \mathbf{x} + \mathbf{n}$ where $\mathbf{n}$ is white noise. In this paper, we do not restrict ourselves to this simplest version of $\mathbf{y}$ and assume that the dependence of $\mathbf{y}$ on $\mathbf{x}$ and $\mathbf{n}$ is arbitrary.

[6]More generally, T can be considered as a set of parameter vectors $\alpha = (\alpha^{(1)}, \ldots, \alpha^{(q)})^T \in C^q \subseteq \mathbb{R}^q$, where C^q is a q-dimensional cube, i.e., $\mathbf{y} = \mathbf{y}(\alpha, \cdot)$ and $\mathbf{x} = \mathbf{x}(\alpha, \cdot)$. One coordinate, say $\alpha^{(1)}$ of α, could be interpreted as time.

set, $K_{Y,j}$, may lead to a *smaller associated error* than that for the processing of the whole set K_Y by a filter which is not specifically adjusted to each particular piece $K_{Y,j}$.

As a result, $\mathcal{F}^{(p-1)}[\mathbf{y}(t,\cdot)]$ represents a special *piecewise interpolation procedure* and, thus, should be attributed with the associated advantages such as, for example, the high accuracy of estimation.

In Section 4.4, this observation is confirmed. In Sections 4.5 and 5, it is also shown that the proposed technique allows us to avoid the difficulties discussed in Section 1.3 above.

3. Description of the problem

3.1. Piecewise linear filter model

Let $\mathcal{F}^{(p-1)} : K_Y \to K_X$ be a filter such that, for each $t \in T$,

$$\mathcal{F}^{(p-1)}[\mathbf{y}(t,\cdot)] = \sum_{j=1}^{p-1} \delta_j \mathcal{F}_j[\mathbf{y}(t,\cdot)], \tag{3.1}$$

where

$$\mathcal{F}_j[\mathbf{y}(t,\cdot)] = \alpha_j + \mathcal{B}_j[\mathbf{y}(t,\cdot)] \quad \text{and} \quad \delta_j = \begin{cases} 1, & \text{if } t_j \le t \le t_{j+1}, \\ 0, & \text{otherwise.} \end{cases} \tag{3.2}$$

Here, $\mathcal{F}_j$ is a sub-filter defined for $t_j \le t \le t_{j+1}$. In (3.2), $\alpha_j = [\alpha_j^{(1)}, \dots, \alpha_j^{(m)}]^T \in \mathbb{R}^m$ and $\mathcal{B}_j : L^2(\Omega, \mathbb{R}^n) \to L^2(\Omega, \mathbb{R}^m)$ is a linear operator given by a matrix $B_j \in \mathbb{R}^{m\times n}$, so that

$$[\mathcal{B}_j(\mathbf{y})](t,\omega) = B_j[\mathbf{y}(t,\omega)].$$

Thus, $\mathcal{F}_j$ is defined by a matrix $F_j \in \mathbb{R}^{m\times n}$ such that

$$F_j[\mathbf{y}(t,\omega)] = \alpha_j + B_j[\mathbf{y}(t,\omega)]. \tag{3.3}$$

Filter $\mathcal{F}^{(p-1)}$ defined by (3.1)–(3.3) is called the *piecewise* filter[7].

3.2. Assumptions

In the known approaches related to filtering of stochastic signals (e.g., see [1]–[13], [23], [25]), it is assumed that covariance matrices formed from the reference signal and observed signal are known or can be estimated.

The assumption used here is similar. The covariance matrices that are assumed to be known or can be estimated, are formed from *selected* signal pairs $\{\mathbf{x}(t_j,\cdot), \mathbf{y}(t_j,\cdot)\}$ with $j = 1,\dots,p$ and p to be a small number[8], $p \ll N$, where N is the number of signals in K_X or K_Y.

---

[7]Hereinafter, we will use a non-curly symbol to denote an operator and associated matrix (e.g., the operator $\mathcal{F}_j : L^2(\Omega,\mathbb{R}^n) \to L^2(\Omega,\mathbb{R}^m)$ and the associated matrix $F_j \in \mathbb{R}^{m\times n}$ are denoted by F_j).

[8]It is worthwhile to note that it is *not* assumed that the covariance matrices are known for *each* signal pair from $K_X \times K_Y$, $\{\mathbf{x}(t,\cdot), \mathbf{y}(t,\cdot)\}$ with $t \in [a, b]$.

3.3. The problem

In (3.1)–(3.3), parameters of the filter $\mathcal{F}^{(p-1)}$, i.e., vector α_j and matrix B_j, for $j = 1,\dots,p-1$, are unknown. Therefore, under the assumptions described in Section 3.2, the problem is to determine α_j and B_j, for $j = 1,\dots,p-1$. The related problem is to estimate an error associated with the filter $\mathcal{F}^{(p-1)}$.

Solutions to the both problems are given in Sections 4.2 and 4.4, respectively. In particular, in the following Section 3.4, interpolation conditions (3.5) and (3.8) are introduced that lead to a determination of α_j and B_j.

3.4. Interpolation conditions

Let us denote

$$\|\mathbf{x}(t_j,\cdot)\|_\Omega^2 = \int_\Omega \|\mathbf{x}(t_j,\omega)\|_2^2 d\mu(\omega) \tag{3.4}$$

where $\|\mathbf{x}(t_j,\omega)\|_2$ is the Euclidean norm of $\mathbf{x}(t_j,\omega) \in \mathbb{R}^m$.

For $t = t_1$, let $\widehat{\mathbf{x}}(t_1,\cdot)$ be an estimate of $\mathbf{x}(t_1,\cdot)$ determined by known methods [1]–[13], [23], [25]. This is *the initial condition* of the proposed technique.

For $j = 1,\dots,p-1$, each sub-filter F_j in (3.2)–(3.3) is defined so that α_j and $\mathcal{B}_j$ satisfy the conditions as follows.

Sub-filter $\mathcal{F}_1$: For $j = 1$, α_1 and $\mathcal{B}_1$ solve

$$\widehat{\mathbf{x}}(t_1,\cdot) = \alpha_1 + \mathcal{B}_1[\mathbf{y}(t_1,\cdot)] \quad \text{and} \quad \min_{\mathcal{B}_1} \|[\mathbf{x}(t_2,\cdot) - \alpha_1] - \mathcal{B}_1[\mathbf{y}(t_2,\cdot)]\|_\Omega^2, \tag{3.5}$$

respectively. Then an estimate of $\mathbf{x}(t,\cdot)$, $\widehat{\mathbf{x}}(t,\cdot)$, for $t \in [t_1,\ t_2]$, is determined as

$$\widehat{\mathbf{x}}(t,\cdot) = \mathcal{F}_1[\mathbf{y}(t,\cdot)] = \alpha_1 + \mathcal{B}_1[\mathbf{y}(t,\cdot)] = \widehat{\mathbf{x}}(t_1,\cdot) + \mathcal{B}_1[\mathbf{y}(t,\cdot) - \mathbf{y}(t_1,\cdot)] \tag{3.6}$$

where α_1 and $\mathcal{B}_1$ satisfy (3.5). In particular, $\alpha_1 = \widehat{\mathbf{x}}(t_1,\cdot) - \mathcal{B}_1[\mathbf{y}(t_1,\cdot)]$ and

$$\widehat{\mathbf{x}}(t_2,\cdot) = \mathcal{F}_1[\mathbf{y}(t_2,\cdot)].$$

Extending this procedure up to $j = k-1$, where $k = 3,\dots,p$, we set the following. Let $\widehat{\mathbf{x}}(t_{k-1},\cdot)$ be an estimate of $\mathbf{x}(t_{k-1},\cdot)$ defined by the preceding steps as

$$\widehat{\mathbf{x}}(t_{k-1},\cdot) = \mathcal{F}_{k-2}[\mathbf{y}(t_{k-1},\cdot)]. \tag{3.7}$$

Then sub-filter $\mathcal{F}_{k-1}$ is defined as follows.

Sub-filter $\mathcal{F}_{k-1}$: For $j = k-1$, α_{k-1} and $\mathcal{B}_{k-1}$ solve

$$\widehat{\mathbf{x}}(t_{k-1},\cdot) = \alpha_{k-1} + \mathcal{B}_{k-1}[\mathbf{y}(t_{k-1},\cdot)] \tag{3.8}$$

$$\text{and} \quad \min_{\mathcal{B}_{k-1}} \|[\mathbf{x}(t_k,\cdot) - \alpha_{k-1}] - \mathcal{B}_{k-1}[\mathbf{y}(t_k,\cdot)]\|_\Omega^2, \tag{3.9}$$

respectively. Then an estimate of $\mathbf{x}(t,\cdot)$, $\widehat{\mathbf{x}}(t,\cdot)$, for $t \in [t_{k-1},\ t_k]$, is determined as

$$\begin{aligned}\widehat{\mathbf{x}}(t,\cdot) &= \mathcal{F}_{k-1}[\mathbf{y}(t,\cdot)] = \alpha_{k-1} + \mathcal{B}_{k-1}[\mathbf{y}(t,\cdot)] \\ &= \widehat{\mathbf{x}}(t_{k-1},\cdot) + \mathcal{B}_1[\mathbf{y}(t,\cdot) - \mathbf{y}(t_{k-1},\cdot)].\end{aligned} \tag{3.10}$$

The conditions (3.5) and (3.8)–(3.9) are motivated by the device of piecewise function interpolation and associated advantages [14].

Filter $\mathcal{F}^{(p-1)}$ of the form (3.1)–(3.2) with α_j and $\mathcal{B}_j$ satisfying (3.5) and (3.8) is called the *piecewise linear interpolation* filter. The pair of signals $\{\mathbf{x}(t_k,\cdot), \mathbf{y}(t_k,\cdot)\}$ associated with time t_k defined by (2.1) is called the *interpolation pair.*

4. Main results

4.1. General device

In accordance with the scheme presented in Sections 3.1 and 3.4 above, an estimate of the reference signal $\mathbf{x}(t,\cdot)$, for any $t \in T = [a,\ b]$, by the *piecewise linear interpolation* filter $\mathcal{F}^{(p-1)}$, is given by

$$\widehat{\mathbf{x}}(t,\cdot) = \mathcal{F}^{(p-1)}[\mathbf{y}(t,\cdot)] = \sum_{j=1}^{p-1} \delta_j \mathcal{F}_j[\mathbf{y}(t,\cdot)], \tag{4.1}$$

where, for each $j = 1,\dots,p-1$, the sub-filter $\mathcal{F}_j$ is given by (3.2), and is defined from the interpolation conditions (3.5) and (3.8).

Below, we show how to determine $\mathcal{F}_j$ to satisfy the conditions (3.5) and (3.8).

4.2. Determination of piecewise linear interpolation filter

Let us denote

$$\mathbf{z}(t_j,t_{j+1},\cdot) = \mathbf{x}(t_{j+1},\cdot) - \widehat{\mathbf{x}}(t_j,\cdot) \quad \text{and} \quad \mathbf{w}(t_j,t_{j+1},\cdot) = \mathbf{y}(t_{j+1},\cdot) - \mathbf{y}(t_j,\cdot). \tag{4.2}$$

We need to represent $\mathbf{z}(t_j,t_{j+1},\cdot)$ and $\mathbf{w}(t_j,t_{j+1},\cdot)$ in terms of their components as follows:

$$\mathbf{z}(t_j,t_{j+1},\cdot) = [\mathbf{z}^{(1)}(t_j,t_{j+1},\cdot),\dots,\mathbf{z}^{(m)}(t_j,t_{j+1},\cdot)]^T$$
$$\text{and} \quad \mathbf{w}(t_j,t_{j+1},\cdot) = [\mathbf{w}^{(1)}(t_j,t_{j+1},\cdot),\dots,\mathbf{w}^{(n)}(t_j,t_{j+1},\cdot)]^T,$$

where $\mathbf{z}^{(j)}(t_j,t_{j+1},\cdot) \in L^2(\Omega,\mathbb{R})$ and $\mathbf{w}^{(i)}(t_j,t_{j+1},\cdot) \in L^2(\Omega,\mathbb{R})$ are random variables, for all $j = 1,\dots,m$.

Then we can introduce the covariance matrix

$$E_{z_j w_j} = \left\{ \left\langle \mathbf{z}^{(i)}(t_j,t_{j+1},\cdot), \mathbf{w}^{(k)}(t_j,t_{j+1},\cdot) \right\rangle \right\}_{i,k=1}^{m,n}, \tag{4.3}$$

where $\langle \mathbf{z}^{(i)}(t_j,t_{j+1},\cdot), \mathbf{w}^{(k)}(t_j,t_{j+1},\cdot) \rangle = \int_\Omega \mathbf{z}^{(i)}(t_j,t_{j+1},\omega)\mathbf{w}^{(k)}(t_j,t_{j+1},\omega)\, d\mu(\omega)$.

Below, $M^\dagger$ is the Moore–Penrose generalized inverse of a matrix M.

Now, we are in a position to establish the main results.

Theorem 4.1. *Let $K_X = \{\mathbf{x}(t,\cdot) \in L^2(\Omega,\mathbb{R}^m) \mid t \in T = [a,\ b]\}$ and $K_Y = \{\mathbf{y}(t,\cdot) \in L^2(\Omega,\mathbb{R}^n) \mid t \in T = [a,\ b]\}$ be sets of reference signals and observed signals, respectively. Let $t_j \in [a,\ b]$, for $j = 1,\dots,p$, be such that*

$$a = t_1 < \dots < t_p = b.$$

For $t = t_1$, let $\widehat{\mathbf{x}}(t_1,\cdot)$ be a known estimate of $\mathbf{x}(t_1,\cdot)$[9]*. Then, for any $t \in [a,\ b]$, the proposed piecewise linear interpolation filter $\mathcal{F}^{(p-1)} : L^2(\Omega,\mathbb{R}^n) \to L^2(\Omega,\mathbb{R}^m)$*

[9]As it has been mentioned in Section 3.4, $\widehat{\mathbf{x}}(t_1,\cdot)$ can be determined by the known methods.

transforming any signal $\mathbf{y}(t,\cdot) \in L^2(\Omega,\mathbb{R}^m)$ *to an estimate of* $\mathbf{x}(t,\cdot)$, $\widehat{\mathbf{x}}(t,\cdot)$, *is given by*

$$\widehat{\mathbf{x}}(t,\cdot) = \mathcal{F}^{(p-1)}[\mathbf{y}(t,\cdot)] = \sum_{j=1}^{p-1} \delta_j \mathcal{F}_j[\mathbf{y}(t,\cdot)] \tag{4.4}$$

where

$$\mathcal{F}_j[\mathbf{y}(t,\cdot)] = \widehat{\mathbf{x}}(t_j,\cdot) + \mathcal{B}_j[\mathbf{y}(t,\cdot) - \mathbf{y}(t_j,\cdot)], \tag{4.5}$$

$$\widehat{\mathbf{x}}(t_j,\cdot) = \mathcal{F}_{j-1}[\mathbf{y}(t_j,\cdot)] \quad (\textit{for } j = 2,\dots,p-1), \tag{4.6}$$

$$B_j = E_{z_j w_j} E^{\dagger}_{w_j w_j} + M_{B_j}[I_n - E_{w_j w_j} E^{\dagger}_{w_j w_j}], \tag{4.7}$$

and where I_n *is the* $n \times n$ *identity matrix and* M_{B_j} *is an* $m \times n$ *arbitrary matrix.*

Proof. It follows from (3.5) and (3.8) that α_j, for $j = 1,\dots,p-1$, is given by

$$\alpha_j = \widehat{\mathbf{x}}(t_j,\omega) - B_j[\mathbf{y}(t_j,\omega)]. \tag{4.8}$$

Further, for α_j given by (4.8),

$$\begin{aligned} &\|[\mathbf{x}(t_{j+1},\cdot) - \alpha_j] - \mathcal{B}_j[\mathbf{y}(t_{j+1},\cdot)]\|^2_\Omega \\ &= \|\mathbf{z}(t_j,t_{j+1},\cdot) - \mathcal{B}_j[\mathbf{w}(t_j,t_{j+1},\cdot))]\|^2_\Omega \qquad (4.9) \\ &= \operatorname{tr}\{E_{z_j z_j} - E_{z_j w_j} B_j^T - B_j E_{w_j z_j} + B_j E_{w_j w_j} B_j^T\} \\ &= \|E^{1/2}_{z_j z_j}\|^2 - \|E_{z_j w_j}(E^{1/2}_{w_j w_j})^{\dagger}\|^2 + \|(B_j - E_{z_j w_j} E^{\dagger}_{w_j w_j}) E^{1/2}_{w_j w_j}\|^2 \\ &= \|E^{1/2}_{z_j z_j}\|^2 - \|E_{z_j w_j}(E^{1/2}_{w_j w_j})^{\dagger}\|^2 + \|E_{z_j w_j}(E^{1/2}_{w_j w_j})^{\dagger} - B_j E^{1/2}_{w_j w_j}\|^2, \qquad (4.10) \end{aligned}$$

where $\|\cdot\|$ is the Frobenius norm. The latter is true because

$$E^{\dagger}_{w_j w_j} E^{1/2}_{w_j w_j} = (E^{1/2}_{w_j w_j})^{\dagger} \quad \text{and} \quad E_{z_j w_j} E^{\dagger}_{w_j w_j} E_{w_j w_j} = E_{z_j w_j} \tag{4.11}$$

by Lemma 24 in [5]. Thus, the second expression in (3.8) is reduced to the problem

$$\min_{B_j} \|E_{z_j w_j}(E^{1/2}_{w_j w_j})^{\dagger} - B_j E^{1/2}_{w_j w_j}\|^2. \tag{4.12}$$

It is known (see, for example, [5], p. 304) that the solution of problem (4.12) is given by (4.7). The equation (4.5) follows from (3.3) and (4.8).

Theorem 4.1 is proven. □

It is worthwhile to observe that, due to an arbitrary matrix M_{B_j} in (4.7), the filter $\mathcal{F}^{(p-1)}$ is not unique. In particular, M_{B_j} can be chosen as the zero matrix $\mathbb{O}$ similarly to the generic optimal linear [5] (which is also not unique by the same reason).

4.3. Numerical realization of filter $\mathcal{F}^{(p-1)}$ and associated algorithm

4.3.1. NUMERICAL REALIZATION. In practice, the set $T = [a,\ b]$ (see Section 2.1) is represented by a finite set $\{\tau_k\}_{k=1}^N$, i.e., $[a,\ b] = [\tau_1, \tau_2, \ldots, \tau_N]$ where $a \le \tau_1 < \tau_2 < \cdots < \tau_N \le b$.

For $k = 1, \ldots, N$, the estimate of $\mathbf{x}(\tau_k, \cdot)$, $\widehat{\mathbf{x}}(\tau_k, \cdot)$, and observed signal $\mathbf{y}(\tau_k, \cdot)$ are represented by $m \times q$ and $n \times q$ matrices

$$\widehat{X}^{(k)} = [\widehat{\mathbf{x}}(\tau_k, \omega_1), \ldots, \widehat{\mathbf{x}}(\tau_k, \omega_q)] \quad \text{and} \quad Y^{(k)} = [\mathbf{y}(\tau_k, \omega_1), \ldots, \mathbf{y}(\tau_k, \omega_q)]. \quad (4.13)$$

The sequence of fixed time-points $\{t_k\}_1^p \subset [a,\ b]$ introduced in (2.1) is such that

$$\tau_1 = t_1 < \cdots < t_p = \tau_N, \quad (4.14)$$

where $t_1 = \tau_{n_0}, \quad t_2 = \tau_{n_0+n_1}, \quad \ldots, \quad t_p = \tau_{n_0+n_1\ldots+n_{p-1}}$, and where $n_0 = 1$ and $n_1, \ldots, n_{p-1}$ are positive integers such that $N = n_0 + n_1 + \cdots + n_{p-1}$.

For $j = 1, \ldots, p$, signal $\mathbf{y}(tj, \cdot)$ associated with t_j in (4.14) is represented by

$$Y_j = [\mathbf{y}(t_j, \omega_1), \ldots, \mathbf{y}(t_j, \omega_N)].$$

4.3.2. ALGORITHM. As it has been mentioned in Section 3.4, it is supposed that, for $t = t_1$, an estimate of X_1, $\widehat{X}_1$, is known and can be determined by the known methods. This is the initial condition of the proposed technique.

On the basis of the results obtained in Sections 3.4 and 4.2, the performance algorithm of the proposed filter consists of the following steps. For $j = 1 \ldots, p$, we write $N_j = n_0 + n_1 + \cdots + n_{j-1}$.

Initial parameters: $Y^{(1)}, \ldots, Y^{(N)}$, $\{t_j\}_{j=1}^p$ (see (4.14)), $\{E_{z_j w_j}\}_{j=1}^p$, $\{E_{w_j w_j}\}_{j=1}^p$ (see (4.2) and (4.3)), $\widehat{X}_1$, $n_0 = 1$ and $M_{B_j} = \mathbb{O}$, for $j = 1, \ldots, p-1$.

(Possible ways to get estimates of $E_{z_j w_j}$ and $E_{w_j w_j}$ are discussed below in Section 4.5.)

Final parameters: $\widehat{X}^{(2)}$, $\widehat{X}^{(3)}$, …, $\widehat{X}^{(N)}$.

Algorithm:

- for $j = 1$ to p do
  begin

$$B_j = E_{z_j w_j} E_{w_j w_j}^{\dagger};$$

  - for $k = N_{j-1} + 1$ to N_j do
    begin

$$\widehat{X}^{(k)} = \widehat{X}_j + B_j (Y^{(k)} - Y_j);$$

    end

  end

4.4. Error analysis

It is natural to expect that the error associated with the piecewise interpolating filter $\mathcal{F}^{(p-1)}$ decreases when $\max\limits_{j=1,\dots,p-1} \Delta t_j$ decreases. Below, in Theorem 4.3, we justify that this observation is true. To this end, first, in the following Theorem 4.2, we establish an estimate of the error associated with the filter F.

Let us introduce the norm by

$$\|\mathbf{x}(t,\cdot)\|^2_{T,\Omega} = \frac{1}{b-a}\int_T \|\mathbf{x}(t,\cdot)\|^2_\Omega dt. \tag{4.15}$$

We also denote $\|\mathbf{x}(t,\omega)\|^2_{T,\Omega} = \|\mathbf{x}(t,\cdot)\|^2_{T,\Omega}$.

Let us suppose that $\mathbf{x}(\cdot,\omega)$ and $\mathbf{y}(\cdot,\omega)$ are Lipschitz continuous signals, i.e., that there exist real non-negative constants λ_j and γ_j, with $j = 1,\dots,p$, such that, for $t \in [t_j,\ t_{j+1}]$,

$$\|\mathbf{x}(t,\omega) - \mathbf{x}(t_j,\omega)\|^2_{T,\ \Omega} \le \lambda_j \Delta t_j \quad \text{and} \quad \|\mathbf{y}(t,\omega) - \mathbf{y}(t_{j+1},\omega)\|^2_{T,\ \Omega} \le \gamma_j \Delta t_j \tag{4.16}$$

where $\Delta t_j = |t_{j+1} - t_j|$.

Theorem 4.2. *Under the conditions* (4.16) *the error associated with the piecewise interpolation filter,* $\|\mathbf{x}(t,\omega) - F^{(p-1)}[\mathbf{y}(t,\omega)]\|^2_{T,\ \Omega}$, *is estimated as follows:*

$$\begin{aligned}&\|\mathbf{x}(t,\omega) - F^{(p-1)}[\mathbf{y}(t,\omega)]\|^2_{T,\ \Omega}\\ &\quad\le \max_{j=1,\dots,p-1} \left[(\lambda_j + \gamma_j\|B_j\|^2)\Delta t_j + \|E^{1/2}_{z_j z_j}\|^2 - \|E_{z_j w_j}(E^{1/2}_{w_j w_j})^\dagger\|^2\right].\end{aligned} \tag{4.17}$$

Proof. For $t \in [t_j,\ t_{j+1}]$ and F_j defined by (4.5)–(4.7),

$$\begin{aligned}&\mathbf{x}(t,\omega) - F[\mathbf{y}(t,\omega)]\\ &\quad= \mathbf{x}(t,\omega) - F_j[\mathbf{y}(t,\omega)] &(4.18)\\ &\quad= \mathbf{x}(t,\omega) - \widehat{\mathbf{x}}(t_j,\omega) + B_j\mathbf{y}(t_j,\omega) - B_j\mathbf{y}(t,\omega)\\ &\quad= [\mathbf{x}(t,\omega) - \mathbf{x}(t_{j+1},\omega)]\\ &\qquad+ \mathbf{z}(t_j,t_{j+1},\omega) - B_j\mathbf{w}(t_j,t_{j+1},\omega) + B_j[\mathbf{y}(t_{j+1},\omega) - \mathbf{y}(t,\omega)]. &(4.19)\end{aligned}$$

Then (4.18) and (4.19) imply

$$\begin{aligned}\|\mathbf{x}(t,\omega) - F[\mathbf{y}(t,\omega)]\|^2_{T,\Omega} &\le \|\mathbf{x}(t,\omega) - \mathbf{x}(t_{j+1},\omega)\|^2_{T,\Omega}\\ &\quad+ \|\mathbf{z}(t_j,t_{j+1},\omega) - B_j\mathbf{w}(t_j,t_{j+1},\omega)\|^2_\Omega\\ &\quad+ \|B_j[\mathbf{y}(t_{j+1},\omega) - \mathbf{y}(t,\omega)]\|^2_{T,\Omega}.\end{aligned} \tag{4.20}$$

It follows from (4.9) and (4.10) that for B_j given by (4.7),

$$\|\mathbf{z}(t_j,t_{j+1},\omega) - B_j\mathbf{w}(t_j,t_{j+1},\omega)\|^2_\Omega = \|E^{1/2}_{z_j z_j}\|^2 - \|E_{z_j w_j}(E^{1/2}_{w_j w_j})^\dagger\|^2. \tag{4.21}$$

Then (4.4)–(4.7), (4.16) and (4.18)–(4.21) imply that for all $t \in [a,\ b]$ and $\omega \in \Omega$, (4.17) is true. □

Further, to show that the error of the reference signal estimate tends to the zero, we need to assume that, for $t \in [t_1,\ t_2]$, the known estimate $\widehat{\mathbf{x}}(t_1,\omega)$ differs from $\mathbf{x}(t,\omega)$ for the value of the order Δt_1, i.e., that, for some constant $c_1 \geq 0$,

$$\|\mathbf{x}(t,\omega) - \widehat{\mathbf{x}}(t_1,\omega)\|_\Omega^2 \leq c_1 \Delta t_1, \quad \text{for } t \in [t_1,\ t_2]. \tag{4.22}$$

Theorem 4.3. *Let the conditions* (4.16) *and* (4.22) *be true. Then the error associated with the piecewise interpolating filter* F, $\|\mathbf{x}(t,\omega) - F^{(p-1)}[\mathbf{y}(t,\omega)]\|_{T,\ \Omega}^2$, *decreases in the following sense:*

$$\|\mathbf{x}(t,\omega) - F^{(p-1)}[\mathbf{y}(t,\omega)]\|_{T,\ \Omega}^2 \to 0 \quad \textit{as} \quad \max_{j=1,\dots,p-1} \Delta t_j \to 0 \quad \textit{and} \quad p \to \infty. \tag{4.23}$$

Proof. The relation (4.15) implies that

$$\|\mathbf{x}(t,\omega) - F[\mathbf{y}(t,\omega)]\|_{T,\Omega}^2 = \frac{1}{b-a} \sum_{j=1}^{p-1} \int_{t_j}^{t_{j+1}} \|\mathbf{x}(t,\omega) - F_j[\mathbf{y}(t,\omega)]\|_\Omega^2 dt, \tag{4.24}$$

where

$$\begin{aligned}
&\|\mathbf{x}(t,\omega) - F_j[\mathbf{y}(t,\omega)]\|_\Omega^2 \\
&\quad = \|\mathbf{x}(t,\omega) - \widehat{\mathbf{x}}(t_j,\omega) + B_j[\mathbf{y}(t_j,\omega) - B_j\mathbf{y}(t,\omega)]\|_\Omega^2 \\
&\quad \leq \|\mathbf{x}(t,\omega) - \mathbf{x}(t_j,\omega)\|_\Omega^2 + \|\mathbf{x}(t_j,\omega) - \widehat{\mathbf{x}}(t_j,\omega)\|_\Omega^2 + \|B_j[\mathbf{y}(t_j,\omega) - B_j\mathbf{y}(t,\omega)]\|_\Omega^2.
\end{aligned}$$

Then

$$\int_{t_j}^{t_{j+1}} \|\mathbf{x}(t,\omega) - F_j[\mathbf{y}(t,\omega)]\|_\Omega^2 dt \tag{4.25}$$

$$\begin{aligned}
&\leq \int_{t_j}^{t_{j+1}} \|\mathbf{x}(t,\omega) - \mathbf{x}(t_j,\omega)\|_\Omega^2 dt + \int_{t_j}^{t_{j+1}} \|\mathbf{x}(t_j,\omega) - \widehat{\mathbf{x}}(t_j,\omega)\|_\Omega^2 dt \\
&\qquad + \|B_j\| \int_{t_j}^{t_{j+1}} \|\mathbf{y}(t_j,\omega) - \mathbf{y}(t,\omega)\|_\Omega^2 dt
\end{aligned}$$

$$\leq \lambda_j(\Delta t_j)^2 + \|\mathbf{x}(t_j,\omega) - \widehat{\mathbf{x}}(t_j,\omega)\|_\Omega^2 \Delta t_j + \|B_j\|\gamma_j(\Delta t_j)^2 \tag{4.26}$$

Let us consider an estimate of $\|\mathbf{x}(t_j,\omega) - \widehat{\mathbf{x}}(t_j,\omega)\|_\Omega^2$, for $j = 1,\dots,p-1$. To this end, let us denote $\Delta t = \max_{j=1,\dots,p-1} \Delta t_j$.

For $j = 1$, i.e., for $t \in [t_1,\ t_2]$,

$$\begin{aligned}
&\|\mathbf{x}(t,\omega) - F_1\mathbf{y}(t,\omega)\|_\Omega^2 \\
&\quad \leq \|\mathbf{x}(t,\omega) - \mathbf{x}(t_1,\omega)\|_\Omega^2 + \|\mathbf{x}(t_1,\omega) - \widehat{\mathbf{x}}(t_1,\omega)\|_\Omega^2 + \|B_1\|\|\mathbf{y}(t_1,\omega) - \mathbf{y}(t,\omega)\|_\Omega^2 \\
&\quad \leq \lambda_1 \Delta t_1 + c_1 \Delta t_1 + \|B_1\|\gamma_1 \Delta t_1 \\
&\quad \leq \beta_1 \Delta t,
\end{aligned}$$

where $\beta_1 = \lambda_1 + c_1 + \|B_1\|\gamma_1$. In particular, the latter implies

$$\|\mathbf{x}(t_2,\omega) - \widehat{\mathbf{x}}(t_2,\omega)\|_\Omega^2 = \|\mathbf{x}(t_2,\omega) - F_1\mathbf{y}(t_2,\omega)\|_\Omega^2 \leq \beta_1 \Delta t$$

For $j = 2$, i.e., for $t \in [t_2,\ t_3]$,

$$\begin{aligned}
&\|\mathbf{x}(t,\omega) - F_2\mathbf{y}(t,\omega)\|_\Omega^2\\
&\quad \le \|\mathbf{x}(t,\omega) - \mathbf{x}(t_2,\omega)\|_\Omega^2 + \|\mathbf{x}(t_2,\omega) - \widehat{\mathbf{x}}(t_2,\omega)\|_\Omega^2 + \|B_2\|\|\mathbf{y}(t_2,\omega) - \mathbf{y}(t,\omega)\|_\Omega^2\\
&\quad \le \lambda_2 \Delta t_2 + \beta_1 \Delta t + \|B_2\|\gamma_2 \Delta t_2\\
&\quad \le \beta_2 \Delta t,
\end{aligned}$$

where $\beta_2 = \lambda_2 + \beta_1 + \|B_2\|\gamma_2$. In particular, then it follows that

$$\|\mathbf{x}(t_3,\omega) - \widehat{\mathbf{x}}(t_3,\omega)\|_\Omega^2 = \|\mathbf{x}(t_3,\omega) - F_2\mathbf{y}(t_3,\omega)\|_\Omega^2 \le \beta_2 \Delta t.$$

On the basis of the above, let us assume that, for $j = k-1$ with $k = 2,\dots,p-1$, i.e., for $t \in [t_{k-1},\ t_k]$,

$$\|\mathbf{x}(t_k,\omega) - \widehat{\mathbf{x}}(t_k,\omega)\|_\Omega^2 = \|\mathbf{x}(t_k,\omega) - F_{k-1}\mathbf{y}(t_k,\omega)\|_\Omega^2 \le \beta_{k-1} \Delta t$$

where β_{k-1} is defined by analogy with β_2.

Then, for $j = k$ with $k = 2,\dots,p-1$, i.e., for $t \in [t_k,\ t_{k+1}]$,

$$\begin{aligned}
&\|\mathbf{x}(t,\omega) - F_k\mathbf{y}(t,\omega)\|_\Omega^2\\
&\quad \le \|\mathbf{x}(t,\omega) - \mathbf{x}(t_k,\omega)\|_\Omega^2 + \|\mathbf{x}(t_k,\omega) - \widehat{\mathbf{x}}(t_k,\omega)\|_\Omega^2 + \|B_k\|\|\mathbf{y}(t_k,\omega) - \mathbf{y}(t,\omega)\|_\Omega^2\\
&\quad \le \lambda_k \Delta t_k + \beta_{k-1} \Delta t + \|B_k\|\gamma_2 \Delta t_k\\
&\quad \le \beta_k \Delta t,
\end{aligned}$$

where $\beta_k = \lambda_k + \beta_{k-1} + \|B_k\|\gamma_k$. Thus, the following is true:

$$\|\mathbf{x}(t_{k+1},\omega) - \widehat{\mathbf{x}}(t_{k+1},\omega)\|_\Omega^2 = \|\mathbf{x}(t_{k+1},\omega) - F_k\mathbf{y}(t_{k+1},\omega)\|_\Omega^2 \le \beta_k \Delta t. \tag{4.27}$$

Therefore, (4.25), (4.26) and (4.27) imply

$$\begin{aligned}
&\int_{t_j}^{t_{j+1}} \|\mathbf{x}(t,\omega) - F_j[\mathbf{y}(t,\omega)]\|_\Omega^2 dt\\
&\qquad \le \lambda_j(\Delta t_j)^2 + \beta_{j-1}(\Delta t_j)^2 + \|B_j\|\gamma_j(\Delta t_j)^2\\
&\qquad \le \eta_j(\Delta t)^2
\end{aligned} \tag{4.28}$$

where $\eta_j = \lambda_j + \beta_{j-1} + \|B_j\|$, and then it follows from (4.24)–(4.26) and (4.28) that for all $t \in [a,\ b]$,

$$\|\mathbf{x}(t,\omega) - F[\mathbf{y}(t,\omega)]\|_{T,\Omega}^2 \le \frac{1}{b-a}\sum_{j=1}^{p-1}\eta_j(\Delta t)^2 = \frac{1}{b-a}\Delta t\sum_{j=1}^{p-1}\eta_j \Delta t. \tag{4.29}$$

Let us now choose $c \in \mathbb{R}$ and $d \in \mathbb{R}$ so that $\Delta t = \dfrac{d-c}{p}$ and partition interval $[c,\ d] \subset \mathbb{R}$ by points $\tau_1,\dots,\tau_p$ so that $c = \tau_1$ and $\tau_j = \tau_1 + j\Delta t$ with $j = 1,\dots,p$. There exists an integrable (bounded) function $\varphi : [c,\ d] \to \mathbb{R}$ such that, for $\xi_j \in (\tau_j,\ \tau_{j+1})$, $\varphi(\xi_j) = \eta_j$. Then

$$\lim_{\Delta t\to\infty}\sum_{j=1}^{p-1}\eta_j \Delta t = \lim_{\Delta t\to\infty}\sum_{j=1}^{p-1}\varphi(\xi_j)\Delta t = \int_c^d \varphi(\tau)d\tau < +\infty. \tag{4.30}$$

Thus,

$$\frac{1}{b-a}\Delta t\sum_{j=1}^{p-1}\eta_j\Delta t \to 0 \quad \text{as } \Delta t \to 0. \tag{4.31}$$

As a result, (4.29)–(4.31) imply (4.23). □

Remark 4.4. We would like to emphasize that the statement of Theorem 4.3 is fulfilled only under assumptions (4.16) and (4.22). At the same time, the assumptions (4.16) and (4.22) are not restrictive from a practical point of view. The condition (4.16) is true for Lipschitz continuous signals $\mathbf{x}$ and $\mathbf{y}$, i.e., for very wide class of signals. The condition (4.22) is achieved by a choosing an appropriate known method (e.g., see [1]–[13], [23], [25]) to find the estimate $\widehat{\mathbf{x}}(t_1,\omega)$ used in the proposed filter $\mathcal{F}^{(p-1)}$ (see (3.5) and Theorem 4.1).

4.5. Some remarks related to the assumptions of the method

As it has been mentioned in Section 3.2, for $j = 1, \ldots, p$, matrices $E_{z_j w_j}$ and $E_{w_j w_j}$ in (4.7) are assumed to be known or can be estimated. Here, p is a chosen number of selected interpolation signal pairs (see Section 3.4). We note that normally p is much smaller than the number of input-output signals $\mathbf{x}(t,\cdot)$ and $\mathbf{y}(t,\cdot)$. Therefore, to estimate any signal $\mathbf{x}(t,\cdot)$ from an arbitrarily large set K_X, only a small number, p, of matrices $E_{z_j w_j}$ and $E_{w_j w_j}$ should be estimated (or be known). This issue has also been discussed in Sections 1.1.1 and 1.1.4.

By the proposed method, $\mathbf{x}(t,\cdot)$ is estimated for $t \in [t_j,\ t_{j+1}]$. While $E_{w_j w_j}$ in (4.7) can be directly estimated from observed signals $\mathbf{y}(t_{j+1},\cdot)$ and $\mathbf{y}(t_j,\cdot)$, an estimate of matrix $E_{z_j w_j}$ depends on the reference signal $\mathbf{x}(t_{j+1},\cdot)$ (see (4.2) and (4.3)) which is unknown (because the estimate is considered for $t \in [t_j,\ t_{j+1}]$).

Some possible approaches to an estimation of matrix $E_{z_j w_j}$ could be as follows.

1. In the general case, when $\mathbf{x}(t,\cdot)$ and $\mathbf{y}(t,\cdot)$ are arbitrary signals as discussed in Section 2.1 above, matrix $E_{z_j w_j}$ can be estimated as proposed, for example, in [27], from samples of z_j and w_j.
2. In the case of incomplete observations, the method proposed in [28, 29] can be used.
3. Let $E_{\hat{z}_j w_j}$ be a matrix obtained from matrix $E_{z_j w_j}$ where the term $\mathbf{x}(t_{j+1},\cdot)$ is replaced by $\widehat{\mathbf{x}}(t,\cdot)$ with $t \in [t_{j-1},\ t_j]$. Since $\widehat{\mathbf{x}}(t,\cdot)$ with $t \in [t_{j-1},\ t_j]$ is known, matrix $E_{\hat{z}_j w_j}$ can be considered as an estimate of $E_{z_j w_j}$.
4. In the important case of an *additive* noise, $E_{z_j w_j}$ can be represented in the explicit form. Indeed, if

$$\mathbf{y}(t,\cdot) = \mathbf{x}(t,\cdot) + \xi(t,\cdot)$$

where $\xi(t,\cdot) \in L^2(\Omega,\mathbb{R}^m)$ is a random noise, then $\mathbf{z}(t_j,t_{j+1},\cdot) = \mathbf{y}(t_{j+1},\cdot) - \xi(t_{j+1},\cdot) - \widehat{\mathbf{x}}(t_j,\cdot)$ and matrix $E_{z_j w_j}$ can be represented as follows:

$$E_{z_j w_j} = E_{(y_{j+1}-\xi_{j+1})(y_{j+1}-y_j)} - E_{\widehat{x}_j(y_{j+1}-y_j)} \tag{4.32}$$

We note that the RHS of (4.32) depends only on observed signals $\mathbf{y}(t_j,\cdot)$, $\mathbf{y}(t_{j+1},\cdot)$, estimated signal $\widehat{\mathbf{x}}(t_j,\cdot)$, and noise $\xi(t_{j+1},\cdot)$, not on the reference signal $\mathbf{x}(t_{j+1},\cdot)$. In particular, in (4.32), the term $E_{\xi_{j+1}(y_{j+1}-y_j)}$ can be estimated as

$$\pm(E[\xi_{j+1}^2])^{1/2} \times (E[(y_{j+1}-y_j)^2])^{1/2}$$

where $E[\xi_{j+1}^2] = \displaystyle\int_\Omega [\xi(t_{j+1},\omega)]^2 \, d\mu(\omega)$. It is motivated by the Holder's inequality for integrals. The second term in (4.32), $E_{\widehat{x}_j(y_{j+1}-y_j)}$, can be estimated from the samples of $\widehat{\mathbf{x}}(t_{j+1},\cdot)$ and $\mathbf{y}(t_{j+1},\cdot)-\mathbf{y}(t_j,\cdot)$.
We also note that the first term in the RHS of (4.32), $E_{(y_{j+1}-\xi_{j+1})(y_{j+1}-y_j)}$, is similar to the related covariance matrix in the Wiener filtering approach [5].
5. Other known ways to estimate $E_{\xi_{j+1}(y_{j+1}-y_j)}$ can be found in [5], Section 5.3.

In general, an estimation of covariance matrices is a special research topic which is not a subject of this paper. The relevant references can be found, for example, in [5, 29].

5. Simulations

5.1. General consideration

In these simulations, in accordance with Section 4.3.1, signal sets K_X and K_Y (see Section 2.1) are given by

$$K_X = \{\mathbf{x}(\tau_1,\cdot),\mathbf{x}(\tau_2,\cdot),\dots,\mathbf{x}(\tau_N,\cdot)\} \quad \text{and} \quad K_Y = \{\mathbf{y}(\tau_1,\cdot),\mathbf{y}(\tau_2,\cdot,\dots,\mathbf{y}(\tau_N,\cdot)\},$$

where, for $k=1,\dots,N$, $\mathbf{x}(\tau_k,\cdot)\in L^2(\Omega,\mathbb{R}^m)$ and $\mathbf{y}(\tau_k,\cdot)\in L^2(\Omega,\mathbb{R}^n)$. In many practical problems (arising, for example, in a DNA analysis the number N is quite large, for instance, $N=\mathcal{O}(10^4)$.

We set $N=141$ and $m=n=116$. Thus, in these simulations, the interval $T=[a,\ b]$ (see Sections 2.1 and 4.3.1) is modelled as 141 points τ_k with $k=1,\dots,141$ so that $[a,\ b]=[\tau_1,\tau_2,\dots,\tau_{141}]$.

The sequence of fixed time-points $\{t_k\}_1^p \subset T$ in (2.1) is now such that

$$\tau_1 = t_1 < \cdots < t_p = \tau_{141}. \tag{5.1}$$

Below, in Examples 1–12, four particular choices of the specific interpolation signals pairs $\{\mathbf{x}(t_k,\cdot),\ \mathbf{y}(t_k,\cdot)\}_1^p$ (introduced in Section 3.4) are considered, for $p=5,8,15$ and 28.

Signals $\mathbf{x}(\tau_k,\cdot)$ and $\mathbf{y}(\tau_k,\cdot)$ have been simulated as digital images represented by 116×256 matrices

$$X^{(k)} = [\mathbf{x}(\tau_k,\omega_1),\dots,\mathbf{x}(\tau_k,\omega_{256})] \quad \text{and} \quad Y^{(k)} = [\mathbf{y}(\tau_k,\omega_1),\dots,\mathbf{y}(\tau_k,\omega_{256})], \tag{5.2}$$

respectively, for $k=1,\dots,141$, so that $X^{(k)}$ represents an image that should be estimated from an observed image $Y^{(k)}$. A column of matrices $X^{(k)}$ and $Y^{(k)}$, $\mathbf{x}(\tau_k,\omega_i)\in\mathbb{R}^{116}$ and $\mathbf{y}(\tau_k,\omega_i)\in\mathbb{R}^{116}$, for $i=1,\dots,256$, represents a realization of signals $\mathbf{x}(\tau_k,\cdot)$ and $\mathbf{y}(\tau_k,\cdot)$, respectively.

Note that $X^{(1)},\dots,X^{(141)}$ are not used in the piecewise linear filter $F^{(p-1)}$ below since they are not supposed to be known. They are represented here for illustration purposes only. In particular, $X^{(1)},\dots,X^{(141)}$ are used to compare their estimates by different filters.

Observed noisy signals $Y^{(1)},\dots,Y^{(141)}$ have been simulated in the form presented by (5.13) in the Examples 1–5 in Sections 5.2 and 5.3. We note that the considered observed *signals are grossly corrupted.*

To estimate the signals $X^{(1)}$, ..., $X^{(141)}$ from the observed signals $Y^{(1)}$, ..., $Y^{(141)}$, the proposed piecewise linear filter $F^{(p-1)}$, the generic optimal linear (GOL) filters [5] and the averaging polynomial filter [12] have been used.

The filters proposed in [12, 13, 22, 23] have not been applied here by the reasons discussed in Section 1. In particular, the filter in [23] cannot be applied to signals represented by $Y^{(1)}$, ..., $Y^{(141)}$ in the form (5.13) below because the associated inverse matrices used in [23] do not exist.

For signals under consideration (given by matrices $X^{(k)}$ and $Y^{(k)}$ with $k = 1,\dots,141$), the filter $F^{(p-1)}$, the generic optimal linear (GOL) filters [5] and the averaging polynomial filter [10, 12] are represented as follows.

(i) *Piecewise linear filter* $F^{(p-1)}$. For $j = 1,\dots,p$, $\{X_j, Y_j\}$ designates an interpolation pair defined similarly to that in Section 3.4. Each X_j and Y_j is associated with t_j in (5.1) so that

$$X_j = [\mathbf{x}(t_j,\omega_1),\dots,\mathbf{x}(t_j,\omega_{256})] \quad \text{and} \quad Y_j = [\mathbf{y}(t_j,\omega_1),\dots,\mathbf{y}(t_j,\omega_{256})].$$

The estimate $\widehat{X}^{(k)}$ of $X^{(k)}$ by the filter $F^{(p-1)}$ is given by

$$\widehat{X}^{(k)} = F^{(p-1)}[Y^{(k)}], \tag{5.3}$$

where, by (4.4)–(4.7) in Section 4.2,

$$F^{(p-1)}[Y^{(k)}] = \sum_{j=1}^{p-1} \delta_j F_j^{(p-1)}[Y^{(k)}], \quad \delta_j = \begin{cases} 1, & \text{if } j \le k \le j+1, \\ 0, & \text{otherwise}, \end{cases} \tag{5.4}$$

$$F_j^{(p-1)}[Y^{(k)}] = \widehat{X}_j + B_j[Y^{(k)} - Y_j], \tag{5.5}$$

$$\widehat{X}_j = F_{j-1}[Y_j], \quad \widehat{X}_1 \text{ is given}, \tag{5.6}$$

$$B_j = E_{z_j w_j}(E_{w_j w_j})^{\dagger}, \tag{5.7}$$

and where $E_{z_j w_j}$ and $E_{w_j w_j}$ are estimates of matrices $E_{z_j w_j}$ and $E_{w_j w_j}$ in (4.7), respectively. In particular, $E_{W_j W_j}$ can be represented in the form

$$E_{w_j w_j} = W_j W_j^T, \quad \text{where } W_j = Y_{j+1} - Y_j. \tag{5.8}$$

Further, matrix $E_{z_j w_j}$ depends on $Z_j = X_{j+1} - \widehat{X}_j$ where X_{j+1} is unknown. Therefore a determination of $E_{z_j w_j}$ is reduced, in fact, to finding an estimate of X_{j+1}. Since it is customary to find $E_{z_j w_j}$ in terms of signal samples [5], $E_{z_j w_j}$ has been presented as

$$E_{z_j w_j} = \widetilde{Z}_j W_j^T, \quad \text{where } \widetilde{Z}_j = \widetilde{X}_{j+1} - \widehat{X}_j \tag{5.9}$$

and $\widetilde{X}_{j+1}$ has been constructed from a sample of X_{j+1} as follows. The sample of X_{j+1} is a 116×178 matrix presented by odd columns of X_{j+1}. Then an estimate of X_{j+1} is chosen as a 116×256 matrix $\widetilde{X}_{j+1}$ where each odd column is a related odd column of X_{j+1}, and each even column is an average of two adjacent columns. The last column in $\widetilde{X}_{j+1}$ is the same as its preceding column.

This way of estimating $E_{z_j w_j}$ was chosen for illustration purposes only. Other related methods have been considered in Section 4.5.

The errors associated with the filter $F^{(p-1)}$ are given by

$$\varepsilon_{k,F}^{(p-1)} = \left\| X^{(k)} - F^{(p-1)}[Y^{(k)}] \right\|_F^2, \quad \text{for } k = 1, \dots, 141. \tag{5.10}$$

(ii) *Generic optimal linear* (*GOL*) *filters* [5]. To each signal $Y^{(k)}$, an individual GOL filter W_k has also been applied, so that W_k estimates $X^{(k)}$ from $Y^{(k)}$ in the form

$$W_k Y^{(k)} = E_{X^{(k)}Y^{(k)}} E_{Y^{(k)}Y^{(k)}}^{\dagger} Y^{(k)},$$

for each $k = 1, \dots, 141$. Thus, the GOL filter W_k requires an estimate of 141 matrices $E_{X^{(k)}Y^{(k)}}$, for each $k = 1, \dots, 141$.

Similarly to matrix $E_{Z_j W_j}$ in the filter $F^{(p-1)}$ above, the matrix $E_{X^{(k)}Y^{(k)}}$ has been estimated from samples of each $X^{(k)}$, $\widetilde{X}^{(k)}$, for each $k = 1, \dots, 141$.

One of the advantages of the proposed filter $F^{(p-1)}$ is that $F^{(p-1)}$ requires a smaller number, p, of samples of X_j, $\widetilde{X}_j$, to be known (where $j = 1, \dots, p$).

The errors associated with filters W_k are given by

$$\epsilon_{k,w} = \|X^{(k)} - W_k Y^{(k)}\|_F^2. \tag{5.11}$$

(iii) *Averaging polynomial filters* [10, 12]. By the methodology in [10], the averaging polynomial filter W is based on the use of the estimates of the covariance matrices, E_{XY} and E_{YY}, in the form

$$E_{XY} = \frac{1}{141} \sum_{k=1}^{141} \widetilde{X}^{(k)} (Y^{(k)})^T \quad \text{and} \quad E_{YY} = \frac{1}{141} \sum_{k=1}^{141} Y^{(k)} (Y^{(k)})^T.$$

Then, for each, $k = 1, \dots, 141$, the estimate of $X^{(k)}$ is given by

$$WY^{(k)} = E_{XY} E_{YY}^{\dagger} Y^{(k)}.$$

The errors associated with the filter W are given by

$$\varepsilon_{kW} = \|X^{(k)} - WY^{(k)}\|_F^2, \quad \text{for } k = 1, \dots, 141. \tag{5.12}$$

5.2. Simulations with signals modelled from images 'plant': application of piecewise interpolation filter and GOL filters

Here, results of simulations for reference signals represented by matrices $X^{(1)}$, $\dots$, $X^{(141)}$ (see (5.2) above) formed from images 'plant'[10] are considered. Typical selected images $X^{(k)}$ are shown in Figure 1.

[10]The database is available in http://sipi.usc.edu/services/database.html.

(a) Signal $X^{(1)}$.

(b) Signal $X^{(55)}$.

(c) Signal $X^{(95)}$.

(d) Signal $X^{(141)}$.

FIGURE 1. Examples of selected signals to be estimated from observed data.

Observed noisy images $Y^{(1)}, \ldots, Y^{(141)}$ have been simulated in the form

$$Y^{(k)} = X^{(k)} \bullet \mathtt{randn}_{(k)} \bullet \mathtt{rand}_{(k)}, \tag{5.13}$$

for each $k = 1, \ldots, 141$. Here, $\bullet$ means the Hadamard product, and $\mathtt{randn}_{(k)}$ and $\mathtt{rand}_{(k)}$ are 116×256 matrices with random entries. The entries of $\mathtt{randn}_{(k)}$ are normally distributed with mean zero, variance one and standard deviation one. The entries of $\mathtt{rand}_{(k)}$ are uniformly distributed in the interval $(0,1)$. A typical example of such images is given in Figure 2 (a).

To demonstrate the effectiveness of the proposed filter $F^{(p-1)}$, sub-filters $F_j^{(p-1)}$ and associated interpolation signal pairs $\{X_j, Y_j\}_{j=1}^p$ have been chosen in four different ways as follows.

Example 1. First, for $p = 5$, the interpolation signal pairs are

$$\{X_1, Y_1\} = \{X^{(1)}, Y^{(1)}\}, \ \{X_2, Y_2\} = \{X^{(35)}, Y^{(35)}\}, \ \{X_3, Y_3\} = \{X^{(70)}, Y^{(70)}\},$$
$$\{X_4, Y_4\} = \{X^{(105)}, Y^{(105)}\}, \quad \{X_5, Y_5\} = \{X^{(141)}, Y^{(141)}\}. \tag{5.14}$$

The error values $\{\varepsilon_{k,F}^{(4)}\}_1^{141}$ associated with filter $F^{(4)}$ are evaluated by (5.10). The graph of $\{\varepsilon_{k,F}^{(4)}\}_1^{141}$ is presented in Figure 3 (a).

Example 2. For $p = 8$, the interpolation signal pairs are

$$\{X_1, Y_1\} = \{X^{(1)}, Y^{(1)}\}, \quad \{X_j, Y_j\} = \{X^{(20(j-1))}, Y^{(20(j-1))}\}, \quad \text{for } j = 2, \ldots, 7;$$
$$\text{and} \quad \{X_8, Y_8\} = \{X^{(141)}, Y^{(141)}\}. \tag{5.15}$$

The error magnitudes $\{\varepsilon_{k,F}^{(7)}\}_1^{141}$ associated with the piecewise interpolation filter $F^{(7)}$ constructed by (5.4)–(5.9) with the interpolation signal pairs given by (5.15) are diagrammatically shown in Figure 3 (b).

It follows from Figure 3 (b) that the errors associated with filter $F^{(7)}$ is less than those of filter $F^{(4)}$. This is a confirmation of Theorem 4.3.

Example 3. Further, for$p = 15$, the interpolation pairs are

$$\{X_1, Y_1\} = \{X^{(1)}, Y^{(1)}\}, \quad \{X_j, Y_j\} = \{X^{(10(j-1))}, Y^{(10(j-1))}\} \quad \text{for } j = 2, \ldots, 14;$$
$$\text{and} \quad \{X_{15}, Y_{15}\} = \{X^{(141)}, Y^{(141)}\}. \tag{5.16}$$

In Figure 3 (c), the errors $\{\varepsilon_{k,F}^{(15)}\}_1^{141}$ associated with the piecewise interpolation filter $F^{(15)}$ are presented. The Figure 3 (c) demonstrates a further confirmation of Theorem 4.3: the errors associated with the piecewise interpolation filter diminishes as p increases.

Example 4. Finally, the number of interpolation signal pairs $\{X_j, Y_j\}_{j=1}^p$ is $p = 29$ so that

$$\{X_1, Y_1\} = \{X^{(1)}, Y^{(1)}\}, \quad \{X_j, Y_j\} = \{X^{(5(j-1))}, \quad Y^{(5(j-1))}\} \quad \text{for } j = 2, \ldots, 28;$$
$$\text{and} \quad \{X_{29}, Y_{29}\} = \{X^{(141)}, Y^{(141)}\}. \tag{5.17}$$

(a) Observed signal $Y^{(95)}$.

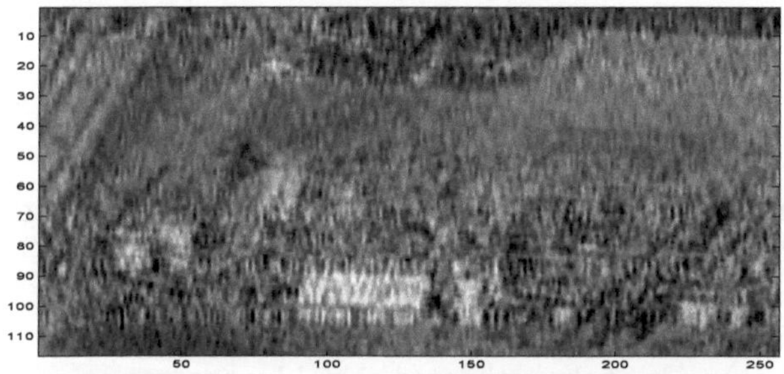

(b) Estimate of $X^{(95)}$ by piecewise filter $F^{(28)}$.

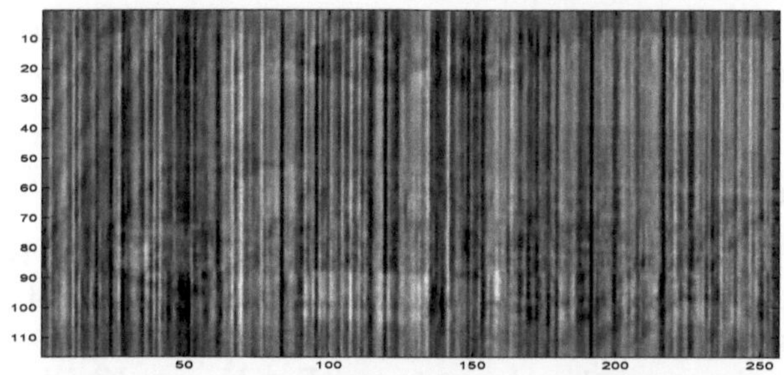

(c) Estimate of $X^{(95)}$ by generic optimal linear (GOL) filters.

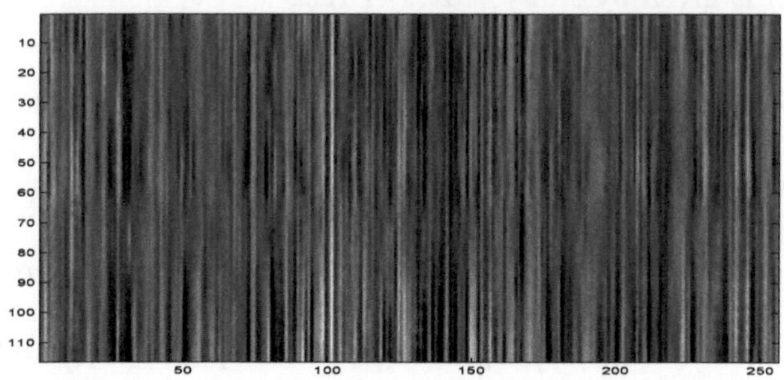

(d) Estimate of $X^{(95)}$ by averaging polynomial filter.

FIGURE 2. Examples of the observed signal and the estimates obtained by different filters.

In this case, when p is grater than in the previous Examples 1–3, the errors $\{\varepsilon_{k,F}^{(29)}\}_1^{141}$ associated with the piecewise interpolation filter $F^{(29)}$ are smaller than those associated with filters $F^{(4)}$, $F^{(8)}$ and $F^{(15)}$ – see Figure 3 (d).

The diagrams of errors associated with the GOL filters [5] are also presented in Figure 3. It follows from Figure 3 that proposed filters $F^{(4)}$, $F^{(8)}$, $F^{(15)}$ and $F^{(29)}$ provide the better accuracy then that of the GOL filters.

At the same time, the filter $F^{(p-1)}$ is easer to implement since it requires less initial information compared to GOL filters, as it has been discussed in Sections 1.1.1 and 1.1.4.

5.3. Results of simulations for averaging polynomial filter [10, 12]

To further illustrate the effectiveness of the proposed piecewise interpolation filter, in this Section, results of simulations for the averaging polynomial filter [10, 12] are presented.

Example 9. The filter [10, 12] applied to signals considered in Section 5.2 gives the associated errors $\{\epsilon_{kW}\}_{k=1}^{141}$ (see (5.12)) represented in Figure 4. For a comparison, the errors associated with the piecewise interpolation filter $F^{(28)}$ and the GOL filters [5] are also given in Figure 4.

A typical example of the estimated signal by the averaging polynomial filter [10, 12] is presented in Figure 2 (d).

Together with Figures 2 and 3, Figure 4 illustrates the advantage of the piecewise interpolation filter.

6. Conclusions

The theory for a new approach to the operator approximation is provided. The approach is motivated, in particular, by the problem of filtering of arbitrarily large sets of stochastic signals K_Y and K_X. An idealistic filter transforming signal sets K_X and K_Y is interpreted as an operator $\mathcal{F} : K_Y \to K_X$. Its approximating operator (filter) is given by $\mathcal{F}^{(p-1)} : K_Y \to K_X$.

Distinctive features of the approach are as follows.

(i) The proposed filter $\mathcal{F}^{(p-1)} : K_Y \to K_X$ is presented in the form of a sum with $p-1$ terms where each term, $\mathcal{F}_j : K_{Y,j} \to K_{X,j}$, is interpreted as a particular sub-filter. Here, $K_{Y,j}$ and $K_{X,j}$ are 'small' pieces of K_Y and K_X, respectively.

(ii) The prime idea is to exploit a priori information only on *few reference signals*, p, from the set K_X that contains $N \gg p$ signals (or even an infinite number of signals) and determine $\mathcal{F}_j$ separately, for each pieces $K_{Y,j}$ and $K_{X,j}$, so that the associated error is minimal. In other words, the filter $\mathcal{F}^{(p-1)}$ is flexible to changes in the sets of observed and reference signals K_Y and K_X, respectively.

(iii) Due to the specific way of determining $\mathcal{F}_j$, the filter $\mathcal{F}^{(p-1)}$ provides a smaller associated error than that for the processing of the whole set K_Y by a filter

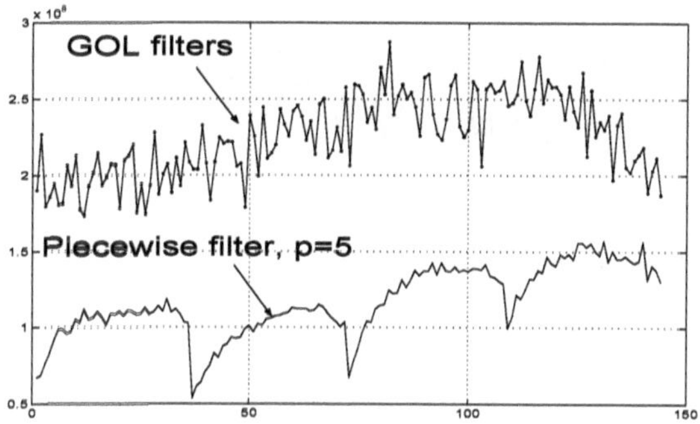

(a)

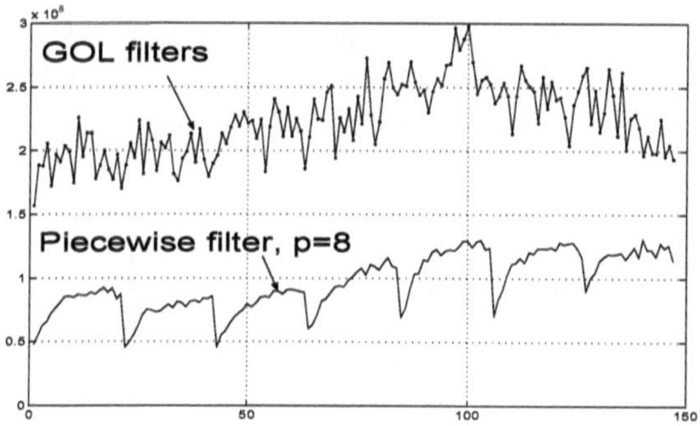

(b)

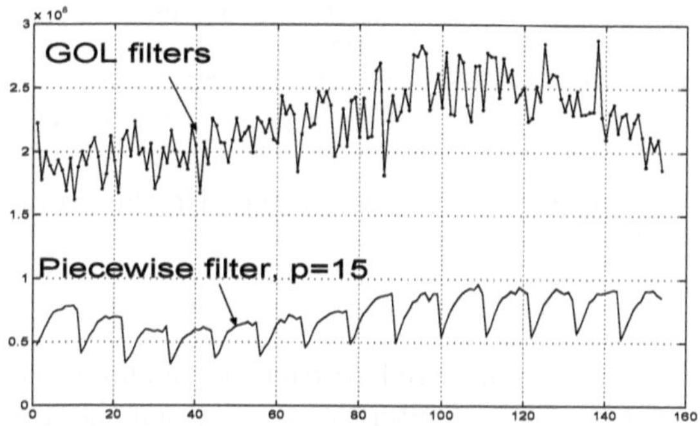

(c)

FIGURE 3. Illustration of the errors associated with the piecewise interpolation filters $F^{(p-1)}$ and the generic optimal linear (GOL) filters [5] applied to signals described in Examples 1–3.

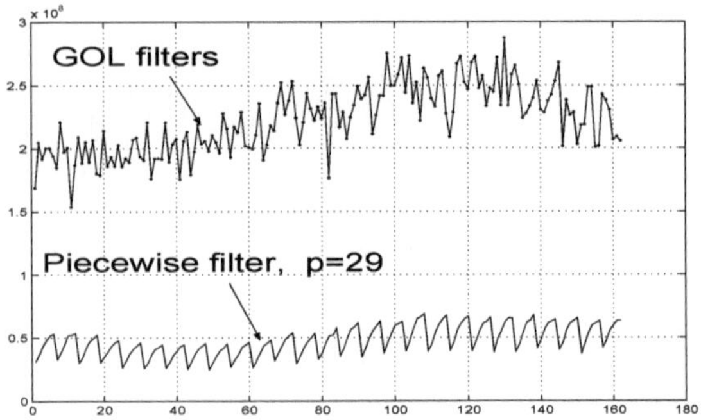

(d)

FIGURE 3. (Cont.) Illustration of the errors associated with the piecewise interpolation filters $F^{(p-1)}$ and the generic optimal linear (GOL) filters [5] applied to signals described in Example 4.

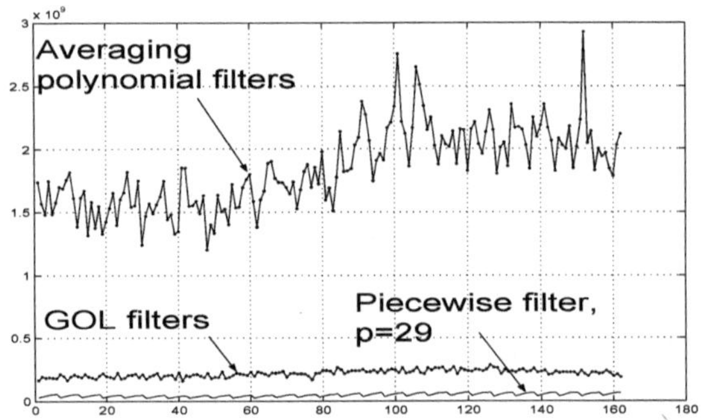

FIGURE 4. Illustration of errors associated with the averaging polynomial filters [10, 12] in Example 5.

which is not specifically adjusted to each particular piece $K_{Y,j}$. Moreover, the error associated with our filter decreases when the number of its terms, $\mathcal{F}_1, \ldots, \mathcal{F}_{p-1}$, increases.

(iv) While the proposed filter $\mathcal{F}^{(p-1)}$ processes arbitrarily large (and even infinite) signal sets, the filter is nevertheless fixed for all signals in the sets.

(v) The filter $\mathcal{F}^{(p-1)}$ is determined in terms of pseudo-inverse matrices so that the filter always exists.

(vi) The computational load associated with the filter $\mathcal{F}^{(p-1)}$ is less than that associated with other known filters applied to the processing of large signal sets.

References

[1] J. Chen, J. Benesty, Y. Huang, and S. Doclo, New Insights Into the Noise Reduction Wiener Filter, *IEEE Trans. on Audio, Speech, and Language Processing*, 14, No. 4, pp. 1218–1234, 2006.

[2] M. Spurbeck and P. Schreier, Causal Wiener filter banks for periodically correlated time series, *Signal Processing*, 87, 6, pp. 1179–1187, 2007.

[3] J.S. Goldstein, I. Reed, and L.L. Scharf, "A Multistage Representation of the Wiener Filter Based on Orthogonal Projections," *IEEE Trans. on Information Theory*, vol. 44, pp. 2943–2959, 1998.

[4] Y. Hua, M. Nikpour, and P. Stoica, "Optimal Reduced-Rank estimation and filtering," *IEEE Trans. on Signal Processing*, vol. 49, pp. 457–469, 2001.

[5] A. Torokhti and P. Howlett, *Computational Methods for Modelling of Nonlinear Systems*, Elsevier, 2007.

[6] E.D. Sontag, *Polynomial Response Maps*, Lecture Notes in Control and Information Sciences, 13, 1979.

[7] S. Chen and S.A. Billings, Representation of non-linear systems: NARMAX model, *Int. J. Control*, vol. 49, no. 3, pp. 1013–1032, 1989.

[8] V.J. Mathews and G.L. Sicuranza, *Polynomial Signal Processing*, J. Wiley & Sons, 2001.

[9] A. Torokhti and P. Howlett, Optimal Transform Formed by a Combination of Nonlinear Operators: The Case of Data Dimensionality Reduction, *IEEE Trans. on Signal Processing*, 54, No. 4, pp. 1431–1444, 2006.

[10] A. Torokhti and P. Howlett, Filtering and Compression for Infinite Sets of Stochastic Signals, *Signal Processing*, 89, pp. 291–304, 2009.

[11] J. Vesma and T. Saramaki, Polynomial-Based Interpolation Filters – Part I: Filter Synthesis, *Circuits, Systems, and Signal Processing*, Volume 26, Number 2, Pages 115–146, 2007.

[12] A. Torokhti and J. Manton, Generic Weighted Filtering of Stochastic Signals, *IEEE Trans. on Signal Processing*, 57, Issue 12, pp. 4675–4685, 2009.

[13] A. Torokhti and S. Miklavcic, Data Compression under Constraints of Causality and Variable Finite Memory, *Signal Processing*, 90 , Issue 10, pp. 2822–2834, 2010.

[14] I. Babuska, U. Banerjee, J.E. Osborn, Generalized finite element methods: main ideas, results, and perspective, *International Journal of Computational Methods*, 1 (1), pp. 67–103, 2004.

[15] S. Kang and L. Chua, A global representation of multidimensional piecewise-linear functions with linear partitions, *IEEE Trans. on Circuits and Systems*, 25 Issue:11, pp. 938–940, 1978.

[16] L.O. Chua and A.-C. Deng, Canonical piecewise-linear representation, *IEEE Trans. on Circuits and Systems*, 35 Issue:1, pp. 101–111, 1988.

[17] J.-N. Lin and R. Unbehauen, Adaptive nonlinear digital filter with canonical piecewise-linear structure, *IEEE Trans. on Circuits and Systems*, 37 Issue:3, pp. 347–353, 1990.

[18] J.-N. Lin and R. Unbehauen, Canonical piecewise-linear approximations, *IEEE Trans. on Circuits and Systems I: Fundamental Theory and Applications*, 39 Issue:8, pp. 697–699, 1992.

[19] S.B. Gelfand and C.S. Ravishankar, A tree-structured piecewise linear adaptive filter, *IEEE Trans. on Inf. Theory*, 39, Issue 6, pp. 1907–1922, 1993.

[20] E.A. Heredia and G.R. Arce, Piecewise linear system modeling based on a continuous threshold decomposition, *IEEE Trans. on Signal Processing*, 44 Issue:6, pp. 1440–1453, 1996.

[21] G. Feng, Robust filtering design of piecewise discrete time linear systems, *IEEE Trans. on Signal Processing*, 53 Issue:2, pp. 599 - 605, 2005.

[22] F. Russo, Technique for image denoising based on adaptive piecewise linear filters and automatic parameter tuning, *IEEE Trans. on Instrumentation and Measurement*, 55, Issue:4, pp. 1362–1367, 2006.

[23] J.E. Cousseau, J.L. Figueroa, S. Werner, T.I. Laakso, Efficient Nonlinear Wiener Model Identification Using a Complex-Valued Simplicial Canonical Piecewise Linear Filter, *IEEE Trans. on Signal Processing*, 55 Issue:5, pp. 1780–1792, 2007.

[24] P. Julian, A. Desages, B. D'Amico, Orthonormal high-level canonical PWL functions with applications to model reduction, *IEEE Trans. on Circuits and Systems I: Fundamental Theory and Applications*, 47 Issue:5, pp. 702–712, 2000.

[25] T. Wigren, Recursive Prediction Error Identification Using the Nonlinear Wiener Model, *Automatica*, 29, 4, pp. 1011–1025, 1993.

[26] G.H. Golub and C.F. van Loan, *Matrix Computations*, Johns Hopkins University Press, Baltimore, 1996.

[27] T. Anderson, *An Introduction to Multivariate Statistical Analysis*, New York, Wiley, 1984.

[28] L.I. Perlovsky and T.L. Marzetta, Estimating a Covariance Matrix from Incomplete Realizations of a Random Vector, *IEEE Trans. on Signal Processing*, 40, pp. 2097–2100, 1992.

[29] O. Ledoit and M. Wolf, A well-conditioned estimator for large-dimensional covariance matrices, *J. Multivariate Analysis* 88, pp. 365–411, 2004.

Anatoli Torokhti
School of Mathematics and Statistics
University of South Australia
1 Mawson Lakes Blv.
SA 5095, Australia
e-mail: anatoli.torokhti@unisa.edu.au

Operator Theory:
Advances and Applications, Vol. 236, 509–516

On Spectral Periodicity for the Sturm–Liouville Problem: Cantor Type Weight, Neumann and Third Type Boundary Conditions

A.A. Vladimirov and I.A. Sheipak

Abstract. We consider second and third boundary conditions for the Sturm–Liouville eigenvalue problem with a generalized derivative of Cantor type function as a weight. The property of spectral periodicity of eigenvalues for some class of boundary conditions is established.

Mathematics Subject Classification (2010). Primary 34L20; Secondary 34B05.

Keywords. Sturm–Liouville operator, self-similar weights, Cantor type functions, spectral periodicity.

1. Introduction

We consider spectral properties of the eigenvalue boundary problem

$$-y'' - \lambda\rho y = 0, \tag{1}$$

$$y'(0) - \gamma_0 y(0) = y'(1) + \gamma_1 y(1) = 0, \tag{2}$$

with self-similar Cantor type measure ρ. The eigenvalue asymptotics for problem (1) with Dirichlet boundary conditions and classic Cantor measure is studied in [1]. The eigenvalue counting function $N(\lambda) := \#\{\lambda_n : \ 0 < \lambda_n \leqslant \lambda\}$ is estimated as $\lambda \to +\infty$: $N(\lambda) \asymp \lambda^{\log_6 2}$.

In [2] the same question for arbitrary singular self-similar measure and Dirichlet boundary conditions is considered. It is proved that the counting function $N(\lambda)$ of eigenvalues of problem (1) with a probability self-similar measure has the asymptotics, as $\lambda \to +\infty$,

$$N(\lambda) = \lambda^D \cdot \big(s(\ln\lambda) + o(1)\big), \tag{3}$$

where the number $D \in (0, 1/2)$ can be determined by the self-similar parameters of the measure ρ and s is some continuous strictly positive periodic function which

The work is supported by RFFI, project N 13-01-00705 11-01-12115-ofi-m-2011.

depends on measure ρ. In a *non-arithmetic* case of self-similarity of measure ρ, function s is known to be a constant. The proof of (3) is based on renewal theory [3]. An analogous technique is applied for studying eigenvalue problems for differential operators on self-similar domains and domains with fractal boundaries ([4], [5], [6]).

The results of [2] were later developed in two directions. First, the class of weights were generalized: in papers [7] and [8] the indefinite weights $\rho \in W_2^{-1}[0,1]$ were considered. In this case weight ρ is a derivative in a distribution sense of a self-similar function $P \in L_2[0,1]$. If P has *positive spectral order* D (in case $P \in L_2[0,1]$ the exponent D can take on any value from the interval $(0,1)$) the counting functions $N_\pm$ of positive and negative eigenvalues of problem (1) have the asymptotics

$$N_\pm(\lambda) = |\lambda|^D \cdot \big(s_\pm(\ln|\lambda|) + o(1)\big) \quad \text{as} \quad |\lambda| \to \infty,$$

where $s_\pm$ are some continuous strictly positive periodical functions which depend on weight ρ. If P has a non-arithmetic type of self-similarity the functions $s_\pm$ are known to be constants (possibly different).

In the second direction we refer to the paper [9], where the boundary spectral problems for differential operators of arbitrary order are studied. In this case weight function is a singular probability measure.

We consider problem (1) under Neumann and third type boundary conditions and show that the boundary conditions have no affect on the main terms of asymptotics (3). However Neumann and third type boundary conditions imply additional properties on eigenvalue behaviour and the main purpose of our paper is to establish the *spectral periodicity* of eigenvalues for problem (1), (2) with *Cantor type self-similar weight* ρ. This property of eigenvalues was obtained in [10] for more narrow class of weights and boundary conditions.

For the proof we apply methods based not on a renewal theory as in [2], [7] and [8], but on the oscillating theory of Sturm–Liouville problems with singular coefficients developed in recent papers [12] and [13]. A similar technique was applied in [14] to obtain some properties of eigenfunction zeroes for a class of fractal Sturm–Liouville operators.

The formal boundary problem (1), (2) is understood as an eigenvalue problem of the linear pencil $T_\rho : \mathbb{C} \to \mathcal{B}(W_2^1[0,1], W_2^{-1}[0,1])$ with quadratic form

$$(\forall\lambda\in\mathbb{C})(\forall y\in W_2^1[0,1]) \quad \langle T_\rho(\lambda)y,y\rangle = \int_0^1 |y'|^2\,dx - \lambda\langle\rho,|y|^2\rangle + \gamma_0|y(0)|^2 + \gamma_1|y(1)|^2.$$

The paper is organized as follows. Section 2 contains the necessary information on Cantor type self-similar functions. In Section 3 we establish the spectral periodicity for some boundary problems with Cantor type weights. Finally, in Section 4 we illustrate the effect of spectral periodicity by computational data for classic Cantor measure as a weight function under certain boundary conditions.

2. Cantor type self-similar functions

Let us consider $\varkappa \in \mathbb{N}$, $\varkappa > 1$ and an arbitrary number $a \in (0, 1/\varkappa)$. We define $b := (1 - \varkappa a)/(\varkappa - 1)$ and the set of points $\alpha_{2k} := k(a+b)$, $\alpha_{2k+1} := \alpha_{2k} + a$ on interval $[0,1]$, where $k \in \{0, \dots, \varkappa - 1\}$.

Definition 2.1. *Function $f \in C[0,1]$ with the properties $f(0) = 0$ and $f(1) = 1$ is a self-similar function of Cantor type with parameters $\varkappa$ and a if it satisfies the self-similarity equation*

$$f(x) = \sum_{k=0}^{\varkappa-1} \left(\frac{1}{\varkappa} f\left(\frac{x - \alpha_{2k}}{a} \right) + \frac{k}{\varkappa} \right) \chi_{[\alpha_{2k}, \alpha_{2k+1}]} + \sum_{k=1}^{\varkappa-1} \frac{k}{\varkappa} \chi_{[\alpha_{2k-1}, \alpha_{2k}]}. \tag{4}$$

The classic Cantor ladder is self-similar with parameters $\varkappa = 2$, $a = 1/3$.

It is known ([7], [15]) that any acceptable set of parameters $\varkappa$ and a defines the unique self-similar function with an arithmetic type of self-similarity. Moreover from [15] it follows that this function is continuous (the right side of (4) defines the compression in $C[0,1]$). The spectral order of a Cantor type function is defined by the equality

$$D = \frac{\ln \varkappa}{\ln \varkappa - \ln a}.$$

The definition of the parameters $\varkappa$ and a implies that $D \in \left(0, \frac{1}{2}\right)$.

It is easy to see that for any self-similar Cantor type function f, $1 - f(1 - \cdot)$ is Cantor type self-similar too with the same parameters $\varkappa$ and a. This implies the truth of the following statement.

Lemma 2.1. *Let $f \in C[0,1]$ be a self-similar Cantor type function. Then for any $x \in [0,1]$ the identity*

$$f(x) = 1 - f(1 - x)$$

is valid.

The quadratic form of problem (1) with Dirichlet boundary conditions and weight $\rho \in \overset{\circ}{W}{}_2^{-1}[0,1]$ is totally determined by the generalized primitive P of the weight. In the case of Neumann boundary conditions, the definition of the weight-distribution by the formula

$$(\forall y \in W_2^1[0,1]) \qquad \langle \rho, y \rangle = -\int_0^1 P\overline{y'}\, dx + P\overline{y}\,\Big|_0^1$$

needs the additional "boundary values" $P(0)$ and $P(1)$.

If the generalized primitive P is continuous we can choose the usual values of the function P at the endpoints. So in this case the quadratic form of the operators $T_\rho(\lambda)$ corresponding to the problem (1), (2), where $\rho = P'$ (the derivative is understood in a distribution sense) looks like

$$\langle T_\rho(\lambda)y, y\rangle = \int_0^1 \{|y'|^2 + \lambda P(|y|^2)'\}\,dx + \gamma_0\,|y(0)|^2 + (\gamma_1 - \lambda)\,|y(1)|^2 \tag{5}$$

and the boundary conditions (2) are understood in a direct sense.

Theorem 2.1. *For any $\lambda \in \mathbb{R}$ and function $y \in \ker T_\rho(\lambda)$ we have $y \in C^1[0,1]$ and*

$$(y' + \lambda P y)' = \lambda P y'. \tag{6}$$

For any $\lambda \in \mathbb{R}$ and function $y \in C^1[0,1]$, satisfying equation (6) and boundary conditions (2), it follows that $y \in \ker T_\rho(\lambda)$.

Proof. From (5) and the equality $\langle T_\rho(\lambda)y, z\rangle = 0$ for any function $z \in \overset{\circ}{W}{}_2^1[0,1]$ it follows that function $y' + \lambda P y \in L_2[0,1]$ coincides a.e. with some absolutely continuous primitive of the function $\lambda P y' \in L_2[0,1]$. The condition $P \in C[0,1]$ implies that function $y' \in L_2[0,1]$ is continuous a.e., which is why we have $y \in C^1[0,1]$ and $y' + \lambda P y \in C^1[0,1]$. Then the equality (5) for arbitrary linear function z implies the boundary conditions (2).

The correctness of the second part of the theorem can be checked by direct calculation. □

For further study we need the following result; the details can be found in [13, Statement 11].

Statement 2.1. *Let $\{\lambda_n\}_{n=0}^\infty$ the sequence of eigenvalues of problem (1), (2) enumerated in increasing order. Then for any $n \in \mathbb{N}$ the eigenvalue λ_n is simple and the corresponding eigenfunction y_n has exactly n different zeroes in open interval $(0,1)$ and $y_n(0) \neq 0$, $y_n(1) \neq 0$.*

From (5) it follows that the variation of the coefficients γ_0 and γ_1 in the boundary conditions (2) leads to the perturbation of each operator $T_\rho(\lambda)$ by some operator with rank equal to 2. Let N_{γ_0,γ_1} and $N_{\gamma_0',\gamma_1'}$ be counting functions of eigenvalues of (1) with different boundary conditions corresponding to the different parameters γ_0, γ_1 and γ_0', γ_1' correspondingly. From general variational theory of self-adjoint operators (see [16]) it follows that for any $\lambda > 0$ the inequality

$$|N_{\gamma_0,\gamma_1}(\lambda) - N_{\gamma_0',\gamma_1'}(\lambda)| \leqslant 2$$

is true.

There is an analogous relationship between counting functions of third type boundary conditions and Dirichlet conditions. The quadratic form for a spectral problem with Dirichlet conditions can be derived by contraction of the quadratic form for the same spectral problem with third type boundary conditions on the space of co-dimension equal to 2. That is why the asymptotic formula (3) doesn't depend on self-adjoint boundary conditions.

3. Spectral periodicity

Theorem 3.1. *Let $\{\lambda_n\}_{n=0}^{\infty}$ be the sequence of increasing eigenvalues of the problem* (1) *with boundary conditions*

$$y'(0) = y'(1) = 0. \tag{7}$$

Then for any $n \in \mathbb{N}$ the equation

$$\lambda_{\varkappa n} = (\varkappa/a)\,\lambda_n \tag{8}$$

is valid.

Proof. Let us fix the eigenfunction y_n corresponding to the eigenvalue λ_n. According to Statement 2.1 y_n doesn't vanish at endpoints of the interval $[0,1]$ and has exactly n different zeroes in the open interval $(0,1)$. Now we construct the new function $z \in C[0,1]$ with the following conditions:

1. For any $k \in \{1,\dots,\varkappa-1\}$ function z is constant on the interval $(\alpha_{2k-1},\alpha_{2k})$.
2. For any $k \in \{0,\dots,\varkappa-1\}$ function z_k of the form

$$z_k(x) := z(\alpha_{2k} + ax)$$

coincides with y_n up to the multiplicative constant.

By Theorem 2.1 the boundary conditions (7) guarantee that $z \in C^1[0,1]$ and $z'(0) = z'(1) = 0$ The self-similarity of P provides that $z' + (\varkappa/a)\,\lambda_n P z$ is absolutely continuous and

$$(z' + (\varkappa/a)\,\lambda_n P z)' = (\varkappa/a)\,\lambda_n P z'.$$

By Theorem 2.1 function z is an eigenfunction of the problem (1), (7) with eigenvalue $(\varkappa/a)\,\lambda_n$. Moreover function z has exactly $\varkappa n$ zeroes in the interval $(0,1)$ – n zeroes in each interval $(\alpha_{2k},\alpha_{2k}+a)$, $k \in \{0,\dots,\varkappa-1\}$. Application of Statement 2.1 completes the proof. □

In other words the construction of the new eigenfunction z now follows. We shift the shrunken copies of origin eigenfunction y_n to the intervals $(\alpha_{2k},\alpha_{2k+1})$, $k = 0,1,\dots\varkappa-1$ and "glue" these copies on interval $(\alpha_{2k-1},\alpha_{2k})$ by constant linear functions.

Theorem 3.2. *Let $\{\lambda_n\}_{n=0}^{\infty}$ be the increasing sequence of the eigenvalues of the spectral problem* (1) *and boundary conditions*

$$by'(0) - 2y(0) = by'(1) + 2y(1) = 0,$$

and $\{\mu_n\}_{n=0}^{\infty}$ be the increasing sequence of the eigenvalues of the same spectral problem (1) *and boundary conditions*

$$by'(0) - 2ay(0) = by'(1) + 2ay(1) = 0.$$

Then for any $n \in \mathbb{N}$ the equation

$$\lambda_{\varkappa(n+1)-1} = (\varkappa/a)\,\mu_n$$

is valid.

n	λ_n	$6\lambda_n$	λ_{2n}
1	$7,0974 \pm 10^{-4}$	$42,584 \pm 10^{-3}$	$42,584 \pm 10^{-3}$
2	$42,584 \pm 10^{-3}$	$255,51 \pm 10^{-2}$	$255,51 \pm 10^{-2}$
3	$61,344 \pm 10^{-3}$	$368,06 \pm 10^{-2}$	$368,06 \pm 10^{-2}$
4	$255,51 \pm 10^{-2}$	$1533,0 \pm 10^{-1}$	$1533,0 \pm 10^{-1}$
5	$272,98 \pm 10^{-2}$	$1637,9 \pm 10^{-1}$	$1637,9 \pm 10^{-1}$
6	$368,06 \pm 10^{-2}$	$2208,4 \pm 10^{-1}$	$2208,4 \pm 10^{-1}$
7	$383,55 \pm 10^{-2}$	$2301,3 \pm 10^{-1}$	$2301,3 \pm 10^{-1}$
8	$1533,0 \pm 10^{-1}$	$9198,2 \pm 10^{-1}$	$9198,2 \pm 10^{-1}$
9	$1548,0 \pm 10^{-1}$	$9288,3 \pm 10^{-1}$	$9288,3 \pm 10^{-1}$

TABLE 1. Estimations of eigenvalues for Neumann boundary conditions, $\varkappa = 2$, $a = 1/3$.

The proof is the same as for Theorem 3.1. The main difference is in construction of new eigenfunction z – we "glue" copies of the origin eigenfunction y by inclined linear functions with zeroes in the centers of the intervals $(\alpha_{2k-1}, \alpha_{2k})$.

Theorem 3.3. *Let $\{\lambda_n\}_{n=0}^{\infty}$ be the increasing sequence of the eigenvalues of the spectral boundary problem* (1), (7), *and let $\{\mu_n\}_{n=0}^{\infty}$ be the increasing sequence of eigenvalues of the spectral problem* (1) *with boundary conditions*

$$y'(0) = by'(1) + 2ay(1) = 0. \tag{9}$$

If $\varkappa$ is even then for any $n \in \mathbb{N}$ the equation

$$\lambda_{\varkappa(n+1/2)} = (\varkappa/a)\,\mu_n$$

is valid.

The proof is the same as for Theorem 3.1. We take into account that for any eigenfunction y_n of the boundary problem (1), (9) function $y_n(1-\cdot)$ is an eigenfunction too.The latter satisfies the equation (1) and boundary conditions

$$by'(0) - 2ay(0) = y'(1) = 0.$$

The new eigenfunction z of the problem (1), (7) is composed from alternating, shrunken copies of y_n and $y_n(1-\cdot)$ which we "glue" by linear functions with zeroes in the centers of the intervals $(\alpha_{2k-1}, \alpha_{2k})$ and by constants on the intervals $(\alpha_{2k+1}, \alpha_{2k+2})$.

4. Examples

All tables in this section illustrate the relationships between eigenvalues of problem (1) where ρ is generalized derivative of a Cantor ladder and different boundary conditions.

n	μ_n	$6\mu_n$	λ_{2n+1}
0	$3,2983 \pm 10^{-4}$	$19,790 \pm 10^{-3}$	$19,790 \pm 10^{-3}$
1	$12,558 \pm 10^{-3}$	$75,349 \pm 10^{-3}$	$75,349 \pm 10^{-3}$
2	$48,946 \pm 10^{-3}$	$293,67 \pm 10^{-2}$	$293,67 \pm 10^{-2}$
3	$65,832 \pm 10^{-3}$	$394,99 \pm 10^{-2}$	$394,99 \pm 10^{-2}$
4	$261,64 \pm 10^{-2}$	$1569,8 \pm 10^{-1}$	$1569,8 \pm 10^{-1}$

TABLE 2. Estimations of eigenvalues of problem (1) with boundary conditions $y'(0) - 2y(0) = 0$, $y'(1) + 2y(1) = 0$, $\varkappa = 2$, $a = 1/3$.

n	μ_n	$6\mu_n$	λ_{2n+1}
0	$1,1829 \pm 10^{-4}$	$7,0974 \pm 10^{-4}$	$7,0974 \pm 10^{-4}$
1	$10,224 \pm 10^{-3}$	$61,344 \pm 10^{-3}$	$61,344 \pm 10^{-3}$
2	$45,497 \pm 10^{-3}$	$272,98 \pm 10^{-2}$	$272,98 \pm 10^{-2}$
3	$63,925 \pm 10^{-3}$	$383,55 \pm 10^{-2}$	$383,55 \pm 10^{-2}$
4	$258,01 \pm 10^{-2}$	$1548,0 \pm 10^{-1}$	$1548,0 \pm 10^{-1}$

TABLE 3. Estimations of eigenvalues of problem (1) with boundary conditions $y'(0) = 0$, $y'(1) + 2y(1) = 0$, $\varkappa = 2$, $a = 1/3$.

Table 1 presents first nine positive eigenvalues of problem (1) with Neumann boundary conditions. It illustrates Theorem 3.1.

Table 2 presents the numerical computation for the eigenvalues of two problems: μ_n are the eigenvalues of the problem (1) under boundary conditions $y'(0) - 2y(0) = y'(1)+2y(1) = 0$; λ_n are the eigenvalues of (1) under boundary conditions $y'(0) - 6y(0) = y'(1) + 6y(1) = 0$. The data illustrates Theorem 3.2.

Table 3 presents a numerical computation for the eigenvalues of another two problems: μ_n are the eigenvalues of problem (1) under boundary conditions $y'(0) = y'(1)+2y(1) = 0$; λ_n are the eigenvalues of (1) and Neumann boundary conditions. The data illustrates Theorem 3.3.

Numerical computation is based on methods described in [17].

References

[1] T. Uno, I. Hong. *Some consideration of asymptotic distribution of eigenvalues for the equation* $\frac{d^2u}{dx^2} + \lambda\rho u = 0$. Japan Math. J., 1959, V. 29, 152–164.

[2] M. Solomyak, E. Verbitsky. *On a spectral problem related to self-similar measures.* Bull. London Math. Soc., 1995, V. 27, N. 3, 242–248.

[3] W. Feller *Introduction to Probability Theory.* V. 2, Wiley, 1967.

[4] T. Fujita. *A fractional dimension, self-similarity and a generalized diffusion operator.* Taniguchi Symp. PMMP. Katata, 1985, 83–90.

[5] J. Kigami, M.L. Lapidus. *Weyl's problem for the spectral distributions of Laplacians on p.c.f. self-similar fractals.* Comm. Math. Phys., 1993, V. 158, 93–125.

[6] M. Levitin, D. Vassiliev. *Spectral asymptotics, renewal theorem, and the Berry conjecture for a class of fractals.* Proc. Lond. Math. Soc., 1996, V. 72, 188–214.

[7] A.A. Vladimirov and I.A. Sheipak. *Self-similar functions in* $L_2[0,1]$ *and the Sturm–Liouville problem with a singular indefinite weight.* Sbornik: Mathematics, 2006, 197, N. 11, 1569–1586, DOI: 10.1070/SM2006v197n11ABEH003813.

[8] A.A. Vladimirov and I.A. Sheipak. *Indefinite Sturm–Liouville Problem for Some Classes of Self-similar Singular Weights.* Proceedings of the Steklov Institute of Mathematics, 2006, Vol. 255, 1–10, DOI: 10.1134/S0081543806040079.

[9] A.I. Nazarov. *Logarithmic small ball asymptotics for some Gaussain processes in* L_2*-norm in respect with sel-similar measure.* Prob. and statistics, Zap. Nauch. Semin. Piter. Otdel. Mat. Inst. Steklov (POMI), **311** (2004), 190–213 (in Russian).

[10] U. Freiberg *Refinement of the spectral asymptotics of generalized Krein Feller operators.* Forum Math., **23** (2009) 427–445.

[11] A.A. Vladimirov and I.A. Sheipak. *Asymptotics of the eigenvalues of the Sturm–Liouville problem with discrete self-similar weight.* Mathematical Notes, 2010, V. 88, N. 5-6, 637–646, DOI: 10.1134/S0001434610110039.

[12] J. Ben Amara and A.A. Shkalikov. *Oscillation Theorems for Sturm–Liouville Problems with Distribution Potentials.* Moscow Univ. Math. Bulletin, 2009, V. 64, N. 3, 132–137.

[13] A.A. Vladimirov. *On the oscillation theory of the Sturm–Liouville problem with singular coefficients.* Computational Mathematics and Mathematical Physics, 2009, V. 49, N. 9, 1535–1546, DOI: 10.1134/S0965542509090085.

[14] E.J. Bird, S.-M. Ngai, A. Teplyaev. *Fractal laplacians on the unit interval.* Ann. Sci. Math. Québec, 2003, V. 27, N. 2, 135–168.

[15] I.A. Sheipak. *On the construction and some properties of self-similar functions in the spaces* $L_p[0,1]$. Mathematical Notes, 2007, V. 81, N. 5-6, 827–839, DOI: 10.1134/S0001434607050306.

[16] P. Lancaster, A. Shkalikov, Qiang Ye. *Strongly definitizable linear pencils in Hilbert space.* Integr. Equat. Oper. Th., 1993, V. 17, 338–360.

[17] A.A. Vladimirov. *Calculating the eigenvalues of the Sturm–Liouville problem with a fractal indefinite weight.* Computational Mathematics and Mathematical Physics, 2007, V. 47, N. 8, 1295–1300, DOI: 10.1134/S0965542507080076.

A.A. Vladimirov and I.A. Sheipak
Dorodnicyn Computing Centre of RAS

and

Moscow Lomonosov State University
e-mail: `vladimi@mech.math.msu.su`
`iasheip@mech.math.msu.su`

Operator Theory:
Advances and Applications, Vol. 236, 517–535

Spectral Analysis and Representations of Solutions of Abstract Integro-differential Equations in Hilbert Space

Victor V. Vlasov and Nadezda A. Rautian

Abstract. We study an abstract integro-differential equations with unbounded operator coefficients in Hilbert space. We obtain the expansion of the strong solutions of such type equations as the exponential series corresponding to the spectra of operator-functions which are the symbols of these equations.

Mathematics Subject Classification (2010). Primary 34D05; Secondary 34C23.

Keywords. Integro-differential equations, Sobolev space, Gurtin–Pipkin heat equation, spectra, operator-function.

1. Introduction

We study integro-differential equations with unbounded operator coefficients in a Hilbert space. Most of these equations is an abstract hyperbolic equations disturbed by terms containing abstract Volterra integral operators. These equations are abstract forms of the Gurtin–Pipkin integro-differential equations (see [21], [20], [39] for more details), which describes the process of heat propagation in media with memory, process of wave propagation in the visco-elastic media (see [22]–[25]) and also arising in the problems of porous media (Darcy law) (see [26], [30], [31], [39])

We formulate the results about correct solvability of the initial-valued problems for these equations in the weighted Sobolev spaces on the positive semi-axis. The proofs of these results are given in previous articles of the authors (see [34]–[37]).

This work was completed with the support of Russian Foundation for Basic Research (grant 11-01-00790), and also Leading Scientific School 7322.2010.1.

We analyse spectral problems for operator-valued functions which are the symbols of these equations. Moreover we study the spectra of the abstract integro-differential equation of Gurtin–Pipkin type. The proofs of these results are also given in previous works of the authors (see [34]–[36]).

For this reason, it is natural and convenient to consider integro-differential equations with unbounded operator coefficients (abstract integro-differential equations), which can be realized as integro-differential partial differential equations with respect to spatial variables when necessary. For the self-adjoin positive operator A considered in what follows we can take, in particular, the operator $A^2y = -y''$, where $x \in (0,\pi)$, $y(0) = y(\pi) = 0$, or the operator $A^2y = -\Delta y$ satisfying the Dirichlet conditions on a bounded domain with sufficiently smooth boundary. At present, there is an extensive literature on abstract integro-differential equations (see, e.g., [1]–[18], [21]–[25], [34]–[37] and the references therein).

The main purpose of our paper is to study the asymptotic behavior of the solutions of evolutionary equations on the base of spectral analysis of its symbols. In accordance with this purpose we obtain the expansion of the strong solutions of the equations mentioned above in exponential series corresponding to spectra of operator-functions which are the symbols of these equations.

We think that these representations of the solutions as the series of exponentials are new and were not obtained earlier for this class of integro-differential equations. The main part of known results deals with the problem of correct solvability and estimates of the solutions. Let us note that papers (see also the bibliography cited there)are devoted to the researching of the integro-differential equations with the main part is an abstract parabolic equation (see [1]–[12]).

The equations with the main part is an abstract hyperbolic equation were studied not so intensively in comparison with the parabolic case. The following papers are closest to this context (see [13]–[18], [21]–[26], [34]–[37]):

Let's emphasize also that important difference of our results from results known earlier (for example) consists in the following fact: we consider the integral operators with kernels that may have the singularities. In earlier known results the restrictions imposed on kernels were stronger. The restrictions imposed on kernels of integral operators for integro-differential equations of parabolic type is essentially weaker (see [1]–[12], [16] for example).

2. Statement of the problem

We study integro-differential equations of the form

$$\frac{d^2u(t)}{dt^2} + K(0)A^2u(t) + \int_0^t K'(t-s)A^2u(s)ds = f(t), \quad t \in \mathbb{R}_+, \tag{2.1}$$

with unbounded operator coefficients in a separable Hilbert space H with the initial conditions

$$u(+0) = \varphi_0, \quad u^{(1)}(+0) = \varphi_1, \tag{2.2}$$

It is assumed that A is a self-adjoint positive operator in H with compact inverse and the scalar function K(t) admits the representation

$$K(t) = \sum_{j=1}^{\infty} \frac{c_j}{\gamma_j} e^{-\gamma_j t}, \tag{2.3}$$

where

$$c_j > 0, \quad \gamma_{j+1} > \gamma_j > 0, \quad j \in \mathbb{N}, \quad \gamma_j \to +\infty \ (j \to +\infty)$$

and

$$K(0) = \sum_{j=1}^{\infty} \frac{c_j}{\gamma_j} < \infty. \tag{2.4}$$

It follows from condition (2.4) that $K'(t) \in L_1(\mathbb{R}_+)$. If we supplement condition (2.4) with the condition

$$\sum_{j=1}^{\infty} c_j < +\infty, \tag{2.5}$$

then the kernel $K'(t)$ will belong to the space $W_1^1(\mathbb{R}_+)$.

Such equations are an abstract form of integro-differential equations describing heat propagation in media with memory (equations of Gurtin–Pipkin type; see [20] and [21], [27]), and sound propagation in viscoelastic media (see [22]–[25]) as well as integro-differential equations arising in homogenization problems for perforated media (the Darcy law; see [26], [28]).

We obtain representations of solutions of problem (2.1), (2.2) in the form of series in exponentials corresponding to points of spectrum of the operator function that is the symbol of (2.1).

3. Definitions and notations

We make the domain $\mathrm{Dom}(A^\beta)$ of the operator A^β, $\beta > 0$ a Hilbert space H_β, by equipping $\mathrm{Dom}(A^\beta)$ with the norm $\|\cdot\|_\beta = \|A^\beta \cdot\|$.

Let $W_{2,\gamma}^n(\mathbb{R}_+, A^n)$ be the Sobolev space of vector functions defined on the half-line $\mathbb{R}_+ = (0, \infty)$ ranging in H and having the finite norm

$$\|u\|_{W_{2,\gamma}^n(\mathbb{R}_+, A^n)} \equiv \left(\int_0^\infty e^{-2\gamma t} \left(\left\| u^{(n)}(t) \right\|_H^2 + \|A^n u(t)\|_H^2 \right) dt \right)^{1/2}, \quad \gamma \geq 0.$$

More details on the spaces $W_{2,\gamma}^n(\mathbb{R}_+, A^n)$ can be found in the monograph [19]. For $n = 0$ we set $W_{2,\gamma}^0(\mathbb{R}_+, A^0) \equiv L_{2,\gamma}(\mathbb{R}_+, H)$.

Definition 3.1. *A vector function u is called a strong solution of problem* (2.1), (2.2) *if it belongs to the space* $W_{2,\gamma}^2(\mathbb{R}_+, A^2)$ *for some* $\gamma \geqslant 0$, *satisfies* (2.1) *a.e. on* $\mathbb{R}_+$, *and satisfies the initial conditions* (2.2).

Let us formulate the results about correct solvability of the problem (2.1), (2.2).

Theorem 3.1. *Let us suppose that $Af(t) \in L_{2,\gamma_2}(\mathbb{R}_+, H)$ or $f'(t) \in L_{2,\gamma_2}(\mathbb{R}_+, H)$, $f(0) = 0$ for some $\gamma_2 \geqslant 0$ and the condition (2.4) is satisfied. Then*

1) *if the condition (2.5) is satisfied and $\varphi_0 \in H_2$, $\varphi_1 \in H_1$, then the problem (2.1), (2.2) has the unique solution in the space $W^2_{2,\gamma}(\mathbb{R}_+, A^2)$ for arbitrary $\gamma > \gamma_2$. Moreover the following estimate*
$$\|u\|_{W^2_{2,\gamma}(\mathbb{R}_+,A^2)} \leq d\left(\|Af(t)\|_{L_{2,\gamma}(\mathbb{R}_+,H)} + \left\|A^2\varphi_0\right\|_H + \|A\varphi_1\|_H\right); \tag{3.1}$$
is valid with constant d independent of f, φ_0, φ_1.
2) *if the condition (2.5) is not satisfied and $\varphi_0 \in H_3$, $\varphi_1 \in H_2$, then the problem (2.1), (2.2) has the unique solution in the space $W^2_{2,\gamma}(\mathbb{R}_+, A^2)$ for arbitrary $\gamma > \gamma_2$. Moreover the following estimate*
$$\|u\|_{W^2_{2,\gamma}(\mathbb{R}_+,A^2)} \leq d\left(\|Af(t)\|_{L_{2,\gamma}(\mathbb{R}_+,H)} + \left\|A^3\varphi_0\right\|_H + \left\|A^2\varphi_1\right\|_H\right). \tag{3.2}$$
is valid with constant d independent of f, φ_0, φ_1.
If $f'(t) \in L_{2,\gamma_2}(\mathbb{R}_+, H)$, $f(0) = 0$ for some $\gamma_2 \geq 0$, then the estimates (3.1), (3.2) are valid, if the term $\|Af(t)\|_{L_{2,\gamma}(\mathbb{R}_+,H)}$ substitute to the term $\|f'(t)\|_{L_{2,\gamma}(\mathbb{R}_+,H)}$.

The proof of Theorem 3.1 is given in the articles [34], [36], [37]. In these cited articles we present the comparison the Theorem 3.1 with the results of L. Pandolfi [21].

Consider the operator function $L(\lambda)$ that is the symbol (an analog of the characteristic quasipolynomial) (2.1):
$$L(\lambda) := \lambda^2 I + \lambda\hat{K}(\lambda)A^2, \quad \hat{K}(\lambda) = \sum_{k=1}^{\infty} \frac{c_k}{\gamma_k(\lambda + \gamma_k)}. \tag{3.3}$$
Here $\hat{K}(\lambda)$ is the Laplace transform of $K(t)$, I and I is the identity operator in H.

Let $\{e_j\}_{j=1}^{\infty}$ be the orthonormal basis consisting of the eigenvectors of A corresponding to the eigenvalues a_j: $Ae_j = a_j e_j$, $j \in \mathbb{N}$. The eigenvalues of A satisfy the strict inequalities $0 < a_1 < a_2 < \cdots < a_n \cdots$; $a_n \to +\infty$, $n \to +\infty$.

Consider the restriction $l_n(\lambda) := (L(\lambda)e_n, e_n) = \lambda^2 + a_n^2\lambda\hat{K}(\lambda)$ of the operator function $L(\lambda)$ to the one-dimensional space spanned by the vector e_n. Thus, we obtain a countable set of meromorphic functions $l_n(\lambda)$, $n \in \mathbb{N}$.

The representations of solutions of problem (2.1), (2.2) in the form of exponential series obtained in the paper depend on the structure and properties of the spectrum of the operator function $L(\lambda)$.

We assume that the following condition is satisfied:
$$\sup_k \{\gamma_k(\gamma_{k+1} - \gamma_k)\} = +\infty. \tag{3.4}$$
Note that condition (3.4) was introduced and used in [32].

Now we formulate the results that describe the structure of the spectra of operator-function $L(\lambda)$.

For real eigenvalues $\lambda_{k,n}$ we have the following chain of inequalities:

Theorem 3.2. *Let us suppose the conditions (2.4) and (3.4) are satisfied. Then the spectrum of the operator-valued function $L(\lambda)$ is the closure of the zeroes set of the functions $\{l_n(\lambda)\}_{n=1}^{\infty}$ that is*

$$\sigma(L) := \overline{\{\lambda_{k,n} | k \in \mathbb{N}, n \in \mathbb{N}\} \cup \{\lambda_n^{\mp} | n \in \mathbb{N}\}}, \tag{3.5}$$

Moreover the real zeroes satisfy the inequalities

$$\cdots - \gamma_{k+1} < x_{k+1} < \lambda_{k+1,n} < -\gamma_k < \cdots < -\gamma_1 < \lambda_{1,n}, \quad k \in \mathbb{N}, \tag{3.6}$$

where x_k are the real zeroes of the function $\lambda \hat{K}(\lambda)$, and $\lambda_{1,n} = x_1 = 0$, $\lambda_{k,n} = x_k + O\left(1/a_n^2\right)$.

Theorem 3.3. *Let us suppose the conditions (2.4), (3.4) and (2.5) are satisfied. Then the conjugate complex zeroes $\lambda_n^{\pm}$, $\lambda_n^{+} = \overline{\lambda_n^{-}}$ of the meromorphic function $l_n(\lambda)$ asymptotically represented in the form*

$$\lambda_n^{\pm} = \pm i \left(\sqrt{K(0)} \cdot a_n + O\left(\frac{1}{a_n}\right)\right) - \frac{1}{2K(0)} \sum_{k=1}^{\infty} c_k + O\left(\frac{1}{a_n^2}\right), \quad a_n \to +\infty. \tag{3.7}$$

Theorem 3.4. *Let us suppose the conditions (2.4) and (3.4) are satisfied, but the condition (2.5) is not satisfied. Then the pair of the conjugate complex zeroes $\lambda_n^{\pm}$, $\lambda_n^{+} = \overline{\lambda_n^{-}}$ of the meromorphic function $l_n(\lambda)$ asymptotically represented in the form*

$$\lambda_n^{\pm} = \pm i\Theta \cdot a_n + \Phi(a_n, \{c_k\}_{k=1}^{\infty}, \{\gamma_k\}_{k=1}^{\infty}), \quad k \in \mathbb{N} \tag{3.8}$$

where $\Theta = \Theta(\{c_k\}_{k=1}^{\infty}, \{\gamma_k\}_{k=1}^{\infty})$ is a positive constant, depending on the sequences $\{c_k\}_{k=1}^{\infty}$, $\{\gamma_k\}_{k=1}^{\infty}$, $\operatorname{Re}\Phi = O(a_n)$, $\operatorname{Im}\Phi = o(a_n)$ while $a_n \to +\infty$ and $\lim_{a_n \to \infty} \operatorname{Re}\Phi(a_n, \{c_k\}_{k=1}^{\infty}, \{\gamma_k\}_{k=1}^{\infty}) = -\infty$.

The proofs of Theorems 3.2–3.4 are given in [34]–[36].

It is relevant to indicate that the structure of spectra of Gurtin–Pipkin equation was also studied in the following papers [32], [33], [40].

In what follows we use the expression $D \lesssim E$ if the following inequality $D \le cE$, is satisfied with some positive constant c. In turn the expression $D \approx E$ means the following inequalities $D \lesssim E \lesssim D$. We use the symbols := =: for introducing new values.

4. Statement of results

Theorem 4.1. *Suppose that $f(t) = 0$ for $t \in \mathbb{R}_+$, the vector function $u(t) \in W_{2,\gamma}^2\left(\mathbb{R}_+, A^2\right)$, $\gamma > 0$ is a strong solution of problem (2.1), (2.2) and conditions*

(2.4) and (3.4) are satisfied. Then, for each $t \in \mathbb{R}_+$ the solution $u(t)$ of problem (2.1), (2.2) can be represented as the following sum of series:

$$u(t) = \frac{1}{\sqrt{2\pi}} \sum_{n=1}^{\infty} \frac{(\varphi_{1n} + \lambda_n^+ \varphi_{0n})\, e^{\lambda_n^+ t} e_n}{l_n'(\lambda_n^+)} + \frac{1}{\sqrt{2\pi}} \sum_{n=1}^{\infty} \frac{(\varphi_{1n} + \lambda_n^- \varphi_{0n})\, e^{\lambda_n^- t} e_n}{l_n'(\lambda_n^-)}$$
$$+ \frac{1}{\sqrt{2\pi}} \sum_{n=1}^{\infty} \left(\sum_{k=1}^{\infty} \frac{(\varphi_{1n} + \lambda_{k,n}\varphi_{0n})\, e^{\lambda_{k,n} t}}{l_n'(\lambda_{k,n})} \right) e_n, \tag{4.1}$$

convergent in the norm of H, where the $\lambda_{k,n}$ are the real zeros of the meromorphic function $l_n(\lambda)$, which satisfy the inequalities

$$\cdots - \gamma_{k+1} < x_{k+1} < \lambda_{k+1,n} < -\gamma_k < \cdots < -\gamma_1 < \lambda_{1,n}, \quad k \in \mathbb{N}, \tag{4.2}$$

and the x_k are the real zeros of the function $\lambda \hat{K}(\lambda)$; moreover,

$$\lambda_{1,n} = x_1 = 0, \quad \lambda_{k,n} = x_k + O\left(1/a_n^2\right) \quad \text{as} \quad a_n \to +\infty,$$

and $\lambda_n^{\pm}$ is a pair of complex conjugate zeros, $\lambda_n^+ = \overline{\lambda_n^-}$, asymptotically representable in the following form:

1. *If condition (2.5) holds, then*

$$\lambda_n^{\pm} = \pm i \left(\sqrt{K(0)} \cdot a_n + O\left(\frac{1}{a_n}\right)\right) - \frac{1}{2K(0)} \sum_{k=1}^{\infty} c_k + O\left(\frac{1}{a_n^2}\right), \quad a_n \to +\infty. \tag{4.3}$$

2. *If condition (2.5) does not hold, then*

$$\lambda_n^{\pm} = \pm i\Theta(\{c_k\}, \{\gamma_k\}) a_n + \Phi(a_n, c_k, \gamma_k), \quad k \in \mathbb{N} \tag{4.4}$$

where $\Theta(\{c_k\}, \{\gamma_k\})\}$ is a positive constant depending on the sequences $\{c_k\}_{k=1}^{\infty}$, $\{\gamma_k\}_{k=1}^{\infty}$,

$$\operatorname{Re}\Phi(a_n, c_k, \gamma_k) = O(a_n), \quad \operatorname{Im}\Phi(a_n, c_k, \gamma_k) = o(a_n) \quad \text{as} \quad a_n \to +\infty,$$

and moreover,

$$\lim_{a_n \to \infty} \operatorname{Re}\Phi(a_n, c_k, \gamma_k) = -\infty.$$

Note that case 2 covers the sequences $\{c_k\}_{k=1}^{\infty}$ and $\{\gamma_k\}_{k=1}^{\infty}$ that have the asymptotic representations

$$c_k = \mathscr{A}/k^{\alpha} + O(1/k^{\alpha+1}), \quad \gamma_k = \mathscr{B}k^{\beta} + O(k^{\beta-1}), \quad k \to +\infty, \quad 0 \le \alpha < 1, \quad \alpha + \beta > 1.$$

Asymptotic formulas for the complex eigenvalues $\lambda_n^{\pm}$ for these sequences are given in [36].

Theorem 4.2. *If $\varphi_0 \in H_2$ and $\varphi_1 \in H_1$, then the series obtained from (4.1) by p-fold termwise differentiation with respect to t for $p = 0, 1, 2$ converges in the space H_{2-p} uniformly with respect to t on any interval $[t_0, T]$, where $0 < t_0 < T < +\infty$. Moreover, for all $t \in [t_0, T]$ one has the estimates*

$$\left\| \sum\nolimits_{n=1}^{\infty} u_n^{(p)}(t) e_n \right\|_{H_{2-p}}^2 \leqslant d(\|A\varphi_1\|^2 + \|A^2\varphi_0\|^2), \quad p = 0, 1, 2,$$

with constant d independent of the vector functions φ_1 and φ_0.

Further, if there are finitely many summands in (2.3) (i.e., $c_j = 0$ for $j > N$, $N \in \mathbb{N}$), then one can set $t_0 = 0$.

Theorem 4.3. *Suppose that $f(t) \in C([0,T],H)$ for arbitrary $T > 0$ and suppose that the vector-function $u(t) \in W^2_{2,\gamma}(\mathbb{R}_+, A^2)$ for some $\gamma > 0$, is a strong solution of problem (2.1), (2.2), conditions (2.4), (3.4) are satisfied, and $\varphi_0 = \varphi_1 = 0$. Then, for each $t \in \mathbb{R}_+$, the solution $u(t)$ of problem (2.1), (2.2) can be represented as the following sum of series:*

$$u(t) = \frac{1}{\sqrt{2\pi}}\sum_{n=1}^{\infty}\omega_n(t,\lambda_n^+)e_n + \frac{1}{\sqrt{2\pi}}\sum_{n=1}^{\infty}\omega_n(t,\lambda_n^-)e_n + \frac{1}{\sqrt{2\pi}}\sum_{n=1}^{\infty}\left(\sum_{k=1}^{\infty}\omega_n(t,\lambda_{kn})\right)e_n, \tag{4.5}$$

converging in the norm of H, where

$$\omega_n(t,\lambda) = \frac{\int_0^t f_n(\tau)e^{\lambda(t-\tau)}d\tau}{l_n'(\lambda)}.$$

Theorem 4.4. *Let the assumptions of Theorem 4.3 be satisfied together with the condition $\sum_{j=1}^{\infty}\gamma_j^{-3/2} < \infty$. Then the series obtained from (4.5) by p-fold term-by-term differentiation with respect to t for $p = 0,1,2$ is convergent in the space H_{2-p} uniformly with respect to t on any interval $[t_0,T]$, where $t_0 < T < +\infty$ ($t_0 = 0$ for $p = 0,1$ and $t_0 > 0$ for $p = 2$); moreover the following estimates hold for all $t \in [t_0,T]$:*

$$\left\|\sum_{n=1}^{\infty}u_n(t)e_n\right\|^2_{H_2} \leqslant d\left\|A^2 f(t)\right\|_{L_{2,\gamma}(R_+,H)}, \tag{4.6}$$

$$\left\|\sum_{n=1}^{\infty}u_n^{(p)}(t)e_n\right\|^2_{H_{2-p}} \tag{4.7}$$
$$\leqslant d\left(\left\|A^{2-p}f^{(p)}(t)\right\|^2_{L_{2,\gamma}(R_+,H)} + \left\|A^{2-p}f(0)\right\|^2_H + (p-1)\left\|f'(0)\right\|^2_H\right), \quad p = 1,2.$$

The series (4.1) and (4.5) are obtained by applying the inverse Laplace transform to the solution of problem (2.1), (2.2) with the use of integration over rectangular contours separating the points γ_j. (The construction of these contours is described in [36].) Here the estimates of the operator function $L^{-1}(\lambda)$ on these contours play an important role. Besides, the proof of Theorems 4.1–4.4 heavily uses the representations (4.2), (4.3) and (4.4) obtained in [36].

In conclusion, note that the solvability of problem (2.1), (2.2) in the Sobolev spaces described above was studied in [36], but the representation of solutions by series in elementary functions was not considered. Further, the solvability of the equation of Gurtin–Pipkin type was studied in [16] in the same spaces and in [21] in different function spaces. In special cases, the localization of the spectrum of the operator function $L(\lambda)$ was studied in the papers [32] and [33]. The papers [35], [18], and [34] contain results on the well-posed solvability and the spectral analysis of equations of the form (2.1).

5. Proofs

Proof of Theorem 4.1. We introduce the following contours

$$\begin{aligned} R_N &:= \frac{\gamma_N + \gamma_{N+1}}{2}, \quad \gamma_N \to \infty \\ \Gamma &= \left\{\Gamma_\gamma \cup \Gamma_{R_N} \cup \Gamma^+ \cup \Gamma^-\right\}, \\ C_{R_N} &:= \left\{\Gamma_{R_N} \cup \Gamma^+ \cup \Gamma^-\right\}, \quad N \in \mathbb{N}, \\ \Gamma_\gamma &= \left\{z = x + iy \in \mathbb{C} | x = \gamma, |y| \leqslant R_N\right\}, \\ \Gamma_{R_N} &= \left\{z = x + iy \in \mathbb{C} | x = -R_N, |y| \leqslant R_N\right\}, \\ \Gamma^\pm &= \left\{z = x + iy \in \mathbb{C} | -R_N \leqslant x \leqslant \gamma, y = \pm R_N\right\}. \end{aligned}$$

According to the conditions of Theorem 4.1, the problem (2.1), (2.2) has the unique solution $u(t) \in W^2_{2,\gamma}\left(\mathbb{R}_+, A^2\right)$, $\gamma > 0$.

The Laplace transform of the strong solution of the equation (2.1) with initial conditions (2.2) has the form

$$\hat{u}(\lambda) = L^{-1}(\lambda)(\varphi_1 + \lambda\varphi_0). \tag{5.1}$$

where operator-function $L(\lambda)$ is the symbol of the equation (2.1). Hence we have

$$\hat{u_n}(\lambda) = \frac{(\varphi_{1n} + \lambda\varphi_{0n})}{l_n(\lambda)}.$$

Using the Paley–Wiener theorem and the Cauchy theorem about residues for sufficiently large $N = N(m)$ $(R_N > a_m\sqrt{S})$ we have

$$\begin{aligned} u(t) &= \sum_{n=1}^{\infty} u_n(t)e_n = \sum_{n=1}^{\infty}\left(\frac{1}{\sqrt{2\pi}}\lim_{R_N\to\infty}\int_{-R_N}^{R_N}\hat{u_n}(\gamma+iy)e^{(\gamma+iy)t}dy\right)e_n \\ &= \sum_{n=1}^{\infty}\left(\frac{1}{\sqrt{2\pi}}\lim_{R_N\to\infty}\int_{-R_N}^{R_N}\frac{(\varphi_{1n}+(\gamma+iy)\varphi_{0n})}{l_n(\gamma+iy)}e^{(\gamma+iy)t}dy\right)e_n \\ &= \frac{1}{\sqrt{2\pi}}\lim_{R_N\to\infty}\lim_{m\to\infty}\int_{-R_N}^{R_N}\left(\sum_{n=1}^{m}\frac{(\varphi_{1n}+(\gamma+iy)\varphi_{0n})}{l_n(\gamma+iy)}e^{(\gamma+iy)t}e_n\right)dy \\ &= \frac{1}{\sqrt{2\pi}}\lim_{m\to\infty}\int_{-R_N}^{R_N}\left(\sum_{n=1}^{m}\frac{(\varphi_{1n}+(\gamma+iy)\varphi_{0n})}{l_n(\gamma+iy)}e^{(\gamma+iy)t}e_n\right)dy \end{aligned}$$

$$= \frac{1}{\sqrt{2\pi}} \lim_{m\to\infty} \left(\sum_{n=1}^{m} \frac{(\varphi_{1n} + \lambda_n^+ \varphi_{0n})\, e^{\lambda_n^+ t} e_n}{l_n'(\lambda_n^+)} \right.$$

$$+ \sum_{n=1}^{m} \frac{(\varphi_{1n} + \lambda_n^- \varphi_{0n})\, e^{\lambda_n^- t} e_n}{l_n'(\lambda_n^-)} + \sum_{n=1}^{m} \left(\sum_{k=1}^{N(m)} \frac{(\varphi_{1n} + \lambda_{k,n}\varphi_{0n})\, e^{\lambda_{k,n} t}}{l_n'(\lambda_{k,n})} \right) e_n$$

$$\left. + \sum_{n=1}^{m} \left(\int_{C_{R_N}} \frac{(\varphi_{1n} + z\varphi_{0n}) e^{zt}}{l_n(z)} dz \right) e_n \right).$$

So for arbitrary $t \in \mathbb{R}_+$ we obtain

$$u(t) = \lim_{m\to\infty} \left(P_{Nm}(t) + \Phi_{Nm}(t) \right),$$

where

$$P_{Nm}(t) := \sum_{n=1}^{m} \frac{(\varphi_{1n} + \lambda_n^+ \varphi_{0n})\, e^{\lambda_n^+ t} e_n}{l_n'(\lambda_n^+)} + \sum_{n=1}^{m} \frac{(\varphi_{1n} + \lambda_n^- \varphi_{0n})\, e^{\lambda_n^- t} e_n}{l_n'(\lambda_n^-)}$$

$$+ \sum_{n=1}^{m} \left(\sum_{k=1}^{N(m)} \frac{(\varphi_{1n} + \lambda_{k,n}\varphi_{0n})\, e^{\lambda_{k,n} t}}{l_n'(\lambda_{k,n})} \right) e_n, \tag{5.2}$$

m, $N(m) \in \mathbb{N}$, $p = 0, 1, 2$, $\lambda_{k,n}$ are real zeroes of meromorphic function $l_n(\lambda)$ satisfying the inequalities (4.2), $\lambda_n^{\pm}$ are complex zeroes of function $l_n(\lambda)$, function

$$\Phi_{Nm}(t) := \sum_{k=1}^{m} \left(\int_{C_{R_N}} \frac{(\varphi_{1n} + z\varphi_{0n})}{l_n(z)} e^{zt} dz \right) e_n.$$

Our main purpose is to prove that $\lim_{m\to\infty} \|\Phi_{Nm}(t)\|_H = 0$. In order to do this we use the following three propositions. The proofs of these propositions have pure technical character and we omit them.

Proposition 5.1. *If* $|\arg z_1 - \arg z_2| < \pi - \delta$, *where* $\delta > 0$. *Then* $|z_1 + z_2|^2 \approx |z_1|^2 + |z_2|^2$, $z_1, z_2 \in \mathbb{C}$.

Statement 5.1. *Suppose that condition* (3.4) *is satisfied. Then the following estimates*

1) $|z^p \cdot l_n^{-1}(z)| \lesssim 1/R_N^{2-p}$ *for* $z \in C_{R_N}$, $p = 0, 1$.
2) $|a_n \cdot l_n^{-1}(z)| \lesssim 1/R_N$ *for* $z \in C_{R_N}$

are valid.

Proposition 5.2. *For some $\mu > 0$ for arbitrary $t > 0$ and $p = 0, 1$ for $z \in C_{R_N}$ the following estimates*

$$\frac{1}{R_N^{2-p}} \int\limits_{-R_N \pm iR_N}^{-\mu \ln R_N \pm iR_N} e^{t\,\mathrm{Re}\, z} d|z| \leqslant \frac{1}{R_N^{2-p}} \left(R_N - \mu \ln R_N\right) e^{-t\mu \ln R_N} \lesssim \frac{1}{R_N^{1-p}} e^{-t\mu \ln R_N},$$

$$\frac{1}{R_N^{2-p}} \int\limits_{-\mu \ln R_N \pm iR_N}^{\gamma \pm iR_N} e^{t\,\mathrm{Re}\, z} d|z| \leqslant \frac{1}{R_N^{2-p}} \left(\gamma + \mu \ln R_N\right) e^{t\gamma},$$

$$\frac{1}{R_N^{2-p}} \int\limits_{-R_N - iR_N}^{-R_N + iR_N} e^{t\,\mathrm{Re}\, z} d|z| = \frac{1}{R_N^{2-p}} \int\limits_{-R_N}^{R_N} e^{t\,\mathrm{Re}(-R_N + iy)} dy = \frac{2}{R_N^{1-p}} e^{-tR_N}$$

are valid.

On the base of Proposition 5.2 for $t > 0$ we have following inequalities

$$\begin{aligned}
&\left| \int\limits_{C_{R_N}} \frac{(\varphi_{1n} + z\varphi_{0n}) e^{zt}}{l_n(z)} dz \right| \\
&\leqslant \frac{|\varphi_{1n}| + R_N |\varphi_{0n}|}{R_N^2} \left(\int\limits_{\gamma - iR_N}^{-R_N - iR_N} + \int\limits_{-R_N - iR_N}^{-R_N + iR_N} + \int\limits_{-R_N + iR_N}^{\gamma + iR_N} \right) e^{t\,\mathrm{Re}\, z} d|z| \\
&\leqslant \left(|\varphi_{1n}| + R_N |\varphi_{0n}|\right) \left(\frac{1}{R_N} e^{-t\mu \ln R_N} + \frac{1}{R_N^2} \left(\gamma + \mu \ln R_N\right) e^{t\gamma} + \frac{2}{R_N} e^{-tR_N} \right) \\
&\lesssim \frac{\ln R_N}{R_N^2} \left(|\varphi_{1n}| + R_N |\varphi_{0n}|\right) e^{t\gamma}.
\end{aligned}$$

Hence we have for $t > 0$

$$\begin{aligned}
\|\Phi_{Nm}(t)\|_H^2 &= \left\| \sum_{k=1}^{m} \left(\int\limits_{C_{R_N}} \frac{\varphi_{1n} + z\varphi_{0n}}{l_n(z)} e^{zt} dz \right) e_n \right\|_H^2 \\
&= \sum_{k=1}^{m} \left| \int\limits_{C_{R_N}} \frac{(\varphi_{1n} + z\varphi_{0n})}{l_n(z)} e^{zt} dz \right|^2 \\
&\lesssim \sum_{k=1}^{m} \frac{\left(|\varphi_{1n}|^2 + R_N^2 |\varphi_{0n}|^2\right) \ln^2 R_N}{R_N^4} e^{2t\gamma} \\
&= e^{2\gamma t} \left(\frac{\|\varphi_1\|_H^2}{R_N^4} + \frac{\|\varphi_0\|_H^2}{R_N^2} \right) \ln^2 R_N \to 0, \quad N \to +\infty,
\end{aligned}$$

So we finish the proof of Theorem 4.1 so as for arbitrary $t \in \mathbb{R}_+$ we have

$$\lim_{m\to\infty} \|\Phi_{Nm}(t)\|_H = \lim_{m\to\infty} \|u(t) - P_{Nm}(t)\|_H = 0. \qquad \square$$

Proof of Theorem 4.2. Let us differentiate the series (4.1) p times for $p = 1,2$ with respect to t and let us show that series obtained thus converges in the space H_{2-p} uniformly with respect to t on any interval $[t_0, T]$, where $0 < t_0 < T < +\infty$. Keeping in the mind asymptotical representation (4.2), (4.3) (4.4) and Proposition 5.1 we obtain the following estimates:

$$\sum_{n=1}^{\infty}\left|\frac{(\varphi_{1n}+\lambda_n^{\pm}\varphi_{0n})(\lambda_n^{\pm})^p a_n^{2-p}}{l_n^{(1)}(\lambda_n^{\pm})}e^{\lambda_n^{\pm}t}\right|^2 = \sum_{n=1}^{\infty}\left|\frac{(\varphi_{1n}+\lambda_n^{\pm}\varphi_{0n})(\lambda_n^{\pm})^p a_n^{2-p}}{2\lambda_n^{\pm}+a_n^2\sum\limits_{j=1}^{\infty}\dfrac{c_j}{(\gamma_j+\lambda_n^{\pm})^2}}e^{\lambda_n^{\pm}t}\right|^2$$

$$\lesssim \sum_{n=1}^{\infty}\frac{\left(|\varphi_{1n}|^2+|\lambda_n^{\pm}|^2|\varphi_{0n}|^2\right)|\lambda_n^{\pm}|^{2p}a_n^{4-2p}}{4|\lambda_n^{\pm}|^2+a_n^4\left|\sum\limits_{j=1}^{\infty}\dfrac{c_j}{(\gamma_j+\lambda_n^{\pm})^2}\right|^2}e^{\operatorname{Re}\lambda_n^{\pm}t}$$

$$\lesssim \sum_{n=1}^{\infty}\frac{\left(|\varphi_{1n}|^2+|\lambda_n^{\pm}|^2|\varphi_{0n}|^2\right)a_n^{4-2p}}{|\lambda_n^{\pm}|^{2-2p}} \lesssim \sum_{n=1}^{\infty}\frac{\left(|\varphi_{1n}|^2+a_n^2|\varphi_{0n}|^2\right)a_n^{4-2p}}{a_n^{2-2p}}$$

$$= \sum_{n=1}^{\infty}(|a_n\varphi_{1n}|^2+|a_n^2\varphi_{0n}|^2) = \|A\varphi_1\|^2+\|A^2\varphi_0\|^2$$

$$\sum_{n=1}^{\infty}\left|\sum_{k=1}^{\infty}\frac{(\varphi_{1n}+\lambda_{kn}\varphi_{0n})(\lambda_{kn})^p a_n^{2-p}}{l_n'(\lambda_{kn})}e^{\lambda_{kn}t}\right|^2$$

$$\leqslant \sum_{n=1}^{\infty}\left(\sum_{k=1}^{\infty}\left|\frac{(\varphi_{1n}+\lambda_{kn}\varphi_{0n})(\lambda_{kn})^p a_n^{2-p}}{2\lambda_{kn}+a_n^2\sum\limits_{j=1}^{\infty}\dfrac{c_j}{(\gamma_j+\lambda_{kn})^2}}e^{\lambda_{kn}t}\right|\right)^2$$

$$\lesssim \sum_{n=1}^{\infty}\left(\sum_{k=1}^{\infty}\frac{(|\varphi_{1n}|+|x_k||\varphi_{0n}|)|x_k|^p a_n^{2-p}}{\left(2|x_k|^2+a_n^4\left|\sum\limits_{j=1}^{\infty}\dfrac{c_j}{(\gamma_j+x_k)^2}\right|^2\right)^{1/2}}e^{x_k t}\right)^2$$

$$\lesssim \sum_{n=1}^{\infty}\left(\sum_{k=1}^{\infty}\frac{(|\varphi_{1n}|+|x_k||\varphi_{0n}|)a_n^{2-p}|x_k|^p}{a_n|x_k|^{1/2}}e^{x_k t}\right)^2$$
$$\lesssim \sum_{n=1}^{\infty}|a_n^{1-p}\varphi_{1n}|^2\left(\sum_{k=1}^{\infty}|x_k|^{p-1/2}e^{x_k t}\right)^2$$
$$+\sum_{n=1}^{\infty}|a_n^{1-p}\varphi_{0n}|^2\left(\sum_{k=1}^{\infty}|x_k|^{p+1/2}e^{x_k t}\right)^2$$
$$\lesssim \|A^{1-p}\varphi_1\|^2+\|A^{1-p}\varphi_0\|^2,$$

where x_k are real zeroes of the function $\lambda\hat{K}(\lambda)$, and $\lambda_{k,n}$ have the asymptotics $\lambda_{1,n}=x_1=0$, $\lambda_{k,n}=x_k+O\left(1/a_n^2\right)$. Using the d'Alambert and Weierstrass theorems we obtain that $\sum_{k=1}^{\infty}|x_k|^{p+1/2}e^{x_k t}$ converges absolutely and uniformly on the semiaxis $t\in[t_0,+\infty)$, $t_0>0$. □

Proof of Theorem 4.3. We consider the problem (2.1), (2.2) with conditions $\varphi_0=\varphi_1=0$. According to Theorem 4.3 the problem (2.1), (2.2) has the unique solution $u(t)\in W_{2,\gamma}^2\left(\mathbb{R}_+,A^2\right)$, $\gamma>0$, which can be represented in the form

$$u(t)=\int_0^t v(t,\tau)d\tau,$$

where vector-function $v(t,\tau)$ is the solution of the following problem

$$\frac{d^2v(t,\tau)}{dt^2}+K(0)A^2v(t,\tau)+\int_\tau^t K^{(1)}(t-s)A^2v(s,\tau)ds=0,\quad t>\tau, \tag{5.3}$$
$$v(\tau,\tau)=0,\quad v_t^{(1)}(\tau,\tau)=f(\tau), \tag{5.4}$$

and $v(t,\tau)=0$ for $t<\tau$.

Let us verify it. We have

$$A^2u(t)=\int_0^t A^2v(t,\tau)d\tau.$$

Using differentiating of the solution $u(t)$ by variable t and taking into account the conditions (5.4) we obtain

$$u_t(t)=v(t,t)+\int_0^t v_t(t,\tau)d\tau=\int_0^t v_t(t,\tau)d\tau,$$

$$u_{tt}(t)=v_t(t,t)+\int_0^t v_{tt}(t,\tau)d\tau=f(t)+\int_0^t v_{tt}(t,\tau)d\tau.$$

Moreover we have

$$\int_0^t K^{(1)}(t-s)A^2\left(\int_0^s v(s,\tau)d\tau\right)ds = \int_0^t\left(\int_\tau^t K^{(1)}(t-s)A^2 v(s,\tau)ds\right)d\tau.$$

The Laplace transform of the strong solution of the problem (5.3), (5.4) has the form

$$\hat{v}(\lambda,\tau) = L^{-1}f(\tau)e^{-\lambda\tau},$$

where operator-function $L(\lambda)$ is the symbol of the equation (5.3) that has the representation (3.3). Hence we have

$$\hat{v_n}(\lambda,\tau) = \frac{f_n(\tau)e^{-\lambda\tau}}{l_n(\lambda)}.$$

Providing the arguments analogously to the proof of the theorem 4.1 for all $t > \tau > 0$ we obtain the following representation for the solution $v(t,\tau)$ of the problem (5.3), (5.4)

$$v(t,\tau) = \lim_{m\to\infty}\left(P_{Nm}(t,\tau) + \Phi_{Nm}(t,\tau)\right),$$

where

$$P_{Nm}(t,\tau) := \sum_{n=1}^m \frac{f_n(\tau)e^{\lambda_n^+(t-\tau)}e_n}{l_n^{(1)}(\lambda_n^+)} + \sum_{n=1}^m \frac{f_n(\tau)e^{\lambda_n^-(t-\tau)}e_n}{l_n^{(1)}(\lambda_n^-)}$$
$$+\sum_{n=1}^m\left(\sum_{k=1}^{N(m)} \frac{f_n(\tau)e^{\lambda_{k,n}(t-\tau)}}{l_n^{(1)}(\lambda_{k,n})}\right)e_n, \tag{5.5}$$

m, $N(m) \in \mathbb{N}$, $p = 0,1,2$, $\lambda_{k,n}$ are the real zeroes of the meromorphic function $l_n(\lambda)$ satisfying the inequalities (4.2), $\lambda_n^{\pm}$ are the complex zeroes of function $l_n(\lambda)$

$$\Phi_{Nm}(t,\tau) := \sum_{k=1}^m\left(\int_{C_{R_N}} \frac{f_n(\tau)e^{z(t-\tau)}}{l_n(z)}dz\right)e_n,$$

Let us prove that $\lim_{m\to\infty}\|\Phi_{Nm}(t,\tau)\|_H = 0$ for arbitrary $T > 0$ and $0 \leqslant \tau \leqslant t \leqslant T$. For $0 \leqslant \tau \leqslant t$ we have

$$\left|\int_{C_{R_N}} \frac{f_n(\tau)e^{z(t-\tau)}}{l_n(z)}dz\right| \leqslant \frac{|f_n(\tau)|}{R_N^2}\left(\int_{\gamma-iR_N}^{-R_N-iR_N} + \int_{-R_N-iR_N}^{-R_N+iR_N} + \int_{-R_N+iR_N}^{\gamma+iR_N}\right)e^{(t-\tau)\operatorname{Re} z}d|z|$$
$$\leqslant |f_n(\tau)|\left(\frac{1}{R_N}e^{-(t-\tau)\mu\ln R_N} + \frac{1}{R_N^2}(\gamma+\mu\ln R_N)e^{(t-\tau)\gamma} + \frac{2}{R_N}e^{-(t-\tau)R_N}\right)$$
$$\lesssim \frac{\ln R_N}{R_N^2}|f_n(\tau)|e^{(t-\tau)\gamma}.$$

Owing to the conditions of Theorem 4.3 vector-function $f(t) \in C([0,T],H)$ for arbitrary $T > 0$. Hence for arbitrary $0 \leqslant \tau \leqslant t \leqslant T$ we have

$$\|\Phi_{Nm}(t,\tau)\|_H^2 = \left\| \sum_{k=1}^{m} \left(\int_{C_{R_N}} \frac{f_n(\tau)e^{z(t-\tau)}}{l_n(z)} dz \right) e_n \right\|_H^2 = \sum_{k=1}^{m} \left| \int_{C_{R_N}} \frac{f_n(\tau)e^{z(t-\tau)}}{l_n(z)} dz \right|^2$$

$$\lesssim \left(\frac{\ln R_N}{R_N^2}\right)^2 \sum_{k=1}^{m} |f_n(\tau)|^2 e^{2(t-\tau)\gamma} \leqslant \left(\frac{\ln R_N}{R_N^2}\right)^2 \|f(\tau)\|_H^2 e^{2(t-\tau)\gamma} \to 0, \quad N \to +\infty.$$

So for arbitrary $0 \leqslant \tau \leqslant t \leqslant T$ we obtain

$$\lim_{m\to\infty} \|\Phi_{Nm}(t,\tau)\|_H = \lim_{m\to\infty} \|v(t,\tau) - P_{Nm}(t,\tau)\|_H = 0,$$

and

$$v(t,\tau) = \frac{1}{\sqrt{2\pi}} \sum_{n=1}^{\infty} \frac{f_n(\tau)e^{\lambda_n^+(t-\tau)}e_n}{l_n^{(1)}(\lambda_n^+)} + \frac{1}{\sqrt{2\pi}} \sum_{n=1}^{\infty} \frac{f_n(\tau)e^{\lambda_n^-(t-\tau)}e_n}{l_n^{(1)}(\lambda_n^-)}$$
$$+ \frac{1}{\sqrt{2\pi}} \sum_{n=1}^{\infty} \left(\sum_{k=1}^{\infty} \frac{f_n(\tau)e^{\lambda_{k,n}(t-\tau)}}{l_n^{(1)}(\lambda_{k,n})} \right) e_n. \tag{5.6}$$

Moreover for $0 \leqslant \tau \leqslant t \leqslant T$ we have

$$\sup_{\tau\in[0,t]} \|\Phi_{Nm}(t,\tau)\|_H \lesssim \frac{\ln R_N}{R_N^2} \sup_{\tau\in[0,t]} \|f(\tau)\|_H \, e^{T\gamma} \to 0, \quad N \to +\infty,$$

and then

$$\lim_{m\to\infty} \sup_{\tau\in[0,t]} \|\Phi_{Nm}(t,\tau)\|_H = \lim_{m\to\infty} \sup_{\tau\in[0,t]} \|v(t,\tau) - P_{Nm}(t,\tau)\|_H = 0.$$

Thus the series (5.6) converges uniformly on $\tau \in [0,t]$ in the space H.

Integrating the series (5.6) by $\tau \in (0,t)$ we obtain the following representation for the solution $u(t)$ of the problem (2.1), (2.2) with zero initial conditions

$$u(t) = \frac{1}{\sqrt{2\pi}} \sum_{n=1}^{\infty} \frac{\left(\int_0^t f_n(\tau)e^{\lambda_n^+(t-\tau)}d\tau\right) e_n}{l_n^{(1)}(\lambda_n^+)} + \frac{1}{\sqrt{2\pi}} \sum_{n=1}^{\infty} \frac{\left(\int_0^t f_n(\tau)e^{\lambda_n^-(t-\tau)}d\tau\right) e_n}{l_n^{(1)}(\lambda_n^-)}$$
$$+ \frac{1}{\sqrt{2\pi}} \sum_{n=1}^{\infty} \left(\sum_{k=1}^{\infty} \frac{\int_0^t f_n(\tau)e^{\lambda_{k,n}(t-\tau)}d\tau}{l_n^{(1)}(\lambda_{k,n})} \right) e_n. \tag{5.7}$$

□

Proof of Theorem 4.4. Let us differentiate p-fold term-by-term the series (5.7) with respect to t for $p = 1, 2$. Then we have

$$\sum_{n=1}^{\infty} u_n^{(p)}(t)e_n = \frac{1}{\sqrt{2\pi}} \sum_{n=1}^{\infty} \omega_n^{(p)}(t,\lambda_n^+)e_n + \frac{1}{\sqrt{2\pi}} \sum_{n=1}^{\infty} \omega_n^{(p)}(t,\lambda_n^-)e_n$$
$$+ \frac{1}{\sqrt{2\pi}} \sum_{n=1}^{\infty} \left(\sum_{k=1}^{\infty} \omega_n^{(p)}(t,\lambda_{kn}) \right) e_n, \tag{5.8}$$

where

$$\omega_n^{(p)}(t,\lambda) = \frac{\left((p-1)f_n^{(1)}(0) + \lambda^{(p-1)} f_n(0)\right) e^{\lambda t} + \int\limits_0^t f_n^{(p)}(\tau) e^{\lambda(t-\tau)} d\tau}{l_n^{(1)}(\lambda)}.$$

Let us show that thus obtained series converges in the space H_{2-p} $(p = 0, 1, 2)$ uniformly with respect to t on any interval $[t_0, T]$, where $t_0 < T < +\infty$ ($t_0 = 0$ for $p = 0, 1$ and $t_0 > 0$ for $p = 2$). Taking in mind the asymptotic representations (4.2), (4.3) and (4.4) we obtain the following estimates:

$$\begin{aligned}
&\sum_{n=1}^{\infty} \left| \frac{a_n^{2-p} \int_0^t f_n^{(p)}(\tau) e^{\lambda_n^{\pm}(t-\tau)} d\tau}{2\lambda_n^{\pm} + a_n^2 \sum_{j=1}^{\infty} \frac{c_j}{(\lambda_n^{\pm} + \gamma_j)^2}} \right|^2 \\
&\leqslant \sum_{n=1}^{\infty} \frac{a_n^{4-2p} \left(\int_0^t |f_n^{(p)}(\tau)| e^{\operatorname{Re}\lambda_n^{\pm}(t-\tau)} d\tau\right)^2}{\left|2\lambda_n^{\pm} + a_n^2 \sum_{j=1}^{\infty} \frac{c_j}{(\lambda_n^{\pm} + \gamma_j)^2}\right|^2} \\
&\lesssim \sum_{n=1}^{\infty} \frac{a_n^{4-2p} \left(\int_0^t |f_n^{(p)}(\tau)|^2 d\tau\right) \left(\int_0^t e^{2\operatorname{Re}\lambda_n^{\pm}(t-\tau)} d\tau\right)}{4|\lambda_n^{\pm}|^2 + a_n^4 \left|\sum_{j=1}^{\infty} \frac{c_j}{(\lambda_n^{\pm} + \gamma_j)^2}\right|^2} \\
&\lesssim \sum_{n=1}^{\infty} \frac{a_n^{4-2p} \left(\int_0^t |f_n^{(p)}(\tau)|^2 d\tau\right) \frac{1}{2\operatorname{Re}\lambda_n^{\pm}} \left(1 - e^{2\operatorname{Re}\lambda_n^{\pm} t}\right)}{|\lambda_n^{\pm}|^2} \\
&\lesssim \sum_{n=1}^{\infty} \frac{a_n^{4-2p} \left(\int_0^t |f_n^{(p)}(\tau)|^2 d\tau\right)}{a_n^2} = \sum_{n=1}^{\infty} \left(\int_0^t |a_n^{1-p} f_n^{(p)}(\tau)|^2 d\tau\right) \\
&= \int_0^t \sum_{n=1}^{\infty} |a_n^{1-p} f_n^{(p)}(\tau)|^2 d\tau = \int_0^t \left\| A^{1-p} f^{(p)}(\tau) \right\|_H^2 d\tau \\
&\leqslant \left\| A^{1-p} f^{(p)}(t) \right\|^2_{L_{2,\gamma}(\mathbb{R}_+, H)} < \infty, \quad p = 0, 1, 2,
\end{aligned}$$

$$\begin{aligned}
&\sum_{n=1}^{\infty} \left| \sum_{k=1}^{\infty} \frac{a_n^{2-p} \int_0^t f_n^{(p)}(\tau) e^{\lambda_{kn}(t-\tau)} d\tau}{2\lambda_{kn} + a_n^2 \sum_{j=1}^{\infty} \frac{c_j}{(\lambda_{kn} + \gamma_j)^2}} \right|^2 \\
&\leqslant \sum_{n=1}^{\infty} \left(\sum_{k=1}^{\infty} \frac{a_n^{2-p} \int_0^t |f_n^{(p)}(\tau)| e^{\lambda_{kn}(t-\tau)} d\tau}{\left|2\lambda_{kn} + a_n^2 \sum_{j=1}^{\infty} \frac{c_j}{(\lambda_{kn} + \gamma_j)^2}\right|} \right)^2
\end{aligned}$$

$$\approx \sum_{n=1}^{\infty}\left(\sum_{k=1}^{\infty}\frac{a_n^{2-p}\int_0^t|f_n^{(p)}(\tau)|e^{\lambda_{kn}(t-\tau)}d\tau}{\left(4|\lambda_{kn}|^2+a_n^4\left|\sum_{j=1}^{\infty}\frac{c_j}{(\lambda_{kn}+\gamma_j)^2}\right|^2\right)^{1/2}}\right)^2$$

$$\approx \sum_{n=1}^{\infty}\left(\sum_{k=1}^{\infty}\frac{a_n^{2-p}\int_0^t|f_n^{(p)}(\tau)|e^{x_k(t-\tau)}d\tau}{\left(4|x_k|^2+a_n^4\left|\sum_{j=1}^{\infty}\frac{c_j}{(x_k+\gamma_j)^2}\right|^2\right)^{1/2}}\right)^2$$

$$\leqslant \sum_{n=1}^{\infty}\left(\sum_{k=1}^{\infty}\frac{a_n^{2-p}\left(\int_0^t|f_n^{(p)}(\tau)|^2d\tau\right)^{1/2}\left(\int_0^t e^{2x_k(t-\tau)}d\tau\right)^{1/2}}{|x_k|}\right)^2$$

$$=\sum_{n=1}^{\infty}\left(\int_0^t|a_n^{2-p}f_n^{(p)}(\tau)|^2d\tau\right)\left(\sum_{k=1}^{\infty}\frac{1}{\sqrt{2}|x_k|^{3/2}}\left(1-e^{2x_kt}\right)^{1/2}\right)^2$$

$$\leqslant \frac{1}{2}\int_0^t\sum_{n=1}^{\infty}|a_n^{2-p}f_n^{(p)}(\tau)|^2d\tau\left(\sum_{k=1}^{\infty}\frac{1}{\gamma_k^{3/2}}\right)^2$$

$$=\frac{1}{2}\left(\sum_{k=1}^{\infty}\frac{1}{\gamma_k^{3/2}}\right)^2\int_0^t\left\|A^{2-p}f^{(p)}(\tau)\right\|_H^2d\tau$$

$$\lesssim \frac{1}{2}\left(\sum_{k=1}^{\infty}\frac{1}{\gamma_k^{3/2}}\right)^2\left\|A^{2-p}f^{(p)}(\tau)\right\|^2_{L_{2,\gamma}(\mathbb{R}_+,H)}<\infty,\quad p=0,1,2.$$

Thus we have the estimate (4.6). Then for $p=1,2$ we have the following chain of inequalities

$$\sum_{n=1}^{\infty}\left|\frac{a_n^{2-p}\left(f_n(0)(\lambda_n^{\pm})^{p-1}+(p-1)f_n^{(1)}(0)\right)e^{\lambda_n^{\pm}t}}{2\lambda_n^{\pm}+a_n^2\sum_{j=1}^{\infty}\frac{c_j}{(\lambda_n^{\pm}+\gamma_j)^2}}\right|^2$$

$$\leqslant \sum_{n=1}^{\infty}\frac{a_n^{4-2p}\left(|f_n(0)|^2(\lambda_n^{\pm})^{2(p-1)}+(p-1)^2|f_n^{(1)}(0)|^2\right)e^{2\operatorname{Re}\lambda_n^{\pm}t}}{4|\lambda_n^{\pm}|^2+a_n^4\left|\sum_{j=1}^{\infty}\frac{c_j}{(\lambda_n^{\pm}+\gamma_j)^2}\right|^2}$$

$$\lesssim \sum_{n=1}^{\infty}\frac{a_n^{4-2p}\left(|f_n(0)|^2a_n^{2(p-1)}+(p-1)^2|f_n^{(1)}(0)|^2\right)}{a_n^2}$$

$$=\sum_{n=1}^{\infty}|f_n(0)|^2+(p-1)^2a_n^{2(1-p)}|f_n^{(1)}(0)|^2\leqslant\|f(0)\|_H^2+(p-1)^2\left\|A^{1-p}f^{(1)}(0)\right\|_H^2.$$

$$
\sum_{n=1}^{\infty}\left|\sum_{k=1}^{\infty}\frac{a_n^{2-p}\left(f_n(0)\lambda_{kn}^{p-1}+(p-1)f_n^{(1)}(0)\right)e^{\lambda_{kn}t}}{2\lambda_{kn}+a_n^2\sum_{j=1}^{\infty}\frac{c_j}{(\lambda_{kn}+\gamma_j)^2}}\right|^2
$$

$$
\leqslant\sum_{n=1}^{\infty}\left(\sum_{k=1}^{\infty}\frac{a_n^{2-p}\left(|f_n(0)||\lambda_{kn}|^{p-1}+(p-1)|f_n^{(1)}(0)|\right)e^{\lambda_{kn}t}}{\left|2\lambda_{kn}+a_n^2\sum_{j=1}^{\infty}\frac{c_j}{(\lambda_{kn}+\gamma_j)^2}\right|}\right)^2
$$

$$
\lesssim\sum_{n=1}^{\infty}\left(\sum_{k=1}^{\infty}\frac{a_n^{2-p}\left(|f_n(0)||x_k|^{p-1}+(p-1)|f_n^{(1)}(0)|\right)e^{x_kt}}{\left(4|x_k|^2+a_n^4\left|\sum_{j=1}^{\infty}\frac{c_j}{(x_k+\gamma_j)^2}\right|^2\right)^{1/2}}\right)^2
$$

$$
\lesssim\sum_{n=1}^{\infty}\left(\sum_{k=1}^{\infty}\frac{a_n^{2-p}\left(|f_n(0)||x_k|^{p-1}+(p-1)|f_n^{(1)}(0)|\right)e^{x_kt}}{|x_k|}\right)^2
$$

$$
\lesssim\sum_{n=1}^{\infty}|a_n^{2-p}f_n(0)|^2\left(\sum_{k=1}^{\infty}|x_k|^{p-2}e^{x_kt}\right)^2+(p-1)^2\sum_{n=1}^{\infty}|f_n^{(1)}(0)|^2\left(\sum_{k=1}^{\infty}|x_k|^{-1}e^{x_kt}\right)^2
$$

$$
\leqslant\left\|A^{2-p}f(0)\right\|_H^2\left(\sum_{k=1}^{\infty}|x_k|^{p-2}e^{x_kt}\right)^2+(p-1)^2\left\|f^{(1)}(0)\right\|_H^2\left(\sum_{k=1}^{\infty}|x_k|^{-1}e^{x_kt}\right)^2.
$$

Using the d'Alambert and Weierstrass theorems we obtain that series $\sum_{k=1}^{\infty}|x_k|^{p-2}e^{x_kt}$ converges absolutely and uniformly on the semiaxis $t\in[t_0,+\infty)$, where $t_0=0$ for $p=1$ and $t_0>0$, $p=2$ so as $\sum_{j=1}^{\infty}\gamma_j^{-3/2}<\infty$ in accordance with conditions of Theorem 4.4. Hence we have the estimate (4.7). □

References

[1] Di Blasio G. *Parabolic Volterra equations of convolution type.* J. Integral Equations Appl. **6** (1994), 479–508.

[2] Di Blasio G., Kunisch K., Sinestari E. *L^2-regularity for parabolic partial integrodifferential equations with delays in the highest order derivatives.* J. Math. Anal. Appl. **102** (1984), 38–57.

[3] Di Blasio G., Kunisch K., Sinestari E. *Stability for abstract linear functional differential equations.* Izrael. J. Mathematics. **50:3** (1985), 231–263.

[4] Kunisch K., Mastinsek M. *Dual semigroups and structural operators for partial differential equations with unbounded operators acting on the delays.* Differ. Integral Equations **3:4** (1990), 733–756.

[5] Kunisch K., Shappacher W. **3:4** *Necessary conditions for partial differential equations with delay to generate l_0-semigroup.* J. Differ. Equations **50** (1983), 49–79.

[6] Wu J. *Semigroup and integral form of class of partial differential equations with infinite delay.* Differ. Integr. Equations **4:6** (1991), 1325–1351.

[7] Wu J. *Theory and applications of partial functional differential equations.* Appl. Math. Sci. New York: Springer-Verlag **119** (1996).

[8] Vlasov V.V. *On the solvability and properties of solutions of functional-differential equations in Hilbert spaces.* Sbornik Mathem. **186:8** (1995), 67–92.

[9] Vlasov V.V. *On the solvability and estimates of solutions of functional-differential equations in Sobolev spaces.* Proc. Steklova Math. Inst. **227** (1999), 109–121.

[10] Vlasov V.V. *On the solvability of abstract parabolic equations with aftereffect.* Dokl. Ross. Akad. Nauk. **415:2** (2007), 151–152.

[11] Miller R.K. *An integrodifferential equation for rigid heat conductors with memory.* J. Math. Anal. Appl. **66** (1978), 313–332.

[12] Miller R.K., Wheeler R.L. *Well-posedness and stability of linear Volterra interodifferential equations in abstract spaces.* Funkcialaj Ekvac. **21** (1978), 279–305.

[13] Vlasov V.V., Shmatov K.I. *On the solvability of delayed hyperbolic equations in Hilbert spaces.* Proc. Steklov Math. Inst. **243** (2003), 127–137.

[14] Medvedev D.A., Vlasov V.V., Wu J. *Solvability and structural properties of abstract neutral functional differential equations.* Functional Differential Equations **15:3-4** (2008), 249–272.

[15] Vlasov V.V., Wu J. *Solvability and Spectral Analysis of Abstract Hyperbolic equations with delay* Functional Differential Equations **16:4** (2009), 751–768.

[16] Vlasov V.V., Medvedev D.A. *Functional-differential equations in Sobolev spaces and related problems of spectral theory* Contemporary Math. Fundamental Directions **30** (2008), 3–173. (English translation Journ. of Math. Sc. **164:5** (2010), 659–841).

[17] Miller R.K. *Volterra Integral Equation in Banach Space* Funkcialaj Ekvac. **18** (1975), 163–194.

[18] V.V. Vlasov, Wu J., G.R. Kabirova *Correct Solvability and Spectral Properties of Abstract Hyperbolic Equations with Aftereffect.* Contemporary Math. Fundamental Directions **35** (2010), 44–59. (English translation in J. Math. Sci., New York, 2010, 170:3 388–404).

[19] *J.L. Lions and E. Magenes Nonhomogeneous Boundary-Value Problems and Applications* Springer-Verlag, Berlin – Heidelberg – New York, 1972.

[20] Gurtin M.E., Pipkin A.C. *Theory of heat conduction with finite wave speed* Arch. Rat. Mech. Anal. **31** (1968), 113–126.

[21] Pandolfi L. *The controllability of the Gurtin–Pipkin equations: a cosine operator approach.* Appl. Math. Optim. **52** (2005), 143–165.

[22] Desch W., Miller R.K. *Exponential stabilization of Volterra Integrodifferential equations in Hilbert space.* J. Differential Equations **70** (1987), 366–389.

[23] Miloslavskii A.I. *Spectral properties of the operator pencil arising in the viscoelasticity* Available from Ukr. VINITI, No. 1229-UK87 (Kharkov) [in Russian].

[24] Miloslavskii A.I. *Spectral Analysis of Small Oscillations of Visco-elastic Hereditary Medium* [in Russian]. Preprint, Harkov, 1989.

[25] Kopachevsky N.D., Kreĭn S.G. *Operator approach to Linear* Problems of Hydrodynamics. Vol. 2. Nonselfadjoint Problems for Viscous Fluids. Berlin: Basel-Boston, 2003.

[26] D.A. Kosmodemyanskiy, A.S. Shamaev *Spectral properties of some problems in mechanics of strongly inhomogeneous media.* Contemporary Math. Fundamental Directions **17** (2006), 88–109 (English translation in Journal of Mathematical Sciences **149:6** (2008), 1679–1700.)

[27] Ivanov S., Pandolfi L. *Heat equations with memory: lack of controllability to the rest.* Journal of Mathematical analysis and applications **355** (2009), 1–11.

[28] Vlasov V.V., Gavrikov A.A., Ivanov S.A., Knyazkov D.U., Samarin V.A., Shamaev A.S. *Spectral properties of combined media.* Modern problems of Mathematics and Mechanics **1** (2009), 134–155.

[29] Vlasov V.V. *On the spaces of the vector-functions holomorphic in the angle* [in Russian]. Preprint. Moscow, 1981.

[30] Zhikov V.V. *On an extension of the method of two-scale convergence and its applications.* Sbornik: Mathematics. **191:7** (2000), 31–72.

[31] Zhikov V.V. *On a method of two-scale convergence.* Tr. Semin. Im. Petrovskogo. **23** (2003), 149–187.

[32] S. Ivanov and T. Sheronova *Spectrum of the Heat Equation with Memory* arXiv: math. CV/ 0912.1818v1.

[33] S.A. Ivanov *Spectrum of the Heat Equation with Memory* arXiv: math. CV/1002.2831

[34] Vlasov V.V., Rautian N.A. *Well-Defined Solvability and Spectral Analysis of Abstract Hyperbolic.* Tr. Semin. Im. Petrovskogo. **28** (2011), 15–46. (English translation in Journal of Mathematical Sciences **179:3** (2011), 390–415.)

[35] Vlasov V.V., Rautian N.A., Shamaev A.S. *Solvability and Spectral Analysis of Integro-Differential Equations Arising in the Theory of Heat Transfer and Acoustics.* Doklady Mathematics **82:2** (2010), 684–687.

[36] Vlasov V.V., Rautian N.A., Shamaev A.S. *Spectral Analysis and Correct solubility of an abstract integrodiferential equations Arising in the Theory of Heat Transfer and Acoustics.* Current Problems in Mathematics: Fundamental Directions. **39** (2011), 36–65 [in Russian]. (English translation in J. Math. Sci., New York, 2013, 190:1 34–65).

[37] Vlasov V.V., Rautian N.A. *Correct solvability of integro-differential equations arising in the theory of heat transfer and acoustics.* Functional Differential Equations **19:1-2** (2012), 207–224.

[38] Lykov A.V. *Problems of Heat and Mass Transfer* [in Russian] Minsk, 1976.

[39] Sandrakov G.V. *Multiphase homogenized diffusion models for problems with several parameters.* Izvestiya: Mathematics. **71:6** (2007), 119–172.

[40] Eremenko A., Ivanov S. *Spectra of the Gurtin–Pipkin Type Equations.* SIAM J. Math. Anal. **43** (2011), 2296–2306.

Victor V. Vlasov and Nadezda A. Rautian
Faculty of Mechanics and Mathematics
Moscow Lomonosov State University
Vorobievi Gori, Moscow, 117234, Russia
e-mail: `vikvvlasov@rambler.ru`
`nrautian@mail.ru`

Operator Theory: Advances and Applications (OT)

This series is devoted to the publication of current research in operator theory, with particular emphasis on applications to classical analysis and the theory of integral equations, as well as to numerical analysis, mathematical physics and mathematical methods in electrical engineering.

Edited by
Joseph A. Ball (Blacksburg, VA, USA), Harry Dym (Rehovot, Israel), Marinus A. Kaashoek (Amsterdam, The Netherlands), Heinz Langer (Vienna, Austria), Christiane Tretter (Bern, Switzerland)

■ **OT 238: Edmunds, D.E. / Evans, W.D.,** Representations of Linear Operators Between Banach Spaces (2013).
ISBN 978-3-0348-0641-1

■ **OT 237: Kaashoek, M.A. / Rodman, L. / Woerdeman, H.J.** (Eds.), Advances in Structured Operator Theory and Related Areas. The Leonid Lerer Anniversary Volume (2013).
ISBN 978-3-0348-0638-1

■ **OT 236: Cepedello Boiso, M. / Hedenmalm, H. / Kaashoek, M.A. / Montes Rodríguez, A. / Treil, S.** (Eds.), Concrete Operators, Spectral Theory, Operators in Harmonic Analysis and Approximation. 22nd International Workshop in Operator Theory and its Applications, Sevilla, July 2011 (IWOTA11) (2013).
ISBN 978-3-0348-0647-3

■ **OT 234/OT235: Eidelman, Y. / Gohberg, I. / Haimovici, I.** (Eds.), Separable Type Representations of Matrices and Fast Algorithms.
Vol. 1. Basics. Completion problems. Multiplication and inversion algorithms (2013).
ISBN 978-3-0348-0605-3
Vol 2. Eigenvalue method (2013).
ISBN 978-3-0348-0611-4

■ **OT 233: Todorov, I.G. / Turowska, L.** (Eds.), Algebraic Methods in Functional Analysis. The Victor Shulman Anniversary Volume (2013).
ISBN 978-3-0348-0501-8

■ **OT 232: Demuth, M. / Kirsch, W.** (Eds.), Mathematical Physics, Spectral Theory and Stochastic Analysis (2013).
ISBN 978-3-0348-0590-2

■ **OT 231: Molahajloo, S. / Pilipović, S. / Toft, J. / Wong, M.W.** (Eds.), Pseudo-Differential Operators, Generalized Functions and Asymptotics (2013).
ISBN 978-3-0348-0584-1

■ **OT 230: Brown, B.M. / Eastham, M.S.P. / Schmidt, K.M.**, Periodic Differential Operators (2013).
ISBN 978-3-0348-0527-8

■ **OT 229: Almeida, A. / Castro, L. / Speck, F.-O.** (Eds.), Advances in Harmonic Analysis and Operator Theory. The Stefan Samko Anniversary Volume (2013).
ISBN 978-3-0348-0515-5

■ **OT 228: Karlovich, Y.I. / Rodino, L. / Silbermann, B. / Spitkovsky, I.M.** (Eds.), Operator Theory, Pseudo-Differential Equations, and Mathematical Physics (2013).
ISBN 978-3-0348-0536-0

■ **OT 227: Janas, J. / Kurasov, P. / Laptev, A. / Naboko, S.** (Eds.), Operator Methods in Mathematical Physics. Conference on Operator Theory, Analysis and Mathematical Physics (OTAMP) 2010, Bedlewo, Poland (2013).
ISBN 978-3-0348-0530-8

■ **OT 226: Alpay, D. / Kirstein, B.**, Interpolation, Schur Functions and Moment Problems II (2012).
ISBN 978-3-0348-0427-1

■ **OT 225: Sakhnovich, L.A.**, Levy Processes, Integral Equations, Statistical Physics: Connections and Interactions (2012).
ISBN 978-3-0348-0355-7

MIX
Papier aus verantwortungsvollen Quellen
Paper from responsible sources
FSC® C105338

If you have any concerns about our products,
you can contact us on
ProductSafety@springernature.com

In case Publisher is established outside the EU,
the EU authorized representative is:
Springer Nature Customer Service Center GmbH
Europaplatz 3, 69115 Heidelberg, Germany

Printed by Libri Plureos GmbH
in Hamburg, Germany